T0348474

TRANSPORT IN BIOLOGICAL MEDIA

TRANSPORT IN BIOLOGICAL MEDIA

Edited by

SID M. BECKER
University of Canterbury, Department of Mechanical Engineering, Christchurch, New Zealand

ANDREY V. KUZNETSOV
North Carolina State University, Department of Mechanical and Aerospace Engineering, Raleigh, NC, USA

AMSTERDAM • BOSTON • HEIDELBERG • LONDON
NEW YORK • OXFORD • PARIS • SAN DIEGO
SAN FRANCISCO • SINGAPORE • SYDNEY • TOKYO
Academic Press is an imprint of Elsevier

Academic Press is an imprint of Elsevier
The Boulevard, Langford Lane, Kidlington, Oxford OX5 1GB, UK
Radarweg 29, PO Box 211, 1000 AE Amsterdam, The Netherlands
225 Wyman Street, Waltham, MA 02451, USA
525 B Street, Suite 1900, San Diego, CA 92101-4495, USA

Copyright © 2013 Elsevier Inc. All rights reserved.

No part of this publication may be reproduced, stored in a retrieval system or transmitted in any form or by any means electronic, mechanical, photocopying, recording or otherwise without the prior written permission of the publisher.

Permissions may be sought directly from Elsevier's Science & Technology Rights Department in Oxford, UK: phone (+44) (0) 1865 843830; fax (+44) (0) 1865 853333; email: permissions@elsevier.com. Alternatively you can submit your request online by visiting the Elsevier website at *http://elsevier.com/locate/permissions*, and selecting *Obtaining permission to use Elsevier material.*

Notice
No responsibility is assumed by the publisher for any injury and/or damage to persons or property as a matter of products liability, negligence or otherwise, or from any use or operation of any methods, products, instructions or ideas contained in the material herein. Because of rapid advances in the medical sciences, in particular, independent verification of diagnoses and drug dosages should be made.

Library of Congress Cataloging-in-Publication Data
A catalog record for this book is available from the Library of Congress.

British Library Cataloguing in Publication Data
A catalogue record for this book is available from the British Library.

ISBN: 978-0-12-415824-5

For information on all **Elsevier** publications
visit our website at **store.elsevier.com**

Contents

Preface xi

1. Modeling Momentum and Mass Transport in Cellular Biological Media: From the Molecular to the Tissue Scale

GEORGE E. KAPELLOS AND TERPSICHORI S. ALEXIOU

1.1 Introduction 2
 1.1.1 Cellular Biological Media 2
 1.1.2 Interplay Between Transport Phenomena, Structure and Function 4
 1.1.3 Modeling of Transport Phenomena Across Multiple Scales 6
 1.1.4 An Engineer's Perspective 6
1.2 Mechanics of Biomolecules, Subcellular Structures and Biological Cells 7
 1.2.1 Biomacromolecules 8
 1.2.2 Subcellular Structures 9
 1.2.3 Biological Cells 11
1.3 Formulation of Balance Laws and Constitutive Equations 12
 1.3.1 Single-Scale, Single-Phase Approaches 12
 1.3.2 Biot's Theory of Poroelasticity 16
 1.3.3 Theory of Interacting Continua 16
 1.3.4 Multiscale Bottom-Up Approaches 18
 1.3.5 Multiscale Computational, Equation-Free Approaches 23
1.4 Calculation of Constitutive Parameters 23
 1.4.1 Generic Framework for Theoretical Calculation 23
 1.4.2 Remarks on the Experimental Determination 24
 1.4.3 Transport Properties of the Extracellular Phase 25
 1.4.4 Mechanical Properties of Biological Cells 25
 1.4.5 Mechanical Properties of Cellular Biological Media 29
 1.4.6 Hydraulic Permeability of Porous Cellular Biological Media 29
 1.4.7 Diffusion Coefficients in Cellular Biological Media 31
1.5 Modeling of Growth and Pattern Formation 32
 1.5.1 Continuum-Based Models 33
 1.5.2 Discrete-Based Models 33
Acknowledgments 36
References 37

2. Thermal Pain in Teeth: Heat Transfer, Thermomechanics and Ion Transport

MIN LIN, SHAOBAO LIU, FENG XU, TIANJIAN LU, BOFENG BAI, AND GUY M. GENIN

2.1 Introduction 42
2.2 Modeling of Thermally Induced Dentinal Fluid Flow 43
 2.2.1 Analysis of Thermomechanics of the Tooth 43
 2.2.2 Analysis of Dentinal Fluid Flow 46
2.3 Modeling of Nociceptor Transduction 47
 2.3.1 Modeling of Shear Stress 47
 2.3.2 Modeling Transduction 49
2.4 Results and Discussion 51
 2.4.1 Tooth Thermomechanics 51
 2.4.2 The Mechanism Underlying Thermally Induced DFF 52
 2.4.3 DFF and Its Implications for Tooth Thermal Pain 53
 2.4.4 The Difference Between Hot and Cold Tooth Pain 54
2.5 Conclusion 56
References 56

3. Drug Release in Biological Tissue

FILIPPO DE MONTE, GIUSEPPE PONTRELLI, AND SID BECKER

3.1 Introduction 61
3.2 Continuum Modeling of Mass Transport in Porous Media 65
 3.2.1 Porosity and Volume-Averaged Variables 66

3.2.2 Permeability, Darcy's Law and the Continuity Equation 68
3.2.3 Tortuosity and Fick's Law 72
3.3 Conservation of Drug Mass 73
3.3.1 Drug Mass Balance in the Fluid Phase 74
3.3.2 Drug Mass Balance in the Solid Phase 77
3.3.3 Governing Equations 78
3.4 Analytical Solutions for Local Mass Non-Equilibrium 78
3.4.1 Nusselt's Solution 79
3.4.2 Schumann's Solution 83
3.4.3 Anzelius's Solution 85
3.4.4 Recent Solutions 86
3.5 Analytical Solutions for Local Mass Equilibrium 87
3.5.1 A Worked Example 87
3.6 Applications of Porous Media to the Drug-Eluting Stent 92
3.6.1 The Fluid Wall Model: The Pure Diffusion Approximation 96
3.6.2 The Fluid Wall Model: The Advection-Reaction-Diffusion Equation 102
3.6.3 The Multi-Layered Wall Model: The Pure Diffusion Approximation 106
3.7 Conclusion 115
References 116

4. Transport of Water and Solutes Across Endothelial Barriers and Tumor Cell Adhesion in the Microcirculation

BINGMEI M. FU AND YANG LIU

4.1 Introduction 120
4.2 Microvascular Transport 122
4.2.1 Transvascular Pathways 122
4.2.2 Transport Coefficients 129
4.2.3 Permeability Measurement 130
4.2.4 Transport Models for Water and Solutes Through Interendothelial Cleft 135
4.2.5 Endothelial Surface Glycocalyx 146
4.2.6 Transport Across Fenestrated Microvessels 152
4.2.7 Transport in Tissue Space (Interstitium) 152
4.3 Modulation of Microvascular Transport 153
4.3.1 Permeability Increase 154
4.3.2 Permeability Decrease 156
4.3.3 Permeability Increase and Decrease 158
4.3.4 Microvascular Hyperpermeability and Tumor Metastasis 158
4.4 Tumor Cell Adhesion in the Microcirculation 162
4.4.1 Tumor Cell Adhesion Under Flow 162
4.4.2 Mathematical Models for Tumor Cell Adhesion in the Microcirculation 163
4.4.3 Model Predictions for Tumor Cell Adhesion in the Microcirculation 166
4.5 Summary and Opportunities for Future Study 170
Acknowledgments 170
References 170

5. Carrier-Mediated Transport Through Biomembranes

RANJAN K. PRADHAN, KALYAN C. VINNAKOTA, DANIEL A. BEARD, AND RANJAN K. DASH

5.1 Introduction 182
5.2 Physicochemical Principles and Kinetic Modeling of Carrier-Mediated Transport 183
5.2.1 Thermodynamics of Solute Transport 183
5.2.2 Kinetic Treatment of a Simple Carrier 183
5.2.3 Kinetic Treatment of a Simple Pore 187
5.2.4 Membrane Potential Dependency of Solute Transport 188
5.3 Experimentally Observable Features of Carrier-Mediated Transport Phenomena 190
5.4 Kinetic Modeling of Mitochondrial Ca^{2+} Uniporter 191
5.4.1 Historical Background 191
5.4.2 Kinetic Scheme for Mitochondrial Ca^{2+} Transport via the Ca^{2+} Uniporter 193
5.4.3 Derivation of Mitochondrial Ca^{2+} Uniporter Flux Expression 195
5.4.4 $\Delta\Psi$ Dependency of Mitochondrial Ca^{2+} Uniporter Flux: Free Energy Barrier Formalism 197
5.4.5 Mitochondrial Ca^{2+} Uniporter Model Parameterization and Simulations 200
5.4.6 $\Delta\Psi$ Dependency of Mitochondrial Ca^{2+} Uniporter Flux: Alternate Formulation 201
5.4.7 Mg^{2+} Inhibition and Pi Regulation of Mitochondrial Ca^{2+} Uniporter Function 204

5.4.8 Kinetic Scheme for Mg^{2+} Inhibition of Mitochondrial Ca^{2+} Uniporter Function 204
5.4.9 Mitochondrial Ca^{2+} Uniporter Model with Mg^{2+} Inhibition 206
5.5 Other Modes of Carrier-Mediated Transport: Antiport and Cotransport 208
5.6 Summary and Conclusion 210
Acknowledgment 211
References 211

6. Blood Flow Through Capillary Networks

C. POZRIKIDIS AND J.M. DAVIS

6.1 Introduction 214
6.2 Equations of Steady Capillary Blood Flow 216
6.2.1 Balances at a Bifurcation 217
6.2.2 Discharge Hematocrit 217
6.2.3 Tube Hematocrit and Hematocrit Ratio 218
6.2.4 Effective Viscosity 218
6.2.5 Cell Partitioning at a Bifurcation 219
6.2.6 Converging Bifurcations 220
6.2.7 Diverging Bifurcation 220
6.2.8 Klitzman and Johnson Cell Partitioning Law 220
6.2.9 General Expression for the Cell Partitioning Law 221
6.2.10 Empirical Cell Partitioning Law 221
6.2.11 Numerical Studies of Suspension Flow Through a Bifurcation 221
6.3 Steady Flow Through Tree Networks 222
6.3.1 Geometrical Construction 222
6.3.2 Numerical Method 223
6.3.3 Results and Discussion 223
6.4 Steady Flow Through Homogeneous Networks 226
6.4.1 Theoretical Model 226
6.4.2 Geometrical Construction 228
6.4.3 Numerical Method 228
6.4.4 Flow Through a Pristine Network 228
6.4.5 Dimensions and Parameters 229
6.4.6 Effective Hydraulic Permeability 229
6.4.7 Results and Discussion 230
6.4.8 Significance of the Bifurcation Law 231
6.4.9 Significance of the Viscosity Correlation 233
6.4.10 Discussion 234
6.5 Equations of Unsteady Blood Flow 235
6.5.1 Unsteady Flow Through a Straight Capillary 235
6.5.2 Circular Capillaries 236
6.5.3 Correlations 237
6.5.4 Balances at Bifurcations 237
6.5.5 Numerical Method 238
6.5.6 Single-Node Dynamics 240
6.6 Unsteady Flow Through Tree Networks 243
6.6.1 Steady Flow for Subcritical Exponents 244
6.6.2 Unsteady Flow for Supercritical Exponents 246
6.6.3 State Space 248
6.6.4 Summary and Discussion 248
6.7 Summary and Outlook 249
References 250

7. Models of Cerebrovascular Perfusion

T. DAVID AND R.G. BROWN

7.1 Introduction 254
7.2 From Arteries to Cells and Back Again (Cerebral Anatomy and Physiology) 254
7.2.1 Cerebral Arterial Structure 254
7.2.2 Circle of Willis 256
7.2.3 Penetrating Arteries and the Cortex 256
7.3 Structure of Arterial Blood Vessels 257
7.4 A Simple Description of Cerebral Autoregulation 258
7.4.1 Organ Autoregulation 258
7.4.2 Local Autoregulation: 'Functional Hyperemia' 258
7.5 Vascular Trees and Their Numerical Simulation 259
7.6 Simple Models of Autoregulated Cerebral Perfusion 259
7.6.1 Blood Flow Through the Circle of Willis 259
7.6.2 Simple Models of Autoregulated Cerebral Perfusion 261
7.7 More Complex Models 264
7.7.1 Arteriolar Models of Autoregulation 264
7.7.2 Perfusion via the Capillary Bed 267
7.8 Conclusions 272
Acknowledgments 272
References 272

8. Mechanobiology of the Arterial Wall

ANNE M. ROBERTSON AND PAUL N. WATTON

8.1 Introduction 276
8.2 Overview of the Arterial Wall 278
8.2.1 Brief Overview of the Architecture of the Arterial Wall 279
8.2.2 Design Requirements for the Arterial Wall 281
8.3 The Extracellular Matrix 283
8.3.1 Collagen and the Arterial Wall 283
8.3.2 Elastin in the Arterial Wall 291
8.4 Vascular Cells 292
8.4.1 Endothelial Cells 293
8.4.2 Vascular Smooth Muscle Cells 295
8.4.3 Fibroblasts 297
8.4.4 Matrix Assembly by Vascular Cells 297
8.5 Architecture of the Arterial Wall 301
8.5.1 Tunica Intima 301
8.5.2 Tunica Media 306
8.5.3 Tunica Adventitia 308
8.6 Constitutive Models for the Arterial Wall 309
8.6.1 Multiple Mechanism Models 311
8.6.2 Isotropic Mechanism 312
8.6.3 Anisotropic Mechanisms: Kinematics of Fiber Recruitment 313
8.6.4 *N*-Fiber Anisotropic Models 314
8.6.5 Anisotropic Models with a Distribution of Fiber Orientations 316
8.6.6 Distributions of Fiber Recruitment Stretch 317
8.6.7 Multi-Mechanism Models: Growth, Remodeling and Damage (GR&D) 319
8.7 Modeling Vascular Disease: Intracranial Aneurysms 324
8.7.1 Background 324
8.7.2 Computational Modeling of Intracranial Aneurysms 326
8.7.3 Example: Fluid-Solid-Growth Model of Aneurysm Evolution 329
8.7.4 Discussion 331
References 333

9. Shear Stress Variation and Plasma Viscosity Effect in Microcirculation

XUEWEN YIN AND JUNFENG ZHANG

9.1 Introduction 350
9.2 Models and Methods 351
9.2.1 The Lattice-Boltzmann Method (LBM) for Fluid Dynamics 351
9.2.2 Red Blood Cell (RBC) Model and Membrane Mechanics 352
9.2.3 Intercellular Aggregation 353
9.2.4 The Immersed-Boundary Method (IBM) for Fluid-Membrane Interaction 354
9.2.5 Viscosity Update Algorithm 355
9.3 Algorithm Validations 356
9.3.1 The Laplace Relationship for Stationary Bubbles 356
9.3.2 The Dispersion Relationship for Capillary Waves 357
9.3.3 The Drag Coefficient for Circular Cylinders 358
9.3.4 RBC Deformation and Rotation in Shear Flows 360
9.4 WSS Variation Induced by Blood Flows in Microvessels 360
9.4.1 Single-File RBC Flows 361
9.4.2 Cell-Free Layer (CFL) and Wall Shear Stress (WSS) Variation in Multiple RBC Flows 368
9.5 Suspending Viscosity Effect 372
9.5.1 Single RBC Rotation in Shear Flows 372
9.5.2 Single RBC Migration in Channel Flows 374
9.5.3 Multiple RBC Flows in Microvessels 376
9.5.4 RBC Motion and Deformation in Bifurcated Microvessels 380
9.6 Summary 386
Acknowledgments 387
References 387

10. Targeted Drug Delivery: Multifunctional Nanoparticles and Direct Micro-Drug Delivery to Tumors

CLEMENT KLEINSTREUER, EMILY CHILDRESS, AND ANDREW KENNEDY

10.1 Introduction 392
10.2 Diagnostic Imaging and Image-Guided Drug Delivery 392
10.2.1 Imaging Techniques 392
10.2.2 Image-Guided Drug Delivery 394
10.3 Free Transport 394
10.3.1 Passive and Active Targeting 395
10.3.2 Nanodrug Carriers 395
10.3.3 Summary 400

10.4 Forced Transport 400
10.5 Direct Transport 401
10.5.1 Implementation of Optimal Targeted Drug Delivery 401
10.5.2 Applications of Optimal Micro-Drug Delivery 404
10.6 Conclusions 413
Acknowledgments 413
References 414

11. Electrotransport Across Membranes in Biological Media: Electrokinetic Theories and Applications in Drug Delivery

S. KEVIN LI, JINSONG HAO, AND MARK LIDDELL

11.1 Introduction 418
11.2 Nernst-Planck Theory and Model Simulation Analyses 419
11.3 Electrotransport Under a Constant Electric Field Across Membrane (Symmetric Conditions) 420
11.3.1 Membrane Flux and the Modified Nernst-Planck Equation 421
11.3.2 Membrane Flux and Transference Number 424
11.3.3 Transport Lag Time 424
11.3.4 Electroosmotic Transport 425
11.4 Electrotransport Under Variable Electric Field Across Membrane (Asymmetric Conditions) 426
11.4.1 The Nernst-Planck Equation with Electroneutrality Approximation 426
11.4.2 Electrotransport Under Asymmetric Conditions with Electroosmosis 430
11.5 Electrotransport Across Multiple Barriers/ Membranes 430
11.5.1 Electrotransport Across Two-Membrane Systems Under Symmetric and Asymmetric Conditions 432
11.5.2 Effects of Membrane Porosity and Applied Voltage Upon Electrotransport Across Two-Membrane Systems 436
11.6 Electrotransport Under Alternating Current 440
11.7 Electropermeabilization Effect 441
11.8 Electrokinetic Methods of Enhanced Transport Across Biological Membranes 443
11.8.1 Transdermal Iontophoresis 444
11.8.2 Transungual Iontophoresis 447
11.8.3 Transscleral Iontophoresis 448
References 450

12. Mass Transfer Phenomena in Electroporation

ALEXANDER GOLBERG AND BORIS RUBINSKY

12.1 Introduction 458
12.2 Electroporation Background and Theory 459
12.3 Applications of Electroporation-Mediated Mass Transport in Biological Systems 463
12.4 Mechanisms of Pulsed Electric Field-Mediated Transport into Cells 464
12.5 Experimental Methods Used to Study Mass Transfer During Electroporation 465
12.6 Mathematical Models Describing Molecular Transport During Reversible Electroporation 467
12.6.1 Physico-Chemical Model for Electroporation 467
12.6.2 Electropermeabilization Model 470
12.6.3 Electrodiffusion Model of DNA Cluster Formation 472
12.6.4 Model of Conductivity Changes During Cell Suspension Electroporation 475
12.6.5 Model of Small Molecule Transport Kinetics Due to Electroporation 477
12.6.6 Two Compartment Pharmacokinetic Model for Molecular Uptake During Electroporation 479
12.6.7 Statistical Model for Cell Electrotransformation 481
12.6.8 A Multiscale Model for Mass Transfer of Drug Molecules in Tissue 481
12.7 Future Needs in Mathematical Modeling of Mass Transport for Electroporation Research 484
References 485

13. Modeling Cell Electroporation and Its Measurable Effects in Tissue

NATAŠA PAVŠELJ, DAMIJAN MIKLAVČIČ, AND SID BECKER

13.1 Introduction – Electroporation 493

13.2 Skin Electroporation 495
13.3 Physical Changes in Biological Tissue Following Electroporation 497
13.3.1 Electrical Conductivity Increase 497
13.3.2 Tissue Heating During Electroporation 499
13.3.3 Molecular Transport During Skin Electroporation 502
13.4 Modeling of Skin Electroporation Transport 504
13.4.1 Modeling Non-Thermal Electroporation (Short Pulses) 506
13.4.2 Modeling Thermal Electroporation (Long Pulse) 507
13.5 Conclusions 514
Acknowledgments 515
References 515

14. Modeling Intracellular Transport in Neurons

ANDREY V. KUZNETSOV

14.1 Introduction 522
14.2 A Model of Axonal Transport Drug Delivery 523
14.3 Effect of Dynein Velocity Distribution on Propagation of Positive Injury Signals in Axons 532
14.4 Simulation of Merging of Viral Concentration Waves in Retrograde Viral Transport in Axons 540
14.5 Conclusions 546
References 547

Index 551

Preface

The purpose of this book is to provide a platform from which emerging and leading researchers are able to present the interdisciplinary modeling strategies and theoretical tools that are used to understand the diverse phenomena associated with transport within biological media.

In order to successfully navigate through the applied field of transport in biological media, the theoretical and computational aspects of transport modeling must be understood. Furthermore, this task requires a fundamental understanding of the nature of the physiology of the biological media in which the transport processes take place. This book seeks to present the reader with the experience and knowledge of an international group of expert academic and clinical researchers that will provide this understanding.

Empirical observations, while critical in identifying the existence and behavior of transport within living tissue, do not always provide direct insight into the physics underlying the phenomena. Similarly, empirically based models, which can predict the outcomes of transport to accurately match controlled experimental conditions, do not need to capture the physical laws which govern the transport.

Recently there has been a push in the field's research community to abandon purely empirical descriptions in favor of mechanistic physics-based conceptualizations of transport phenomena. It is the goal of the editors as well as the expert contributing authors of this book to promote an understanding of the physics, the modeling, and the biological physiology in order to help bridge the gap between biology and transport modeling.

Chapter 1 reviews the modeling strategies of transport at the cellular and subcellular scales. A technical review of the physiological considerations is presented in the context of transport modeling at these length scales in order to allow the reader to understand the rationale behind the choice of model.

Chapter 2 details the multiscale considerations between dentinal fluid flow and the neurological receptor transduction. To do this the illustrative example of the nociceptive mechanoreceptor response to thermally induced stimuli within the tooth is considered. Modeling strategies of capturing the local physiology and the multiscale interactions are detailed.

Chapter 3 provides a review of the exact solution development for the modeling of molecular transport in composite and porous perfuse tissue. The development of the governing equations and boundary conditions are derived from first principles as well as the current understanding of the underlying physics. Porous media models involving the solid-fluid space are developed and applied to specific applications of transport through the arterial wall.

Chapter 4 details the transport of water and solute across variety of endothelial barriers. The physiology of tissue specific to the microvessel environments is presented specifically to explain the choices of models used to represent the physics mathematically. The chapter closes with a detailed discussion of the effect of the presence of tumor cells within the microcirculatory flow field.

Chapter 5 considers the nanoscale transport across biological membranes in which the electro-chemical forces contribute significantly to transport. In order to accomplish this, a detailed introductory review of the physiology of the molecular scale make-up of common biomebranes is presented as is the current understanding of the fundamental physics governing the electrochemical transport of ions. Advanced methods of modeling the delivery across bio-membranes are presented in detail.

Chapter 6 discusses modeling strategies that capture the flow of blood within the fractal geometry of microcapillary networks. The physical implications of the common assumptions used to represent these extremely complex structures are discussed in depth, providing an indispensible link between theory, physics, and computational modeling. A detailed discussion of the unsteady effects associated with the interaction between the microscale geometry and pulsatile flow provides further insight into the physical understanding of these phenomena.

Chapter 7 continues with the investigation of vascular networks, focusing on the specific application to the dynamic vasculature within the brain. A detailed discussion of the multiscale physiology representing the brain's anatomy is formulated in the context of the theoretical model development. A review of the current models used to represent transport to the cerebral tissue is provided.

Chapter 8 discusses the mechanics of the arterial wall. This is a necessary component to better understand the nature of the conjugate interaction between the arterial wall's mechanical response and the flow field's distribution of momentum which is communicated through the local wall shear stress. This chapter provides an extraordinarily detailed and insightful introduction into the physiology of the arterial wall as a considerable review of the current modeling strategies that capture the dynamic mechanical response of this soft tissue.

Chapter 9 presents the conjugate perspective of the preceding chapter by analyzing the development of shear stresses within the microcirculatory flow field. The effect of the presence of individual elastic red blood cells on the flow is investigated in detail. The Lattice Boltzmann Method is used and an intensive introduction to this modeling tool is provided as well. The calculation of local variations in viscosity and the interpretation of the results are provided from a physics-based perspective.

Chapter 10 provides a novel look at the manner in which the prediction of the arterial flow field can be used with imaging data to develop patient-specific treatments. A review of the current understanding of the contributing methods is provided.

Chapter 11 considers the electrokinetic effects on transport across biological membranes at very low voltages. This chapter reviews the field's understanding of the physics underlying the electrically assisted diffusion as well as the flow. Comparison of current theories and modeling strategies is presented in a detailed review that provides the reader not only with a compendium of models, but also the conditions under which individual models are appropriate.

Chapter 12 discusses the transport at the cellular scale associated with electroporation. The physics and empirical observations associated with exposing a cell to an intense electrical field are introduced in a review of the literature. Eight specific modeling strategies are provided in order to describe the nanoscale transport associated with this phenomenon.

Chapter 13 follows with an in-depth analysis of transport-associated electroporation at the level of tissue. The empirical observations of skin electroporation are reviewed and categorized according to the underlying physics. A review of existing transdermal transport models is provided and the implications of each model type are discussed.

Chapter 14 discusses intracellular transport in neurons. The theoretical model development is provided in three applications: the utilization of axonal transport for targeted drug delivery, the problem of how injured axons measure the distance to the lesion site, and viral trafficking in axons by means of retrograde axonal transport. Specific modeling strategies are discussed in detail, and analytical solutions are presented.

This book helps to bridge the gap between physiology, biology, and transport modeling. We believe that the reader will find this book to be an indispensable resource in navigating through the field of modeling transport in living media.

Sid M. Becker
University of Canterbury
Mechanical Engineering Department
Christchurch, New Zealand

Andrey V. Kuznetsov
North Carolina State University,
Department of Mechanical and
AerospaceEngineering
Raleigh, NC, USA

CHAPTER

1

Modeling Momentum and Mass Transport in Cellular Biological Media: From the Molecular to the Tissue Scale

George E. Kapellos and Terpsichori S. Alexiou

Laboratory of Transport Phenomena & Physicochemical Hydrodynamics, Department of Chemical Engineering, University of Patras, Patras, Greece

Dedicated to Alkiviades C. Payatakes.

OUTLINE

1.1 Introduction 2
1.1.1 Cellular Biological Media 2
1.1.2 Interplay Between Transport Phenomena, Structure and Function 4
1.1.3 Modeling of Transport Phenomena Across Multiple Scales 6
1.1.4 An Engineer's Perspective 6

1.2 Mechanics of Biomolecules, Subcellular Structures and Biological Cells 7
1.2.1 Biomacromolecules 8
1.2.2 Subcellular Structures 9
1.2.2.1 Glycocalyx 9
1.2.2.2 Cell Membrane 10
1.2.2.3 Intracellular Matrix 11
1.2.3 Biological Cells 11

1.3 Formulation of Balance Laws and Constitutive Equations 12
1.3.1 Single-Scale, Single-Phase Approaches 12
1.3.2 Biot's Theory of Poroelasticity 16
1.3.3 Theory of Interacting Continua 16
1.3.4 Multiscale Bottom-Up Approaches 18
1.3.4.1 Asymptotic Spatial Homogenization 19
1.3.4.2 Spatial Averaging Method 21
1.3.5 Multiscale Computational, Equation-Free Approaches 23

1.4 Calculation of Constitutive Parameters 23
1.4.1 Generic Framework for Theoretical Calculation 23

Transport in Biological Media
http://dx.doi.org/10.1016/B978-0-12-415824-5.00001-1

© 2013 Elsevier Inc. All rights reserved.

1.4.2 *Remarks on the Experimental Determination* 24
1.4.3 *Transport Properties of the Extracellular Phase* 25
1.4.4 *Mechanical Properties of Biological Cells* 25
1.4.5 *Mechanical Properties of Cellular Biological Media* 29
1.4.6 *Hydraulic Permeability of Porous Cellular Biological Media* 29
1.4.7 *Diffusion Coefficients in Cellular Biological Media* 31

1.5 Modeling of Growth and Pattern Formation 32
1.5.1 *Continuum-Based Models* 33
1.5.2 *Discrete-Based Models* 35

Acknowledgments 36

References 37

1.1 INTRODUCTION

Momentum and mass transport in mammalian tissues, microbial biofilms and similar biological materials are of great importance in quite diverse fields of science and technology, including medicine, biology, biotechnology and tissue and environmental engineering. In particular, the analysis of transport phenomena in such materials is encountered in many natural processes and technological applications, such as the *in vitro* construction of artificial tissues as well as the *in vivo* regeneration of native ones, the biodegradation of organic contaminants by microbial biofilms in porous media, the settling and consolidation of suspended microbial flocs in solid–liquid separation processes, the formation of microbial cakes on membrane filtration units and the delivery of chemotherapeutic agents for the eradication of malignant tumors and biofilm infections.

1.1.1 Cellular Biological Media

Animal and plant tissues, microbial flocs and biofilms, artificial tissues and cell-entrapping gels can be considered, under a unified perspective, as cellular biological media. In this regard, *cellular biological media*[1] are defined as multiphase complex systems, which consist of biological cells along with their extracellular matrix (ECM), and exhibit a dynamically evolving and highly organized hierarchical structure. The ECM is produced and actively secreted by cells. Its composition and architecture depends on the cell type and also, on epigenetic factors, such as the available nutrients and applied mechanical forces. For instance, the microbial

[1] The term *cellular biological media* was coined by Alkiviades C. Payatakes to describe collectively biofilms, tissues, etc. during a group meeting with G.E.K. & T.S.A. on September, 2005. The original intention was to develop simple models for the calculation of transport coefficients in three-phase biological systems, with application mainly to biofilms. Yet, it was soon realized that an integrated, multiscale theoretical framework and a unified perspective on these biomaterials were both missing. Critical feedback from colleagues and reviewers fueled our motivation to work towards this direction.

TABLE 1.1 Components of Some Mammalian Tissues

Tissue	Cells	ECM key constituents	Network of vessels	Reference
Cartilage	Chondrocytes	Collagen type I, fibronectin, hyaluronan, aggrecan, chondroadherin	No	*Wilson et al.* [99]
Bone	Osteocytes, osteoblasts, osteoclasts	Collagen type I, hydroxyapatite, fibronectin, osteocalcin, osteonectin	Yes	*Sikavitsas et al.* [85]
Brain	Nerve cells (neurons and glial cells)	Lecticans, tenascin, versican	Yes	*Ruoslahti* [80]
Muscle	Muscle cells	Collagen type I, decorin, fibromodulin, depican	Yes	*Gillies et al.* [28]
Skin	Fibroblasts, keratinocytes, adipocytes	Collagen type I, keratin, cadherin, laminin	Yes	*Wang et al.* [95]
Tendons	Tenoblasts, tenocytes, synovial cells	Collagen type I, elastin, decorin, aggrecan, biglycan, tenascin	No	*James et al.* [36]

ECM is composed mainly of polysaccharides (e.g., alginates, xanthan, cellulose) and, to a lesser extent, of proteins, lipids and even nucleic acids, with the latter probably originating from cell lysis. In mammalian tissues, the ECM contains proteins (e.g., collagen, fibrin, elastin, laminin), glycosaminoglycans (GAGs), and in some cases inorganic minerals (e.g., calcium phosphates in bone) (see also Table 1.1). The extracellular aqueous solution exists either as fluid *bound* by physico-chemical forces to ECM components (e.g., heparan sulfate and other GAGs), or as *free fluid* moving through the cellular biological medium. Finally, some cellular biological media might contain a network of vessels with a tree-like structure that serves to overcome flow and mass transport limitations.

Despite the existence of significant differences in the detailed composition, structure and function of different types of cellular biological media, they all exhibit a *hierarchical structure*, which evolves over time and presents a high degree of spatial organization at every characteristic length-scale. For example, Fig. 1.1 illustrates three fundamental scales of the structural hierarchy of a microbial biofilm. The characteristic lengths which are associated with these scales of observation are defined by:

1. The diameter of the extracellular polymeric fibers (several nm, Π-scale).
2. The size of a single microbial cell (a few μm, K-scale).
3. The size of the overall biofilm (from tenths of μm to several mm, B-scale).

As another example, one may identify five characteristic levels of the structural organization in bone [49]. In this context, the term *scale* denotes the spatial dimension of an object or process characterized by both *extent* and *resolution*. The term extent denotes the overall size of the observable area or volume, and the term resolution denotes the size of the smallest discernible element of area or volume. Both extent and resolution depend on the instrument which is used to make the observation.

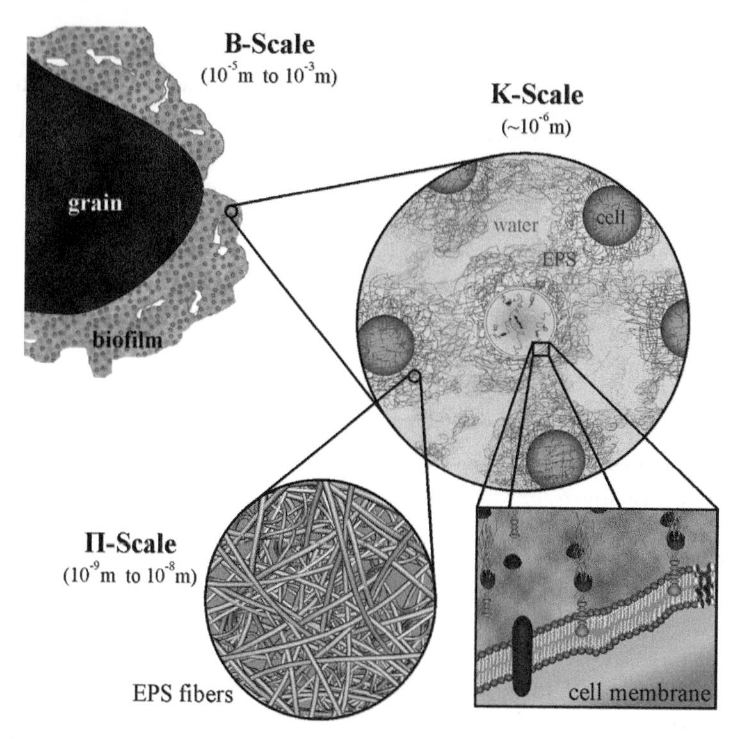

FIGURE 1.1 Schematic representation of the hierarchical structure of a microbial biofilm. Reprinted from *Kapellos et al.* [40] with permission from Elsevier.

The development of sharpened experimental tools, such as magnetic resonance imaging, atomic force microscopy, confocal laser microscopy, computed micro-tomography, optical tweezers and others, has substantially advanced our understanding of the complex physical and biological mechanisms associated with the structure, function and mechanics of cellular biological media. These advanced instruments offer a precise observation window in space–time, and thus enable the tracking of phenomena and behavior associated with single cells or even single biomacromolecules. For example, *Liu et al.* [57] combined atomic force with fluorescence microscopy to study the mechanical properties of individual fibrin fibers. Interestingly, they found that fibrin fibers can be stretched to nearly three times their normal length without losing elasticity and up to six times before rupturing.

1.1.2 Interplay Between Transport Phenomena, Structure and Function

Transport phenomena, manifested as mechanical loading, interstitial flow, and solute transport, are intrinsically and inextricably coupled to the structure and function of cellular biological media. Over a century ago, *Julius Wolff* hypothesized that bone adapts its internal architecture and external shape in response to applied mechanical loads, and that these adaptations are in accordance with specific mathematical laws. While this statement (usually referred to as *Wolff's law*) has been roundly criticized in respect of its theoretical background and domain of validity [79], it is nevertheless among the first examples that clearly highlight the important role of force in the force-form-function paradigm for cellular biological media. Subsequently, significant light has been shed on the interrelations between bone remodeling and applied loads. For instance, bone deformation induces the flow of ionic aqueous solution in the canalicular network, which in turn results in an electrical (streaming) potential. This combined electromechanical stimulation triggers an increase in the mass of bone [1,38].

A more recent development is the experimental work by *Sikavitsas et al.* [84], which clearly shows that the fluid shear strongly affects the *in vitro* osteogenic activity of osteoblasts. Specifically, by increasing the viscosity and keeping constant the flow rate of the feed solution, they were able to increase the fluid shear while maintaining a practically constant mass transfer flux in a perfusion bioreactor. Under these conditions, they observed that the increased shear causes increased deposition and more uniform spatial distribution of a mineralized matrix. Nowadays, cutting-edge research is focused on understanding the mechanisms through which the bone cells sense and translate different physical forces to the corresponding intracellular signals, a process known as mechanotransduction (e.g., [85]). Similar observations with regard to the effects of applied forces have been made for other cellular biological media as well. Examples include the adaptation of the composition, internal architecture and overall thickness of arterial walls in response to changes of the intraluminal fluid shear and pressure [29], the electromechanical couplings in cartilage (i.e., deformation-induced streaming potential/current and current-generated mechanical stress [24]), the regulation of the metabolic activity in *E. coli* aggregates by forced interstitial flow [23], among many others.

Mass transport mechanisms are also of paramount importance for the development of structure and function in cellular biological media. First and foremost, efficient transport of nutrients, wastes, signals, growth factors and other biochemical agents are crucial for life sustenance. Limitations in the supply of vital nutrients can lead either to adaptation or to necrosis. For example, the rat brain adapts to prolonged exposure to 'mild' hypoxia (i.e., low oxygen supply) by increasing the number of blood capillaries per tissue volume, and thus decreasing the intercapillary distance and ensuring that oxygen reaches all cells [50]. In the interstitial space, the primary mass transfer mechanism is molecular diffusion, which in several cases is augmented by bulk convection [89]. Among the first contributions highlighting the role of flow-enhanced solute transport in tissues is the work of *Swabb et al.* [88], who showed that the relative importance of convective versus diffusive mass transfer depends strongly on the GAG content of the tissue and the size of solute molecules. Larger solutes and lower GAG content result in convection dominated mass transfer. In a similar vein, *Piekarski and Munro* [71] suggested that the cyclic mechanical loading in bone causes significant pulsatile flow in the canalicular network, which in turn results in enhanced mass transfer. Perhaps, the most remarkable and intriguing mechanisms of mass transport are the protein-mediated mechanisms of transmembrane and intracellular transport, which can take place either spontaneously (facilitated passive transport) or at the cost of biochemical energy (active transport). Examples of such mechanisms are the transmembrane passage of water through channel proteins (aquaporins), the intracellular transport of oxygen by mobile carrier proteins (e.g., hemoglobin, myoglobin), and the transport of vesicles and other large molecular structures by motor proteins (kinesins, dyneins) that move along microtubules of the cytoskeleton (e.g., [82]).

1.1.3 Modeling of Transport Phenomena Across Multiple Scales

Mathematical modeling and computational simulation are indispensable tools for the study of the complex systems under consideration. These tools can be used to postulate and test hypotheses concerning the underlying mechanisms, to design and interpret experiments and, ultimately, to predict the behavior of the system. A landmark of mathematical modeling of biological systems is Turing's paper on morphogenesis [91]. Based on the idea that pattern formation in biological systems is dictated by the concentration of specific biochemical agents (morphogens), Turing described the spatiotemporal distribution of a morphogen using diffusion reaction equations. Later, the chemotactic migration of biological cells in response to a chemical concentration gradient was also mathematically described using the diffusion–reaction equations (e.g., the *Keller–Segel* model [45]). The first models accounting for the coupling between fluid flow and matrix deformation in tissues appeared early in the latter half of the 20th century and were based on Biot's theory of poroelasticity [65] and mixture theory [63]. Subsequently, these theories have been further developed and refined in order to account for finite deformations, electrokinetics and other features of the physical systems (see, for instance, [41] and references therein). Recently, upscaling methods, such as volume averaging and homogenization, were used to connect the macroscale description to the microscale structure and the function of cellular biological media (e.g., [40,43,52,101]). These works were focused primarily on verifying and defining the validity domain of the equations used in poroelasticity and mixture theories, and also on providing a means for the theoretical calculation of constitutive parameters.

In many practical applications, a sufficient description of the transport processes in cellular biological media is achieved at the macroscopic scale. In those cases, it would be useful to know how and to what extent finer-scale phenomena (molecular, cellular) affect the behavior that is observed and measured at the macroscopic scale. This is usually referred to as an *upscaling* problem. An integrated theoretical analysis of such a problem requires dealing with two fundamental issues. The first issue regards the formulation of balance equations and constitutive relations at the macroscopic scale. The second issue regards the proper connection of the parameters, which appear in the constitutive equations, with geometrical and physicochemical parameters that pertain to the phenomena taking place at the molecular and cellular scales. These issues are discussed in detail in this chapter.

Nowadays, some applications (e.g., localized drug delivery) require effective regulation of certain aspects of the molecular or cellular scale phenomena while manipulating macroscopic variables, such as solute concentration and mechanical stresses. This is usually referred to as a *downscaling* problem. Multiscale theoretical frameworks are required in order to simultaneously handle both types of problems (upscaling and downscaling). In principle, multiscale approaches should incorporate appropriate *structural* and *process* models at every fundamental structural level, as well as consistent cross-scale interfacing between the component models [94].

1.1.4 An Engineer's Perspective

The analysis of transport phenomena in cellular biological media requires a multidisciplinary approach, integrating knowledge from mathematics, biology, biophysics, biochemistry along with engineering principles. The fact that this is a daunting problem requiring considerations across multiple scales of length, time and force was exquisitely captured by *E.N. Lightfoot* a few decades ago. Specifically, in the preface of his book on 'Transport phenomena and living systems' [54], he wrote:

'... truly faithful models of any interesting biological system are impossible. To obtain useful models one must set rather specific goals and work toward them by systematic alternation of theory and experiment.'

Later, at the Dahlem conference on biofilms [55] he expressed a 'minority opinion' stating that:

'the complexity of biofilms and the need to model their behavior at many different time and length scales, and from a multidisciplinary point of view, make the elaboration of a hierarchic transport modeling strategy imperative'.

These statements soundly express the point of view of the authors in this chapter as well.

The objective of this chapter is to present a comprehensive review of the theoretical modeling of momentum and mass transport in cellular biological media. In the spirit of a bottom-up approach, the chapter begins with a concise exposition of the state of the art on the mechanics of individual biomacromolecules, subcellular structures and biological cells that build up the cellular biological medium (Section 1.2). Thereafter, the chapter's focus is on the formulation of balance laws and constitutive equations (Section 1.3) as well as on the experimental and theoretical calculation of constitutive parameters at the scale of the cellular biological medium (Section 1.4). The conceptual framework and the mathematical foundation is presented for various theoretical approaches, including single-scale–single-phase models, the theory of interacting continua, upscaling methods (spatial averaging, homogenization) and multiscale equation-free approaches. Finally, an introduction on modeling growth and pattern formation is given in Section 1.5. This chapter is a significant extension to our recent review article [41].

1.2 MECHANICS OF BIOMOLECULES, SUBCELLULAR STRUCTURES AND BIOLOGICAL CELLS

"Πάντων γὰρ ὅσα πλείω μέρη ἔχει καὶ μὴ ἔστιν οἷον σωρὸς τὸ πᾶν ἀλλ' ἔστι τι τὸ ὅλον παρὰ τὰ μόρια, ἔστι τι αἴτιον"[2] (Aristotle, Metaphysics [1045a.10]).

This statement by Aristotle, which is commonly used in the abridged version of '*[there is some cause for which] the whole is more than the sum of its parts*', is true for cellular biological media as well. In this regard, the observation and quantification of the mechanical behavior of isolated, individual cells and biomolecules *in vitro*, albeit very useful, is not sufficient to infer the mechanics of a living cellular biological medium. Beyond the fundamental understanding of the nature and behavior of individual constituent parts, it is also necessary to unravel the fundamental mechanisms that govern the interactions between them. Therefore, extending

[2] A rough translation is: 'In all things which consist of many parts and are not a mere heap, but the whole is something more than its parts, there is some cause'.

Aristotle's thoughts to the systems under consideration, the *interactions* between biomolecules, supramolecular structures and cells (occurring within and across multiple scales) are the *cause* for the emergent behavior of a cellular biological medium. Slow but steady progress is being made towards the understanding of such interactions. In this section, we present briefly some recent advances in understanding the mechanics of the components of a cellular biological medium, beginning with individual biomolecules and moving up to subcellular structures and cells. This is presented with an emphasis on models of their interactions.

1.2.1 Biomacromolecules

The three-dimensional molecular structure of proteins, nucleic acids and other biomolecules significantly affects their biological functionality and, in turn, the related cellular functions. Quite accurate reconstructions of the molecular structure of biomolecules can be experimentally obtained with X-ray crystallography and NMR spectroscopy. Thereafter, the structural data can be used as input parameters into the computational molecular dynamics methods, which provide predictions for the motion and dynamics of individual molecules and of molecular groups.

Within the framework of the molecular dynamics method, the time evolution of a system of N interacting particles (atoms or groups of atoms) can be tracked through the numerical integration of Newton's equations of motion:

$$m_i \frac{d^2 \mathbf{r}_i}{dt^2} = \mathbf{F}_i(t) \quad \text{for } i = 1, 2, \ldots, N \tag{1.1}$$

where m_i is the mass, $\mathbf{r}_i$ is the position vector and $\mathbf{F}_i$ is the total force acting on the ith particle at the time instant t. It is customary to assume that $\mathbf{F}_i = -\nabla U_i$, where U_i is the so-called potential function, which measures the energy of interaction between the ith particle and the other particles of the system [74]. The potential function obtains contributions from bonded interactions (e.g., stretching and bending between two atoms in a molecule) and non-bonded interactions (e.g., van der Waals and Coulomb). The vast amount of information about the position and velocity of each of the particles making up a molecular system can be filtered properly in the context of statistical mechanics in order to establish estimates for average state variables (e.g., pressure, chemical potential), as well as transport and mechanical properties (e.g., mass diffusivity, fluid viscosity).

An interesting example is the protein actin, which is a major structural component of the eukaryotic cytoskeleton. Briefly, the actin monomer (G-actin) is a globular protein, which is activated by binding ATP (adenosine-5'-triphosphate) and forms oligomers and filaments (F-actin). Hydrolysis of the bound ATP to ADP (adenosine diphosphate) leads to destabilization and depolymerization of the filaments. A specific region in the three-dimensional structure of actin, which is called a D-loop and serves as binding site for the enzyme DNase I, undergoes a transition from a *loop* in ATP-bound G-actin to a short α-*helix* in ADP-bound G-actin. It has been proposed that this loop-to-helix conformational transition might be responsible for the destabilization of F-actin after ATP hydrolysis. To this end, *Chu and Voth* [14] performed molecular dynamics simulations for short filaments of actin and found that the helical conformation of the D-loop results in wider, shorter, more disordered filaments with almost half the persistence length[3] as compared to filaments with the loop conformation of the D-loop.

With currently available computer power, a molecular dynamics simulation of a system with dimensions of a few nanometers is possible at times of several hundreds of nanoseconds. The study of larger molecular systems for longer time-spans has been achieved with hybrid multiscale methods, and also with so-called coarse-grained methods (e.g., Brownian Dynamics). For example, *Fyta et al.* [25] introduced a hybrid multiscale method for the computer-aided simulation of fluid–biomolecule interactions, which combines molecular dynamics simulation for the polymer dynamics with Lattice Boltzmann simulation for the fluid dynamics. Two additional contributions were included in the force acting on each particle (Eq. 1.1): a deterministic frictional force proportional to the relative velocity of the fluid with respect to the polymer, and a stochastic force in the form of Gaussian noise that accounts for the random fluctuations of the fluid. They applied this approach to the translocation of DNA through a nanopore, and found that the hydrodynamic interactions decrease the translocation time.

1.2.2 Subcellular Structures

1.2.2.1 Glycocalyx

The glycocalyx is a highly-hydrated fibrous meshwork of carbohydrates that projects out and covers the membrane of endothelial cells, many bacteria and other cells. Proteoglycans and glycoproteins are generic structural components of a glycocalyx, but the precise biochemical composition and structure is determined by the specific cell type and the prevailing mechanical and physicochemical conditions. It has been proposed that the glycocalyx serves several important functions. For example, the bacterial glycocalyx mediates cell attachment, retains humidity during exposure to dry environments, protects against molecular and cellular antibacterial agents (antibiotics, surfactants, bacteriophages, phagocytes) and other vital functions [15]. Likewise, the endothelial glycocalyx regulates the vascular permeability, modulates the interactions between blood and endothelial cells and transmits physical forces to the cytoskeleton of endothelial cells [75].

Of particular interest is the function of the endothelial glycocalyx as a transducer of fluid shearing forces [97]. Specifically, it has been long known that fluid shear stress strongly affects the morphology and function of endothelial cells. For instance, fluid shear modulates the production of the biochemical signal nitric oxide, which 'informs' the muscle cells around a blood vessel to contract or relax thus constricting or widening the vessel, so as to regulate the flow of blood. However, the presence of the glycocalyx results in negligible shear acting on the outer surface of the endothelial membrane, and therefore gives rise to the question of how are these physical forces transmitted across the glycocalyx. To this end, *Weinbaum et al.* [97] developed a simple model, which combines the Brinkman equation for flow through the porous glycocalyx with a discrete representation of core proteins as cantilever beams. With

[3] The persistence length is defined as the distance between successive points of change in the tangent direction of a macromolecular chain. It is a measure of the stiffness of the macromolecule and is related to Young's modulus for the bulk (macroscopic) material.

this model, they analyzed several glycocalyx-mediated interactions, and proposed that bush-like formations of core proteins in the glycocalyx serve as *lever arms*, which amplify and convert fluid shearing stresses at the blood–glycocalyx interface to deformations at the glycocalyx–cytoskeleton interface.

1.2.2.2 Cell Membrane

The plasma membrane consists of phospholipids and proteins, and its primary role is to regulate the rate of transport of chemical substances between the extracellular and the intracellular regions. Moreover, it modulates the cell shape in combination with the intracellular matrix. At the molecular scale of observation, the plasma membrane exhibits some very interesting properties [35]:

- Dynamic rearrangement of proteins and phospholipids via lateral diffusion,
- Spatial heterogeneity with regard to molecular structure and functionality,
- Asymmetric structure with varying composition and arrangement of proteins at the two opposite faces of the membrane.

The exact mechanism of transport across the cell membrane depends on the chemical species being transported and the cell type. Small and/or non-polar molecules (such as oxygen, carbon dioxide, urea, etc.) can diffuse spontaneously through the lipid bilayer down their concentration gradient (simple passive diffusion). On the other hand, the transport of large and/or polar molecules as well as ions (e.g., potassium, glucose, amino acids) through the lipid bilayer is prohibited by hydrophobic and steric interactions and is achieved only through the expenditure of biochemical energy (active transport). Certain proteins facilitate the transport of specific molecules, either by providing a narrow channel for the molecule to pass through (channel proteins, also referred to as permeases or porins), or by binding to the molecule on one side of the membrane and releasing it on the other side (carrier proteins).

Most types of bacterial cells exhibit additional polymeric structures beyond their plasma membrane (excluding the glycocalyx and ECM). In particular, the so-called Gram-positive bacteria have a thick peptidoglycan mesh tethered to their plasma membrane by polysaccharide chains (teichoic acid). On the other hand, the plasma membrane of Gram-negative bacteria is surrounded by a thin peptidoglycan layer and a second (outer) plasma membrane. *Mycobacteria* constitute an intriguing example with regard to their extra-membranous structure, because they do not strictly fit into either the Gram-positive or the Gram-negative category. Specifically, the plasma membrane of mycobacteria is surrounded by a thick mesh of peptidoglycans, which in turn, is enclosed by an additional peripheral layer that consists mainly of complexes between polysaccharides (arabinogalactan) and long-chain fatty acids (mycolic acids) [64]. This unique feature has been associated with the virulence and inherent drug resistance of mycobacteria, and is accordingly considered a crucial factor for rational drug design. Along this line, *Hong and Hopfinger* [34] performed molecular dynamics simulations of the motion of thirteen drugs in tightly packed chains of mycolic acid. The selected drugs covered a wide range of molecular size, shape, degree of lipophilicity and activity against *M. tuberculosis*. Their results highlighted the important role of molecular shape. For instance, a comparison of the drugs clofazimine and thiocarlide, which have similar molecular weight, shows that the bulky-shaped clofazimine diffuses at a slower rate than the flat-shaped thiocarlide.

1.2.2.3 Intracellular Matrix

The cytoskeleton is a dynamic, three-dimensional, interconnected network of filamentous proteins embedded within the cytoplasm of eukaryotic cells. The fundamental functions of the cytoskeleton are involved in modulating the shape of the cell, providing mechanical strength and integrity, enabling the movement of cells and facilitating the intracellular transport of supramolecular structures, vesicles and even organelles. Three main types of structural units comprise the cytoskeleton, each one associated with a specific protein [22]:

i. *Microfilaments*, which are single-helical strands of 7–9 nm in diameter and are mainly composed of G-actin subunits.
ii. *Intermediate filaments*, which are composite twisted-together strands, 8–12 nm in diameter and composed of keratins, vimentins and other keratin-like subunits.
iii. *Microtubules*, which are hollow cylindrical structures of about 24 nm in diameter and composed of α- and β-tubulin.

Although bacteria were long considered to lack a cytoskeleton, research over the last fifteen years [92] has provided evidence for the existence of prokaryotic homologues of all the major structural components in the eukaryotic cytoskeleton (e.g., the protein *MreB* is considered as a homologue of actin).

Several recent experimental studies have been concerned with the *in vitro* reconstruction and mechanical characterization of actin networks. For example, *Lieleg et al.* [53] investigated the effect of a crosslinking agent, α-actinin, on the emergent pattern formation of actin networks *in vitro*. Interestingly, the resulting networks exhibited a remarkable polymorphism depending on the concentration ratio of α-actinin to actin. At low concentration ratios of α-actinin/actin, networks of entangled filaments were formed. As the concentration ratio of α-actinin/actin increased, the degree of crosslinking also increased, and bundles of filaments appeared in the network. Above a threshold value of this concentration ratio, a highly heterogeneous pattern was formed with star-like clusters of actin filaments. In addition, these authors proposed that three-dimensional networks of Timoshenko beams could be used for the theoretical description of the mechanical behavior of actin networks.

1.2.3 Biological Cells

The experimentally observed mechanical response of biological cells has been found to range from solid-like to fluid-like, depending primarily on features of the cytoskeleton and of the membrane of the cell under consideration, as well as on the time scale of observation [44]. For time scales spanning a few milliseconds up to several tenths of seconds, a cell behaves as an elastic material, as for example during instantaneous probing with the cantilever of an atomic force microscope. At longer time scales, creep behavior might be observed, as for example during micropipette aspiration experiments [56]. Both micro-continuum models and physically-based discrete models have been used to describe individual cell mechanics.

Perhaps more than for any other cell, the mechanics of the red blood cell has been studied extensively, primarily because of its structural simplicity. It is commonly characterized as a 'bag of hemoglobin' because of its high hemoglobin content (a mobile carrier protein for oxygen) and its lack of most intracellular organelles (mitochondria, nucleus and other).

The membrane and the cytoskeleton, which is composed of the protein spectrin, largely determine the mechanical behavior of red blood cells. An interesting phenomenon associated with the red blood cell is the so-called Fahraeus–Lindqvist effect. Specifically, for blood vessels characterized by diameters within the range of approximately 30 to 300 μm, the apparent viscosity of whole blood (including both plasma and cells) decreases with decreasing vessel diameter until it reaches a plateau, after which it increases again for capillaries with size less than 10 μm [73].

This phenomenon, which has been observed both *in vitro* and *in vivo*, is explained as follows. In large blood vessels, the red blood cell performs a three-dimensional tumbling motion characterized by significant viscous dissipation. As the blood vessels narrow, the hydrodynamic interactions between red cells, plasma and vessel walls constrain the motion of red cells and, eventually, force them to move along a single file line, leaving a cell-free layer of plasma between the cell path and the vessel wall. This structural transition is accompanied by an overall decreased resistance to flow of whole blood in the vessel. Several theoretical models have captured certain aspects of this phenomenon. For example, *Pozrikidis* [72] investigated numerically the motion of a file of red blood cells through a straight vessel using a boundary integral method, for which each individual cell was modeled as a hyperelastic continuum capsule. The model predicted the experimentally observed plateau in the apparent viscosity for flows when the vessel diameters were comparable to the size of the red cell. In an alternative approach, *Dzwinel et al.* [19] developed a discrete particle model to simulate the motion of red cells and plasma within a deformable vessel. Interestingly, for narrow vessels, their simulations also reproduced the spontaneous migration of red cells toward the center of the vessel and the presence of the plasma layer near the vessel wall.

1.3 FORMULATION OF BALANCE LAWS AND CONSTITUTIVE EQUATIONS

The complex nature of cellular biological media sets up a challenge for the theoretical formulation of mass and momentum conservation laws along with the necessary constitutive relations at the scale of the cellular biological medium. Information stemming from the multiphasic–multicomponent composition, the dynamically evolving hierarchical structure and a mechanical behavior that falls somewhere in the spectrum between fluid-like (blood, lymph, etc.) and solid-like (bone, muscle, skin, etc.), must be analyzed and filtered up to some extent in order to develop reliable and useful models of the real phenomena. In the following subsections, we present the conceptual framework and sample equations for several different approaches that have been used to tackle down this problem (Fig. 1.2).

1.3.1 Single-Scale, Single-Phase Approaches

One commonly used approach is to disregard the finer spatial scales of the structure and consider the cellular biological medium as a *monophasic* continuum at the macroscopic scale (e.g., the B-scale in Fig. 1.1). In this case, the mechanical behavior of the bulk material (and the

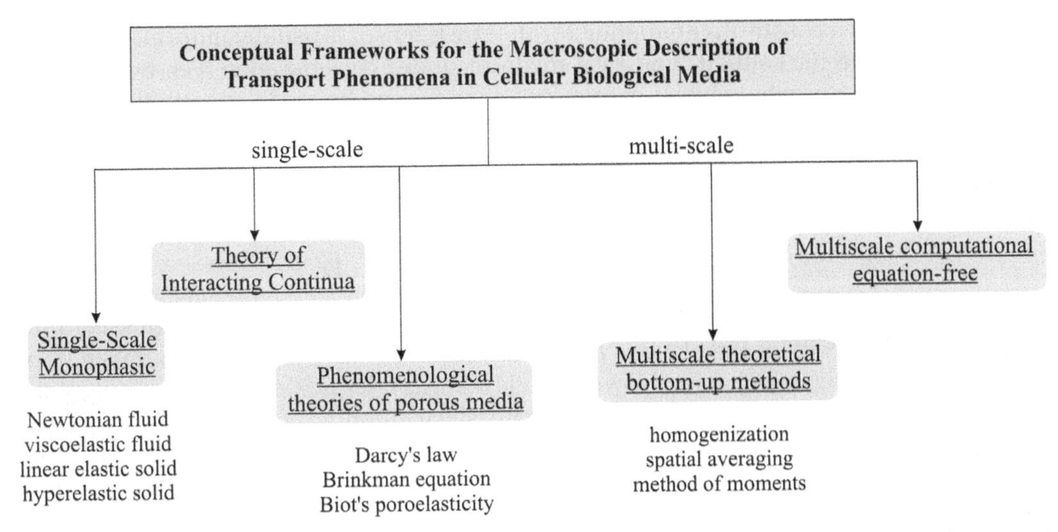

FIGURE 1.2 Overview of the conceptual frameworks for the macroscopic description of transport phenomena in cellular biological media.

transport of chemical species within it) may be described by conservation laws in the standard (differential) form for single-phase materials. For instance, conservation of mass:

$$\frac{\partial \rho}{\partial t} + \nabla \cdot (\rho \mathbf{v}) = 0 \tag{1.2}$$

conservation of linear momentum:

$$\frac{\partial \rho \mathbf{v}}{\partial t} + \nabla \cdot (\rho \mathbf{v}\mathbf{v}) = \nabla \cdot \boldsymbol{\sigma} + \mathbf{f}_b \tag{1.3}$$

conservation of solute:

$$\frac{\partial C_A}{\partial t} + \nabla \cdot \mathbf{J}_A = R_A \tag{1.4}$$

Here, ρ is the density, $\mathbf{v}$ is the velocity, $\boldsymbol{\sigma}$ is the Cauchy stress tensor, $\mathbf{f}_b$ is a body force for the cellular biological medium and C_A is the solute concentration, $\mathbf{J}_A$ is the solute flux, R_A is a reaction rate for the Ath chemical substance. Typically, the solute flux might include contributions from diffusion caused by the concentration gradient (or, more strictly, the gradient in chemical potential), from convection caused by bulk motion, from an imposed electrical field and other forces. For example:

$$\mathbf{J}_A = \lambda_{Ac} C_A \mathbf{v} - D_A \nabla C_A - \frac{z_A F_c}{R_g T} D_A C_A \nabla \mathcal{V}_{el} \tag{1.5}$$

where D_A is the diffusion coefficient, λ_{Ac} is the convection hindrance factor, z_A is the charge of the Ath chemical substance, F_c is the Faraday constant, R_g is the ideal gas constant, T is the absolute temperature and $\mathcal{V}_{el}$ the electrical potential.

The definition of a constitutive equation for the stress tensor is usually more elaborate and depends on the specific cellular biological medium under consideration. For example, the rheological behavior of whole blood in arteries and veins can be modeled with the constitutive equation for an incompressible generalized Newtonian fluid:

$$\boldsymbol{\sigma} = -p\mathbf{I} + 2\eta\,(\dot{\gamma})\,\mathbf{D} \tag{1.6}$$

$$\mathbf{D} = \frac{1}{2}\left[\nabla\mathbf{v} + (\nabla\mathbf{v})^T\right]; \quad \dot{\gamma} = \sqrt{2\mathbf{D}:\mathbf{D}} \tag{1.7}$$

where p is the pressure, $\mathbf{D}$ is the rate of strain tensor and η is the viscosity. At high shear rates ($\dot{\gamma} > 50\ \mathrm{s}^{-1}$) the viscosity of the blood is practically constant (ca. 4×10^{-3} Pa s), thus retrieving the simple Newtonian behavior. However, at low shear rates ($<50\,\mathrm{s}^{-1}$) the apparent viscosity of the blood decreases with increasing shear rate (shear-thinning effect), while at extremely low shear rates it might even exhibit a yield point [98]. Attempts to capture the non-Newtonian behavior of blood at low shear rates have been made, for example using the empirical correlation Carreau-Yasuda [27]:

$$\frac{\eta - \eta_\infty}{\eta_0 - \eta_\infty} = \left[1 + \left(\lambda_\eta \dot{\gamma}\right)^\alpha\right]^{(n-1)/\alpha} \tag{1.8}$$

where $\eta_0, \eta_\infty, \lambda_\eta, \alpha$ and n are system-specific parameters.

On the other end of the spectrum of material behavior, bone is a characteristic example of the application of the linear elasticity theory. The constitutive equation for the stress tensor is:

$$\boldsymbol{\sigma} = \lambda_s tr(\boldsymbol{\varepsilon})\mathbf{I} + 2\mu_s\boldsymbol{\varepsilon} \tag{1.9}$$

$$\boldsymbol{\varepsilon} = \frac{1}{2}\left[\nabla\mathbf{u} + (\nabla\mathbf{u})^T\right] \tag{1.10}$$

where $\mathbf{u}$ is the displacement, $\boldsymbol{\varepsilon}$ is the strain tensor, λ_s is the first Lamé constant and μ_s is the shear modulus for the elastic material. The representation of bone as a linear elastic solid has some usefulness for analyzing the stress distribution under various load bearing conditions and determining potential sites for fracture development. Nonetheless, there exists a significant amount of *free fluid* that can flow in the porous microstructure of bone. It was the recognition of the importance of interstitial fluid flow in bone remodeling processes that motivated the development of *biphasic* models, which are presented in the following subsections.

On the other hand, for instance, in muscular tissues most of the extracellular fluid is *bound* by extracellular polymers (mainly GAGs) and, therefore, a monophasic macroscopic description seems reasonable. The theory of nonlinear hyperelasticity, which describes an elastic solid undergoing large deformations, has found significant acceptance and success in its application to muscle and other soft tissues. In general, a solid material retains memory of a reference configuration to which it returns when the applied loads are removed. This essential information can be incorporated into formulated constitutive equations via the definition of an appropriate function that measures the energy stored in the deformed material. For example, let Ω_0 denote a reference stress-free configuration of the material, and Ω denote the current (deformed) configuration. Further, let $\mathbf{x} = \chi\ (\mathbf{X})$ denote the mapping from a point $\mathbf{X}$ in the reference configuration to a point **x** in the deformed configuration. The following (second order) tensors are defined:

$$\mathbf{F} \equiv \frac{\partial \chi(\mathbf{X})}{\partial \mathbf{X}} = \nabla_x \chi(\mathbf{X}) \tag{1.11}$$

$$\mathbf{C} = \mathbf{F}^T \cdot \mathbf{F}, \quad \mathbf{B} = \mathbf{F} \cdot \mathbf{F}^T, \quad \mathbf{E} = \frac{1}{2}(\mathbf{C} - \mathbf{I}) \tag{1.12}$$

Here, **F** is the deformation gradient tensor, **E** is the Green–Lagrange strain tensor, **I** is the second order unit tensor, **C** and **B** are the right and left Cauchy–Green strain tensors, respectively. Now, the basic idea is to consider that there exists a Helmholtz free energy function Ψ, which can be expressed in terms of some measure of the deformation [32,33]. A possible choice is to use the invariants of the left Cauchy–Green tensor:

$$I_1 = tr(\mathbf{B}) \tag{1.13a}$$

$$I_2 = \frac{1}{2}\left[(tr(\mathbf{B}))^2 - tr(\mathbf{B} \cdot \mathbf{B})\right] \tag{1.13b}$$

$$I_3 = \det(\mathbf{B}) = (\det(\mathbf{F}))^2 \tag{1.13c}$$

Several commonly used expressions for $\Psi(I_1, I_2, I_3)$ are given in Table 1.2. The Helmholtz free energy is related to the Cauchy stress tensor $\boldsymbol{\sigma}$ as follows:

$$\boldsymbol{\sigma} = J^{-1}\mathbf{F} \cdot \mathbf{S} \cdot \mathbf{F}^T \tag{1.14}$$

$$\mathbf{S} = 2\frac{\partial \Psi}{\partial \mathbf{C}} = \frac{\partial \Psi}{\partial \mathbf{E}} \tag{1.15}$$

where $J = \det(\mathbf{F})$ and **S** is the second Piola–Kirchhoff stress tensor. For isotropic materials, the Cauchy stress tensor can be expressed in terms of the left Cauchy–Green tensor [33, Eq. 6.34, p. 217] as follows:

$$\boldsymbol{\sigma} = 2J^{-1}\left[I_3\frac{\partial \Psi}{\partial I_3}\mathbf{I} + \left(\frac{\partial \Psi}{\partial I_1} + I_1\frac{\partial \Psi}{\partial I_2}\right)\mathbf{B} - \frac{\partial \Psi}{\partial I_2}\mathbf{B} \cdot \mathbf{B}\right] \tag{1.16}$$

TABLE 1.2 Some Common Expressions for the Helmholtz Free Energy of Isotropic Hyperelastic Materials (μ_s Is the Shear Modulus, and C_k Denotes a Material Parameter – Different for Each Model)

Expression for the free energy	Comments
$\Psi(I_1) = C_1(I_1 - 3)$	neo-Hookean model
$\Psi(I_1, I_2) = C_1(I_1 - 3) + C_2(I_2 - 3)$	Mooney-Rivlin model
$\Psi(I_1) = C_1(I_1 - 3) + C_2(I_1 - 3)^2 + C_3(I_1 - 3)^3$	Yeoh model
$\Psi(I_1) = \mu_s \sum_k \frac{C_k}{n^{k-1}}\left(I_1^k - 3^k\right)$	Arruda-Boyce model n: material parameter (number of links per chain)
$\Psi(\lambda_1, \lambda_2, \lambda_3) = \sum_k \frac{\mu_s}{\alpha_k}\left(\lambda_1^{\alpha_k} + \lambda_2^{\alpha_k} + \lambda_3^{\alpha_k} - 3\right)$	Ogden model λ_i: principal stretches

1.3.2 Biot's Theory of Poroelasticity

Motivated by the problem of soil subsidence, *Biot* [8] formulated a phenomenological theory on the consolidation of deformable porous media saturated with a fluid. In that pioneering work, he considered the porous medium as homogeneous and isotropic at the macroscale, and treated the solid matrix as a linearly elastic solid and the fluid, which occupies the pore space, as incompressible and Newtonian. Later, he extended his theory by taking into account the effects of anisotropy, viscoelasticity for the solid matrix and inertia for the fluid (a summary of his work can be found in [9]). Biot's theory of poroelasticity has been used to predict the mechanics of bone, brain, solid tumors and polymeric hydrogels (see [41] and references therein).

A cellular biological medium can be considered as a poroelastic material consisting of an elastic solid skeleton (cells and ECM) and a pore structure that is well interconnected and occupied by the extracellular free fluid. In the context of Biot's theory of poroelasticity, momentum transport is described by the following equations [4,8]:

$$\nabla \cdot \boldsymbol{\sigma}_{eff} = \mathbf{0} \tag{1.17}$$

$$\mathbf{v} = -\frac{k_{eff}}{\mu_f} \nabla p \tag{1.18}$$

$$\frac{\partial \theta}{\partial t} = -\nabla \cdot \mathbf{v} \tag{1.19}$$

with:

$$\boldsymbol{\sigma}_{eff} = \mu_{eff} \left[\nabla \mathbf{u} + (\nabla \mathbf{u})^T \right] + \lambda_{eff} e \mathbf{I} - \alpha p \mathbf{I} \tag{1.20}$$

$$\theta = \alpha e + p/M \tag{1.21}$$

where $\boldsymbol{\sigma}_{eff}$ is the 'effective' stress tensor for the entire material (fluid plus solid), $\mathbf{u}$ is the displacement of the material, p is the fluid pressure, $e = \nabla \cdot \mathbf{u}$ is the material dilatation, $\mathbf{v}$ is the fluid velocity, μ_{eff} and λ_{eff} are the Lamé parameters, μ_f is the fluid viscosity, k_{eff} is the permeability of the material, θ is the variation of the fluid volume fraction (*'variation in water content'*) and α, M are phenomenological coefficients (also known as Biot or Biot–Willis coefficients). The coefficient α expresses the ratio of the variation of pore volume to the dilatation of the solid, under constant total volume conditions. Usually, the fluid and the solid matrix are considered to be intrinsically incompressible and, thus, the values $\alpha = 1$ and $1/M = 0$ are used [4].

1.3.3 Theory of Interacting Continua

The theory of interacting continua, or mixture theory, was developed for the mathematical description of transport phenomena in multicomponent systems at a scale of observation with resolution much larger than the characteristic length of the individual constituents.

The structure of the system at finer spatial scales is not taken into account explicitly. *Truesdell* [90] developed the mathematical foundation of the theory on the conceptual basis that the constituents of the mixture can be modeled as superimposed, interacting continua. A material point is assigned for each constituent at every point in space that is occupied by the mixture. The governing conservation laws for each constituent of the mixture contain the usual terms that appear in monophasic formulations, and an additional term which accounts for the interaction between the reference constituent and the other constituents of the mixture. A thorough review of the theory of mixtures is given, for instance, in [6].

The theory was introduced to the field of biomaterials by *Mow et al.* [63] as a model for the mechanical behavior of cartilage. These authors showed that a biphasic mixture theory (linearly elastic solid, Newtonian fluid) captures very well the experimentally observed behavior of articular cartilage under confined compression. In order to achieve this agreement between theory and experiment, it was necessary to express the permeability of the cartilage as a function of the solid dilatation. Another key finding of their work is that the tissue scale viscoelastic behavior (creep, stress relaxation) can be observed as a result of the relative motion between the two constituents, even if none of the constituents are viscoelastic. Thereafter, the theory has been extended and used widely to predict the mechanics of cartilage, endothelial glycocalyx, tumors, hydrogels, soft tissues in general and filter cakes (see [41] and references therein). In the framework of mixture theory, a significant effort has been made to incorporate electrokinetic phenomena, the dependence of the permeability on solid dilatation, finite solid deformation and viscoelastic behavior. As an example, the equations of the triphasic mixture theory of *Lai et al.* are presented below [48,87]:

$$\frac{\partial(\rho_\alpha \varphi_\alpha)}{\partial t} + \nabla \cdot (\rho_\alpha \varphi_\alpha \mathbf{v}_\alpha) = 0 \quad (\alpha = s, w, +, -) \tag{1.22}$$

$$\nabla \cdot \boldsymbol{\sigma}_{mix} = \mathbf{0} \tag{1.23}$$

$$-\rho_\alpha \varphi_\alpha \nabla \mu_\alpha + \sum_\omega f_{\alpha\omega} (\mathbf{v}_\omega - \mathbf{v}_\alpha) = 0 \quad (\alpha = w, +, -) \tag{1.24}$$

$$\nabla \cdot \mathbf{i}_e = 0 \tag{1.25}$$

$$\varphi_s = \frac{\varphi_{s0}}{1 + tr(\boldsymbol{\varepsilon})}; \quad \varphi_s + \varphi_w + \varphi_+ + \varphi_- = 1 \tag{1.26}$$

with:

$$\boldsymbol{\sigma}_{mix} = -p\mathbf{I} - T_c\mathbf{I} + \lambda_s tr(\boldsymbol{\varepsilon}) + 2\mu_s \boldsymbol{\varepsilon}; \quad \boldsymbol{\varepsilon} = \frac{1}{2}\left[\nabla \mathbf{u} + (\nabla \mathbf{u})^T\right] \tag{1.27}$$

$$\mathbf{i}_e = \varphi_w F_c \left[C_+ (\mathbf{v}_+ - \mathbf{v}_s) - C_- (\mathbf{v}_- - \mathbf{v}_s)\right] \tag{1.28}$$

$$\mu_w = \mu_{w0} + \left[p - R_g T \phi_{osm} (C_+ + C_-) + B_w tr(\boldsymbol{\varepsilon})\right] / \rho_w \tag{1.29}$$

$$\mu_\alpha = \mu_{\alpha 0} + (R_g T/M_\alpha) \ln(\gamma_\alpha C_\alpha) + z_\alpha F_c V_{el}/M_\alpha \quad (\alpha = +, -) \tag{1.30}$$

$$T_c = \frac{\alpha_0 c_0^F}{1 + tr(\boldsymbol{\varepsilon})/\varphi_{w0}} \exp\left[-\kappa \left(\frac{\gamma_+ \gamma_-}{\gamma_+^* \gamma_-^*} C_+ C_- \right)^{1/2} \right] \tag{1.31}$$

Here, φ_α is the volume fraction, $\mathbf{v}_\alpha$ is the velocity, ρ_α is the density, μ_α is the electrochemical potential, C_α is the molar concentration of the αth constituent, $\boldsymbol{\sigma}_{mix}$ is the total stress tensor of the mixture, $\mathbf{i}_e$ is the electric current, p is the fluid pressure, $\mathbf{u}$ is the displacement, T_c is the chemical expansion stress, $f_{\alpha\omega}$ is the friction coefficient between the αth and ωth constituents per unit tissue volume ($f_{\alpha\omega} = f_{\omega\alpha}$), μ_s and λ_s are the Lamé parameters for the solid, ϕ_{osm} is an osmotic coefficient, z_α is the charge, M_α is the molecular weight, γ_α is the activity coefficient in the tissue, γ_α^* is the activity coefficient in the external bath of the αth ion, V_{el} is the electric potential, c_0^F is the fixed charge density and κ, B_w, α_0 are material coefficients. Note that the subscript '0' denotes reference to the stress-free configuration of the tissue.

1.3.4 Multiscale Bottom-Up Approaches

A robust approach that accounts for both the multiphasic and hierarchical features of cellular biological media requires the development of governing equations for the coarser scale of observation (tissue scale) by starting from 'well established' first principles on finer scales of the system (cellular or molecular scale). For this purpose, one can implement mathematical procedures, including the asymptotic homogenization method and volume averaging methods, among others, which were initially applied to the analysis of transport phenomena and mechanics of porous or composite materials, in the context of continuum mechanics.

These upscaling methods offer two significant advantages over single-scale approaches. First, one can obtain precise correspondence between theoretical variables defined at different spatial scales, as well as between theoretical variables and the respective experimentally measured quantities (via the use of appropriate weight functions in the averaging procedure). Second, a framework is provided for the calculation and correlation of constitutive parameters defined at the coarse spatial scale with the structure of and the phenomena occurring at finer spatial scales. Thereby, a direct comparison between theory and experiment can be accomplished without the use of adjustable parameters.

In any case, the derived upscaled equations should satisfy two prerequisites. First, these must contain only *measurable* average variables (e.g., concentration, velocity, displacement) and a minimum number of *physically meaningful* constitutive parameters (e.g., diffusion coefficient, elastic moduli, hydraulic permeability). Second, the spatiotemporal evolution of the average variables, which is determined from the solution of the upscaled equations, should match the corresponding field that would be obtained if a complete description at the molecular scale was feasible. This prerequisite is usually verified through comparison with data from well-controlled laboratory or computational experiments.

A relatively limited amount of work has been done regarding the analysis of transport processes in cellular biological media using upscaling methods. The first applications of the volume averaging [66] and homogenization [10] methods appeared in the 1980s and were focused on diffusion. Later, the asymptotic homogenization method was used to derive macroscopic equations, and also to correlate the apparent elastic moduli of bone with its microstructure [16,31]. More recently, these upscaling methods have been used for systems with 2 scales/2 phases [52,101] and also 3 scales/3 phases [40,43].

In the following paragraphs, we outline the conceptual basis and give examples for two upscaling methods: the asymptotic homogenization and the spatial averaging methods. Both methods begin with the formulation of the governing equations at the microscale. For example, let ψ_α denote a microscale variable of interest, which is associated with the αth phase. The spatiotemporal evolution of ψ_α can be described mathematically by:

$$\mathcal{T}_m \{\psi_\alpha(\mathbf{x}, t)\} = \mathcal{F}_m(\mathbf{x}, t) \tag{1.32}$$

where $\mathcal{T}_m$ denotes a generic (linear differential) operator, $\mathcal{F}_m$ denotes a generic source term, t is time, $\mathbf{x} = \mathbf{x}^*/\ell_\alpha^m$ is a dimensionless spatial variable with resolution at the microscale and ℓ_α^m is a characteristic length associated with the αth phase at the microscale. Further, let L^M be a characteristic length associated with a well-separated coarser scale of observation ($L^M >> \ell_\alpha^m$). In many cases, one is encountered with the solution of Eq. (1.32) in a domain with *extent* $\mathcal{O}\left(L^M\right)$ and *resolution* $\mathcal{O}\left(\ell_\alpha^m\right)$, which poses a formidable task even with current computational resources. Therefore, it is desired to establish an 'equivalent' mathematical description in terms of averaged variables and of resolution $\mathcal{O}\left(L^M\right)$.

1.3.4.1 Asymptotic Spatial Homogenization

In the context of the homogenization method, the microscale structure is assumed to be periodic. Let $\mathcal{Y} \subset \mathbb{R}^3$ denote a representative periodic unit cell. The general idea is to express the dependent variables in terms of some 'small' (perturbation) parameters and multiple independent spatial variables, each one associated with a different scale [81]. The perturbation parameters are defined as the ratios of characteristic lengths. Subsequently, by means of standard perturbation analysis, the microscale problem can be decomposed into a set of subproblems, which by the application of an averaging operator produce the upscaled equations. A two-scale problem involves a single perturbation parameter $\epsilon = \ell_\alpha^m/L^M << 1$ and two dimensionless spatial variables $\mathbf{x}^m = \mathbf{x}^*/\ell_\alpha^m$ and $\mathbf{x}^M = \mathbf{x}^*/L^M$, which are associated with the microscale and macroscale, respectively. Furthermore, it is assumed that the microscale variable $\psi_\alpha(\mathbf{x})$ can be approximated by an expansion in powers of ϵ as follows:

$$\psi_\alpha(\mathbf{x}) \to \psi_\alpha^\epsilon(\mathbf{x}^M, \mathbf{x}^m; \epsilon) = \psi_\alpha^{(0)}(\mathbf{x}^M, \mathbf{x}^m) + \epsilon\psi_\alpha^{(1)}(\mathbf{x}^M, \mathbf{x}^m) + \epsilon^2\psi_\alpha^{(2)}(\mathbf{x}^M, \mathbf{x}^M) + \cdots \tag{1.33}$$

where $\psi_\alpha^{(i)}(\mathbf{x}^M, \mathbf{x}^m)$ are $\mathcal{Y}$-periodic functions. Upon this consideration, the derivatives must be replaced accordingly:

$$\frac{\partial}{\partial x_i} \to \frac{\partial}{\partial x_i^m} + \epsilon\frac{\partial}{\partial x_i^M} \tag{1.34}$$

Introduction of the expressions given above in Eq. (1.32) results in a polynomial in ϵ, in which the coefficients must be zero. Thus, a set of subproblems is obtained:

$$\mathcal{O}\left(\epsilon^0\right): \quad \mathcal{T}_m^{(0,0)}\left\{\psi_\alpha^{(0)}(\mathbf{x}^M, \mathbf{x}^m)\right\} = \mathcal{F}_m^{(0)}(\mathbf{x}^m, t) \tag{1.35a}$$

$$\mathcal{O}\left(\epsilon^{1}\right): \qquad \mathcal{T}_{m}^{(1,1)}\left\{\psi_{\alpha}^{(1)}(\mathbf{x}^{M},\mathbf{x}^{m})\right\} + \mathcal{T}_{m}^{(1,0)}\left\{\psi_{\alpha}^{(0)}(\mathbf{x}^{M},\mathbf{x}^{m})\right\} = \mathcal{F}_{m}^{(1)}(\mathbf{x}^{m},t) \tag{1.35b}$$

$$\mathcal{O}\left(\epsilon^{n}\right): \sum_{i=0}^{n}\mathcal{T}_{m}^{(n,i)}\left\{\psi_{\alpha}^{(i)}(\mathbf{x}^{M},\mathbf{x}^{m})\right\} = \mathcal{F}_{m}^{(n)}(\mathbf{x}^{m},t) \quad n=2,3,\ldots \tag{1.35c}$$

Typically, on the basis that $\epsilon << 1$, it suffices to limit the examination of the subproblems to first-order analysis. Averaging both sides of Eq. (1.35b) over the volume of $\mathcal{Y}$ (i.e., with respect to the $\mathbf{x}^{m}$ spatial variable) results in an upscaled equation in terms of an average dependent variable $\{\psi_{\alpha}\}(\mathbf{x}^{M})$, and parameters that can be obtained from the solution of Eq. (1.32) in a representative periodic cell.

A recent example is the application of the asymptotic homogenization method to the study of electro-hydromechanical couplings in cortical bone, by *Lemaire et al.* [52]. They obtained a set of upscaled equations, which can be expressed in compact form as follows:

$$\nabla\cdot\boldsymbol{\sigma}_{tot} = \mathbf{0} \tag{1.36}$$

$$\mathbf{v} = -\mathbf{K}_{P}\cdot\nabla p_{b} - \mathbf{K}_{C}\cdot\nabla C_{b} - \mathbf{K}_{E}\cdot\nabla\overline{V}_{b} \tag{1.37}$$

$$\nabla\cdot\mathbf{v} = -\boldsymbol{\beta}^{*} : \frac{\partial\boldsymbol{\varepsilon}}{\partial t} + \delta^{*}p_{b} + q_{e}^{*} \tag{1.38}$$

$$\frac{\partial\varphi_{f}}{\partial t} + \nabla\cdot\mathbf{v} + \varphi_{f}\nabla\cdot\left(\frac{\partial\mathbf{u}}{\partial t}\right) = 0 \tag{1.39}$$

$$\frac{\partial\left(\varphi_{f}\Phi_{i}^{*}C_{b}\right)}{\partial t} = \nabla\cdot\left[\mathbf{D}_{i}^{*}\cdot\left(\nabla C_{b} + z_{i}C_{b}\nabla\overline{V}_{b}\right)\right] \quad \text{for } i=+,- \tag{1.40}$$

with:

$$\boldsymbol{\sigma}_{tot} = \mathbb{C}^{*} : \boldsymbol{\varepsilon} - \boldsymbol{\alpha}^{*}p_{b} + \boldsymbol{\tau}_{e}^{*}; \quad \boldsymbol{\varepsilon} = \frac{1}{2}\left[\nabla\mathbf{u} + (\nabla\mathbf{u})^{T}\right] \tag{1.41}$$

Here, $\boldsymbol{\sigma}_{tot}$ is the effective Cauchy stress tensor for the bone, $\mathbf{v}$ is the bulk fluid velocity, p_b is the bulk pressure, $\mathbf{u}$ is the bulk displacement, C_b is the bulk salinity, $\overline{V}_b = F_cV_b/R_gT$ is the bulk streaming potential, φ_f is the porosity, $\mathbb{C}^{*}$ is the effective elastic tensor (fourth order), $\boldsymbol{\tau}_e^{*}$ is an electrochemical stress tensor (second order), q_e^{*} is an electrochemical source term, Φ_i^{*}, z_i and $\mathbf{D}_i^{*}$ is a double-layer factor, the charge and the effective diffusion tensor of ith ionic species, respectively. The quantities $\boldsymbol{\tau}_e^{*}, q_e^{*}, \Phi_i^{*}$ must be calculated from the solution of specific closure problems in the context of a representative periodic unit cell. Moreover, the effective parameters $\mathbb{C}^{*}, \mathbf{D}_i^{*}, \boldsymbol{\alpha}^{*}, \boldsymbol{\beta}^{*}, \mathbf{K}_P, \mathbf{K}_C, \mathbf{K}_E, \delta^{*}$ can also be estimated theoretically from respective closure problems. The interested reader is directed to the paper of *Lemaire et al.* [52] for details.

1.3.4.2 *Spatial Averaging Method*

In the context of the spatial averaging method, a *point* on a coarse scale of observation is considered as the centroid of a volume element with a spatial resolution at a finer scale. If the volume element is sufficiently large to include all the phases present at the finer scale and much smaller than the extent of the coarse scale, then it is referred to as a representative elementary volume (REV). The general idea is to define a macroscale quantity by averaging the spatial distribution of its microscale counterpart over a REV with the help of an appropriate weight function. This approach was introduced, in a different context, in the averaging methods developed by *Anderson and Jackson* [2] in their study of momentum transport in fluidized beds, and by *Marle* [59,60] in his study on transport phenomena in porous media. Specifically, in Marle's formulation, the finer-scale quantity is defined as a generalized function (distribution) over the entire three-dimensional space, and the corresponding spatial average quantity is defined by the convolution product of the finer-scale distribution with a weight function. Thereafter, upscaled equations are obtained by applying the averaging operator to the microscale equations.

Let $m_w(\mathbf{x})$ denote a normalized, positive function with compact support that coincides with the REV. Further, let the generalized function corresponding to the microscale quantity ψ_α be denoted by $\bar{\psi}_\alpha$ and, for instance, defined as:

$$\bar{\psi}_\alpha = \psi_\alpha H_\alpha + \psi_{\alpha\omega}\delta_{\alpha\omega} \tag{1.42}$$

where H_α is a (Heaviside) function associated with the αth phase, and $\delta_{\alpha\omega}$ is the surface Dirac delta distribution concentrated on the $\alpha\omega$-interface [43]. The *superficial* and *intrinsic* spatial-averages of ψ_α at a macroscale point $\mathbf{x}^M$ are defined as:

$$\psi_\alpha = \left(m_w{*}\overline{\psi}_\alpha\right)(\mathbf{x}^M, t); \quad \varphi_\alpha\{\psi_\alpha\}^\alpha = \left(m_w{*}\overline{\psi}_\alpha\right)(\mathbf{x}^M, t) \tag{1.43}$$

respectively. Here, the symbol $*$ denotes convolution product and φ_α is the volume fraction of the αth phase. The two averages are related as $\{\psi_\alpha\} = \varphi_\alpha\{\psi_\alpha\}^\alpha$. The main steps involved in the derivation of upscaled equations can be represented in a generic scheme as follows:

$$\begin{aligned} \mathcal{T}_m\left\{\psi_\alpha(\mathbf{x}^m, t)\right\} = \mathcal{F}_m(\mathbf{x}^m, t) \overset{\bar{\psi}_\alpha}{\rightarrow} \overline{\mathcal{T}}_m\left\{\bar{\psi}_\alpha(\mathbf{x}^m, t)\right\} = \overline{\mathcal{F}}_m(\mathbf{x}^m, t) \\ \overset{m_w*}{\rightarrow} \mathcal{T}_M\left\{\{\psi_\alpha\}(\mathbf{x}^M, t)\right\} = \mathcal{F}_M(\mathbf{x}^M, t) \end{aligned} \tag{1.44}$$

In the first step, the microscale equations are expressed in terms of generalized functions, and thereafter the convolution product with the weight function is formed. The weight function serves as a mathematical representation of the influence of the measuring process. The practical importance of using a weight function in the spatial averaging method lies in the fact that it allows one to obtain a precise correspondence between the theoretical dependent variables and the corresponding experimentally measured quantities [5,17]. In addition, the use of weight functions allows one to relate experimental measurements obtained with different techniques, so as to acquire more precise information about the physical system under consideration.

Recently, *Kapellos et al.* [40,43] applied the spatial averaging method to the study of diffusive mass transfer and interstitial flow in deformable, porous cellular biological media exhibiting three characteristic length scales, with specific emphasis on microbial biofilms. The set of upscaled equations so obtained is:

$$\nabla \cdot \{\mathbf{v}_f\} = 0 \tag{1.45}$$

$$\mathbf{0} = \nabla \cdot \{\boldsymbol{\sigma}_f\} + \mathbf{F}_{s \to f} \tag{1.46}$$

$$\mathbf{0} = \nabla \cdot \{\boldsymbol{\sigma}_s\} - \mathbf{F}_{s \to f} \tag{1.47}$$

$$\frac{\partial}{\partial t}\left(K_{A\beta}\{C_{Af}\}^f\right) + \nabla \cdot \{\mathbf{J}_{A\beta}\} = \{R_{A\beta}\} \tag{1.48}$$

with:

$$\{\boldsymbol{\sigma}_f\} = -\varepsilon_\beta \{p_f\}^f \mathbf{I} + \mu_f \left[\nabla\{\mathbf{v}_f\} + \left(\nabla\{\mathbf{v}_f\}\right)^T\right] \tag{1.49}$$

$$\{\boldsymbol{\sigma}_s\} = \mu_s \left[\nabla\{\mathbf{u}_s\}^s + \left(\nabla\{\mathbf{u}_s\}^s\right)^T\right] + \lambda_s \mathbf{I}\left(\nabla \cdot \{\mathbf{u}_s\}^s\right) + (\textit{extra stresses}) \tag{1.50}$$

$$\mathbf{F}_{s \to f} = -\nabla \varepsilon_\beta \cdot \left\{-\{p_f\}^f \mathbf{I} + \mu_f \left[\nabla\{\mathbf{v}_f\}^f + \left(\nabla\{\mathbf{v}_f\}^f\right)^T\right]\right\} - \mu_f \varepsilon_\beta^2 \mathbf{K}_\beta^{-1} \cdot \{\mathbf{v}_f\}^f \tag{1.51}$$

$$\{\mathbf{J}_{A\beta}\} = -\mathbf{D}_{A\beta} \cdot \nabla\{C_{Af}\}^f \tag{1.52}$$

Here, the subscripts f, s, β denote the interstitial free fluid, the solid skeleton (cells and ECM) and the overall cellular biological medium, respectively. Further, $\{\sigma_\alpha\}$ is the superficial-average Cauchy stress tensor for the αth phase, $\{\mathbf{v}_f\}$ is the superficial-average velocity, $\{p_f\}^f$ is the intrinsic average pressure, μ_f is the dynamic viscosity of the interstitial fluid, $\{\mathbf{u}_s\}^s$ is the intrinsic average displacement, μ_s and λ_s are the effective Lamé parameters for the solid skeleton, $\{C_{Af}\}^f$ is the intrinsic average concentration of the Ath chemical species (per volume of interstitial fluid), ε_β is the volume fraction of interstitial free fluid, $\mathbf{K}_\beta$ is the hydraulic permeability of the cellular biological medium, $\mathbf{D}_{A\beta}$ is the effective diffusivity tensor and $K_{A\beta}$ is the equilibrium partition coefficient of the Ath species (between the cellular biological medium and an external bath made of interstitial free fluid). The effective parameters $\mathbf{K}_\beta$, $\mathbf{D}_{A\beta}$ can be calculated from the solution of specific closure problems in the context of a representative unit cell [40,43]. Typically, periodic unit cells are used for this purpose. However, recently *Kapellos et al.* [40] developed a consistent closure problem in the context of a non-periodic unit cell, which is based on an effective-medium approximation (also referred to as self-consistent or window method in the literature). The developed models for these parameters are presented in Section 1.4. Note that for uniform weighting, the incompressibility of the fluid imposes that $\nabla \varepsilon_\beta \cdot \left(\nabla\{\mathbf{v}_f\}^f\right)^T = 0$ and expression 1.51 should be simplified accordingly.

1.3.5 Multiscale Computational, Equation-Free Approaches

In many cases of practical interest, the behavior of a cellular biological medium cannot be captured satisfactorily over a wide range of working conditions using well established, *simple* constitutive equations. Single-scale methods usually rely on heuristic approaches, which stem from the modeler's experience along with trial and error, so as to extend the validity range of existing constitutive models. Further, the implementation of standard upscaling methods, such as homogenization and volume averaging, becomes a formidable task if there exist material components which exhibit nonlinear mechanical behavior at the finer spatial scale or, if there does not exist a large separation of characteristic length scales (which is necessary in order to perform significant simplifications in the calculus).

Over the last decade, a new line of analysis has been developed for the study of complex systems with hierarchical structure: the so-called *equation-free approach* [46]. The main idea of the equation-free approach is to circumvent the formulation of macroscopic constitutive equations and calculate the quantities of interest at the macroscale by performing appropriately initialized computational simulations at the microscale. A representative elementary volume, which contains a geometric representation of the structure at the microscale, is assigned to every macroscale point or, in practice, to every element of the macroscale computational mesh. Regarding the conceptual framework of the equation-free approach, the establishment of direct handshaking for bidirectional information transfer between two scales is of crucial importance. There is no golden rule for the successful cross-scale linkage. Local information from the macroscale must be properly projected to the microscale and used to define appropriate boundary and initial conditions (*downscaling*). Then, in the context of the REV, computational simulation of the process is performed and the distributions of microscale quantities are properly averaged to obtain the values of the corresponding macroscale quantities at that point or element (*upscaling*). The upscaling step is achieved using theorems established in the context of the homogenization or volume averaging methods. In the field of biomechanics, equation-free approaches have been developed to predict the mechanical response of tissue constructs [11] and collagen networks [86].

1.4 CALCULATION OF CONSTITUTIVE PARAMETERS

The calculation of the parameters that appear in constitutive equations, which describe the behavior of a material at the macroscopic scale, can be performed either experimentally or theoretically. Both approaches constitute a challenge for cellular biological media. In this section, a few general points are presented, alongside some specific results for several material parameters (elastic moduli, hydraulic permeability, diffusivity).

1.4.1 Generic Framework for Theoretical Calculation

The theoretical calculation of a constitutive parameter requires the development of:

1. A *structural model*, which provides the internal geometry and topology of the cellular biological medium with sufficiently high spatial resolution.

2. A *process model*, which mathematically describes transport processes in the context of the structural model.
3. A *closure scheme*, which mathematically links the value of the constitutive parameter with the structural and process models used for the finer spatial scale.

With regard to the structural model, there exist three different approaches. The simplest approach is to consider the cellular biological medium as an ensemble of (geometric) unit cells. Each unit cell represents the structure in the vicinity of a few biological cells. A more realistic approach is to represent the cellular biological medium as a regular or random array of biological cells. Finally, the most promising approach is to use digitized images of real cellular biological media. These images can be obtained using three-dimensional reconstruction from two-dimensional serial sections, confocal laser scanning microscopy (CLSM), X-ray computed micro-tomography (micro-CT), or any other suitable visualization technique. This approach has been used for the prediction of bone mechanical properties and is gradually gaining ground for other types of cellular biological media as well.

With regard to the process model, there exist two different approaches. In the first of these, a constituent phase is treated as a *continuum* at the finer spatial scale and transport processes are described with a set of partial differential equations, which can be solved analytically or numerically depending on the complexity of the adopted structural model. In the second approach, a constituent phase is treated as a population of interacting *agents* (fluid and solid particles, mass-and-spring elements, molecules, etc.) and transport processes are described using computer-aided simulations in the context of statistical mechanics (such as spring network analysis, molecular dynamics or other). With regard to the closure scheme, several different approaches have been developed in the literature, including heuristic REV analysis, effective-medium approaches, asymptotic homogenization and volume averaging (see [40,41] and references therein).

1.4.2 Remarks on the Experimental Determination

In the laboratory setting, one can measure the fundamental dimensions: length, mass and time, as well as certain other physical quantities (force, temperature, etc.). Constitutive parameters *cannot be measured*. Constitutive parameters *can be estimated* empirically by fitting the outcome of an appropriate theoretical model to the corresponding measurements from well-controlled experiments. This can easily be overlooked in practice. The appropriateness of the employed model for the material behavior of a cellular biological medium is of major significance. It is also very important to ensure that the fundamental hypotheses of the model are not violated by unintended external influences in experimental setup. For example, it has been recently found that the large variability (over an order of magnitude) in reported values of Young's modulus for biofilms was caused, in part, by a misinterpretation of the theoretical variables [3].

The degree of experimental complexity is another important issue. For most composite and porous materials, the experimental determination of transport and mechanical properties is a standard practice, which is even demonstrated at the introductory level of undergraduate courses. Yet, for a number of reasons, the experimental determination remains a formidable challenge for most cellular biological media. First of all, it is usually difficult to obtain a tissue sample with prescribed size and shape, in accordance with an experimental protocol (e.g., for tensile testing of tendons, ligaments). In the case of biofilms, the situation is even more complicated because the biofilm must somehow be grown within the measuring device. Clearly the task can rapidly become overwhelming in complexity with regard to additional features like heterogeneity, anisotropy along with the conformational changes, which occur at every level of the hierarchical structure during material testing.

Finally, an issue of crucial importance is to ensure that the determined constitutive parameters are independent of the specific procedure, measuring device and specimen size used in the experiments. This issue is also frequently overlooked. The influence of these factors should be minimized by the appropriate choice of experimental method. Furthermore, these factors should be taken into account explicitly in the formulation of the theoretical model of the process. A number of approaches have been taken to resolve this issue. A common straightforward approach is to perform experiments with samples of various sizes and shapes in order to ensure that the determined parameters are independent of the sample dimensions. Another approach is to obtain estimates for a specific parameter with different experimental procedures. In this case, the consistency of the underlying models must be ensured. Furthermore, the influence of the measuring process can be represented mathematically using weight functions, which allow one to obtain a precise correspondence between theoretical dependent variables and the respective experimentally measured quantities. Yet, the determination of appropriate weight functions is not an easy task in itself. Finally, a robust approach is to perform computer-aided simulations of the experimental test (virtual experiments). Some noteworthy examples include the simulated tissue compression test [93], the simulated cell aspiration test [18] and the simulated tissue infusion test [62], among others.

1.4.3 Transport Properties of the Extracellular Phase

The architecture of the ECM is a key determinant of the mechanical and transport properties of the extracellular phase and, consequently, of the cellular biological medium. Other parameters, such as the surface charge [21], and the cohesion/bonding between different components (e.g., for collagen fibers and GAGs [83]) are also important. Experimental data and models for the mechanical properties of the ECM are relatively scarce (e.g., [77,83]). However, theoretical estimates for the hydraulic permeability can be obtained from models that have been developed for flow through fibrous porous media (see Table 1.3 and [41]). Furthermore, theoretical estimates for the diffusion coefficient of a chemical substance in the extracellular space can be obtained from models that have been developed for diffusion in hydrogels (see Table 1.4 and [40]).

1.4.4 Mechanical Properties of Biological Cells

The mechanical properties of cells are typically determined experimentally using appropriate techniques, such as micropipette aspiration, cytoindentation, shearing flow, optical tweezers and other. For example, Fig. 1.3 presents an approach for shear-flow testing of adherent cells. An important point of concern is related to the state of the biological cell

TABLE 1.3 Theoretical Models for the Hydraulic Permeability of Fibrous Porous Materials (r_π Is the Fiber Radius and φ_σ Is the Fiber Volume Fraction). Reprinted from *Kapellos et al.* [41] (see References therein)

Model	Equation	Comments
Jackson and James	$\frac{k_\pi}{r_\pi^2} = \frac{3}{20\varphi_\sigma}\left[-\ln\varphi_\sigma - 0.931\right]$	Combination of analytical solutions
Davies	$\frac{k_\pi}{r_\pi^2} = \frac{1}{16\varphi_\sigma^{3/2}(1+56\varphi_\sigma^3)}$	Fit to a large number of data for the flow of air through fibrous materials
Johnston	$\frac{k_\pi}{r_\pi^2} = 0.01064\left[\frac{(1-\varphi_\sigma)(2-\varphi_\sigma)}{\varphi_\sigma}\right]^2$	Fit to experimental data for flow through disordered fibrous materials
Koponen et al.	$\frac{k_\pi}{r_\pi^2} = \frac{5.55}{\exp(10.1\varphi_\sigma)-1}$	Fit to data from the simulation of flow in disordered fibrous structures with the Lattice Boltzmann method
Clague et al.	$\frac{k_\pi}{r_\pi^2} = c_1\left[\frac{1}{2}\sqrt{\pi/\varphi_\sigma} - 1\right]^2 \exp(c_2\varphi_\sigma)$	Fit to data from flow simulations in periodic arrays with the Lattice Boltzmann
		$c_1 = 0.50941$; $c_2 = -1.8042$ (square)
		$c_1 = 0.71407$; $c_2 = -0.51854$ (random)
Happel	$\frac{k_{\pi\perp}}{r_\pi^2} = \frac{1}{8\varphi_\sigma}\left[-\ln\varphi_\sigma - \frac{1-\varphi_\sigma^2}{1+\varphi_\sigma^2}\right]$	Cylinder in cell model. Analytical solution ($\perp$ denotes flow normal to the fiber axis)
Happel	$\frac{k_{\pi//}}{r_\pi^2} = \frac{1}{4\varphi_\sigma}\left[-\ln\varphi_\sigma - \frac{3}{2} + 2\varphi_\sigma - \frac{\varphi_\sigma^2}{2}\right]$	Cylinder in cell model. Analytical solution (// denotes flow parallel to the fiber axis)
Kuwabara	$\frac{k_{\pi\perp}}{r_\pi^2} = \frac{1}{8\varphi_\sigma}\left[-\ln\varphi_\sigma - \frac{3}{2} + 2\varphi_\sigma\right]$	Cylinder in cell model. Analytical solution
Sangani and Acrivos	$\frac{k_{\pi\perp}}{r_\pi^2} = \frac{1}{8\varphi_\sigma}\left[-\ln\varphi_\sigma - 1.476 + 2\varphi_\sigma - 1.774\varphi_\sigma^2 + 4.076\varphi_\sigma^3\right]$	Periodic square array
Sangani and Acrivos	$\frac{k_{\pi\perp}}{r_\pi^2} = \frac{1}{8\varphi_\sigma}\left[-\ln\varphi_\sigma - 1.490 + 2\varphi_\sigma - 0.5\varphi_\sigma^2\right]$	Periodic hexagonal array
Drummond and Tahir	$\frac{k_{\pi//}}{r_\pi^2} = \frac{1}{4\varphi_\sigma}\left[-\ln\varphi_\sigma + K_{DT} + 2\varphi_\sigma - \frac{\varphi_\sigma^2}{2}\right]$	$K_{DT} = -1.476$ (square array)
		$K_{DT} = -1.498$ (triangular array)
		$K_{DT} = -1.354$ (hexagonal array)
Drummond and Tahir	$\frac{k_{\pi\perp}}{r_\pi^2} = \frac{1}{8\varphi_\sigma}\left[-\ln\varphi_\sigma - 1.476 + 2\varphi_\sigma - 1.774\varphi_\sigma^2 + 4.078\varphi_\sigma^3\right]$	Periodic square array

(Continued)

TABLE 1.3 (Continued)

Model	Equation	Comments
Gebart	$\frac{k_{\pi\perp}}{r_\pi^2} = (16/9\pi c_1)\left[\sqrt{\varphi_{\sigma,\max}/\varphi_\sigma} - 1\right]^{5/2}$	Analytical solution for periodic arrays based on lubrication approximation $(c_1, \varphi_{\sigma,\max}) = \left(\sqrt{2}, \pi/4\right)$ (square) $(c_1, \varphi_{\sigma,\max}) = \left(\sqrt{6}, \pi/\sqrt{12}\right)$ (hexagonal)
Gebart	$\frac{k_{\pi//}}{r_\pi^2} = \frac{8}{c_2}\frac{(1-\varphi_\sigma)^3}{\varphi_\sigma^2}$	$c_2 = 57$ square arrangement $c_2 = 53$ hexagonal arrangement

TABLE 1.4 Theoretical Models for the Diffusion Coefficient in Hydrogels ($D_{A\pi}$: Diffusion Coefficient in Hydrogel, $D_{A\upsilon}$: Diffusion Coefficient in Solvent, r_π: Fiber Radius, φ_σ: Fiber Volume Fraction, r_s: Solute Size, $\overline{M}$: Number Average Molecular Weight (MW))

Source	Equation	Comments
Ogston et al. [67]	$\frac{D_{A\pi}}{D_{A\upsilon}} = \exp\left[-\left(1 + \frac{r_s}{r_\pi}\right)\sqrt{\varphi_\sigma}\right]$	
Peppas and Reinhart [68]	$\frac{D_{A\pi}}{D_{A\upsilon}} = k_1\left(\frac{\overline{M}_c - \overline{M}_c^*}{\overline{M}_n - \overline{M}_c^*}\right)\exp\left[-k_2 r_s^2\left(\frac{\varphi_\sigma}{1-\varphi_\sigma}\right)\right]$	k_1, k_2: system-specific parameters $\overline{M}_c$: cross-linked MW $\overline{M}_n$: uncross-linked MW $\overline{M}_c^*$: critical MW
Johnson et al. [37]	$\frac{D_{A\pi}}{D_{A\upsilon}} = \exp\left[-0.84\alpha^{1.09}\right]\left[1 + \frac{r_s}{\sqrt{k_\pi}} + \frac{1}{9}\frac{r_s^2}{k_\pi}\right]^{-1}$	k_π: hydrogel permeability $\alpha = \varphi_\sigma\,(1 + r_s/r_\pi)^2$
Phillips [69]	$\frac{D_{A\pi}}{D_{A\upsilon}} = \exp\left[-0.84\alpha^{1.09}\right]\exp\left[-\pi\varphi_\sigma^b\right]$	$b = -0.174\ln(59.6 r_s/r_\pi)$

during mechanical testing, which has been identified as a potential source of variability in experimentally estimated parameters [56]. For example, the obtained values might depend on whether the cell is living or dead, attached to a solid substrate or suspended in a fluid, squeezed or stretched and so on. Discrepancies between such cases may be attributed primarily to the cytoskeleton, which for a living cell responds actively to mechanical stimulation. In this regard, different experimental methods can be employed in a complementary fashion in order to deduce more complete information about the mechanical behavior of a specific cell over a wide range of mechanical loading conditions. Of course, the theoretical models underlying each experimental protocol must be taken into consideration. Standard

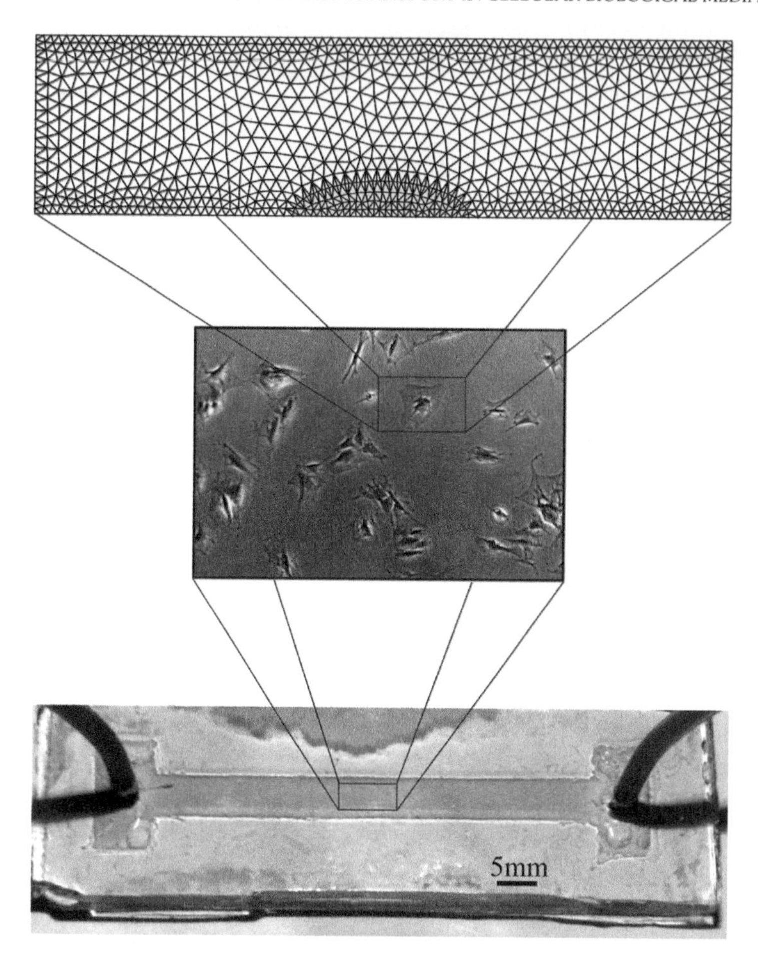

FIGURE 1.3 Experimental setup for shear-flow testing of adherent biological cells. Bottom: flow channel; Middle: rat osteocytes attached to the lower surface of the flow channel; Top: unstructured mesh for FEM analysis of fluid-cell interactions (G.E. Kapellos, unpublished data).

continuum-based models of cell mechanics, which are usually employed to interpret experimental data, include [56]:

- The *liquid drop model*, which represents the biological cell as a composite core–shell particle consisting of a liquid core enclosed by an elastic shell of constant surface tension.
- The *elastic solid model*, which represents the biological cell as a continuum elastic body either linear or nonlinear (hyperelastic).
- The *'standard linear solid' model*, which represents the biological cell as a continuum viscoelastic body, and is based on phenomenological descriptions of the spring-and-dashpot type for cell mechanics.
- The *biphasic mixture model*, which represents the biological cell as a biphasic continuum body consisting of interacting fluid (cytoplasm) and solid (cytoskeleton) constituents.

1.4.5 Mechanical Properties of Cellular Biological Media

In general, the mechanical properties of a cellular biological medium depend on the intrinsic material properties, the characteristic dimensions, the shape, the interconnectivity and the spatial arrangement of the component material phases (cells, ECM, pores). Considerable work, both in theory and experiment, has been done for the estimation of elastic properties of tissues (especially of bone and cartilage [41]). Most of those studies have demonstrated the prominent effect of the volume fractions. In particular, an increase in the volume fraction of solids (cells and ECM) results in an increase of the elastic and shear moduli. Several theoretical models have also been developed for anisotropic elastic tissues (see, for instance, the review in [102] with application to bone). More recently, a number of experimental studies provide parameters for biphasic [62] and nonlinear hyperelastic models [26].

1.4.6 Hydraulic Permeability of Porous Cellular Biological Media

The hydraulic permeability is considered as a measure of the flow resistance, which is exerted by the solid matrix (cells, ECM) on the free fluid flowing through a cellular biological medium, and depends strongly on the detailed geometry and topology of the pore structure (i.e., the space allocated to free fluid). Typically, the hydraulic permeability of a cellular biological medium is obtained experimentally, with suitable experimental methods being different for various types of cellular biological media. In practice it is very difficult (if feasible at all) to follow the conformational changes that occur at every level of the hierarchical structure during the experimental test, and thus it is customary to correlate the hydraulic permeability with other macroscopic quantities, such as the dilatation/strain of the solid matrix, and the pressure gradient of the extracellular fluid (see [41] and references therein). Moreover, rather limited work has been done for the theoretical calculation of the hydraulic permeability and its correlation with microstructural features. For example, *Beno et al.* [7] developed a model for the hydraulic permeability of cortical bone, and concluded that the dimensions of canaliculi and the geometry of the GAG matrix (interfiber spacing and fiber radius) that partially occupies the canaliculi both have a strong effect on the permeability. In addition, they suggested that the shape and size of the osteocyte lacuna, along with the three-dimensional distribution of canaliculi, determine the degree of anisotropy of the permeability.

Recently, *Kapellos et al.* [43] developed a theoretical model of the hydraulic permeability tensor of cellular biological media, which is consistent with upscaled equations for interstitial flow obtained via the spatial averaging method. They solved numerically the model equations in the context of a three phase, periodic spherocylinder-in-cell model, which accounts for salient geometric features of microbial aggregates and biofilms at the cellular scale. The model was validated against previous theoretical data for sphere packings and experimental data for mycelial cakes. Some novel and interesting results were obtained with regard to the dependence of the permeability tensor on the degree of structural anisotropy arising from the shape, orientation and spatial arrangement of cells. For example, Fig. 1.4 shows the dimensionless principal components of the permeability tensor as functions of the aspect ratio $R_{ECS} = l_z/l_y$, where l_z is the *head-to-head* distance and l_y is the *body-to-body* distance. In the case where $R_{ECS} << 1$, the head-to-head spacing is very small in comparison to the body-to-body spacing, and the cells form *nematia* in space (Fig. 1.4B1). In this case, the longitudinal

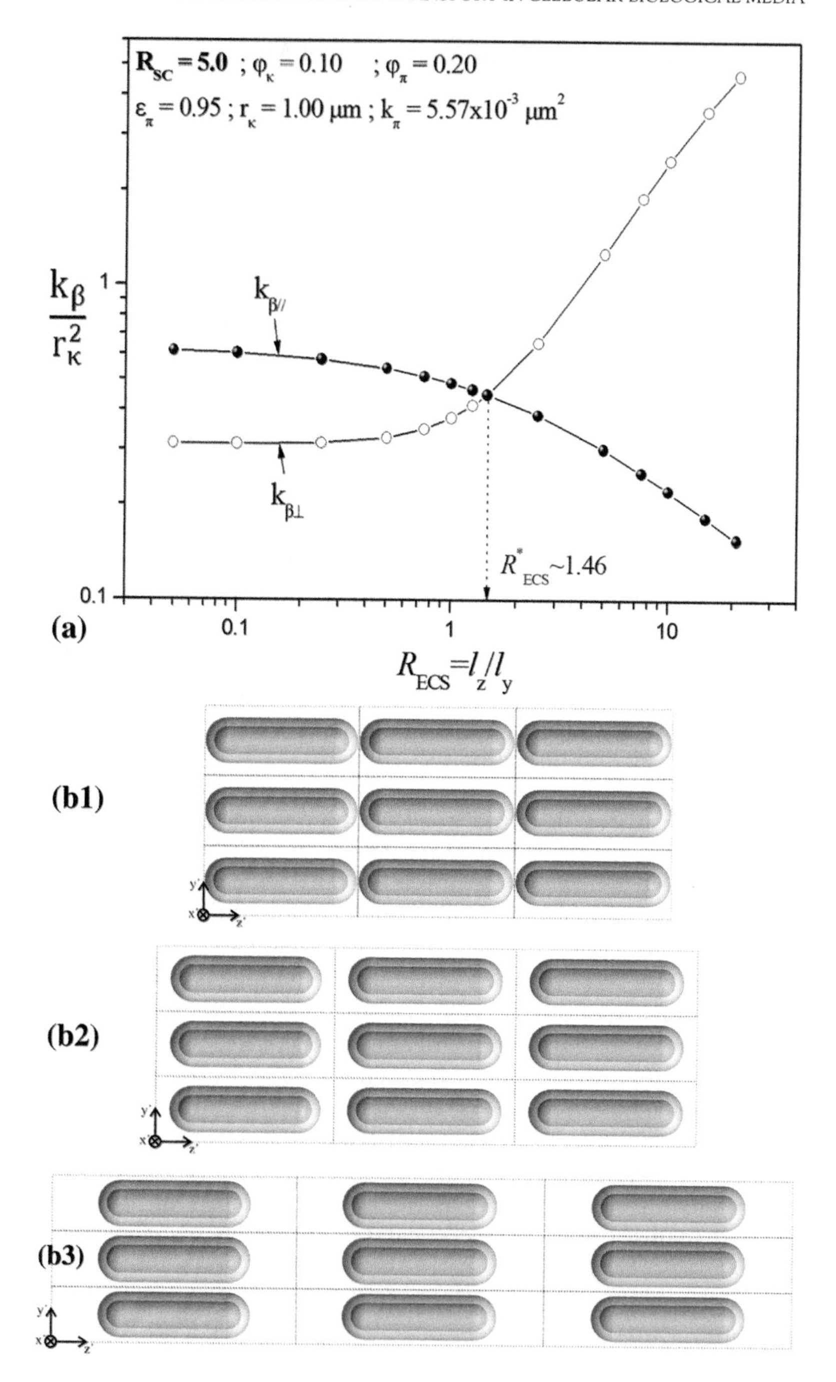

FIGURE 1.4 Effect of cell arrangement on the hydraulic permeability tensor. (a) Dimensionless principal components of the permeability tensor as functions of $R_{ECS}=l_z/l_y$ (l_z: *head-to-head* distance; l_y: *body-to-body* distance). (b) Schematic illustration of cell arrays for: (b1) $R_{ECS} = 0.1$ (*nematic alignment*), (b2) $R_{ECS} = 1.46$ (*isotropic permeability tensor*), and (b3) $R_{ECS} = 10.0$ (*stack formation*). Reprinted from *Kapellos et al.* [43] with permission from Elsevier.

component of the permeability tensor ($k_{\beta//}$) is much larger than the transverse component ($k_{\beta\perp}$), as expected on the basis of previous studies of fluid flow past infinitely long cylinders. As the value of R_{ECS} increases, the longitudinal permeability component decreases and the transverse component increases monotonically. For $R_{ECS} >> 1$, the head-to-head spacing is very large in comparison to the body-to-body spacing, and the cells form *stacks* (Fig. 1.4B3). The longitudinal component of the permeability tensor is much smaller than the transverse component as shown in Fig. 1.4A. Furthermore, there exists a critical value, R^*_{ECS}, for which the principal components of the hydraulic permeability tensor are equal to each other; that is, the flow process is isotropic in a geometrically anisotropic structure.

1.4.7 Diffusion Coefficients in Cellular Biological Media

Extensive experimental and theoretical work has been done for the estimation of the diffusion coefficient of a chemical substance in a cellular biological medium. A recent review on theoretical models can be found in [40, Section 1.3]. The recent model of *Kapellos et al.* [40] is presented here in some detail. In particular, they developed a three-phase, effective-medium model for the calculation of the diffusion coefficient, consistent with the upscaled equations for diffusive mass transport obtained via the spatial averaging method. On the cellular scale of observation, the structural model accounts for three phases: biological cells (κ -phase), porous ECM (π-phase) and large interstitial pores filled with free fluid (υ-phase). The equations of the process model were solved analytically and the following simple expression was obtained for the diffusion coefficient:

$$\frac{D_{A\beta}}{D_{A\upsilon}} = \frac{(1-\delta)\lambda_\kappa \Phi_{11} + (\xi_\mu + 1)\lambda_\pi \Phi_{12}}{(1-\delta)\lambda_\kappa \Phi_{21} + (\xi_\mu + 1)\lambda_\pi \Phi_{22}} \tag{1.53}$$

with:

$$\Phi_{11} = 2\varphi_\upsilon \varphi_\pi + 3(1-\varphi_\upsilon)\left[\varphi_\pi + 3(1-\delta)\varphi_\kappa\right]\lambda_\pi + \varphi_\upsilon(\varphi_\pi + 3\varphi_\kappa)\lambda_\pi \tag{1.54a}$$

$$\Phi_{12} = 2\varphi_\upsilon(2\varphi_\pi + 3\varphi_\kappa) + 2\varphi_\pi(3 - 2\varphi_\upsilon)\lambda_\pi \tag{1.54b}$$

$$\Phi_{21} = \varphi_\pi(3-\varphi_\upsilon) + \varphi_\upsilon(\varphi_\pi + 3\varphi_\kappa)\lambda_\pi \tag{1.54c}$$

$$\Phi_{22} = (3-\varphi_\upsilon)(2\varphi_\pi + 3\varphi_\kappa) + 2\varphi_\upsilon\varphi_\pi\lambda_\pi \tag{1.54d}$$

$$\lambda_\kappa = K_{A,\pi/\upsilon} K_{A,\kappa/\pi} D_{A\kappa}/D_{A\upsilon}, \quad \lambda_\pi = D_{A\pi}/D_{A\upsilon}, \quad \xi_\mu = \frac{K_{A,\kappa/\pi} D_{A\kappa}}{r_\kappa J_{A\mu}} \tag{1.54e}$$

Here, $D_{A\beta}$ is the *effective* diffusion coefficient in the cellular biological medium, $D_{A\upsilon}$ is the diffusion coefficient in large interstitial pores filled with free fluid, $D_{A\pi}$ is the effective diffusion coefficient in the porous ECM, $D_{A\kappa}$ is the diffusion coefficient in the cellular phase, φ_α is the volume fraction of the αth phase ($\alpha = \kappa, \pi, \upsilon$), $K_{A,\alpha/\omega}$ is the equilibrium partition coefficient between the phases α and ω, $J_{A\mu}$ is the membrane permeability, $\delta = \delta_\mu / r_\kappa$ is the dimensionless membrane thickness and r_κ is the cell size. Useful expressions can be obtained for limiting cases of certain parameters of the system. For instance, if the extracellular space is occupied by a single phase, say the π-phase, and the membrane thickness is neglected ($\delta << 1$), then by setting $\lambda_\pi = 1$ (i.e., $D_{A\pi} = D_{A\upsilon}$) and $\varphi_\upsilon = 0$ in Eq. (1.53), one obtains:

$$\frac{D_{A\beta}}{D_{A\pi}} = \frac{3\lambda_\kappa - 2\varphi_\pi(\lambda_\kappa - 1) + 2\varphi_\pi \xi_\mu}{3 + \varphi_\pi(\lambda_\kappa - 1) + (3 - \varphi_\pi)\xi_\mu} \tag{1.55}$$

Further, if the resistance across the membrane is negligible (i.e., $\xi_\mu = 0$), then:

$$\frac{D_{A\beta}}{D_{A\pi}} = \frac{3\lambda_\kappa - 2\varphi_\pi(\lambda_\kappa - 1)}{3 + \varphi_\pi(\lambda_\kappa - 1)} \tag{1.56}$$

This expression is identical to the well-known Maxwell's result for electrical conduction in two-phase systems [61, Eq. 17, p. 440]. Importantly, Eq. (1.53) has predicted the qualitative trend as well as the quantitative variability of a large number of published experimental data on the diffusion coefficient of oxygen in cell-entrapping gels, microbial flocs, biofilms and mammalian tissues.

1.5 MODELING OF GROWTH AND PATTERN FORMATION

Biological cells possess the fascinating ability to take up an *unstructured* aqueous solution of specific organic compounds and, provided that certain other conditions are met, create a *well-structured* living tissue through a complex and highly dynamic process. Following a reductionist approach, cellular and molecular biology have made significant progress towards unraveling fundamental mechanisms that control cellular functions as well as the interactions of cells with their environment. Yet, the abundant information from biological observations must be properly integrated into mathematical models and computer simulators in order to better understand the emergence and evolution of patterns during the development of a cellular biological medium.

Notwithstanding the significant differences in the detailed mechanisms associated with specific cellular biological media, there exists a simple general picture of the involved fundamental processes that has served as a basis for modeling tissue formation from cells. The starting point is set by the converging migration and attachment of a few cells to a favorable niche, in terms of nutrient availability, temperature and other conditions. Following cell proliferation and ECM synthesis, a cell aggregate is formed where the cell to cell communication (via physical contacts and chemical signals) enables the emergence of coordinated actions. As the cellular biological medium obtains a critical size, the build-up of mechanical stresses along with mass and flow limitations come into play and may lead to responses of varying complexity depending on the cell type. For example, most mammalian and plant tissues might cease growth in order to achieve homeostatic equilibrium (preservation of cellular health and functionality). On the other hand, tumors and biofilms might continue growing, and thus induce the formation of internal necrotic spots, dispersive migration of cells, and other. The complexity increases with the consideration of multiple cell types.

Depending on the consideration of the material nature of the cellular biological medium, mathematical models of cell growth can be classified in two fundamentally different classes: (a) continuum-based and (b) discrete-based models. An outline of each class is given in the following paragraphs.

1.5.1 Continuum-Based Models

Theoretical models, which represent a cellular biological medium as a *continuum body*, are formulated in terms of spatial average quantities (e.g., cell number density) that are assumed to be continuous functions of space and time. Features of the system at finer spatial scales can be taken into account only implicitly and, in principle, this approach is reasonable for systems with large populations of cells. The process of growth is mathematically described using standard forms of mass and momentum balances for either monophasic or multiphasic (mixture-type) materials. An issue of primary difficulty involves the definition of constitutive equations for mass and momentum fluxes in relation to cell proliferation and ECM formation.

In a first approach, tissue growth can be modeled as a diffusion/chemotaxis process (e.g., [20]). Any additional effects of bulk motion are disregarded and the momentum balance is not used. One step further, convective bulk motion can be considered in conjunction with an *a priori* hypothesis for the direction of growth, and thus the momentum balance is circumvented again. For instance, it has been considered that tumor spheroids expand along the radial direction (e.g., [12]) and biofilms normal to the substrate (e.g., [96]). In the most general case of three-dimensional volumetric growth, the definition of a growth-associated stress tensor is required. Various approaches have been proposed to confront this issue in the literature, yet none has gained uniform acceptance. For example, based upon the reasoning that a tissue undergoes continuous and permanent deformation under the action of the growth-force, without ever returning to a 'reference' configuration (i.e., fluid-like behavior), the growth-associated stress tensor can be approximated using the constitutive relation for a highly viscous (Newtonian) fluid (e.g., [58]). From a solid mechanics perspective in the context of finite elasticity, it has been proposed to model tissue growth as a process that alters the 'reference' state of the material. Mathematically, this can be achieved by a multiplicative decomposition of the deformation gradient tensor into elastic and growth parts ($\mathbf{F} = \mathbf{F}_e \cdot \mathbf{F}_g$). The growth contribution ($\mathbf{F}_g$) creates an intermediate new reference state, which needs to be further modified by the elastic contribution ($\mathbf{F}_e$) so as to ensure that the final configuration is a continuous body, albeit one that might be exhibiting residual stresses [76]. Intuitively, the updated reference state of a growing elastic tissue could be viewed in analogy to an unstable intermediate compound that is formed during a chemical reaction.

1.5.2 Discrete-Based Models

In computer simulations, which represent the cellular biological medium as a population of *discrete entities*, the macroscopically observed form and function emerge from the interactions between entities and their environment. An entity might represent a single biological cell, or a coarse-grained particle made up of several cells and perhaps also ECM. Typically, each entity is assigned with three vectors: position, velocity and a *state* vector that contains information about other properties, such as volume, age, metabolic activity, etc. The behavior of each entity is defined by a *set of locally-applied rules* that are based on information about the entity and its neighborhood. If multiple cell types are involved, then a different set of rules can be assigned for each cell type. The rules might be deterministic or stochastic and are defined on the basis of:

- Fundamental physical laws (e.g., mass balance for the cell).
- Experimental observations from the biological system under consideration (e.g., if the volume of a cell reaches a critical value, then the cell is divided in two daughter cells).
- Work hypotheses based on experience from other multi-particle physical systems (e.g., cells push their neighbors along a path of least resistance).

There are two general methods of discrete-based modeling:

1. The method of *cellular automata.*
2. The method of *individual-based modeling* (also referred to as agent- or particle-based).

Cellular automata are defined as rule-based dynamic systems, which are discrete both in space and time. The computational domain is discretized with a simple or complex grid and every grid-cell might be empty or contain one or more entities (cells and/or ECM particles). The exact position of an entity in the grid-cell is not defined. In individual-based models, each entity moves in space (continuous or discrete) and interacts with other entities and the environment in pursuit of specific objectives (e.g., maximization of nutrient availability, minimization of shear stress).

In practice, there is some overlap between these two methods. Specifically, a method of individual entities moving in a tessellated domain is equivalent to cellular automata, provided that a common set of rules/objectives is applied in both. In such a case the difference is subtle and mainly philosophical. In the cellular automaton method, the focus is on grid-cells (e.g., the Xth grid-cell contains the Yth entity). On the other hand, in the individual-based method the focus is on the individual entity (e.g., the Yth entity is in the Xth grid-cell). Of course, the two viewpoints have significant differences in terms of computational program structure. In any case, the most important point of differentiation between discrete-based models is the designated set of rules.

It should be mentioned that over the last decade *hybrid approaches* have been developed, which combine continuum-based models (e.g., for solute transport) with discrete-based models (e.g., for cell proliferation) (see, for instance, [39,42]). The rationale for such approaches is the disparity in characteristic sizes between solute molecules, biological cells, and other components in a cellular biological medium. Hybrid approaches have shown a strong potential for unraveling the dynamics of growth and pattern formation of cellular biological media. For example, Fig. 1.5 shows experimentally observed and computer-simulated sample configurations of biofilms in a microfluidic pore network. In the recent literature, the majority of models and computer simulators address the growth of cancerous tumors and microbial biofilms, mainly because of the availability of relevant experimental data. A comprehensive review on modeling tumor growth can be found in [78], while some key features and findings of selected discrete-based models of biofilm formation have been assembled here in Table 1.5.

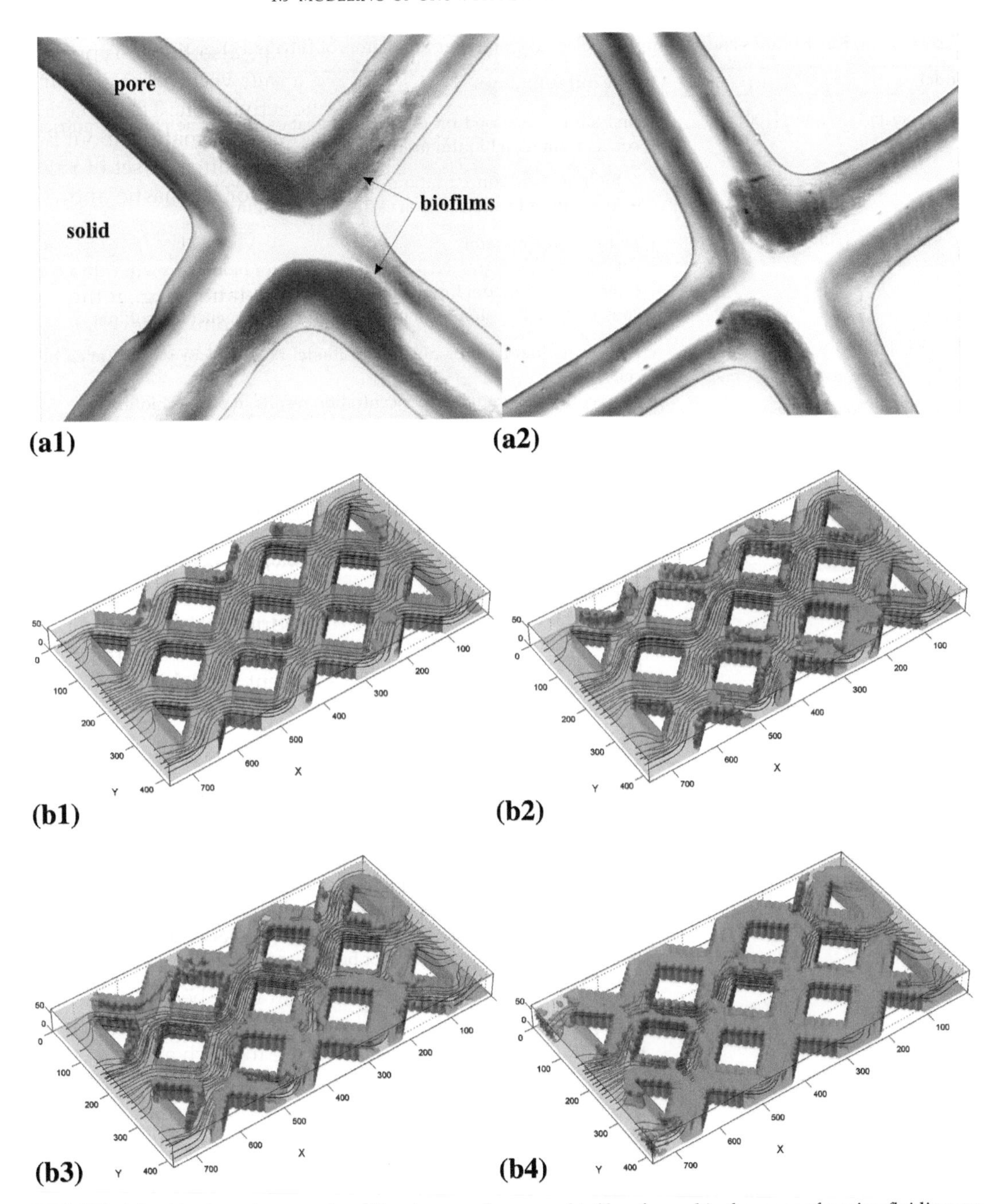

FIGURE 1.5 (a) Microphotographs of *Pseudomonas fluorescens* biofilms formed in the pores of a microfluidic pore network. (b) Representative snapshots from a computer simulation of the evolution of biofilms in a pore network (*Kapellos et al.* [42]).

TABLE 1.5 Key Features and Findings of Selected Computer Simulators of Biofilm Formation

Model	Key features/findings
Wimpenny and Colasanti [100]	• first simulator based on cellular automata showed the effect of nutrient concentration on biofilm morphology: – high nutrient concentration → compact biofilms – low nutrient concentration → dendritic biofilms
Picioreanu et al. [70]	• first hybrid simulator • combined coarse-grained cellular automata for biofilm growth with a continuum-based model for nutrient diffusion–reaction • showed that diffusion-limitation results in heterogeneous biofilms
Kreft et al. [47]	• first individual-based simulator to model the behavior of each microbial cell • showed that low nutrient concentration results in porous biofilms
Hermanowicz [30]	• coarse-grained cellular automata • first introduced the concept of 'least mechanical resistance' to model the growth-induced deformation of biofilms • accounted for detachment (albeit, heuristically – without solving flow equations) • showed that detachment results in more compact biofilms
Chang et al. [13]	• cellular automata model the behavior of each microbial cell • first used Brownian Dynamics simulation for nutrient transport • showed that the production of a growth-inhibitor by the cells results in heterogeneous and porous biofilms
Laspidou and Rittmann [51]	• coarse-grained cellular automata • first distinguished between active/inert biomass • accounted for biofilm consolidation (albeit, heuristically – without solving momentum equations) • showed that consolidation increases heterogeneity in biofilms • observed stratification of activity in biofilms
Kapellos et al. [39]	• first hybrid simulator for biofilm growth in porous media • first modeled explicitly fluid-biofilm interactions (flow through the biofilm; shear-induced biofilm detachment) • combined continuum-based approaches for fluid flow and solute transport and discrete-based approaches for biofilm growth and detachment • showed the importance of detachment/re-entrainment/re-attachment for biofilm migration in porous media • observed multiple biofilm structures (compact, heterogeneous, porous, threads, etc.)

Acknowledgments

G.E.K. is grateful to the Human Frontier Science Program for granting him with a short-term fellowship award (ST-000234/2009), which allowed him to receive initial training and perform experimental research on bone cell and tissue culture *in vitro* at the Mikos laboratory, Department of Bioengineering, Rice University, Houston, USA (in 2010). This chapter has benefited greatly from in-depth discussions with Professor Antonios Mikos and his students, especially Dr. Richard Thibault. Thanks are also expressed to Professor Stavros Pavlou of Patras University for providing useful comments on a first draft of this chapter.

References

[1] J.C. Anderson, C. Eriksson, Piezoelectric properties of dry and wet bone, Nature 227 (1970) 491–492.

[2] T.B. Anderson, R. Jackson, A fluid mechanical description of fluidized beds, Industrial & Engineering Chemistry Fundamentals 6 (4) (1967) 527–539.

[3] N. Aravas, C.S. Laspidou, On the calculation of the elastic modulus of a biofilm streamer, Biotechnology and Bioengineering 101 (1) (2008) 196–200.

[4] P. Basser, Interstitial pressure, volume, and flow during infusion into brain tissue, Microvascular Research 44 (1992) 143–165.

[5] P. Baveye, G. Sposito, The operational significance of the continuum hypothesis in the theory of water movement through soils and aquifers, Water Resources Research 20 (5) (1984) 521–530.

[6] A. Bedford, D.S. Drumheller, Theory of immiscible and structured mixtures, International Journal of Engineering Science 21 (1983) 863–960.

[7] T. Beno, Y.-J. Yoon, S.C. Cowin, S.P. Fritton, Estimation of bone permeability using accurate microstructural measurements, Journal of Biomechanics 39 (2006) 2378–2387.

[8] M.A. Biot, General theory of three-dimensional consolidation, Journal of Applied Physics 12 (1941) 155–164.

[9] M.A. Biot, Mechanics of deformation and acoustic propagation in porous media, Journal of Applied Physics 33 (4) (1962) 1482–1498.

[10] J.J. Blum, G. Lawler, M. Reed, I. Shin, Effect of cytoskeletal geometry on intracellular diffusion, Biophysical Journal 56 (1989) 995–1005.

[11] R.G.M. Breuls, B.G. Sengers, C.W.J. Oomens, C.V.C. Bouten, F.P.T. Baaijens, Predicting local cell deformations in engineered tissue constructs: a multilevel finite element approach, Journal of Biomechanical Engineering 124 (2002) 198–207.

[12] J.J. Casciari, S.V. Sotirchos, R.M. Sutherland, Mathematical modelling of microenvironment and growth in EMT6/Ro multicellular tumor spheroids, Cell Proliferation 25 (1992) 1–22.

[13] I. Chang, E.S. Gilbert, N. Eliashberg, J.D. Keasling, A three-dimensional, stochastic simulation of biofilm growth and transport-related factors that affect structure, Microbiology 149 (2003) 2859–2871.

[14] J.-W. Chu, G.A. Voth, Allostery of actin filaments: molecular dynamics simulations and coarse-grained analysis, Proceedings of the National Academy of Sciences of the United States of America 102 (37) (2005) 13111–13116.

[15] J.W. Costerton, R.T. Irvin, K.-J. Cheng, The bacterial glycocalyx in nature and disease, Annual Review of Microbiology 35 (1981) 299–324.

[16] J.M. Crolet, B. Aoubiza, A. Meunier, Compact bone: numerical simulation of mechanical characteristics, Journal of Biomechanics 26 (6) (1993) 677–687.

[17] J.H. Cushman, On unifying the concepts of scale, instrumentation, and stochastics in the development of multiphase transport theory, Water Resources Research 20 (11) (1984) 1668–1676.

[18] D.E. Discher, D.H. Boal, S.K. Boey, Simulations of the erythrocyte cytoskeleton at large deformation. II. Micropipette aspiration, Biophysical Journal 75 (1998) 1584–1597.

[19] W. Dzwinel, K. Boryczko, D.A. Yuen, A discrete-particle model of blood dynamics in capillary vessels, Journal of Colloid and Interface Science 258 (2003) 163–173.

[20] H.J. Eberl, D.F. Parker, M.C.M. van Loosdrecht, A new deterministic spatiotemporal continuum model for biofilm development, Journal of Theoretical Medicine 3 (3) (2001) 161–176.

[21] S.R. Eisenberg, A.J. Grodzinsky, Electrokinetic micromodel of extracellular matrix and other polyelectrolyte networks, Physicochemical Hydrodynamics 10 (4) (1988) 517–539.

[22] D.A. Fletcher, R.D. Mullins, Cell mechanics and the cytoskeleton, Nature 463 (2010) 485–492.

[23] J.D. Fowler, C.R. Robertson, Metabolic behavior of immobilized aggregates of *Escherichia coli* under conditions of varying mechanical stress, Applied and Environmental Microbiology 57 (1) (1991) 93–101.

[24] E.H. Frank, A.J. Grodzinsky, Cartilage electromechanics – I. Electrokinetic transduction and the effects of electrolyte pH and ionic strength, Journal of Biomechanics 20 (6) (1987) 615–627.

[25] M.G. Fyta, S. Melchionna, E. Kaxiras, S. Succi, Multiscale coupling of molecular dynamics and hydrodynamics: application to DNA translocation through a nanopore, SIAM Multiscale Modeling and Simulation 5 (4) (2006) 1156–1173.

[26] Z. Gao, K. Lister, J.P. Desai, Constitutive modeling of liver tissue: experiment and theory, Annals of Biomedical Engineering 38 (2) (2010) 505–516.

[27] F.J.H. Gijsen, F.N. van de Vosse, J.D. Janssen, The influence of the non-Newtonian properties of blood on the flow in large arteries: steady flow in a carotid bifurcation model, Journal of Biomechanics 32 (1999) 601–608.
[28] A.R. Gillies, R.L. Lieber, Structure and function of the skeletal muscle extracellular matrix, Muscle & Nerve 44 (2011) 318–331.
[29] S. Glagov, R. Vito, D.P. Giddens, C.K. Zarins, Micro-architecture and composition of artery walls: relationship to location, diameter and the distribution of mechanical stress, Journal of Hypertension 10 (1992) S101–S104.
[30] S.W. Hermanowicz, A simple 2D biofilm model yields a variety of morphological features, Mathematical Biosciences 169 (2001) 1–14.
[31] S.J. Hollister, D.P. Fyhrie, K.J. Jepsen, S.A. Goldstein, Application of homogenization theory to the study of trabecular bone mechanics, Journal of Biomechanics 24 (9) (1991) 825–839.
[32] G.A. Holzapfel, T.C. Gasser, R.W. Ogden, A new constitutive framework for arterial wall mechanics and a comparative study of material models, Journal of Elasticity 61 (2000) 1–48.
[33] G.A. Holzapfel, Nonlinear Solid Mechanics: A Continuum Approach for Engineering, John Wiley & Sons Ltd., Chichester, 2000.
[34] X. Hong, A.J. Hopfinger, Molecular modeling and simulation of *Mycobacterium tuberculosis* cell wall permeability, Biomacromolecules 5 (2004) 1066–1077.
[35] K. Jacobson, E.D. Sheets, R. Simson, Revisiting the fluid mosaic model of membranes, Science 268 (1995) 1441–1442.
[36] R. James, G. Kesturu, G. Balian, A.B. Chhabra, Tendon: biology, biomechanics, repair, growth factors, and evolving treatment options, Journal of Hand Surgery 33A (2008) 102–112.
[37] E.M. Johnson, D.A. Berk, R.K. Jain, W.M. Deen, Hindered diffusion in agarose gels: test of effective medium model, Biophysical Journal 70 (1996) 1017–1026.
[38] M.W. Johnson, D.A. Chakkalakal, R.A. Harper, J.L. Katz, S.W. Rouhana, Fluid flow in bone *in vitro*, Journal of Biomechanics 15 (11) (1982) 881–885.
[39] G.E. Kapellos, T.S. Alexiou, A.C. Payatakes, Hierarchical simulator of biofilm growth and dynamics in granular porous materials, Advances in Water Resources 30 (2007) 1648–1667.
[40] G.E. Kapellos, T.S. Alexiou, A.C. Payatakes, A multiscale theoretical model for diffusive mass transfer in cellular biological media, Mathematical Biosciences 210 (1) (2007) 177–237.
[41] G.E. Kapellos, T.S. Alexiou, A.C. Payatakes, Theoretical modeling of fluid flow through cellular biological media: an overview, Mathematical Biosciences 225 (2) (2010) 83–93.
[42] G.E. Kapellos, T.S. Alexiou, S. Pavlou, Hierarchical hybrid simulation of biofilm growth dynamics in 3D porous media in: Papadrakakis et al.(Eds.), Computational Methods for Coupled Problems in Science and Engineering IV, CIMNE, Spain, 2011., pp. 710–720.
[43] G.E. Kapellos, T.S. Alexiou, A.C. Payatakes, A multiscale theoretical model for fluid flow in cellular biological media, International Journal of Engineering Science 51 (2012) 241–271.
[44] K.E. Kasza, A.C. Rowat, J. Liu, T.E. Angelini, C.P. Brangwynne, G.H. Koenderink, D.A. Weitz, The cell as a material, Current Opinion in Cell Biology 19 (2007) 101–107.
[45] E.F. Keller, L.A. Segel, Initiation of slime mold aggregation viewed as an instability, Journal of Theoretical Biology 26 (1970) 399–415.
[46] I.G. Kevrekidis, C.W. Gear, G. Hummer, Equation-free: the computer-aided analysis of complex multiscale systems, AIChE Journal 50 (7) (2004) 1346–1355.
[47] J.-U. Kreft, G. Booth, J.W.T. Wimpenny, BacSim, a simulator for individual-based modeling of bacterial colony growth, Microbiology 144 (1998) 3275–3287.
[48] W.M. Lai, J.S. Hou, V.C. Mow, A triphasic theory for the swelling and deformation behaviors of articular cartilage, Journal of Biomechanical Engineering 113 (1991) 245–258.
[49] R. Lakes, Materials with structural hierarchy, Nature 361 (1993) 511–515.
[50] J.C. LaManna, J.C. Chavez, P. Pichiule, Structural and functional to hypoxia in the rat brain, The Journal of Experimental Biology 207 (2004) 3163–3169.
[51] C.S. Laspidou, B.E. Rittmann, Modeling the development of biofilm density including active bacteria, inert biomass, and extracellular polymeric substances, Water Research 38 (2004) 3349–3361.
[52] T. Lemaire, E. Capiez-Lernout, J. Kaiser, S. Naili, E. Rohan, V. Sansalone, A multiscale theoretical investigation of electric measurements in living bone. Piezoelectricity and electrokinetics, Bulletin of Mathematical Biology 73 (2011) 2649–2677.

[53] O. Lieleg, K.M. Schmoller, C.J. Cyron, Y. Luan, W.A. Wall, A.R. Bausch, Structural polymorphism in heterogeneous cytoskeletal networks, Soft Matter 5 (2009) 1796–1803.

[54] E.N. Lightfoot, Transport Phenomena and Living Systems. Biomedical Aspects of Momentum and Mass Transport, John Wiley & Sons Inc., USA, 1974.

[55] E.N. Lightfoot, Minority opinion. Biofilms as dynamic transport–reaction systems, in: W.G. Characklis, P.A. Wilderer (Eds.), Structure and Function of Biofilms, John Wiley & Sons, Great Britain, 1989., pp. 193–198.

[56] C.T. Lim, E.H. Zhou, S.T. Quek, Mechanical models for living cells – a review, Journal of Biomechanics 39 (2006) 195–216.

[57] W. Liu, L.M. Jawerth, E.A. Sparks, M.R. Falvo, R.R. Hantgan, R. Superfine, S.T. Lord, M. Guthold, Fibrin fibers have extraordinary extensibility and elasticity, Science 313 (2006) 634.

[58] S.R. Lubkin, T. Jackson, Multiphase mechanics of capsule formation in tumors, Journal of Biomechanical Engineering 124 (2002) 237–243.

[59] C.M. Marle, Écoulements monophasiques en milieu poreux, Revue de l' Institut Français du Pétrole XXII (10) (1967) 1471–1509.

[60] C.M. Marle, On macroscopic equations governing multiphase flow with diffusion and chemical reactions in porous media, International Journal of Engineering Science 20 (5) (1982) 643–662.

[61] J.C. Maxwell, A Treatise on Electricity and Magnetism vol. 1, Dover Publications, 1954.

[62] M. Milosevic, S.J. Lunt, E. Leung, J. Skliarenko, P. Shaw, A. Fyles, R.P. Hill, Interstitial permeability and elasticity in human cervix cancer, Microvascular Research 75 (2008) 381–390.

[63] V.C. Mow, S.C. Kuei, W.M. Lai, C.G. Armstrong, Biphasic creep and stress relaxation of articular cartilage in compression: theory and experiments, Journal of Biomechanical Engineering 102 (1980) 73–84.

[64] M. Niederweis, O. Danilchanka, J. Huff, C. Hoffmann, H. Engelhardt, Mycobacterial outer membranes: in search of proteins, Trends in Microbiology 18 (3) (2010) 109–116.

[65] J.L. Nowinski, C.F. Davis, A model of the human skull as a poroelastic spherical shell subjected to a quasistatic load, Mathematical Biosciences 8 (1970) 397–416.

[66] J.A. Ochoa, S. Whitaker, P. Stroeve, Determination of cell membrane permeability in concentrated cell ensembles, Biophysical Journal 52 (1987) 763–774.

[67] A.G. Ogston, B.N. Preston, J.D. Wells, On the transport of compact particles through solutions of chain-polymers, Proceedings of the Royal Society of London A 333 (1973) 297–316.

[68] N.A. Peppas, C.T. Reinhart, Solute diffusion in swollen membranes 1. A new theory, Journal of Membrane Science 15 (1983) 275–287.

[69] R.J. Phillips, A hydrodynamic model for hindered diffusion of proteins and micelles in hydrogels, Biophysical Journal 79 (2000) 3350–3354.

[70] C. Picioreanu, M.C.M. van Loosdrecht, J.J. Heijnen, Mathematical modeling of biofilm structure with a hybrid differential-discrete cellular automaton approach, Biotechnology Bioengineering 58 (1) (1998) 101–116.

[71] K. Piekarski, M. Munro, Transport mechanism operating between blood supply and osteocytes in long bones, Nature 269 (1977) 80–82.

[72] C. Pozrikidis, Axisymmetric motion of a file of red blood cells through capillaries, Physics of Fluids 17 (2005) 031503.

[73] A.R. Pries, T.W. Secomb, P. Gaehtgens, Biophysical aspects of blood flow in the microvasculature, Cardiovascular Research 32 (1996) 654–667.

[74] A. Redondo, R. LeSar, Modeling and simulation of biomaterials, Annual Review of Materials Research 34 (2004) 279–314.

[75] S. Reitsma, D.W. Slaaf, H. Vink, M.A.M.J. van Zandvoort, M.G.A. oude Egbrink, The endothelial glycocalyx: composition, functions, and visualization, Pflügers Archiv European Journal of Physiology 454 (2007) 345–359.

[76] E.K. Rodriguez, A. Hoger, A.D. McCulloch, Stress-dependent finite growth in soft elastic tissues, Journal of Biomechanics 27 (4) (1994) 455–467.

[77] B.A. Roeder, K. Kokini, J.E. Sturgis, J.P. Robinson, S.L. Voytik-Harbin, Tensile mechanical properties of three-dimensional type I collagen extracellular matrices with varied microstructure, Journal of Biomechanical Engineering 124 (2002) 214–222.

[78] T. Roose, S.J. Chapman, P.K. Maini, Mathematical models of avascular tumor growth, SIAM Review 49 (2) (2007) 179–208.

[79] C. Ruff, B. Holt, E. Trinkaus, Who's afraid of the big bad Wolff? 'Wolff's law' and bone functional adaptation, American Journal of Physical Anthropology 129 (2006) 484–498.
[80] E. Ruoslahti, Brain extracellular matrix, Glycobiology 6 (5) (1996) 489–492.
[81] E. Sanchez-Palencia, Non-Homogeneous Media and Vibration Theory, Springer-Verlag, Berlin, 1980.
[82] M. Schliwa, G. Woehlke, Molecular motors, Nature 422 (2003) 759–765.
[83] M.H. Schwartz, P.H. Leo, J.L. Lewis, A microstructural model for the elastic response of articular cartilage, Journal of Biomechanics 27 (7) (1994) 865–873.
[84] V.I. Sikavitsas, G.N. Bancroft, H.L. Holtorf, J.A. Jansen, A.G. Mikos, Mineralized matrix deposition by marrow stromal osteoblasts in 3D perfusion culture increases with increasing fluid shear forces, Proceedings of the National Academy of Sciences of the United States of America 100 (25) (2003) 14683–14688.
[85] V.I. Sikavitsas, J.S. Temenoff, A.G. Mikos, Biomaterials and bone mechanotransduction, Biomaterials 22 (2001) 2581–2593.
[86] T. Stylianopoulos, V.H. Barocas, Volume-averaging theory for the study of the mechanics of collagen networks, Computer Methods in Applied Mechanics and Engineering 196 (2007) 2981–2990.
[87] D.N. Sun, W.Y. Gu, X.E. Guo, W.M. Lai, V.C. Mow, A mixed finite element formulation of triphasic mechano-electrochemical theory for charged, hydrated biological soft tissues, International Journal for Numerical Methods in Engineering 45 (1999) 1375–1402.
[88] E.A. Swabb, J. Wei, P.M. Gullino, Diffusion and convection in normal and neoplastic tissues, Cancer Research 34 (1974) 2815–2822.
[89] M.A. Swartz, M.E. Fleury, Interstitial flow and its effects in soft tissues, Annual Review of Biomedical Engineering 9 (2007) 229–256.
[90] C.T. Truesdell, On the foundations of mechanics and energetics, in: Continuum Mechanics, The Rational Mechanics of Materials, vol. II, Gordon & Breach, NY, 1965., pp. 293–305.
[91] A.M. Turing, The chemical basis of morphogenesis, Philosophical Transactions of the Royal Society of London, Series B, Biological Sciences 237 (641) (1952) 37–72.
[92] F. van den Ent, L.A. Amos, J. Löwe, Prokaryotic origin of the actin cytoskeleton, Nature 413 (2001) 39–44.
[93] B. van Rietbergen, H. Weinans, R. Huiskes, A. Odgaard, A new method to determine trabecular bone elastic properties and loading using micromechanical finite-element models, Journal of Biomechanics 28 (1) (1995) 69–81.
[94] D.G. Vlachos, A review of multiscale analysis: examples from systems biology, materials engineering, and other fluid–surface interacting systems, Advances in Chemical Engineering 30 (2005) 1–61.
[95] T.-W. Wang, J.-S. Sun, Y.-C. Huang, H.-C. Wu, L.-T. Chen, F.-H. Lin, Skin basement membrane and extracellular matrix proteins characterization and quantification by real time RT-PCR, Biomaterials 27 (2006) 5059–5068.
[96] O. Wanner, W. Gujer, A multispecies biofilm model, Biotechnology and Bioengineering 28 (3) (1986) 314–328.
[97] S. Weinbaum, X. Zhang, Y. Han, H. Vink, S.C. Cowin, Mechanotransduction and flow across the endothelial glycocalyx, Proceedings of the National Academy of Sciences of the United States of America 100 (13) (2003) 7988–7995.
[98] R.L. Whitmore, The flow behavior of blood in the circulation, Nature 215 (1967) 123–126.
[99] R. Wilson, A.F. Diseberg, L. Gordon, S. Zivkovic, L. Tatarczuch, E.J. Mackie, J.J. Gorman, J.F. Bateman, Comprehensive profiling of cartilage extracellular matrix formation and maturation using sequential extraction and label-free quantitative proteomics, Molecular & Cellular Proteomics 9 (2010) 1296–1313.
[100] J.W.T. Wimpenny, R. Colasanti, A unifying hypothesis for the structure of microbial biofilms based on cellular automaton models, FEMS Microbiology Ecology 22 (1997) 1–16.
[101] B.D. Wood, S. Whitaker, Diffusion and reaction in biofilms, Chemical Engineering Science 53 (3) (1998) 397–425.
[102] P.K. Zysset, A review of morphology-elasticity relationships in human trabecular bone: theories and experiments, Journal of Biomechanics 36 (2003) 1469–1485.

CHAPTER

2

Thermal Pain in Teeth: Heat Transfer, Thermomechanics and Ion Transport

Min Lin[a], ShaoBao Liu[b], Feng Xu[a], TianJian Lu[b], BoFeng Bai[c], and Guy M. Genin[d]

[a]The Key Laboratory of Biomedical Information Engineering of the Ministry of Education, School of Life Science and Technology, Xi'an Jiaotong University, and Biomedical Engineering and Biomechanics Center, Xi'an Jiaotong University, Xi'an, P.R. China, [b]Biomedical Engineering and Biomechanics Center, Xi'an Jiaotong University, Xi'an, P.R. China, [c]State Key Laboratory of Multiphase Flow in Power Engineering, Xi'an Jiaotong University; Xi'an, P.R. China, [d]Department of Mechanical Engineering and Materials Science, Washington University in St. Louis, St. Louis, MO, USA

OUTLINE

2.1 Introduction 42

2.2 Modeling of Thermally Induced Dentinal Fluid Flow 43
- *2.2.1 Analysis of Thermomechanics of the Tooth* 43
- *2.2.2 Analysis of Dentinal Fluid Flow* 46

2.3 Modeling of Nociceptor Transduction 47
- *2.3.1 Modeling of Shear Stress* 47
- *2.3.2 Modeling Transduction* 49

2.4 Results and Discussion 51
- *2.4.1 Tooth Thermomechanics* 51
- *2.4.2 The Mechanism Underlying Thermally Induced DFF* 52
- *2.4.3 DFF and Its Implications for Tooth Thermal Pain* 53
- *2.4.4 The Difference Between Hot and Cold Tooth Pain* 54

2.5 Conclusion 56

References 56

© 2013 Elsevier Inc. All rights reserved.

http://dx.doi.org/10.1016/B978-0-12-415824-5.00002-3

2.1 INTRODUCTION

The tooth is a sensory tissue, and its sensory responses to various stimuli have been studied for decades [1–5]. Microstructure studies have shown that dentinal microtubules (DMTs) radiate from the pulp wall to the surrounding cementum or dentine-enamel junction (DEJ) [6] (Fig. 2.1). Most of the DMTs contain non-myelinated terminal fibrils and odontoblastic processes (OPs, extension of odontoblast) which are situated in an environment filled with dentinal fluid [1]. This dental microstructure has favored the 'hydrodynamic theory' for explaining tooth pain sensation under thermal stimulation, although several other theories have also been proposed (e.g., the odontoblastic transduction theory, the neural theory) [7]. The 'hydrodynamic theory' assumes that the pain sensation from teeth is attributed to the stimulation of nociceptive mechanoreceptors (nociceptors) as a consequence of dentinal fluid flow (DFF) within DMTs. Direct evidence to support the hydrodynamic theory has been obtained by simultaneously recording the DFF through DMTs and the corresponding intradental neural discharge evoked in the cat tooth [1,8].

A wide range of external noxious stimuli for inducing thermal pain in the tooth (e.g., thermal, mechanical, dental restorative processes) have been found to cause either an inward (toward the pulp chamber) or outward (away from the pulp chamber) DFF in DMTs [9,10]. Nevertheless, the relationship between DFF and the associate intradental neural discharge patterns remains unknown. It has also been found that the transduction of pain in teeth (human and cats, *in vivo*) is more sensitive to the outward DFF than to the inward DFF [1,8,11], indicating that the associated mechanoreceptors may not be equally sensitive to the inward and outward flows. However, the underlying mechanism remains unknown. Besides, the major difference between the effects of hot and cold stimuli has been identified: the former causes inward fluid movement while the latter causes outward fluid movement [1,8,12]. Cold stimuli were found to evoke more rapid transient pain sensations, while hot stimuli generally

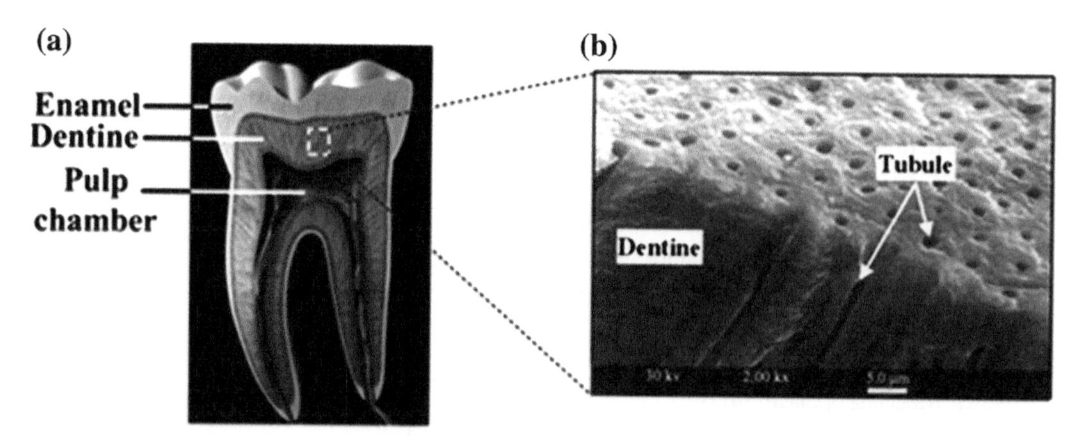

FIGURE 2.1 Tooth structures: (a) cut-away image of human tooth illustrating composite layers (www.3dscience.com); (b) SEM image of dentine showing solid dentine material and microtubules running perpendicularly from pulpal wall toward dentine-enamel junction [36].

induce a dull pain which lasts longer than cold-induced pain [13]. The different pain responses have been further confirmed by *in vivo* experimental electrophysiological data (from dogs) [14]. Hence, a better understanding of the DFF and the associated intradental nerve response could provide an insight into the mechanisms for the different tooth responses to cold and hot stimuli.

This chapter describes the development of a thermomechanical model of teeth to uncover the intradental neural discharge behaviors under hot or cold stimulation. The effects of DFF velocity and direction on the shear stress experienced by nerve terminals were analyzed using a computational fluid dynamics (CFD) model. After that, a modified Hodgkin-Huxley (H-H) model was introduced to simulate the ion transport and neural discharge of intradental nociceptors. These models were validated by comparing the modeling results with the experimental observations of *Andrew & Matthews* [1]. Mechanistic insights into the difference between hot and cold tooth pain are provided based on the developed models. The models developed here, together with advances in a number of areas related to pain research, could lead to the development of tools for novel and more effective analgesics.

2.2 MODELING OF THERMALLY INDUCED DENTINAL FLUID FLOW

2.2.1 Analysis of Thermomechanics of the Tooth

The tooth is modeled as a one-dimensional (1-D) layered structure (Fig. 2.2a). Such a simplification will not affect the temperature gradient inducing bending behaviors of dentine in tooth [15].

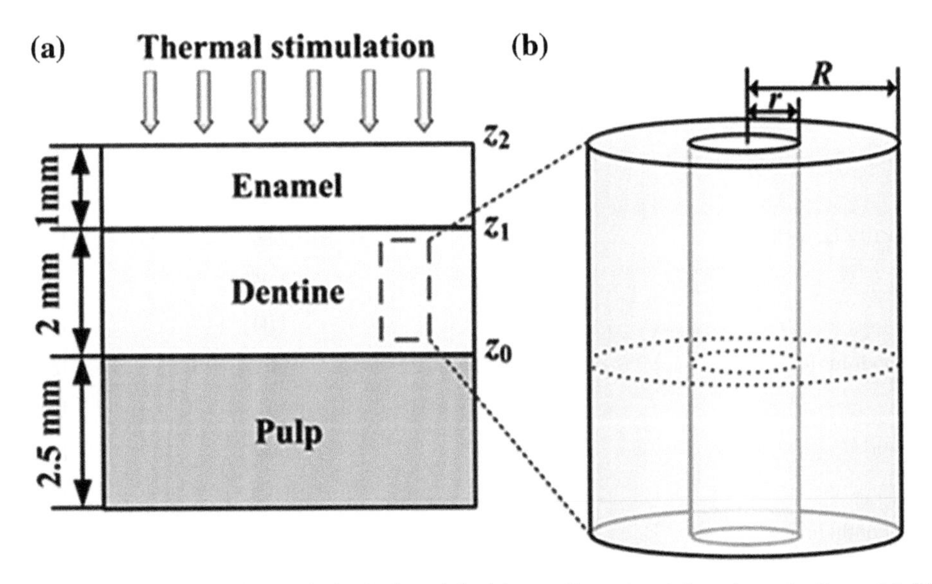

FIGURE 2.2 Illustration of idealized physical models: (a) one-dimensional three-layer tooth model; (b) single dentine microtubule.

A 1-D Fourier heat transfer equation is employed for studying the heat transfer in the layered model:

$$\rho_i c_i \frac{\partial T(z,t)}{\partial t} = k_i \frac{\partial^2 T(z,t)}{\partial z^2} \tag{2.1}$$

in which i is the index of sub-layers ($i = 1, 2, 3$ represents the enamel, dentine and pulp layers, respectively); T is the temperature (dependent on location z (m) and time t(s)); ρ_i (kg m^{-3}), k_i (W m^{-1} K^{-1}) and c_i (J kg^{-1} K^{-1}) are the mass density, thermal conductivity and specific heat of sub-layer i, respectively. The physical properties of tooth structures are listed in Table 2.1. The third boundary condition is applied to the enamel surface, and the convective heat transfer coefficient is 500 W/(m2•K) during hot or cold water stimulation. After the thermal stimulation (5 s duration) was removed, the enamel was exposed to natural convection with a heat transfer coefficient ~10 W/(m2•K). Initial body temperature was 37°C. The temperature at the bottom of the pulp layer was assumed to remain unchanged, at 37°C, during and after thermal stimulation; in other words, the first boundary conditions is applied to the bottom of pulp layer.

The pulp layer is a liquid layer, acting only as a heat transfer medium without restraining the solid structure (e.g., enamel, dentine) from thermal deformation [16]. Hence the thermal stress in the pulp layer is assumed to be negligible in the 1-D layered model.

TABLE 2.1 Physical Properties of the Tooth

Property	Tooth Component	Values	Ref.
k Thermal conductivity [W m^{-1} K^{-1}]	Enamel	0.81	[17]
	Dentine	0.48	[17]
	Pulp†	0.63	[18]
c_p Specific heat [$\times 10^{-3}$ J kg^{-1} K^{-1}]	Enamel	0.71	[19]
	Dentine	1.59	[19]
	Pulp†	4.2	[18]
ρ Density [$\times 10^{-3}$ kg m^{-3}]	Enamel	2.80	[19]
	Dentine	1.96	[19]
	Pulp†	1.00	[18]
E Young's modulus [GPa]	Enamel	94	[20]
	Dentine	20	[20]
n Poisson ratio	Enamel	0.30	[21]
	Dentine	0.25	[20]
λ Coefficient of thermal expansion	Enamel	1.696	[22]
[$\times 10^{-5}$ K^{-1}]	Dentine	1.059	[22]

†*Values taken from water following de Vree et al. [37].*

To model the application of thermal stimulation to the enamel surface, a one-dimensional three-layer model (see Fig. 2.2(a)) was employed. The in-plane thermal stresses in the enamel and dentine layers [23] are as follows:

$$\sigma_e(z,t) = \overline{E}_e(1+\nu_e)\left\{\begin{array}{l} -\overline{\lambda}_e\Delta T + \left[\begin{array}{l}(a'_{11}+a'_{12})\left(\int_{z_1}^{z_2}\overline{E}_e\overline{\lambda}_e\Delta T dz + \int_{z_0}^{z_1}\overline{E}_d\overline{\lambda}_d\Delta T dz\right) + \\ (b'_{11}+b'_{12})\left(\int_{z_1}^{z_2}\overline{E}_e\overline{\lambda}_e\Delta T z dz + \int_{z_0}^{z_1}\overline{E}_d\overline{\lambda}_d\Delta T z dz\right)\end{array}\right] \\ +z\left[\begin{array}{l}(b'_{11}+b'_{12})\left(\int_{z_1}^{z_2}\overline{E}_e\overline{\lambda}_e\Delta T dz + \int_{z_0}^{z_1}\overline{E}_d\overline{\lambda}_d\Delta T dz\right) + \\ (d'_{11}+d'_{12})\left(\int_{z_1}^{z_2}\overline{E}_e\overline{\lambda}_e\Delta T z dz + \int_{z_0}^{z_1}\overline{E}_d\overline{\lambda}_d\Delta T z dz\right)\end{array}\right]\end{array}\right\}$$

$$z_2 \leqslant z < z_1 \quad \text{Enamel layer} \tag{2.2a}$$

$$\sigma_d(z,t) = \overline{E}_d(1+\nu_d)\left\{\begin{array}{l} -\overline{\lambda}_d\Delta T + \left[\begin{array}{l}(a'_{11}+a'_{12})\left(\int_{z_1}^{z_2}\overline{E}_e\overline{\lambda}_e\Delta T dz + \int_{z_0}^{z_1}\overline{E}_d\overline{\lambda}_d\Delta T dz\right) + \\ (b'_{11}+b'_{12})\left(\int_{z_1}^{z_2}\overline{E}_e\overline{\lambda}_e\Delta T z dz + \int_{z_0}^{z_1}\overline{E}_d\overline{\lambda}_d\Delta T z dz\right)\end{array}\right] \\ +z\left[\begin{array}{l}(b'_{11}+b'_{12})\left(\int_{z_1}^{z_2}\overline{E}_e\overline{\lambda}_e\Delta T dz + \int_{z_0}^{z_1}\overline{E}_d\overline{\lambda}_d\Delta T dz\right) + \\ (d'_{11}+d'_{12})\left(\int_{z_1}^{z_2}\overline{E}_e\overline{\lambda}_e\Delta T z dz + \int_{z_0}^{z_1}\overline{E}_d\overline{\lambda}_d\Delta T z dz\right)\end{array}\right]\end{array}\right\}$$

$$z_1 \leqslant z < z_0 \quad \text{Dentine layer} \tag{2.2b}$$

where $\Delta T = T(z,t) - T_0$, and T_0 is the initial body temperature. $\overline{E_i} = E_i/(1-v^2), \overline{\lambda}_i = (1+v)\lambda_i$. E (Pa), v and λ (K^{-1}) are the Young's modulus, Poisson ratio and coefficient of thermal expansion of dentine, respectively. The in-plane extensional, coupling and bending stiffnesses of the overall laminate of structure are governed respectively by the following [24]:

$$\left.\begin{array}{l} A_{ij} = \sum_{k=1}^{2}(\overline{Q}_{ij})_k(z_k - z_{k-1}) \\ B_{ij} = \frac{1}{2}\sum_{k=1}^{2}(\overline{Q}_{ij})_k(z_k^2 - z_{k-1}^2) \\ D_{ij} = \frac{1}{3}\sum_{k=1}^{2}(\overline{Q}_{ij})_k(z_k^3 - z_{k-1}^3)\end{array}\right\}(i,j = 1,2,6) \tag{2.3}$$

in which k is the index of sub-layers ($k=1,2$ represents the enamel and dentine layers, respectively). A_{ij}, B_{ij}, C_{ij} are separately assembled into the elements of stiffness matrices [A], [B] and [D]. The elements $a'_{11}, a'_{12}, b'_{11}, b'_{12}, d'_{11}, d'_{12}$ can be determined by [24]:

$$\left.\begin{array}{ll} a'_{11} = [A^{-1}B(D-BA^{-1}B)^{-1}BA^{-1}]_{11} & a'_{12} = [A^{-1}B(D-BA^{-1}B)^{-1}BA^{-1}]_{12} \\ b'_{11} = [-(A^{-1}B)(D-BA^{-1}B)^{-1}]_{11} & b'_{12} = [-(A^{-1}B)(D-BA^{-1}B)^{-1}]_{12} \\ d'_{11} = [(D-BA^{-1}B)^{-1}]_{11} & d'_{12} = [(D-BA^{-1}B)^{-1}]_{12}\end{array}\right\} \tag{2.4}$$

$[\overline{Q}]_k$ is the stiffness matrix of the layered structure, defined by:

$$\left[\overline{Q}\right]_k = \begin{bmatrix} \frac{E_k}{1-v_k^2} & \frac{\nu_k E_k}{1-v_k^2} & 0 \\ \frac{\nu_k E_k}{1-v_k^2} & \frac{E_k}{1-v_k^2} & 0 \\ 0 & 0 & \frac{E_k}{2(1+v_k)}\end{bmatrix} \tag{2.5}$$

2.2.2 Analysis of Dentinal Fluid Flow

The distributions of temperature $T(z, t)$ and thermal stress $\sigma(z, t)$ in the dentine layer could be obtained by thermomechanics analysis in the 1-D layered model. The DMT is modeled as a cylinder with inner and outer radii of r and R (Fig. 2.2b). The analysis assumes that the outer surface of the cylinder (through the thickness of dentine layer) is subjected to thermal stress in the radial direction. The displacement around the cylinder can be given by [3]:

$$u_\rho = \frac{1}{E}\left[\frac{(1-v_d)\left(r^2 p_i - R^2 p_o\right)}{R^2 - r^2}\rho + \frac{(1+v_d)a^2b^2\left(p_i - p_o\right)}{r^2 - R^2}\frac{1}{\rho}\right] \tag{2.6}$$

where p_i represents the stress on the inner wall and p_o represents the stress on the outer wall.

Here, we assume that $p_i=0$ and $p_o=\sigma$. Given that in general R>3r, and setting $E = \frac{E}{1-v_d^2}$ and $v_d = \frac{v_d}{1-v_d}$, the displacement of DMT wall can be expressed as:

$$u_r \approx \frac{2}{E}(1+v_d)(1-v_d)r\sigma(z,t) \tag{2.7}$$

where υ_d is the Poisson's ratio for the dentine layer. The volume change of DMT is given by:

$$V(t) = \int_{z_0}^{z_1} [\pi(u_r(z,t)+r)^2 - \pi r^2]dz \tag{2.8}$$

Fluid flow velocity at the pulpal end of DMT caused by deformation of DMT is calculated as

$$u'(\mathrm{t}) = \frac{V'(t)}{\pi r^2} \tag{2.9}$$

Thermal expansion/contraction of the dentinal fluid causes a volume change in the dentinal fluid, which is governed by:

$$\delta V(t) = -\int_{z_0}^{z_1} 3(T(z,t) - T_0)\alpha\pi r^2 dz \tag{2.10}$$

where α is the coefficient of thermal expansion of the dentinal fluid. The fluid flow velocity due to this volume change in the dentinal fluid is:

$$u''(\mathrm{t}) = \frac{\delta V'(t)}{\pi r^2} \tag{2.11}$$

When considering the deformation in both DMT and dentinal fluid, the DFF velocity is:

$$u(\mathrm{t}) = \frac{V' + \delta V'(t)}{\pi r^2} \tag{2.12}$$

2.3 MODELING OF NOCICEPTOR TRANSDUCTION

2.3.1 Modeling of Shear Stress

The DMT innervation system consists of dentinal fluid, non-myelinated sensory nerve fibrils and odontoblastic process (OP). A schematic representation of this system is shown in Fig. 2.3a–b. There is only one terminal fibril and one OP inside a DMT in most cases [25]. Outward DFF can cause slight movement of odontoblasts toward the DMT (Fig. 2.3a), while inward flow will move the odontoblasts in the opposite direction (Figure 2.3b) [1]. The diameter of the OP decreases along the direction from the pulpal wall to the DEJ [25]. The values of physiological parameters of the tooth innervation system are listed in Table 2.2. The diameter of the OP was assumed to change linearly along its longitudinal direction, with consideration that the maximum diameter d_{op} is larger than 1 μm [6]. To model the effect of OP movement on the shear stress on the terminal bead (TB), we assume here that the OP displacement (OPD) in the DFF direction changes linearly with flow velocity. The physical model for the fluid dynamics simulation of the inward flow is shown in Fig. 2.3b.

Based on the *in vivo* structure of the dentinal microtubule innervation system described above and the symmetrical structure of the terminal bead and odontoblastic process in the longitudinally sectioned plane (along their axes), the three-dimensional (3-D) structure of

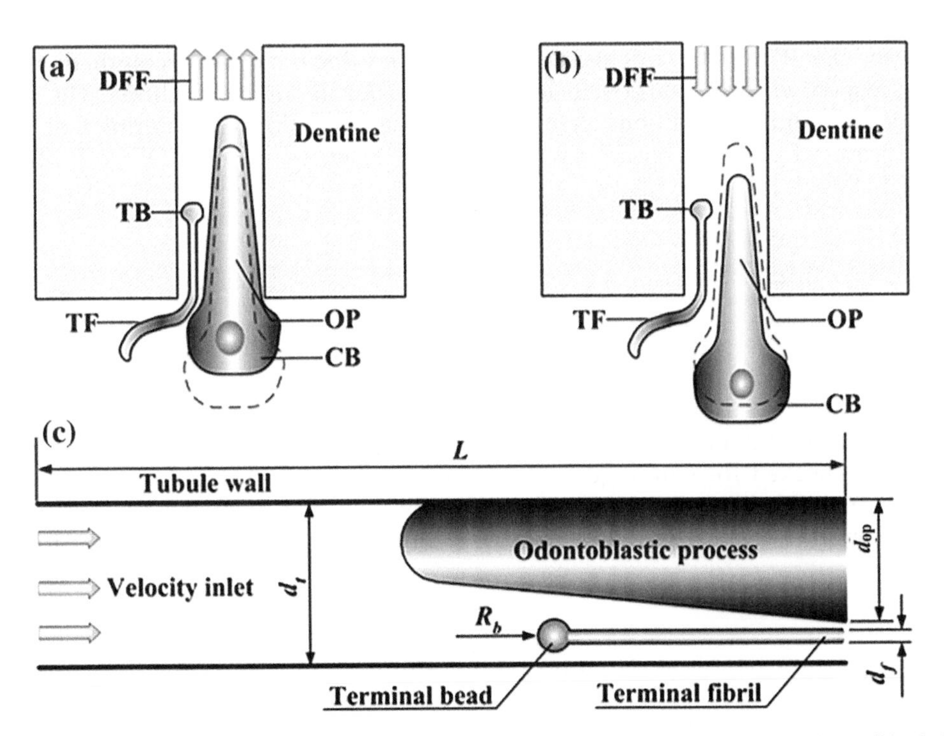

FIGURE 2.3 Physiological relevant structures: (a) a slightly outward displacement of OP and its cell body (CB) in response to the outward flow; (b) a slightly inward flow of the OP in response to the inward flow; (c) the physically realistic model for fluid dynamics simulation (inward flow).

TABLE 2.2 Physiological Parameters of the Tooth's Innervation System

Parameters	Value	Ref.
DMT (cat canine) diameter, d_t (μm)	0.73	[1]
Fluid viscosity, μ (Pa·s)	1.55×10^{-3}	[26]
Fluid density, ρ (kg/m3)	1010	[26]
Terminal fibril diameter, d_f (μm)	0.1	[27]
Terminal bead diameter, d_b (μm)	0.2	[28]

fluid flow through the dentinal microtubule innervation system was simplified to a two-dimensional (2-D) model. Since the focus of the present research was on the τ_{MSS}, this simplification provides a reasonable approximation for the numerical simulation of fluid dynamics [5]. The steady-state Navier-Stokes equations employed to model the shear stress experienced by TB, are expressed as:

$$\nabla \bullet v = 0 \tag{2.13}$$

$$\rho\,(v \bullet \nabla)\,v = -\nabla p + \mu \nabla^2 v \tag{2.14}$$

where v (m/s), p (Pa), ρ (kg/m^3) and μ (Pa·s) are the velocity vector, pressure, density and viscosity, respectively. Zero fluid velocity is assumed for the initial condition. The constant fluid velocity boundary conditions were applied in the simulation. Considering that ~30% of the tubules are in free communication with the pulp [29], we estimate the DFF velocities in an individual DMT as [5]:

$$V = \frac{10^6 V_c}{0.3N\pi\left(\frac{d_d}{2}\right)^2} \tag{2.15}$$

Where V_c (nl s^{-1} mm^{-2}) is the recorded flow velocity taken from literature [1]; the number of DMT exposed by cutting 1 mm from the tip of a cat's canine tooth was averaged at $N \approx 2249/\text{mm}^2$ [1] and the mean diameter of the DMT was $d_d \approx 0.73$ μm [1]. The simulation results were checked for convergence to ensure no significant dependence on element size. Since the local channel diameter (the gap between TB and OP) is less than 1 μm, the slip boundary effect on the simulation results should be considered by using the following Eq. [4]:

$$\frac{\tau_{\text{slip}}}{\tau_{\text{non - slip}}} = \frac{1}{1 + (\delta/h)} \tag{2.16}$$

where τ_{slip} (Pa) and $\tau_{\text{non-slip}}$ (Pa) are the wall shear stresses when slip and non-slip boundary conditions are applied, respectively; δ (μm) is the slip length at the wall (~0.1 μm); h (μm) is the distance between two parallel walls (e.g., the local channel diameter of ~0.12 μm).

2.3.2 Modeling Transduction

Nociceptors are the receptors for pain sensations. When stimulated by noxious stimuli, the nociceptors will mediate the selective passage of ions across ion channels on the cell membrane generating action potentials [30]. The ion channels can generally be activated by three stimuli (e.g., mechanical, thermal and/or chemical stimuli) and three different transmembrane ion currents are induced accordingly. Since the ion channels are arranged in parallel in the membranes of nociceptors, I_{st} ($\mu A/cm^2$) (denoting the total stimuli-induced current) can be calculated as the sum of the three currents:

$$I_{st} = I_{mech} + I_{heat} + I_{chem} \tag{2.17}$$

where I_{mech}, I_{heat} and I_{chem} are the currents generated by the opening of the mechanically, thermally and chemically gated ion channels, respectively (all in $\mu A/cm^2$). In the present case, the intradental nerve terminals are stimulated by the shear stress, and only mechanical-gated ion channels are taken into account for the generation of the stimulus-induced ion current.

It is reported that the mechanically-gated ion current is exponentially proportional to the mechanical stimulation [31]; the current is calculated as [5]:

$$I_{st} = \left(\left[C_{h1} \exp\left(\frac{(\tau_{MSS} - \tau_{thr})/\tau_{thr}}{C_{h2}} \right) + C_{h3} \right] + I_{shift} \right) \times \mathrm{H}\left(\tau_{MSS} - \tau_{thr}\right) \tag{2.18}$$

where I_{st} is the evocated current; C_{h1}, C_{h2} and C_{h3} are constants; $H(x)$ is the Heaviside function responsible for the threshold process and I_{shift} ($\mu A/cm^2$) is the shift current [5]. The constants C_{h1}, C_{h2} and C_{h3} are set to be 2.0 $\mu A/cm^2$, 2.0 $\mu A/cm^2$ and -1.0 $\mu A/cm^2$, respectively [5].

All neurons are found to behave in a quantitatively similar way to that described by the Hodgkin-Huxley (H-H) model [32]. Hence, a modified H-H model has been proposed to introduce more than one K^+ channel into the modeling of the frequency modulation of nociceptors [33,34]. The intradental nociceptor transduction is modeled using the modified H-H model as [5]:

$$C_{mem} \frac{dV_{mem}}{dt} = I_{st} + I_{Na} + I_K + I_L + I_{K2} \tag{2.19}$$

where V_{mem} is the membrane potential (mV); t (ms) is the neural discharging time; C_{mem} ($\mu F/cm^2$) is the membrane capacity per unit area; I_{Na}, I_K and I_L are the sodium (Na^+), K^+ and leakage currents ($\mu A/cm^2$), respectively and I_{K2} is the fast, transient K^+ current. I_{Na}, I_K, I_L and I_{K2} are given by [33]:

$$I_{Na} = g_{Na} m^3 h \left(V_{Na} - V_{mem}\right) \tag{2.20}$$

$$I_K = g_K n^4 \left(V_K - V_{mem}\right) \tag{2.21}$$

$$I_L = g_L \left(V_L - V_{mem}\right) \tag{2.22}$$

$$I_{K2} = g_A A^3 B \left(V_{K2} - V_{mem}\right) \tag{2.23}$$

where m, n and h are gating variables; A and B are factors having the same functional significance as factors m and h; V_{Na}, V_K, V_L and V_{K2} are the reversal potentials (mV) for the Na^+, K^+, leakage and fast transient K^+ currents, respectively, and g_{Na}, g_K, g_L and g_A are the maximal ionic conductances (mS/cm^2) through Na^+, K^+, leakage and the fast transient K^+ currents, respectively. The ion currents are driven by electrical potential difference, conductance or permeability coefficient [32]. The conductance of the ion current can be regulated by voltage dependent activation and inactivation variables (gating variables), given as [32]:

$$\frac{dx}{dt} = \frac{x_\infty(V_{\text{mem}}) - x}{\tau_x(V_{\text{mem}})} \quad \text{or} \quad \frac{dx}{dt} = \alpha_x(1-x) - \beta_x x \tag{2.24}$$

where x can be any one of the three gating variables m, n or h and the initial values for m, n and h are 0.06, 0.33 and 0.83, respectively. $\tau_x = x_{fac}[1/(\alpha_x + \beta_x)]$ and $x_\infty = \alpha_x/(\alpha_x + \beta_x)$, α_x and β_x are rate constants (in sec^{-1}), which can be approximated from voltage clamp experiments [32], and are given as:

$$\alpha_n = -0.01(V_{\text{mem}} + 50)/(\exp[-(V_{\text{mem}} + 50)/10] - 1) \tag{2.25}$$

$$\beta_n = 0.125\exp[-(V_{\text{mem}} + 60)/80] \tag{2.26}$$

$$\alpha_m = -0.1(V_{\text{mem}} + 35)/(\exp[-(V_{\text{mem}} + 35)/10] - 1) \tag{2.27}$$

$$\beta_m = 4\exp[-(V_{\text{mem}} + 60)/18] \tag{2.28}$$

$$\alpha_h = 0.07\exp[-(V_{\text{mem}} + 60)/20] \tag{2.29}$$

$$\beta_h = 1/(\exp[-(V_{\text{mem}} + 30)/10] + 1) \tag{2.30}$$

The factors A and B are determined according to [35]:

$$\tau_A\frac{dA}{dt} + A = A_\infty, \quad A_\infty = \left(0.0761\frac{\exp[(V_{\text{mem}} + 94.22)/31.84]}{1 + \exp[(V_{\text{mem}} + 1.17)/28.93]}\right)^{1/3} \tag{2.31}$$

$$\tau_A = A_{\text{fac}}\left(0.3632 + \frac{1.158}{1 + \exp[(V_{\text{mem}} + 55.96)/20.12]}\right) \tag{2.32}$$

$$\tau_B\frac{dB}{dt} + B = B_\infty, \quad B_\infty = \left(\frac{1}{1 + \exp[(V_{\text{mem}} + 53.3)/14.54]}\right)^4 \tag{2.33}$$

$$\tau_B = B_{\text{fac}}\left(1.24 + \frac{2.678}{1 + \exp[(V_{\text{mem}} + 50)/16.027]}\right) \tag{2.34}$$

No data on intradental nociceptors has been reported, the data used for the current model are adopted from those for a squid axon [33]: $g_A = 47.7$ mS/cm^2, $C_{mem} = 2.8$ μF/cm^2, $g_{Na} = 120$ mS/cm^2, $g_K = 36$ mS/cm^2, $g_L = 0.3$ mS/cm^2, $A_{fac} = B_{fac} = 7.0$, $m_{fac} = H_{fac} = 0.263$ and

$n_{fac} = 2.63$. $V_{Na} = 57.19$ mV and $V_K = V_{K2} = -78.78$ mV [27]; the reversal potential of the leakage current, $V_L = -63.79$ mV, is obtained by adjusting V_L so that the equilibrium membrane potential is achieved [24,25].

2.4 RESULTS AND DISCUSSION

2.4.1 Tooth Thermomechanics

Upon cold stimulation and followed by natural convection on the enamel surface, the temperature and thermal stress distributions within the tooth have been simulated, and the results are shown in Fig. 2.4a–b. The temperature change has not yet reached the inner layer at the initial stage of cooling (Fig. 2.4a). Cold stimulation causes tensile stresses (positive in

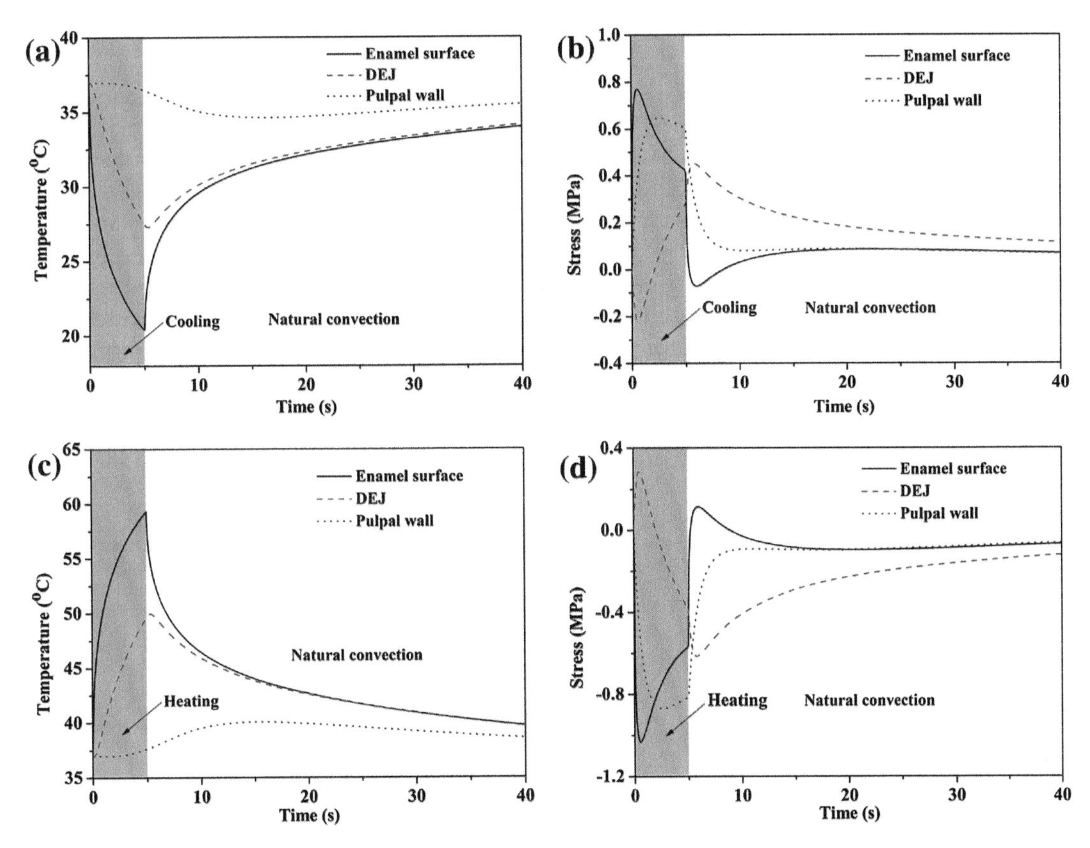

FIGURE 2.4 Simulated temperature and thermal stress change as a function of time at enamel surface, dentine-enamel junction (DEJ) and pulpal wall during and after application of 5°C cold water (a, b) and 80°C hot water (c, d) on enamel surface. After thermal stimulation (5 s duration) was removed, the enamel was exposed to natural convection, namely, cooling with ambient temperature 25°C with heat transfer coefficient ~10 W/(m^2 K). Initial body temperature was 37°C. Temperature at bottom of pulp layer was assumed to remain unchanged (37°C) during and after thermal stimulation [3].

value) at the enamel surface, and compressive stresses (negative in value) at the DEJ (Fig. 2.4b). Values of both the tensile and compress stresses reach their maximum at $t \approx 1$ s, agreeing with the finite elements analysis of *Llyod et al.* [37]. Cold stimulation induces contraction of the outer layer, resulting in flexure of the layered tooth structure. Such deformation stretches the pulpal wall, causing the rapid development of tensile stresses there (Fig. 2.4b). As temperature change reaches deeper into the inner layer, the tensile stresses at the pulpal wall decrease, causing thermal contraction of the structure, counteracting the initial flexure. These deformation characteristics are consistent with the experimental observations by *Linsuwanont et al.* [15]. When hot stimulation is applied to the enamel surface, the present model simulations show the opposite trend in the changes in both temperature and stresses to that seen under cold stimulation (Fig. 2.4c–d).

2.4.2 The Mechanism Underlying Thermally Induced DFF

DFF caused by the thermal stimulation of exposed dentine (without an enamel layer) has been simulated, and the result is shown in Fig. 2.5a. Significant differences can be observed when comparing the simulated fluid flow velocity caused by dentine tubule thermal deformation only (dashed line, Fig. 2.5a) with that caused by thermal expansion/contraction of dentinal fluid only (dotted line, Fig. 2.5a). An initially rapid inward flow under heating and outward flow under cooling is seen in the case of dentine tubule thermal deformation. In contrast, a thermal expansion/contraction of dentinal fluid causes a slow response of fluid flow at the initial stage of heating or cooling. In addition, the change in DFF as induced by expansion/contraction of dentinal fluid at the interval of heating and cooling is opposite to that induced by dentinal tubule deformation. The DFF simulated by considering both dentinal tubule deformation and dentinal fluid expansion/contraction (solid line, Fig. 2.5a) agrees qualitatively with experimental observations by *Andrew et al.* [1]. The results of Fig. 2.5a–b

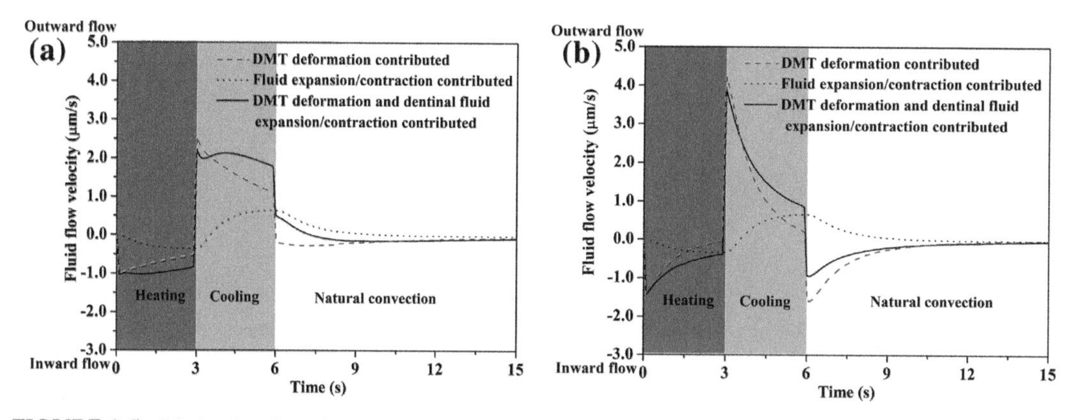

FIGURE 2.5 Mechanism for initiation of dentinal fluid flow induced by thermal stimulation on (a) exposed dentine surface and (b) intact teeth surface (enamel surface). Thermal boundary was identical to experimental condition [1], namely, heated by 55°C hot water for 3 s, followed by 5°C cold water for 3 s, and thereafter natural convection on exposed dentine surface (ambient temperature 25°C) [3].

imply that the initiation of DFF could be attributed to both dentinal tubule deformation and dentinal fluid expansion/contraction. Furthermore, dentinal tubule deformations are responsible for the initially rapid response of fluid flow during and after thermal stimulation. Compared to Fig. 2.5a, Fig. 2.5b shows similar changes in DFF when thermal stimulation was applied on the enamel surface, though the magnitude of the fluid flow velocity differs. Such similarity indicates that the mechanism of dentinal fluid flow remains the same when thermal stimulation is applied either to the enamel surface or the dentine surface.

2.4.3 DFF and Its Implications for Tooth Thermal Pain

A rapid transient tooth pain response has been widely reported in daily life and dentistry under cold stimulation [13,38,39]. *In vivo* studies have shown that after a short latency (<1 s) of cold stimulation (0∼5°C), intradental neurons respond with an initially high-frequency discharge [1,40,41]. The discharge rate decreased [40] and ceased within 4 s [41] though the same cold stimulus was still present. The mechanism underlying the dynamic neural discharge patterns remains unclear. The variation in DFF velocity during thermal stimulation may offer some mechanistic insight into the characteristics of this cold stimulation-induced neural discharge pattern. Figure 2.6 (dashed line) shows a high rate of outward DFF at the initial stage (<1 s) of cold stimulation, and a decrease in velocity as time elapses. The high rate

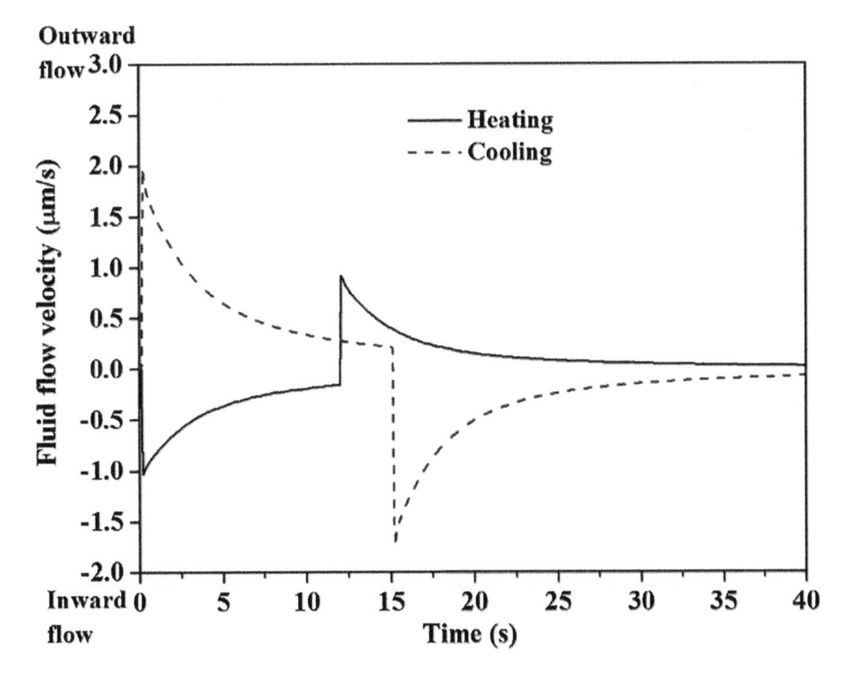

FIGURE 2.6 Simulated DFF flow velocity as a function of time during thermal stimulation. Cooling: 5°C, 15 s duration, re-warming: 37°C. Heating: 55°C, 12 s duration, re-warming: 37°C.

of DFF after a short period of cold stimulation may induce large shear stresses, resulting in mechanical activation of nerve terminals and a consequent neural discharge [1,6,11,42]. The rapid decrease in DFF velocity will dramatically reduce the shear stress, corresponding to a decrease and cessation of neural discharge.

When hot stimulation (55°C) is applied, a relatively long latency (>10 s) [40,41] in neural responses can be observed. No neural discharge can be detected [40,41] although an initially rapid response in inward DFF appears (Fig. 2.6, solid line). These findings do not contradict with those for cold stimulation. The intradental neurons are much 'less sensitive' to inward DFF than outward DFF, hence a higher rate of inward DFF is required to evoke an intradental nerve response [1,11]. A neural discharge after a long latency should be associated with the activation of thermo-sensitive nociceptors due to excessive temperature rise [41].

2.4.4 The Difference Between Hot and Cold Tooth Pain

The mechanical threshold for pulpal nociceptors (mechanoreceptors) is calculated to be about 90 Pa [5]. To verify the model, fluid velocities have been adopted from the literature [1] and the corresponding shear stresses have been calculated, and are listed in Table 2.3.

Figure 2.7 shows the relationship between the neural discharge rate (over 5 s) and the flow velocity. The experimental observations show that the nociceptive receptors respond in a significantly different manner to the DFF in different flow directions [1]. The neural discharge rate increases progressively as the outward flow velocity increases, above a threshold. In contrast, the associated receptors are much less sensitive to inward flow. The simulated results are in good agreement with the experimental results. The numerical simulations theoretically reveal that the OPD accounts for the difference in the responses of the intradental nerve to the inward and outward flows.

Cold stimuli (0~5°C) induce outward flow velocities ranging between 531.2 and 849.9 μm/s [1,39], while the range is approximately 354.1–779.1 μm/s [1] for inward flow responding to hot stimuli (55°C). The inward and outward flow directions and their corresponding fluid velocity magnitudes are inconsistent with those employed as boundary

TABLE 2.3 Fluid Flow Velocity and the Simulated Shear Stress

Fluid Flow Velocity (μm/s)	Simulated Shear Stress (Pa)
−984	99.4
−850	92.6
−473	57.8
−111	17.4
460	90.4
611	152.4
791	227.7
1011	351.2

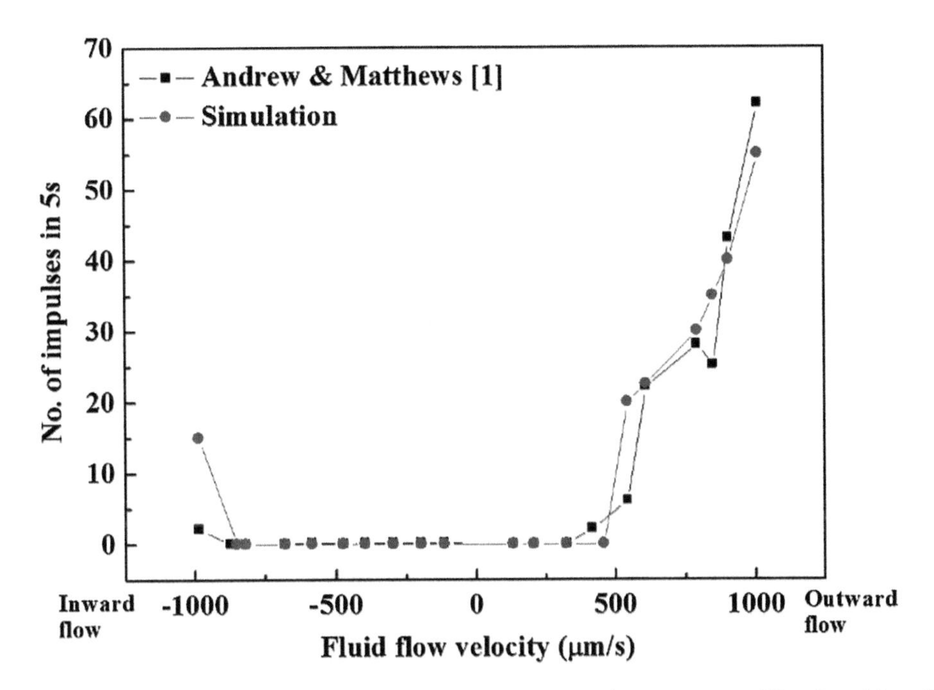

FIGURE 2.7 Comparison of frequency response between experimental measurements [1] and model predictions. Note that cold stimulation (0 ~ 5°C) is reported to cause outward flow velocities ranging between 531.2 ~ 849.9 μm/s [1], while hot stimulation (~55°C) causes inward flow velocities ranging between 354.1 ~ 779.1 μm/s [1].

conditions in the present work (Fig. 2.7). Although fluid velocities were employed as the boundary conditions in modeling the TB MSS and the subsequent neural discharge, the difference between hot and cold sensations still can be revealed.

In the case of cold stimuli (0 ~ 5°C), a short latency (~1 s) of the neural response can be observed [40,41]. During that stage, the local temperature (where the TB is located) is still far from being able to activate the thermal receptors [41]. Therefore, it seems unlikely that the rapid response to the cold stimulation originates from thermo-sensitive receptors. Note that fluid flow could be detected before a temperature change in DEJ, and the latency of the initiation of the DFF (<1 s) [1,12] (induced by either hot or cold stimuli) corresponds to the latency of the neural response. In addition, the flow velocity induced by cold stimuli may easily exceed the threshold [1,39] for activating the mechanoreceptors (Fig. 2.7). Therefore, the initial stage of cold-induced tooth pain (sharp, shooting pain) may involve the activation of mechano-sensitive receptors by the DFF. It should be mentioned that tooth pain (dull, burning pain) after a long latency (~30 s) of cold stimulation may be attributed to the activation of cold-sensitive nociceptors [43,44].

A neural response can only be detected after 10 s of hot stimuli (55°C) [40,41]. Prior to this, no neural discharge signal can be detected [1,40,41]. This does not contradict with the conclusion that the DFF may evoke the neural response, since hot stimuli cannot initiate the high rate of DFF [1] needed to activate the mechano-sensitive receptors. It is possible that after such a long latency, the temperature around the thermally sensitive receptors reaches the threshold [41], triggers the receptors and causes pain sensation [40,41].

2.5 CONCLUSION

A thermomechanical model has been developed to simulate thermally-induced DFF in a single DMT. The results of the simulation using the proposed models indicate that an initially rapid outward DFF may cause large shear stresses upon intradental nerve terminals, causing a neural discharge after short period of cold stimulation. By coupling the CFD model with a modified Hodgkin-Huxley (H-H) model, we provide some mechanistic insight into long-standing questions regarding the phenomenon that cold stimuli seem to cause higher pain intensity than do hot stimuli, when applied to teeth.

References

[1] D. Andrew, B. Matthews, Displacement of the contents of dentinal tubules and sensory transduction in intradental nerves of the cat, Journal of Physiology (London) 529 (2000) 791–802.
[2] M. Brännström, The hydrodynamic theory of dentinal pain: sensation in preparations, caries, and the dentinal crack syndrome, Journal of Endodontics 12 (1986) 453–457.
[3] M. Lin, S.B. Liu, L. Niu, F. Xu, T.J. Lu, Analysis of thermal-induced dentinal fluid flow and its implications in dental thermal pain, Archives of Oral Biology 56 (2011) 846–854.
[4] M. Lin, Z.Y. Luo, B.F. Bai, F. Xu, T.J. Lu, Fluid dynamics analysis of shear stress on nerve endings in dentinal microtubule: a quantitative interpretation of hydrodynamic theory for tooth pain, Journal of Mechanics in Medicine and Biology 11 (2011) 205–219.
[5] M. Lin, Z.Y. Luo, B.F. Bai, F. Xu, T.J. Lu, Fluid mechanics in dentinal microtubules provides mechanistic insights into the difference between hot and cold dental pain, PLoS ONE 6 (2011) e18068.
[6] D.H. Pashley, Dynamics of the pulpo-dentin complex, Critical Reviews in Oral Biology and Medicine 7 (1996) 104–133.
[7] B.J. Sessle, Invited review: the neurobiology of facial and dental pain: present knowledge, future directions, Journal of Dental Research 66 (1987) 962–981.
[8] N. Vongsavan, B. Matthews, The relationship between the discharge of intradental nerves and the rate of fluid flow through dentine in the cat, Archives of Oral Biology 52 (2007) 640–647.
[9] S.-Y. Kim, J. Ferracane, H.-Y. Kim, I.-B. Lee, Real-time measurement of dentinal fluid flow during amalgam and composite restoration, Journal of Dentistry 38 (2010) 343–351.
[10] M. Lin, F. Xu, T.J. Lu, B.F. Bai, A review of heat transfer in human tooth – experimental characterization and mathematical modeling, Dental Materials 26 (2010) 501–513.
[11] P. Charoenlarp, S. Wanachantararak, N. Vongsavan, B. Matthews, Pain and the rate of dentinal fluid flow produced by hydrostatic pressure stimulation of exposed dentine in man, Archives of Oral Biology 52 (2007) 625–631.
[12] P. Linsuwanont, J.E.A. Palamara, H.H. Messer, An investigation of thermal stimulation in intact teeth, Archives of Oral Biology 52 (2007) 218–227.
[13] M. Brännström, G. Johnson, Movements of the dentine and pulp liquids on application of thermal stimuli. An in vitro study, Acta Odontologica Scandinavica 28 (1970) 59–70.
[14] B. Matthews, Cold-sensitive and heat-sensitive nerves in teeth, Journal of Dental Research 47 (1968) 974–975.
[15] P. Linsuwanont, A. Versluis, J.E. Palamara, H.H. Messer, Thermal stimulation causes tooth deformation: a possible alternative to the hydrodynamic theory? Archives of Oral Biology 53 (2008) 261–272.
[16] M. Lin, Q.D. Liu, F. Xu, B.F. Bai, T.J. Lu, In vitro investigation of heat transfer in human tooth, in: C. Quan (Ed.), Fourth International Conference on Experimental Mechanics, Spie-Int Soc Optical Engineering, Bellingham, 2010.
[17] M. Lin, Q.D. Liu, T. Kim, F. Xu, B.F. Bai, T.J. Lu, A new method for characterization of thermal properties of human enamel and dentine: influence of microstructure, Infrared Physics and Technology 53 (2010) 457–463.
[18] J.H. de Vree, T.A. Spierings, A.J. Plasschaert, A simulation model for transient thermal analysis of restored teeth, Journal of Dental Research 62 (1983) 756–759.

[19] W.S. Brown, W.A. Dewey, H.R. Jacobs, Thermal properties of teeth, Journal of Dental Research 49 (1970) 752–755.
[20] H.H. Xu, D.T. Smith, S. Jahanmir, E. Romberg, J.R. Kelly, V.P. Thompson, E.D. Rekow, Indentation damage and mechanical properties of human enamel and dentin, Journal of Dental Research 77 (1998) 472–480.
[21] D.N. Fenner, P.B. Robinson, P.M.Y. Cheung, Three-dimensional finite element analysis of thermal shock in a premolar with a composite resin MOD restoration, Medical Engineering and Physics 20 (1998) 269–275.
[22] H.C. Xu, W.Y. Liu, T. Wang, Measurement of thermal expansion coefficient of human teeth, Australian Dental Journal 34 (1989) 530–535.
[23] F. Xu, T. Wen, K.A. Seffen, T.J. Lu, Biothermomechanics of skin tissue, Journal of the Mechanics and Physics of Solids 56 (2008) 1852–1884.
[24] F. Xu, T.J. Lu, Skin biothermomechanics: modeling and experimental characterization, Advances in Applied Mechanics 43 (2009) 147–248.
[25] C. Carda, A. Peydro, Ultrastructural patterns of human dentinal tubules, odontoblasts processes and nerve fibers, Tissue and Cell 38 (2006) 141–150.
[26] G. Berggren, M. Brännström, The rate of flow in dentinal tubules due to capillary attraction, Journal of Dental Research 44 (1965) 408–415.
[27] G.R. Holland, B. Matthews, P.P. Robinson, An electrophysiological and morphological study of the innervation and reinnervation of cat dentine, The Journal of Physiology 386 (1987) 31–43.
[28] M.R. Byers, Dental sensory receptors, International Review of Neurobiology 25 (1984) 39–94.
[29] D.H. Pashley, M.I. Livingston, O.W. Reeder, J.A. Horner, Effects of the degree of tubule occlusion on the permeability of human dentine, in vitro, Archives of Oral Biology 23 (1978) 1127–1133.
[30] E.W. McCleskey, M.S. Gold, Ion channels of nociception, Annual Review of Physiology 61 (1999) 835–856.
[31] R. Francois, J.D. Liam, N.W. John, Kinetic properties of mechanically activated currents in spinal sensory neurons, The Journal of Physiology 588 (2010) 301–314.
[32] A.L. Hodgkin, A.F. Huxley, A quantitative description of membrane current and its application to conduction and excitation in nerve, The Journal of Physiology 117 (1952) 500–544.
[33] F. Xu, T. Wen, T.J. Lu, K.A. Seffen, Modeling of nociceptor transduction in skin thermal pain sensation, Journal of Biomechanical Engineering 130 (2008) 041013.
[34] F. Xu, T.J. Lu, K.A. Seffen, Skin thermal pain modeling – a holistic method, Journal of Thermal Biology 33 (2008) 223–237.
[35] J.A. Connor, D. Walkter, R. McKown, Neural repetitive firing modifications of the Hodgkin-Huxley axon suggested by experimental results from crustacean axons, Biophysical Journal 18 (1977) 81–102.
[36] A.L. Hodgkin, The Conduction of Nervous Impulses, Liverpool University Press, Liverpool, 1964.
[37] B.A. Llyod, M.B. McGinley, W.S. Brown, Thermal stress in teeth, Journal of Dental Research 57 (1978) 571–582.
[38] M.K.C. Mengel, A.E. Stiefenhofer, E. Jyväsjärvi, K.-D. Kniffki, Pain sensation during cold stimulation of the teeth: differential reflection of A[delta] and C fiber activity? Pain 55 (1993) 159–169.
[39] W. Chidchuangchai, N. Vongsavan, B. Matthews, Sensory transduction mechanisms responsible for pain caused by cold stimulation of dentine in man, Archives of Oral Biology 52 (2007) 154–160.
[40] E. Jyväsjärvi, K.D. Kniffki, Cold stimulation of teeth: a comparison between the responses of cat intradental A delta and C fibres and human sensation, The Journal of Physiology 391 (1987) 193–207.
[41] B. Matthews, Responses of intradental nerves to electrical and thermal stimulation of teeth in dogs, Journal of Physiology (London) 264 (1977) 641–664.
[42] M. Brännström, A. Astroem, A study on the mechanism of pain elicited from the dentin, Journal of Dental Research 43 (1964) 619–625.
[43] M.N. Naylor, Studies on the mechanism of sensation to cold stimulation of human dentine, in: D.J. Anderson (Ed.), Sensory mechanisms in dentine, Oxford Press, 1963, pp. 80–87.
[44] Chul-Kyu Park, Mi Sun Kim, Zhi Fang, Hai Ying Li, Sung Jun Jung, Se-Young Choi, Sung Joong Lee, Kyungpyo Park, J.S. Kim, S.B. Oh, Functional expression of thermo-transient receptor potential channels in dental primary afferent neurons: implication for tooth pain, Journal of Biological Chemistry 281 (2006) 17304–17311.

CHAPTER

3

Drug Release in Biological Tissues

Filippo de Monte[a], Giuseppe Pontrelli[b], and Sid Becker[c]

[a]Department of Industrial and Information Engineering and Economics, University of L'Aquila, L'Aquila, Italy,

[b]Istituto per le Applicazioni del Calcolo (IAC), CNR, Rome, Italy,

[c]University of Canterbury, Department of Mechanical Engineering, Christchurch, New Zealand

OUTLINE

Nomenclature 60

Greek Symbols 60

Acronyms 61

Subscripts 61

Superscripts 61

3.1 Introduction 61

3.2 Continuum Modeling of Mass Transport in Porous Media 65

3.2.1 Porosity and Volume-Averaged Variables 66

3.2.1.1 Averaged Concentration 67

3.2.2 Permeability, Darcy's Law and the Continuity Equation 68

3.2.2.1 The Continuity Equation 69

3.2.2.2 Extended Continuity Equation 71

3.2.3 Tortuosity and Fick's Law 72

3.3 Conservation of Drug Mass 73

3.3.1 Drug Mass Balance in the Fluid Phase 74

3.3.2 Drug Mass Balance in the Solid Phase 77

3.3.3 Governing Equations 78

3.4 Analytical Solutions for Local Mass Non-Equilibrium 78

3.4.1 Nusselt's Solution 79

3.4.2 Schumann's Solution 83

3.4.3 Anzelius's Solution 85

3.4.4 Recent Solutions 86

3.5 Analytical Solutions for Local Mass Equilibrium 87

3.5.1 A Worked Example 87

© 2013 Elsevier Inc. All rights reserved.

http://dx.doi.org/10.1016/B978-0-12-415824-5.00003-5

3.6 Applications of Porous Media to the Drug-Eluting Stent 92
3.6.1 The Fluid Wall Model: The Pure Diffusion Approximation 95
3.6.1.1 General Physiological and Mathematical Description 96
3.6.1.2 Concentration Solutions and Results 99
3.6.2 The Fluid Wall Model: The Advection-Reaction-Diffusion Equation 102
3.6.2.1 General Physiological and Mathematical Description 103
3.6.2.2 Concentration Solutions and Results 104
3.6.3 The Multi-Layered Wall Model: The Pure Diffusion Approximation 108
3.6.3.1 General Physiological and Mathematical Description 108
3.6.3.2 Concentration Solutions and Results 111
3.7 Conclusion 116
References 116

Nomenclature

A	cross-sectional area (m^2)
Bi	Biot number
c	concentration ($kg\ m^{-3}$)
$\langle c \rangle$	volume-averaged concentration ($kg\ m^{-3}$)
D	effective diffusivity ($m^2\ s^{-1}$)
f	retardation coefficient
G_X	One-dimensional Green function along $x(m^{-1})$
j	specific mass flux due to a concentration gradient ($kg\ s^{-1}\ m^{-2}$)
$\hbar$	mass transfer coefficient ($m\ s^{-1}$)
k	partition coefficient
K	porous medium permeability (m^2)
l	length (m)
m	mass (kg)
$\dot{m}$	mass flow rate ($kg\ s^{-1}$)
M	mass per unita area ($kg\ m^{-2}$)
p	pressure (Pa)
P	membrane permeability ($m\ s^{-1}$)
Pe	Péclet number
r_h	hydraulic radius (m)
t	time (s)
u	velocity along $x(m\ s^{-1})$
V	volume (m^3)
x,y,z	Cartesian space coordinates (m)

Greek Symbols

β	reaction rate coefficient (s^{-1}). Also, β denotes eigenvalues in Subsection 3.5.1.
γ	diffusivity ratio

ε	volumetric porosity
ε_k	available volume fraction, $k\varepsilon$
λ	tortuosity. Also, λ denotes eigenvalues in Section 3.6.
μ	dynamic viscosity ($kg\ m^{-1}\ s^{-1}$)
ρ	density ($kg\ m^{-3}$)
σ	available volume fraction ratio
ϕ	dimensionless membrane permeability
Φ	Thiele Modulus

Acronyms

DES	Drug-Eluting Stent
EOS	Equation of State
LME	Local Mass Equilibrium
LMNE	Local Mass Non-Equilibrium
SOV	Separation of Variables

Subscripts

c	continuum
d	drug
rev	representative elementary volume

Superscripts

f	fluid phase
s	solid phase

3.1 INTRODUCTION

Problems involving the release of a drug from a polymeric gel matrix into biological tissues arise in a number of scientific and bioengineering disciplines. Important technological areas include drug-eluting stents for the prevention of restenosis [1–6], therapeutic contact lenses to increase the ocular bioavailability of ophthalmic drugs [7,8] and dermal and transdermal drug delivery [9–11].

The first application, concerning the process of restenosis after stent implantation for the treatment of coronary artery disease, is a result of a complex interplay of several implant-induced biological processes. Stent-based drug delivery has been shown to inhibit several of these biological processes, thereby preventing restenosis. A drug-eluting stent (DES) consists of a metallic stent platform with a polymeric coating that encapsulates a therapeutic drug, as shown in Fig. 3.1. Before the advent of DESs, stents were primarily used as scaffolds for keeping the arterial lumen opened, and their design involved interaction between vascular biology and engineering. As DES has come of age, the picture has become much more complex and multi-disciplinary. It now involves an interplay between vascular biology, polymer chemistry, pharmacology [1–4] and engineering [5,6]. In addition to being designed for structural integrity, deliverability to the lesion site and conformability with the arterial wall upon deployment, the stent also has to be optimized for drug delivery [3].

FIGURE 3.1 Drug-eluting coronary stent. It is coated with a drug that inhibits cell growth that could reclose a propped-open artery.

As far as the second application is concerned, diseases of the anterior segment of the eye are mostly treated by topical ocular administration in the inferior fornix of the conjunctiva. However, this procedure is extremely inefficient because when a drop (50 to 100 μl per drop) is instilled into the eye, the ophthalmic drug has a short residence time in the conjunctival sac of less than 5 min, and only 1–5% of the applied drug penetrates the cornea and reaches the intraocular tissues. The biovailability tends to be low and depends on the precorneal fluids dynamics, how well the drug binds to tear proteins, conjunctival drug absorption, tears turn over, resistance to corneal penetration, nasolachrymal drainage, metabolic degradation and non-productive absorption. The absorption and the efficacy of the instilled drug can be increased by altering its formulation and/or by changing the local conditions. For this reason, in recent years many researchers have proposed the use of therapeutic contact lenses to increase the ocular bioavailability of ophthalmic drugs [7] as illustrated in Fig. 3.2. The first attempt to increase the residence time of an ophthalmic drug involved the use of soaked contact lenses. The lens is hydrated and then placed onto the cornea where it releases the drug until an equilibrium is reached between drug concentration in the contact lens and that in the conjunctival sac. The maximum drug loading is limited by the solubility of the drug in the polymeric matrix and the delivery period of time is still very short. However, from a medical point of view, the central question is to be able to predict the concentration of the drug in the anterior chamber of the eye [8]. In this case, mathematical models and numerical simulations are the only available tools to make such predictions.

Transdermal drug delivery (Fig. 3.3) has made an important contribution to medical practice, but has still to fully achieve its potential as an alternative to oral delivery and hypodermic injections. While much time and effort have been devoted to understanding the Fickian principles of drug diffusion across the skin membrane, one area of research that has been neglected has been transfollicular drug delivery. First-generation transdermal delivery systems

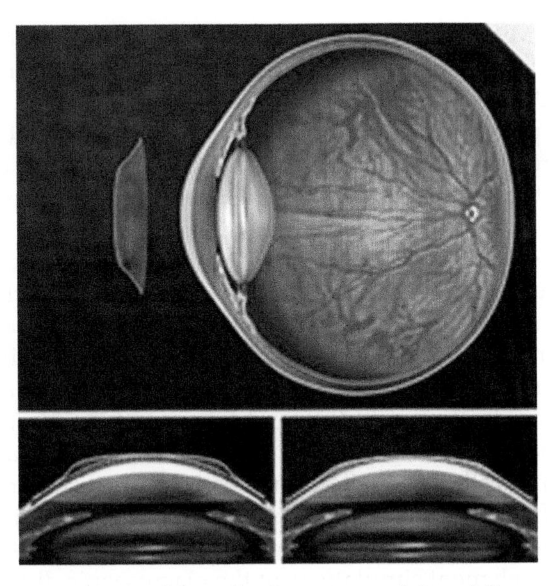

FIGURE 3.2 Therapeutic contact lens used for reshaping cornea and drug delivery.

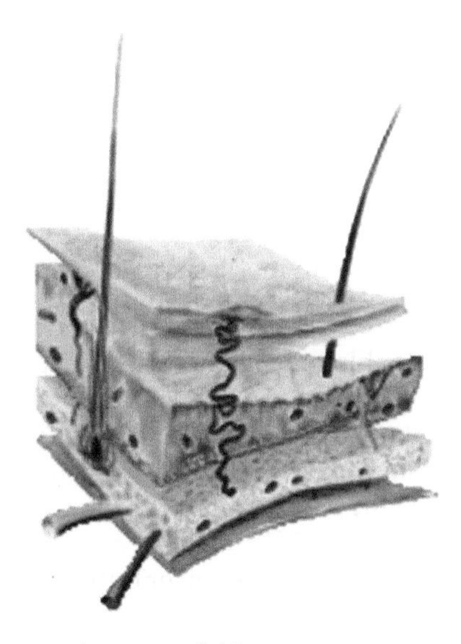

FIGURE 3.3 A depiction of the composite nature of skin.

have continued their steady increase in the clinical, where they are used for the delivery of small, lipophilic, low-dose drugs. Second-generation delivery systems using chemical enhancers, noncavitational ultrasound and iontophoresis have also resulted in clinical products; the ability of iontophoresis to control delivery rates in real time provides added functionality. Third-generation delivery systems target their effects to the skin's barrier layer, the stratum corneum, using microneedles, thermal ablation, microdermabrasion, electroporation and cavitational ultrasound. Microneedles and thermal ablation are currently progressing through clinical trials for delivery of macromolecules and vaccines, such as insulin, parathyroid hormone and influenza vaccine. Using these novel second- and third-generation enhancement strategies, transdermal delivery is poised to significantly increase its impact on medicine [10].

Several studies address the modeling of transdermal diffusion of drugs to better understand the permeation of molecules through the skin, especially the stratum corneum, which forms the main permeation barrier to percutaneous permeation [9]. In order to ensure the reproducibility and predictability of drug permeation through the skin and into the body, a quantitative understanding of the permeation barrier properties of the stratum corneum is crucial. Multiscale frameworks for modeling the multicomponent transdermal diffusion of molecules have been proposed in the literature [11]. The problem is in general divided into smaller problems of increasing length scale: microscopic, mesoscopic and macroscopic. First, the microscopic diffusion coefficient in the lipid bilayers of the stratum corneum is found through molecular dynamics (MD) simulations. Then, a homogenization procedure is performed over a model unit cell of the heterogeneous stratum corneum, resulting in effective diffusion parameters. These effective parameters are the macroscopic diffusion coefficients for the homogeneous medium that is 'equivalent' to the heterogeneous stratum corneum, and thus can be used in finite element simulations of the macroscopic diffusion process.

All these applications share the same modeling framework: the delivery of a drug from a polymeric matrix in contact with a biological tissue for therapeutic purposes. Such a tissue can be treated as a porous medium [12] as it is composed of dispersed cells separated by connective voids which allow nutrients, drugs, minerals, etc. to reach all cells within the tissue. Mass transport of these substances in many biological and medical applications is achieved by advection and diffusion within the tissue which may be enhanced by interstitial flows. Polymeric gels can be treated as porous media too. To obtain a sound understanding of the drug delivery process, mathematical modeling can be divided into two categories, as follows:

1. modeling of drug elution from the polymeric gel; and
2. modeling of transport of the drug in the biological tissue.

Both these models provide valuable insights to the engineers and scientists to assist in the design of drug delivery systems. They are strictly coupled, and can help the system designer to achieve a particular release rate and release time, as well a desired drug distribution and tissue concentration. For modeling purposes, it is important to identify the dominant physicochemical processes by the relevant parameters and take them into account when developing the mathematical model (see Sections 3.2 and 3.3). For example, in the coating of DESs, we have two phases (solid and fluid) whose drug concentrations are different, so generating the so-called *local mass non-equilibrium* (LMNE) involves a complex and challenging mathematical treatment [13] (see Section 3.4). However, the simplifying assumption of *local mass equilibrium* (LME) can sometimes be used [14] and this is detailed in Section 3.5. Only one phase, the fluid phase, will be considered when the drug diffuses within biological tissues.

In the second part of this chapter (Section 3.6), DES application has been selected to show the use of a hierarchy of models of increasing complexity. The first subsection 3.6.1 considers the fluid wall model for the arterial wall by using the Pure Diffusion Approximation. Then, this is extended in the second subsection 3.6.2 by making use of the Diffusion-Reaction-Advection Equation. Finally, the third subsection 3.6.3 shows the multi-layered wall model for the artery by again applying the Pure Diffusion Approximation.

3.2 CONTINUUM MODELING OF MASS TRANSPORT IN POROUS MEDIA

Consider a porous medium whose fluid and solid phases are binary mixtures, that is, mixtures of only two components; for example, when a DES is considered, these are drug (solute) and plasma (solvent) for the fluid phase, and drug (solute) and polymeric gel (solvent) for the solid phase.

To describe the mass transport of drug in each of these two phases, two different scales, microscopic and macroscopic (continuum), may be used. In the microscopic approach, the typical microscopic governing equations have to be written for each phase and then solved, either numerically or analytically. Specifically, for each phase the physics are captured by:

- the fluid phase: continuity and momentum equations, conservation of drug mass, boundary and initial conditions;
- the solid phase: conservation of drug mass and related boundary and initial conditions.

This approach would allow us to derive the drug concentration in both phases as a function of the space and time coordinates. These concentrations are defined as:

$$c_d^f \triangleq \frac{dm_d^f}{dV^f}, \quad c_d^s \triangleq \frac{dm_d^s}{dV^s} \tag{3.1}$$

where dm_d^f and dm_d^s are the elemental masses of drug contained into the fluid and solid elemental volumes dV^f and dV^s, respectively, as shown in Fig. 3.4.

Note also that $(x^f, y^f, z^f) \in V^f$, where x^f, y^f and z^f are the Cartesian coordinates of a point P^f within the fluid-phase. Similarly, c_d^s depends on $(x^s, y^s, z^s) \in V^s$, where x^s, y^s and z^s are the space coordinates of a point P^s in the solid-phase.

The microscopic approach, however, requires knowledge of the specific and local geometric shape of the individual pore structure networks, but these are in general not known. This is the motivation behind the continuum approach, which assumes that a porous medium may be represented as a globally homogeneous material by appropriately defining average variables over a sufficiently large volume, 'representative elementary volume' (r.e.v., for short), as depicted in Fig. 3.4. Hence, when dealing with the continuum approach (macroscopic), an elemental volume dV_c (where the subscript c denotes continuum) is greater than V_{rev}.

The length scale l of the r.e.v. is much larger than the pore scale given by the average size δ of the pores, but considerably smaller than the length scale L over which macroscopic changes of physical quantities, such as drug concentration and fluid velocity, have to be considered. Hence, the continuum approach requires that $\delta \ll l \ll L \Rightarrow L$ has to be at least two orders of magnitude larger than δ as in Section 8.3 of Ref. [15]. In biological tissues, it happens that $\delta < 0.1\ \mu m, l \sim 1\ \mu m$, and $L \sim 10\ \mu m$ to 1 cm, where L is chosen to be the characteristic linear dimension of the biological medium, for example the size of tissues or the distance between adjacent blood vessels. As a matter of fact, typical values found in the literature for the geometrical dimensions of arteries indicate a thickness L of only 2 μm for endothelium and IEL layers, but in this case the average size δ of the pores can be considered sufficiently less than

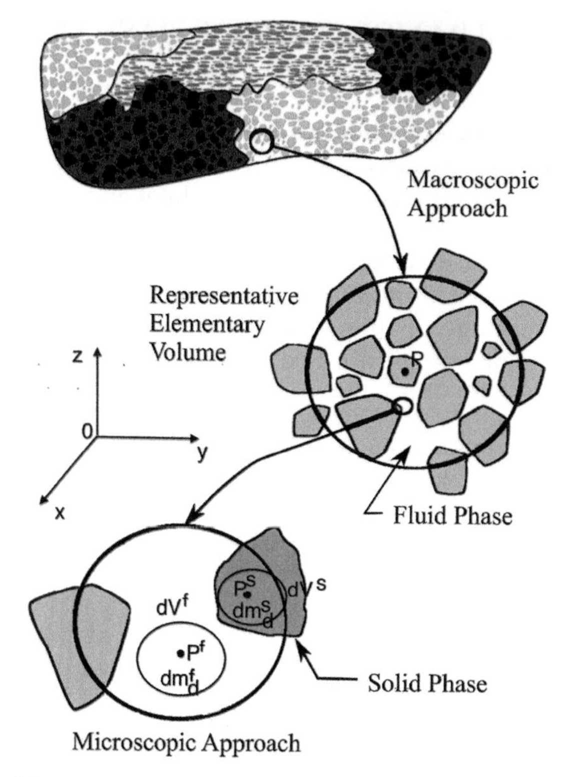

FIGURE 3.4 Schematic of the main scales in a porous medium with the representative elementary volume (r.e.v.).

0.1 μm. In the polymeric gel coating of DESs, $\delta < 0.1\ \mu m$ and $L \sim 5\ \mu m$ [5]. Therefore, solute transport in coatings and biological tissues can be analyzed by using the continuum approach.

When the continuum modeling is used, the details of the pore structures are neglected and macroscopic geometrical and physical variables, such as porosity, permeability and tortuosity, have to be defined. The porosity is strictly geometrical and is defined in Section 3.2.1 alongside the volume-averaged variables. The permeability appears in the momentum equation, and when the inertial contributions are negligible, it provides a linear relationship between pressure drop and flow rate, as indicated by the Darcy law (Section 3.2.2). Finally, Fick's law still applies for diffusion in porous media but is affected by another strictly geometrical variable, the tortuosity, as discussed in Section 3.2.3.

3.2.1 Porosity and Volume-Averaged Variables

The porosity ε (dimensionless), also called the 'volumetric' porosity, is the ratio of the void volume V^f contained in the porous medium and occupied by the fluid to the total volume V of the same medium. We have:

$$\varepsilon = \frac{V^f}{V} = \frac{V^f}{V^f + V^s} \tag{3.2}$$

(less than 1) where V^s is the volume of the solid part, represented by biological cells or polymeric matrix of the coating, as displayed in Fig. 3.1.

The porosity does not give any information about the interconnectedness of the void (the pores). In general pores may be penetrable or isolated, where the penetrable pores can be further classified as either passing or nonpassing as in Section 8.2 of Ref. [15]. Nonpassing pores can be considered as part of the solid matrix when dealing with transport analysis and an 'effective' porosity (ratio of connected void to total volume) can likely be introduced.

In fluid-saturated porous media (of interest in this chapter) that can be modeled as nondeformable, homogeneous, and isotropic, the volumetric porosity ε is the same as the so-called 'surface' porosity; that is, the fraction of void area (or free-flow area, A^f) to total area A of a typical cross section [16,17]. Hence, in these porous structures, all the void volume is connected; that is, all the pores are penetrable and passing.

It is relevant to observe that not all the penetrable and passing pores are accessible to solutes. In fact, it can happen that a pore is inaccessible to a solute if the solute molecule is larger than the pore, or if the pore is large enough but is surrounded by pores that are smaller than the solutes. The portion of void volume that is accessible to a solute is called 'available volume'. The ratio of the available volume to the void volume is defined as the 'partition coefficient' of the solute and is denoted by $k(\leq 1)$. Then, the product of k and ε is defined as the 'available volume fraction' ε_k, which provides the ratio of the available volume to the total volume of the porous structure, with $\varepsilon_k \leq \varepsilon$.

3.2.1.1 Averaged Concentration

There are two different ways of averaging over a volume. One is based on the volume of each phase contained in r.e.v., that is, V^f_{rev} for the fluid phase (which is a portion of the r.e.v., i.e., ε) and V^s_{rev} for the solid phase (which is the complementary portion of the r.e.v., i.e., $1-\varepsilon$). Another is based on the total volume of the r.e.v. (incorporating both fluid and solid material), given by $V_{rev} = V^f_{rev} + V^s_{rev}$.

For example, we can take a volume average of c^f_d as defined by Eq. (3.1) with respect to the corresponding phase volume or over the total volume (r.e.v.). Thus we have, respectively,

$$\left\langle c^f_d \right\rangle^f \triangleq \frac{1}{V^f_{rev}} \underbrace{\int_{V^f_{rev}} c^f_d \, dV^f_{rev}}_{m^f_d} \tag{3.3a}$$

$$\left\langle c^f_d \right\rangle \triangleq \frac{1}{V_{rev}} \int_{V_{rev}} c^f_d \, dV_{rev} = \frac{1}{V_{rev}} \underbrace{\int_{V^f_{rev}} c^f_d \, dV^f_{rev}}_{m^f_d} \tag{3.3b}$$

where $\left\langle c^f_d \right\rangle$ and $\left\langle c^f_d \right\rangle^f$ depend on x, y and z, which are the Cartesian coordinates of the centroid P of the r.e.v. (see Fig. 3.4), and m^f_d is the mass of drug contained in the fluid phase volume V^f_{rev}.

Equation (3.3a) gives the so-called intrinsic volume-averaged concentration of c^f_d, while Eq. (3.3b) yields its volume-averaged concentration [13,14,17]. Comparing the above two equations yields $\left\langle c^f_d \right\rangle = \varepsilon \left\langle c^f_d \right\rangle^f$. In addition, if not all the penetrable and passing pores are accessible to the drug, we have $\left\langle c_d^f \right\rangle = \varepsilon_k \left\langle c_d^f \right\rangle^f$. Also, the averaging operation of c_d^f performed through the above integrals provides the value of the drug concentration in the fluid phase at the centroid P, which can fall in the fluid or solid phase. In the case of Fig. 3.4, it falls in the solid phase. Also, it is assumed that the result of averaging over a volume is independent of the size of the r.e.v.

Similarly, we can take an average of c_d^s with respect to the solid phase volume V_{rev}^s (intrinsic average) or over the total volume V_{rev}. We obtain, respectively,

$$\left\langle c_d^s \right\rangle^s \triangleq \frac{1}{V_{rev}^s} \underbrace{\int_{V_{rev}^s} c_d^s \, dV_{rev}^s}_{m_d^s} \tag{3.4a}$$

$$\left\langle c_d^s \right\rangle \triangleq \frac{1}{V_{rev}} \int_{V_{rev}} c_d^s \, dV_{rev} = \frac{1}{V_{rev}} \underbrace{\int_{V_{rev}^s} c_d^s \, dV_{rev}^s}_{m_d^s} \tag{3.4b}$$

where $\left\langle c_d^s \right\rangle$ and $\left\langle c_d^s \right\rangle^s$ depend on x, y and z and m_d^s is the mass of drug contained in the solid phase of volume V_{rev}^s. Also, $\left\langle c_d^s \right\rangle = (1 - \varepsilon) \left\langle c_d^s \right\rangle^s$, or $\left\langle c_d^s \right\rangle = (1 - \varepsilon_k) \left\langle c_d^s \right\rangle^s$.

The averaging operation of c_d^s yields the value of drug concentration in the solid phase at the centroid P of the r.e.v., where a drug concentration in the fluid phase also exists. Therefore, each spatial point of the porous medium simultaneously contains two phases: a fluid phase with a volume fraction of ε (or ε_k) and a solid phase with a volume fraction of $1 - \varepsilon$ (or $1 - \varepsilon_k$). Similarly, we can get the spatial average of other variables involved in the mass transport phenomenon, such as fluid velocity, pressure, temperature and so on.

3.2.2 Permeability, Darcy's Law and the Continuity Equation

The permeability K is a measure of the flow conductivity in the porous medium. In flows for which the viscous effects dominate the inertial interactions (very low Reynolds number, probably around 5), the pressure gradient and the flow rate are linearly related. This flow regime is characteristic of many transport mechanisms in biological media. The permeability appears in Darcy's law (which is a derivation of the momentum balance in porous media) as the constant of linearity between the average flow velocity u and the pressure gradient (driving potential). In one dimension along x, Darcy model is expressed by

$$u = -\frac{K}{\mu} \frac{\partial p}{\partial x} \tag{3.5}$$

where p and μ are the fluid pressure and dynamic viscosity, respectively. The permeability has dimensions of (length)^2 and depends only on the geometry of the medium. The ratio K/μ is defined as the 'hydraulic conductivity'; for this reason, the permeability K is also called the 'specific hydraulic permeability'.

As the Darcy model ignores boundary effects on the flow, advanced models such as Forchheimer's equation and Brinkman's equation are available in the porous medium literature.

The former is applicable for large flow velocities, while the latter accounts for the boundary effects. The reader may refer to the studies [12] in Section 8.3 of Ref. [15] and Section 1.5 of Ref. [17].

The flow velocity u appearing in Eq. (3.5) is actually a spatial averaged velocity taken with respect to the total volume (r.e.v.), that is, $\langle u \rangle$. In general it depends on the space x and time t coordinates, and is termed the 'Darcy velocity', though various names are also used in the literature, such as filtration velocity or seepage velocity [16,17]. This velocity is related to the intrinsic volume-averaged velocity $\langle u \rangle^f$ through the Dupuit-Forchheimer relationship, $\langle u \rangle = \varepsilon \langle u \rangle^f$ [16]. We may also have $\langle u \rangle = \varepsilon_k \langle u \rangle^f$.

It is relevant to note that in the Darcy model, Eq. (3.5), u (effect) is an average velocity over the total volume, while p in the pressure gradient (cause) is an 'intrinsic' quantity, that is, a quantity whose average is based on the fluid phase volume, $\langle p \rangle^f$. For the sake of simplicity and brevity, in this chapter we will use u, u^f and p in place of, $\langle u \rangle$, $\langle u \rangle^f$ and $\langle p \rangle^f$, respectively.

Then, as we have two unknowns, u and p, and only Darcy's law, we need another equation to close the system. We can make use of the conservation of fluid mass in one dimension along x, as indicated in the next two subsections.

3.2.2.1 The Continuity Equation

Consider a 1-D differential volume element in the macroscopic flow field, $dV_c^f = A^f dx$, as depicted in Fig. 3.5. It is greater than V_{rev}^f and, hence, other than dV^f of Fig. 3.4 and Eq. (3.1) related to the microscopic approach. Writing a mass balance for this elemental volume yields the continuity equation along x. This balance may be stated as:

$$\begin{pmatrix}\text{net rate of mass flow entering}\\ \text{volume element in } x \text{ direction}\end{pmatrix} = \begin{pmatrix}\text{rate of increase of the fluid}\\ \text{mass in volume element}\end{pmatrix} \tag{3.6}$$

Both terms in the above statement can be evaluated as follows.

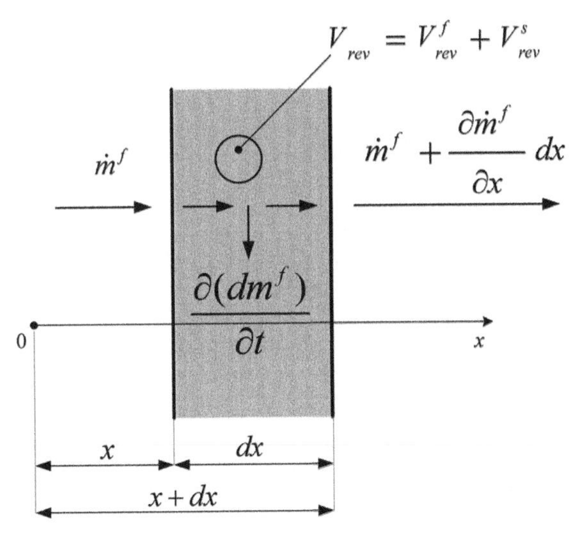

FIGURE 3.5 Nomenclature for the derivation of the continuity equation in a porous medium using the continuum approach.

First, let ρ^f be the fluid density, which in general depends on both temperature and pressure. It is defined as dm^f/dV_c^f, where dm^f is the elemental fluid mass contained into the fluid elemental volume dV_c^f. If $\dot{m}^f = \rho^f u^f A^f$ is the mass flow rate into the element in the x direction through the cross-sectional free-flow area A^f at x, then $\dot{m}^f + (\partial \dot{m}^f/\partial x)dx$ is the mass flow rate leaving the element along the same x direction through the surface A^f at $x + dx$. The net rate of mass flow into the element is the difference between the entering and leaving flow rates, given by:

$$\begin{pmatrix}\text{net rate of mass flow entering} \\ \text{volume element in } x \text{ direction}\end{pmatrix} = -\frac{\partial \dot{m}^f}{\partial x}dx = -A^f\frac{\partial(\rho^f u^f)}{\partial x}dx \tag{3.7}$$

noting that A^f is independent of x.

Second, the fluid mass content dm^f of the volume element dV_c^f is $\rho^f A^f dx$. Then, the rate of increase of this mass content is obtained by taking its partial derivative with respect to time, that is, $[\partial(\rho^f A^f)/\partial t]dx$. Therefore, we can write:

$$\begin{pmatrix}\text{rate of increase of the fluid} \\ \text{mass in volume element}\end{pmatrix} = A^f\frac{\partial \rho^f}{\partial t}dx \tag{3.8}$$

noting that A^f is also independent of t.

Substituting Eqs. (3.7) and (3.8) into Eq. (3.6), we obtain the continuity equation at a macroscopic level (continuum approach):

$$\varepsilon_k\frac{\partial \rho^f}{\partial t} + \frac{\partial(\rho^f u)}{\partial x} = 0 \tag{3.9}$$

where we have used $u^f = u/\varepsilon_k$, that is, $\langle u\rangle^f = \langle u\rangle/\varepsilon_k$; also, the fluid density ρ^f is an 'intrinsic' quantity, i.e., $\langle\rho^f\rangle^f$, and is related to the fluid pressure p appearing in Eq. (3.5) through the equation of state (EOS) of the fluid under consideration. For instance, for an ideal gas, we have $\rho^f = p/RT^f$, where T^f is the 'intrinsic' volume-averaged gas temperature, i.e., $\langle T^f\rangle^f$, and R its characteristic constant. If T^f is also unknown, we need another equation, that is, the conservation-of-energy equation in the fluid phase, as indicated in Section 3.3.3.

Note that the continuity equation defined by Eq. (3.9) is valid for pure fluids. If the fluid that saturates the porous structure is a mixture of two or more chemical species, for example drug (solute) and plasma (solvent), the fluid density ρ^f is the fluid mixture density. Also, u^f refers to the fluid velocity that is in general different from the velocities of either chemical species contained in the fluid (see Section 3.3.1).

Alternatively, Eq. (3.9) can be obtained by first writing the microscopic continuity equation for the fluid phase and, then by integrating it over the total volume V_{rev}. When integrating it, well-established theorems such as the transport and averaging theorems have to be properly used, as shown in [13] and Section 3.5 of Ref. [17].

Equations (3.5) and (3.9), along with the initial and boundary conditions for fluid velocity and pressure, are the governing equations for the transient, one-dimensional flow of a fluid through a porous medium. For an incompressible, isothermal fluid, ρ^f is uniform and constant, and Eq. (3.9) reduces to $\partial u/\partial x = 0 \Rightarrow u$ is independent of the space coordinate x.

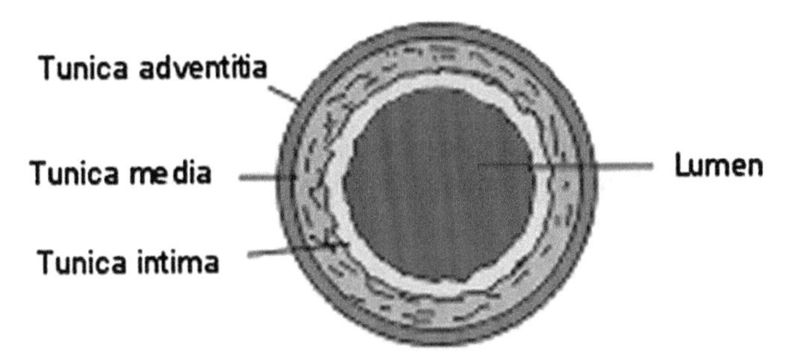

FIGURE 3.6 A normal elastic artery. The walls of arteries are composed of three main concentric zones: Tunica adventitia (outer), tunica media (middle), and tunica intima (inner layer).

Then, integrating the Darcy model given by Eq. (3.5) between $x = 0$ (pressure p_1) and $x = l$ (pressure p_2) yields the following simple velocity profile:

$$u = \frac{p_1 - p_2}{\mu l / K} \tag{3.10}$$

Hence, the velocity is proportional to the applied pressure difference and inversely proportional to the quantity $\mu l/K$, which may be seen as a hydraulic resistance to the fluid flow. In human medium sized arteries (see Fig. 3.6), the pressure difference between lumen and adventitia, under normal conditions, does not exceed 100 mmHg. As $K = 2 \cdot 10^{-14}\ cm^2$ represents the average permeability of an arterial wall with a thickness of 220 μm, and $\mu = 0.72 \cdot 10^{-2}\ g\ cm^{-1}\ s^{-1}$ is the viscosity of plasma, the magnitude of the Darcy velocity given by Eq. (3.10) is about 10^{-6} cm s^{-1}.

It is important to observe that Eq. (3.9) is valid if there is no fluid production (source) or fluid consumption (sink) in the medium. However, sources and sinks can be present in biological (porous) tissues because of fluid exchanges between the interstitial space and the blood or lymph vessels, as discussed in next section.

3.2.2.2 *Extended Continuity Equation*

An extension of Eq. (3.7) for biological media is:

$$\begin{aligned} &\begin{pmatrix}\text{net rate of mass flow entering} \\ \text{volume element in } x \text{ direction}\end{pmatrix} + \begin{pmatrix}\text{fluid mass rate produced} \\ \text{into the elemental volume}\end{pmatrix} \\ &- \begin{pmatrix}\text{fluid mass rate consumed} \\ \text{into the elemental volume}\end{pmatrix} = \begin{pmatrix}\text{rate of increase of the fluid} \\ \text{mass in volume element}\end{pmatrix} \end{aligned} \tag{3.11}$$

If $g_b^f = d\dot{m}_b^f / dV^f$ is the rate of mass fluid flow per unit volume of fluid phase from blood vessels into the interstitial fluid space (units of $kg\ s^{-1}\ m^{-3}$), we have:

$$\begin{pmatrix}\text{fluid mass rate produced} \\ \text{into the elemental volume}\end{pmatrix} = g_b^f A^f dx \tag{3.12}$$

Similarly, if $g_l^f = d\dot{m}_l^f/dV^f$ is the rate of mass fluid flow per unit volume of fluid phase from the interstitial fluid space into lymph vessels (in $kg\ s^{-1}\ m^{-3}$), we can write:

$$\begin{pmatrix}\text{fluid mass rate consumed}\\ \text{into the elemental volume}\end{pmatrix} = g_l^f A^f dx \tag{3.13}$$

Substituting Eqs. (3.7), (3.8), (3.12) and (3.13) into Eq. (3.11), after some algebraic manipulation, we get:

$$\varepsilon_k \frac{\partial \rho^f}{\partial t} + \frac{\partial(\rho^f u)}{\partial x} = \varepsilon_k g_b^f - \varepsilon_k g_l^f \tag{3.14}$$

where the subscripts b and l refer to the blood and lymph vessels, respectively.

We recall that, in the above equation, ρ^f, g_b^f and g_l^f are average quantities per unit volume of the fluid phase, whereas u is an average quantity per unit volume of the biological tissue (porous medium). For an incompressible, isothermal fluid, this equation reduces to:

$$\frac{\partial u}{\partial x} = \varepsilon_k \left(g_b^f\right)' - \varepsilon_k \left(g_l^f\right)' \tag{3.15}$$

where $(g_b^f)'$ and $(g_l^f)'$ are the rates of volumetric fluid flow per unit volume of fluid phase, respectively. Their values are determined by using Starling's law (see Chap. 9 of Ref. [15]).

3.2.3 Tortuosity and Fick's Law

The tortuosity, λ, is an important characteristic for the combination of the fluid and the geometry of the porous medium. It accounts for the motion of fluid molecules that follow tortuous pathways in the void space and is defined on p. 430 of Ref. [15] as:

$$\lambda = \left(\frac{L_{\min}}{L}\right)^2 \tag{3.16}$$

where $L_{\min}$ is the shortest path length (measured through connected pores) between any two points of the fluid in a porous medium, and L is the straight-line distance between the same two points.

Equation (3.16) states that the tortuosity is dimensionless, and is always greater than or equal to unity. Along with other factors, it affects the solute transport by diffusion in porous media. Within the fluid and solid phases of the porous structure Fick's law still applies for diffusion, but the diffusivity has to be appropriately revisited in order to account for the tortuosity. For this purpose, we recall that for a binary fluid mixture of two components, for example drug (solute) and plasma (solvent), Fick's model is expressed, in the absence of the porous medium and in one dimension (along x), by:

$$j_d^f = -D_{d0}^f \frac{\partial c_d^f}{\partial x} \tag{3.17}$$

where j_d^f is the mass flux of drug per unit area in the fluid binary mixture (superscript f). Also, D_{d0}^f is the diffusivity or diffusion coefficient of the drug in the same fluid mixture, where the subscript 0 denotes the absence of the porous medium.

Hence, Fick's equation linearly relates the drug mass flux by diffusion to the concentration gradient (driving potential) through a physical property called the diffusivity. In the presence of the porous structure, Eq. (3.17) is still valid, but the diffusion of solute is characterized by the so-called 'effective' diffusivity or 'effective' diffusion coefficient, D_d^f. This is related to the former, D_{d0}^f, through the tortuosity of pathways for diffusion and an additional viscosity function f_η accounting for local boundaries and local viscosity as follows [12]:

$$D_d^f = \frac{D_{d0}^f}{(\lambda f_\eta)^2} \tag{3.18}$$

where D_d^f is less than D_{d0}^f.

The quantity $(\lambda f_\eta)^2$ represents the hindrance to 'flow diffusion' inside the pores. As the tortuosity generally increases with a decrease in porosity, a decrease in the latter significantly reduces the effective mass diffusivity of a drug in the fluid mixture, and hence, the drug flux j_d^f.

When applying Fick's model to the fluid phase of a porous medium, we use the local concentration c_d^f, as defined by Eq. (3.1), provided a microscopic approach is used. But if the treatment deals with a continuum approach (see Section 3.2), we have to account for a volume-averaged concentration of drug at point x and time t. This average is to be taken over the volume occupied by the fluid, as indicated by Eq. (3.3a). Hence, the term c_d^f appearing in Eq. (3.17) is actually $\left\langle c_d^f \right\rangle^f = \left\langle c_d^f \right\rangle / \varepsilon_k$, where $\left\langle c_d^f \right\rangle$ is defined by Eq. (3.3b). Also, the effective diffusivity D_d^f of Eq. (3.18) is averaged over the volume of the fluid phase, that is, $\left\langle D_d^f \right\rangle^f = \left\langle D_d^f \right\rangle / \varepsilon_k$, where $\left\langle D_d^f \right\rangle$ is averaged over the total volume (r.e.v.).

Similarly, when the solid phase of the porous structure is a binary mixture, we can apply Fick's model. In one dimension (along x), it is given by:

$$j_d^s = -D_d^s \frac{\partial c_d^s}{\partial x} \tag{3.19}$$

where j_d^s is the flux of drug per unit area in the solid binary mixture, c_d^s is an intrinsic volume-averaged concentration as defined by Eq. (3.4a), i.e., $\langle c_d^s \rangle^s = \langle c_d^s \rangle / (1 - \varepsilon_k)$, and D_d^s is the 'effective' diffusivity of the drug in the solid mixture averaged over the solid phase volume, namely $\langle D_d^s \rangle^s = \langle D_d^s \rangle / (1 - \varepsilon_k)$. Then, in some way D_d^s is related to the diffusivity D_{d0}^s in absence of the porous structure.

3.3 CONSERVATION OF DRUG MASS

The fluid that saturates the porous structure can be a mixture of two or more chemical species. For the purposes of this chapter, it suffices to consider a binary mixture of only two chemical species; one being the plasma and the other being the drug. However, as the concentration of the drug is several orders of magnitude less than that of plasma, our attention focuses only on the transport of the drug.

In addition, as the drug can also be contained in the solid phase of the porous medium, such as in the coating of DESs or therapeutic contact lenses, a rigorous analysis of mass transfer involves solving two coupled partial differential equations expressing the conservation of drug mass, one for the fluid and the other for the porous material, which represent the so-called fluid and solid phases, respectively.

In this case, the porous material gives up the drug to the fluid flowing through it when its concentration in the fluid phase c_d^f is less than that c_d^s in the solid phase. This model may be denoted as 'two-phase model', and must not be confused with the well-known notation of 'two-phase flow' (or 'two-phase fluid flow'), which indicates that two miscible fluids (a liquid and a gas, or two liquids/gases) share the void volume of the porous structure [17]. In the current case, the void is saturated by a single fluid ('single-phase flow'), which is a mixture of two components.

As there are two coupled partial differential equations expressing the drug conservation, one in the fluid phase and the other in the solid phase, the solution will be represented by two concentration functions of the same substance, c_d^f and c_d^s, respectively, which are intrinsic volume-averaged concentrations.

3.3.1 Drug Mass Balance in the Fluid Phase

Let us assume a one-dimensional (1-D) model where the fluid containing the substance of interest flows uniformly with velocity u^f in the direction of the x axis, which in general depends on the space x and time t coordinates. The fluid is supposed to be well mixed, so that the concentration of the only substance can be considered uniform in any plane perpendicular to the direction of flow.

The concentration distribution $c_d^f(x,t)$ in the fluid flow field is governed by the conservation of drug mass, which can be derived by writing a drug mass balance for a 1-D differential volume element in the flow field, $dV_c^f = A^f dx$, as shown in Fig. 3.7. This balance may be stated as:

$$\begin{aligned}&\begin{pmatrix}\text{net rate of drug mass}\\ \text{entering volume element}\end{pmatrix}+\begin{pmatrix}\text{net rate of drug mass diffusion}\\ \text{into the elemental volume}\end{pmatrix}\\ &+\begin{pmatrix}\text{drug consumption}\\ \text{rate in element}\end{pmatrix}+\begin{pmatrix}\text{drug mass rate produced by chemical}\\ \text{reaction into the elemental volume}\end{pmatrix}\\ &+\begin{pmatrix}\text{mass transfer rate of}\\ \text{drug with solid-phase}\end{pmatrix}=\begin{pmatrix}\text{rate of increase of drug}\\ \text{mass in element}\end{pmatrix}\end{aligned} \tag{3.20}$$

Each term in the above statement can be evaluated as given in the following paragraphs. The fluid flowing through the porous medium is considered a constant property. First, let u_d^f be the drug (solute) velocity in the fluid plasma (solvent), which is not necessarily equal to solvent velocity u_p^f (the subscript p denotes plasma). Solutes are actually more hindered by porous structures than solvents are. This phenomenon is taken into account by the so-called 'retardation coefficient', f, which expresses the ratio of the drug (solute) velocity u_d^f in the fluid to the solvent velocity u_p^f (this coefficient is paid little attention in biological tissues). If we

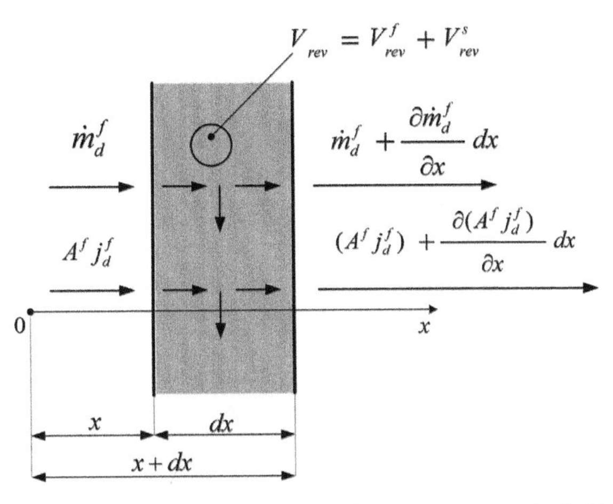

FIGURE 3.7 Nomenclature for the derivation of the drug balance equation in the fluid phase of a porous medium.

assume that the solvent velocity is equal to the fluid velocity, u^f, we can write $f = u^f_d/u^f$ (with the value of f being between zero and unity). This assumption is reasonable in dilute systems since, in these systems, the fluid velocity u^f may be considered independent of the solute concentration and, hence, can be derived independently, using the momentum and continuity equations given in Section 3.2.2.

Then, if $\dot{m}^f_d = c^f_d u^f_d A^f$ is the drug mass flux into the element in the x direction through the cross-sectional free-flow area A^f at x, then $\dot{m}^f_d + (\partial \dot{m}^f_d/\partial x)dx$ is the rate of drug mass flow leaving the element along the same x direction through the surface at $x + dx$. The net rate of drug mass into the element is the difference between the entering and leaving flow rates, given by:

$$\begin{pmatrix}\text{net rate of drug mass}\\ \text{entering volume element}\end{pmatrix} = -\frac{\partial \dot{m}^f_d}{\partial x}dx = -A^f\frac{\partial (fu^f c^f_d)}{\partial x}dx \tag{3.21}$$

where we have assumed that $u^f_p \approx u^f$, as for dilute systems, and noting that A^f is independent of x.

Equation (3.21) says that the convective (or advective) transport of a drug across the tissue is due to its velocity $u^f_d \approx fu^f$, where f accounts for the reduction of the convective influence due to the collisions of large molecules of drug with the solid matrix of the porous structure. We also recall that the flow velocity u^f is related to the Darcy velocity u through the relationship $u^f = u/(k\varepsilon)$.

Second, referring to the nomenclature shown in Fig. 3.7, the diffusion contribution is represented by:

$$\begin{pmatrix}\text{net rate of drug mass diffusion}\\ \text{into the elemental volume}\end{pmatrix} = -\frac{\partial (A^f j^f_d)}{\partial x}dx = A^f D^f_d \frac{\partial^2 c^f_d}{\partial x^2}dx \tag{3.22}$$

for which the diffusive flux of drug j^f_d per unit free-flow area A^f is given by Fick's law (see Section 3.2.3), and noting that A^f is independent of x.

Third, the drug consumption rate in the differential volume element dV_c^f may be approximated by a linear reaction having $\beta_d^f > 0$ as an effective first-order consumption rate coefficient. We can write:

$$\begin{pmatrix}\text{drug consumption}\\ \text{rate in element}\end{pmatrix} = -\beta_d^f c_d^f A^f dx \tag{3.23}$$

Fourth, if $\dot{c}_d^f = d\dot{m}_d^f / dV_c^f$ is the mass of substance produced by a chemical reaction per unit time and per unit volume of fluid phase, we have:

$$\begin{pmatrix}\text{drug mass rate produced by chemical}\\ \text{reaction into the elemental volume}\end{pmatrix} = \dot{c}_d^f A^f dx \tag{3.24}$$

Fifth, account must be taken of local mass transfer between solid and fluid phases whose exchange surface into the elemental volume is $P_w dx$, with P_w wetted perimeter. We can write:

$$\begin{pmatrix}\text{mass transfer rate of}\\ \text{drug with solid-phase}\end{pmatrix} = \hbar P_w \left(c_d^s - c_d^f\right) dx \tag{3.25}$$

where $\hbar$ is the solid-fluid local mass transfer coefficient.

Sixth, the mass content of the volume element dV_c^f in terms of drug is $c_d^f A^f dx$. Then, the rate of increase of this mass content is obtained by taking its partial derivative with respect to time, that is, $[\partial(c_d^f A^f)/\partial t]dx$. Therefore, we can write:

$$\begin{pmatrix}\text{rate of increase of drug}\\ \text{mass in element}\end{pmatrix} = A^f \frac{\partial c_d^f}{\partial t} dx \tag{3.26}$$

noting that A^f is also independent of t.

Finally, Eqs. (3.21)–(3.26) are introduced into Eq. (3.20) expressing the conservation of drug in the fluid phase. By some algebra, the fluid phase concentration equation satisfied by the intrinsic volume-averaged concentrations of drug $c_d^f = \left\langle c_d^f \right\rangle^f$, and $c_d^s = \left\langle c_d^s \right\rangle^s$, defined by Eqs. (3.3a) and (3.4a), respectively, is found to be:

$$\underbrace{\varepsilon_k \frac{\partial c_d^f}{\partial t}}_{\text{mass storage}} = \underbrace{\varepsilon_k D_d^f \frac{\partial^2 c_d^f}{\partial x^2}}_{\text{diffusive term}} - \underbrace{\frac{\partial\left(f u c_d^f\right)}{\partial x}}_{\text{advective term}} - \underbrace{\varepsilon_k \beta_d^f c_d^f}_{\text{reaction term}} + \underbrace{\frac{\varepsilon_k \hbar}{r_h}\left(c_d^s - c_d^f\right)}_{\text{local mass transfer}} + \underbrace{\varepsilon_k \dot{c}_d^f}_{\text{mass source}} \tag{3.27a}$$

where r_h is the hydraulic radius (free-flow area A^f/wetted perimeter P_w).

Note, for instance, that $\varepsilon_k \dot{c}_d^f$ is the drug produced by a chemical reaction per unit time and per unit volume of porous medium, while $\dot{c}_d^f$ is the same quantity but per unit volume of fluid phase (i.e., intrinsic). Hence, in Eq. (3.27a), $c_d^f, c_d^s, \beta_d^f, D_d^f, f$ and $\dot{c}_d^f$ are volume-averaged quantities per unit volume of the fluid phase (intrinsic), whereas u is an average quantity per unit volume of the tissue (porous medium).

The fluid phase concentration equation listed before may also be rewritten in terms of the volume-averaged drug concentrations $\bar{c}_d^f = \left\langle c_d^f \right\rangle$ and $\bar{c}_d^s = \left\langle c_d^s \right\rangle$ defined by Eqs. (3.3b) and (3.4b), respectively. In such a case, we have:

$$\underbrace{\frac{\partial \bar{c}_d^f}{\partial t}}_{\text{mass storage}} = \underbrace{D_d^f \frac{\partial^2 \bar{c}_d^f}{\partial x^2}}_{\text{diffusive term}} - \underbrace{\frac{1}{\varepsilon_k}\frac{\partial (f u \bar{c}_d^f)}{\partial x}}_{\text{advective term}} - \underbrace{\beta_d^f \bar{c}_d^f}_{\text{reaction term}} + \underbrace{\frac{\hbar}{r_h}\left(\frac{\varepsilon_k}{1-\varepsilon_k}\bar{c}_d^s - \bar{c}_d^f\right)}_{\text{local mass transfer}} + \underbrace{\dot{\bar{c}}_d^f}_{\text{mass source}} \tag{3.27b}$$

where we have used $\left\langle c_d^f \right\rangle = \varepsilon_k \left\langle c_d^f \right\rangle^f$ and $\left\langle c_d^s \right\rangle = (1-\varepsilon_k)\left\langle c_d^s \right\rangle^s$ (see Section 3.2.1.1).

In Eq. (3.27b), β_d^f, D_d^f, and f are intrinsic volume-averaged quantities, that is, per unit volume of the fluid phase, whereas $\bar{c}_d^f, \bar{c}_d^s, \dot{\bar{c}}_d^f$ and u are average quantities per unit volume of the tissue (porous medium).

Alternatively, Eq. (3.27) can be obtained by first writing the microscopic drug balance equation for the fluid phase, and then integrating it over V_{rev}. The integrating operation requires the application of well-established theorems such as the transport and averaging theorems ([13] and Section 3.5 of [17]).

3.3.2 Drug Mass Balance in the Solid Phase

In a similar manner, by writing a drug mass balance for a differential volume element in the solid, $dV_c^s = A^s dx$, the conservation equation of substance in the solid matrix phase for the intrinsic volume-averaged drug concentrations $c_d^s = \left\langle c_d^s \right\rangle^s$ and $c_d^f = \left\langle c_d^f \right\rangle^f$ is:

$$\underbrace{(1-\varepsilon_k)\frac{\partial c_d^s}{\partial t}}_{\text{mass storage}} = \underbrace{(1-\varepsilon_k) D_d^s \frac{\partial^2 c_d^s}{\partial x^2}}_{\text{diffusive term}} - \underbrace{(1-\varepsilon_k)\beta_d^s c_d^s}_{\text{reaction term}} + \underbrace{\frac{\varepsilon_k \hbar}{r_h}\left(c_d^f - c_d^s\right)}_{\text{local mass transfer}} + \underbrace{(1-\varepsilon_k)\dot{c}_d^s}_{\text{mass source}} \tag{3.28a}$$

where the advective term is absent; also, D_d^s is the effective mass diffusivity of the considered substance in the solid phase and β_d^s (positive) is a linear consumption rate coefficient.

The above solid phase concentration equation may also be rewritten in terms of the volume-averaged concentrations of drug $\bar{c}_d^s = \left\langle c_d^s \right\rangle$ and $\bar{c}_d^f = \left\langle c_d^f \right\rangle$ as follows:

$$\underbrace{\frac{\partial \bar{c}_d^s}{\partial t}}_{\text{mass storage}} = \underbrace{D_d^s \frac{\partial^2 \bar{c}_d^s}{\partial x^2}}_{\text{diffusive term}} - \underbrace{\beta_d^s \bar{c}_d^s}_{\text{reaction term}} + \underbrace{\frac{\hbar}{r_h}\left(\bar{c}_d^f - \frac{\varepsilon_k}{1-\varepsilon_k}\bar{c}_d^s\right)}_{\text{local mass transfer}} + \underbrace{\dot{\bar{c}}_d^s}_{\text{mass source}} \tag{3.28b}$$

where we have used $\left\langle c_d^s \right\rangle = (1-\varepsilon_k)\left\langle c_d^s \right\rangle^s$ and $\left\langle c_d^f \right\rangle = \varepsilon_k \left\langle c_d^f \right\rangle^f$ (see Section 3.2.2).

For complete solubility of the drug in the fluid, we have $\varepsilon_k = \varepsilon$ in both Eqs. (3.28).

3.3.3 Governing Equations

Eqs. (3.5), (3.9), (3.27a) and (3.28a) are the governing equations for transient, one-dimensional flow of a fluid through a porous medium, where the fluid that saturates the porous structure is a mixture of two chemical species; for example, drug (solute) and plasma (solvent). The EOS of the fluid needs to be known, as well as the initial and boundary conditions for the unknown variables, namely c_d^f, c_d^s, u and p.

If the temperature of the fluid mixture T^f appearing in EOS is unknown, we also need the conservation-of-energy equation in the fluid phase, which is in turn related to the solid phase energy equation of the porous medium having T^s as unknown (see Subsection 15.2.3 of Ref. [16]). However, for the purposes of this chapter, the mass transfer process will be considered isothermal, and hence the energy balance equations do not play any role.

It is noteworthy to point out that the four partial differential equations (3.5), (3.9), (3.27a) and (3.28a) that govern the conservation of momentum and the conservation of fluid mass and of drug mass are not coupled. This approximation is only valid for dilute systems which, as pointed out in section 3.3.1, are of interest here. First, we can derive the velocity u and pressure p fields by solving the 1-D flow problem described by momentum and continuity equations. Concerning this, we recall that for an incompressible, isothermal fluid, the Darcy velocity u is simply given by Eq. (3.10). Then, once the fluid velocity u appearing in Eq. (3.27a) is known, we can solve the two coupled partial differential Eqs. (3.27a) and (3.28a) expressing the LMNE [13]. This solution will be discussed in the next section where, as we have mixtures of only two components, for simplicity the subscript d will be done away with.

3.4 ANALYTICAL SOLUTIONS FOR LOCAL MASS NON-EQUILIBRIUM

The mathematical difficulties of analytically solving the two coupled partial equations, one for the fluid and the other for the porous material (solid), seem to be insuperable. However, for some special cases, analytical solutions are available in the literature. They come from the theory of heat transfer and will now be briefly revisited and discussed.

Before proceeding to describe them, we note that all the classical special cases (due to Nusselt, Schumann and Anzelius) neglect the diffusive and reaction terms as well as the heat (here, mass) sources or sinks. In addition, as depicted in Fig. 3.8, the porous medium is considered as semi-infinite ($x \geq 0$), the fluid velocity (here, solute velocity fu) uniform and constant, and the boundary and initial conditions for mass transfer are:

$$c^f(x = 0, t) = C_{x0}^f \tag{3.29a}$$

$$c^s(x, t = 0) = F^s(x) \tag{3.29b}$$

Hence, the boundary condition at $x = 0$ is known only for the drug concentration in the fluid phase. In fact, that in the solid phase cannot be assigned since it comes out as a consequence of the local transfer of drug mass between the two phases, as shown in the next subsections. Similarly, the initial condition $t = 0$ is given only for the drug concentration in the solid phase.

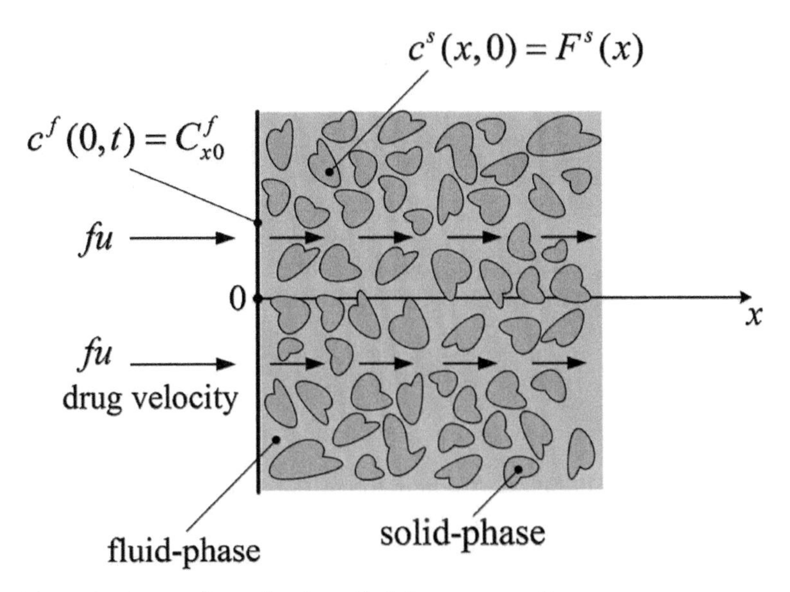

FIGURE 3.8 Schematic of a one-dimensional semi-infinite porous medium.

The one for the fluid phase derives from the interaction with the other phase, as will be shown afterwards.

3.4.1 Nusselt's Solution

Nusselt considered the problem as part of a heat regenerator theory (see Subsection 35–6b of Ref. [18]). For our purposes, with concentration in place of temperature and mass transfer instead of heat transfer, by using the simplifying assumptions stated before, Eqs. (3.27a) and (3.28a) reduce, respectively, to:

$$-r_h \frac{1}{\varepsilon_k} fu \frac{\partial c^f}{\partial x} = \hbar(c^f - c^s) \quad (x > 0; t \geq 0) \tag{3.30a}$$

$$r_h \frac{1 - \varepsilon_k}{\varepsilon_k} \frac{\partial c^s}{\partial t} = \hbar(c^f - c^s) \quad (x \geq 0; t > 0) \tag{3.30b}$$

where the drug storage in the fluid phase is neglected.

In Nusselt's former treatment, in fact, the heat storage in the fluid was neglected, which can be done only when the u velocity is very large. However, Nusselt made the general assumption that the solid temperature (here concentration) is an arbitrary function of x at $t = 0$, as shown by Eq. (3.29b).

Equation (3.30a) is not valid at $x = 0$, where the boundary condition Eq. (3.29a) applies, but is valid at the initial time $t = 0$ as the time partial derivative, $\partial/\partial t$, is absent in the fluid phase equation. For $t = 0$, it reduces to a nonhomogeneous, ordinary, first-order differential equation with constant coefficients, whose solution is:

$$c^f(x,0) = \left[C^f_{x0} + \frac{\hbar}{r_h}\frac{\varepsilon_k}{fu}\int_{\theta=0}^{x} F^s(\theta)\exp\left(\frac{\hbar}{r_h}\frac{\varepsilon_k}{fu}\theta\right)d\theta\right]\exp\left(-\frac{\hbar}{r_h}\frac{\varepsilon_k}{fu}x\right) \tag{3.31a}$$

where the constant of integration has been determined by using Eq. (3.29a).

Equation (3.31a) is to be considered as a second initial condition for $t = 0$, the first being Eq. (3.29b). For the special case of $F^s(x) = C^s_{t0}$ (i.e., uniform), Eq. (3.31a) reduces to:

$$c^f(x,0) = C^s_{t0} + \left(C^f_{x0} - C^s_{t0}\right)\exp\left(-\frac{\hbar}{r_h}\frac{\varepsilon_k}{fu}x\right) = C^f_{x0} + \left(C^s_{t0} - C^f_{x0}\right)\left[1 - \exp\left(-\frac{\hbar}{r_h}\frac{\varepsilon_k}{fu}x\right)\right] \tag{3.31b}$$

Similarly, Eq. (3.30b) is not valid at $t = 0$, where the initial condition Eq. (3.29b) applies, but is valid at the boundary condition $x = 0$ as the space partial derivative, $\partial/\partial x$, is absent in the solid phase equation. For $x = 0$, it becomes a nonhomogeneous, ordinary, first-order differential equation with constant coefficients. Its solution is:

$$c^s(0,t) = C^f_{x0} + \left[F^s(0) - C^f_{x0}\right]\exp\left(-\frac{\hbar}{r_h}\frac{\varepsilon_k}{1-\varepsilon_k}t\right) \tag{3.32a}$$

where the constant of integration has been determined by using Eq. (3.29b).

Equation (3.32a) can be considered as a second boundary condition for $x = 0$, the first being expressed by Eq. (3.29a). For the special case of $F^s(x) = C^s_{t0}$, Eq. (3.32a) becomes:

$$c^s(0,t) = C^f_{x0} + \left(C^s_{t0} - C^f_{x0}\right)\exp\left(-\frac{\hbar}{r_h}\frac{\varepsilon_k}{1-\varepsilon_k}t\right) \tag{3.32b}$$

The solution for the drug concentration in the fluid phase is given in an integral form as (see Eq. (35–52) on p. 277 of Ref. [18]):

$$\begin{aligned} c^f(x,t) = {} & \exp\left[-\frac{\hbar\varepsilon_k}{r_h}\left(\frac{x}{fu} + \frac{t}{1-\varepsilon_k}\right)\right]\cdot\left\{C^f_{x0}J_0\left(2i\frac{\hbar\varepsilon_k}{r_h}\sqrt{\frac{xt}{(1-\varepsilon_k)fu}}\right)\right. \\ & + \frac{\hbar\varepsilon_k}{r_h fu}\int_{\theta=0}^{x}\exp\left(\frac{\hbar\varepsilon_k\theta}{r_h fu}\right)F^s(\theta)J_0\left[2i\frac{\hbar\varepsilon_k}{r_h}\sqrt{\frac{t(x-\theta)}{(1-\varepsilon_k)fu}}\right]d\theta \\ & \left. + C^f_{x0}\frac{\hbar}{r_h}\frac{\varepsilon_k}{1-\varepsilon_k}\int_{\theta=0}^{t}\exp\left(\frac{\hbar}{r_h}\frac{\varepsilon_k\theta}{1-\varepsilon_k}\right)J_0\left[2i\frac{\hbar\varepsilon_k}{r_h}\sqrt{\frac{x(t-\theta)}{(1-\varepsilon_k)fu}}\right]d\theta\right\} \end{aligned} \tag{3.33a}$$

where $J_0(z)$ is the Bessel function of the first kind of zero order as in Chap. 9 of Ref. [19], which becomes real for an imaginary argument. In detail, $J_0(0) = 1$ and $J_0(iz) = I_0(z)$, where $I_0(z)$ is the modified Bessel function of the first kind of zero order.

The drug concentration in the solid phase can be derived either by substituting c^f given by Eq. (3.33a) in Eq. (3.30a), after differentiation, or in Eq. (3.30b), after integration. Choosing the former yields:

$$\begin{aligned}
c^s(x,t) =& \exp\left(-\frac{\hbar}{r_h}\frac{\varepsilon_k t}{1-\varepsilon_k}\right)F^s(x) - \exp\left[-\frac{\hbar\varepsilon_k}{r_h}\left(\frac{x}{fu}+\frac{t}{1-\varepsilon_k}\right)\right] \\
&\cdot\left\{C^f_{x0}\sqrt{\frac{fu}{1-\varepsilon_k}\frac{t}{x}}iJ_1\left(2i\frac{\hbar\varepsilon_k}{r_h}\sqrt{\frac{xt}{(1-\varepsilon_k)fu}}\right)\right. \\
&+\frac{\hbar\varepsilon_k}{r_h}\int_{\theta=0}^{x}\exp\left(\frac{\hbar\varepsilon_k\theta}{r_h fu}\right)F^s(\theta)\sqrt{\frac{1}{(1-\varepsilon_k)fu}\frac{t}{(x-\theta)}}iJ_1\left[2i\frac{\hbar\varepsilon_k}{r_h}\sqrt{\frac{t(x-\theta)}{(1-\varepsilon_k)fu}}\right]d\theta \\
&\left.+C^f_{x0}\frac{\hbar\varepsilon_k}{r_h}\frac{fu}{1-\varepsilon_k}\int_{\theta=0}^{t}\exp\left(\frac{\hbar}{r_h}\frac{\varepsilon_k\theta}{1-\varepsilon_k}\right)\sqrt{\frac{1}{(1-\varepsilon_k)fu}\frac{(t-\theta)}{x}}iJ_1\left[2i\frac{\hbar\varepsilon_k}{r_h}\sqrt{\frac{x(t-\theta)}{(1-\varepsilon_k)fu}}\right]d\theta\right\}
\end{aligned} \tag{3.33b}$$

where $J_1(z)$ is the Bessel function of the first kind of first order [19] which, when multiplied by the imaginary unit i, becomes real for an imaginary argument. In detail, $iJ_1(iz) = -I_1(z)$, where $I_1(z)$ is the modified Bessel function of the first kind of first order.

The solution development in the fluid phase is clearly described in [20]. Nusselt did not solve the integrals appearing in the two equations listed before. It can be proved that, for $t = 0$, the former reduces to Eq. (3.31a) and, hence, to Eq. (3.31b) when $F^s(x) = C^s_{t0}$. By some algebra, for $x = 0$, the latter reduces to Eq. (3.32a) and, hence, to Eq. (3.32b) when $F^s(x) = C^s_{t0}$.

For the special case of $F^s(x) = C^s_{t0}$, the governing Eqs. (3.29)–(3.30) may conveniently be rewritten in a dimensionless form by using the following variables:

$$\tilde{x} = \frac{\hbar}{r_h}\frac{\varepsilon_k}{fu}x \quad \tilde{t} = \frac{\hbar}{r_h}\frac{\varepsilon_k}{1-\varepsilon_k}t \quad \tilde{c} = \frac{c - C^f_{x0}}{C^s_{t0} - C^f_{x0}} \tag{3.34}$$

where the $r_h/\hbar$ ratio (having units of s) can be considered as a time constant characterizing the local transfer of drug between the solid and fluid phases.

In dimensionless form, the defining equations become:

$$-\frac{\partial\tilde{c}^f}{\partial\tilde{x}} = (\tilde{c}^f - \tilde{c}^s) \quad (\tilde{x} > 0; \tilde{t} \geq 0) \tag{3.35a}$$

$$\frac{\partial\tilde{c}^s}{\partial\tilde{t}} = (\tilde{c}^f - \tilde{c}^s) \quad (\tilde{x} \geq 0; \tilde{t} > 0) \tag{3.35b}$$

$$\tilde{c}^f(\tilde{x} = 0, \tilde{t}) = 0 \tag{3.35c}$$

$$\tilde{c}^s(\tilde{x}, \tilde{t} = 0) = 1 \tag{3.35d}$$

whose concentration solutions may be derived from Eqs. (3.33a) and (3.33b) as, respectively:

$$\tilde{c}^f(\tilde{x},\tilde{t}) = \exp[-(\tilde{x}+\tilde{t})] \cdot \int_{\tilde{\theta}=0}^{\tilde{x}} \exp(\tilde{\theta}) I_0 \left[2\sqrt{\tilde{t}(\tilde{x}-\tilde{\theta})}\right] d\tilde{\theta} \tag{3.36a}$$

$$\tilde{c}^s(\tilde{x},\tilde{t}) = \exp(-\tilde{t}) + \exp\left[-(\tilde{x}+\tilde{t})\right] \cdot \int_{\tilde{\theta}=0}^{\tilde{x}} \exp(\tilde{\theta}) \sqrt{\frac{\tilde{t}}{\tilde{x}-\tilde{\theta}}} I_1 \left[2\sqrt{\tilde{t}(\tilde{x}-\tilde{\theta})}\right] d\tilde{\theta} \tag{3.36b}$$

It should be noted that the evaluation of the analytical solutions of Eq. (3.36) requires numerical integration. This was done using an adaptive Gauss-Kronod quadrature scheme in Matlab ambient. Plots of the solutions are given in Fig. 3.9, for which the dimensionless

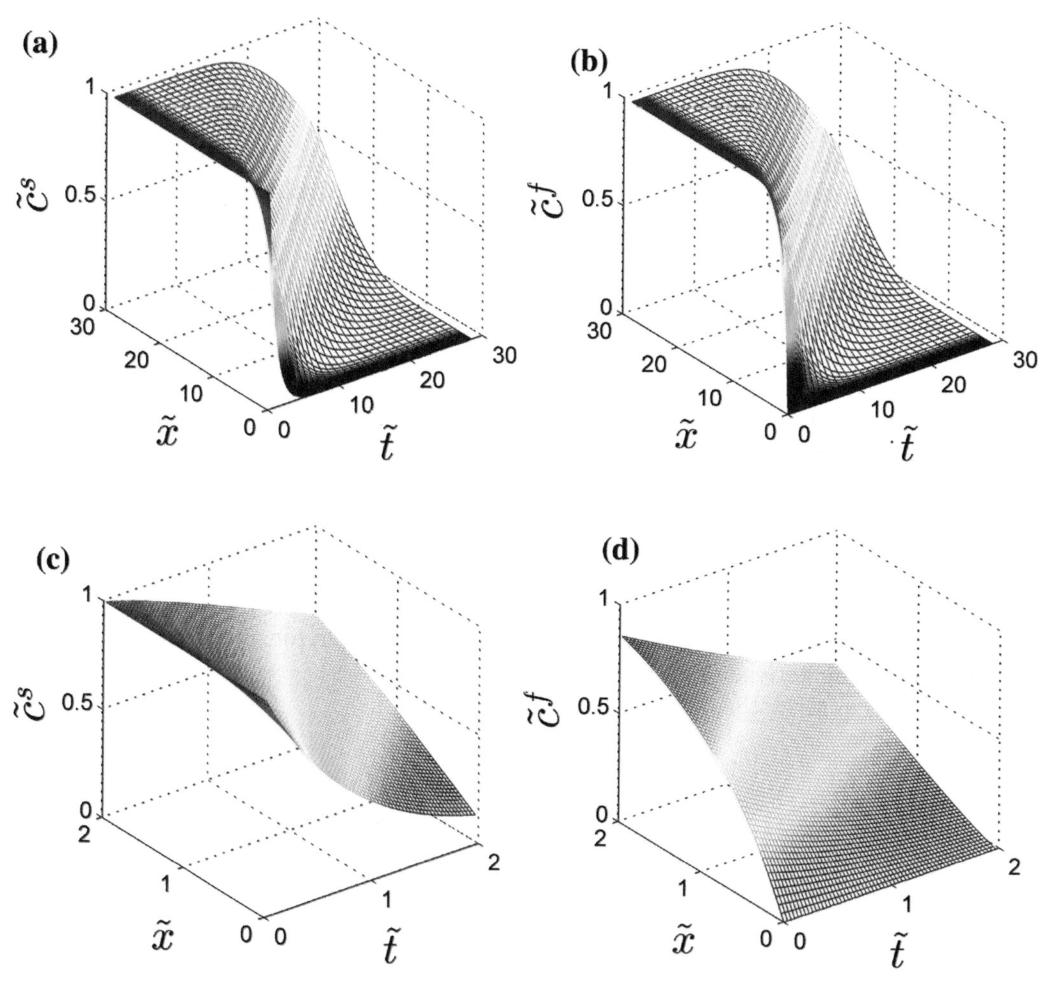

FIGURE 3.9 Dimensionless concentration profiles as function of space and time: (a) solid phase; (b) fluid phase; (c) solid phase for small values of locations and times, and (d) fluid phase for small values of locations and times.

solid phase concentration, cases (a) and (c), rapidly diminishes at small values of $\tilde{x}$ when the time increases, while the further locations take much longer to feel the influence of the fluid within the porous media. In particular, at the boundary surface $\tilde{x} = 0$, we have an exponential decay as indicated by Eq. (3.31b) that, in dimensionless form, becomes $\tilde{c}^s(0, \tilde{t}) = \exp(-\tilde{t})$. The non-dimensional fluid phase concentration, cases (b) and (d), has zero value at the boundary $\tilde{x} = 0$, and then increases with $\tilde{x}$ as drug is released from the solid phase. For $\tilde{t} = 0$, it increases exponentially as given by Eq. (3.31b) that, in dimensionless form, becomes $\tilde{c}^f(\tilde{x}, 0) = 1 - \exp(-\tilde{x})$. It is interesting to note that both solutions are similar for large values of $\tilde{x}$ and $\tilde{t}$, cases a) and b), which makes sense physically. The LMNE is evident only for small values of $\tilde{x}$ and $\tilde{t}$, cases (c) and (d).

3.4.2 Schumann's Solution

It is worth mentioning that Schumann considered the problem as part of a porous medium theory (see Subsection 35–6c of Ref. [18,21]). His goal was in fact to quantify the heating up of a porous prism through which a liquid was flowing. For the purposes that are of interest here, concentration is put in place of temperature and mass transfer instead of heat transfer, so by using the simplifying assumptions stated in Section 3.4, Eqs. (3.27a) and (3.28a) reduce, respectively, to:

$$\varepsilon_k \frac{\partial c^f}{\partial t} + fu \frac{\partial c^f}{\partial x} = -\frac{\hbar \varepsilon_k}{r_h}(c^f - c^s) \quad (x > 0; t > 0) \tag{3.37a}$$

$$r_h \frac{1 - \varepsilon_k}{\varepsilon_k} \frac{\partial c^s}{\partial t} = \hbar(c^f - c^s) \quad (x \geq 0; t > 0) \tag{3.37b}$$

where the term $\partial/\partial t$ in the fluid phase (indicating drug storage) is now taken into account.

According to Schumann's former treatment where the solid temperature was assumed to be uniform at $t = 0$, the initial condition Eq. (3.29b) has to be replaced by:

$$c^s(x, t = 0) = C^s_{t0} \tag{3.38}$$

Note that Eq. (3.37b) is also valid at the boundary condition $x = 0$ in accordance with the absence of the space partial derivative, $\partial/\partial x$. For $x = 0$, it reduces to a nonhomogeneous, ordinary differential equation of the first order with constant coefficients. Its solution is given by Eq. (3.32b), which can be considered as a second boundary condition for $x = 0$, the first being Eq. (3.29a).

On the contrary, Eq. (3.37a) does not apply at the initial time $t = 0$ consistently to the presence of the time partial derivative, $\partial/\partial t$. Hence, an initial condition for $c^f(x, t)$ has to be assigned as:

$$c^f(x, t = 0) = C^s_{t0} \tag{3.39}$$

Equations (3.38) and (3.39) state that there exists a LME between the two phases at $t = 0$. As the diffusive term is absent in both the drug balance equations, namely Eqs. (3.37a) and

(3.37b), the above equilibrium at $t = 0$ keeps up to $t = x/(fu)$. The quantity $x/(fu)$ is the time that it takes for a drug particle in the fluid phase flowing through the semi-infinite porous structure ($x \geq 0$) of Fig. 3.8 to reach the location x of interest with a velocity of fu.

The solution for the drug concentration in both phases is given in an integral form as (see Eqs. (35–65) and (35–66) on p. 280 of Ref. [18]):

$$\frac{c^f(x,t) - C^s_{t0}}{C^f_{x0} - C^s_{t0}} = \exp\left(-\frac{\hbar}{r_h}\frac{\varepsilon_k}{fu}x\right) \cdot \left\{\exp\left[-\frac{\hbar}{r_h}\frac{\varepsilon_k}{1-\varepsilon_k}\left(t - \frac{x}{fu}\right)\right] J_0\left[2i\frac{\hbar\varepsilon_k}{r_h}\sqrt{\frac{x}{fu}\frac{1}{1-\varepsilon_k}\left(t-\frac{x}{fu}\right)}\right] + \frac{\hbar}{r_h}\frac{\varepsilon_k}{1-\varepsilon_k}\int_{\theta=0}^{t-\frac{x}{fu}} \exp\left(-\frac{\hbar}{r_h}\frac{\varepsilon_k}{1-\varepsilon_k}\theta\right) J_0\left(2i\frac{\hbar\varepsilon_k}{r_h}\sqrt{\frac{x}{fu}\frac{\theta}{1-\varepsilon_k}}\right) d\theta\right\} \tag{3.40a}$$

$$\frac{c^s(x,t) - C^s_{t0}}{C^f_{x0} - C^s_{t0}} = \frac{\hbar}{r_h}\frac{\varepsilon_k}{1-\varepsilon_k}\exp\left(-\frac{\hbar}{r_h}\frac{\varepsilon_k}{fu}x\right) \cdot \int_{\theta=0}^{t-\frac{x}{fu}} \exp\left(-\frac{\hbar}{r_h}\frac{\varepsilon_k}{1-\varepsilon_k}\theta\right) J_0\left(2i\frac{\hbar\varepsilon_k}{r_h}\sqrt{\frac{x}{fu}\frac{\theta}{1-\varepsilon_k}}\right) d\theta \tag{3.40b}$$

For a given location x, the above equations apply only for $t \geq x/(fu)$. For $0 \leq t < x/(fu)$, we have $c^f(x,t) = c^s(x,t) = C^s_{t0}$. For $t = x/(fu)$, they reduce, respectively, to:

$$\frac{c^f(x) - C^s_{t0}}{C^f_{x0} - C^s_{t0}} = \exp\left(-\frac{\hbar}{r_h}\frac{\varepsilon_k}{fu}x\right) \quad \frac{c^s(x) - C^s_{t0}}{C^f_{x0} - C^s_{t0}} = 0 \tag{3.41}$$

For a given time t, Eqs. (3.40a) and (3.40b) apply only for $0 \leq x \leq (fu)t$. For $x = (fu)t$, they reduce, respectively, to:

$$\frac{c^f(t) - C^s_{t0}}{C^f_{x0} - C^s_{t0}} = \exp\left(-\frac{\hbar\varepsilon_k}{r_h}t\right) \quad \frac{c^s(t) - C^s_{t0}}{C^f_{x0} - C^s_{t0}} = 0 \tag{3.42}$$

For $x > (fu)t$, we have a LME, i.e., $c^f(x,t) = c^s(x,t) = C^s_{t0}$, where $(fu)t$ is the distance traveled (from the boundary surface $x = 0$) by a drug particle flowing in the fluid phase of the porous medium with a fu velocity during the time t of interest.

After some lengthy algebraic manipulations, Schumann was able to solve the integrals of Eqs. (3.40a) and (3.40b), arriving at two infinite series which he solved explicitly (more details can be found in Subsection 35–6.c of Ref. [18,21]).

Note that the drug concentration in the solid phase can alternatively be derived by substituting c^f given by Eq. (3.40a) in Eq. (3.37a), and performing two differentiations, or in Eq. (3.37b), and performing one integration. In addition, combining Eqs. (3.40a) and (3.40b) yields the following drug concentration difference between the two phases:

$$\frac{c^s(x,t) - c^f(x,t)}{C^s_{t0} - C^f_{x0}} = I_0\left[2\frac{\hbar\varepsilon_k}{r_h}\sqrt{\frac{x}{fu}\frac{1}{1-\varepsilon_k}\left(t - \frac{x}{fu}\right)}\right] \cdot \exp\left[-\frac{\hbar}{r_h}\frac{\varepsilon_k}{1-\varepsilon_k}\left(t - \frac{x}{fu}\right)\right] \quad (t \geq x/(fu)) \tag{3.43}$$

where we have used $J_0(iz) = I_0(z)$.

By using the dimensionless variables as defined by Eq. (3.34), we have that Eq. (3.43) becomes:

$$\tilde{c}^s(\tilde{x},\tilde{t}) - \tilde{c}^f(\tilde{x},\tilde{t}) = I_0\left[2\sqrt{\tilde{x}\left(\tilde{t} - \frac{\tilde{x}}{1-\varepsilon_k}\right)}\right] \cdot \exp\left[-\left(\tilde{t} - \frac{\tilde{x}}{1-\varepsilon_k}\right)\right] \quad (\tilde{t} \geq \tilde{x}/(1-\varepsilon_k)) \tag{3.44}$$

Hence, the above dimensionless drug concentration difference depends not only on the non-dimensional space and time but also on the available volume fraction ε_k.

3.4.3 Anzelius's Solution

Anzelius's work concerned the heating of a fluid flowing through a hot porous matrix. The partial differential equations he used are exactly the same as the ones used by Schumann; that is, Eqs. (3.37a) and (3.37b), with boundary and initial conditions given by Eqs. (3.29a), (3.38) and (3.39), respectively. The solution is given in an integral form as (see Eqs. (16) and (17) on p. 394 of Ref. [22]):

$$\frac{c^f(x,t) - C^s_{t0}}{C^f_{x0} - C^s_{t0}} = \exp\left(-\frac{\hbar}{r_h}\frac{\varepsilon_k}{fu}x\right) \cdot \left\{1 + \frac{\hbar\varepsilon_k}{r_h}\sqrt{\frac{1}{1-\varepsilon_k}\frac{x}{fu}} \int_{\theta=0}^{t-\frac{x}{fu}} \exp\left(-\frac{\hbar}{r_h}\frac{\varepsilon_k}{1-\varepsilon_k}\theta\right)\frac{1}{\sqrt{\theta}} I_1\left(2\frac{\hbar\varepsilon_k}{r_h}\sqrt{\frac{1}{1-\varepsilon_k}\frac{x}{fu}\theta}\right)d\theta\right\} \tag{3.45a}$$

$$\frac{c^s(x,t) - C^s_{t0}}{C^f_{x0} - C^s_{t0}} = \frac{\hbar}{r_h}\frac{\varepsilon_k}{1-\varepsilon_k}\exp\left(-\frac{\hbar}{r_h}\frac{\varepsilon_k}{fu}x\right) \cdot \int_{\theta=0}^{t-\frac{x}{fu}} \exp\left(-\frac{\hbar}{r_h}\frac{\varepsilon_k}{1-\varepsilon_k}\theta\right) I_0\left(2\frac{\hbar\varepsilon_k}{r_h}\sqrt{\frac{x}{fu}\frac{\theta}{1-\varepsilon_k}}\right)d\theta \tag{3.45b}$$

For a given location x, the above equations apply only for $t \geq x/(fu)$. For a given time t, they apply only for $x \leq (fu)t$. Also, note that Eq. (3.45a) derived by Anzelius is exactly the same as Eq. (3.40a) obtained by Schumann. In fact, by using the relation $I_0(z) = J_0(iz)$, Eq. (3.40a) becomes:

$$\frac{c^f(x,t) - C^s_{t0}}{C^f_{x0} - C^s_{t0}} = \left\{ \exp\left[-\frac{\hbar}{r_h}\frac{\varepsilon_k}{1-\varepsilon_k}\left(t - \frac{x}{fu}\right)\right] \cdot I_0\left[2\frac{\hbar\varepsilon_k}{r_h}\sqrt{\frac{x}{fu}\frac{1}{1-\varepsilon_k}\left(t-\frac{x}{fu}\right)}\right] \right.$$
$$\left. + \frac{\hbar}{r_h}\frac{\varepsilon_k}{1-\varepsilon_k}\int_{\theta=0}^{t-\frac{x}{fu}} \exp\left(-\frac{\hbar}{r_h}\frac{\varepsilon_k}{1-\varepsilon_k}\theta\right) I_0\left(2\frac{\hbar\varepsilon_k}{r_h}\sqrt{\frac{x}{fu}\frac{\theta}{1-\varepsilon_k}}\right) d\theta \right\}$$
$$\cdot \exp\left(-\frac{\hbar}{r_h}\frac{\varepsilon_k}{fu}x\right) \tag{3.46}$$

By integrating the integral on the RHS of the above equation by parts, and recalling that $dI_0(z)/dz = I_1(z)$ (see p. 376, Eq. (9.6.27) of Ref. [19]), we obtain exactly Eq. (3.45a). Similarly, as $I_0(z) = J_0(iz)$, Eq. (3.45b) is exactly the same as Eq. (3.40b).

Then, by using the dimensionless variables as defined by Eq. (3.34), Eqs. (3.45a) and (3.45b) become, respectively:

$$\tilde{c}^f(\tilde{x},\tilde{t}) = 1 - \exp(-\tilde{x}) \cdot \left[1 + \sqrt{\tilde{x}}\int_{\tilde{\theta}=0}^{\tilde{t}-\frac{\tilde{x}}{1-\varepsilon_k}} \exp(-\tilde{\theta})\frac{1}{\sqrt{\tilde{\theta}}} I_1\left(2\sqrt{\tilde{x}\tilde{\theta}}\right) d\tilde{\theta}\right] \quad (\tilde{t} \geq \tilde{x}/(1-\varepsilon_k)) \tag{3.47a}$$

$$\tilde{c}^s(\tilde{x},\tilde{t}) = 1 - \exp(-\tilde{x}) \cdot \int_{\tilde{\theta}=0}^{\tilde{t}-\frac{\tilde{x}}{1-\varepsilon_k}} \exp(-\tilde{\theta}) \cdot I_0\left(2\sqrt{\tilde{x}\tilde{\theta}}\right) d\tilde{\theta} \quad (\tilde{t} \geq \tilde{x}/(1-\varepsilon_k)) \tag{3.47b}$$

3.4.4 Recent Solutions

Recent analytical solutions and physical analysis for the Schumann model (based on a perturbation technique) were obtained by Kuznetsov in Refs. [23–25]. In particular, these papers demonstrated that the wave describing a temperature difference between the fluid and solid phases (in terms of the present paper, a concentration difference) forms a wave that is localized in space and propagates in the direction of the flow. As the wave propagates, it spreads out and its amplitude decreases. This new physical behavior of the temperature (or concentration) difference was first shown in the above papers.

In addition, following *Tzou*'s approach [26], *Haji-Sheikh et al.* were able to obtain analytical solutions using Green's functions for both fluid and solid phases of a porous medium subject to rapid transient heating (or cooling) [27,28]. In such a case, in fact, the local thermal equilibrium (LTE) hypothesis is usually not valid. A criterion for LTE based on the magnitude of the Sparrow number and on the rate of change of the heat input was developed and proposed. Its extension to LME is subject of future research.

3.5 ANALYTICAL SOLUTIONS FOR LOCAL MASS EQUILIBRIUM

In many cases, it is acceptable to assume a LME so that $c^f = c^s = c$. Then, assuming that the drug velocity fu in the fluid phase is uniform and constant, summing up Eqs. (3.27a) and (3.28a) yields:

$$\frac{\partial c}{\partial t} = \overline{D}\frac{\partial^2 c}{\partial x^2} - (fu)\frac{\partial c}{\partial x} - \bar{\beta}c + \bar{\dot{c}} \tag{3.48}$$

where $\overline{D}, \overline{\beta}$ and $\bar{\dot{c}}$ are, respectively, the overall effective diffusivity, the overall consumption rate coefficient and the overall mass production rate, all of them per unit volume of porous medium. They may be taken as:

$$\overline{D} = (1 - \varepsilon_k)D^s + \varepsilon_k D^f \tag{3.49a}$$

$$\bar{\beta} = (1 - \varepsilon_k)\beta^s + \varepsilon_k \beta^f \tag{3.49b}$$

$$\bar{\dot{c}} = (1 - \varepsilon_k)\dot{c}^s + \varepsilon_k \dot{c}^f \tag{3.49c}$$

Hence, by assuming that at any point in the solid-fluid porous matrix the two concentrations are identical, the analysis can significantly be simplified. In particular, when there is no motion and reaction through the pores of the solid structure, we have pure mass diffusion. This means that the mathematical methods developed for diffusion in solids [22,29,30] apply to porous media saturated with stagnant fluid with no reaction. However, if the convection and reaction terms are present in the LME-based Eq. (3.48), by some appropriate transformations the problem can still be reduced to a pure diffusive one, as shown in the next subsection.

3.5.1 A Worked Example

A finite polymeric matrix of available volume fraction ε_k and thickness l contains drug initially at a uniform concentration C^s_{t0}, as depicted in Fig. 3.10. Plasma at the same drug concentration C^s_{t0} flows through the matrix with velocity $u_p \approx u$ (averaged over r.e.v.), where u is the velocity of the binary mixture of plasma and drug (for dilute systems, $u_p \approx u$). Therefore, the initial conditions for the solid and fluid phases are, respectively:

$$(1 - \varepsilon_k)c^s(x, t = 0) = (1 - \varepsilon_k)C^s_{t0} \quad (0 < x < l) \tag{3.50a}$$

$$\varepsilon_k c^f(x, t = 0) = \varepsilon_k C^s_{t0} \quad (0 < x < l) \tag{3.50b}$$

For $t > 0$, the concentration of drug at the plasma inlet $x = 0$ the concentration of drug is kept at C^f_{x0}. At the opposite boundary surface $x = l$ a highly resistive drug barrier allows the plasma to pass while restricting the drug's transport, effectively acting as a no-flux boundary condition. This is depicted in Fig. 3.10. The boundary conditions for the solid and fluid phases at $x = 0$ and $x = l$ are, respectively:

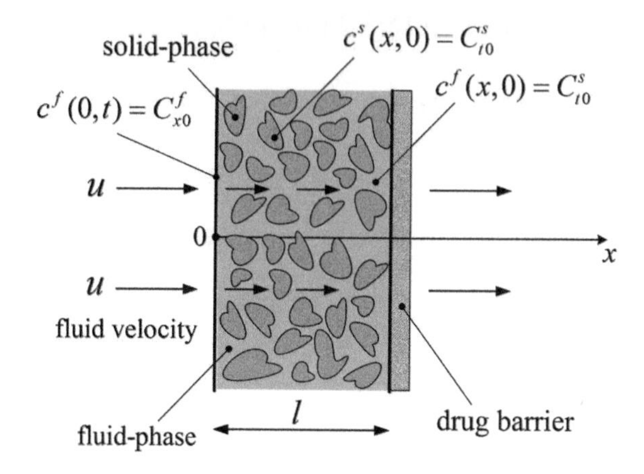

FIGURE 3.10 Schematic of a one-dimensional finite polymeric matrix (porous medium) with a drug barrier at $x = l$.

- $x = 0$

$$\hbar_0(1-\varepsilon_k)\left[C^f_{x0} - c^s(x=0,t)\right] = -D^s(1-\varepsilon_k)\left(\frac{\partial c^s}{\partial x}\right)_{x=0} \quad (t > 0) \tag{3.51a}$$

$$\varepsilon_k c^f(x=0,t) = \varepsilon_k C^f_{x0} \quad (t > 0) \tag{3.51b}$$

- $x = l$

$$-D^s(1-\varepsilon_k)\left(\frac{\partial c^s}{\partial x}\right)_{x=l} = 0 \quad (t > 0) \tag{3.52a}$$

$$-D^f\varepsilon_k\left(\frac{\partial c^f}{\partial x}\right)_{x=l} = 0 \quad (t > 0) \tag{3.52b}$$

where $\hbar_0$ is the mass transfer coefficient at $x = 0$.

Equations (3.50) state that there exists a LME between solid and fluid phases at $t = 0$. On the contrary, Eqs. (3.51) say that there exists no mass equilibrium at $x = 0$. However, if we assume that $\hbar_0$ is very large ($\hbar_0 \to \infty$), we have mass equilibrium at $x = 0$ too, and Eq. (3.51a) reduces to:

$$(1-\varepsilon_k)c^s(x=0,t) = (1-\varepsilon_k)C^f_{x0} \quad (t > 0) \tag{3.53}$$

Now, for the sake of simplicity, the approximation of LME is made not only at $x = 0$, but at any location x in the solid-fluid porous matrix. Consequently, Eq. (3.48), expressing the conservation of drug in the whole porous medium, holds. When mass generation and reaction terms are absent, it reduces to:

$$\frac{\partial \theta}{\partial t} = \overline{D}\frac{\partial^2 \theta}{\partial x^2} - (fu)\frac{\partial \theta}{\partial x} \quad (0 < x < l; t > 0) \tag{3.54}$$

where the reduced concentration $\theta(x,t) = c(x,t) - C^s_{t0}$ has been used. Looking ahead in the analysis, in fact, it makes the initial concentration zero, hence simplifying the derivation of the solution, as will be shown in the next paragraph.

Also, the boundary and initial conditions collapse into:

$$\theta(0,t) = \theta_0 \quad \left(\frac{\partial c}{\partial x}\right)_{x=l} = 0 \quad (t > 0) \tag{3.55a}$$

$$\theta(x, t = 0) = 0 \quad (0 < x < l) \tag{3.55b}$$

where $\theta_0 = C^f_{x0} - C^s_{t0}$.

Solution

The above defining Eqs. (3.56a)–(3.56c) are transformed using the following relation ([31] and p. 91 of [30]):

$$\theta(x,t) = w(x,t) \cdot \exp\left[\frac{fu}{2\overline{D}}x - \frac{(fu)^2}{4\overline{D}}t\right] \tag{3.56}$$

where $w(x,t)$ is a new dependent variable.

Substituting the above equation in the governing Eqs. (3.54)–(3.55), we obtain:

$$\frac{\partial w}{\partial t} = \overline{D}\frac{\partial^2 w}{\partial x^2} \quad (0 < x < l; t > 0) \tag{3.57a}$$

$$w(0,t) = \theta_0 \exp\left[\frac{(fu)^2}{4\overline{D}}t\right] \quad (t > 0) \tag{3.57b}$$

$$-\overline{D}\left(\frac{\partial w}{\partial x}\right)_{x=l} = \hbar_{le} w(l,t) \quad (t > 0) \tag{3.57c}$$

$$w(x,0) = 0 \quad (0 < x < l) \tag{3.57d}$$

where $\hbar_{le} = fu/2$ is the effective mass transfer coefficient at $x = l$. Also, as in Eq. (3.57c) we have $\overline{D}$ defined by Eq. (3.49a) in place of D^s, the plasma seems to play an active role in this boundary condition (see p. 167 of [32]).

The concentration solution to Eqs. (3.57a)–(3.57d) may be obtained by using Green's function (GF) solution equation (see p. 66 of [30]):

$$w(x,t) = -\overline{D}\theta_0 \int_{\tau=0}^{t} \exp\left[\frac{(fu)^2}{4\overline{D}}\tau\right] \cdot \left[-\frac{\partial}{\partial x'} G_{X13}(x,x',t-\tau)\right]_{x'=0} d\tau \tag{3.58}$$

where the large-cotime form of $G_{X13}(x,x',t-\tau)$ is on p. 590 of Ref. [30]:

$$G_{X13}(x,x',t-\tau) = \frac{2}{l}\sum_{n=1}^{\infty} \frac{\beta_n^2 + Bi_{le}^2}{\beta_n^2 + Bi_{le}^2 + Bi_{le}} \sin\left(\beta_n \frac{x}{l}\right) \sin\left(\beta_n \frac{x'}{l}\right) \exp\left[-\frac{\beta_n^2 \overline{D}}{l^2}(t-\tau)\right] \tag{3.59}$$

(The subscript X indicates mass diffusion along the x-axis; 1 and 3 denote the boundary conditions of the 1st and 3rd kind at $x = 0$ and $x = l$, respectively. More details can be found in Chap. 2 of Ref. [30] for the notation system devised by *Beck et al.*).

The β_n is the nth dimensionless eigenvalue of the following eigencondition: $\beta_n \cot \beta_n = -Bi_{le}$, where $Bi_{le} = \bar{h}_{le} l/\overline{D}$ is the effective Biot number at $x = l$. The β_n eigenvalues may be computed by using explicit approximate equations with six-digit accuracy developed by *Haji-Sheikh and Beck* [33]. Substituting the large-cotime GF (3.59) in Eq. (3.58) and integrating gives:

$$w(x,t) = 2\theta_0 \sum_{n=1}^{\infty} \frac{\beta_n^2 + Bi_{le}^2}{\beta_n^2 + Bi_{le}^2 + Bi_{le}} \frac{\beta_n}{(fPe/2)^2 + \beta_n^2} \sin\left(\beta_n \frac{x}{l}\right) \cdot \left\{\exp\left[\frac{(fu)^2}{4\overline{D}}t\right] - \exp\left(-\frac{\beta_n^2 \overline{D}}{l^2}t\right)\right\} \tag{3.60}$$

where $Pe = ul/\overline{D}$ is the Péclet number associated with the fluid.

Once $w(x,t)$ is obtained, the transformation (3.56) can be inverted to find the concentration of drug. Then, we have:

$$\begin{aligned} \theta(x,t) = {} & 2\theta_0 \exp\left(f\frac{Pe}{2}\frac{x}{l}\right) \cdot \sum_{n=1}^{\infty} \frac{\beta_n^2 + Bi_{le}^2}{\beta_n^2 + Bi_{le}^2 + Bi_{le}} \frac{\beta_n}{(fPe/2)^2 + \beta_n^2} \sin\left(\beta_n \frac{x}{l}\right) \\ & - 2\theta_0 \exp\left(f\frac{Pe}{2}\frac{x}{l}\right) \cdot \sum_{n=1}^{\infty} \frac{\beta_n^2 + Bi_{le}^2}{\beta_n^2 + Bi_{le}^2 + Bi_{le}} \frac{\beta_n}{(fPe/2)^2 + \beta_n^2} \sin\left(\beta_n \frac{x}{l}\right) \\ & \cdot \exp\left\{-\left[\left(f\frac{Pe}{2}\right)^2 + \beta_n^2\right]\frac{\overline{D}t}{l^2}\right\} \end{aligned} \tag{3.61}$$

where the steady state part (first term on the RHS) has an algebraic convergence (very slow).

Using the following algebraic identity (see App. B, p. 267 of Ref. [34]):

$$\begin{aligned} & 2\sum_{n=1}^{\infty} \frac{\beta_n^2 + Bi_{le}^2}{\beta_n^2 + Bi_{le}^2 + Bi_{le}} \frac{\beta_n}{(fPe/2)^2 + \beta_n^2} \sin\left(\beta_n \frac{x}{l}\right) \\ & = \frac{(fPe + 2Bi_{le})e^{-f\frac{Pe}{2}\frac{x}{l}} + (fPe - 2Bi_{le}) \cdot \exp\left[-f\frac{Pe}{2}\left(2 - \frac{x}{l}\right)\right]}{(fPe + 2Bi_{le}) + (fPe - 2Bi_{le}) \cdot \exp\left(-fPe\right)}, \end{aligned} \tag{3.62}$$

and also noting that $Bi_{le} = \hbar_{le} l/\overline{D} = ful/2\overline{D} = fPe/2$, Eq. (3.61) becomes:

$$\theta(x,t) = \theta_0 - 2\theta_0 \exp\left(f\frac{Pe}{2}\frac{x}{l}\right) \cdot \sum_{n=1}^{\infty} \frac{\beta_n}{\beta_n^2 + (fPe/2)^2 + fPe/2} \sin\left(\beta_n \frac{x}{l}\right) \cdot \exp\left\{-\left[\left(f\frac{Pe}{2}\right)^2 + \beta_n^2\right]\frac{\overline{D}t}{l^2}\right\} \tag{3.63}$$

which reduces to the well-known solution given in the literature when $u = 0$ ($\Rightarrow Pe = 0$). This is an application of the 'symbolic' intrinsic verification (see Chap. 5 of Ref. [30,35]). In a dimensionless form, we have:

$$\tilde{c}(\tilde{x},\tilde{t}) = 1 - 2\exp\left(f\frac{Pe}{2}\tilde{x}\right) \cdot \sum_{n=1}^{\infty} \frac{\beta_n}{\beta_n^2 + (fPe/2)^2 + fPe/2} \sin(\beta_n \tilde{x}) \cdot \exp\left\{-\left[\left(f\frac{Pe}{2}\right)^2 + \beta_n^2\right]\tilde{t}\right\} \tag{3.64}$$

where:

$$\tilde{x} = \frac{x}{l} \quad \tilde{t} = \frac{\overline{D}t}{l^2} \quad \tilde{c} = \frac{\theta}{\theta_0} = \frac{c - C_{t0}^s}{C_{x0}^f - C_{t0}^s} \tag{3.65}$$

The plots in Fig. 3.11 depict the evaluation of the solution to the dimensionless solute concentration as a function of dimensionless position and time for different Péclet numbers with a retardation coefficient $f = 0.5$. At very short times, the Péclet number has no noticeable influence on the distribution, and the presence of convective influences at early times is only noticeable for $Pe \sim 10$. At longer times, even minor convective influences are noticeable at locations far from the source.

Next, to quantify the influence of Pe on transient behavior, the dimensionless drug mass per unit area $\tilde{M}$ contained within the domain has been plotted as a function of time for $f = 0.5$, as shown in Fig. 3.12. The dimensionless drug mass may be calculated as:

$$\tilde{M}(\tilde{t}) = \int_{\tilde{x}=0}^{1} \tilde{c}(\tilde{x},\tilde{t})d\tilde{x} \tag{3.66}$$

where $\tilde{M}$ is defined as:

$$\tilde{M} = \frac{M - C_{t0}^s l}{\theta_0 l} = \frac{M - C_{t0}^s l}{(C_{x0}^f - C_{t0}^s)l} \tag{3.67}$$

At $Pe = 0$, the mass is slowly transported by diffusion. As convective influences increase (with increasing values of Pe), the rate at which the mass is transported also increases. Comparing the times at which the steady state concentration is approached for $Pe = 0$ around and $Pe = 10$ it is evident that even a small convective contribution will greatly speed up the drug transfer.

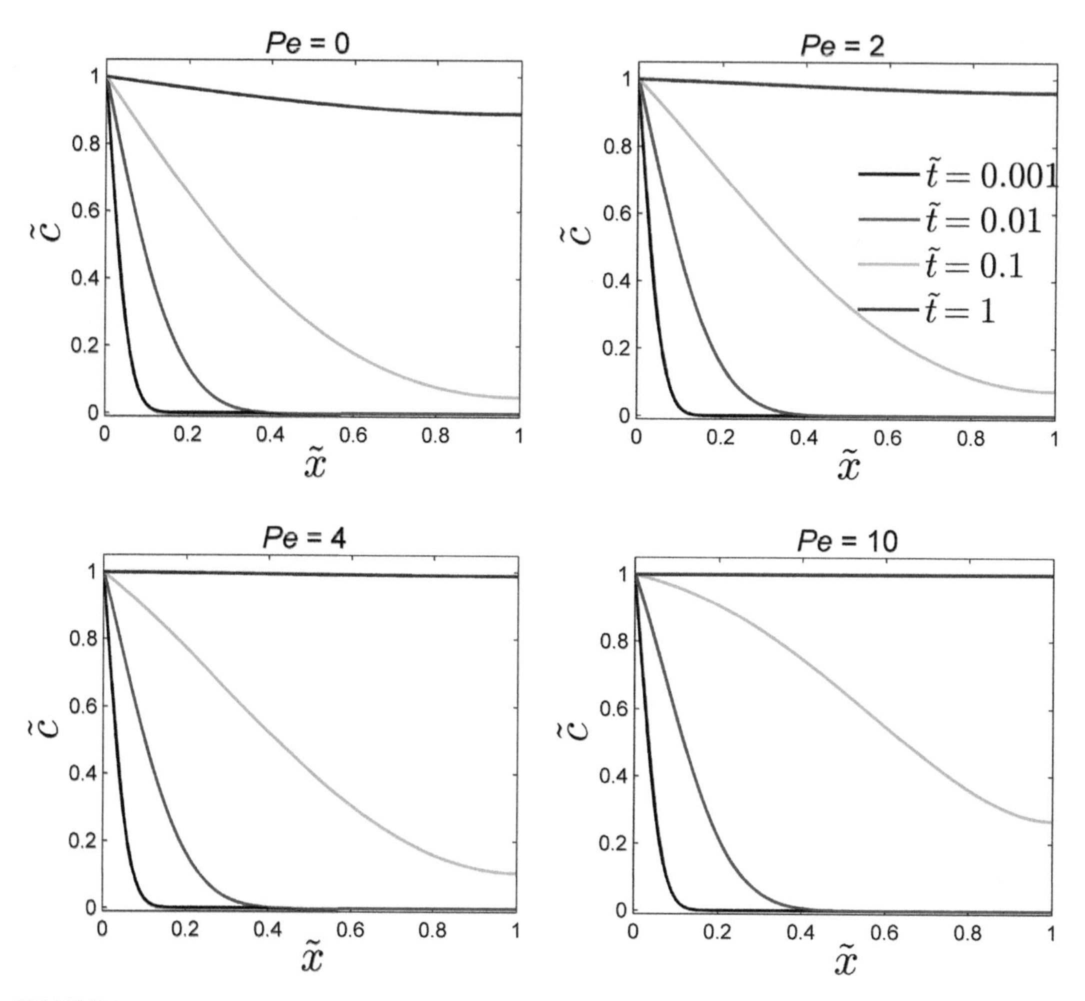

FIGURE 3.11 Transient concentration solutions for 4 different Péclet numbers.

3.6 APPLICATIONS OF POROUS MEDIA TO THE DRUG-ELUTING STENT

This section describes a very common procedure that involves the application of porous media to the previously discussed LME theory. The subject of this particular application involves what is referred to as the drug-eluting stent (DES).

As was already discussed in Section 3.1, the alteration of blood flows due to the narrowing (stenosis) or occlusion of an artery is one of the most common occurrences in cardiovascular diseases. Among the various available medical treatments, a well-established technique regards the insertion into the artery of a wired scaffold, or stent. The stent is designed to

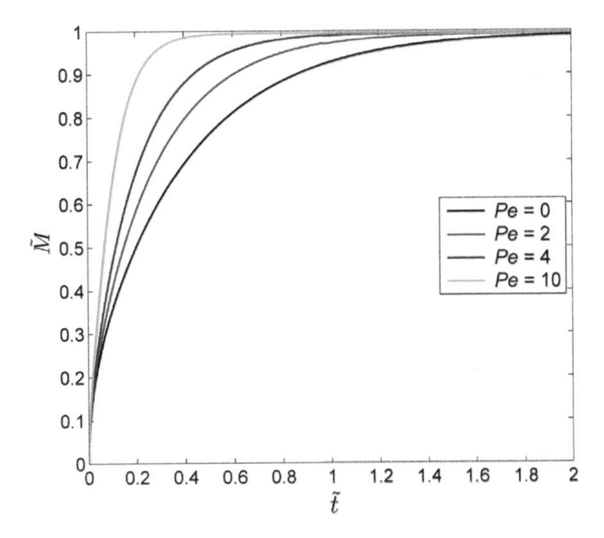

FIGURE 3.12 Drug mass versus time with Péclet number as a parameter.

mechanically provide the structural stability required to hold open the injured vessel in order to restore the correct blood flow. A typical design of the scaffold itself is provided in Fig. 3.13, and this metallic mesh is placed directly into the inner lining of the injured vessel. A challenge to the long term effectiveness of this treatment, however, has been reported, due to the tendency for tissue to grow around the scaffold's wire structure over time, thus recreating the blockage and re-occluding the lumen [4]. A solution to this is chemical inhibition of this growth by a local and controlled delivery of drugs. To ensure that the delivery is localized, the drug is contained within a very thin, porous, polymeric layer coating of the stent's surface. This design is known as the drug-eluting stent (DES), and its application has been targeted at healing the vascular tissues, and at preventing possible restenosis by virtue of its anti-proliferative action against smooth muscle cells. The application of the DES is an emerging technology that combines mechanical support of restricted lumen with local drug delivery [3,4], and although different configurations exist, a typical DES consists of the metallic strut, coated with one or more biocompatible polymeric layers containing the therapeutic agent to be delivered [36,37].

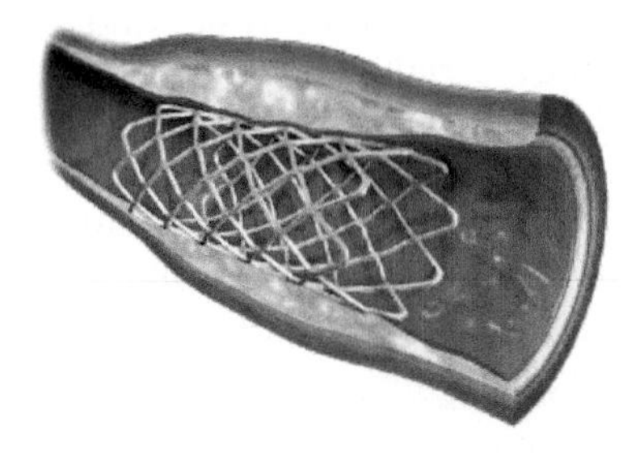

FIGURE 3.13 Metallic mesh of a stent implanted in a stenotic artery.

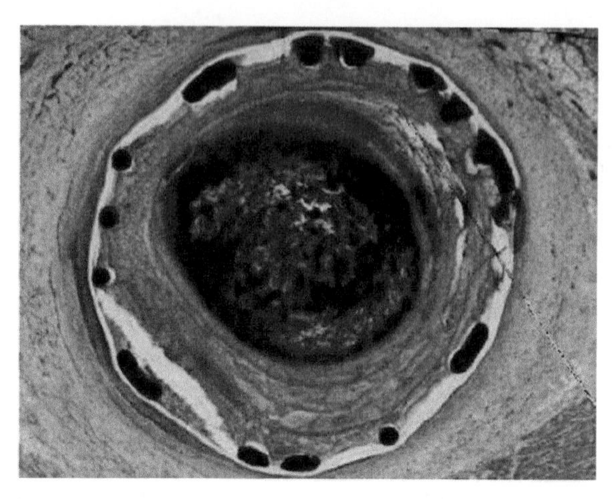

FIGURE 3.14 Section of a stented artery with the struts (in black) embedded into the wall.

Recently, polymeric gels have become much more prominent in their use as drug carrier devices in many biotechnological applications [38]. When designing these porous media, a drug-filled gel is stored within a solid polymer matrix, which enables the diffusive delivery of the drug to the surrounding tissue. With this in mind, the growth-inhibiting drug is loaded into a polymeric gel matrix that is added to the metallic struts of the stent. This allows slow, controlled, drug release after the stent has been implanted. In typical designs, the stent struts are uniformly covered by the polymer, and this is referred to as a coated stent. In an alternative design, honeycombed strut elements are inlaid with the polymer gel (stent with drug reservoirs, shown in Fig. 3.14) [39].

It is important to understand the dynamic nature of the delivery, because, while the drug concentration should meet some minimum requirement in order to achieve its desired effect, exceeding a critical threshold concentration may produce a toxic effect resulting in the destruction of tissue [40]. The drug is released through the vessel wall by both convective and diffusive processes. However, the theoretical understanding of the local transport is further complicated by the potential of metabolism of the drug by the surrounding tissue. Furthermore, it is likely that transport is locally impeded by partitioning at the interface of different tissue types that make up the arterial wall. The dynamic design of DES is aimed at the prolonged release of the drug and controlled delivery to the surrounding tissue, and mathematical models have been proposed to simulate the complex physics underlying the transient transport of the drug from the stent and into the arterial wall.

Drug release and transport depend on many of the geometric and physical parameters related to the drug, the stent, and the tissue making up the arterial wall [41]. Computational modeling has been shown to provide remarkable insight into the physical factors that influence the delivery process. These include the geometrical design of the stent, the mechanical characteristics of the materials and the chemical properties of the drug [42]. These effects play out over different time and space scales and, in order to illuminate the underlying physics associated with the drug's deposition, integrative biomechanical methods are required [43]. Recently, a number of mathematical models have been developed that address the fundamental problem of the mass release from DES through the arterial wall [5,6,44,45]. Unconventional approaches of drug delivery, such as those based on the endoluminal gel paving technology, have been recently proposed [46]. Multi-physics studies incorporating the principles of solid and fluid mechanics are able to capture both the mechanical expansion of the strut as well as the DES elution properties [47]. Difficulties in coupling different geometrical scales have been reported [48], and solutions have recently been proposed that incorporate multiscale modeling strategies [49].

Mathematical descriptions of the system can be categorized according to the level of complexity of the arterial wall description and, according to *Prosi et al.* [48], there are three cases into which these models fall.

- The simplest and least descriptive of these is the wall-free model which describes the entire arterial wall through boundary conditions. While this straightforward approach is easily implemented, unfortunately all information involving any spatial variation drug concentration within the arterial wall is lost.
- The fluid wall model approximates the wall structure as a single homogeneous layer. In many studies, the complex composite nature of the multi-layered structure making up the arterial wall is disregarded and, instead, porous media theory is used to depict the physical domain by a homogenous, spatially averaged representation. This approach has been used with great success in numerical and analytical studies of DES behavior [5,50]. Although the fluid wall model is more accurate than the wall-free model, its inherent integrative approach to the description of the arterial wall is unable to capture the role that discontinuities within the tissue microstructure can play in the drug release mechanism.
- The motivation of the multi-layered wall model is based on the knowledge that the actual anatomy of the arterial wall is known to be composed of multiple layers, each with its own distinct structural and chemical properties [50,51]. Furthermore, it is well accepted that a more accurate description of the drug delivery requires a realistic depiction of the wall structure. The multi-layered model is the most complete model, and takes into account the heterogeneous nature of the composite structure of the arterial wall: that each layer has its own distinct properties, and thus that the drug's behavior in each layer is distinct as well. However, due to its composite nature, the multi-layered wall has a high potential for producing very complicated and algebraically involved solutions.

The choice between the fluid wall model and the multi-layered wall model is made by considering the tradeoff between the required level of detail of the solution, the availability of data describing the physical parameter values and the acceptable degree of solution complexity. The multi-layer model is more comprehensive than the fluid wall model because it can provide a detailed description of the inter-tissue drug behavior. However, it should come as no surprise that with each additional layer the complexity of the solution increases. Furthermore, in order to present an accurate portrayal of the drug's behavior as it passes through the multi-layered domain, knowledge of the parameter values that characterize the physical properties of each layer must be available. For this reason, the analytical solutions of both the fluid wall model and the multi-layered model are provided in the next subsections.

These solutions have been derived under the simplifying assumption of one-dimensional transport of the drug. In addition, as the drug is initially contained only in the solid polymeric matrix of the coating, whose pores (channels) are occupied by the plasma after DES implantation, at $t = 0$ we have $c^s(x, t = 0) = C^s_{t0}$ and $c^f(x, t = 0) = 0$. Strictly speaking, it is not acceptable to assume a LME, so that $c^f = c^s = c$ in Eqs. (3.27a) and (3.28a), as shown in 3.5. However, to simplify the analysis, it is here assumed that at $t = 0$ the drug mass is instantaneously transferred to the fluid phase from the solid matrix and, then, released into the arterial wall. Hence, $c^f(x, t = 0) = C^s_{t0}$ and the solid gel can be completely neglected. This simplifying assumption is removed in [52] where a two-phase model is considered.

3.6.1 The Fluid Wall Model: The Pure Diffusion Approximation

The fluid wall model eliminates some of the inherent complexities associated with the composite nature of the arterial wall by using a single layer whose homogenous physical properties represent the combined effects of the multiple layers of the actual artery [39,48].

This simplistic model will be extended in the next sections to include more general effects and wall structure, but its description provides the methodological framework used. The analytical studies [5] use the fluid wall model approximation in conjunction with porous media representations of both the coating and the arterial wall in order to successfully provide a parametric analysis. The solution of that study, which is summarized in this section, provides an excellent tool for analysis in order to develop a better understand the underlying physics of DES delivery.

3.6.1.1 *General Physiological and Mathematical Description*

The conceptual physical space of the fluid wall model of DES consists of two layers. The stent is coated with a polymeric gel that is represented by Layer 0 (of thickness l_0). The stent is embedded into the arterial wall (Layer 1 of thickness l_1), as illustrated in Fig. 3.15. Because the fluid wall approximation is being applied, the complex multi-layered structure of the arterial wall can be considered to behave homogeneously with averaged properties inside Layer 1. Both the coating and the arterial wall are treated as porous media [12].

Because most of the mass transport process occurs along the direction normal to the two layers (radial direction), we restrict our study to a simplified one-dimensional model. In particular, we consider the transient transport along a coordinate axis that crosses the metallic strut, the coating, and the arterial wall. The wall thickness is very small relative to the arterial radius, and thus the Cartesian coordinate system is used (see Fig. 3.15). Taking the perspective that applies to either the luminal or the adventitial side, the drug transport route originates within the stent's porous coating, and progresses into the arterial wall, so that the positive coordinate axis direction is oriented away from the stent surface and into the arterial wall.

A volume-averaged concentration of drug in the fluid phase $\bar{c}^f = \langle c^f \rangle$ is considered in place of the intrinsic volume-averaged concentration $c^f = \langle c^f \rangle^f$, that is related to the previous quantity by $c^f = \bar{c}^f / \varepsilon_k$ (Section 3.2.1.1). For the sake of brevity, $\bar{c}^f$ will henceforth simply be denoted by c. Because it is anticipated that the drug's behavior is distinct in each of the system's two porous layers, the drug behavior may be considered in terms of two distinct components: $c_0(x, t)$ which corresponds to the drug concentration in the polymeric gel of Layer 0, and $c_1(x, t)$ which corresponds to the drug concentration in the arterial wall of Layer 1.

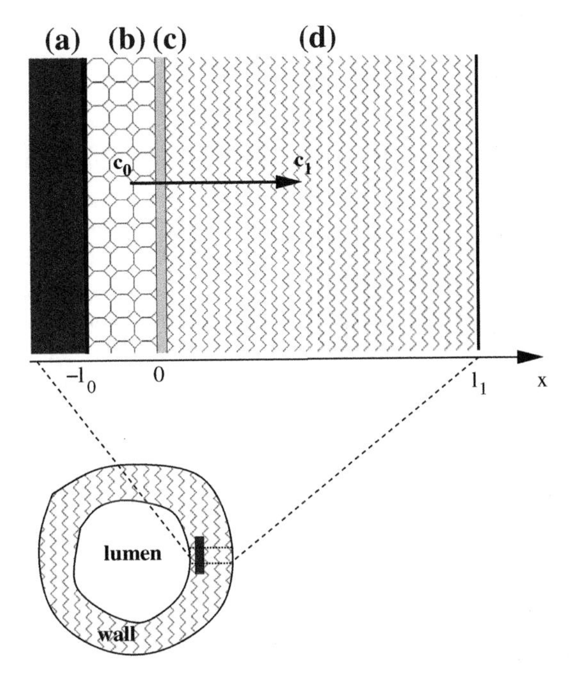

FIGURE 3.15 Cross section of a stented artery with a zoomed area near the wall that shows the metallic mesh and the two-layer medium at the adventitial side described by the model (3.68)–(3.70): (a) stent strut, (b) coating, (c) topcoat, (d) arterial wall. Due to an initial difference of concentration, drug is released in the arterial wall from (b) to (d) through the permeable membrane (c). An analogous two-layer pattern is present on the opposite side of the strut, referring to the drug dynamics towards the lumen (lumenal side).

As was already said, initially ($t = 0$) the drug mass is instantaneously transferred to the fluid phase from the solid matrix, and then released into the arterial wall. Hence, $c_0(x, t = 0) = C_0$ where C_0 is some uniform concentration value. Since the strut is impermeable to the drug, no mass flux passes through the boundary surface corresponding to the stent-coating interface at $x = -l_0$. Moreover, it is assumed that the plasma velocity through the pores of the polymeric gel matrix of the coating is very low after DES implantation, so that the drug transport in Layer 0 is governed by pure diffusion. Bearing in mind Eq. (3.27b), the dynamics of the drug in the fluid phase of the coating can be described by the following differential equations:

$$\frac{\partial c_0}{\partial t} = D_0 \frac{\partial^2 c_0}{\partial x^2} \quad (-l_0 < x < 0; t > 0) \tag{3.68a}$$

$$\left(\frac{\partial c_0}{\partial x}\right)_{x=-l_0} = 0 \quad (t > 0) \tag{3.68b}$$

$$c_0(x, t = 0) = C_0 \quad (-l_0 < x < 0) \tag{3.68c}$$

where D_0 is the effective diffusion coefficient of the drug in the fluid phase of the polymeric coating. Also, we recall that it is a volume-averaged quantity per unit volume of the fluid phase (intrinsic quantity), that is, $D_0 = \left\langle D^f_{d0} \right\rangle^f$, where the subscript d denotes drug while 0 indicates coating (Layer 0).

As implied in the discussion of the stent coating's initial condition, an equally valid assumption can be that initially no drug is present in the tissue of the arterial wall. A zero concentration can be prescribed at the domain boundaries of the artery on both the luminal side and the adventitial side. This is justified as follows. Because the arterial wall of the adventitial side is much thicker than the strut coating ($l_1 \gg l_0$), the concentration c_1 at $x = l_1$ does not change with time, so that its initial zero value is preserved at that location. In the case of the lumenal side, the position $x = l_1$ corresponds to the interface between the arterial wall and the blood stream. Because of the extremely high mass transfer coefficient of the blood stream (washout), it is reasonable to assume that the concentration at this point is zero.

It should be noted that in both cases, this homogeneous boundary condition of the first kind at $x = l_1$ will result in a fraction of drug being transported out of the arterial wall and into the tissues adjacent to the adventitia, or on the luminal side, a fraction of the drug will be dispersed into the lumen side blood stream. Once it has been transported out of the arterial wall, these fractions of the drug are considered as 'lost', because that component of the drug is no longer available to treat the tissues of the arterial wall.

In the wall (Layer 1), the drug dynamics are described by the following equation and related boundary/initial conditions:

$$\frac{\partial c_1}{\partial t} = D_1 \frac{\partial^2 c_1}{\partial x^2} \quad (0 < x < l_1;\, t > 0) \tag{3.69a}$$

$$c_1(x = l_1, t) = 0 \quad (t > 0) \tag{3.69b}$$

$$c_1(x, t = 0) = 0 \quad (0 < x < l_1) \tag{3.69c}$$

where D_1 is the effective diffusion coefficient of the drug in the fluid phase of the arterial wall.

This representation of the transport is extremely simple, primarily because it neglects any advection that would be associated with the flow of plasma within the wall tissue. This equation also neglects any uptake of the drug by the arterial lining. Similar assumptions are described in [5] and, while this has been shown to provide reasonable results, in the next subsection (Section 3.6.2) we develop a more comprehensive model including both advection and drug reaction.

To close the system of Eqs. (3.68) and (3.69), the conditions at the interface $x = 0$ (the so-called inner boundary conditions) have to be assigned. One of them is obtained by imposing continuity of the mass flux at the interface:

$$D_0 \left(\frac{\partial c_0}{\partial x} \right)_{x=0} = D_1 \left(\frac{\partial c_1}{\partial x} \right)_{x=0} \quad (t > 0) \tag{3.70a}$$

Now consider that, in order to slow down the drug release rate, a permeable membrane (called the topcoat) of permeability P is used to cover the surface of the drug-filled coating of Layer 0. In our model, the topcoat is located at the interface ($x = 0$) between the coating and the arterial wall, as shown in Fig. 3.15. The topcoat is sufficiently thin (even relative to Layer 0) that its interfacial resistance can be represented by a jump discontinuity in concentration at the interface. To do this, the mass transfer through the topcoat can be described using the second Kedem-Katchalsky equation (see Subsection 9.3.3 of [15,39,48,53]). Thus, the continuous flux of mass passing across the membrane normally to the coating may be expressed by:

$$-D_0\left(\frac{\partial c_0}{\partial x}\right)_{x=0} = P\left[\frac{c_0(0,t)}{\varepsilon_{k,0}} - \frac{c_1(0,t)}{\varepsilon_{k,1}}\right] \quad (t>0) \tag{3.70b}$$

A generalized form of the above equation, including the effect of the boundary layers formed on either side of the membrane on the volume and solute fluxes, was derived in [54].

3.6.1.2 *Concentration Solutions and Results*

Using the following dimensionless transformations:

$$\begin{gathered}\tilde{x} = \frac{x}{l_1}, \quad \tilde{t} = \frac{D_1}{l_1^2}t, \quad \tilde{c}_0 = \frac{c_0}{C_0}, \quad \tilde{c}_1 = \frac{c_1}{C_0}\\ \gamma_0 = \frac{D_0}{D_1}, \quad \tilde{l}_0 = \frac{l_0}{l_1}, \quad \phi = \frac{Pl_1}{D_1\varepsilon_{k,1}}, \quad \sigma_0 = \frac{\varepsilon_{k,0}}{\varepsilon_{k,1}}\end{gathered} \tag{3.71}$$

and by re-setting the variables:

$$\tilde{x} \to x, \quad \tilde{t} \to t, \quad \tilde{c}_0 \to c_0, \quad \tilde{c}_1 \to c_1, \quad \tilde{l}_0 \to l_0 \tag{3.72}$$

the governing Eqs. (3.68)–(3.70) given in the previous subsection may be represented in a dimensionless form as:

$$\frac{\partial c_0}{\partial t} = \gamma_0\frac{\partial^2 c_0}{\partial x^2} \quad (-l_0 < x < 0;\, t>0) \tag{3.73a}$$

$$\left(\frac{\partial c_0}{\partial x}\right)_{x=-l_0} = 0 \quad (t>0) \tag{3.73b}$$

$$\gamma_0\left(\frac{\partial c_0}{\partial x}\right)_{x=0} = \left(\frac{\partial c_1}{\partial x}\right)_{x=0} \quad (t>0) \tag{3.73c}$$

$$c_0(x, t=0) = 1 \quad (-l_0 < x < 0) \tag{3.73d}$$

and:

$$\frac{\partial c_1}{\partial t} = \frac{\partial^2 c_1}{\partial x^2} \quad (0 < x < 1;\, t>0) \tag{3.74a}$$

$$-\gamma_0\left(\frac{\partial c_0}{\partial x}\right)_{x=0} = \phi\left[\frac{c_0(0,t)}{\sigma_0} - c_1(0,t)\right] \quad (t>0) \tag{3.74b}$$

$$c_1(x=1,t)=0 \quad (t>0) \tag{3.74c}$$

$$c_1(x,t=0)=0 \quad (0<x<1) \tag{3.74d}$$

The initial boundary value problem of Eqs. (3.73) and (3.74) is similar to the problem of heat diffusion in a two-layer slab. The analytical solution to this problem has been developed for the transient diffusion of mass associated with the DES in the study [5] by using the Separation of Variables (SOV) method. The solution is presented as follows.

The non-dimensional concentrations are represented by the Fourier series solutions:

$$c_0(x,t)=\sum_{m=1}^{\infty} A_m X_{0m}(x)\exp\left(-\gamma_0\lambda_{0m}^2 t\right) \quad (-l_0\le x\le 0;\, t\ge 0) \tag{3.75a}$$

$$c_1(x,t)=\sum_{m=1}^{\infty} A_m X_{1m}(x)\exp\left(-\lambda_{1m}^2 t\right) \quad (0\le x\le 1;\, t\ge 0) \tag{3.75b}$$

where the two layers' eigenvalues λ_{0m} and λ_{1m} are related by the relation $\lambda_{0m}=\lambda_{1m}/\sqrt{\gamma_0}$.

Also, the corresponding eigenfunctions are represented by:

$$X_{0m}(x)=a_{0m}\cos\left(\lambda_{0m}x\right)+b_{0m}\sin\left(\lambda_{0m}x\right) \quad (-l_0\le x\le 0) \tag{3.76a}$$

$$X_{1m}(x)=a_{1m}\cos\left(\lambda_{1m}x\right)+\sin\left(\lambda_{1m}x\right) \quad (0\le x\le 1) \tag{3.76b}$$

for which:

$$a_{0m}=-\sigma_0\left[\frac{\lambda_{1m}}{\phi}+\tan\left(\lambda_{1m}\right)\right] \quad b_{0m}=\frac{1}{\sqrt{\gamma_0}} \quad a_{1m}=-\tan\left(\lambda_{1m}\right) \tag{3.77}$$

Each of the terms of the series of solution of Eq. (3.75) has an eigenvalue associated with it. The closure of the components of the solution, Eqs. (3.76)–(3.77), requires the value of each of the 'm' eigenvalues ($m=1,2,\ldots$) of Layer 1, λ_{1m}. These are evaluated numerically as roots of the transcendental equation (eigencondition), $a_0\tan\left(\lambda_0 l_0\right)+b_0=0$, that is:

$$\underbrace{\sigma_0\left[\frac{\lambda_1}{\phi}+\tan\left(\lambda_1\right)\right]}_{-a_0}\tan\left(\frac{l_0}{\sqrt{\gamma_0}}\lambda_1\right)-\underbrace{\frac{1}{\sqrt{\gamma_0}}}_{b_0}=0 \tag{3.78}$$

Then, by appying the initial condition, the constant of integration appearing in Eqs. (3.75) is evaluated as $A_m=-b_{0m}/(N_m\lambda_{0m})$, where the dimensionless norm N_m coming from the orthogonality property of the eigenfunctions is given by (see App. 5 of Ref. [5]):

$$N_m=\frac{1}{2}\left[\left(a_{0m}^2+\frac{1}{\gamma_0}\right)l_0-\frac{a_{0m}}{\lambda_{1m}}+\sigma_0\left(a_{1m}^2+1-\frac{a_{1m}}{\lambda_{1m}}\right)\right] \tag{3.79}$$

For a discussion of the determination of the eigenvalues and the sensitivity of the solution to the number of terms in Eqs. (3.75), the reader is encouraged to consider the study [5]. In that study, dimensionless parameter values are used which are made up of the dimensional parameter values that are representative of a typical DES. These parameters have been chosen on a physical basis and in agreement with the typical scales in DES and data in literature for the arterial wall and heparin drug in the coating layer ([40,42,55]). The resulting parameter values:

$$\phi = 0.234, \quad \sigma_0 = 0.164, \quad l_0 = 0.05, \quad \gamma_0 = 0.0014 \tag{3.80}$$

are used to develop the solution and to evaluate the eigenvalues that are the roots of Eq. (3.79). The concentration profiles for three values of dimensionless time are displayed in Fig. 3.16. The drug elutes from the coating to the wall, with its concentration decaying over time. Because of the different properties of the two layers, at early times the concentrations exhibit very steep gradients near $x = 0$, which disappears at later times.

To present a more globalized, transient depiction of the process, that study introduces the concept of the instantaneous dimensionless drug mass per unit area, $\widetilde{M}_i = M_i/C_0 l_1$, present in both the coating ($i = 0$) and the arterial wall ($i = 1$) layers. By re-setting $\widetilde{M}_i \to M_i$, these masses are defined as:

$$M_0(t) = \int_{-l_0}^{0} c_0(x,t)dx = \sum_{m=1}^{\infty} A_m^2 N_m \exp(-\lambda_{1m}^2 t) \tag{3.81a}$$

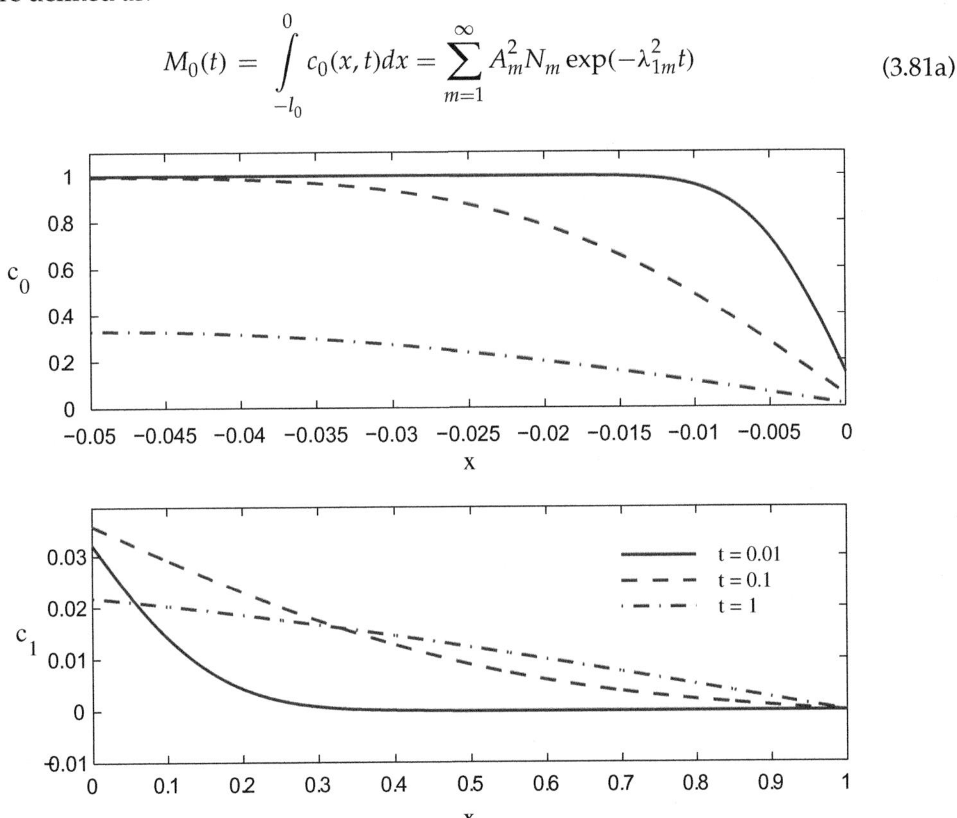

FIGURE 3.16 Drug concentration profiles in the coating (above) and in the wall (below) for three different times (note the different space scales).

$$M_1(t) = \int_0^1 c_1(x,t)dx = \sum_{m=1}^{\infty} A_m \frac{\cos(\lambda_{1m}) - 1}{\lambda_{1m}\cos(\lambda_{1m})} \exp(-\lambda_{1m}^2 t) \tag{3.81b}$$

Figure 3.17 shows that the drug mass in the coating layer, M_0, decreases monotonically, while the mass in the wall, M_1, first increases to a maximum M_1^* at time t^*, then decreases to zero at the same rate as M_0. Since the drug is absorbed at $x = 1$, the total mass is not preserved and tends to zero at very large times.

In principle, the drug diffusion occurs from the coating towards both the lumen (inner wall) as well as towards the adventitia (outer wall). Therefore an inner two-layer system (lumenal side) and an outer two-layer system (adventitial side) should be distinguished. The relative importance of the transport in either direction depends on the penetration depth of the stent. The adventitial side is generally larger in size and deemed more relevant from a clinical point of view. However, it should be noted that the methodology and solution formulation presented in this section are general and apply to transport in both the inner (lumenal) and outer (adventitial) sides of the stent.

3.6.2 The Fluid Wall Model: The Advection-Reaction-Diffusion Equation

The previous solution does not consider the effects of the plasma flow through the tissues composing the arterial wall. While this assumption has been shown to hold using scale analysis [5] as well as numerical models [39], the description of the fluid wall model of the

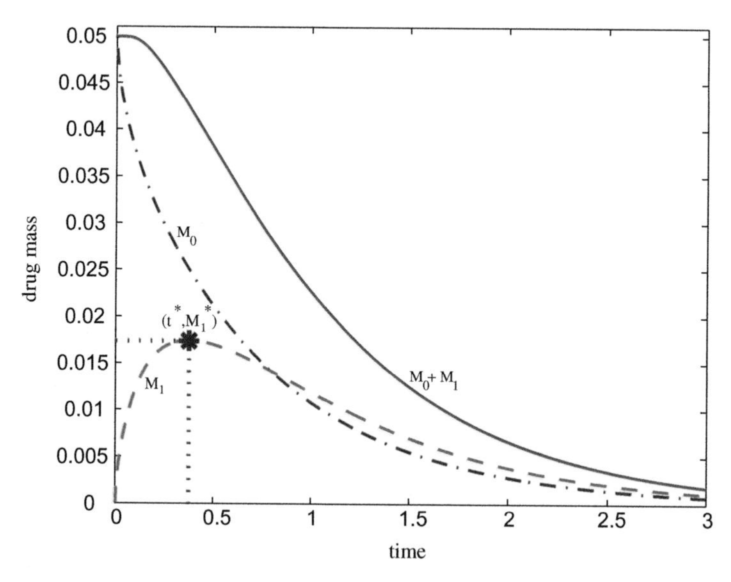

FIGURE 3.17 Dimensionless drug mass in the coating (- · - · -), in the wall (- - -) and total mass (—) as function of time. In the coating, drug mass M_0 is monotonically decreasing, while in the wall there is a characteristic time t^* at which the drug reaches a maximum peak M_1^*. Due to the absorption at the wall side, drug is not preserved and vanishes at a time large enough.

DES is not complete without considering this phenomenon. This section develops the governing equations, boundary and interface conditions and presents the solution of the drug deposition from the DES that includes the advective contribution within the arterial wall as well as the uptake of the drug by the arterial wall tissue. Referring once more to Fig. 3.15, and recalling that the adventitial side is generally larger in size and deemed more relevant from a clinical point of view, the following discussion will only consider the transport from the stent to the outer arterial wall (adventitial side).

3.6.2.1 *General Physiological and Mathematical Description*

Following the coordinate system presented in Fig. 3.15, we first consider the solute within the stent's thin porous coating. Initially, the drug is contained only in the fluid phase of the stent thin porous coating, and the initial concentration is represented by a potentially non-uniform distribution, $C_0 \cdot F_0(x)$, where C_0 is a constant and $F_0(x)$ is an arbitrary dimensionless function such that $0 \leq F_0(x) \leq 1$. Again, the strut is impermeable to the drug, so no mass passes through the boundary surface corresponding to the stent-coating interface at $x = -l_0$. Transport in Layer 0 is governed by pure diffusion because the velocity of plasma penetrating into the stent's coating is very low. Thus, the dynamics of the drug in the fluid phase of the coating (Layer 0) is still described by Eqs. (3.68a)–(3.68c), and only the initial condition is now different.

In the wall (Layer 1), bearing in mind Eq. (3.27b), the drug dynamics are described by the following advection-reaction-diffusion equation:

$$\frac{\partial c_1}{\partial t} = D_1 \frac{\partial^2 c_1}{\partial x^2} - 2\delta_1 \frac{\partial c_1}{\partial x} - \beta_1 c_1 \quad (0 < x < l_1; t > 0) \tag{3.82}$$

where the advective component is represented by the parameter $2\delta_1 = f_1 u_1 / \varepsilon_{k,l}$.

The quantity $f_1 u_1$ is the drug velocity in the arterial wall, which has been considered to be uniform and constant. In particular, the term f_1 is the retardation coefficient defined in Section 3.3.1 (also known as the hindrance coefficient) and u_1 is the plasma velocity. For an isothermal and incompressible fluid in a porous medium, the local flow velocity u_1 may be calculated by Eq. (3.10). Note that the advective component is sometimes represented as $2\delta_1 = (1 - v_1)u_1$, where $v_1 = 1 - f_1/\varepsilon_{k,l}$ is known as the Stavernan filtration coefficient [56].

The last term on the left-hand side of Eq. (3.82) represents the drug reaction inside the media layer of the arterial wall. Here, it is approximated by a linear reaction having $\beta(> 0)$ as an effective first-order consumption rate coefficient. The boundary and initial conditions associated with Eq. (3.82) are still defined by Eqs. (3.69b) and (3.69c), respectively. To close the mass transfer system of defining equations, the conditions at the interface ($x = 0$) will now be assigned. One of them is obtained by imposing continuity of the mass flux across the two layers as expressed by Eq. (3.70a). For the current purposes, it is rewritten as:

$$D_0 \left(\frac{\partial c_0}{\partial x} \right)_{x=0} = D_1 \left(\frac{\partial c_1}{\partial x} \right)_{x=0} - 2\delta_1 c_1(0, t) \quad (t > 0) \tag{3.83}$$

The interfacial resistance due to the transport inhibiting influence of the topcoat coating of Layer 0 is again represented by the second Kedem-Katchalsky equation, i.e., Eq. (3.70b).

3.6.2.2 *Concentration Solutions and Results*

In order to represent the system in non-dimensional terms, the same dimensionless variables previously listed in Eq. (3.71) can be used. When this is done, the convection and reaction terms are represented by:

$$Pe = \frac{\delta_1 l_1}{D_1} \quad \Phi = \sqrt{\frac{\beta_1 l_1^2}{D_1}} \tag{3.84}$$

The Péclet number, *Pe*, expresses the ratio of the transport of the drug by advection to the diffusion of the drug within the arterial wall. The Thiele modulus, Φ, is a dimensionless number characterizing the ratio of the reaction rate to the diffusion rate [53].

The dimensionless variables can then be rewritten as in Eq. (3.72) to simplify the solution presentation. In order to resolve the solution at the interface, the following transformation of Layer 1 is performed which was suggested by *Özişik* (see Eq. (2–168) on p. 90 of Ref. [57]):

$$c_1(x,t) = w_1(x,t)\exp[Pe\cdot x - (Pe^2 + \Phi^2)t] \tag{3.85}$$

The governing equations of the current problem may be represented in dimensionless form as:

$$\frac{\partial c_0}{\partial t} = \gamma_0 \frac{\partial^2 c_0}{\partial x^2} \quad (-l_0 < x < 0;\, t > 0) \tag{3.86a}$$

$$\left(\frac{\partial c_0}{\partial x}\right)_{x=-l_0} = 0 \quad (t > 0) \tag{3.86b}$$

$$\gamma_0\left(\frac{\partial c_0}{\partial x}\right)_{x=0} = \left[\left(\frac{\partial w_1}{\partial x}\right)_{x=0} - Pe\cdot w_1(0,t)\right]\exp\left[-(Pe^2+\Phi^2)t\right] \quad (t>0) \tag{3.86c}$$

$$c_0(x,t=0) = F_0(x) \quad (-l_0 < x < 0) \tag{3.86d}$$

and:

$$\frac{\partial w_1}{\partial t} = \frac{\partial^2 w_1}{\partial x^2} \quad (0 < x < 1;\, t > 0) \tag{3.87a}$$

$$-\gamma_0\left(\frac{\partial c_0}{\partial x}\right)_{x=0} = \phi\left\{\frac{c_0(0,t)}{\sigma_0} - w_1(0,t)\exp[-(Pe^2+\Phi^2)t]\right\} \quad (t>0) \tag{3.87b}$$

$$w_1(x=1,t) = 0 \quad (t>0) \tag{3.87c}$$

$$w_1(x,t=0) = 0 \quad (0 < x < 1) \tag{3.87d}$$

The analytical solution to the above initial boundary value problem has been developed for the transient diffusion-advection of mass associated with the DES in the study [31] by using the SOV method. The solution is presented as follows.

The non-dimensional concentrations are represented by the Fourier series solutions:

$$c_0(x,t) = \sum_{m=1}^{\infty} A_m X_{0m}(x) \exp(-\gamma_0 \lambda_{0m}^2 t) \quad (-l_0 \le x \le 0; t \ge 0) \tag{3.88a}$$

$$c_1(x,t) = \exp[Pe \cdot x] \sum_{m=1}^{\infty} A_m X_{1m}(x) \exp\left[-(\lambda_{1m}^2 + Pe^2 + \Phi^2)t\right] \quad (0 \le x \le 1; t \ge 0) \tag{3.88b}$$

where the two layers' eigenvalues are related by the relationship $\gamma_0 \lambda_{0m}^2 = \lambda_{1m}^2 + Pe^2 + \Phi^2$, while the eigenfunctions are still represented by Eqs. (3.76) but with a different expression for their coefficients. We have:

$$a_{0m} = -\sigma_0 \left(1 + \frac{Pe}{\phi}\right) \left[\frac{\lambda_{1m}}{\phi + Pe} + \tan(\lambda_{1m})\right] \tag{3.89a}$$

$$b_{0m} = \frac{1}{\sqrt{\gamma_0}} \frac{\lambda_{1m} + Pe \tan(\lambda_{1m})}{\sqrt{\lambda_{1m}^2 + Pe^2 + \Phi^2}} \tag{3.89b}$$

$$a_{1m} = -\tan(\lambda_{1m}) \tag{3.89c}$$

where the eigenvalues λ_{1m} of Layer 1 may be determined by computing the roots of the transcendental equation (eigencondition), $a_0 \tan(\lambda_0 l_0) + b_0 = 0$, that is:

$$\underbrace{\sigma_0 \left(1 + \frac{Pe}{\phi}\right) \left[\frac{\lambda_1}{\phi + Pe} + \tan(\lambda_1)\right]}_{-a_0} \tan\left(\frac{l_0}{\sqrt{\gamma_0}} \sqrt{\lambda_1^2 + Pe^2 + \Phi^2}\right) - \underbrace{\frac{1}{\sqrt{\gamma_0}} \frac{\lambda_1 + Pe \tan(\lambda_1)}{\sqrt{\lambda_1^2 + Pe^2 + \Phi^2}}}_{b_0} = 0 \tag{3.90}$$

In this way each of the 'm' roots of Eq. (3.90) are used to develop a corresponding term of the series of the solution, Eqs. (3.88). These roots are in general real, but imaginary roots are possible too. Note that, when $Pe = 0$ and $\Phi = 0$, Eq. (3.90) reduces exactly to Eq. (3.78). Similarly, the coefficients defined by Eqs. (3.89) reduce to the ones given by Eqs. (3.77).

In Eq. (3.88) the constant of integration, A_m, is evaluated as:

$$A_m = \frac{1}{N_m} \int_{x=-l_0}^{0} F_0(x) X_{om}(x) dx \tag{3.91}$$

which for the special case of a Layer 0 uniform initial distribution, $F_0(x) = 1$, may be expressed as $A_m = -b_{0m}/(N_m\lambda_{0m})$. The normalizing constant (norm), N_m, is defined by:

$$N_m = \frac{1}{2}\left[(a_{0m}^2 + b_{0m}^2)l_0 - \frac{a_{0m}b_{0m}}{\lambda_{0m}} + \sigma_0\left(a_{1m}^2 + 1 - \frac{a_{1m}}{\lambda_{1m}}\right)\right] \tag{3.92}$$

In the study in [31], a parametric investigation is conducted regarding the effects of the advection of the drug within the arterial wall, which in this solution is represented by the Péclet number, *Pe*. The study also examines the relative role of the drug reaction which is represented in the Thiele modulus, Φ. Using the dimensionless parameter values listed in Eq. (3.81), the concentration solution given by Eqs. (3.88) is plotted for different values of *Pe* and Φ. (In such a case, an imaginary eigenvalue was found.)

To show the influence of the filtration velocity on the drug release, a value $\delta_1 \cong 10^{-6} cm/s$ is considered, which reflects measurements presented in the studies in [47,58]. Simulations for four values of *Pe* in a compatible range have been carried out to show the trend of the solution. The effects on the drug transport within the arterial wall are plotted in Fig. 3.18. It is clear that introducing even a very small advective term lowers the concentration curves at all times. This is because the convection velocity sweeps the drug away from the wall, where it is dispersed. At intermediate and later instants, the profiles may appear bulged and therefore a more uniform concentration is guaranteed. A critical value of *Pe* exists; beyond it the solution no longer represents a physical reality.

The importance of the reaction term depends on the drug used, on the specific tissue and on individual factors. However, the presence of a reactive term acts as a sink for concentration. A typical value for the consumption rate coefficient is $\beta_1 \simeq 10^{-4} s^{-1}$ [59]. Consequently, $\Phi = \sqrt{0.15}$ and the trend of the concentration at three increasing values of Φ is depicted in Fig. 3.19. Increasing the value of Φ accelerates the drug consumption and diminishes the concentration, but preserves the shapes at fixed times. A negligible variation with Φ is reported at early times. At later instants, the concentration profiles flatten and decrease linearly. In conclusion, the effect on varying *Pe* is similar, though more sensitive, to that on Φ, and they combine when both coexist.

3.6.3 The Multi-Layered Wall Model: The Pure Diffusion Approximation

In this section, a multi-layered extension of pure diffusion model in Section 3.6.1 is developed. Following [50,60], an idealized model of wall consisting of four layers (namely endothelium, intima, internal elastic lamina and media) is proposed. Each layer is treated as a macroscopically homogeneous porous medium with its own distinct diffusion coefficient. A further simplification involves the endothelium that is covered by a thin ciliate layer called the *endothelial surface layer* or *glycocalyx* and this is composed of a sequence of long chain macromolecules and proteins [51]. As the drug transport properties through the glycocalyx are unknown, it has been included in the endothelium layer for simplicity. The stent coating is assumed to be a thin porous slab in imperfect contact with the endothelium due to the presence of a topcoat.

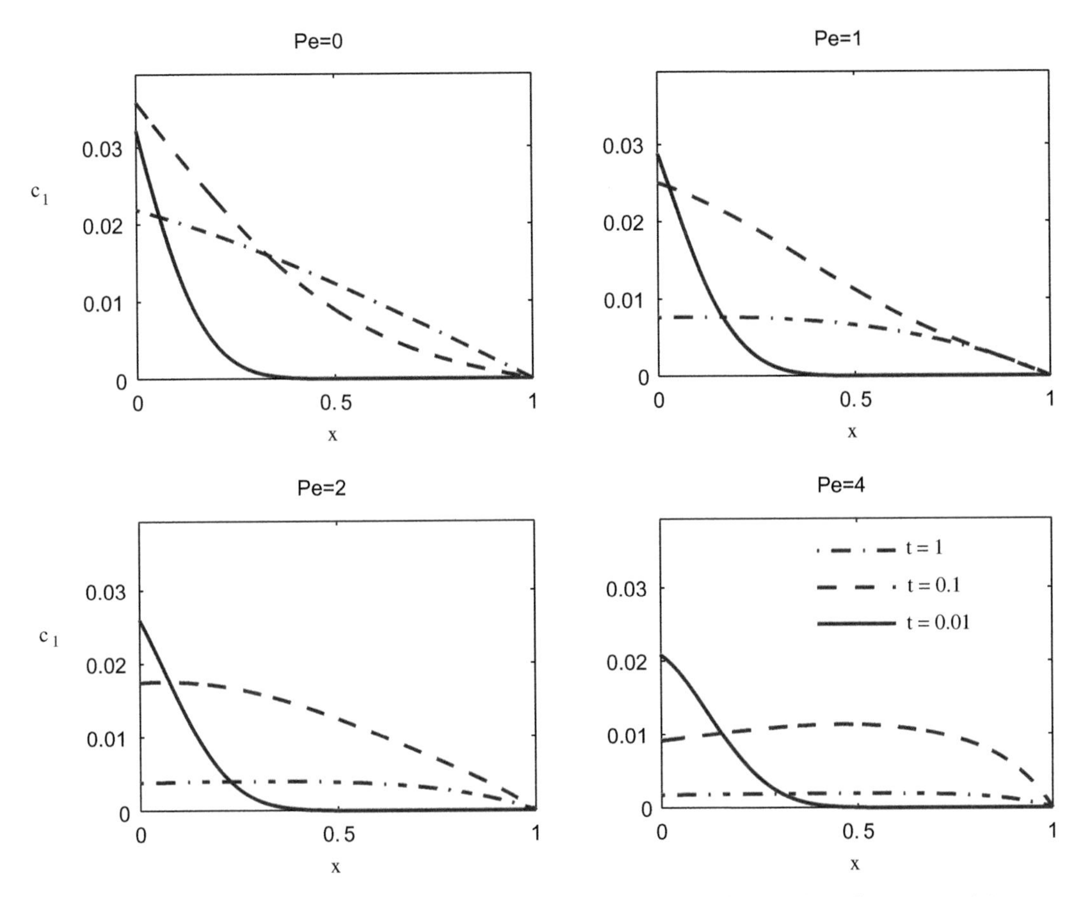

FIGURE 3.18 Wall concentration decay profiles at varying *Pe* (with $\Phi = 0$) for different locations and times.

3.6.3.1 *General Physiological and Mathematical Description*

The adventitia and the surrounding tissues form the outmost wall layer and have a sufficiently large extent to be considered as semi-infinite. The classical SOV method leads to a Sturm-Liouville problem with discontinuous coefficients and severe spectral irregularities. Drug concentration in each layer at various times is given in the form of a Fourier series by using dimensionless parameters which control the transfer mechanism across the layered wall. For generality, the problem is developed such that there is any number of arterial tissue layers.

First, consider a thin layer (of thickness l_0) of gel containing a drug that coats a stent which is embedded into the arterial wall [44]. Again we restrict our study to a simplified 1-D model and consider only the transport from the stent into the adventitial side arterial wall. In this way, the drug's behavior is considered along an outward pointing line crossing the metallic strut, the coating and the layers of the arterial wall. In a general 1-D framework, let us

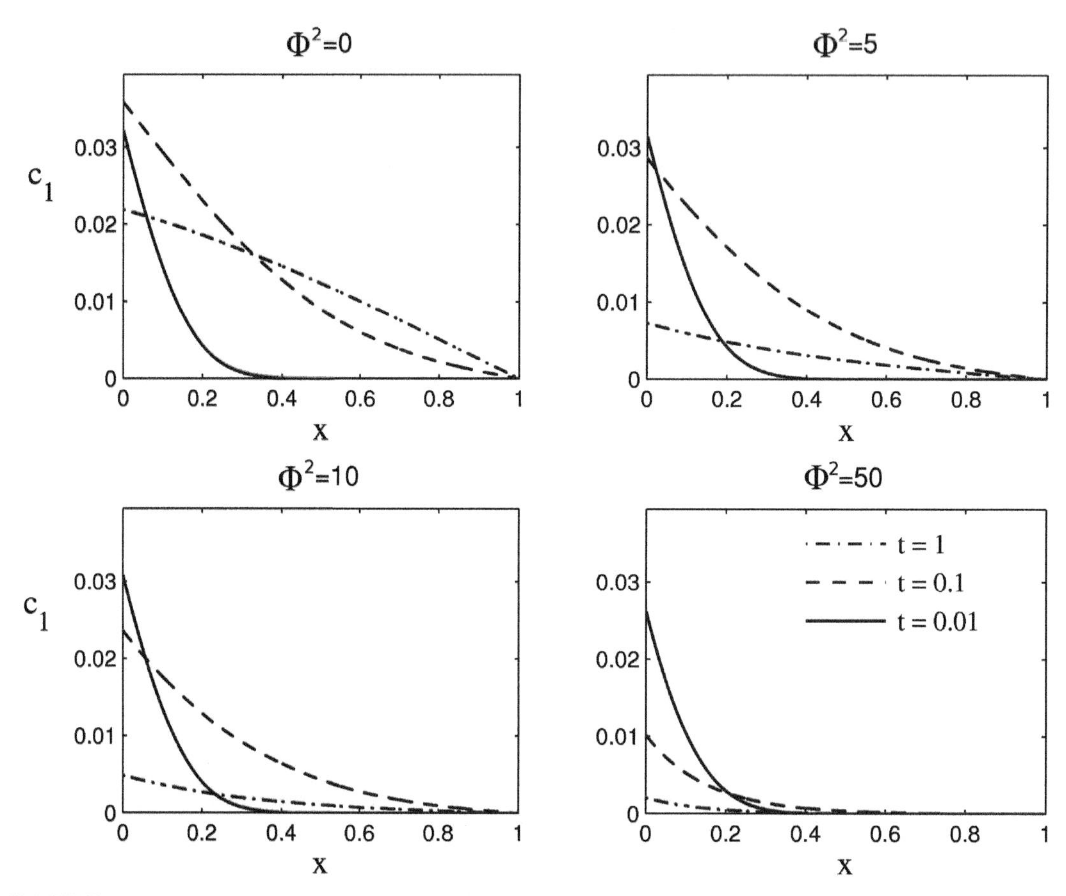

FIGURE 3.19 Wall concentration decay profiles at varying Φ (with $Pe=0$) for different locations and times.

consider a set of intervals $[x_{i-1}, x_i]$ for $i = 0, 1, 2, \ldots, n$ each having thickness $l_i = x_i - x_{i-1}$ so that the coating (layer 0) and the wall layers (layers $i = 1, 2, \ldots, n$) are conceptually represented as in Fig. 3.20 where, without loss of generality, $x_0 = 0$ has been assumed to be the coating-wall interface.

At the initial time ($t=0$), the drug is contained only in the plasma of the coating and it is distributed with a maximum, possibly non-uniform, concentration $C_0 \cdot F_0(x)$, with $0 \leq F_0(x) \leq 1$. Since the metallic strut is impermeable to the drug, no mass flux passes through the boundary surface $x = -l_0$. The dynamics of the drug in the coating (Layer 0) is described by a 1-D diffusion equation, and related boundary-initial conditions, as defined by Eqs. (3.68), where only the initial condition is now different, being non-uniform.

As reported in [50], the convection-reaction terms are only relevant in the media layer, whereas in the other layers these effects can be neglected. However, it has been shown in Section 3.6.2 that a more general model including convection-reaction terms can be reduced to a pure diffusive system by a variable transformation, namely Eq. (3.85). For this reason,

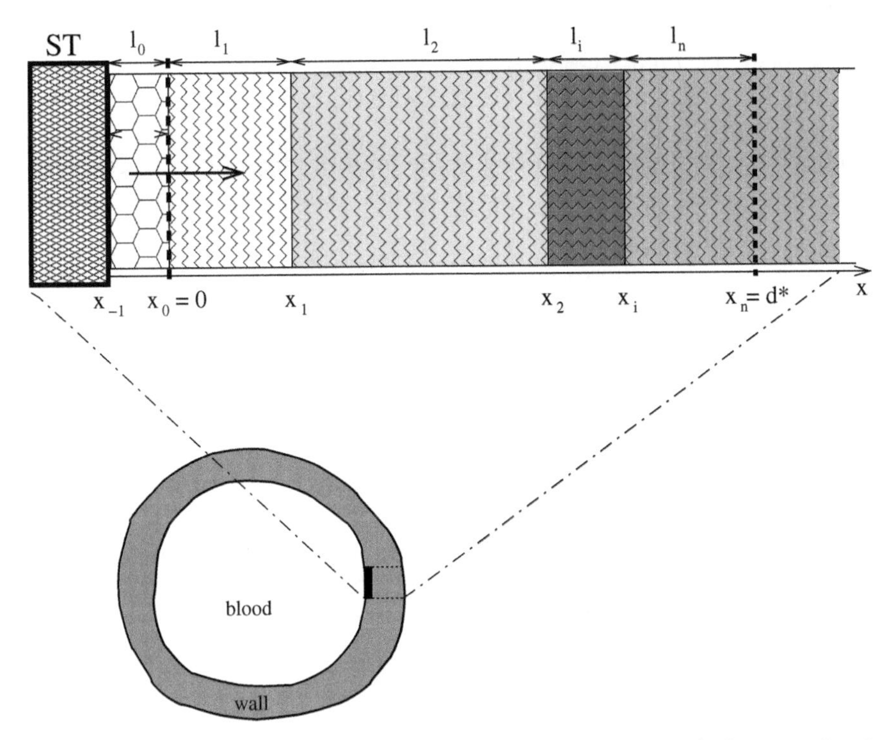

FIGURE 3.20 A sketch of the layered wall. The 1-D wall model is defined along the line normal to the strut stent surface and extends with a sequence of n contiguous layers $[x_{i-1}; x_i]$, $i = 1, 2, \ldots, n$ from the polymer coating interface $x_0 = 0$ up to the wall bound x_n estimated by the penetration distance d^*. ST indicates the metallic stent strut bearing the coating (figure not in scale).

within the $i = 1, 2, \ldots, n$ layers of the arterial wall, the drug dynamics are described by the following diffusion equation and related initial condition:

$$\frac{\partial c_i}{\partial t} = D_i \frac{\partial^2 c_i}{\partial x^2} \quad (x_{i-1} < x < x_i; t > 0) \tag{3.93a}$$

$$c_i(x, t = 0) = 0 \quad (x_{i-1} < x < x_i) \tag{3.93b}$$

where D_i is the effective diffusivity of the drug in the plasma of the ith layer of the arterial wall and c_i is the ith drug concentration.

At the interfaces between the layers of the arterial wall, the continuity of flux and the continuity of concentration relate the solutions of the $i = 1, 2, \ldots, n$ layers to one another:

$$D_i \left(\frac{\partial c}{\partial t}\right)_{x=x_i} = D_{i+1} \left(\frac{\partial c}{\partial x}\right)_{x=x_i} \quad (i = 1, 2, \ldots, n-1; t > 0) \tag{3.94a}$$

$$\frac{c_i(x_i, t)}{\varepsilon_{k,i}} = \frac{c_{i+1}(x_i, t)}{\varepsilon_{k,i+1}} \quad (i = 1, 2, \ldots, n-1; t > 0) \tag{3.94b}$$

As in the previous examples, the topcoat is used on the coating to slow the drug release into the arterial wall. This effect appears in the interface conditions between the coating (Layer 0) and the first tissue layer of the arterial wall (Layer 1), as indicated by Eqs. (3.70a) and (3.70b).

Finally, a boundary condition is imposed at the limit of the outermost tissue layer of the adventitia, x_n. Some controversy occurs when measuring this wall bound. Different values of the thickness are in fact given in the literature, depending on the arterial size; these generally lie in the range 100 to 200 μm for a medium sized artery [39,40,45,50]. Actually, as the arterial wall is embedded in the surrounding tissues, the drug transport does not suddenly stop at the outer limit of the adventitia but proceeds forward into the external tissues, and penetrates to a depth that depends on the time at which the process is observed. In principle, the exact range of the domain encompassing the diffusion process cannot be estimated *a priori*, and any truncation of the domain is rather arbitrary, if made on physiological considerations only. As a matter of fact, instead of arbitrarily guessing the correct wall thickness at which concentration and mass flux vanish asymptotically, a more rigorous approach may be made. This is done by modeling the outer boundary as a semi-infinite medium that is in perfect contact with the arterial media and having uniform properties (equal to those of the last layer, i.e., the adventitia) so that $x_n \to \infty$. The concentration and mass flux as $x \to \infty$ are finite (boundary condition of 0th kind or of Beck type; see Chap. 2 of Ref. [30]):

$$c_n(x \to \infty, t) = \text{finite}, \quad \left(\frac{\partial c_n}{\partial x}\right)_{x \to \infty} = \text{finite} \tag{3.95}$$

For computational purposes, it is possible to model the outmost layer within a bounded domain instead. This, in fact, is what has been suggested in [61], by using the concept of penetration distance d^*. Clearly the value of d^* increases with time and, in the case of a composite layered system, the penetration depth depends on the thickness and the diffusivities of each of the arterial wall layers. Also, it increases on the accuracy desired, for example 10^{-p}, with $p = 1, 2, \ldots, 10$. To estimate the penetration distance with which one can determine (with errors less than 10^{-p}) the outer bound of the outer arterial wall layer, x_n, the following estimate is provided in [60] as:

$$d^* \approx \sqrt{10pD_nt} + \sum_{j=1}^{n-1} l_j \left(1 - \delta_{ij}\right) \left(1 - \prod_{s=j}^{i-1} \sqrt{\frac{D_{s+1}}{D_s}}\right) \tag{3.96}$$

where δ_{ij} is the Kronecker delta.

Thus, without losing any significant accuracy, the nth semi-infinite layer can be truncated at the penetration distance, $x_n = d^*$, and the 0th kind boundary conditions (3.95) are replaced by:

$$c_n(x_n = d^*, t) = 0, \quad \left(\frac{\partial c_n}{\partial x}\right)_{x_n = d^*} = 0 \tag{3.97}$$

within an accuracy of 10^{-p}. In other words, setting condition (3.97) guarantees that the concentration and the mass flux vanish at d^* with a mass loss comparable with 10^{-p} at a given time. In the outmost nth layer, both the above conditions hold, but the absorbing condition $c_n = 0$ is expected to be more realistic, since the vasa vasorum of the adventitia are continually replenished with fresh blood and sweep away any residual drug [45].

3.6.3.2 *Concentration Solutions and Results*

Using the following dimensionless variables:

$$\tilde{x} = \frac{x}{d^*}, \quad \tilde{t} = \frac{\mathrm{Dmax}}{(d^*)^2}t, \quad \tilde{c}_i = \frac{c_i}{\mathrm{C}_0}, \quad \tilde{l}_i = \frac{l_i}{d^*}$$
$$\gamma_i = \frac{D_i}{\mathrm{Dmax}}, \quad \phi = \frac{Pd^*}{\mathrm{Dmax}\,(\varepsilon_k)\mathrm{max}}, \quad \sigma_i = \frac{(\varepsilon_k)_i}{(\varepsilon_k)\mathrm{max}} \tag{3.98}$$

where the subscript *max* refers to the maximum value across the $n+1$ layers, and then by means of the change of variables of Eq. (3.72) (here adding $\tilde{l}_i \to l_i$), the drug balance equations are recast in a dimensionless form as:

$$\frac{\partial c_i}{\partial t} = \gamma_i \frac{\partial^2 c_i}{\partial x^2} \quad (x_{i-1} < x < x_i; t > 0; i = 0, 1, \ldots, n) \tag{3.99a}$$

with the following non-dimensional initial conditions:

$$c_0(x, t = 0) = F_0(x) \quad (-l_0 = -x_{-1} < x < x_0 = 0) \tag{3.99b}$$

$$c_i(x, t = 0) = 0 \quad (x_{i-1} < x < x_i; i = 1, 2, \ldots, n) \tag{3.99c}$$

and with the following dimensionless interface and boundary conditions:

$$\left(\frac{\partial c_0}{\partial x}\right)_{x=-l_0} = 0 \quad (t > 0) \tag{3.99d}$$

$$\gamma_0 \left(\frac{\partial c_0}{\partial x}\right)_{x=0} = \gamma_1 \left(\frac{\partial c_1}{\partial x}\right)_{x=0} \quad (t > 0) \tag{3.99e}$$

$$-\gamma_0 \left(\frac{\partial c_0}{\partial x}\right)_{x=0} = \phi \left[\frac{c_0(0, t)}{\sigma_0} - \frac{c_1(0, t)}{\sigma_1}\right] \quad (t > 0) \tag{3.99f}$$

$$\gamma_i \left(\frac{\partial c_i}{\partial x}\right)_{x=x_i} = \gamma_{i+1} \left(\frac{\partial c_{i+1}}{\partial x}\right)_{x=x_i} \quad (i = 1, 2, \ldots, n-1; t > 0) \tag{3.99g}$$

$$\frac{c_i(x_i, t)}{\sigma_i} = \frac{c_{i+1}(x_i, t)}{\sigma_{i+1}} \quad (i = 1, 2, \ldots, n-1; t > 0) \tag{3.99h}$$

$$c_n(x = x_n = 1, t) = 0 \quad (t > 0) \tag{3.99i}$$

The development of the solution of Eq. (3.99a) according to the conditions (3.99b)–(3.99i) is presented in the study [61] in which the SOV method is still used to arrive at the following series solution:

$$c_i(x,t) = \sum_{m=1}^{\infty} A_m X_{im}(x) \exp\left(-\gamma_i \lambda_{im}^2 t\right) \quad (x_{i-1} \le x \le x_i;\, t \ge 0;\, i = 0,1,\ldots,n) \tag{3.100}$$

where the n layer's eigenvalues are related among them by the relationship $\lambda_{0m}\sqrt{\lambda_0} = \lambda_{im}\sqrt{\lambda_i}$, with $i = 1,2,\ldots,n$. A different method of solution based on a generalized integral transform is given in [62].

For Layer 0 the eigenfunction is represented by the expression:

$$X_{om}(x) = -\frac{1}{\tan(\lambda_{0m} l_0)} \cos(\lambda_{om} x) + \sin(\lambda_{om} x) \tag{3.101a}$$

The eigenfunctions of layers $i = 1,2,\ldots,m$ are

$$X_{im}(x) = a_{im} \cos(\lambda_{im} x) + b_{im} \sin(\lambda_{im} x) \tag{3.101b}$$

for which $a_{i,m}$ and $b_{i,m}$ are constants associated with series term 'm' and Layer i. In detail, the constants of Layer 1 are:

$$a_{1m} = \frac{\sigma_1}{\sigma_0} X_{om}(0) + \frac{\gamma_0 \sigma_1}{\phi} \left(\frac{dX_{0m}}{dx}\right)_{x=0} \tag{3.102a}$$

$$b_{1m} = \frac{1}{\lambda_{1m}} \frac{\gamma_0}{\gamma_1} \left(\frac{dX_{0m}}{dx}\right)_{x=0} \tag{3.102b}$$

and the constants of the layers $i = 1,2,\ldots,n$ are:

$$a_{im} = \frac{\sigma_i}{\sigma_{i-1}} \cos\left(\lambda_{im} x_{i-1}\right) X_{i-1m}(x_{i-1}) - \frac{1}{\lambda_{im}} \frac{\gamma_{i-1}}{\gamma_i} \sin(\lambda_{im} x_{i-1}) \left(\frac{dX_{i-1m}}{dx}\right)_{x=x_{i-1}} \tag{3.103a}$$

$$b_{im} = \frac{\sigma_i}{\sigma_{i-1}} \sin\left(\lambda_{im} x_{i-1}\right) X_{i-1m}(x_{i-1}) + \frac{1}{\lambda_{im}} \frac{\gamma_{i-1}}{\gamma_i} \cos(\lambda_{im} x_{i-1}) \left(\frac{dX_{i-1m}}{dx}\right)_{x=x_{i-1}} \tag{3.103b}$$

In Eq. (3.100), the constant A_m may still be evaluated by using Eq. (3.91) as the initial drug concentration is different from zero only in the coating. For the special case of a Layer 0 uniform initial distribution, $F_0(x) = 1$, it may be taken as $A_m = -1/(\sigma_0 N_m \lambda_{om})$, where the norm, N_m, is defined by:

$$N_m = \sum_{i=0}^{n} \frac{1}{2\sigma_i} \left\{ x X_{im}^2 + \frac{1}{\lambda_{im}^2} \left[x \left(\frac{dX_{im}}{dx}\right)^2 - X_{im} \frac{dX_{im}}{dx} \right] \right\} \Bigg|_{x_{i-1}}^{x_i} \tag{3.104}$$

And to close the solution set, the eigenvalues of Layer 0, λ_{0m}, may be determined by evaluating the roots of the transcendental equation (multi-layer eigencondition):

$$a_{nm} \cos(\lambda_{nm}) + b_{nm} \sin(\lambda_{nm}) = 0 \tag{3.105}$$

where the coefficients a_{nm} and b_{nm} are given by Eqs. (3.103a) and (3.103b), respectively, for $i = n$, and $\lambda_{nm} = \lambda_{0m}\sqrt{\gamma_0/\gamma_n}$.

Once the drug concentration is known, the dimensionless drug mass per unit area, $\tilde{M}_i = M_i/(C_0 d^*)$, contained within each layer ($i = 0, 1, \ldots, n$), may be taken as (re-setting $\tilde{M}_i \to M_i$):

$$M_0(t) = \int_{-l_0}^{0} c_0(x,t)\,dx = -\sum_{m=1}^{\infty} \frac{A_m}{\lambda_{0,m}} \exp\left(-\gamma_0 \lambda_{0,m}^2 t\right) \tag{3.106a}$$

$$M_i(t) = \int_{x_{i-1}}^{x_i} c_i(x,t)\,dx = \sum_{m=1}^{\infty} A_m \frac{\exp\left(-\gamma_i \lambda_{im}^2 t\right)}{\lambda_{im}^2} \left[\left(\frac{dX_{im}}{dx}\right)_{x_{i-1}} - \left(\frac{dX_{im}}{dx}\right)_{x_i}\right] \quad (i = 1, 2, \ldots, n) \tag{3.106b}$$

In the study [61] the multi-layered wall model is developed in which the stent coating and five layers of arterial wall are represented using the geometric and physical parameter values listed in Table 3.1, with a uniform initial concentration within the stent coating, and a membrane permeability value of $P = 10^{-6}$ cm^2 s^{-1}.

The dimensionless parameter groups are scaled by the penetration depth d^*, and so it is evaluated from Eq. (3.96) with an accuracy of p=2, i.e., 1%. See the study [61] for a detailed account of the relationship between penetration depth, time and accuracy. The results of that study are shown in Fig. 3.21, in which the dimensionless concentration profiles within each of the layers are plotted at representative times.

The drug is retained differently in each layer, which receives mass from the inner and transmits to the outer, in a cascade sequence, and that are completely damped out at the distance d^* which represents the outer domain boundary. It is interesting to note that the levels of concentration in layer 2 (intima) are nearly constant and can be higher than in the others at intermediate times. This is in agreement with the higher diffusivity D_2 and relatively small layer thickness l_2.

TABLE 3.1 The parameters used in the simulations for the coating and the wall layers. The penetration distance d^* estimates the wall bound, provides the thickness l_5 of the external layer and depends on the maximum simulated time.

	Coating (0)	Endothelium (1)	Intima (2)	IEL (3)	Media (4)	Adventitia (5)
$l_i = x_i - x_{i-1}$(cm)	5×10^{-4}	2×10^{-4}	10^{-3}	2×10^{-4}	2×10^{-2}	$d^* - x_4$
D_i(cm^2/s)	10^{-10}	8×10^{-9}	7.7×10^{-8}	4.2×10^{-8}	7.7×10^{-8}	12×10^{-8}
ϵ_i	0.1	5×10^{-4}	0.61	4×10^{-3}	0.61	0.85
k_i	1	1	1	1	1	1

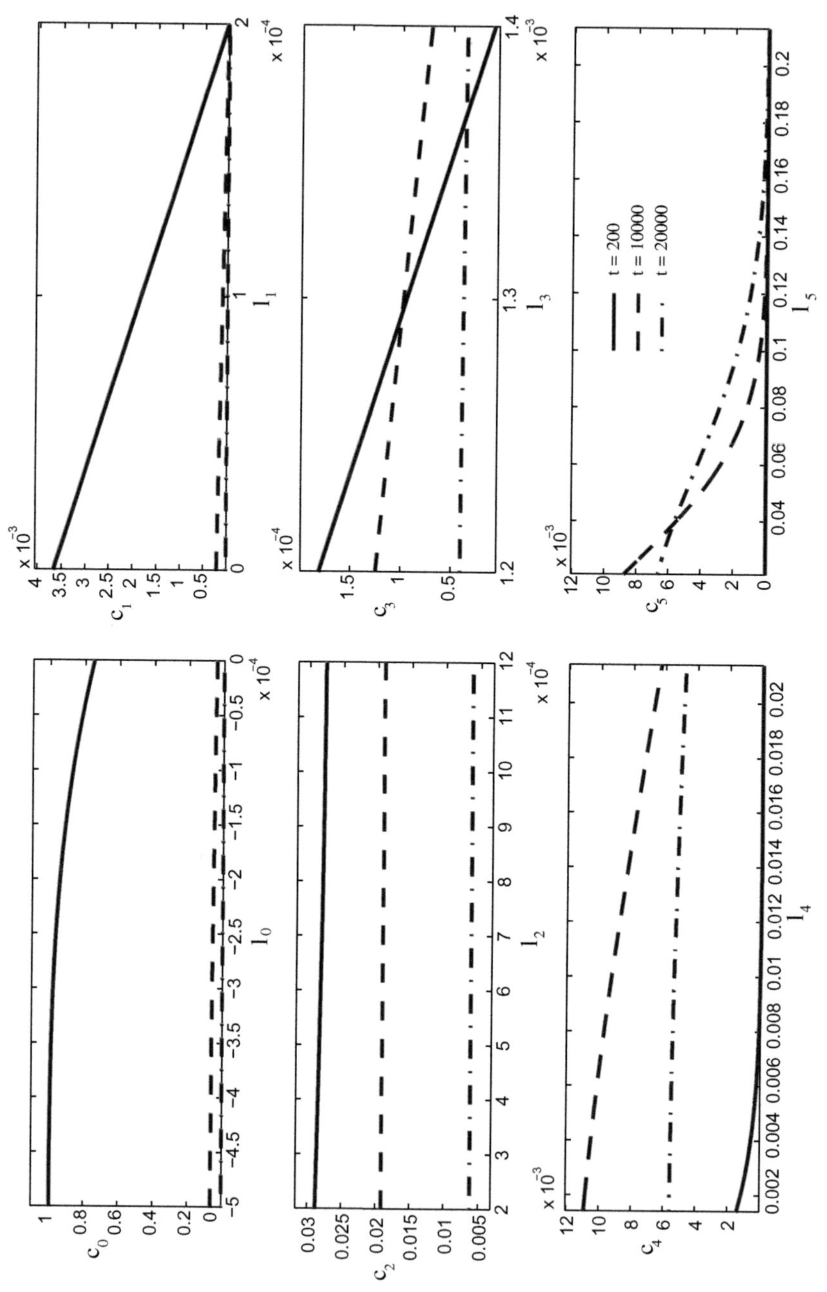

FIGURE 3.21 Concentration profiles in the six layers at three different times (in s) (note the different scale for coordinates and concentrations).

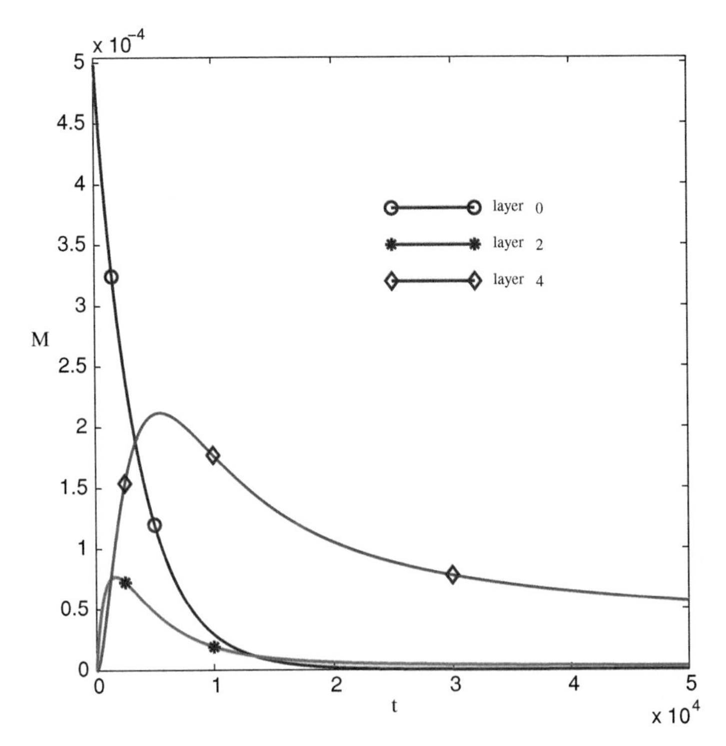

FIGURE 3.22 Dimensionless mass in the coating (layer 0), endothelium (layer 2) and media (layer 4) vs. time (in s). In the coating, mass is monotonically decreasing, while in the others there is a characteristic time at which the drug reaches a peak.

Due to the diffusive coefficient and to the porosity, the mass decreases exponentially in the coating and in layer (1) (innermost layers). However, in layers 2–5 the concentration experiences an initial build-up and then decays exponentially. This is very evident in the dimensionless mass contained in each layer, which is depicted in Fig. 3.22. In the outmost layer (5) the mass accumulates as the time proceeds. The simulation points out that the time of peak mass in the intima (layer 2) is at 1640 s ($\approx$ 27 min), in the media (layer 4) is 5460 s ($\approx$ 1 h: 30 min). The thin layers 1 and 3 retain a negligible mass due to their thickness, and the media is completely emptied after a time of about 57 days. At that time, all the mass is transferred to the external wall layer, with a slight mass loss. However, the therapeutic effects of DES is limited in the endothelium-media, while the residual drug in the outmost layer is considered lost.

3.7 CONCLUSION

The transport of drugs into biological media is a complex phenomenon that occurs over a range of time and length scales. Modeling of transport requires a deeper understanding not only of the solute and drug characteristics, but also of the physical structure of the biological media. The porous media representation of the biological tissue in which the drug is carried allows for a continuum scale analysis of the interaction between drug and tissue, which may be based on relatively simple fundamental conservation laws. These lead to the ability to apply the analytical solutions of Nusselt, Schumann and Anzelius to the transient behavior of drug in a homogenous medium.

When, as often is the case, the biological tissue is not macroscopically homogenous, but rather is constituted from different discrete layers, a composite porous media representation of the tissue is likely to be used. An obvious and important application of the composite structure representation is the transport of drug from a stent through the wall of a blood vessel. Analytical solutions of different representations have been provided that allow for variability in application including the fluid wall model with pure diffusion, the fluid wall model with diffusion, advection and drug reaction and the diffusive multi-layered wall model.

References

[1] J.E. Sousa, P.W. Serruys, M.A. Costa, New frontiers in cardiology, drug-eluting stents: Part I. Circulation 107 (2003) 2274–2279.

[2] J.E. Sousa, P.W. Serruys, M.A. Costa, New frontiers in cardiology, drug-eluting stents: Part II. Circulation 107 (2003) 2382–2389.

[3] H. Hara, M. Nakamura, J.C. Palmaz, R.S. Schwartz, Role of stent design and coatings on restenosis and thrombosis, Advanced Drug Delivery Reviews 58 (2006) 377–386.

[4] W.H. Maisel, Unanswered questions – drug-eluting stents and the risk of late thrombosis, New England Journal of Medicine 356 (2007) 981–984.

[5] G. Pontrelli, F. de Monte, Mass diffusion through two-layer porous media: An application to the drug-eluting stent, International Journal of Heat and Mass Transfer 50 (2007) 3658–3669.

[6] J.M. Weiler, E.M. Sparrow, R. Ramazani, Mass transfer by advection and diffusion from a drug-eluting stent, International Journal of Heat and Mass Transfer 55 (2012) 1–7.

[7] N.I. Border, P.E. Buchwald, Ophthalmic drug design based on the metabolic activity of the eye: soft drugs and chemical delivery systems, AAPS Journal 7 (2005) 820–833.

[8] R. Avtar, D. Tandon, Modeling the drug transport in the anterior segment of the eye, European Journal of Pharmaceutical Sciences 35 (2008) 175–182.

[9] R. Manitz, W. Lucht, K. Strehmel, R. Weiner, R. Neubert, On mathematical modeling of dermal and transdermal drug delivery, Journal of Pharmaceutical Sciences 87 (1998) 873–879.

[10] M.R. Prausnitz, R. Langer, Transdermal drug delivery, Nature Biotechnology 26 (2008) 1261–1268.

[11] J.E. Rim, P.M. Pinsky, W.W. van Osdol, Multiscale modeling framework of transdermal drug delivery, Annals of Biomedical Engineering 37 (2009) 1217–1229.

[12] A.R.A. Khaled, K. Vafai, The role of porous media in modeling flow and heat transfer in biological tissues, International Journal of Heat and Mass Transfer 46 (2003) 4989–5003.

[13] Y. Davit, G. Debenest, B.D. Wood, M. Quintard, Modeling non-equilibrium mass transport in biologically reactive porous media, Advances in Water Resources 33 (2010) 1075–1093.

[14] F. Golfier, B.D. Wood, L. Orgogozo, M. Quintard, M. Bues, Biofilms in porous media: development of macroscopic transport equations via volume averaging with closure for local mass equilibrium conditions, Advances in Water Resources 32 (2009) 463–485.

[15] G.A. Truskey, F. Yuan, D.F. Katz, Transport Phenomena in Biological Systems, second ed., Pearson Prentice Hall Bioengineering, Lebanon, Indiana, USA, 2009.

[16] A. Bejan, Porous media, in: A. Bejan, A.D. Kraus (Ed.), Heat Transfer Handbook, Wiley, New York, 2003, pp. 1131–1180.

[17] D.A. Nield, A. Bejan, Convection in Porous Media, third ed., Springer, New York, 2006.

[18] M. Jakob, Heat Transfer, vol. 2, John Wiley & Sons, New York, 1957.

[19] M. Abramowitz, I.A. Stegun, Handbook of mathematical functions, in: Applied Mathematics Series, vol. 55, National Bureau of Standards, 1964.

[20] A.A. Rabah, S. Kabelac, A simplified solution of the regenerator periodic problem: the case for air conditioning, Forsch Ingenieurwes 74 (2012) 207–214.

[21] T.E.W. Schumann, Heat transfer: liquid flowing through a porous prism, Journal of the Franklin 208 (1929) 405–416.

[22] H.S. Carslaw, J.C. Jaeger, Conduction of Heat in Solids, second ed., Clarendon Press, London, 1959.

[23] A. Kuznetsov, An investigation of a wave of temperature difference between solid and fluid phases in a porous packed-bed, International Journal of Heat and Mass Transfer 37 (1994) 3030–3033.

[24] A. Kuznetsov, A perturbation solution for a nonthermal equilibrium fluid flow through a three-dimensional sensible heat storage packed bed, ASME Journal of Heat Transfer 118 (1996) 508–510.

[25] A. Kuznetsov, A perturbation solution for heating a rectangular sensible heat storage packed bed with a constant temperature at the walls, International Journal of Heat and Mass Transfer 40 (1997) 1001–1006.

[26] D.K. Tzou, Macro- to Microscale Heat Transfer, Taylor and Francis, New York, 1997.

[27] W.J. Minkowycz, A. Haji-Sheikh, K. Vafai, On departure from local thermal equilibrium in porous media due to a rapidly changing heat source: the Sparrow number, International Journal of Heat and Mass Transfer 42 (1999) 3373–3385.

[28] A. Haji-Sheikh, W.J. Minkowycz, Heat transfer analysis under local thermal non-equilibrium conditions, in: P. Vadàsz (Ed.), Emerging Topics in Heat and Mass Transfer in Porous Media From Bioengineering and Microelectronics to Nanotechnology, Springer, Dordrecht, 2008, 39–62.

[29] J. Crank, The Mathematics of Diffusion, second ed., Clarendon Press, Oxford, 1979.

[30] K.D. Cole, J.V. Beck, A. Haji-Sheikh, B. Litkouhi, Heat Conduction Using Green's Functions, second ed., CRC Press, Boca Raton, 2011.

[31] G. Pontrelli, F. de Monte, Modeling of mass dynamics in arterial drug-eluting stents, Journal of Porous Media 12 (2009) 19–28.

[32] L.M. Jiji, Heat Conduction, third ed., Springer-Verlag, Berlin, 2009.

[33] A. Haji-Sheikh, J.V. Beck, An efficient method of computing eigenvalues in heat conduction, Numerical Heat Transfer (Part B-Fundamentals) 38 (2000) 133–156.

[34] J.V. Beck, K.D. Cole, Improving convergence of summations in heat conduction, International Journal of Heat and Mass Transfer 50 (2007) 257–268.

[35] J.V. Beck, R. McMasters, K.J. Dowding, D.E. Amos, Intrinsic verification methods in linear heat conduction, International Journal of Heat and Mass Transfer 49 (2006) 2984–2994.

[36] S. Prabhu, S. Hossainy, Modeling of degradation and drug release from a biodegradable stent coating, Journal of Biomedical Materials Research, Part A 80A (2007) 732–741.

[37] S. Hossainy, S. Prabhu, A mathematical model for predicting drug release from a biodurable drug-eluting stent coating, Journal of Biomedical Materials Research, Part A 87A (2008) 487–493.

[38] S. Baek, A.R. Srinivasa, Modeling of the pH-sensitive behavior of a ionic gel in the presence of diffusion, International Journal of Non-Linear Mechanics 39 (2004) 1301–1318.

[39] P. Zunino, Multidimensional pharmacokinetic models applied to the design of drug-eluting stents, Cardiovascular Engineering: An International Journal 4 (2004) 181–191.

[40] D.V. Sakharov, L.V. Kalachev, D.C. Rijken, Numerical simulation of local pharmacokinetics of a drug after intravascular delivery with an eluting stent, Journal of Drug Targeting 10 (2002) 507–513.

[41] M.C. Delfour, A. Garon, V. Longo, Modeling and design of coated stents to optimize the effect of the dose, Siam Journal on Applied Mathematics 65 (2005) 858–881.

[42] C.J. Creel, M.A. Lovich, E.R. Edelman, Arterial paclitaxel distribution and deposition, Circulation Research 86 (2000) 879–884.

[43] G.A. Ateshian, M.H. Friedman, Integrative biomechanics: a paradigm for clinical applications of fundamental mechanics, Journal of Biomechanics 42 (2009) 1444–1451.

[44] G. Vairo, M. Cioffi, R. Cottone, G. Dubini, F. Migliavacca, Drug release from coronary eluting stents: a multidomain approach, Journal of Biomechanics 43 (2010) 1580–1589.

[45] R. Mongrain, I. Faik, R.L. Leask, J. Rodes-Cabau, E. Larose, O.F. Bertrand, Effects of diffusion coefficients and struts apposition using numerical simulations for drug eluting coronary stents, ASME Journal of Biomechanical Engineering 129 (2007) 733–742.

[46] M. Grassi, G. Pontrelli, L. Teresi, G. Grassi, L. Comel, A. Ferluga, L. Galasso, Novel design of drug delivery in stented arteries: a numerical comparative study, Mathematical Biosciences and Engineering 6 (2009) 493–508.

[47] F. Migliavacca, F. Gervaso, M. Prosi, P. Zunino, S. Minisini, L. Formaggia, G. Dubini, Expansion and drug elution model of a coronary stent, Computer Methods in Biomechanics and Biomedical Engineering 10 (2007) 493–508.

[48] M. Prosi, P. Zunino, K. Perktold, A. Quarteroni, Mathematical and numerical models for transfer of low-density lipoproteins through the arterial walls: a new methodology for the model set up with applications to the study of disturbed lumenal flow, Journal of Biomechanics 38 (2005) 903–917.

[49] C. Vergara, P. Zunino, Multiscale boundary conditions for drug release from cardiovascular stents, Multiscale Modeling & Simulation 7 (2008) 565–588.

[50] N. Yang, K. Vafai, Low-density lipoprotein (ldl) transport in an artery – a simplified analytical solution, International Journal of Heat and Mass Transfer 51 (2008) 497–505.

[51] M. Khakpour, K. Vafai, Critical assessment of arterial transport models, International Journal of Heat and Mass Transfer 51 (2008) 807–822.

[52] G. Pontrelli, A. Di Mascio, F. de Monte, Local mass non-equilibrium in arterial drug-eluting stents, WAMS 2012, The International Workshop on Applied Modeling and Simulation, Rome, Italy, 24–27 September, 2012.

[53] A. Kargol, M. Kargol, S. Przestalski, The Kedem-Katchalsky equations as applied for describing substance transport across biological membranes, Cellular and Molecular Biology Letters 2 (1997) 117–124.

[54] A. Kargol, Modified Kedem-Katchalsky equations and their applications, Journal of Membrane Science 174 (2000) 43–53.

[55] C.W. Hwang, D. Wu, E.R. Edelman, Physiological transport forces govern drug distribution for stent-based delivery, Circulation 104 (2001) 600–605.

[56] M. Khakpour, K. Vafai, Effects of gender-related geometrical characteristics of aortailiac bifurcation on hemodynamics and macromolecule concentration distribution, International Journal of Heat and Mass Transfer 51 (2008) 5542–5551.

[57] M.N. Özişik, Heat Conduction, second ed., John Wiley & Sons, New York, 1993.

[58] G. Meyer, R. Merval, A. Tedgui, Effects of pressure-induced stretch and convection on low-density lipoprotein and albumin uptake in the rabbit aortic wall, Circulation Research 79 (1996) 532–540.

[59] L. Ai, K. Vafai, A coupling model for macromolecule transport in a stenosed arterial wall, International Journal of Heat and Mass Transfer 49 (2006) 1568–1591.

[60] M. Khakpour, K. Vafai, A comprehensive analytical solution of macromolecular transport within an artery, International Journal of Heat and Mass Transfer 51 (2008) 2905–2913.

[61] G. Pontrelli, F. de Monte, A multi-layer porous wall model for coronary drug-eluting stents, International Journal of Heat and Mass Transfer 53 (2010) 3629–3637.

[62] C. Liu, W.P. Ball, J.H. Hellis, An analytical solution to the one-dimensional solute advection-dispersion equation in multi-layer porous media, Transport in Porous Media 30 (1998) 25–43.

CHAPTER

4

Transport of Water and Solutes Across Endothelial Barriers and Tumor Cell Adhesion in the Microcirculation

Bingmei M. Fu[a,*] *and Yang Liu*[b]

[a]Department of Biomedical Engineering, The City College of the City University of New York, New York, NY, USA

[b]Department of Mechanical Engineering, The Hong Kong Polytechnic University, Kowloon, Hong Kong, PR China

OUTLINE

4.1 Introduction 120

4.2 Microvascular Transport 122

4.2.1 Transvascular Pathways 122

4.2.1.1 Transport Across Peripheral Microvessels 122

4.2.1.2 Transport Across the Blood-Brain Barrier 124

4.2.2 Transport Coefficients 129

4.2.3 Permeability Measurement 130

4.2.3.1 Permeability Measured in a Single Perfused Microvessel 130

4.2.3.2 Permeability Measured in Cultured Endothelial Cell Monolayers 135

4.2.4 Transport Models for Water and Solutes Through Interendothelial Cleft 135

4.2.4.1 1-D Models 135

4.2.4.2 3-D Models 141

4.2.5 Endothelial Surface Glycocalyx 146

4.2.5.1 Composition, Thickness and Structure of Endothelial Surface Glycocalyx (ESG) 146

*Corresponding author.

© 2013 Elsevier Inc. All rights reserved.

http://dx.doi.org/10.1016/B978-0-12-415824-5.00004-7

4.2.5.2 Role of Endothelial Surface Glycocalyx 148
4.2.5.3 Effect of Charge Carried by Endothelial Surface Glycocalyx 151
4.2.6 Transport Across Fenestrated Microvessels 152
4.2.7 Transport in Tissue Space (Interstitium) 152

4.3 Modulation of Microvascular Transport 153
4.3.1 Permeability Increase 154
4.3.2 Permeability Decrease 156
4.3.3 Permeability Increase and Decrease 158
4.3.4 Microvascular Hyperpermeability and Tumor Metastasis 158
4.3.4.1 VEGF Effects on Microvascular Integrity and on Tumor Cell Adhesion 158
4.3.4.2 Integrin Signaling and Tumor Metastasis 161

4.4 Tumor Cell Adhesion in the Microcirculation 162
4.4.1 Tumor Cell Adhesion Under Flow 162
4.4.2 Mathematical Models for Tumor Cell Adhesion in the Microcirculation 163
4.4.2.1 General Cell Adhesion Models 163
4.4.2.2 Bell's Model 165
4.4.2.3 Dembo et al.'s Model 165
4.4.3 Model Predictions for Tumor Cell Adhesion in the Microcirculation 166
4.4.3.1 Effect of Curvature 166
4.4.3.2 Effect of Wall Shear Stress 168

4.5 Summary and Opportunities for Future Study 170

Acknowledgments 170

References 170

4.1 INTRODUCTION

The microvascular bed is the primary location where water and nutrients are exchanged between circulating blood and body tissues. Usually, these microvessels have a diameter ranging from 5 to 50 μm, categorized as arterioles, capillaries and post-capillary venules. The microvessels in all human body tissues (except the brain) are called peripheral microvessels, whose wall consists mainly of endothelial cells (Fig. 4.1). Under normal conditions, the gap (or cleft) between endothelial cells (interendothelial cleft, Fig. 4.2) is widely believed to be the principal pathway for the transport of water and hydrophilic solutes (such as glucose, amino acids, vitamins, hormones) across the capillary wall [122,128,172,229]. Direct and indirect evidence indicates that there are junctional strands with discontinuous leakages [8,37,38,52,229] inside the interendothelial cleft, and a glycocalyx layer [3,4,142,199,225,235] at the endothelial surface (Fig. 4.2). The transport of proteins or other macromolecules is thought to occur through vesicle shuttle mechanisms [181]. In disease, larger gaps may be formed in the microvessel endothelium to allow the passage of plasma proteins and cells such as blood cells and tumor cells. Microvascular permeability is a quantitative measure of how permeable the microvessel wall is to all kinds of substances including water, solutes and cells. Under healthy conditions, the microvessel wall maintains a normal permeability to water and small solutes

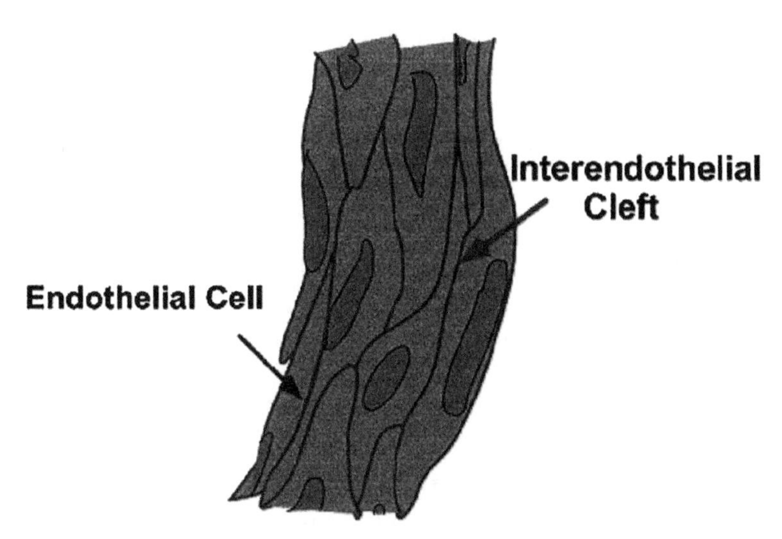

FIGURE 4.1 A sketch for a rat mesenteric post-capillary venule of a diameter ~30 μm. Its wall consists of endothelial cells. The darker areas represent the endothelial nuclei. The gap (or cleft) between adjacent endothelial cells is often called interendothelial cleft.

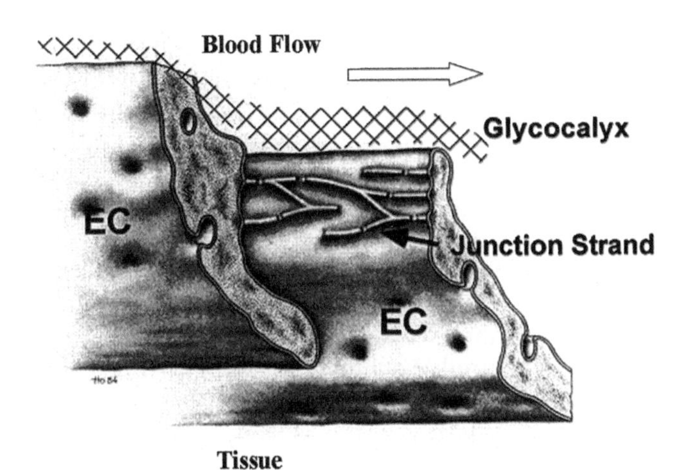

FIGURE 4.2 Ultrastructural organization of junction strands in the interendothelial cleft and the surface glycocalyx. Revised from [37].

for material exchange during metabolic processes. However, in disease, the integrity of the vessel wall structure can be destroyed and much larger solutes, such as proteins and cells, can move through the wall. It is the transvascular pathways at the vessel wall and their structural barriers that determine and regulate the microvascular permeability. Therefore, the current basic understanding about transvascular pathways will be introduced first in this chapter.

4.2 MICROVASCULAR TRANSPORT

4.2.1 Transvascular Pathways

4.2.1.1 Transport Across Peripheral Microvessels

The endothelial cells lining microvessel walls provide the rate-limiting barrier to extravasation of plasma components of all sizes – from electrolytes to proteins. As noted earlier, the microvessels in all mammalian tissues except the brain are named peripheral microvessels. At present, the results of electron microscopy indicate that there are four primary pathways observed in the peripheral microvessel wall: intercellular clefts, transcellular pores, vesicles and fenestrae (Fig. 4.3). Microvessels of different types and in different tissues may have different primary transvascular pathways. Under different physiological and pathological conditions, the primary pathway of a given microvessel can change [181].

4.2.1.1.1 INTERENDOTHELIAL (INTERCELLULAR) CLEFT

The cleft between adjacent endothelial cells is widely believed to be the principal pathway for the transport of water and hydrophilic solutes through the microvessel wall under normal physiological conditions (Fig. 4.2, revised from [37]). The interendothelial cleft is also suggested to be the pathway for the transport of high molecular weight plasma proteins, leukocytes and tumor cells across microvessel walls in disease states. Direct and indirect evidence (summarized in [151]) indicates that there are tight junction strands with discontinuous leakages and fiber matrix components (glycocalyx layer) [58] at the endothelial surface. These structural components of the microvessel wall form the barrier between the blood stream and body tissues, which maintains the normal microvessel permeability to water and solutes. Variations in permeability are caused by the changes in these structural components.

The molecular basis for the passage of molecules at the level of the breaks in tight junctions is more likely to be the localized absence of cell-cell contacts, with a corresponding loss of a closely regulated molecular sieve, as suggested by *Fu et al.* [88] and *Michel and Curry* [151]. Thus the junction break/surface matrix model suggests independent mechanisms to regulate the permeability properties of the microvessel wall. The junction break size and frequency are likely to involve regulation of cell-cell attachment via occludin and other junction proteins, including the cadherin's associated junctions [72] (Fig. 4.4). On the other hand, the regulation of the density and organization of the glycocalyx is likely to involve interaction between the

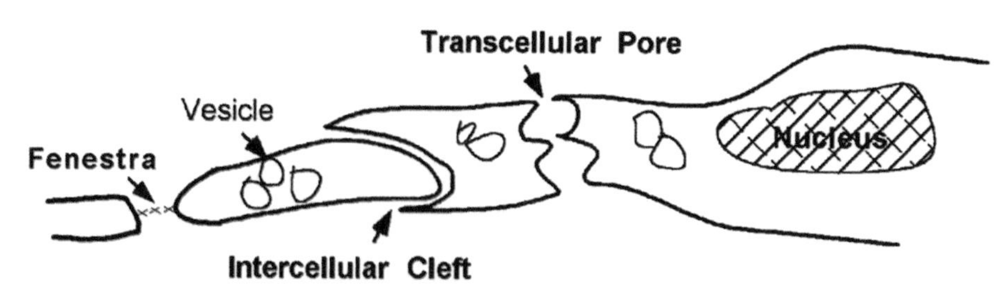

FIGURE 4.3 Schematic depiction of 4 types of typical transvascular pathways in the peripheral microvessel wall.

molecules forming the cell surface with those forming the cytoskeleton, and with circulating plasma proteins. Some of the cellular mechanisms underlying these interactions are reviewed in [72,151]. Under physiological and pathological conditions, microvessel permeability can be regulated acutely and chronically by mechanisms that are currently in the process of being understood.

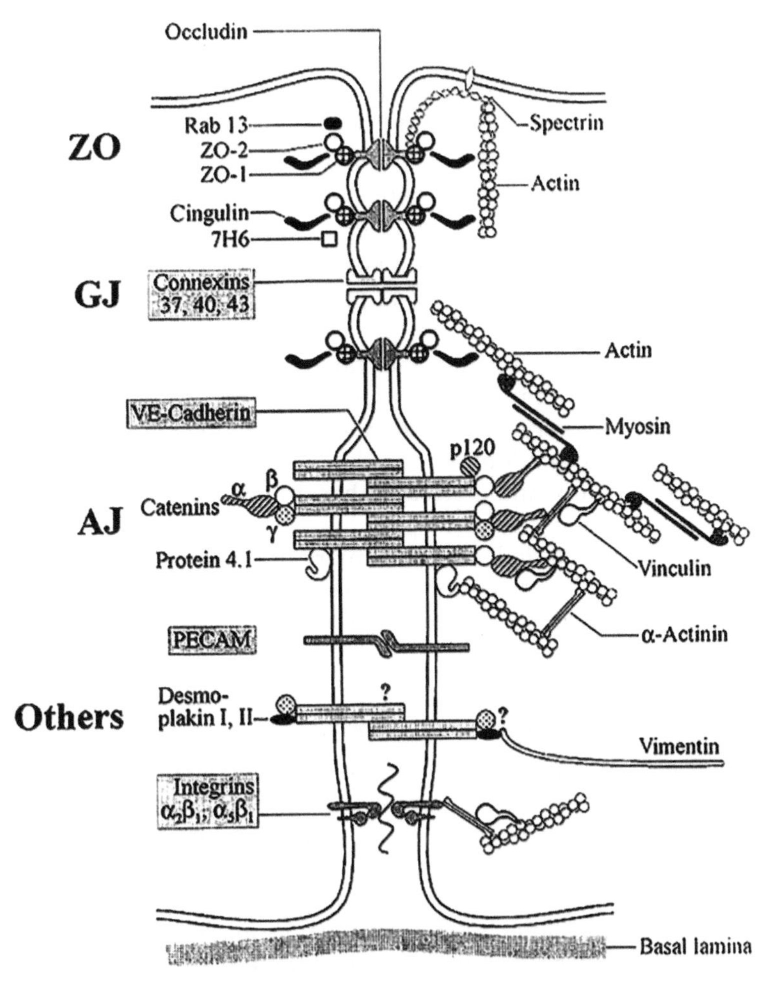

FIGURE 4.4 Model of molecular machinery associated with junction between endothelial cells and junction-associated filament system. Details of these interactions are an area of active investigation, and figures of this type are constantly updated (from [72,151]). In this figure an idealized arrangement of junction components is shown with tight junction (ZO, zonula occludins [154]) spatially separated from adherence junction (AJ). The tight junction is formed by integral membrane adhesion protein occludins. Peripheral membrane proteins associated with tight junction include ZO-1, ZO-2, cingulin, antigen 7H6 and a small GTP-binding protein Rab 13. The main site of attachment of junction-associated actin filaments to plasma membrane appears to be adherence junction (AJ). The channel-like proteins, connexins, form the gap junction (GJ) between adjacent cells for communications.

A serial section electron microscopy study on frog and rat mesenteric capillaries by *Adamson et al.* [6] demonstrated that the junction strand was interrupted by infrequent breaks that, on average, were 150 nm long, spaced 2–4 μm apart along the strand, and which accounted for up to 10% of the length of the strand under control conditions. At these breaks, the space between adjacent endothelial cells (average ~20 nm) was as wide as that in regions of the cleft between adjacent cells with no strands. The luminal surface of endothelial cells (ECs) lining the vasculature is coated with a glycocalyx of membrane-bound macromolecules comprised of sulfated proteoglycans, hyaluronic acids, sialic acids, glycoproteins, and plasma proteins that adhere to this surface matrix [180,210]. The thickness of this endothelial surface glycocalyx (ESG) has been observed to range from less than 100 nm up to 1 μm for the microvessels in different tissues and species by using different preparation and observing methods [98,140,151,203,219,223,235]. Although the ESG thickness varies, its density and organization have been reported to be the same among different tissues and species. The glycocalyx fiber radius is ~6 nm and the gap spacing between fibers is ~8 nm [17,203].

Vesicles: Cytoplasmic vesicular exchange, which behaves like a shuttle bus, is the major pathway for transport of plasma proteins and large molecules under normal physiological conditions [181].
Fenestrae: Fenestrae usually exist in the fenestrated microvessel (e.g., the vessel in the kidney) instead of in the continuous microvessel endothelium. In some fenestrated microvessels, there exists a very thin membrane (~25 nm thick) diaphragm, which covers the fenestra. The fenestrated endothelia have higher hydraulic conductivities and are more permeable to small ions and molecules than are the continuous endothelia. However, their permeabilities to plasma proteins are about the same [181].
Transcellular Pores: In response to local tissue injury or inflammation, additional transport pathways for large molecules may be opened (transcellular pores) and existing pathways may be made less restrictive. The response is complex, and varies among different animals, organs and tissues [152].

4.2.1.2 *Transport Across the Blood-Brain Barrier*

The most complicated organ in our body is the brain. It contains 100 billion neurons and 1 trillion glial cells (non-nerve supporting cells in the brain including astrocytes, oligodendrocytes, microglia and ependymal cells). Along with a tremendous number of blood vessels, these cells and their surrounding extracellular matrix form a highly complex, though well organized 3-D interconnecting array. The movement of ions across the neuronal membrane through voltage-gated channels conducts the information along the neuron axon at a speed of up to 400 km/h. The release of neurotransmitters into the synaptic space between adjacent nerve cells mediates the communication between neurons. In order to perform its highly complicated tasks, the brain needs a substantial amount of energy to maintain electrical gradients across neuronal membranes and consequently requires a sufficient supply of oxygen and nutrients. Although it only accounts for 2% of the body weight, the brain uses 20% of the blood supply. This blood is delivered through a complex network of blood vessels that extend for >650 km and passes a surface area of ~20 m^2. The mean distance between adjacent capillaries is ~40 μm, which allows almost instantaneous equilibration in the brain tissue surrounding the microvessels of small solutes such as glucose, amino acids, vitamins, oxygen, etc. However, unlike peripheral microvessels in other organs, where there is a relatively free small solute exchange between blood and tissue, the microvessels in the brain (cerebral microvessels) constrain the movement of molecules between blood and the brain tissue [1,173]. While this unique characteristic provides a natural defense against toxins circulating in the blood, it also prevents the successful delivery of therapeutic agents to the brain.

4.2.1.2.1 THE BLOOD-BRAIN BARRIER (BBB)

The wall of the microvessels in the brain (cerebral microvessels) has a special name; the blood-brain barrier (BBB), which distinguishes it from the wall of peripheral microvessels. The BBB is a unique dynamic regulatory interface between the cerebral circulation and the brain tissue, and it is essential for maintaining the micro-environment within the brain. No other body organ so absolutely depends on a constant internal micro-environment as does the brain. In brain tissue, the extracellular concentrations of amino acids and ions such as Na^+, K^+ and Ca^{2+} must be retained in very narrow ranges [173]. If the brain is exposed to large variations in the concentrations of these molecules, neurons would not function properly because some amino acids serve as neurotransmitters and certain ions modify the threshold for neuronal firing. The BBB also protects the central nervous system (CNS) from blood-borne neuroactive solutes, such as glutamate, glycine, norepinephrine, epinephrine and peptide hormones [202], which can increase with physiological changes (e.g., diet and stress) and pathological changes (e.g., injury and disease). In addition, the BBB plays a key role in facilitating the brain's uptake of essential nutrients like glucose, hormones and vitamins, and larger molecules like insulin, leptin and iron; all of which aid in sustaining brain growth and metabolism [239].

The term 'blood-brain barrier' was coined by *Lewandowsky* in 1900, when he demonstrated that neurotoxic agents affected brain function only when directly injected into the brain but not when injected into the systemic circulation [134]. The first experimental observation of this vascular barrier between the cerebral circulation and the CNS dates back to the 1880s, when Paul Ehrlich discovered that certain water-soluble dyes, like trypan blue, when injected into the systemic circulation, were rapidly taken up by all organs except the brain and spinal cord [76]. Ehrlich interpreted these observations as a lack of the affinity of the CNS for the dyes. However, subsequent experiments performed by Edwin Goldmann, an associate of Ehrlich, demonstrated that the same dyes, when injected directly into the CNS, stained all types of cells in the brain tissue but no other tissues in the rest of the body [103]. It took an additional 70 years before this barrier was localized to cerebral microvascular endothelial cells, in electron-microscopic studies performed by *Reese and Karnovsky* [179]. Although the concept of the BBB has continued to be refined over the past few decades, the recent understanding of the basic structure of the BBB is built on the general framework established by their studies in the late 1960s [179]. More specifically, the BBB exists primarily as a selective diffusion barrier at the level of cerebral capillary endothelium.

A conceptual depiction of the anatomical structure of the blood-brain barrier is shown in Fig. 4.5a. For comparison, the cross-sectional view of a peripheral microvessel (a typical microvessel in non-brain organs) is also shown in Fig. 4.5b. For both peripheral microvessels and the BBB, the circumference of the microvessel lumen is surrounded by endothelial cells,

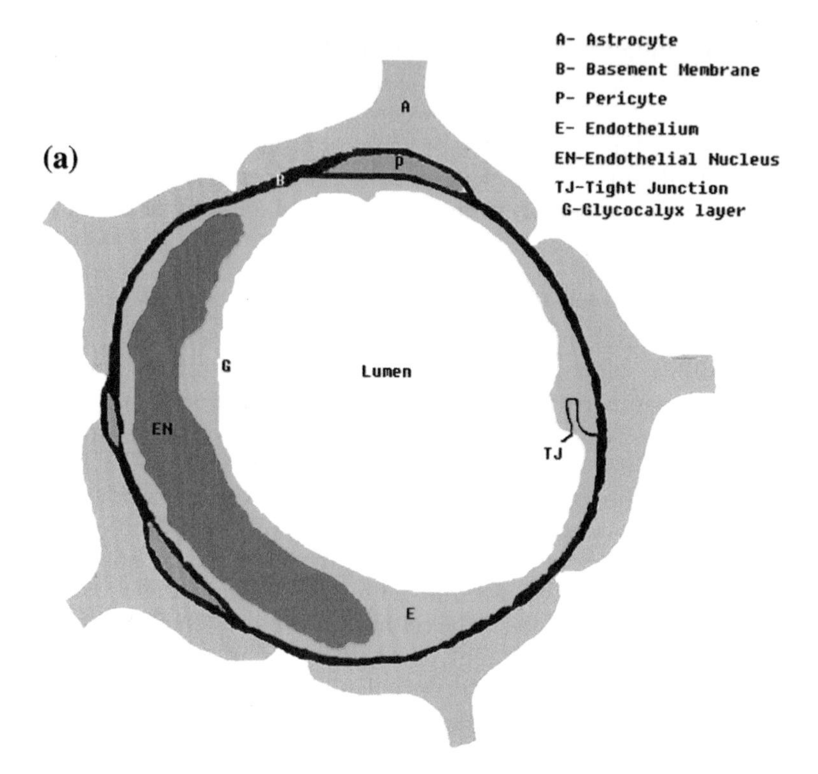

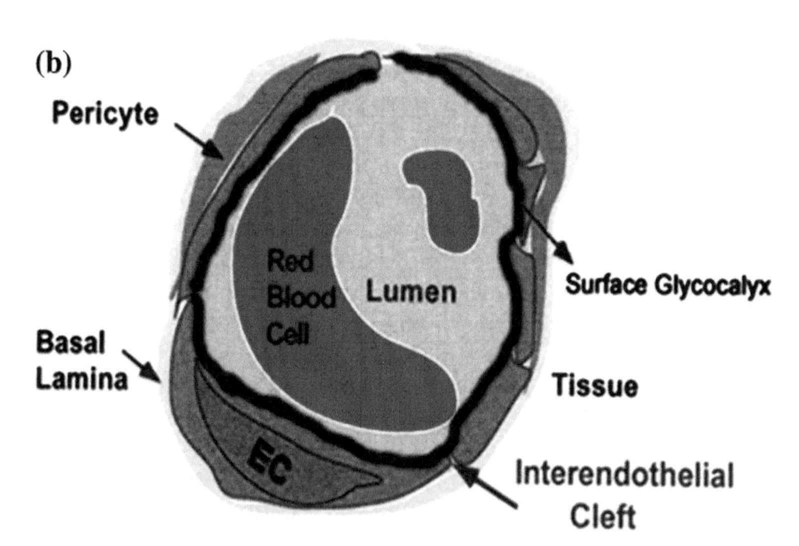

FIGURE 4.5 (a) Shows a cross-sectional view of the blood-brain barrier, formed by endothelial cells with tight junctions at their ends, the pericytes, the basement membrane and the astrocyte processes. Based on the electron microscopic study in [78]. In comparison, (b) shows that of the wall of a peripheral microvessel (in non-brain tissues).

the opposing membranes of which are connected by tight junctions. At the luminal surface of the endothelial cell, there is a rather uniform fluffy glycocalyx layer [4,203,218,223], which is mainly composed of heparan sulfate proteoglycan, chondroitin sulfate proteoglycan and hyaluronic [113]. This mucopolysaccharide structure is highly hydrated in electrolytic solution and contains large numbers of solid-bound fixed negative charges due to the polyanionic nature of its constituents abundant in glycoproteins, acidic oligosaccharides, terminal sialic acids, proteoglycan and glycosaminoglycans aggregates [210]. Pericytes attach to the abluminal membrane of the endothelium at irregular intervals. In a peripheral microvessel, there is a loose and irregular basal lamina (or basement membrane) surrounding the pericytes. In contrast, in the BBB, pericytes and endothelial cells are ensheathed by a very uniform basement membrane of 20–40 nm thickness, which is composed of collagen type IV, heparin sulfate proteoglycans, laminin, fibronectin and other extracellular matrix proteins [78]. The basal lamina is contiguous with the plasma membranes of astrocyte end-feet that wrap almost the entire abluminal surface of the endothelium [173]. The blood-brain barrier (BBB) has unique structures that function to protect the central nervous system (CNS). In addition to a tighter junction of the microvessel endothelium, there is a uniform and narrow matrix-like basement membrane layer (20–40 nm), sandwiched between the vessel wall and the astrocyte processes ensheathing the cerebral microvessel (Fig. 4.5a).

In addition to the anatomical structures, the BBB differs from the peripheral microvessels in the following ways. The mitochondrial content of the endothelial cells forming the BBB is greater than that of such cells in all non-neural tissues. It is suggested that this larger metabolic work capacity may be used to maintain the unique structural characteristics of the BBB, or/and by metabolic pumps that may require energy to maintain the differences in composition of the cerebral circulation and the brain tissue [164]. The BBB has a high electrical resistance, much less fenestration, and more intensive junctions, which are responsible for restricting the paracellular passage of water and polar solutes from the peripheral circulation into the CNS [39,110]. Between adjacent endothelial membranes, there are junctional complexes which include adherens junctions (AJs), tight junctions (TJs) and possibly gap junctions [198]. The structure of the junction complexes between endothelial cells is similar to that shown in Fig. 4.4 [2,126]. Both AJs and TJs act to restrict paracellular transport across the endothelium, while gap junctions mediate intercellular communication. AJs are ubiquitous in the vasculature and their primary component is vascular endothelial (VE)-cadherin. They basically mediate the adhesion of endothelial cells to each other and contact inhibition during vascular growth and remodeling. Although the disruption of AJs at the BBB can lead to increased permeability, it is the TJ that is the major junction which confers the low paracellular permeability and high electrical resistance [187]. The tight junction complex includes two classes of trans-membrane molecules: occludins and claudins. The trans-membrane proteins from adjacent endothelia cells interact with each other, and form seals in the spaces between adjacent endothelial cells. The cytoplasmic tails of the trans-membrane proteins are linked to the actin cytoskeleton via a number of accessory proteins, such as members of the zonula occludens family, ZO-1, ZO-2 and ZO-3.

A number of grafting and cell culture studies have suggested that the ability of cerebral endothelial cells to form the BBB is not intrinsic to these cells, but that the cellular milieu of the brain somehow induces the barrier property in the blood vessels. It is believed that all components of the BBB are essential for maintaining its functionality and stability. Pericytes seem to play a key role in angiogenesis, structural integrity and maturation of cerebral microvessels [20]. The extracellular matrix of the basal lamina appears to serve as an anchor for the endothelial layer via interaction of laminin and other matrix proteins with endothelial integrin receptors [118]. It has been suggested that astrocytes are critical in the development and/or maintenance of the unique features of the BBB. Additionally, astrocytes may act as messengers to (or in) conjunction with neurons in the moment-to-moment regulation of the BBB's permeability [20].

4.2.1.2.2 TRANSPORT PATHWAYS ACROSS THE BLOOD-BRAIN BARRIER

The BBB endothelial cells differ from those in peripheral microvessels by more intensive tight junctions, sparse pinocytic vesicular transport and much less fenestrations. The transport of substances from the capillary blood into the brain tissue depends on the molecular size, lipid solubility, binding to specific transporters and electrical charge [157]. Figure 4.6 summarizes the transport routes across the BBB [162]. Compared to the peripheral microvessel wall, the additional structure of the BBB and tighter endothelial junctions greatly restrict the transport of hydrophilic molecules through the gaps between the cells, i.e., the paracellular pathway of the BBB, route A in Fig. 4.6. In contrast, small hydrophobic molecules such as O_2 and CO_2 diffuse freely across plasma membranes following their concentration gradients, i.e., the transcellular lipophilic diffusion pathway, route C in Fig. 4.6. The permeability of the BBB to most molecules can be estimated on the basis of their octanol/water partition coefficients [189]. For example, diphenhydramine (Benadryl), which has a high partition coefficient and can easily cross the BBB, whereas water-soluble loratadine (Claritin) is not able to penetrate the BBB and has little effect on the CNS [123].

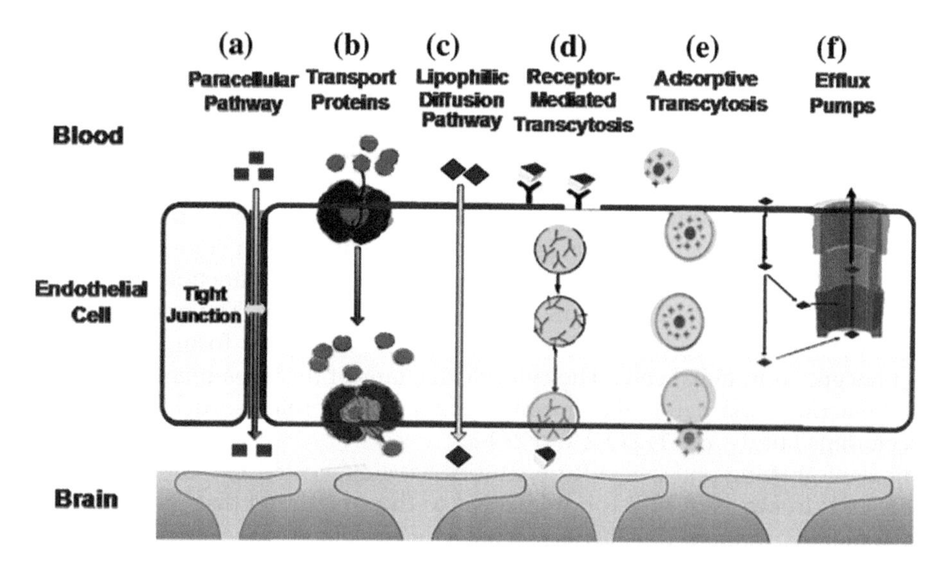

FIGURE 4.6 Transport pathways across the brain microvascular endothelial cell. Modified from [162].

However, the octanol/water partition coefficients do not completely reflect the permeability of the BBB to solutes. Some solutes with low partition coefficients that easily enter into the CNS generally cross the BBB by active or facilitated transport mechanisms, which rely on ion channels, specific transporters, energy-dependent pumps and a limited amount of receptor-mediated transcytosis. Glucose, amino acids and small intermediate metabolites, for example, are ushered into brain tissue via facilitated transport mediated by specific transport proteins (route B in Fig. 4.6), whereas larger molecules, such as insulin, transferrin, low density lipoprotein and other plasma proteins, are carried across the BBB via receptor-mediated (route D) or adsorptive transcytosis (route E). Some small molecules with high octanol/water partition coefficients are observed to poorly penetrate the BBB. Recent studies suggested that these molecules are actively pumped back into the blood by efflux systems (route F in Fig. 4.6). These efflux systems greatly limit drug delivery across the BBB. For instance, P-glycoprotein (P-gp), which is a member of the adenosine triphosphate-binding cassette family of exporters, has been demonstrated to be a potent energy-dependent transporter. P-gp contributes greatly to the efflux of xenobiotics from brain to blood, and has increasingly been recognized to have a protective role, being responsible for impeding the delivery of therapeutic agents [193]. The organic anion transporters and glutathione-dependent multidrug resistance-associated proteins (MRP) also contribute to the efflux of organic anions from the CNS, and many drugs with the BBB permeabilities that are lower than predicted are the substrates for these efflux proteins [2,27,162,173]. While the brain endothelium is the major barrier interface, the transport activity of the surrounding pericytes [197], basement membrane and astrocyte foot processes (Fig. 4.5a) [230] also contribute to the BBB barrier function under physiological conditions, and may act as a substitute defense if the primary barrier at the endothelium is compromised [136].

4.2.2 Transport Coefficients

The aforementioned ultrastructural study using electron microscopy, together with other permeability studies, have shown that the microvessel wall (the paracellular pathway or interendothelial cleft) behaves as a passive membrane for water and hydrophilic solute transport [53,148]. The membrane transport properties are often described by Kedem-Katchalsky equations derived from the theory of irreversible thermodynamics:

$$J_s = PRT\Delta C + (1 - \sigma_f)CJ_v \tag{4.1}$$

$$J_v = L_p(\Delta p - \sigma_d RT\Delta C) \tag{4.2}$$

where J_s and J_v are the solute and total volumetric fluxes; ΔC and Δp are the concentration and pressure differences across the membrane. L_p, the hydraulic conductivity, describes the membrane permeability to water. P, the diffusive permeability, describes the permeability to solutes. σ_f is the solvent drag or ultrafiltration coefficient which describes the retardation of solutes due to membrane restriction, and σ_d, the reflection coefficient, describes the selectivity of membrane to solutes. In many transport processes, σ_f is equal to σ_d [55] and thus we often use σ, the reflection coefficient, to represent both of them. R is the universal gas constant and T is the absolute temperature.

4.2.3 Permeability Measurement

All of the permeability measurements have been interpreted in terms of L_p, P and σ, which are measured experimentally in intact whole organisms (including human subjects), on perfused tissues and organs, on single perfused microvessels and on monolayers of cultured microvascular endothelial cells. Different experimental preparations have their advantages and disadvantages. Although measurements made on the intact regional circulation of an animal subject (usually using radioactive isotope labeled tracers) suffer from uncertainties surrounding the exchange surface area of the microvessel wall and the values of the transvascular differences in pressure and concentration, they usually involve minimal interference with the microvessels themselves. These studies can provide valuable information concerning microvascular exchange under basal conditions. At the other end of the scale are measurements on single perfused vessels. The Landis technique has been used to measure the hydraulic conductivity L_p and reflection coefficient σ. Quantitative fluorescence microscope photometry is used to measure solute diffusive permeability P. Both of these techniques are described in detail in [61]. The surface area of the microvessel can be measured directly, as also can the difference in pressure and concentration across the vessel walls. The disadvantages of the single vessel preparation are: (1) they interfere directly with the vessels, and (2) they are usually restricted to a small number of convenient vessel types (e.g., mesenteric vessels on a two-dimensional translucent tissue). Direct interference with a vessel, whether by exposure to light or micromanipulation, might be expected to increase permeability. However, this concern was allayed when it was shown that L_p [60] and P to potassium ions [55] in single muscle capillaries were similar to values based on indirect measurements on the intact muscle microcirculation.

Although the rapid growth of endothelial cell biology is largely a result of experiments on cultured endothelial cells *in vitro* (in dishes), there are serious limitations to the use of monolayers of cultured endothelial cells for gaining direct information about vascular permeability. For example, the most widely reported permeability measurements on monolayers of cultured endothelium are permeability to serum albumin, which has a mean value in the range of 10^{-6} cm/s [11]. In general, the *in vitro* permeability to albumin is 2 to 10 times larger than that from the *in vivo* (in live animals) measurement. Estimates of reflection coefficients σ of cultured monolayers of endothelial cells to macromolecules are too low for plasma proteins to exert a significant osmotic pressure across them [216]. These results indicate that monolayers of cultured endothelial cells do not completely reflect the permeability characteristics of microvascular endothelium *in vivo*. For this reason, we restrict discussion of permeability properties and the values of permeability coefficients to measurements mainly made on single vessels.

4.2.3.1 Permeability Measured in a Single Perfused Microvessel

4.2.3.1.1 HYDRAULIC CONDUCTIVITY AND REFLECTION COEFFICIENT

Figure 4.7 shows the results of the permeability measured in a single perfused microvessel in thin tissues with rich microvessels, e.g., rat mesentery, rat and mouse cremaster muscles. Figure 4.7a demonstrates the Landis technique, which has been widely used to measure the microvascular hydraulic conductivity L_p and solute reflection coefficient σ. A single microvessel in a thin tissue of a live animal is cannulated with a micropipette A at the upstream side. The micropipette A is connected to a water manometer and is prefilled with 1% BSA mammalian Ringer (a plasma substitute) and red blood cells (RBCs) from another animal. The RBCs

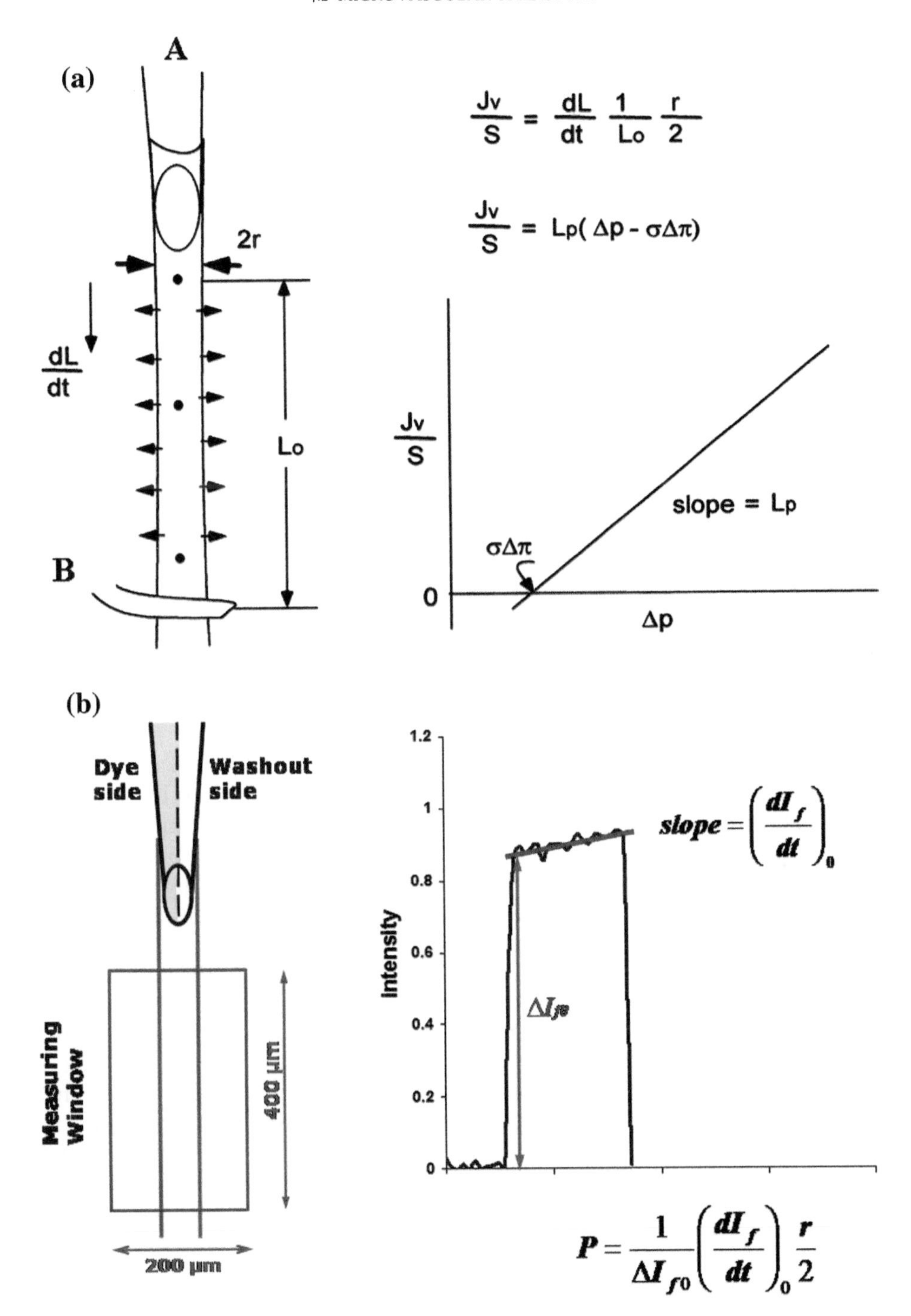

FIGURE 4.7 (a) Measurement of microvascular hydraulic conductivity L_p and reflection coefficient σ to a solute in a single microvessel by the Landis technique. (b) Measurement of microvascular solute permeability in a single microvessel by quantitative fluorescence microscopy or photometry.

are used as the flow markers. The pressure from the water manometer can drive the RBCs and solution in the micropipette A into the vessel. If the downstream side of the microvessel is occluded by another pipette B, there will be no axial flow in the vessel. However, if the vessel wall is permeable to water, there will be a radial flow Jv across the wall. Due to this Jv out of the vessel, we can see the downstream movement of RBCs. A video camera and a recorder can be used to record the movement of an RBC, which can be used to determine the movement velocity of the RBCs, dL/dt. From this movement velocity, the length of the water column, L_0, which is the distance between the initial position of the RBC and the occlusion site B, along with the radius of the vessel, we can calculate the water flux Jv/S out of the microvessel. Here the flux surface area is $S = 2\pi r L_0$ (see Fig. 4.7a). On the other hand, the flux can be theoretically represented as:

$$Jv/S = L_p(\Delta p - \sigma \Delta \pi) \tag{4.3}$$

Here L_p is the microvascular hydraulic conductivity; Δp, $\Delta \pi$ are the hydrostatic and osmotic pressure differences across the vessel wall, respectively. In general, since the p and π in the tissue space outside the vessel are negligible, Δp can be approximated as the driving pressure from the water manometer and $\Delta \pi$ as RTC_L, respectively. Here C_L is the solute concentration in the vessel lumen. In this way, L_p can be determined from the slope of the Jv/S vs. Δp curve shown in Fig. 4.7a. When Δp balances out $\sigma \Delta \pi$, we have no water flux across the vessel wall and thus at the intercept of the Jv/S vs. Δp curve and the abscissa, the reflection coefficient of the vessel wall to a solute $\sigma = \Delta p / \Delta \pi$.

4.2.3.1.2 SOLUTE PERMEABILITY

Figure 4.7b is a sketch of a solute permeability measurement in a single microvessel. Quantitative fluorescence photometry or image microscopy is often used to measure solute permeability. Instead of using a single lumen micropipette, a double lumen θ micropipette (the cross-section of the pipette looks like a Greek letter θ) is used to cannulate the microvessel. One lumen is filled with the washout solution (1% BSA mammalian Ringer) and the other filled with the same, with the addition of fluorescently labeled solutes. Each lumen is connected to a water manometer. Perfusion with the washout for 10 to 20 seconds establishes a baseline. When the perfusion is switched to the lumen containing the test solute (dye), the microvessel lumen is filled with fluorescent solute in about 0.3 sec, producing ΔI_{f0}, the initial total fluorescence intensity in the measuring window of $200 \times 400\,\mu m$ on top of the vessel. With continued perfusion, the measured fluorescence intensity I_f increases, indicating further transport of the solute out of the microvessel and into the surrounding tissue. The initial solute flux into the tissue is measured from $(dI_f/dt)_0$, the slope of the total fluorescence intensity I_f vs. time curve. After 10–30 seconds the microvessel is reperfused with the washout solution. There is a step decrease in the fluorescence intensity as test solute is washed out of the microvessel lumen, and a subsequent return of the measured intensity to control level as solute diffuses from the tissue back into the microvessel lumen. The solute permeability P is calculated by:

$$P = \frac{1}{\Delta I_{f0}} \left(\frac{dI_f}{dt} \right)_0 \frac{r}{2} \tag{4.4}$$

where r is the microvessel radius.

4.2.3.1.3 MEASUREMENT OF THE BLOOD-BRAIN BARRIER SOLUTE PERMEABILITY *IN VIVO*

To measure the solute permeability of the blood-brain barrier, instead of using a micropipette to perfuse the fluorescently labeled solutes into a single microvessel, the solutes are perfused into brain microvessels by a syringe pump via the left or right carotid artery. The method was described in detail in [237]. Briefly, after anesthesia, the rat skull is exposed by cutting away the skin and connective tissue. A section of right or left frontoparietal bone, approximately 5 mm × 5 mm, is ground with a high-speed micro-grinder until it becomes thin enough to allow pial vessels to be seen. The thinned section of rat brain is observed under the microscope, and images are recorded at the same time as a solution containing fluorescently labeled test solutes is introduced into the cerebral circulation via the ipsilateral carotid artery by a syringe pump at a constant perfusion rate of 3 ml/min. We can switch the dye perfusion to a clear solution without the dye to wash away the dye in the vessel. In this way, we can repeat our measurement on the same vessel under various conditions. The images are obtained and then transferred to an image acquisition and analysis workstation for later permeability measurement. A similar method, shown in Fig. 4.7b, is used to determine the BBB solute permeability from the collected images. In general, the rat BBB solute permeability is ~1/10 of that of mesenteric microvessels for solutes of radius ranging from 0.45 nm to 3.5 nm (sodium fluorescein to albumin).

Figure 4.8 shows the measured permeability of frog mesenteric microvessels to hydrophilic solutes as a function of solute size ranging from potassium to albumin. The values of permeability have been plotted on a logarithmic scale, and it can be seen that the decline of P is maintained until the molecular radius reaches around 3.6 nm (the radius of bovine serum

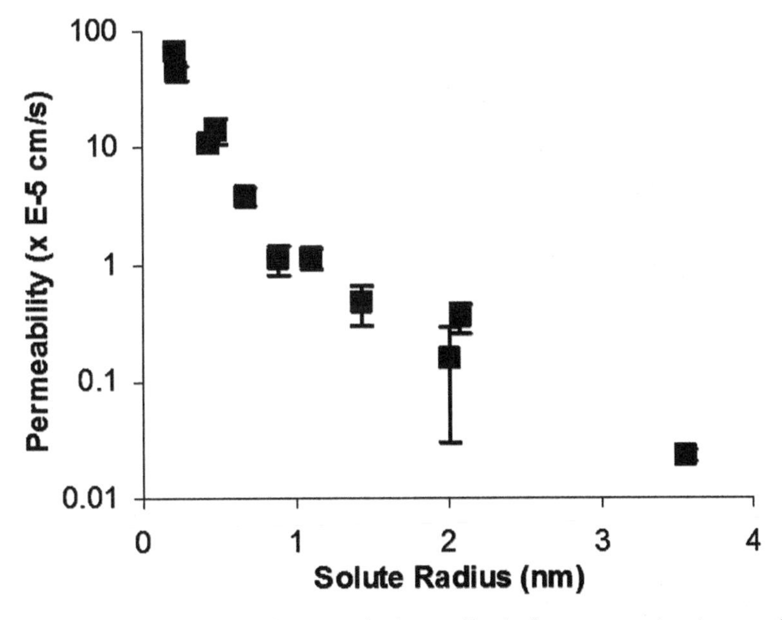

FIGURE 4.8 Solute permeability P as a function of solute radius in frog mesenteric microvessels. Permeability has been plotted on a logarithmic scale (data are from various sources summarized in *Fu et al.* [88]).

albumin). These permeability values are for normal (undisturbed) true capillaries or post capillary venules. The hydraulic conductivity for these vessels ranges from 2 to 4 × 10^{-7} cm/s/cmH_2O [57]. In general, for transport across microvessels, solutes such as ions and vitamins, which have radius of 1 nm or less, can freely transfer across the surface glycocalyx. They are categorized as small solutes. Molecules as large as albumin that hardly transfer across the glycocalyx layer are called large solutes. Those molecules of radius in between 1 nm and 3 nm are called intermediate-sized solutes. Molecules larger than albumin seem not to cross the microvessel wall in the frog mesentery. Values of permeability for molecules larger than serum albumin appear to decrease much less rapidly with increasing molecular size in mammalian skeletal muscle [181], suggesting that either the pathways or mechanisms concerned in the transport of macromolecules differ from those for smaller solutes. Figure 4.8 describes relations between solute permeability and molecular size in microvessels in frog mesentery. Similar relations are found in other types of microvessels such as in mammalian mesentery and skeletal muscle [94,151] as well as in the mammalian brain [237].

Although the absolute values of permeability to water and to the smallest molecules may vary by several orders of magnitude, the microvessels in different tissues have rather similar values in the reflection coefficient, σ, to macromolecules. The wide range of values of hydraulic conductivity (water permeability) L_p and the relatively constant values of σ to serum albumin in different types of microvessel are shown in Fig. 4.9. Each point in this diagram represents the mean value of L_p and σ to albumin for a different type of microvessel or microvascular bed [94]. The L_p values have been plotted on a logarithmic scale so that values covering three orders of magnitude can be displayed. It is seen that there is no correlation between σ to albumin and L_p. The conclusion from Fig. 4.9 is that variations in L_p in different

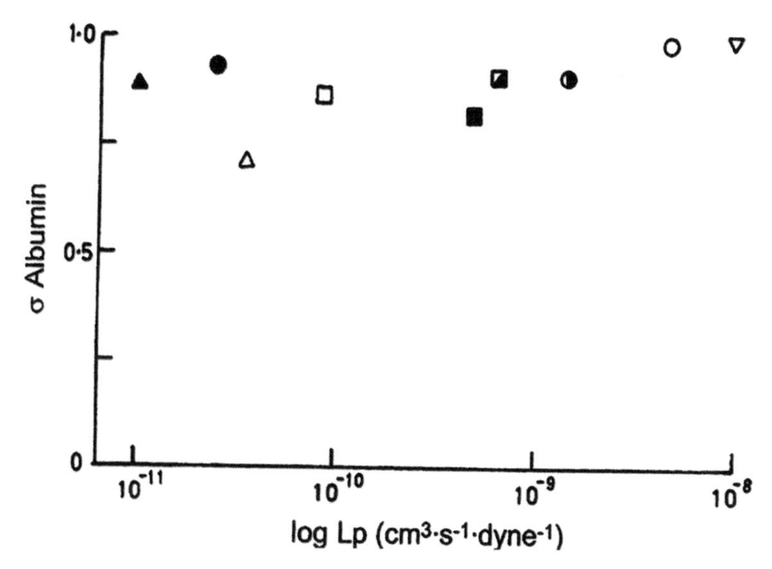

FIGURE 4.9 Reflection coefficient to serum albumin (σ) and hydraulic conductivity (L_p) in different microvascular beds. Each point represents mean value of σ and L_p for microvessels (continuous or fenestrated) in a particular tissue: ▲, cat hindlimb; ●, rat hindlimb; Δ, dog lung; □, dog heart; ■, frog mesentery; ◪, rabbit salivary gland; ◑, dog small intestine; ○, dog glomerulus; ∇, rat glomerulus (from [151]).

microvessels are not accompanied by variations in their leakiness to macromolecules. This phenomenon could be explained if there is a common molecular sieve to serum albumin existing for all types of vessels. The most likely candidate for the sieve is the surface glycocalyx layer in the luminal side of the microvessel, which was observed by using ruthenium red and Alcian blue [142], by using cationized ferritin tracers in electron microscopy of frog mesenteric microvessels [4], and also *in vivo* fluorescent microscopy for hamster cremaster capillaries [223].

The evidence for the surface glycocalyx layer being the molecular sieve for macromolecules is also provided by comparing the L_p and P data for fenestrated and continuous microvessels. The walls of the vessels with high values of L_p ($>10^{-6}$ cm/s/cmH_2O) are fenestrated endothelium, whereas those of vessels with lower L_p values are continuous endothelium. Different pathways are primarily responsible for the transport of fluid and hydrophilic solutes through these two types of endothelia. In fenestrated endothelia, this pathway is through the fenestrae, whereas in continuous endothelium, it is through the intercellular cleft (see Fig. 4.3). It is, therefore, very surprising that the clear differences in morphology are not accompanied by a qualitative change in the properties of the permeability coefficients. Fenestration increases the L_p and the P to small hydrophilic solutes without changing the L_p/P or σ to albumin. Thus the molecular sieving characteristics of microvascular walls appear to be common to both fenestrated and continuous endothelium. *Curry and Michel* [62] suggested that it was luminal glycocalyx that acted as the molecular sieve in both types of vessels.

4.2.3.2 *Permeability Measured in Cultured Endothelial Cell Monolayers*

Although the monolayers of cultured endothelial cells do not completely reflect the permeability characteristics of microvascular endothelium *in vivo*, they are the most accessible and convenient models for studying the molecular mechanisms by which the microvascular permeability is regulated. The techniques for measuring endothelial monolayer permeability to water and solutes are described in [15,40,136].

4.2.4 Transport Models for Water and Solutes Through Interendothelial Cleft

4.2.4.1 *1-D Models*

Prior to the late 1980s, there were two major one-dimensional theories: the pore-slit and the fiber matrix theory (Fig. 4.10), which attempted to correlate the interendothelial cleft structure with the large amount of experimental data for L_p, P and σ. In microvessels with continuous endothelium, the principal pathway for water and solutes lies between the endothelial cells through the interendothelial cleft. The 1-D pore-slit models were developed in terms of the ultrastructure of the cleft (Fig. 4.2).

4.2.4.1.1 PORE-SLIT MODEL

In pore-slit theory, the permeability properties of the microvessel wall can be described in terms of flow through water-filled cylindrical pores or rectangular slits through the vessel wall. The characteristic Reynolds number for the flow is in the order of 10^{-8} for a healthy microvessel and 10^{-7} to 10^{-6} for a stimulated (or leaky) vessel. A Poiseuille type viscous flow

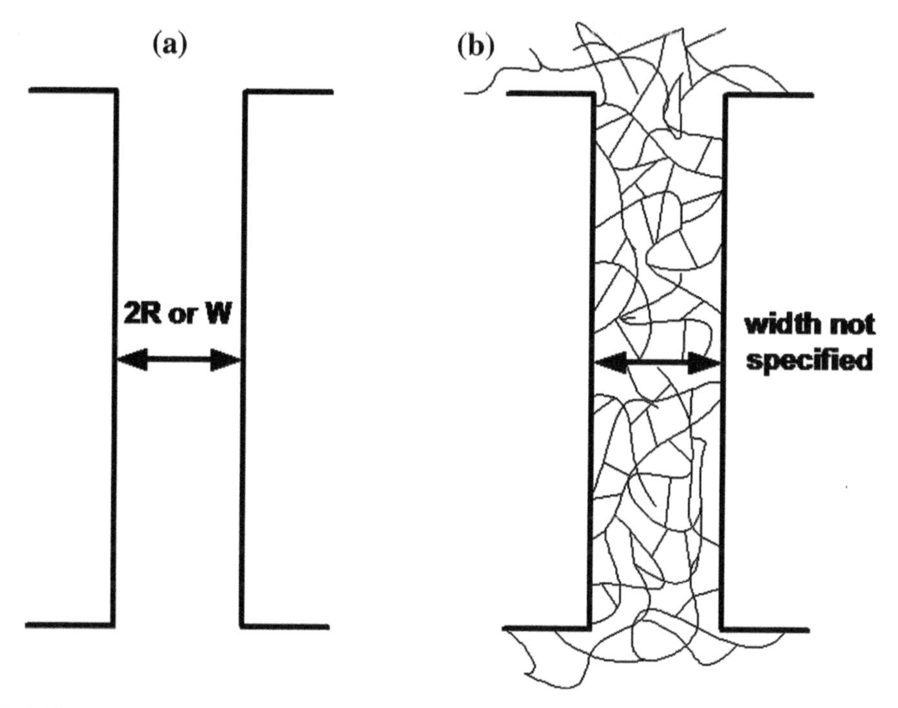

FIGURE 4.10 1-D models of the interendothelial cleft. (a) Pore-slit model, (b) fiber matrix model. Revised from [151].

was thus assumed in the pore/slit to describe water flow due to the narrow channel with an aspect ratio >20. The resistance to solute diffusion was described in terms of the additional drag on a spherical molecule moving within the pore relative to movement in free solution, and the selectivity of the membrane in terms of steric exclusion at the pore entrance [55,84,149,212,229].

The 1-D model for hydraulic conductivity L_p and solute permeability P is, for a cylindrical pore:

$$L_p = \frac{N\pi R^4}{8\mu L} \quad P = N\pi R^2 D_{pore} \frac{\varphi_{pore}}{L} \tag{4.5}$$

and for a rectangular slit:

$$L_p = \frac{L_{jt} f W^3}{12\mu L} \quad P = L_{jt} f W D_{slit} \frac{\varphi_{slit}}{L} \tag{4.6}$$

Here N is the number of pores per unit surface area of the microvessel wall, R is the pore radius, L_{jt} is the total length of the cleft per unit vessel wall surface area, f is the fraction of the slit which is open, L is the thickness of the vessel wall or the depth of the cleft measured from the lumen to the tissue, μ is the water viscosity at experimental temperatures and the parameters D_{pore} and D_{slit} are the restricted (or effective) solute diffusion coefficients within the pore or slit. The parameter $\varphi_{pore/slit}$ is the solute partition coefficient at the pore/slit entrance. For a cylindrical pore of radius R:

$$\begin{aligned}\frac{D_{pore}}{D_{free}} &= 1 - 2.10444\alpha + 2.08877\alpha^3 - 0.094813\alpha^5 - 1.372\alpha^6 \\ \varphi_{pore} &= (1-\alpha)^2 \\ \alpha &= \frac{a}{R}\end{aligned} \tag{4.7}$$

For a long slit of width W:

$$\begin{aligned}\frac{D_{slit}}{D_{free}} &= 1 - 1.004\beta + 0.418\beta^3 + 0.210\beta^4 - 0.1696\beta^5 \\ \varphi_{slit} &= 1-\beta \\ \beta &= \frac{2a}{W}\end{aligned} \tag{4.8}$$

Here a is the solute radius and D_{free} is the free solute diffusion coefficient in aqueous solutions. Expressions in Eqs. (4.7) and (4.8) are polynomial approximations from the numerical solutions for the drug coefficients for the movement of spheres through liquid-filled cylindrical pores [169] and slits [72]. They are good approximation when α or $\beta \leq 0.6$ [55].

The osmotic reflection coefficient of a membrane σ is a measure of the selectivity of the membrane to a particular solute that depends only on the pore size, not the number of pores or the membrane thickness. σ is given by [55]:

$$\sigma = (1-\varphi)^2 \tag{4.9}$$

When there are several pathways in parallel, the membrane reflection coefficient is the sum of the individual coefficients weighted by the fractional contribution of each pathway to the membrane hydraulic conductivity [151]. However, when there are several membranes in series (here we show two in a series as an example), the overall reflection coefficient σ^T is given by [124]:

$$\begin{aligned}\sigma^T &= \frac{P^T}{P^{(1)}}\sigma^{(1)} + \frac{P^T}{P^{(2)}}\sigma^{(2)} \\ P^T &= \frac{P^{(1)}P^{(2)}}{P^{(1)} + P^{(2)}}\end{aligned} \tag{4.10}$$

where $P^{(1)}$, $P^{(2)}$ are solute permeabilities of membranes 1 and 2 and $\sigma^{(1)}$ and $\sigma^{(2)}$ are corresponding reflection coefficients.

4.2.4.1.2 FIBER MATRIX MODEL

The principal hypothesis used to describe the molecular filter of the transvascular pathway is the fiber matrix theory. On the luminal side of the cleft, the presence of a glycocalyx layer on the endothelial cell surface was first described from staining experiments using ruthelium red and Alcian blue for cell surface glycoprotein [142], but there have been more recent studies using different techniques [235]. These experiments suggested that the layer extended into the outer regions of the intercellular cleft. Electron micrographs of microvessels perfused with solutions containing native ferritin suggested that, while the luminal contents had been adequately fixed, the ferritin concentration was greatly reduced close to the luminal surface of the endothelial cells. Quantitative evidence that ferritin was excluded from the endothelial surface was reported in [49,141], which strengthened the idea that the glycocalyx could act as a barrier to macromolecule diffusion. More accurate estimates of the possible thickness of the endothelial cell glycocalyx were provided by *Adamson and Clough* [4] in frog mesenteric capillaries. Using cationized ferritin, they visualized the outer surface of the glycocalyx that was up to 100 nm from the endothelial cell surface when the vessel was perfused with plasma. These observations were consistent with the hypothesis that plasma proteins were absorbed to the endothelial cell glycocalyx and form part of the structure comprising the molecular filter at the cell surface [55,149,190,192]. *Adamson* [3] also demonstrated that enzymatic removal of the glycocalyx, using pronase, increased the hydraulic conductivity of frog mesenteric capillaries by 2.5-fold. The investigator used transmission electron microscopy to carefully examine the microvessels treated with pronase. No morphological features, such as fenestrations, transendothelial channels, or intercellular gaps associated with inflammation, were found, which might account for the increase in hydraulic conductivity. However, the general protease (pronase) may increase L_p by altering more subtle structural components of the microvessel wall that determine microvessel permeability.

Although the nature of the fibers associated with the endothelial cell surface and the cleft entrance is not well understood, the side chains of glycosaminoglycans that are likely to form part of the cell glycocalyx have a characteristic molecular radius close to 0.6 nm. Absorption of plasma proteins like albumin into the side chains of glycosaminoglycans would form a fiber matrix with uniform gap spacing of roughly the diameter of albumin (~7 nm) between adjacent fibers. Regularly arranged electron densities have been demonstrated in this region by *Schultze and Firth* [194], and these could represent fibers of a molecular filter. Using auto-correlation techniques in a recent electron microscopy study, *Squire et al.* [203] revealed a quasi-periodic structure in the surface glycocalyx layer. Their major findings are that the fibers which form the main part of the sieving matrix project from the endothelial surface as a series of bush-like structures, with the effective diameter of the 'branches' of each of bush being 10–12 nm. The sieving elements are pore-like channels, 7–8 nm in diameter, situated between the core proteins forming the glycoproteins on the cell surface, and aligned normal to the endothelial cell surface. These structures have similar sieving properties [6] to the periodic array assumed in [88]. Using the stochastic theory of *Ogston et al.* [163], *Curry and Michel* [62] described the solute partition coefficient φ and the restricted (or effective) solute diffusion coefficient D_{fiber} in terms of the fraction of the matrix volume occupied by fiber S_f and the fiber radius r_f. The partition coefficient φ is defined as the space available to a solute of radius a relative to the space available to water ($a=0$). The restricted solute diffusion coefficient D_{fiber} accounts for the resistance to the solute diffusion due to the existence of the fiber matrix. For a random fiber arrangement, they are expressed as:

$$\varphi = \exp\left[-(1-\varepsilon)\left(\frac{2a}{r_f}+\frac{a^2}{r_f^2}\right)\right]$$
$$\frac{D_{fiber}}{D_{free}} = \exp\left[-(1-\varepsilon)^{0.5}\left(1+\frac{a}{r_f}\right)\right] \tag{4.11}$$

For an ordered fiber arrangement:

$$\begin{aligned} \varphi &= 1 - S_f\left(1 + \frac{a}{r_f}\right)^2 \\ \frac{D_{fiber}}{D_{free}} &= 1 - \left[(1-\varepsilon)^{0.5}\left(1 + \frac{2a}{\pi^{0.5} r_f}\right)\right] \end{aligned} \tag{4.12}$$

where a is the solute radius and $\varepsilon = 1 - S_f$ is void volume of the fiber matrix.

In fact, the above expressions for the effective diffusivity of a solute only consider the steric exclusion of solutes by the fiber array; they do not include hydrodynamic interactions between the fibers and the diffusing solute, which are important when the solute size is comparable to the gap spacing between fibers. Using two approaches, *Philips et al.* [175,176] calculated the effects of hydrodynamic interactions on the hindered transport of solid spherical macromolecules in ordered or disordered fibrous media. One approach was a rigorous 'Stokesian-dynamics' method [33] or generalized Taylor dispersion theory [35], which can calculate local hydrodynamic coefficients at any position in a fibrous bed. But detailed information about the fiber configuration needs to be given. Another approach was an effective medium theory based on Brinkman's equation. Comparing calculated results with the experimental data for the transport of several proteins in hyaluronic acid solutions, these authors found that the use of Brinkman's equation gave good agreement with the more rigorous methods for a homogeneous fiber matrix.

Based on the solution for the flow around parallel square array of infinite cylindrical fibers using hydrodynamic theory, *Tsay et al.* and *Weinbaum et al.* [215,226] found the expressions for φ and D_{fiber} for a confined periodic fiber array in a rectangular slit:

$$\begin{aligned} \varphi &= \frac{1 - b_1 S_f (1 + a/r_f)^2}{1 + b_1 S_f (1 + a/r_f)^2} \\ \frac{D_{fiber}}{D_{free}} &= \frac{D_{slit}}{D_{free}}\left[1 + \frac{a}{K_p{}^{0.5}} + \frac{a^2}{3K_p}\right]^{-1} \end{aligned} \tag{4.13}$$

where D_{slit} is the restricted (or effective) solute diffusion coefficient in a slit, b_1 is the coefficient of the leading term in a doubly periodic Wierstrasse expansion series used in *Tsay et al.* [215], K_p is the Darcy hydraulic conductance for an unbounded fiber array. For a 2-D square fiber array, K_p is given by [215]:

$$K_p = 0.0572a^2\left(\frac{\Delta}{a}\right)^{2.377} \tag{4.14}$$

where Δ is the gap spacing between fibers. $\Delta = r_f[(\pi/S_f)^{0.5} - 2]$. Equation (4.14) is a highly accurate approximation to the exact solution for the 2-D array in *Sangani and Acrivos* [188]. For a 2-D random array, a Carman-Kozeny approximation for K_p is [57]:

$$K_p = \frac{(1 - S_f)^3}{S_f^2}\left(\frac{a^2}{4C}\right) \tag{4.15}$$

where C is a fiber density correction factor. When the fibers are circular cylinders perpendicular to the flow, *Happel* [109] obtained the following approximation for C:

$$C = \frac{2(1-S_f)^3}{S_f}\left[\ln\left(\frac{1}{S_f}\right) - \frac{(1-S_f^2)}{(1+S_f^2)}\right]^{-1} \tag{4.16}$$

When the fiber is confined in a slit of width W, the effective Darcy permeability K_{eff} was related to the value of unbounded K_p by the relation:

$$K_{eff} = K_p\left[1 - \frac{\tanh[(W/2)/\sqrt{K_p}]}{(W/2)/\sqrt{K_p}}\right] \tag{4.17}$$

For a fiber matrix with partition coefficient φ, the reflection coefficient σ is the same as in pore-slit model [35] of Eq. (4.9), $\sigma = (1-\phi)^2$.

The hydraulic conductivity L_p and solute permeability P of a fibrous membrane are [151]:

$$\begin{aligned} L_p &= \frac{A_{fiber}}{L}\frac{K_p}{\mu} \\ P &= \frac{A_{fiber}}{L}\frac{D_{fiber}}{D_{free}}\varphi \end{aligned} \tag{4.18}$$

Here A_{fiber} is the area of fiber filled pathway. Other parameters are the same as in previous section. However, for a confined fiber array in a rectangular slit of width W, K_p in the expression for L_p should be replaced by K_{eff} shown earlier and μ is replaced by an effective μ_{eff}:

$$\mu_{eff} = \mu\frac{[(W/2)/\sqrt{K_p}]^3}{3\left[(W/2)/\sqrt{K_p} - \tanh[(W/2)/\sqrt{K_p}]\right]} \tag{4.19}$$

Although the above described 1-D pore-slit and fiber matrix theories are unable to successfully explain the large body of experimental data for L_p, P and σ, they provide a useful starting point to evaluate the possible cellular and molecular structures that actually determine the permeability properties of the microvessel walls. Results from the fiber matrix theory are also applied in recent 3-D models.

A simplified model of the endothelial surface glycocalyx (ESG) has been used by *Squire et al.* [203], *Sugihara-Seki* [207] and *Sugihara-Seki et al.* [208], in which the core proteins in the ESG were assumed to have a circular cylindrical shape and to be aligned in parallel to form a hexagonal arrangement based on recent detailed structural analyses of the ESG. They analyzed the motion of solute and solvent to estimate the filtration reflection coefficient as well as the diffusive permeability of the ESG. Later, *Zhang et al.* [238] studied osmotic flow through the ESG using a method developed by *Anderson and Malone* [13] for osmotic flow in porous membranes. Instead of a rigorous treatment of the hexagonal geometry of the cylinders, they adopted an approximation in which the geometry is replaced by an equivalent fluid annulus around each cylinder and estimated the osmotic reflection coefficient of the ESG. Further, *Akinaga et al.* [10] examined the charge effect on the osmotic flow for membranes with circular cylindrical pores by extending the formulation of osmotic flow developed by *Anderson and Malone* [13].

4.2.4.2 3-D Models

4.2.4.2.1 MODELS FOR THE PARACELLULAR PATHWAY OF THE PERIPHERAL MICROVESSEL WALL

Until 1984, 1-D models were based on random section electron microscopy. *Bundigaard* [37] was the first to attempt to reconstruct the 3-D junction strand ultrastructure using serial section electron microscopy. In his study, rat heart capillaries were analyzed using conventional 40–60 nm thin and 12 nm ultra thin, serial section electron microscopy. Large pores of 10–20 nm width and 20–80 nm length and small pores of 4–5 nm width and 5–30 nm length were observed. He sketched the latter pores as short discontinuities in the junction strands. In contrast to *Bundgaard*'s study, *Ward et al.* [225] examined the 3-D features of the junction strands of rat cardiac capillaries by using a goniometric tilting technique. After considering the tilting effects, they claimed that more than 70% of the random thin sections of the junction strands were actually open, and concluded that the pathway for water and small and intermediate-sized solutes was not formed by the interruption in continuous lines of membrane fusion, but by continuous junctional regions with an approximate opening width of 5 nm.

Based on the study of *Bundgaard* [37] and *Ward et al.* [225], the studies of *Tsay et al.* [215] and *Weinbaum et al.* [226] proposed a basic 3-D model for the interendothelial cleft. In their model, junctional pores were of three types:

a. A frequent circular pore of 5.5 nm radius,
b. A restricted rectangular slit of 44–88 nm length and 8 nm width, and
c. A large infrequent pore of 44–88 nm length and 22 nm width, which is the same gap width of the wide part of the cleft.

The principal predictions of this model are:

a. That infrequent larger breaks are most likely required to account for the measured L_p and the P to small and intermediate-sized solutes of radius from 0.5 to 2.0 nm,
b. That these large breaks must be accompanied by a sieving matrix only partially occupying the depth of the cleft at the luminal surface,
c. That neither junctional pore, restricted slit or fiber matrix models can by themselves satisfy the permeability and selectivity data, and
d. That 1-D models are a poor description of a cleft with infrequent large breaks since the solute will be confined to small wake-like regions on the downstream side of the junction strand discontinuities and thus not fill the wide part of the cleft.

The prediction in *Weinbaum et al.* [226] as to the likely geometry of the large pores in the junction strand was confirmed by an serial section electron microscopic study on frog mesenteric capillaries [8]. These serial reconstructions revealed rather long breaks, typically 150 nm wide, and the same gap width as the wide part of the cleft. The spacing between adjacent breaks is from 2140 to 4450 nm with an average of 2460 nm. A continuous narrow slit of roughly 2 nm width, which runs along the junctional strand, was also suggested based on goniometric tilting of their sections. The ~2 nm continuous slit was suggested by *Michel and Curry* [151] to be formed by the separation of the outer membrane leaflet due to the snug interlock loops of occludin molecules, provided that these loops from adjacent cells remain entirely extracellular. The 64 kDa transmembrane protein, occludin, was identified to be associated with the tight junction strands [96].

Evidence for a sieving matrix at the endothelial surface, the observation of surface glycocalyx at the luminal surface of the microvessel wall, was also provided by several studies described in the previous section [3,4,142,88]. According to these new experimental results, a modified, combined junction-orifice-fiber entrance layer model, which included a large orifice-like junctional break, a finite region of fiber matrix components at the entrance of the cleft and very small pores or slits in the continuous part of the junction strand, was developed by *Fu et al.* [88]. Figure 4.11 shows their 3-D model for the interendothelial cleft.

This combined junction-orifice-fiber entrance layer model predicted that for the measured hydraulic conductivity L_p to be achieved, the fiber layer must be confined to a relatively narrow region at the entrance to the cleft, where it serves as the primary molecular filter provided that the fiber matrix forms an ordered array and that the junction strand contains at least two types of pores, infrequent 150×20 nm large orifice breaks and a continuous about 2 nm narrow slit or closely spaced 1.5 nm radius circular pores. This model also provided an excellent fit for the hydraulic conductivity L_p and the diffusive permeability (P_d) data for solutes of size ranging from potassium to albumin for frog mesenteric capillaries (see Fig. 4.12). Due to the similarity in morphological wall structure of microvessels in different tissues [151], this 3-D model can be easily adapted to explain the permeability data in other types of microvessels such as the blood-brain barrier [135,136,237].

Fu et al., in another work [87], described a new approach to exploring junction strand structure. Instead of observing the junctional strand structure directly, they attempted to construct a detailed picture of the junctional strand as a transport barrier from a combined three-dimensional theoretical and experimental analysis of the diffusive wake formed by the spreading of low molecular weight tracers (such as lanthanum and sodium fluorescein) on the downstream side of the junction strand discontinuities. Important additional evidence in support of the small pore system proposed in [88] was obtained from time dependent studies of the penetration of lanthanum in the cleft on the abluminal side of the junction strand, confocal microscopic measurements of the spread of sodium fluorescein in the tissue surrounding a perfused microvessel and a theoretical model which describes the time dependent labeling behavior observed in these experiments [87]. The time dependent diffusion wake model of *Fu et al.* [87] provided a new interpretation of labeled tracer studies to define the permeability pathways for low molecular weight tracers which depend on the time dependent filling of the extravascular space. Another experimental difficulty is that the existence of fiber matrix components cannot be directly visualized in the wide part of the cleft, especially in *in vivo* experiments. For this reason, it has not been clear whether the junctional barrier or the fiber matrix is the primary structure in determining the selectivity of the capillary wall. In *Fu et al.* [85] a time dependent convective-diffusion wake model for high molecular weight tracers was proposed to design experiments that can test for the location of the molecular filter. Horseradish peroxidase (Stokes radius = 3 nm) was the test solute in their model. For this size of molecule, most of the tracer concentration difference between the lumen and the tissue was developed across the matrix layer. This observation may explain the common finding in electron microscopy that no large tracer is found in the interendothelial cleft. The result indicates that failure to find tracer in the cleft does not necessarily mean that the tracer does not cross the cleft pathway, but that, under the experimental

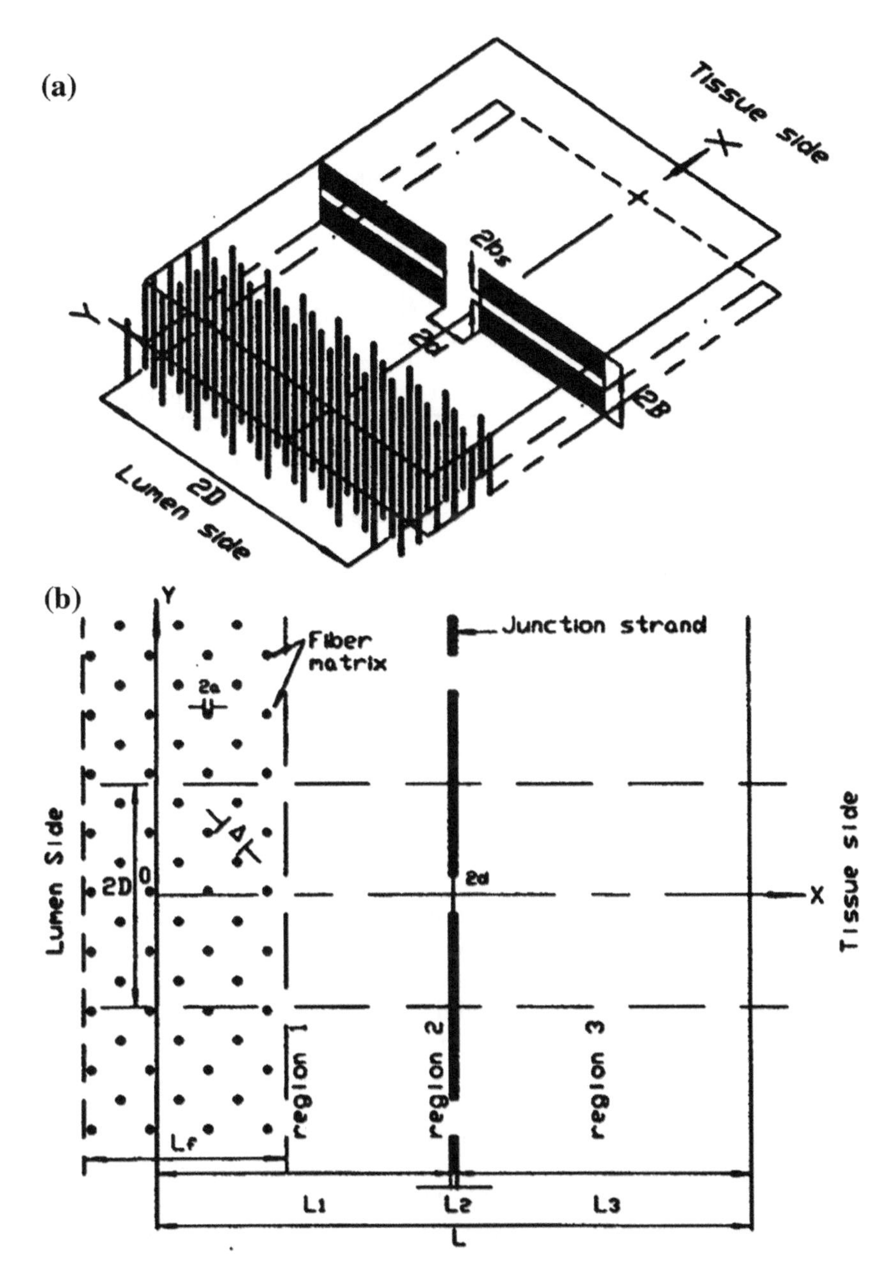

FIGURE 4.11 (a) Three-dimensional sketch of the junction-orifice-matrix entrance layer model for the interendothelial cleft. $2B$ is the width of the cleft. Large junction breaks observed in [8] are $2d \times 2B$, while the small continuous slit in the junction strand is $2b_s$. (b) Plane view of the model. Junction strand with periodic openings lies parallel to the luminal front of the cleft. L_2, depth of pores in junction strand. L_1 and L_3, depths between junction strand and luminal and abluminal fronts of the cleft, respectively. $2D$, distance between adjacent large junction breaks. At the entrance of the cleft on luminal side, glycocalyx is represented by a periodic square array of cylindrical fibers. a, radius of these fibers, Δ, gap spacing between fibers, and L_f, thickness of entrance fiber layer (from [85,87–89]).

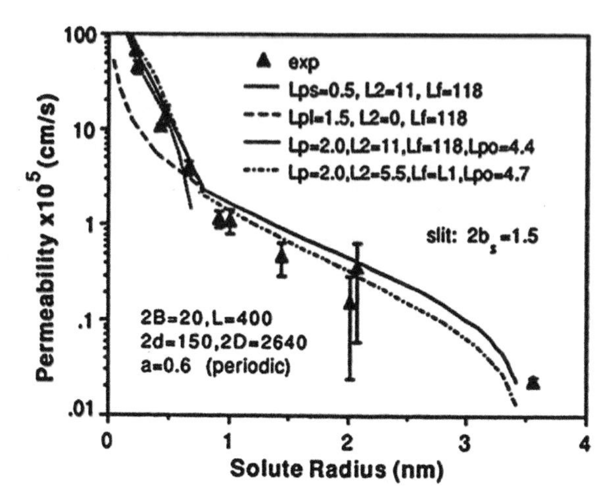

FIGURE 4.12 Comparison of measured permeability data for frog mesenteric microvessels with predictions from the 3-D junction-orifice-fiber matrix model [88]. The dimension for all the geometric parameters in the plot is nm. a is the fiber radius. The gap spacing between fibers $\Delta = 7$ nm. *Lp* is the hydraulic conductivity when the fiber is present and *Lpo* is that when the fiber is absent. *Lps* is the hydraulic conductivity contributed from the small pores and *Lpl* is that from the large pores. *Lp* in cm/s/cmH_2O.

conditions of most tracer experiments, the tracer concentration downstream from the matrix is too low to be detected.

4.2.4.2.2 MODELS FOR THE PARACELLULAR PATHWAY OF THE BLOOD-BRAIN BARRIER

Transport across the BBB includes both paracellular and transcellular pathways [174]. While large molecules cross the BBB through transcellular pathways, water and small hydrophilic solutes cross the BBB through the paracellular pathway [110]. The paracellular pathway of the BBB is formed by the endothelial surface glycocalyx, tight junction openings, the basement membrane (BM) filled with extracellular matrix and the openings between adjacent astrocyte foot processes (Fig. 4.5a). In addition to the endothelial tight junctions, the BM and the astrocyte foot processes provide a significant resistance to water and solute transport across the BBB.

Breakdown and increased permeability of the BBB are widely observed in many brain diseases such as stroke, traumatic head injury, brain edema, Alzheimer's disease, AIDS, brain cancer, meningitis, etc. [19,21,22,25,42,71,95]. Although numerous biochemical factors are found to be responsible for the breakdown of the BBB in disease, the quantitative understanding of how these factors affect the structural components of the BBB to induce leakage is poor. On the other hand, to design therapeutic drugs with better transport properties across the BBB relies greatly on this understanding. Therefore, it is important to investigate how the structural components in the paracellular pathway of the BBB affect its permeability to water and solutes through mathematical modeling.

Extended from a previous three-dimensional model for studying the transport across the peripheral microvessel wall with endothelium only [47,88], *Li et al.* [136] developed a new model for the transport across the BBB, which included the BM and wrapping astrocyte foot processes. The simplified model geometry is shown in Fig. 4.13. This is an enlarged view of the part near the tight junction shown in Fig. 4.5a. At the luminal side, there is an endothelial surface glycocalyx layer with a thickness of L_f from 100–400 nm under normal physiological conditions [4,194,203,223]. Between adjacent endothelial cells, there is an interendothelial cleft with a length of L ~500 nm and a width of $2B$ ~20 nm [8,194]. In the interendothelial cleft, there is a L_{jun} (~10 nm) thick junction strand with a continuous slit-like opening of width $2B_s$, which varies depending on the location of the cerebral microvessels (~1–10 nm). The distance between the junction strand and luminal front of the cleft is L_1. At the tissue side of the cleft, a BM separates the endothelium and the astrocyte foot processes. The thickness of

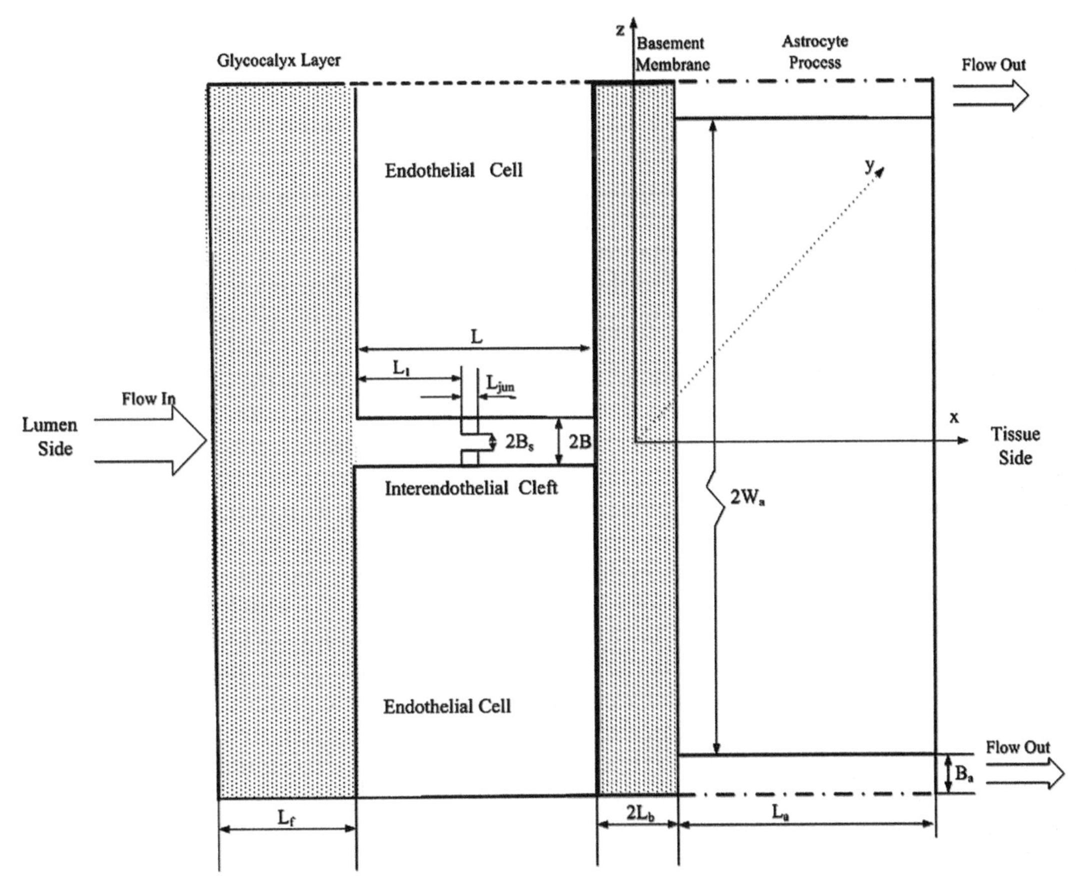

FIGURE 4.13 Model geometry for the paracellular pathway of the BBB (not to scale). The thickness of the endothelial surface glycocalyx layer is L_f. The interendothelial cleft has a length of L and a width of $2B$. The length of the tight junction strand in the interendothelial cleft is L_{jun}. The width of the small continuous slit in the junction strand is $2B_s$. The distance between the junction strand and luminal front of the cleft is L_1. The width of the basement membrane is $2L_b$ and the length of the astrocyte foot processes is $2W_a$. The cleft between astrocyte foot processes has a length of L_a and a width of $2B_a$. The surface glycocalyx layer and the endothelial cells are defined as the Endothelium only while the BBB is defined to include the endothelium, the basement membrane and the astrocytes. Redrawn from [136].

the BM is $2L_b$ (20–40 nm) and the length of the astrocyte foot processes is $2W_a$ (~5000 nm). Between adjacent astrocyte foot processes, there is a cleft with length L_a (~1000 nm) and width $2B_a$ (20–2000 nm). The anatomical parameters for the BBB structural components were obtained from electron microscopy studies in the literature.

Unlike the peripheral microvessel wall, the endothelium of the BBB has negligible large discontinuous breaks in the junction strand of the interendothelial cleft and the small slit in the junction strand is assumed to be continuous [110]. As a result, the cross-sectional BBB geometry is the same along the axial direction (y direction in Fig. 4.13) and thus the model could be simplified to 2-D (in x, z plane). It could be further simplified to a unidirectional flow in each region due to very narrow clefts and the BM. In addition, the curvatures of the BM and the endothelium can be neglected because their widths are much smaller than the diameter of the microvessel. The fluid flow in the cleft regions of the BBB was approximated by the Poiseuille flow while that in the ESG and BM were obtained by the Darcy and Brinkman flows, respectively. Diffusion equations in each region were solved for the solute transport. After solving for the pressure, water velocity and solute concentration profiles, the hydraulic conductivity, L_p, and solute permeability, P, can be calculated.

Figure 4.14a shows the model predictions for L_p as a function of tight junction opening B_s when the BM has different fiber densities. The parameter K_b is the Darcy permeability in the BM. When the fiber density in the BM is the same as that in the ESG, $K_b = 3.16\ \text{nm}^2$. The green line in Fig. 4.14a shows the case of peripheral microvessels with only endothelium. When B_s increases from 0.5 nm to 2 nm, L_p will increase by ~20-fold. In contrast, when the endothelium is wrapped by the BM and the astrocytes as for the BBB, increase in B_s from 0.5 nm to 2 nm only induces a 5-fold increase in L_p when the fiber density in the BM is the same as that in the ESG (dash-dot-dash line). If the fiber density in the BM is 10 times that in the ESG, the increase in L_p (solid line) is only 1.6-fold, while if the fiber density in the BM is 1/10 of that in the ESG, the increase is 12-fold in L_p (dashed line). Even at a large B_s of 5 nm, when the BM is filled with the same density fibers as in the ESG, the BBB permeability is only 17% of that of the endothelium. This percentage can be as low as 2% if the fiber density in the BM is 10 times of that in the ESG. Figure 4.14b shows the model predictions for L_p as a function of the ESG thickness L_f. The green line is for the case of endothelium only, while the solid line is for that of the BBB. We can see that a decrease in L_f from 400 to 0 nm increases L_p by 3-fold in the case of endothelium only, while in the case of the BBB, the increase is only 25% in L_p with the protection of the BM and the astrocytes. Similar results are predicted for the solute permeability [136]. These results indicate that the BM and astrocytes of the BBB provide a great protection to the CNS under both physiological and pathological conditions. However, these unique structures also impede the delivery of drugs to the brain through the BBB.

4.2.5 Endothelial Surface Glycocalyx

4.2.5.1 *Composition, Thickness and Structure of Endothelial Surface Glycocalyx (ESG)*

The surface of endothelial cells (ECs) is decorated with a wide variety of membrane-bound macromolecules, which constitute the ESG. From glycoproteins bearing acidic oligosaccharides and terminal sialic acids (SA), to proteoglycans along with their associated glycosaminoglycan (GAG) side chains, the polyanionic nature of these bound constituents imparts to it

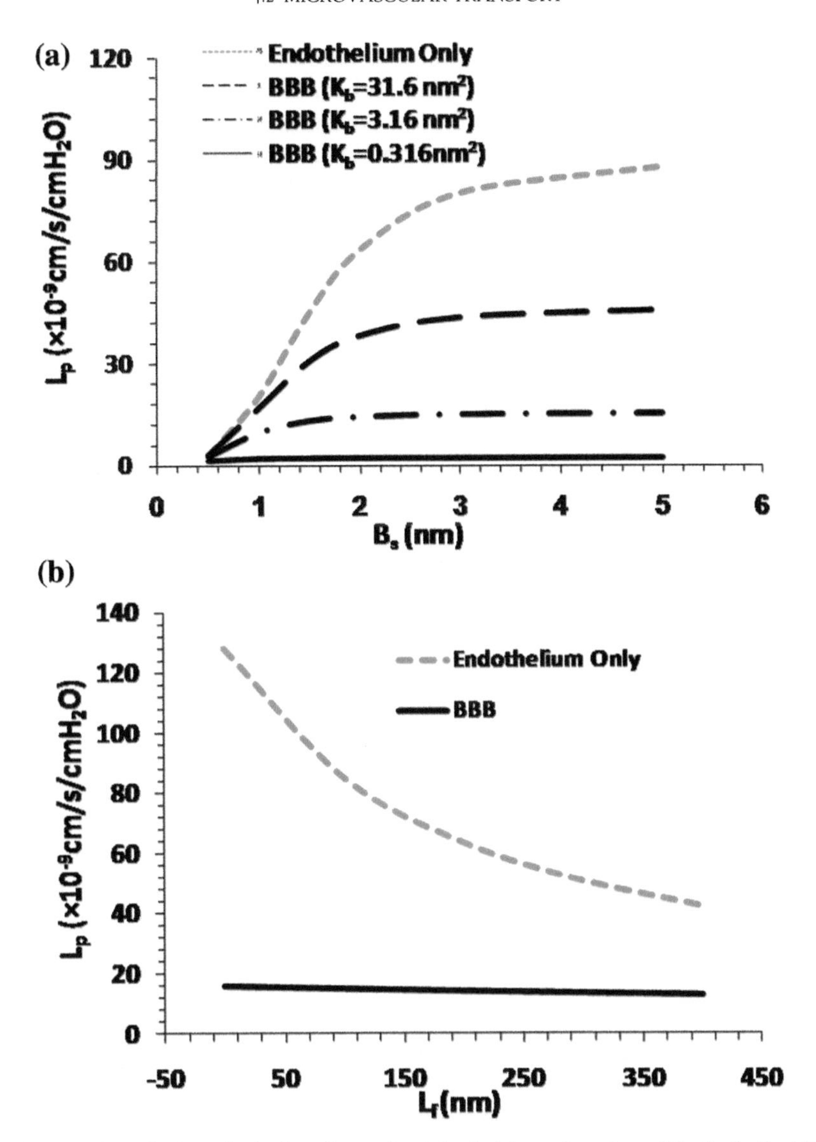

FIGURE 4.14 Model predictions for hydraulic conductivity L_p (a) as a function of B_s, the half width of the small slit in the junction strand under two cases: when considering transport across the endothelium only (Endothelium only, green line), and when considering transport across the entire BBB (BBB). In the BBB case, three different fiber densities were considered for the basement membrane: the same as the fiber density in the surface glycocalyx layer (K_b=3.16 nm^2, the dash-dot-dash line), ten times lower (K_b=31.6 nm^2, the dashed line) and higher (K_b=0.316 nm^2, the solid line); (b) as a function of the surface glycocalyx layer thickness L_f. Redrawn from [136].

a net negative charge. Under physiological conditions, an extended endothelial surface layer arises from the association of components of the ESG with blood-borne molecules [4,178]. Plasma proteins, enzymes, enzyme inhibitors, growth factors and cytokines, through cationic sites in their structure, as well as cationic amino acids, cations and water, all associate with this matrix of biopolyelectrolytes [30,167]. ECs actively regulate the content and physicochemical properties of GAGs on their surface by having high rates of continuous metabolic turnover, which allow adaptation to changes in the local environment [16,170,222]. GAGs are linear polydisperse heteropolysaccharides, characterized by distinct disaccharide unit repeats [119]. Specific combinations of these give rise to different GAG families, such as heparan sulfate (HS), chondroitin/dermatan sulfate (CS) and hyaluronic acid or hyaluronan (HA) found on ECs [166]. Proteoglycans are proteins that contain specific sites where sulfated GAGs are covalently attached [119,180,210]. This ESG plays an important role in regulating vascular permeability, attenuating interactions between circulating blood cells and the ECs, as well as sensing hydrodynamic changes in the blood flow [50,59,63,161,178,180,211,231]. Damage to and modification of the endothelial SGL has been found in many diseases, such as diabetes, ischemia, myocardial edema, chronic infectious diseases and atherosclerosis [26,43,180,219,221,227]. Due to its crucial role in maintaining vascular functions, the structure and composition of the ESG has been widely studied since the 1960s [4,108,142,166,203,223]. The thickness of the ESG has been observed to range from less than 100 nm to 1 μm for the microvessels in different tissues and species, in studies using different preparation and observing methods [98,140,151,203,223,235]. Although the ESG thickness varies, its density and organization have been reported to be the same among different tissues and species. The glycocalyx fiber radius is ~6 nm and gap spacing between fibers ~8 nm [17,203].

4.2.5.2 *Role of Endothelial Surface Glycocalyx*

4.2.5.2.1 MOLECULAR SIEVE

The model in Fig. 4.11 [88] provides a quantitative description of flows through a real endothelial barrier (frog mesenteric capillary) in terms of directly measured geometry of the junctional strand and reasonable estimates of glycocalyx structure. A convective-diffusion model for macromolecule (albumin) transport [116] demonstrates that the primary effect of the fiber matrix at the cleft entrance is to modify the two-dimensional spread of water and solutes through the cleft. In the absence of matrix ($L_f=0$ in Fig. 4.11), the pressure and albumin concentration fall symmetrically about a break in the junctional strand located halfway between the lumen and the tissue (not shown here). However, in the presence of the matrix, Fig. 4.15 shows that more than 90% of the drop occurs across the surface fiber matrix layer when the thickness $L_f=150$ nm [116]. In contrast to the resistance of the matrix to water and large solutes such as albumin (radius = 3.6 nm), the matrix acting as the macromolecular filter offers little resistance to small solutes of less than 1 nm radius [88].

4.2.5.2.2 BUILDING OF STARLING FORCE ACROSS THE CAPILLARY WALL

Nearly every contemporary physiology text has explained Starling's hypothesis in terms of a classic Landis-Starling diagram, in which there is net filtration in the capillaries on the arterial side, and a nearly equal reabsorption on the venous side due to osmotic forces, leaving a small net positive filtration that accounts for the lymph flow. This widely accepted view was first seriously challenged in the provocative review by *Levick* [131]. Levick shows that if one uses the latest measurements of the local average interstitial oncotic pressure in calculating the isogravimetric pressure opposing filtration, one finds that in nearly every tissue (except the renal capillaries and the gut mucosa) the pressure in the postcapillary venules

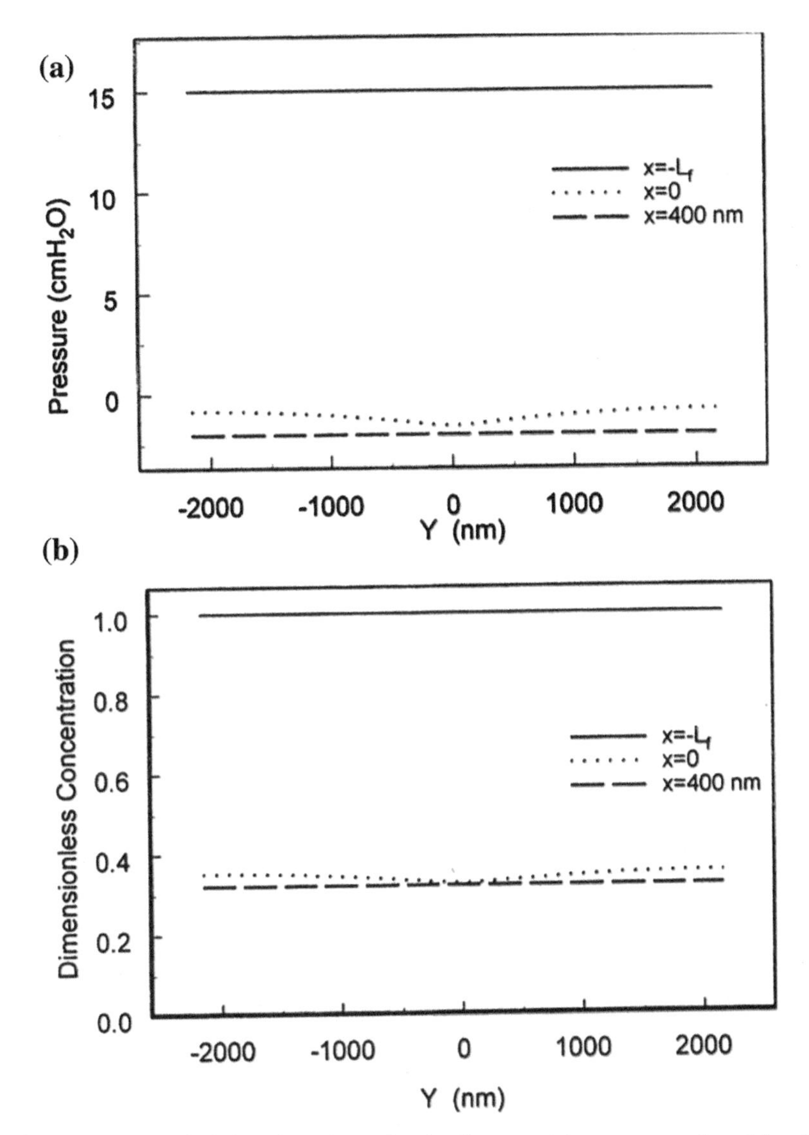

FIGURE 4.15 Effect of endothelial surface glycocalyx (or fiber matrix) on pressure drop (a) and concentration drop (b) from the luminal to abluminal side of an intercellular cleft centered on a discontinuity in junction strand. $x=-L_f$ is the luminal surface, $L_f=150$ nm is the fiber layer thickness. $x=0$ is the cleft entrance and $x=400$ nm is the cleft exit (see Figure 4.11). y is distance along strand [116].

significantly exceeds the isogravimetric pressure and there should be net filtration rather than reabsorption. This fundamental paradox was the focus of a recent review by *Michel* [150], which summarized our latest understanding of microvascular fluid exchange.

Based on arguments presented in *Michel* [150], *Hu and Weibaum* [116] have proposed a new hypothesis which suggests that the oncotic force occurs primarily across the protein sieving layer at the endothelial surface, the surface glycocalyx. In contrast, the filtration pressure drop occurs across both this protein sieving layer and across the interendothelial cleft with its junction strands. This model showed that the paradox described by *Levick* [131] can be resolved if the oncotic force across the surface matrix layer is determined by the local difference in protein concentration between the plasma and the fluid on the tissue side of the matrix layer, rather than the interstitial fluid in the tissue itself. The results from the convective-diffusion model in *Hu and Weinbaum* [116] show that coupling of water flow to albumin flux on the tissue side of the matrix could give rise to a non-uniform distribution of albumin concentration and a corresponding non-uniform distribution of effective osmotic pressure. A similar model for oncotic pressures opposing filtration across rat microvessels [6] further confirms the hypothesis that colloid osmotic forces opposing filtration across non-fenestrated continuous capillaries are developed across the endothelial glycocalyx, and that the oncotic pressure of interstitial fluid does not directly determine fluid balance across microvascular endothelium.

4.2.5.2.3 MECHANOSENSOR

The primary evidence that supports a major role for the ESG in mechanotransduction comes from experiments in which enzymes were used to selectively degrade specific components of the ESG, followed by a reassessment of function, or alternatively from the use of bathing solutions without plasma proteins, where the ESG is compromised [4] and structural organization examined in response to flow shear stress (FSS). *Florian et al.* [80] used the enzyme heparinase III to selectively degrade the HS component of bovine aortic endothelial cell GAGs *in vitro*, and observed that the substantial production of NO induced over 3 hours by steady or oscillatory FSS (20 or 10 ± 15 dyne/cm^2), could be completely inhibited by an enzyme dose that removed only 46% of the fluorescence intensity associated with a HS antibody. The enzyme did not degrade CS and displayed negligible protease activity. It was also demonstrated that receptor mediated NO induction by bradykinin and histamine were not affected by the enzymatic treatment, demonstrating that eNOS activity was not impaired directly by the enzyme. In an earlier study, the enzyme neuraminidase was used to remove SA residues from saline-perfused rabbit mesenteric arteries, and it was observed that flow-dependent vasodilation was abolished by a 30-minute enzymatic pretreatment [177]. Because flow-dependent vasodilation is mediated by NO release in many arteries, this study suggested that SA also contributes to FSS-induced production of NO.

Similarly, *Hecker et al.* [112] showed that when intact segments of rabbit femoral arteries were pre-treated with neuraminidase, FSS-induced NO production was inhibited. They also demonstrated that the same enzyme treatment had no effect on another hallmark response of ECs, the FSS-induced production of prostacyclin (PGI_2) [83]. This study illustrated the fact that there are multiple mechanisms of mechanotransduction and not a single mechanotransducer. In a more recent study, the enzyme hyaluronidase was used in isolated canine femoral arteries to degrade HA from the ESG, and a significant inhibition of FSS-induced NO production was demonstrated [156].

A recent *in vitro* study using bovine aortic endothelial cells showed that the enzyme chondroitinase, employed to selectively degrade CS, did not inhibit the characteristic FSS-induced NO production, but treatment with neuraminidase, hyaluronidase or heparinase completely blocked the response [168]. This study, and the earlier one by *Florian et al.* [80], illustrated the specificity of the ESG GAG components in mechanotransduction. Because the glycocalyx is a complex, multicomponent chemical structure, the results of these experiments are subject to several interpretations. The proteoglycans of the ESG can be linked to both the 'decentralized' and 'centralized' mechanisms of mechanotransduction put forth by *Davies* [66]. Syndecans that contain both HS and CS have an established association with the cytoskeleton [214], and through it, they can decentralize the signal by distributing it to multiple sites within the cell (i.e., nucleus, organelles, focal adhesions, intercellular junctions). Significantly, the platelet-endothelial cell adhesion molecule (PECAM-1) associates with the cytoskeleton through catenins, and has been linked to shear-induced eNOS activation [73,104]. In terms of centralized transduction, it is noteworthy that glypicans that contain HS, but not CS, are linked to caveolae where eNOS resides along with many other signaling molecules [210]. The observation that depletion of HS, but not CS, inhibits shear-induced NO production, favors a glypican-caveolae-eNOS mechanism. It is also important to note that hyaluronic acid binds to its CD44 receptor that is localized in caveolae [81,200]. This provides a link between HA and shear-induced NO. The role of sialic acid that is removed by neuraminidase is less clear, but it is known that CD44 can have oligosaccharides (that are capped by SA) attached to it [81]. Of course, all of the ESG components that have been investigated provide net negative charges to the surface layer that enhance hydration and extension of the multicomponent structure in aqueous media. Loss of charge through enzyme degradation could lead to partial collapse of the integrated structure and reduction of fluid shear sensing [213]. *Pahakis et al.* [168] also showed that none of the four ESG-degrading enzymes used had an inhibitory effect on FSS-induced PGI_2 production. This surprising observation suggests that the transduction machinery for PGI_2 resides in a location distinct from the NO machinery.

4.2.5.3 *Effect of Charge Carried by Endothelial Surface Glycocalyx*

Due to the composition of the ESG, it carries a negative charge which would affect the permeability and selectivity of the microvessel wall to water and solutes. Previously, a simple 1-D Donnan-type model had been proposed to describe the effect of charge on microvessel permeability [56,64,117,151]. It was based on a Donnan equilibrium distribution of ions, which exists as a result of retention of negative charges on the capillary membrane [32,34,67–69]. Later, an electrochemical model was proposed by *Damiano and Stace* [65] for the transport of charged molecules through the capillary glycocalyx without considering transport through the cleft region. To investigate the charge selectivity on microvessel permeability, *Fu et al.* [89] extended the 3-D junction-orifice-fiber matrix model developed by *Fu et al.* [88] for the interendothelial cleft to include a negatively charged glycocalyx layer at the entrance of the interendothelial cleft. Both electrostatic and steric exclusions on charged solutes are considered at the interfaces of the glycocalyx layer between the vessel lumen and between the endothelial cleft. The effect of electrostatic interactions between charged solutes and the matrix on solute transport is also described within the glycocalyx layer. Their model can successfully explain the observations in [151]. Recently, an electrodiffusion-filtration model was developed to describe the transport of negatively charged macromolecules, bovine serum albumin, across venular microvessels in frog mesentery [47]. A very interesting prediction is that the convective component of albumin transport is greatly diminished by the presence of a negatively charged glycocalyx. Most recently, *Li et al.* [135] have developed a model for the charge effect of the ESG and the basement membrane between the endothelium and astrocyte foot processes on the transport across the blood-brain barrier.

Bhalla and Deen [31] studied the effects of charge on osmotic reflection coefficients of macromolecules in porous membranes. *Sugihara-Seki et al.* [209] proposed an electrostatic model to predict the effects of surface charge on the osmotic reflection coefficient of charged spherical solute across the ESG, based on the combination of low-Reynolds-number hydrodynamics and a continuum description of the electric double layers. The ESG was assumed to consist of identical circular cylinders with a fixed surface charge, aligned parallel to each other so as to form an ordered hexagonal arrangement. Their model predicts that the charge of the ESG contributes significantly to the microvessel reflection coefficient to albumin, which was reported in [151].

4.2.6 Transport Across Fenestrated Microvessels

The role of a matrix layer associated with fenestrae (see Fig. 4.3) has been analyzed in detail by *Levick and Smaje* [133]. The size and frequency of fenestration varies from 50 to 60 nm radius in intestinal mucosa, synovium and submandibular gland with a density of 2–5 per μm^2 to 60–88 nm radius and densities of 20–30 per μm^2 in the renal vasculature (excluding the glomerulus). *Levick and Smaje* [133] estimated that, for the typical thickness of the diaphragm in fenestrations of 5 nm, the values of exchange area per path length (A_{fen}/L) (7,000–17,000 cm^{-1}) are close to two orders of magnitude higher than required to account for measured small solute permeabilities and hydraulic conductivities of fenestrated capillaries [182]. *Levick and Smaje* [133] therefore concluded that most of the resistance to water and solute movement across fenestrations must lie within fiber matrix structures that are more than an order of magnitude thicker than the diaphragm. The diaphragm, therefore, might support an overlying glycocalyx and underlying basement membrane containing a mixture of thick and thin fibers [184,185].

Levick [131] also suggested that, in general, the endothelial barriers should be described as a cellular layer sandwiched between surface glycocalyx and an interstitial matrix. In the case of an endothelial barrier with relatively low resistance, the interstitial resistance may be large enough to form a significant fraction of the resistance to water and solute movement between blood and lymph. For example, in synovial joint, *Levick* [131] has estimated that the capillary wall and interstitium contribute approximately equally to the resistance to fluid flow. A similar analysis may apply to the endothelium when gaps form between endothelial cells or through the cells in the presence of inflammatory mediators. A detailed review of flow through the interstitium and fiber matrix structures in relation to the chemical composition of the interstitium is given by *Levick* [132].

4.2.7 Transport in Tissue Space (Interstitium)

The relationship between the structural components of the microvessel wall and its permeability coefficients to solutes of different sizes has been systematically studied in single perfused microvessels [5,54,61,85–91]. However, much less use has been made of single perfused capillaries to investigate, systematically, the relationship between the composition and physical-chemical properties of the interstitium and the interstitial solute diffusion coefficients for probes with a range of molecular sizes. *Fu et al.* [93] extended a method recently developed in [7] to measure both solute capillary permeability coefficients and solute tissue diffusion coefficients from the rate of tissue solute accumulation and the radial concentration gradients measured around individually perfused microvessels in frog mesentery. The unique feature of the experimental design was the use of a confocal microscope to scan the tissue and the microvessel lumen to measure the local concentration of test solute within a series of tissue slices that formed the entire depth of the mesentery. This approach overcomes many limitations of previous attempts to study local capillary and tissue transport, because the relative solute concentration within small volume elements in the tissue is measured directly, allowing the solute concentration profiles at every depth within the tissue to be analyzed. Thus the concentration difference of the test solute across the microvessel wall is measured directly. Further, the method enables comparison of tissue diffusion coefficients calculated using the detailed tissue concentration profiles with those based on average tissue concentrations estimated from the two dimensional projection of the concentration gradients around microvessels as recorded when a video camera focused on the tissue is used to investigate the spread of dye [29,46,82].

The measurements for small and intermediate-sized solute movement in the interstitium of frog mesentery presented some interesting features. At 20°C, the free diffusion coefficient of the small solute sodium fluorescein ($D_{free}^{sodium\ fluorescein}$) is $5.40 \times 10^{-6}\,cm^2/s$ [87], and the free diffusion coefficient of the intermediate-sized solute FITC-α-lactalbumin ($D_{free}^{FITC\text{-}\alpha\text{-}lactalbumin}$) is $1.07 \times 10^{-6}\,cm^2/s$ [5]. In *Fu et al.* [87], the corresponding effective interstitial diffusion coefficients were $D_t^{sodium\ fluorescein} = 30\%\ D_{free}^{sodium\ fluorescein}$ [7,87,93]. $D_t^{FITC\text{-}\alpha\text{-}lactalbumin} = 27\%\ D_{free}^{FITC\text{-}\alpha\text{-}lactalbumin}$. The ratio of $D_t^{sodium\ fluorescein}$ to $D_t^{FITC\text{-}\alpha\text{-}lactalbumin}$, 5.6, is about 1/4 of the ratio of the corrected capillary wall permeability $P^{sodium\ fluorescein}$ to $P^{FITC\text{-}\alpha\text{-}lactalbumin}$ in paired measurements reported in [86]. This result indicates that the structural components in the interendothelial cleft of the microvessel wall are more selective than those in the mesenteric interstitium. Once an α-lactalbumin molecule crosses the microvessel wall, it can diffuse through the interstitial structures, which reduce its free diffusion coefficient to the same extent as for a solute as small as sodium fluorescein. The structural components of the interendothelial cleft are plasma membranes composing the cleft wall, the surface glycocalyx of endothelial cells and the tight junction strands inside the cleft. The structural elements of the interstitium are collagens, elastins, polysaccharides (e.g., hyaluronate, glycosaminoglycans) and plasma proteins extravasated from the microvessel [18,131,132]. *Fu et al.*'s results confirm the hypothesis that, under the conditions of their experiments, the spaces between the fibers in the endothelial cell glycocalyx form the primary size selective structure for solutes up to the size of α-lactalbumin between the lumen and the tissue. In other words, the glycocalyx has smaller inter-fiber distances and is likely to be more highly organized than fibers in the interstitium. Why does the interstitial transport not strongly depend on the molecular weight while the permeability of the microvessel wall to small ions and serum albumin can be different by three orders of magnitude in frog mesentery? While this question is far from being answered, one possible explanation is that the microvessel wall must maintain large differences in plasma protein osmotic pressure and these large differences can only be established by a barrier which is highly selective between small solutes and large solutes.

4.3 MODULATION OF MICROVASCULAR TRANSPORT

The molecular basis for the passage of molecules at the level of the breaks in tight junctions is more likely to be the localized absence of cell-cell contacts with corresponding loss of a closely regulated molecular sieve as suggested by *Weinbaum* [226], *Fu et al.* [88] and *Michel and Curry* [151,154]. Thus the junction break-surface matrix model suggests independent mechanisms to regulate the permeability properties of the microvessel wall. The junction break size and frequency are likely to involve regulation of cell-cell attachment via occludin and other junction proteins, including the cadherin-associated junctions (Fig. 4.4, from [151]). On the other hand, the regulation of glycocalyx density and organization is likely to involve interaction of the molecules forming the cell surface with cytoskeleton, and with circulating plasma proteins. Some of the cellular mechanisms underlying these interactions are reviewed in [14,151]. Under physiological and pathological conditions, microvessel permeability can be regulated acutely and chronically by mechanisms that are underway to being understood.

4.3.1 Permeability Increase

Increased microvessel permeability is usually defined to be the large increase in permeability to fluid and plasma proteins that occurs in acutely and chronically inflamed tissues. Generalized small increases in microvessel permeability have been said to occur in a number of systemic diseases (e.g., diabetes, hypertension and rheumatoid arthritis). In addition, there are physiological variations in microvessel permeability. Studies on amphibians and on rats have shown that microvessel permeability is increased by atrial natriuretic peptide (ANP). Working on single perfused frog mesenteric capillaries, *Meyer and Huxley* [147] have shown that ANP increases hydraulic conductivity by a mechanism that raises particular guanylate cyclase in the endothelium. *Renkin and Tucker* [183] reported the expansion of plasma protein from the vascular to the interstitial compartment by a mechanism that is dependent on ANP. Because it does not appear to be accompanied by a fall in the reflection coefficient to albumin, *Renkin and Tucker* [183] suggested that it might involve either an increase in the porous area of the capillary wall without loss of molecular selectivity, or an increase in the transport of macromolecules by vesicles (Fig. 4.3).

Another type of physiological stimulus, wall shear stress, has also been reported to increase microvessel permeability [151,236]. *Kajimura et al.* [121] used a microperfusion technique to vary the flow velocity in frog mesenteric microvessels and demonstrated that permeability to potassium ions was indeed increased as flow velocity increased. In a very different preparation, *Pallone et al.* [171] and *Turner and Pallone* [217] have shown that the permeability of isolated perfused descending vasa recta to small hydrophilic solutes increases with increasing perfusion rate. If permeability is generally flow dependent, it has important physiological implications particularly for the transport of small hydrophilic solutes to and from the tissues. In muscle, for example, it could account for the large increase in glucose uptake that occurs during exercises without having to invoke large increases in the number of capillaries perfused within the tissue.

The pattern of increased microvessel permeability in inflammation varies between different inflammatory mediators. After mild thermal or chemical injury to the skin, there was an initial increase in permeability lasting for 15–30 min, after which leakage was reduced for a short period before increasing for a second sustained phase that might last for several hours [228]. Exogenously administered mediators such as histamine, serotonin, bradykinin and cytokines have effects that are similar to the initial phase of inflammation. Figure 4.16 shows the typical pattern of permeability increase by these mediators. In the figure, the effect of a cytokine, vascular endothelial growth factor (VEGF), on water and small solute permeability

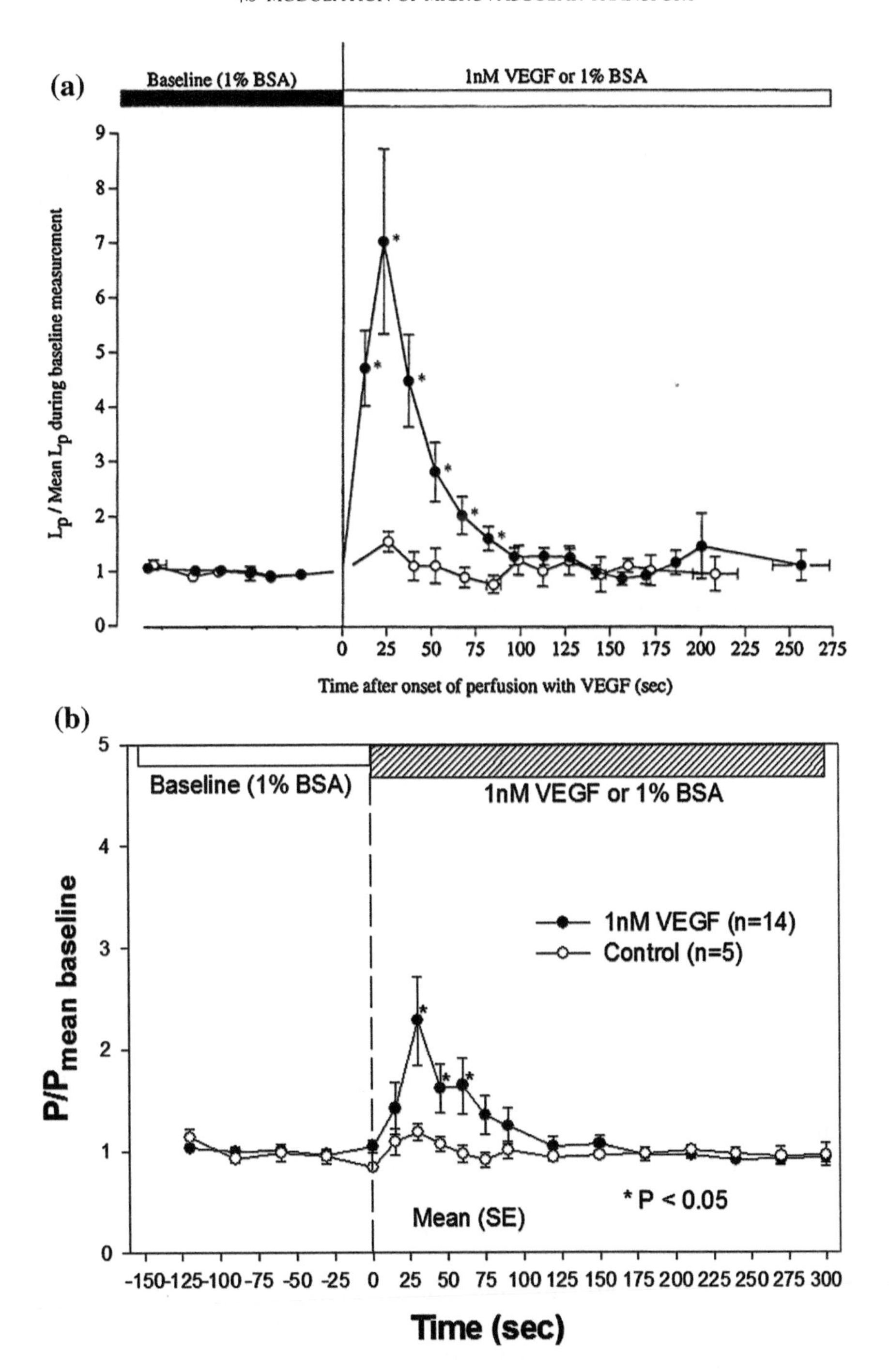

FIGURE 4.16 Vascular Endothelial Grow Factor (VEGF) effect on (a) hydraulic conductivity L_p [23], and (b) small solute sodium fluorescein permeability [91] of frog mesenteric microvessels.

of frog mesenteric microvessels is shown. Exposure to 1 nM VEGF transiently increased L_p within 30 seconds (to 7.8-fold of baseline values) and returned to control within 2 min [23]. Another study by *Fu and Shen* [91] of the response pattern of the small solute sodium fluorescein (MW = 376, radius = 0.45 nm) showed that the permeability $P^{sodium\ fluorescein}$ to VEGF was similar to that of L_p to VEGF and the peak at ~30 seconds is 2.3-fold that of the baseline value. Based on these measured permeability data, the theoretical model [91] predicted that the most likely structural change in the interendothelial cleft by VEGF is ~2.5-fold transient increase in its opening width ($2B$) and partial degradation of surface glycocalyx layer (L_f) (Fig. 4.11).

Using light microscopy and carbon labeling, *Majno et al.* [144] showed that the increased permeability induced by histamine and serotonin was confined to the venules and did not involve the true capillaries. Subsequent electron microscopy by *Majno and Palade* [143] revealed that leakage from the venules was associated with the development of openings or gaps in the venular endothelia. Subsequent work in a number of different laboratories confirmed that histamine, serotonin, bradykinin and many other mediators opened gaps in the endothelium of the postcapillary venules, but did not appear to influence the ultrastructure of the capillaries. The later phase of increased permeability after thermal or chemical injury does involve both capillaries and venules. This appears to involve the development of gaps or openings in the endothelium [51]. More recently, it has been shown that increased capillary (as well as venular) permeability occurs as a result of contact with dead tissue due to endogenous mediators from dying tissue [120]. Gaps and fenestrae in capillary endothelia are also induced by VEGF [186].

4.3.2 Permeability Decrease

Many studies have found that increased intracellular adenosine 3′,5′-cyclic monophosphate (cAMP) can block the inflammatory response in a variety of experimental models. Although the mechanism is not well understood, numerous studies have confirmed that elevated cAMP modulates the paracellular permeability by an increase in tight junction number or complexity of the intercellular cleft [9]. *Adamson et al.* [9] demonstrated that elevation of endothelial cell intracellular cAMP levels by simultaneous adenylate cyclase activation (forskolin) and phosphodiesterase (PDE IV) inhibition (rolipram) reduced capillary hydraulic permeability (L_p) to 43% of baseline values within 20 min (Fig. 4.17a). At the same time, using electron microscopy, they also found that the number of tight junction strands increased from a mean of 1.7 to a mean of 2.2 per cleft (Fig. 4.17b). More junction strands in the cleft will increase the tortuosity of the cleft pathway for water and solutes. *Fu et al.* [86], in their parallel study on the effect of cAMP on solute permeability, indicated that in 20 min, the permeability for the intermediate-sized solute α-lactalbumin (radius = 2.01 nm) $P^{\alpha\text{-}lactalbumin}$ and for the small solute sodium fluorescein (radius = 0.45 nm), P^{sf} decreased to 64% and 67% of the baseline value, respectively.

To investigate the nano-micro-structural mechanisms of decreasing microvascular permeability induced by the enhancement of intraendothelial cAMP levels, *Fu and Chen* [90] extended the previous analytical model developed by *Fu et al.* [88] for the interendothelial cleft to include multiple junction strands in the cleft and an interface between the surface glycocalyx layer and the cleft entrance. Based on the electron microscopic observations of

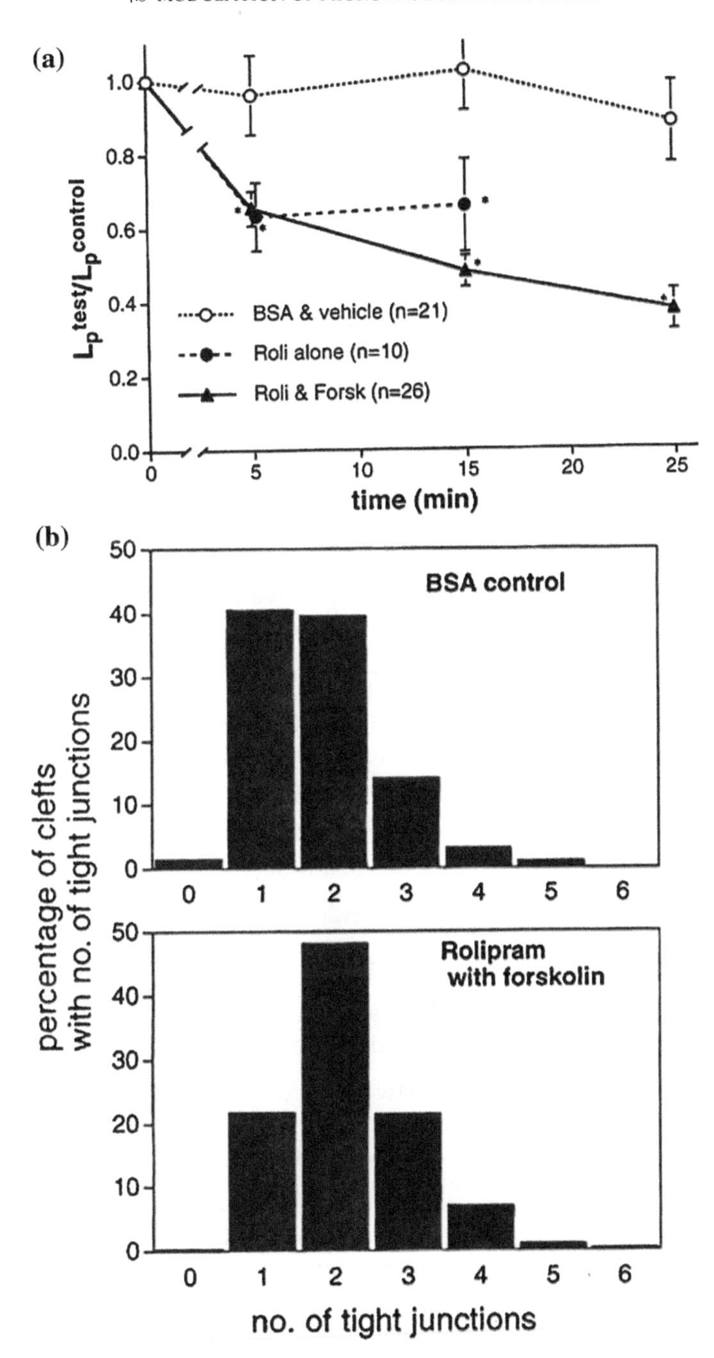

FIGURE 4.17 cAMP effect on microvessel hydraulic conductivity L_p and number of tight junction strands in the interendothelial cleft. (a) Ratio of test L_p to control L_p as a function of time after perfusion with either rolipram alone or rolipram with forskolin induced a fall in L_p. (b) Distribution showing variation in number of tight junctions per endothelial cleft before and after perfusion of rolipram and forskolin (from [9]).

Adamson et al. [9] that elevation of intracellular cAMP levels would increase number of tight junction strands, a numerical method was applied to test the case in which there are two junction strands in the cleft, and large discontinuous breaks and a small continuous slit in each strand. Results from this two-junction-strand and two-pore model can successfully account for the experimental data for the decreased permeability to water [9], small and intermediate-sized solutes [86] by cAMP.

4.3.3 Permeability Increase and Decrease

To investigate the structural mechanisms by which elevation of the intraendothelial cAMP levels abolishes/attenuates the transient increase in microvascular permeability by vascular endothelial growth factor (VEGF), *Fu et al.* [94] examined the cAMP effect on VEGF-induced hyperpermeability to a small solute, sodium fluorescein (Stokes radius = 0.45 nm) $P^{\text{sodium fluorescein}}$, an intermediate-sized solute, α-lactalbumin (Stokes radius = 2.01 nm) $P^{\alpha\text{-lactalbumin}}$, and a large solute, albumin (BSA, Stokes radius = 3.5 nm) P^{BSA} on individually perfused microvessels of frog mesenteries. After 20 min pretreatment with 2 mM of the cAMP analogue, 8-bromo-cAMP, the initial increase by 1 nM VEGF was completely abolished in $P^{\text{sodium fluorescein}}$ (from a peak increase of 2.6 ± 0.37 times control with VEGF alone to 0.96 ± 0.07 times control with VEGF and cAMP), in $P^{\alpha\text{-lactalbumin}}$ (from a peak increase of 2.7 ± 0.33 times control with VEGF alone to 0.76 ± 0.07 times control with VEGF and cAMP), and in P^{BSA} (from a peak increase of 6.5 ± 1.0 times control with VEGF alone to 0.97 ± 0.08 times control with VEGF and cAMP) (Fig. 4.18). Based on these measured data, the prediction from our mathematical models suggested that the increase in the number of tight junction strands in the cleft between endothelial cells forming the microvessel wall is one of the mechanisms for the abolishment of VEGF-induced hyperpermeability by cAMP (Fig. 4.19).

4.3.4 Microvascular Hyperpermeability and Tumor Metastasis

4.3.4.1 VEGF Effects on Microvascular Integrity and on Tumor Cell Adhesion

Vascular endothelial growth factors (VEGFs) are a family of cytokines that act to increase the delivery of nutrients to tissue by three distinct mechanisms:

a. Endothelial cell growth, migration and new blood vessel formation (angiogenesis) [74];
b. Increased blood flow (by vasodilatation) [24]; and
c. Increased vascular permeability to water and solutes [23,74,91,92,114,186,232].

Combining *in vivo* permeability measurement and a mathematical model for the inter-endothelial transport, *Fu et al.* [91] predicted that acute effects of VEGF on microvascular integrity are widened gap opening of the interendothelial cleft and partial degradation of the ESG. Longer term effects of VEGF include the formation of gaps between adjacent endothelial cells in venular microvessels [146], vesiculovascuolar organelle pathways [79], transcellular pores [79,153], and fenestra [79,186].

Previous studies have found that many cancer cells express VEGF to a high degree [130], while the microvascular endothelium has abundant VEGF receptors including VEGFR2 (KDR/Flk-1) [160]. VEGFR2 has been implicated in normal and pathological vascular

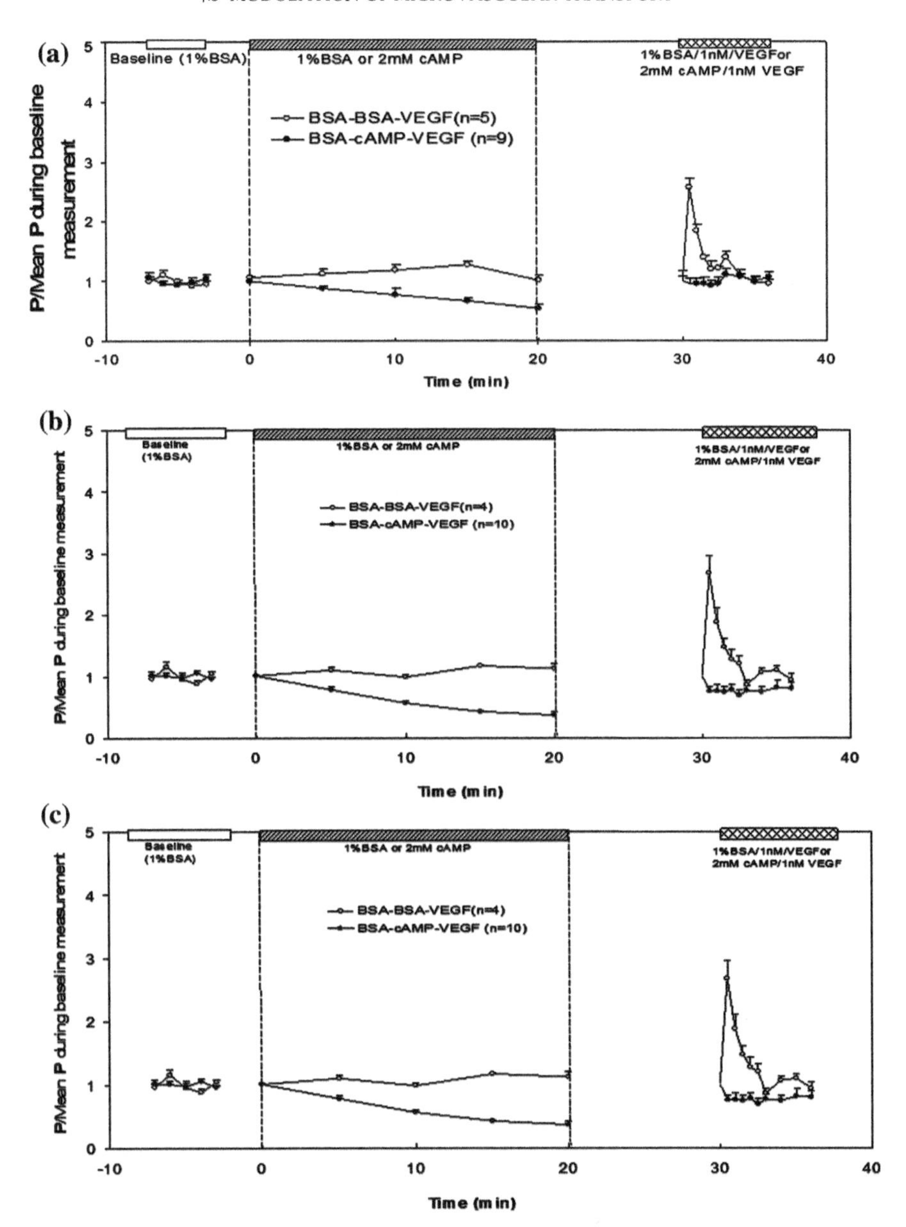

FIGURE 4.18 Effect of cAMP on VEGF-induced hyperpermeability in frog mesenteric microvessels for (a) sodium fluorescein, (b) α-lactalbumin, and (c) BSA (bovine serum albumin). Mean$\pm$SE P relative to baseline plotted as a function of time. In the matched control group (○, BSA-BSA-VEGF), baseline P was first measured with Ringer perfusate containing 1% BSA (bovine serum albumin, 10 mg/ml), then P was measured in a sham experiment of reperfusion with control solution for ~20 min, and finally P was measured under the treatment of 1 nM VEGF for ~5 min. In the test group (•, BSA-cAMP-VEGF), baseline P was first measured with perfusate containing 1% BSA, then P was measured in the test experiment of reperfusion with the same solution, with the addition of 2 mM 8-bromo-cAMP for ~20 min, and finally P was measured with 1 nM VEGF and 2 mM 8-bromo-cAMP for ~5 min. $^{*}p<0.05$ compared with the baseline (from [94]).

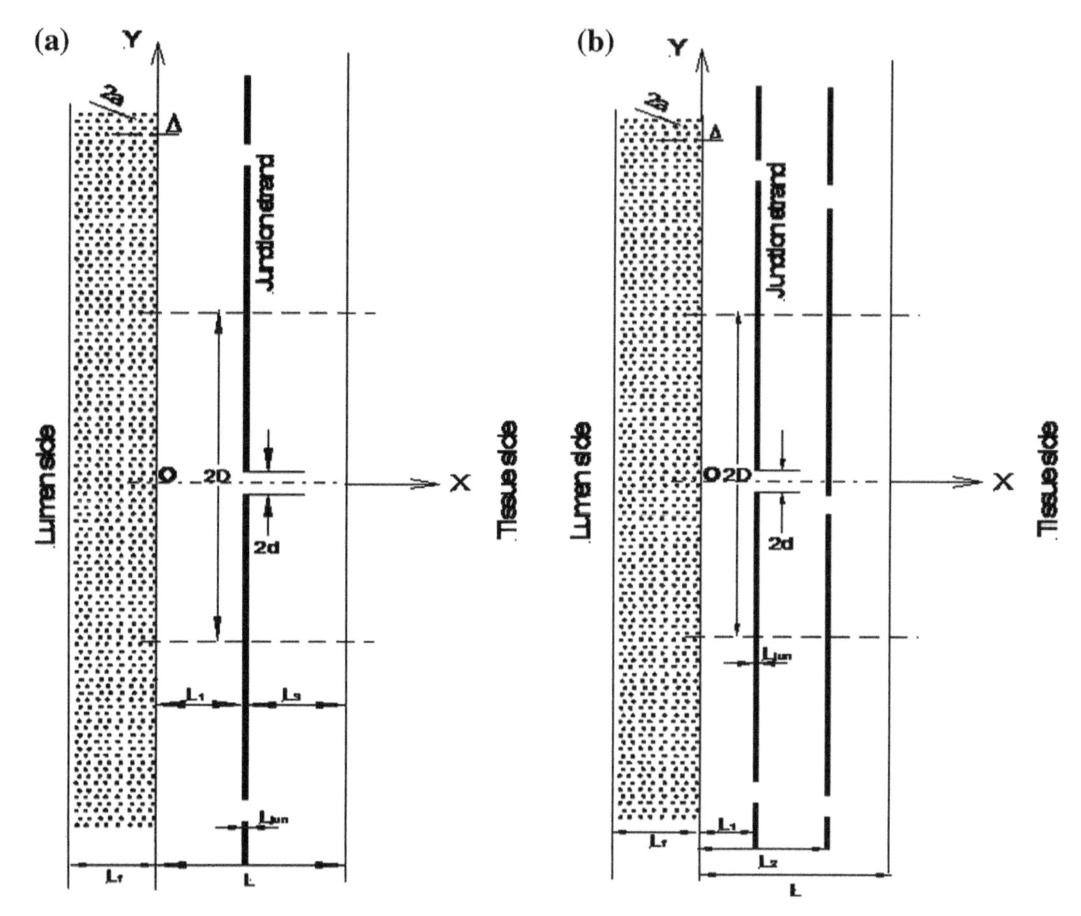

FIGURE 4.19 (a) Plane view of the model of the interendothelial cleft in frog mesenteric microvessels (revised from [88]). The junction strand with periodic breaks lies parallel to luminal front. L is the total depth of the cleft (~400 nm), L_{jun} is the thickness of the junction strand (~10 nm) and L_1 and L_3 are depths between the junction strand and luminal and abluminal fronts, respectively. L_f (~100 nm) is the thickness of fiber matrix at cleft entrance. The distance between two adjacent breaks in a junction strand is $2D$ (~2500 nm), and $2d$ (~150 nm) is the width of large junction breaks. At the entrance of the cleft on the luminal side, surface glycocalyx structures are represented by a periodic square array of cylindrical fibers. The radius of these fibers is a, and the gap spacing between fibers is Δ. (b) The model for explaining cAMP effect on microvessel permeability [94]. There are two junction strands in the cleft under cAMP influence compared to one strand under normal conditions.

endothelial cell biology [165]. Recently, it has been shown that ectopic administration of VEGF enhances the adhesion and transmigration of human breast cancer MDA-MB-231 cells across a monolayer of human brain microvascular endothelial cells under a static condition *in vitro* [130]. In addition, VEGF enhances the adhesion of malignant MDA-MB-435 cells and ErbB2-transformed mouse mammary carcinomas to intact rat mesenteric microvessels under flow *in vivo* [196] (Fig. 4.20).

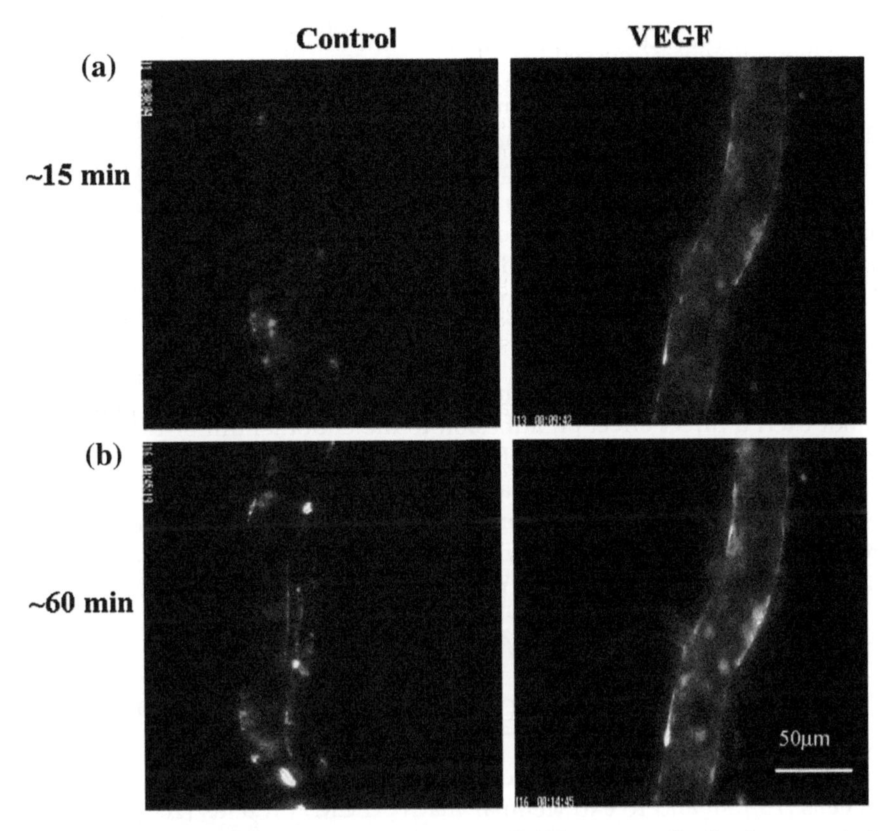

FIGURE 4.20 Photomicrographs showing *in vivo* MDA-MB-435s tumor cell adhesion to a single perfused microvessel under control with 1% BSA Ringer perfusate and under treatment with 1 nM VEGF perfusate after (a) ~15 min and (b) ~60 min perfusion. The perfusion velocity is ~1000 μm/s, which is the mean normal flow velocity in post-capillary venules. Bright spots indicate adherent tumor cells labeled with fluorescence. From [196].

4.3.4.2 *Integrin Signaling and Tumor Metastasis*

Although the non-specific trapping due to the friction between the tumor cells and the narrow part of microvasculature is found to be responsible for the initial tumor cell arrest [99,102,125,158], cell adhesion molecules are required for the adhesion in larger microvessels and transmigration [36,77,99–101,115,130,137,159,191,196]. The integrins are a family of signaling and cell adhesion receptors which attach cells to the extracellular matrix (ECM) and in some cases to other cells, and cooperate with growth factor and cytokine receptors to regulate cell behavior. Signals elicited by integrins enable tumor cells to survive, proliferate, migrate independently of positional constrains [105] and adhere [77]. The α6β4 integrin is a laminin-5 receptor and was originally described as a 'tumor-specific' protein, because of its apparent upregulation in multiple metastatic tumor types [100]. The β4 integrin is unique among integrins, because the cytoplasmic portion of the β4 subunit is 1,017 amino acids long, and possesses distinctive adhesive and signaling functions [100]. Upon binding of the ectodomain of β4 to the basement membrane protein laminin-5, the cytoplasmic portion of β4 interacts with the keratin cytoskeleton to promote the assembly of hemidesmosomal adhesions [138]. In addition, β4 activates intracellular signaling autonomously as well as by associating with multiple receptor tyrosine kinases (RTKs), including the EGFR, ErbB2, Met, and Ron [105,145,155]. Deletion of the β4 signaling domain delayed mammary tumor onset and inhibited primary tumor growth. The tumors arising in mutant mice were significantly more differentiated histologically compared to control tumors. In addition, primary tumor cells expressing signaling-defective β4 displayed a reduced proliferative rate and invasive ability, and underwent apoptosis when deprived of matrix adhesion. Finally, upon injection in the tail vein of nude mice, the mammary tumor cells expressing mutant β4 exhibited a reduced ability to metastasize to the lung [106].

Most recently, *Fan et al.* [77] examined the adhesion of ErbB2-transformed mammary tumor cells to mouse brain microvascular endothelial monolayer. They found that integrin β4 signaling does not exert a direct effect on adhesion to the endothelium or the underlying basement membrane. Rather, it enhances ErbB2-dependent expression of VEGF by tumor cells. VEGF in turn partially disrupts the tight and adherens junctions that maintain the adhesion between endothelial cells, enabling tumor cells to intercalate between endothelial cells and extend projections reaching the underlying exposed basement membrane, and enabling adhesion to occur between cell adhesion molecules (e.g., integrins) and ECM proteins (e.g., laminins).

4.4 TUMOR CELL ADHESION IN THE MICROCIRCULATION

In vitro static adhesion assays have been utilized to investigate tumor cell adhesion to endothelial cells [75,130] and to extracellular matrix (ECM) proteins [204]. Tumor cell adhesion has also been investigated using flow chambers [48,101,201] or artificial blood vessels [36] to address flow effects. Direct injection of tumor cells into the circulation has enabled the observation of tumor cell metastasis in target organs after sacrificing the animals [191], while intravital microscopy has been used to observe the interactions between circulating tumor cells and the microvasculature both *in vivo* and *ex vivo* [12,102,107,158,159,205].

4.4.1 Tumor Cell Adhesion Under Flow

Tumor cell extravasation is a dynamic process in which tumor cell adhesion to the vascular endothelium and transendothelial migration occurs under flow conditions [206]. Therefore, the geometry of the microvasculature and local hydrodynamic factors, along with the cell adhesion molecules at the tumor cell and endothelial cell should play crucial roles in tumor cell adhesion and extravasation. Tumor cells are exposed to flow while (a) circulating from the primary tumor, (b) arresting on downstream vascular endothelium and (c) transmigrating into the secondary target organ. Investigations of the role of shear flow in tumor cell adhesion and extravasation will contribute to the understanding of the complex process of tumor metastasis. Tumor cell extravasation would normally occur in the microvasculature where shear forces are relatively low (like in post-capillary venules) although of sufficient magnitudes to activate cell surface receptors and alter vascular cell function. During tumor cell extravasation there are significant changes in the structure and function of both tumor and endothelial cells.

For example a significant rearrangement of the cell cytoskeleton is required in both the tumor cells during migration [129] and in the endothelial cells as the barrier function is altered [220]. The extravasation of tumor cells also induces endothelial cell remodeling [15].

In an *in vitro* flow chamber study, *Slattery et al.* [201] found that the shear rate, rather than the shear stress, plays a more significant role in PMN (polymorphonuclear neutrophil)-facilitated melanoma adhesion and extravasation. β2 integrin/ICAM-1 adhesion mechanisms were examined, and the results indicated that LFA (lymphocyte function-associated)-1 and Mac-1 (CD11b/CD18) cooperate to mediate the PMN-EC (endothelial cell)-melanoma interactions under shear conditions. In addition, endogenously produced IL-8 contributes to PMN-facilitated melanoma arrest on the EC through the CXC chemokine receptors 1 and 2 (CXCR1 and CXCR2) on PMN [137,201].

To investigate tumor cell adhesion in a well-controlled *in vivo* system, *Shen et al.* [196] and *Yan et al.* [234] used intravital video microscopy to measure the adhesion rate of malignant MDA-MB-435 and 231 cells in straight and curved post-capillary venules on rat mesentery. A straight or curved microvessel was cannulated and perfused with tumor cells by a glass micropipette at a velocity of ~1 mm/s, which is the mean normal blood flow velocity in this type of vessels. At less than 10 min after perfusion, there was a significant difference in cell adhesion to the straight and curved vessel walls. At 60 min post perfusion, the averaged adhesion rate in the curved vessels was ~1.5-fold of that in the straight vessels. In 51 curved segments, 45% of cell adhesion was initiated at the inner side, 25% at outer side, and 30% at both sides of the curved vessels. To investigate the mechanical mechanism by which tumor cells prefer adhering at curved sites, *Yan et al.* [234] performed a computational study in which the fluid dynamics was determined by the lattice Boltzmann method (LBM) and the tumor cell dynamics were governed by Newton's law of translation and rotation. The details of this multi-scale modeling are summarized below.

4.4.2 Mathematical Models for Tumor Cell Adhesion in the Microcirculation

4.4.2.1 General Cell Adhesion Models

Extensive biophysical studies of cell adhesion have led to the development of a number of mathematical models. The construction and application of these models have demonstrated that it is possible to analyze cellular and molecular processes by highly quantitative approaches [240]. *Hammer and Apte* [111] proposed a mathematical model to simulate the interaction of a single cell with a ligand-coated surface under flow. This model can simulate the effect of many parameters on adhesion, such as the number of receptors on microvilli tips, the density of ligands, the rates of reaction between receptors and ligands, the stiffness of the resulting receptor-ligand springs, the response of springs to strains and the magnitude of the bulk hydrodynamic stresses. The model can successfully recreate the entire range of expected and observed adhesive phenomena, from completely unencumbered motion, to rolling, to transient attachment and to firm adhesion. Moreover, this model can generate meaningful statistical measures of adhesion, including the mean and variance in velocity, rate constants for cell attachment and detachment and the frequency of adhesion. *King and Hammer* [127] used the completed double-layer boundary integral equation method to study the adhesive interactions between multiple rigid particles and a planar boundary in a viscous fluid. The simulation results revealed a mechanism for the capture of free-stream cells when an initial cell has adhered to provide a nucleation site. *Wang et al.* [224] developed a population balance model for cell aggregation and adhesion process in a non-uniform shear flow, and they carried out a Monte Carlo simulation based on the model for the heterotypic cell-cell collision and adhesion to a substrate under dynamic shear forces. *Shao and Xu* [195] numerically studied the adhesion between a microvillus-bearing cell and a ligand-coated substrate, also by using the Monte Carlo method. They found that most of the adhesion was mediated by a single bond if the total adhesion frequency was less than 20%.

Mathematical models of cell adhesion relate the forward and reverse reaction rates for receptor-ligand bonds. These reaction rates laws for cell adhesion have been defined as 'adhesive dynamics models', which can be used to couple the effect of receptor-ligand bonds on cell adhesion. Figure 4.21 shows a schematic view of the adhesive dynamics model for the cell. In this model, the cell adhesion molecules on the surface of a circulating cell are defined as receptors, and those on the surface of endothelial cells are defined as ligands. Once the distance between receptor and ligand is smaller than the critical length H_c, it has the chance to form receptor-ligand bonds. Interactions between receptors and ligands are realized by the ideal adhesive springs, and the spring forces are calculated via the compression or expansion of these springs. This dynamic process relates the bond association and dissociation rate of adhesive dynamics models. The following are the two typical cell dynamic models.

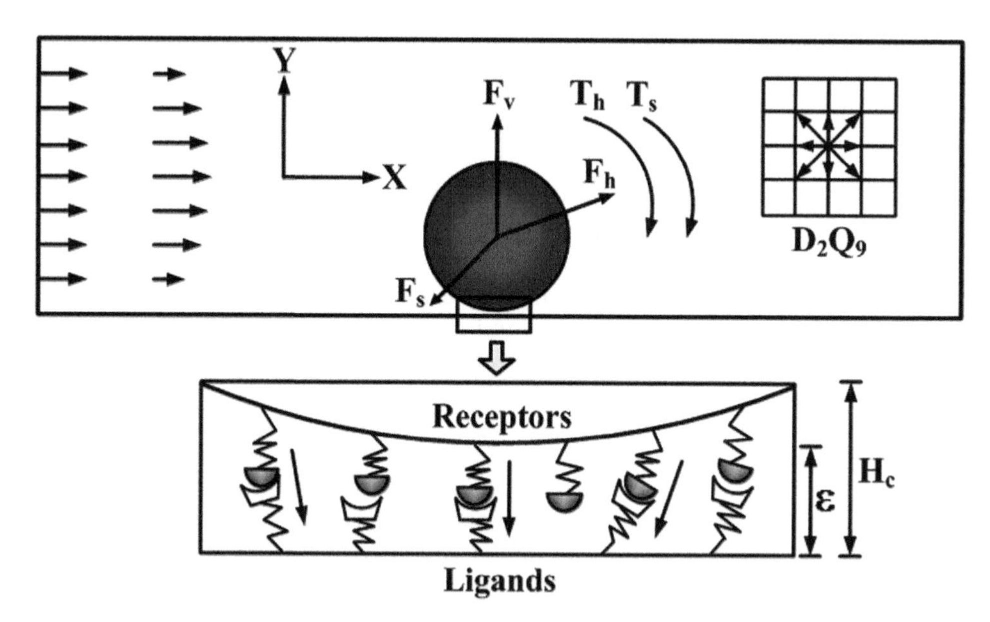

FIGURE 4.21 Schematic diagram of the adhesive dynamics model. $\vec{F}_h$ is the hydrodynamic force, which can be calculated by the momentum exchange method, $\vec{F}_v$ is the repulsive van der Waals force, which can be derived by the Derjaguin approximation, $\vec{F}_s$ is the total spring force that is contributed by the adhesive receptor-ligand bonds and T_h and T_s are the torques induced by the hydrodynamic force and spring force, respectively. ε is the bond length. From *Yan et al.* [233,234].

4.4.2.2 *Bell's Model*

This model [28] was confirmed to be a good approximation for different states of cell adhesion in the straight microvessels, such as no adhesion, rolling, landing and firm adhesion [41]. In this model, the association rate of the bond is 84 s^{-1}, which is a reasonable value that extensive simulations have shown can properly recreate experimental values for velocity and dynamics of rolling in the straight microvessels [45]. For the dissociation rate of the bond, *Bell* [28] adapted the kinetic theory of the strength of solids and proposed a constitutive relation between dissociation rate and force. Therefore, the bond association rate k_f and bond dissociation rate k_r are:

$$k_f = 84 \tag{4.20}$$

$$k_r = k_r^0 \exp\left(\frac{\gamma f}{k_b T}\right) \tag{4.21}$$

where k_b is the Boltzmann constant, T the absolute temperature, k_r^0 the unstressed dissociation rate and γ the reactive compliance that describes the degree to which force facilitates bond breakage. Both k_r^0 and γ are functional properties of adhesion molecules. f is the spring force of each bond, which can be obtained according to the Hooke's law:

$$f = \sigma(\chi - \lambda) \tag{4.22}$$

where σ is the spring constant, χ is the distance between the end points of receptor and ligand and λ is the equilibrium bond length.

4.4.2.3 *Dembo et al.'s Model*

Dembo et al. [70] modeled a piece of membrane with immobile discrete bonds and allowed the membrane to detach. They did this by letting the applied tension exceed the bond stress. This model can be used to predict the critical membrane tension required for detachment, and the resulting peeling velocities of the membrane. The main contribution of this model is the expression for the rate constants as a function of distance between the membranes. *Dembo et al.* [70] demonstrated the reasonable, thermodynamically consistent rate expressions relating the bond association rate k_f and bond dissociation rate k_r to X as:

$$k_f = k_f^0 \exp\left(-\frac{\sigma_{ts}(\chi - \lambda)^2}{2k_b T}\right) \tag{4.23}$$

$$k_r = k_r^0 \exp\left(-\frac{(\sigma - \sigma_{ts})(\chi - \lambda)^2}{2k_b T}\right) \tag{4.24}$$

where k_f^0 and k_r^0 are the reaction rate constants when the spring is at its equilibrium length, and σ and σ_{ts} is the spring constant and 'transition state' spring constant, respectively.

Once the forward association rate and the reverse dissociation rate of the bond are known, the appropriate expressions for the probability of formation and breakage of the bond tethers in a time step dt can be obtained, as formulated by *Chang and Hammer* [44]:

$$P_f = 1 - \exp\left(-k_f \cdot dt\right) \tag{4.25}$$

$$P_r = 1 - \exp\left(-k_r \cdot dt\right) \tag{4.26}$$

where P_f is the probability of forming a bond, and P_r is the probability of breaking a bond in a time interval dt. Recently, these adhesive dynamic models were adapted to describe tumor cell adhesion in the microcirculation on the basis of experimental observations. The major results are presented below.

4.4.3 Model Predictions for Tumor Cell Adhesion in the Microcirculation

4.4.3.1 *Effect of Curvature*

It has been found that both the circulating blood cells and tumor cells prefer to adhere to curved microvessels than straight ones [139,235]. To study the effect of vessel curvature on tumor cell adhesion, *Yan et al.* [234] carried out a numerical simulation using the scheme proposed by *Hammer and Apte* [111], which was based on *Dembo et al.*'s model [70]. They compared the variation of cell velocity and rotational velocity, calculated the force acting on the cell and computed the bonds formed during the cell migration process for straight and curved microvessels. A higher number of bonds means there is a greater opportunity for a cell to adhere to the vessel wall. Comparisons are run for different wall geometries: a straight vessel, a vessel with a positive wall curvature (inner side of the curved vessel) and a vessel with a negative wall curvature (outer side of the curved vessel). For a single cell in a straight vessel, after off and on interactions between the circulating cell and the endothelial cells forming the vessel wall, steady bonds are formed throughout the journey; for a single cell in a curved vessel, the cell adhesion only takes place at the positive curvature wall, once leaving the positive curvature wall, the cell moves freely, and there is no adhesion in the negative curvature wall at all. To compare the probability of cell adhesion between straight and curved microvessels, they calculated the statistics of bond number, the probability of each bond number occurring and the ratio of these two probabilities, as shown in Fig. 4.22. It is found that, for smaller bond numbers, the probability of adhesion in the straight vessel is larger than that in the curved vessel, and the turning point is at 5 bonds with a probability of 25.5%. For the higher bond numbers, i.e., larger than 5, the adhesion probability in the curved vessel is obviously higher than that in the straight vessel, and the larger the bond number, the higher the adhesion probability in the curved vessel. It is understandable that the final cell adhesion depends on the number of bonds: the more bonds form simultaneously, the higher the probability that the cell will adhere to the vessel wall. Therefore vessel curvature has a significant influence on the cell adhesion.

Blood flow always involves multiple cells, and interactions between the cells would affect cell adhesion and migration. For two cells in a straight vessel, due to the interaction between cells, the number of bonds is more than twice that of the single cell case. For two cells in a curved vessel, the number of bonds increases significantly as well, the number of bonds changes dramatically during the whole process, and similarly to the single cell case, the bonds form in the negative curvature wall. A statistical comparison of bond numbers, shown in Fig. 4.23, indicates that, similarly to the single cell case, at lower bond numbers the straight

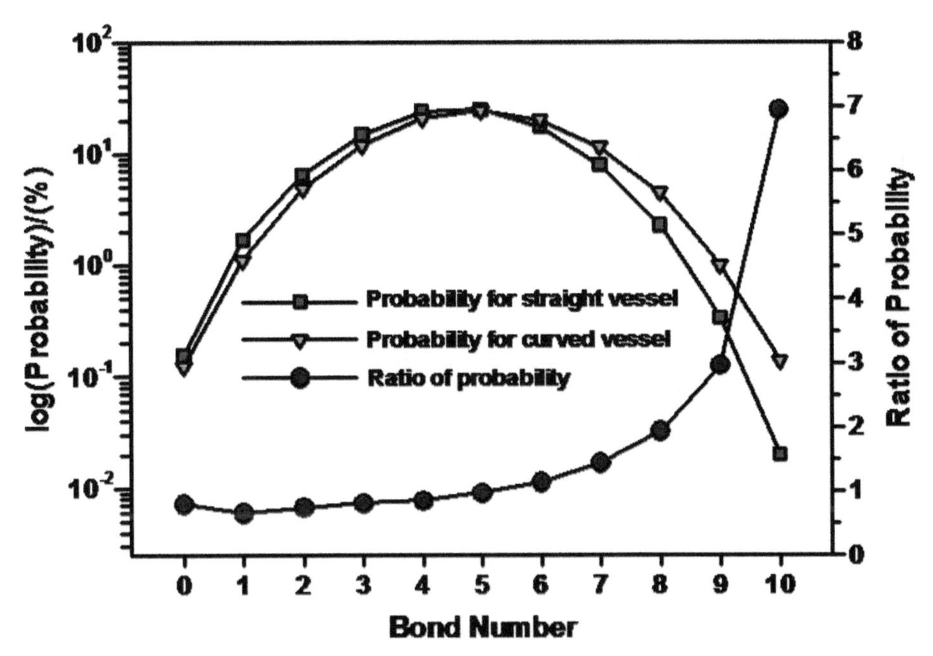

FIGURE 4.22 Comparison of bond formation probabilities and the ratio of the probabilities between straight and curved microvessels for a single cell in the vessel. From *Yan et al.* [234].

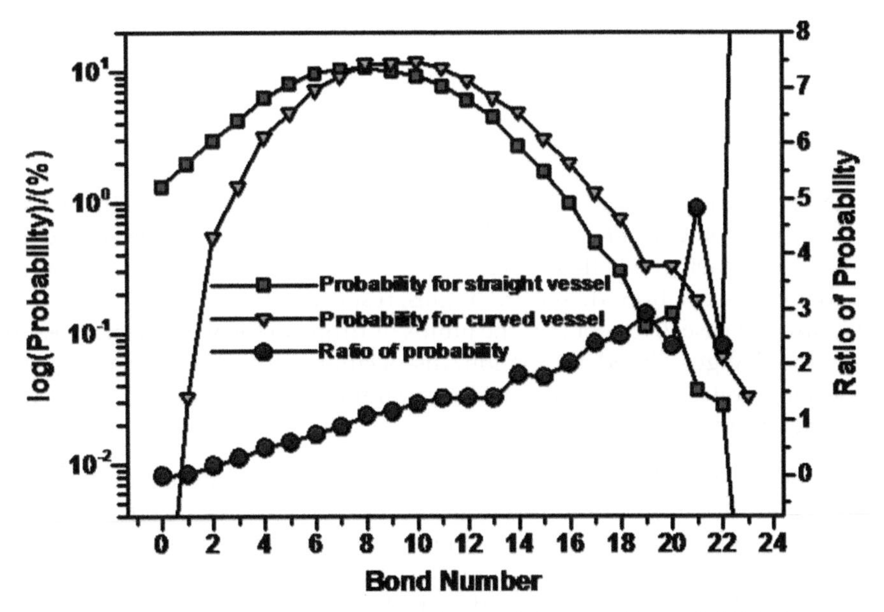

FIGURE 4.23 Comparison of bond formation probabilities and the ratio of the probabilities between straight and curved microvessels for two cells in the vessel. From *Yan et al.* [234].

vessel is more likely to form bonds. The turning point is at 8 bonds. When the bond number is larger than 8, the curved vessel has more chances to form bonds between a cell and the endothelial cells at the vessel wall. At a higher bond number, e.g., 21 bonds or more, the bond forming probability of the curved vessel can be five times that of the straight vessel. The final cell adhesion is actually dependent on simultaneous bond formation. The higher the bond number, the more likely cell adhesion occurs. Therefore, the cell, particularly when interacting with other cells, has a higher probability if adhering to the endothelial cells in a curved vessel than in a straight vessel.

4.4.3.2 *Effect of Wall Shear Stress*

At the curved sites there are rather complicated distributions of wall shear stress, and the wall shear stress or its variation may activate receptor-ligand bond forming, which appears to render these sites prone to catching cells. Based on experimental observation, *Yan et al.* [234] revised Bell's model by integrating the effect of wall shear stress and its gradient, respectively. The models are:

Case 1: The bond association/dissociation rates are related to shear stress:

$$k_f = k_f^n \cdot \left(\frac{\tau}{\tau_0}\right)^{k_1} \quad \text{and} \quad k_r = k_r^n \cdot \left(\frac{\tau}{\tau_0}\right)^{k_2} \tag{4.27}$$

Case 2: The bond association/dissociation rates are related to shear stress gradients:

$$k_f = k_f^n \cdot \exp\left(k_3 \cdot \frac{d\tau}{dl}\right) \tag{4.28}$$

and

$$k_r = k_r^n \cdot \exp\left(k_4 \cdot \frac{d\tau}{dl}\right) \tag{4.29}$$

where τ and τ_0 are the wall shear stresses along the curved vessel and along the straight vessel, respectively, and $d\tau/dl$ is the wall shear stress gradient along the curved vessel. k_1, k_2 and k_3, k_4 are coefficients that represent the sensitivity of wall shear stress and its gradient to bond association/dissociation rates, respectively.
Case 3: Identical to case 2, except that the jumps or drops in the wall shear stress gradient can trigger the change of bond association/dissociation rates. Once triggered, the association/dissociation rates will keep the maximum/minimum value until the next wall shear stress gradient jump or drop occurs.

Figure 4.24 shows the cell trajectory for these three cases: a denser trajectory indicates a slower motion and the coarser trajectory a faster motion. For case 1, the denser trajectory occurs between A_b and B_b due to the centrifugal effect, and the coarser trajectory happens between C_b and D_b, indicating a faster cell motion due to the decrease in k_f and the increase in k_r, both of which result from the drop of wall shear stress. For case 2, the denser trajectory occurs near the conjunction B_b, indicating a slower cell motion there due to the stronger

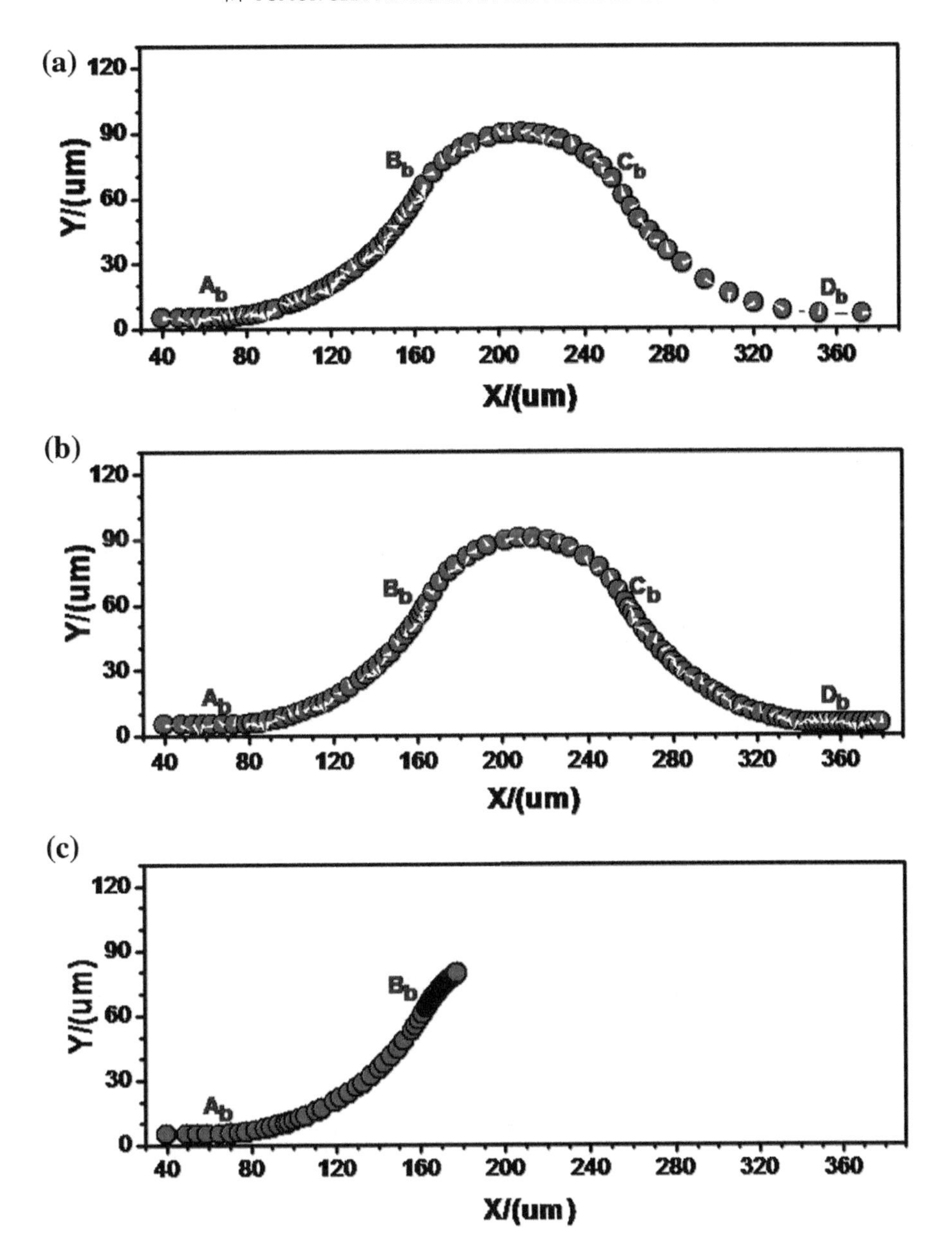

FIGURE 4.24 The history of the trajectory of the tumor cell along the inner wall of a curved vessel, (a) case 1, (b) case 2, and (c) case 3. From *Yan et al.* [234].

adhesive effect caused by the large jump in the wall shear stress gradient at B_b. Another dense trajectory is observed near the conjunction C_b, where the shear stress gradient undergoes a sudden drop. This slower cell motion is not due to the adhesive effect but the centrifugal effect. A coarser trajectory exists in the positive curvature vessel (between B_b and C_b), indicating a faster cell motion due to the centrifugal effect. For case 3, when the cell approaches the conjunction B_b, the cell moves more and more slowly: this is represented by a black band in the cell trajectory. Generally, a positive wall shear stress/gradient jump enhances tumor cell adhesion, whereas a negative wall shear stress/gradient jump would weaken tumor cell adhesion. The wall shear stress/gradient, over a threshold, made a significant contribution to tumor cell adhesion by activating or inactivating cell adhesion molecules. These results explain why tumor cell adhesion preferentially occurs in the positive curvature of curved microvessels with very low Reynolds number (in the order of 10^{-2}) laminar flow.

4.5 SUMMARY AND OPPORTUNITIES FOR FUTURE STUDY

Although transport across endothelium is a classical problem that has been investigated for more than fifty years, the fundamental question related to structure-function of microvessel walls and properties of cells forming the wall still remains unanswered. With help from mathematical models providing more accurate interpretations and predictions, new experimental techniques involving fluorescence, atomic force and electron microscopy and new developments in molecular biology and biochemistry will lead to more fascinating discoveries in this field.

One problem that has not been investigated thoroughly is the selectivity of the endothelium to solutes of various sizes, shapes and charges (especially in the glycocalyx layer). By employing molecular dynamics simulations, new models are expected to elucidate the interactions between solutes and the fiber matrix as well as between solutes when the solute size is very close to the gap spacing between fibers and when the solution is no longer dilute.

Another problem is the development of models for dynamic water and solute transport through multi-transendothelial pathways including intercellular, transcellular, fenestrae and vesicle routes. They are important for predicting malfunctions in the transendothelium process during disease. The third problem is to create transendothelial models for cells such as leukocytes and cancer cells. Cell transport is crucial in many physiological and pathological processes, including the inflammatory response and tumor metastasis.

Acknowledgments

We thank the support from the NSF CBET 0754158 and NIH SC1CA153325-01, U54CA137788-01 and RO1HL094889, and Hong Kong Research Grants Council of the Government of the HKSAR PolyU 5238/08E.

References

[1] N.J. Abbott, Comparative physiology of the blood-brain barrier, in: M.W.B. Bradbury (Ed.), Physiology and Pharmacology of the Blood-Brain Barrier, Springer, Heidelber, 1992., pp. 371–398.

[2] N.J. Abbott, A.K. Patabendige, D.E. Dolman, S.R. Yusof, D.J. Begley, Structure and function of the blood-brain barrier, Neurobiology of Disease 37 (2010) 13–25.

[3] R.H. Adamson, Permeability of frog mesenteric capillaries after partial pronase digestion of the endothelial glycocalyx, Journal of Physiology 428 (1990) 1–13.

[4] R.H. Adamson, G. Clough, Plasma proteins modify the endothelial cell glycocalyx of frog mesenteric microvessel, Journal of Physiology 445 (1992) 473–486.

[5] R.H. Adamson, V.H. Huxley, F.E. Curry, Single capillary permeability to proteins having similar size but different charge, American Journal of Physiology 254 (1988) H304–H312.

[6] R.H. Adamson, J.F. Lenz, X. Zhang, G.N. Adamson, S. Weinbaum, F.E. Curry, Oncotic pressures opposing filtration across non-fenestrated rat microvessels, Journal of Physiology 557 (2004) 889–907.

[7] R.H. Adamson, J.F. Lenz, F.E. Curry, Quantitative laser scanning confocal microscopy on single capillaries: permeability measurement, Microcirculation 1 (1994) 251–265.

[8] R.H. Adamson, C.C. Michel, Pathways through the inter-cellular clefts of frog mesenteric capillaries, Journal of Physiology 466 (1993) 303–327.

[9] R.H. Adamson, B. Liu, G. Nilson-Fry, L.L. Rubin, F.E. Curry, Microvascular permeability and number of tight junctions are modulated by cyclic AMP, American Journal of Physiology 274 (43) (1998) H1885–H1894.

[10] T. Akinaga, M. Sugihara-Seki, T. Itano, Electrical charge effect on osmotic flow through pores, Journal of the Physical Society of Japan 77 (2008) 53–401.

[11] S.M. Albelda, P.M. Sampson, F.R. Haseiton, J.M. Mcniff, S.N. Mueller, S.K. Williams, A.P. Fishman, E.M. Levine, Permeability characteristics of cultured endothelial cell monolayers, Journal of Applied Physiology 64 (1988) 308–322.

[12] A.B. Al-Mehdi, K. Tozawa, A.B. Fisher, L. Shientag, A. Lee, R.J. Muschel, Intravascular origin of metastasis from the proliferation of endothelium-attached tumor cells: a new model for metastasis, Nature-Medicine 6 (2000) 100–102.

[13] J.L. Anderson, D.M. Malone, Mechanism of osmotic flow in porous membranes, Biophysical Journal 14 (1974) 957–982.

[14] J.M. Anderson, C.M. Vaniallie, Tight junctions and the molecular basis for regulation of paracellular permeability, American Journal of Physiology 269 (32) (1995) G467–G475.

[15] D.A. Antonetti, E.B. Wolpert, L. DeMaio, N.S. Harhaj, R.C. Scaduto Jr, Hydrocortisone decreases retinal endothelial cell water and solute flux coincident with increased content and decreased phosphorylation of occludin, Journal of Neurochemistry 80 (2002) 667–677.

[16] T. Arisaka, M. Mitsumata, M. Kawasumi, T. Tohjima, S. Hirose, Y. Yoshida, Effects of shear stress on glycosaminoglycan synthesis in vascular endothelial cells, Annuals of New York Academy of Science 748 (1995) 543–554.

[17] K.P. Arkill, C. Knupp, C.C. Michel, C.R. Neal, K. Qvortrup, J. Rostgaard, J.M. Squire, Similar endothelial glycocalyx structures in microvessels from a range of mammalian tissues: evidence for a common filtering mechanism? Biophysical Journal 101 (2011) 1046–1056.

[18] K. Auckland, R.K. Reed, Interstitial-lymphatic mechanisms in the control of extracellular fluid volume, Physiological Review 73 (1993) 1–78.

[19] S.A. Baldwin, I. Fugaccia, D.R. Brown, L.V. Brown, S.W. Scheff, Blood-brain barrier breach following cortical contusion in the rat, Journal of Neurosurgery 85 (3) (1996) 476–481.

[20] P. Ballabh, A. Braun, M. Nedergaard, The blood-brain barrier: an overview: structure, regulation, and clinical implications, Neurobiology of Disease 16 (1) (2004) 1–13.

[21] P. Barzo, A. Marmarou, P. Fatouros, F. Corwin, J. Dunbar, Magnetic resonance imaging-monitored acute blood-brain barrier changes in experimental traumatic brain injury, Journal of Neurosurgery 85 (6) (1996) 1113–1121.

[22] M.K. Baskaya, A.M. Rao, A. Dogan, D. Donaldson, R.J. Dempsey, The biphasic opening of the blood-brain barrier in the cortex and hippocampus after traumatic brain injury in rats, Neuroscience Letter 226 (1) (1997) 33–36.

[23] D.O. Bates, F.E. Curry, Vascular endothelial growth factor increases hydraulic conductivity of isolated perfused microvessels, American Journal of Physiology 271 (40) (1996) H2520–H2528.

[24] D.O. Bates, R.I. Heald, F.E. Curry, B. Williams, Vascular endothelial growth factor increases Rana vascular permeability and compliance by different signaling pathways, Journal of Physiology 533 (1) (2001) 263–272.

[25] A. Beaumont, A. Marmarou, K. Hayasaki, P. Barzo, P. Fatouros, F. Corwin, C. Marmarou, J. Dunbar, The permissive nature of blood-brain barrier (BBB) opening in edema formation following traumatic brain injury, Acta Neurochirurgica. Supplement 76 (2000) 125–129.

[26] B.F. Becker, Endothelial glycocalyx and coronary vascular permeability: the fringe benefit, Basic Research in Cardiology 105 (2010) 687–701.

[27] D.J. Begley, Structure and function of the blood-brain barrier, in: E. Touitou, B. Barry (Eds.), Enhancement in Drug Delivery, CRC Press, Boca Raton, 2007, pp. 571–690.

[28] G.I. Bell, Models for the specific adhesion of cells to cells, Science 200 (1978) 618–627.

[29] D.A. Berk, F. Yuan, M. Leunig, R.K. Jain, Direct *in vivo* measurement of targeted binding in a human tumor xenograft, Proceeding of National Academy of Science United States of America 94 (1997) 1785–1790.

[30] M. Bernfield, M. Gotte, P.W. Park, O. Reizes, M.L. Fitzgerald, J. Lincecum, M. Zako, Functions of cell surface heparan sulfate proteoglycans, Annual Review of Biochemistry 68 (1999) 729–777.

[31] G. Bhalla, W.M. Deen, Effects of charge on osmotic reflection coefficients of macromolecules in porous membranes, Journal of Colloid Interface Science 333 (2009) 363–372.

[32] M.P. Bohrer, W.M. Deen, C.R. Robertson, J.L. Troy, B.M. Brenner, Influence of molecular configuration on the passage of macromolecules across the glomerular capillary wall, Journal of General Physiology 74 (1979) 583–593.

[33] J.F. Brandy, G. Bossis, Stokesian dynamics, Annual Review of Fluid Mechanics 20 (1988) 111–132.

[34] B.M. Brenner, C.B. Bohrer, W.M. Deen, Determinants of glomerular perselectivity: insights derived from observations *in vivo*, Kidney International 12 (1977) 229–237.

[35] H. Brenner, P.M. Adler, Dispersion resulting from flow through spatially periodic media: II. Surface and intraparticle transport, Philosophical Transactions of the Royal Society of London, Series A 307 (1982) 149–158.

[36] W. Brenner, P. Langer, F. Oesch, C.J. Edgell, R.J. Wieser, Tumor cell-endothelium adhesion in an artificial venule, Annals of Biochemistry 225 (1995) 213–219.

[37] M. Bundgaard, The three-dimensional organization of tight junctions in a capillary endothelium revealed by serial-section electron microscopy, Journal of Ultrastructural Research 88 (1984) 1–17.

[38] M. Bundgaard, J. Frokjaer-Jensen, Functional aspects of the ultrastructure of terminal blood vessels: a qualitative study on consecutive segments of the frog mesenteric microvasculature, Microvascular Research 23 (1982) 1–30.

[39] A.M. Butt, H.C. Jones, N.J. Abbott, Electrical resistance across the blood-brain barrier in anaesthetized rats: a developmental study, Journal of Physiology 429 (1990) 47–62.

[40] L.M. Cancel, A. Fitting, J.M. Tarbell, *In vitro* study of LDL transport under pressurized (convective) conditions, American Journal of Physiology 293 (2007) H126–H132.

[41] K.E. Caputo, D.A. Hammer, Effect of microvillus deformability on leukocyte adhesion explored using adhesive dynamics simulations, Biophysical Journal 89 (2005) 187–200.

[42] I. Cernak, R. Vink, D.N. Zapple, M.I. Cruz, F. Ahmed, T. Chang, S.T. Fricke, A.I. Faden, The pathobiology of moderate diffuse traumatic brain injury as identified using a new experimental model of injury in rats, Neurobiology of Disease 17 (1) (2004) 29–43.

[43] D. Chappell, Sevoflurane reduces leukocyte and platelet adhesion after ischemia-reperfusion by protecting the endothelial glycocalyx, Anesthesiology 115 (2011) 483–491.

[44] K.C. Chang, D.A. Hammer, Influence of direction and type of applied force on the detachment of macromolecularly-bound particles from surfaces, Langmuir 12 (1996) 2271–2282.

[45] K.C. Chang, D.F.J. Tees, D.A. Hammer, The state diagram for cell adhesion under flow: leukocyte rolling and firm adhesion, Proceeding of National Academy of Science United States of America 12 (2000) 2271–2282.

[46] S.R. Chary, R.K. Jain, Direct measurement of interstitial convection and diffusion of albumin in normal and neoplastic tissues by fluorescence photobleaching, Proceeding of National Academy of Science United States of America 86 (1989) 5385–5389.

[47] B. Chen, B.M. Fu, An electrodiffusion-filtration model for effects of surface glycocalyx on microvessel permeability to macromolecules, ASME Journal of Biomechanical Engineering 126 (2004) 614–624.

[48] R. Chotard-Ghodsnia, O. Haddad, A. Leyrat, A. Drochon, C. Verdier, A. Duperray, Morphological analysis of tumor cell/endothelial cell interactions under shear flow, Journal of Biomechanics 40 (2007) 335–344.

[49] G. Clough, C.C. Michel, Quantitative comparisons of hydraulic permeability and endothelial intercellular cleft dimensions in single frog capillaries, Journal of Physiology 405 (1988) 563–576.

[50] A.A. Constantinescu, Endothelial cell glycocalyx modulates immobilization of leukocytes at the endothelial surface, Arteriosclerosis, Thrombosis, and Vascular Biology 23 (2003) 1541–1547.

[51] R.S. Cotran, G. Majino, The delayed and prolonged vascular leakage in inflammation. I. Topography of the leading vessels after thermal injury, American Journal of Physiology 45 (1964) 228–261.

[52] C. Crone, in: P.F. Baker (Ed.), Recent Advances in Physiology, Churchill Livingstone, London, 1984.

[53] C. Crone, D.G. Levitt, Capillary permeability to small solutes, in: Handbook of Physiology. The Cardiovascular System. Microcirculation, sect. 2, vol. IV, pt. 1, American Physiology Society, Bethesda, MD, 1984, pp. 411–466 (Chapter 10).

[54] F.E. Curry, Permeability coefficients of the capillary wall to low molecular weight hydrophilic solutes measured in single perfused capillaries of frog mesentery, Microvascular Research 17 (1979) 290–308.

[55] F.E. Curry, Mechanics and thermodynamics of transcapillary exchange, in: Handbook of Physiology. The Cardiovascular System. Microcirculation, sect. 2, vol. IV, pt. 1, American Physiology Society, Bethesda, MD, 1984, pp. 309–374 (Chapter 8).

[56] F.E. Curry, The effect of albumin on the structure of the molecular filter at the capillary wall, Federation Proceeding 44 (1985) 2610–2613.

[57] F.E. Curry, Determinants of capillary permeability: a review of mechanisms based on single capillary studies in the frog, Circulation Research 59 (1986) 367–380.

[58] F.E. Curry, R.H. Adamson, Endothelial glycocalyx: permeability barrier and mechanosensor engineering, Annals of Biomedical Engineering 40 (2012) 828–839.

[59] F.E. Curry, R.H. Adamson, Vascular permeability modulation at the cell, microvessel, or whole organ level: towards closing gaps in our knowledge, Cardiovascular Research 87 (2010) 218–229.

[60] F.E. Curry, J. Frokjar-Jensen, Water flow across the walls of single muscle capillaries in the frog, *Rana pipiens*, Journal of Physiology 350 (1984) 293–307.

[61] F.E. Curry, V.H. Huxley, I.H. Sarelius, Techniques in the microcirculation: measurement of permeability pressure and flow, Cardiovascular Physiology 309 (1983) 1–34.

[62] F.E. Curry, C.C. Michel, A fiber matrix model of capillary permeability, Microvascular Research 20 (1980) 96–99.

[63] F.E. Curry, T. Noll, Spotlight on microvascular permeability, Cardiovascular Research 87 (2010) 195–197.

[64] F.E. Curry, J.C. Rutledge, J.F. Lenz, Modulation of microvessel wall charge by plasma glycoprotein orosomucoid, American Journal of Physiology 257 (26) (1989) H1354–H1359.

[65] E.R. Damiano, T.M. Stace, A mechano-electrochemical model of radial deformation of the capillary glycocalyx, Biophysical Journal 82 (2002) 1153–1175.

[66] P.F. Davies, Flow-mediated endothelial mechanotransduction, Physiological Review 75 (1995) 519–560.

[67] W.M. Deen, Hindered transport of large molecules in liquid-filled pores, AIChE Journal 33 (1987) 1409–1423.

[68] W.M. Deen, S. Behrooz, J.M. Jamieson, Theoretical model for glomerular filtration of charged solutes, American Journal of Physiology 238 (1980) F126–F139.

[69] W.M. Deen, M.P. Bohrere, C.R. Robertson, B.M. Brenner, Determinants of the transglomerular passage of macromolecules, Federal Proceeding 36 (1977) 2614–2618.

[70] M. Dembo, D.C. Torney, K. Saxman, D.A. Hammer, The reaction-limited kinetics of membrane-to-surface adhesion and detachment, Philosophical Transactions of the Royal Society of London, Series A 234 (1988) 55–83.

[71] W.D. Dietrich, O. Alonso, M. Halley, Early microvascular and neuronal consequences of traumatic brain injury: a light and electron microscopic study in rats, Journal of Neurotrauma 11 (3) (1994) 289–301.

[72] D. Drenckhahn, W. Ness, The endothelial contractile cytoskeleton, in: G.V.R. Born, C.J. Schwartz (Eds.), Vascular Endothelium: Physiology, Pathology and Therapeutic Opportunities, Schattauer, Stuttgart, Germany, 1997, pp. 1–15.

[73] N. Dusserre, N. L'Heureux, K.S. Bell, PECAM-1 interacts with nitric oxide synthase in human endothelial cells: implication for flow-induced nitric oxide synthase activation, Arteriosclerosis, Thrombosis, and Vascular Biology 24 (2004) 1796–1802.

[74] H.F. Dvorak, L.F. Brown, M. Detmar, A.M. Dvorak, Vascular permeability factor/vascular endothelial growth factor, microvascular hyperpermeability, and angiogenesis, American Journal of Pathology 146 (1995) 1029–1039.

[75] S. Earley, G.E. Plopper, Disruption of focal adhesion kinase slows transendothelial migration of AU-565 breast cancer cells, Biochemical and Biophysical Research Communications 350 (2006) 405–412.

[76] P. Ehrlich, Das sauerstufbudurfnis des organismus, Hirschwald, Berlin, 1885.

[77] J. Fan, B. Cai, M. Zeng, Y. Hao, F.G. Giancotti, B.M. Fu, Integrin β4 signaling promotes mammary tumor cell adhesion to brain microvascular endothelium by inducing ErbB2-medicated secretion of VEGF, Annals of Biomedical Engineering 39 (8) (2011) 2223–2241.

[78] E. Farkas, P.G.M. Luiten, Cerebral microvascular pathology in aging and Alzheimer's disease, Progress in Neurobiology 64 (2001) 575–611.
[79] D. Feng, J.A. Nagy, K. Payne, I. Hammel, H.F. Dvorak, A.M. Dvorak, Pathways of macromolecular extravasation across microvascular endothelium in response to VPF/VEGF and other vasoactive mediators, Microcirculation 6 (1) (1999) 23–44.
[80] J.A. Florian, J.R. Kosky, K. Ainslie, Z. Pang, R.O. Dull, J.M. Tarbell, Heparan sulfate proteoglycan is a mechanosensor on endothelial cells, Circulation Research 93 (2003) (2003) e136–e142.
[81] C. Forster-Horvath, L. Meszaros, E. Raso, Expression of CD44v3 protein in human endothelial cells *in vitro* and in tumoral microvessels *in vivo*, Microvascular Research 68 (2004) 110–118.
[82] J. Fox, F. Galey, H. Wayland, Action of histamine on the mesenteric microvasculature, Microvascular Research 19 (1980) 108–126.
[83] J.A. Frangos, S.G. Eskin, L.V. McIntire, C.L. Ives, Flow effects on prostacyclin production by cultured human endothelial cells, Science 227 (1985) 1477–1479.
[84] M.H. Friedman, Principles and Models of Biological Transport, Springer-Verlag, 1986 (Chapter 1)
[85] B. Fu, F.E. Curry, R.H. Adamson, S. Weinbaum, A model for interpreting the tracer labeling of interendothelial clefts, Annals of Biomedical Engineering 25 (1997) 375–397.
[86] B. Fu, R.H. Adamson, F.E. Curry, Test of a two pathway model for small solute exchange across the capillary wall, American Journal of Physiology 274 (43) (1998) H2062–H2073.
[87] B. Fu, F.E. Curry, S. Weinbaum, A diffusion wake model for tracer ultrastructure-permeability studies in microvessels, American Journal of Physiology 269 (38) (1995) H2124–H2140.
[88] B. Fu, F.E. Curry, R.Y. Tsay, S. Weinbaum, A junction-orifice-fiber entrance layer model for capillary permeability: application to frog mesenteric capillaries, Journal of Biomechanical Engineering 116 (1994) 502–513.
[89] B.M. Fu, B. Chen, W. Chen, An electrodiffusion model for effects of surface glycocalyx layer on microvessel solute permeability, American Journal of Physiology 284 (2003) H1240–H1250.
[90] B.M. Fu, B. Chen, A model for the modulation of microvessel permeability by junction strands, Journal of Biomechanical Engineering 125 (2003) 620–627.
[91] B.M. Fu, S. Shen, Structural mechanisms of vascular endothelial growth factor (VEGF) on microvessel permeability, American Journal of Physiology 284 (6) (2003) H2124–H2135.
[92] B.M. Fu, S. Shen, Acute VEGF effect on solution permeability of mammalian microvessels *in vivo*, Microvascular Research 68 (1) (2004) 51–62.
[93] B.M. Fu, R.H. Adamson, F.E. Curry, Determination of microvessel permeability and tissue diffusion coefficient by laser scanning confocal microscopy, ASME Journal of Biomechanical Engineering 127 (2) (2005) 270–278.
[94] B.M. Fu, S. Shen, B. Chen, Structural mechanisms in the abolishment of VEGF-induced microvascular hyperpermeability by cAMP, ASME Journal of Biomechanical Engineering 128 (3) (2006) 313–328.
[95] K. Fukuda, H. Tanno, Y. Okimura, M. Nakamura, A. Yamaura, The blood-brain barrier disruption to circulating proteins in the early period after fluid percussion brain injury in rats, Journal of Neurotrauma 12 (3) (1995) 315–324.
[96] M. Furuse, T. Hirase, M. Itoh, I.A. Nagafuch, S. Yone-Mura, S. Tsukita, Occludin: a novel integral membrane protein localizing at tight junctions, Journal of Cell Biology 123 (1993) 1777–1788.
[97] P. Ganatos, S. Weinbaum, J. Fischbarg, J. Liebovitch, A hydrodynamic theory for determining the membrane coefficients for the passage of spherical molecules through an intercellular cleft, Advanced Bioengineering 3 (1981) 193–196.
[98] L. Gao, H.H. Lipowsky, Composition of the endothelial glycocalyx and its relation to its thickness and diffusion of small solutes, Microvascular Research 80 (2010) 394–401.
[99] P. Gassmann, M.L. Kang, S.T. Mees, J. Haier, *In vivo* tumor cell adhesion in the pulmonary microvasculature is exclusively mediated by tumor cell-endothelial cell interaction, BMC Cancer 10 (2010) 177–189.
[100] F.G. Giancotti, Targeting integrin beta4 for cancer and anti-angiogenic therapy, Trends Pharmacological Science 28 (2007) 506–511.
[101] R. Giavazzi, M. Foppolo, R. Dossi, A. Remuzzi, Rolling and adhesion of human tumor cells on vascular endothelium under physiological flow conditions, Journal of Clinical Investigation 92 (1993) 3038–3044.
[102] O.V. Glinskii, V.H. Huxley, G.V. Glinsky, K.J. Pienta, A. Raz, V.V. Glinsky, Mechanical entrapment is insufficient and intercellular adhesion is essential for metastatic cell arrest in distant organs, Neoplasia 7 (5) (2005) 522–527.
[103] E. Goldmann, Vitalfarbung am zentralnervensystem, Abh. K. Preuss. Akad. Wiss. Phys. Med. 1 (1913) 1–60.

[104] R. Govers, L. Bevers, P. de Bree, T.J. Rabelink, Endothelial nitric oxide synthase activity is linked to its presence at cell-cell contacts, Biochemical Journal 361 (2002) 193–201.

[105] W. Guo, F.G. Giancotti, Integrin signaling during tumor progression, Nature Reviews Molecular Cell Biology 5 (2004) 816–826.

[106] W. Guo, J. Pylayeva, A. Pepe, T. Yoshioka, W.J. Muller, G. Inghirami, F.G. Giancotti, β4 integrin amplifies ErbB2 signaling to promote mammary tumorigenesis, Cell 126 (3) (2006) 489–502.

[107] J. Haier, T. Korb, B. Hotz, H.U. Spiegel, N. Senninger, An intravital model to monitor steps of metastatic tumor cell adhesion within the hepatic microcirculation, Journal of Gastrointestinal Surgury 7 (2003) 507–514.

[108] K.A. Haldenby, D.C. Chappell, C.P. Winlove, K.H. Parker, J.A. Firth, Focal and regional variations in the composition of the glycocalyx of large vessel endothelium, Journal of Vascular Research 31 (1994) 2–9.

[109] J. Happel, Viscous flow relative to arrays of cylinders, AIChE Journal 5 (1959) 174–177.

[110] B.T. Hawkins, T.P. Davis, The blood-brain barrier/neurovascular unit in health and disease, Pharmacological Review 57 (2) (2005) 173–185.

[111] D.A. Hammer, S.M. Apte, Simulation of cell rolling and adhesion on surfaces in shear flow: general results and analysis of selectin-mediated neutrophil adhesion, Biophysical Journal 63 (1992) 35–57.

[112] M. Hecker, A. Mulsch, E. Bassenge, R. Busse, Vasoconstriction and increased flow: two principal mechanisms of shear stress-dependent endothelial autacoid release, American Journal of Physiology 265 (1993) H828–H833.

[113] C.B. Henry, B.R. Duling, Permeation of the luminal capillary glycocalyx is determined by hyaluronan, American Journal of Physiology 277 (2) (1999) H508–H514.

[114] S. Hippenstiel, M. Krull, A. Ikemann, W. Risau, M. Clauss, N. Suttorp, VEGF induces hyperpermeability by a direct action on endothelial-cells, American Journal of Physiology 18 (1998) L678–L684.

[115] J.D. Hood, D.A. Cheresh, Role of integrins in cell invasion and migration, Nature Review Cancer 2 (2002) 91–100.

[116] X. Hu, S. Weinbaum, A new view of Starling's Hypothesis at the microstructural level, Microvascular Research 58 (1999) 281–304.

[117] V.H. Huxley, F.E. Curry, M.R. Powers, B. Thipakorn, Differential action of plasma and albumin on transcapillary exchange of anionic solute, American Journal of Physiology 264 (33) (1993) H1428–H1437.

[118] R.O. Hynes, Integrins: versatility, modulation, and signaling in cell adhesion, Cell 69 (1) (1992) 11–25.

[119] R.L. Jackson, S.J. Busch, A.D. Cardin, Glycosaminoglycans: molecular properties, protein interactions, and role in physiological processes, Physiological Review 71 (1991) 481–539.

[120] I. Joris, H.F. Cuenoud, G.V. Doern, J.M. Underwood, G. Majno, Capillary leakage in inflammation: a study by vascular labeling, American Journal of Pathology 137 (1990) 1353–1363.

[121] M. Kajimura, S.D. Head, C.C. Michel, The effects of flow on the transport of potassium ions through the walls of single perfused frog mesenteric capillaries, Journal of Physiology 511 (1998) 707–718.

[122] M.J. Karnovsky, The ultrastructural basis of capillary permeability studied with peroxidase as a tracer, Journal of Cell Biology 35 (1967) 213–236.

[123] G.G. Kay, The effects of antihistamines on cognition and performance, The Journal of Allergy and Clinical Immunology 105 (6, Pt2) (2000) S622–S657.

[124] O. Kedem, G. Katchalsky, Permeability of composite membranes, Transactions of the Faraday Society 59 (1963) 1931–1953.

[125] Y. Kienast, L. von Baumgarten, M. Fuhrmann, W.E. Klinkert, R. Goldbrunner, J. Herms, F. Winkle, Real-time imaging reveals the single steps of brain metastasis formation, Nature Medicine 16 (1) (2010) 116–122.

[126] J.H. Kim, J.A. Park, S.W. Lee, W.J. Kim, Y.S. Yu, K.W. Kim, Blood-neural barrier: intercellular communication at glio-vascular interface, Journal of Biochemistry and Molecular Biology 39 (4) (2006) 339–345.

[127] M.R. King, D.A. Hammer, Multiparticle adhesive dynamics: hydrodynamic recruitment of rolling leukocytes, Proceedings of the National Academy of Sciences of the United States of America 98 (2001) 14919–14924.

[128] E.M. Landis, J.R. Pappenheimer, Exchange of substances through the capillary walls, Handbook of Physiology. Circulation, Am. Physiol. Soc., Washington, DC, 1963, pp. 961–1034.

[129] D.A. Lauffenburger, A.F. Horwitz, Cell migration: a physically integrated molecular process, Cell 84 (1996) 359–369.

[130] T.H. Lee, H.K. Avraham, S. Jiang, S. Avraham, Vascular endothelial growth factor modulates the transendothelial migration of MDA-MB-231 breast cancer cells through regulation of brain microvascular endothelial cell permeability, Journal of Biological Chemistry 278 (2003) 5277–5284.

[131] J.R. Levick, Capillary filtration-absorption balance reconsidered in light of dynamic extravascular factors, Experimental Physiology 76 (1991) 825–857.

[132] J.R. Levick, Flow through interstitium and other fibrous matrices, Quarterly Journal of Experimental Physiology 72 (1987) 409–437.
[133] J.R. Levick, L.H. Smaje, An analysis of the permeability of a fenestra, Microvascular Research 33 (1987) 233–256.
[134] M. Lewandowsky, Zur lehre von der cerebrospinalflussigkeit, Zeitschrift Fur Klinische Medizin 40 (1900) 480–494.
[135] G. Li, B.M. Fu, An electro-diffusion model for the blood-brain barrier permeability to charged molecules, ASME Journal of Biomechanical Engineering 133 (2) (2011) 0210.
[136] G. Li, W. Yuan, B.M. Fu, A model for water and solute transport across the blood-brain barrier, Journal of Biomechanics 43 (11) (2010) 2133–2140.
[137] S. Liang, M.J. Slattery, C. Dong, Shear stress and shear rate differentially affect the multi-step process of leukocyte-facilitated melanoma adhesion, Experimental Cell Research 310 (2) (2005) 282–292.
[138] S.H. Litjens, J.M. de Pereda, A. Sonnenberg, Current insights into the formation and breakdown of hemidesmosomes, Trends in Cell Biology 16 (2006) 376–383.
[139] Q. Liu, D. Mirc, B.M. Fu, Mechanical mechanisms of thrombosis in intact bent microvessels of rat mesentery, Journal of Biomechanics 41 (2008) 2726–2734.
[140] D.S. Long, Microviscometry reveals reduced blood viscosity and altered shear rate and shear stress profiles in microvessels after hemodilution, Proceedings of the National Academy of Sciences of the United States of America 101 (2004) 10060–10065.
[141] M.F. Loudon, C.C. Michel, I.F. White, The labeling of vesicles in frog endothelial cells with ferritin, Journal of Physiology 296 (1979) 97–112.
[142] J.R. Luft, Fine structure of capillary and endocapillary layer as revealed by ruthenium red, Federation Proceedings 2 (1966) 1773–1783.
[143] G. Majno, G.E. Palade, Studies of inflammation. I. The effect of histamine and serotonin on vascular permeability: an electron microscopy study, The Journal of Biophysical and Biochemical Cytology 11 (1961) 571–605.
[144] G. Majno, G.E. Palade, G. Schoefl, Studies on inflammation. II. The site of action of histamine and serotonin along the vascular tree: a topographic study, The Journal of Biophysical and Biochemical Cytology 11 (1961) 607–626.
[145] F. Mainiero, C. Murgia, K.K. Wary, A.M. Curatola, A. Pepe, M. Blumemberg, J.K. Westwick, C.J. Der, F.G. Giancotti, The coupling of alpha6beta4 integrin to Ras-MAP kinase pathways mediated by Shc controls keratinocyte proliferation, The EMBO Journal 16 (9) (1997) 2365–2375.
[146] D.M. McDonald, G. Thurston, P. Baluk, Endothelial gaps as sites for plasma leakage in inflammation, Microcirculation 6 (1) (1999) 7–22.
[147] D. Meyer, V.H. Huxley, Differential sensitivity of exchange vessel hydraulic conductivity to atrial natriuretic peptide, American Journal of Physiology 258 (27) (1990) H521–H528.
[148] C.C. Michel, Fluid movements through capillary walls, in: Handbook of Physiology. The Cardiovascular System. Microcirculation, sect 2, vol. IV, pt. I, Am. Physiol. Soc., Bethesda, MD, 1984, pp. 375–409 (Chapter 9).
[149] C.C. Michel, Capillary exchange, in: D.W. Seldin, G. Giebisch (Eds.), The Kidney: Physiology and Path Physiology, Raven, New York, 1992, pp. 61–91.
[150] C.C. Michel, Starling: the formulation of his hypothesis of microvascular fluid exchange and its significance after 100 years, Experimental Physiology 82 (1997) 1–30.
[151] C.C. Michel, F.E. Curry, Microvascular permeability, Physiological Reviews 79 (3) (1999) 703–761.
[152] C.C. Michel, C.R. Neal, Pathways through microvascular: endothelium or normal and increased permeability, in: G.V.R. Born, C.J. Schwartz (Eds.), Vascular Endothelium: Physiology, Pathology and Therapeutic Opportunities, Schattauer, Stuttgart, Germany, 1997, pp. 37–48.
[153] C.C. Michel, C.R. Neal, Openings through endothelial cells associated with increased microvascular permeability, Microcirculation 6 (1) (1999) 45–62.
[154] L.L. Mitic, J.M. Anderson, Molecular structure of tight junctions, Annual Review of Physiology 60 (1998) 121–142.
[155] M.M. Moasser, A. Basso, S.D. Averbuch, N. Rosen, The tyrosine kinase inhibitor ZD1839 ('Iressa') inhibits HER2-driven signaling and suppresses the growth of HER2-overexpressing tumor cells, Cancer Research 61 (2001) 7184–7188.
[156] S. Mochizuki, H. Vink, O. Hiramatsu, T. Kajita, F. Shigeto, J.A. Spaan, F. Kajiya, Role of hyaluronic acid glycosaminoglycans in shear-induced endothelium-derived nitric oxide release, American Journal of Physiology 285 (2003) H722–H726.

[157] D.M. Moody, The blood-brain barrier and blood-cerebral spinal fluid barrier, Seminars in Cardiothoracic and Vascular Anesthesia 10 (2) (2006) 128–131.

[158] O.R.F. Mook, J. Marle, H. Vreeling-Sindelarova, R. Jongens, W.M. Frederiks, C.J.K. Noorden, Visualisation of early events in tumor formation of eGFP-transfected rat colon cancer cells in liver, Hepatology 38 (2003) 295–304.

[159] V.L. Morris, E.E. Schmidt, I.C. MacDonald, A.C. Groom, A.F. Chambers, Sequential steps in hematogenous metastasis of cancer cells studied by *in vivo* videomicroscopy, Invasion and Metastasis 17 (6) (1997) 281–296.

[160] D. Mukhopadhyay, J.A. Nagy, E.J. Manseau, H.F. Dvorak, Vascular permeability factor/vascular endothelial growth factor-mediated signaling in mouse mesentery vascular endothelium, Cancer Research 58 (6) (1998) 1278–1284.

[161] A.W. Mulivor, H.H. Lipowsky, Inhibition of glycan shedding and leukocyte-endothelial adhesion in postcapillary venules by suppression of matrixmetalloprotease activity with doxycycline, Microcirculation 16 (2009) 657–666.

[162] E.A. Neuwelt, Mechanisms of disease: the blood-brain barrier, Neurosurgery 54 (1) (2004) 131–142.

[163] A.G. Ogston, B.N. Preston, J.D. Wells, On the transport of compact particles through solutions of chain-polymers, Proceedings of the Royal Society of London, Series A A333 (1973) 297–316.

[164] W.H. Oldendorf, M.E. Cornford, W.J. Brown, The large apparent work capability of the blood-brain barrier: a study of the mitochondrial content of capillary endothelial cells in brain and other tissues of the rat, Annals of Neurology 1 (5) (1977) 409–417.

[165] A.K. Olsson, A. Dimberg, J. Kreuger, L. Claesson-Welsh, VEGF receptor signaling – in control of vascular function, Nature Reviews Molecular Cell Biology 7 (5) (2006) 359–371.

[166] A. Oohira, T.N. Wight, P. Bornstein, Sulfated proteoglycans synthesized by vascular endothelial cells in culture, Journal of Biological Chemistry 258 (1983) 2014–2021.

[167] K. Osterloh, U. Ewert, A.R. Pries, Interaction of albumin with the endothelial cell surface, American Journal of Physiology 283 (2002) H398–H405.

[168] M.Y. Pahakis, J.R. Kosky, R.O. Dull, J.M. Tarbell, The role of endothelial glycocalyx components in mechanotransduction of fluid shear stress, Biochemical and Biophysical Research Communications 355 (2007) 228–233.

[169] P.L. Paine, P. Scherr, Drag coefficients for the movement of rigid spheres through liquid-filled cylindrical pores, Biophysical Journal 15 (1975) 1087–1091.

[170] L. Paka, Y. Kako, J.C. Obunike, S. Pillarisetti, Apolipoprotein E containing high density lipoprotein stimulates endothelial production of heparan sulfate rich in biologically active heparin-like domains. A potential mechanism for the anti-atherogenic actions of vascular apolipoproteine, Journal of Biological Chemistry 274 (1999) 4816–4823.

[171] T.L. Pallone, J. Work, R.L. Myers, R.L. Jamison, Transport of sodium and urea in outer medullary descending vasa recta, The Journal of Clinical Investigation 93 (1994) 212–222.

[172] J.R. Pappenhammer, E.M. Renkin, J.M. Borrero, Filtration, diffusion and molecular sieving through peripheral capillary membranes: a contribution to the pore theory of capillary permeability, American Journal of Physiology 167 (1951) 13–46.

[173] W.M. Pardridge, CNS drug design based on principles of blood-brain barrier transport, Journal of Neurochemistry 70 (1998) 1781–1821.

[174] W.M. Pardridge, Blood-brain barrier drug targeting: the future of brain drug development, Molecular Interventions 3 (2) (2003) 90–105.

[175] R.J. Phillips, W.M. Deen, J.F. Brady, Hindered transport of spherical macromolecules in fibrous membranes and gels, AIChE Journal 35 (11) (1989) 1761–1769.

[176] R.J. Phillips, W.M. Deen, J.F. Brady, Hindered transport in fibrous membranes and gels: effect of solute size and fiber configuration, Journal of Colloid and Interface Science 139 (2) (1990) 363–373.

[177] U. Pohl, K. Herlan, A. Huang, E. Bassenge, EDRF-mediated shear-induced dilation opposes myogenic vasoconstriction in small rabbit arteries, American Journal of Physiology 261 (1991) H2016–H2023.

[178] A.R. Pries, The endothelial surface layer, Pflugers Archiv: European Journal of Physiology 440 (2000) 653–666.

[179] T.S. Reese, M.J. Karnovsky, Fine structural localization of a blood-brain barrier to exogenous peroxidase, Journal of Cell Biology 34 (1) (1967) 207–217.

[180] S. Reitsma, The endothelial glycocalyx: composition, functions, and visualization, Pflugers Archiv: European Journal of Physiology 454 (2007) 345–359.

[181] E.M. Renkin, Transport pathways and processes, in: N. Simionescu, M. Simionescu (Eds.), Endothelial Cell Biology, Plenum, New York, 1988, pp. 51–68.

[182] E.M. Renkin, F.E. Curry, Transport of water and solutes across capillary endothelium, in: D.C. Tosteson, G. Giebisch, H.H. Ussing (Eds.), Membrane Transport in Biology, Springer-Verlag, New York, 1978, pp. 1–45.

[183] E.M. Renkin, V.L. Tucker, Atrial natriuretic peptide as a regulator of transvascular fluid balance, News in Physiological Sciences 11 (1996) 138–143.

[184] H.G. Rennke, R.S. Cotran, M.A. Venkatachalam, Role of molecular charge in glomerular permeability. Tracer studies with cationized ferritins, Journal of Cell Biology 67 (1975) 638–646.

[185] H.G. Rennke, Y. Patel, M.A. Venkatachalam, Glomerular filtration of proteins: clearances of anionic, neutral, and cationic horseradish peroxidase in the rat, Kidney International 13 (1978) 324–328.

[186] W.G. Roberts, G.E. Palade, Increased microvascular permeability and endothelial fenestration induced by vascular endothelial growth factor, Journal of Cell Science 108 (1995) 2369–2379.

[187] I.A. Romero, K. Radewicz, E. Jubin, C.C. Michel, J. Greenwood, P.O. Couraud, P. Adamson, Changes in cytoskeletal and tight junctional proteins correlate with decreased permeability induced by dexamethasone in cultured rat brain endothelial cells, Neuroscience Letters 344 (2) (2003) 112–116.

[188] A.S. Sangani, A. Acrivos, Slow flow past periodic arrays of cylinders with application to heat transfer, International Journal of Multiphase Flow 8 (1982) 193–206.

[189] R.J. Sawchuk, W.F. Elmquist, Microdialysis in the study of drug transporters in the CNS, Advanced Drug Delivery Reviews 45 (2000) 295–307.

[190] E.E. Schbeeberger, R.D. Lynch, B.A. Neary, Interaction of native and chemically modified albumin with pulmonary microvascular endothelium, American Journal of Physiology. Lung Cellular and Molecular Physiology 258 (2) (1990) L89–L98.

[191] K. Schluter, P. Gassmann, A. Enns, T. Korb, A. Hemping-Bovenkerk, J. Holzen, J. Haier, Organ-specific metastatic tumor cell adhesion and extravasation of colon carcinoma cells with different metastatic potential, American Journal of Pathology 169 (2006) 1064–1073.

[192] E.E. Schneeberger, M. Hamelin, Interaction of circulating proteins with pulmonary endothelial glycocalyx and its effect on endothelial permeability, American Journal of Physiology 247 (1984) H206–H217.

[193] E.G. Schuetz, A.H. Schinkel, M.V. Relling, J.D. Schuetz, P-glycoprotein: a major determinant of rifampicin-inducible expression of cytochrome P4503A in mice and humans, Proceedings of the National Academy of Sciences of the United States of America 93 (9) (1996) 4001–4005.

[194] C. Schulze, J.A. Firth, The interendothelial junction in myocardial capillaries: evidence for the existence of regularly spaced, cleft-spanning structures, Journal of Cell Science 101 (1992) 647–655.

[195] J.Y. Shao, G. Xu, The adhesion between a microvillus-bearing cell and a ligand-coated substrate: a Monte Carlo study, Annals of Biomedical Engineering 35 (2007) 397–407.

[196] S. Shen, J. Fan, B. Cai, Y. Lv, M. Zeng, Y. Hao, F.G. Giancotti, B.M. Fu, Vascular endothelial growth factor enhances mammary cancer cell adhesion to endothelium *in vivo*, Journal of Experimental Physiology 95 (2010) 369–379.

[197] S. Shimizu, A novel approach to the diagnosis and management of meralgia paresthetica, Neurosurgery 63 (4) (2008) E820–E832.

[198] M. Simard, G. Arcuino, T. Takano, Q.S. Liu, M. Nedergaard, Signaling at the gliovascular interface, The Journal of Neuroscience 23 (27) (2003) 9254–9262.

[199] M. Sinionescu, N. Sinionescu, Ultrastructure of the microvessel wall: functional correlations, in: Handbook of Physiology. The Cardiovascular System. Microcirculation, sect. 2, vol. IV, pt. 1, Am. Physiol. Soc., Bethesda, MD, 1984, pp. 41–101 (Chapter 3).

[200] P.A. Singleton, L.Y. Bourguignon, CD44 interaction with ankyrin and IP3 receptor in lipid rafts promotes hyaluronan-mediated Ca2+ signaling leading to nitric oxide production and endothelial cell adhesion and proliferation, Experimental Cell Research 295 (2004) 102–118.

[201] M.J. Slattery, S. Liang, C. Dong, Distinct role of hydrodynamic shear in leukocyte-facilitated tumor cell extravasation, American Journal of Physiology 288 (2005) C831–C839.

[202] Q.R. Smith, Transport of glutamate and other amino acids at the blood-brain barrier, Journal of Nutrition 130 (4S Suppl.) (2000) 1016S–1022S.

[203] J.M. Squire, M. Chew, G. Nneji, C. Neal, J. Barry, C.C. Michel, Quasi-periodic substructure in the microvessel endothelial glycocalyx: a possible explanation for molecular filtering? Journal of Structural Biology 136 (2001) 239–255.

[204] L. Spinardi, S. Einheber, T. Cullen, T.A. Milner, F.G. Giancotti, A recombinant tail-less integrin β_4 subunit disrupts hemidesmosomes, but does not suppress $\alpha_6\beta_4$-mediated cell adhesion to laminins, Journal of Cell Biology 129 (1995) 473–487.

[205] M. Steinbauer, M. Guba, G. Cernaianu, G. Köhl, M. Cetto, L.A. Kunz-Schugart, E.K. Gcissler, W. Falk, K.W. Jauch, GFP-transfected tumor cells are useful in examining early metastasis *in vivo*, but immune reaction precludes long-term development studies in immunocompetent mice, Clinical & Experimental Metastasis 20 (2003) 135–141.

[206] P.S. Steeg, D. Theodorescu, Metastasis: a therapeutic target for cancer, Nature Clinical Practice Oncology 5 (4) (2008) 206–219.

[207] M. Sugihara-Seki, Transport of spheres suspended in the fluid flowing between hexagonally arranged cylinders, Journal of Fluid Mechanics 551 (2006) 309–321.

[208] M. Sugihara-Seki, T. Akinaga, T. Itano, Flow across microvessel walls through the endothelial surface glycocalyx and the interendothelial cleft, Journal of Fluid Mechanics 601 (2008) 229–252.

[209] M. Sugihara-Seki, T. Akinaga, T. Itano, Effects of electric charge on osmotic flow across periodically arranged circular cylinders, Journal of Fluid Mechanics 664 (2010) 174–192.

[210] J.M. Tarbell, M.Y. Pahakis, Mechanotransduction and the glycocalyx, Journal of Internal Medicine 259 (2006) 339–350.

[211] J.M. Tarbell, Shear stress and the endothelial transport barrier, Cardiovascular Research 87 (2010) 320–330.

[212] A.E. Taylor, D.N. Granger, Exchange of macromolecules across the microcirculation, in: Handbook of Physiology. The Cardiovascular System. Microcirculation, sect. 2, vol. IV, pt. 2, Am. Physiol. Soc., Bethesda, MD, 1984, pp. 467–520 (Chapter 11).

[213] M.M. Thi, J.M. Tarbell, S. Weinbaum, D.C. Spray, The role of the glycocalyx in reorganization of the actin cytoskeleton under fluid shear stress: a 'bumper-car' model, Proceedings of the National Academy of Sciences of the United States of America 101 (2004) 16483–16488.

[214] E. Tkachenko, J.M. Rhodes, M. Simons, Syndecans: new kids on the signaling block, Circulation Research 96 (2005) 488–500.

[215] R. Tsay, S. Weinbaum, Viscous flow in a channel with periodic cross-bridging fibers of arbitrary aspect ratio and spacing, Journal of Fluid Mechanics 226 (1991) 125–148.

[216] M. Turner, Effects of proteins on the permeability of monolayers of cultured bovine arterial endothelium, The Journal of Physiology (London) 449 (1992) 21–35.

[217] M.R. Turner, T.L. Pallone, Hydraulic and diffusional permeabilities of isolated outer medullary descending vasa recta from the rat, American Journal of Physiology. Heart and Circulatory Physiology 272 (41) (1997) H392–H400.

[218] M. Ueno, H. Sakamoto, Y.J. Liao, M. Onodera, C.L. Huang, H. Miyanaka, T. Nakagawa, Blood-brain barrier disruption in the hypothalamus of young adult spontaneously hypertensive rats, Histochemistry and Cell Biology 122 (2) (2004) 131–137.

[219] B.M. van den Berg, The endothelial glycocalyx protects against myocardial edema, Circulation Research 92 (2003) 592–594.

[220] V.W. van Hinsbergh, G.P. Nieuw Amerongen, Intracellular signaling involved in modulating human endothelial barrier function, Journal of Anatomy 200 (2002) 549–560.

[221] J.W. VanTeeffelen, Agonist-induced impairment of glycocalyx exclusion properties: contribution to coronary effects of adenosine, Cardiovascular Research 87 (2010) 311–319.

[222] P. Vijayagopal, S.R. Srinivasan, E.R. Dalferes, Jr B. Radhakrishnamurthy, G.S. Berenson, Effect of low-density lipoproteins on the synthesis and secretion of proteoglycans by human endothelial cells in culture, The Biochemical Journal 255 (1988) 639–646.

[223] H. Vink, B.R. Duling, Identification of distinct luminal domains for macromolecules, erythrocytes, and leukocytes within mammalian capillaries, Circulation Research 79 (1996) 581–589.

[224] J.K. Wang, M.J. Slattery, M.H. Hoskins, S.L. Liang, C. Dong, Q. Du, Monte Carlo simulation of heterotypic cell aggregation in nonlinear shear flow, Mathematical Biosciences and Engineering 3 (2006) 683–696.

[225] B.J. Ward, K.F. Bauman, J.A. Firth, Interendothelial junctions of cardiac capillaries in rats: their structure and penne-ability properties, Cell and Tissue Research 252 (1988) 57–66.

[226] S. Weinbaum, R. Tsay, F.E. Curry, A three-dimensional junction-pore-matrix model for capillary permeability, Microvascular Research 44 (1992) 85–111.

[227] S. Weinbaum, The structure and function of the endothelial glycocalyx layer, Annual Review of Biomedical Engineering 9 (2007) 121–167.

[228] D.L. Wilhelm, Chemical mediators, in: B.W. Zweifach, L. Grant, R.T. McCluskey (Eds.), The Inflammatory Process, Academic, New York, 1973, pp. 261–301.
[229] S.L. Wissig, Identification of the small pore in muscle capillaries, Acta Physiologica Scandinavica. Supplementum 463 (1979) 33–44.
[230] K. Wolburg-Buchholz, A.F. Mack, E. Steiner, F. Pfeiffer, B. Engelhardt, H. Wolburg, Loss of astrocyte polarity marks blood-brain barrier impairment during experimental autoimmune encephalomyelitis, Acta Neuropathologica 18 (2) (2009) 219–233.
[231] Q. Wu, From red cells to snowboarding: a new concept for a train track, Physical Review Letters 93 (2004) 194–501.
[232] H.M. Wu, Q. Huang, Y. Yuan, H.J. Grange, VEGF induces NO dependent hyperpermeability in coronary venules, American Journal of Physiology 40 (1996) H2735–H2739.
[233] W.W. Yan, B. Cai, Y. Liu, B.M. Fu, Effects of wall shear stress and its gradient on tumor cell adhesion in curved microvessels, Biomechanics and Modeling in Mechanobiology 11 (2012) 641–653.
[234] W.W. Yan, Y. Liu, B.M. Fu, Effects of curvature and cell-cell interaction on cell adhesion in microvessels, Biomechanics and Modeling in Mechanobiology 9 (2010) 629–640.
[235] W.Y. Yen, B. Cai, M. Zeng, J.M. Tarbell, B.M. Fu, Quantification of the endothelial surface glycocalyx on rat and mouse blood vessels, Microvascular Research 83 (3) (2012) 337–346.
[236] Y. Yuan, H.J. Granger, C. Zawieja, W.M. Chilain, Flow modulates coronary venular permeability by a nitric oxide-related mechanism, American Journal of Physiology. Heart and Circulatory Physiology 263 (32) (1992) H641–H646.
[237] W. Yuan, Y. Lv, M. Zeng, B.M. Fu, Non-invasive method for the measurement of solute permeability of rat pial microvessels, Microvascular Research 77 (2009) 166–173.
[238] X. Zhang, F.E. Curry, S. Weinbaum, Mechanism of osmotic flow in a periodic fiber array, American Journal of Physiology 290 (2006) H844–H852.
[239] Y. Zhang, W.M. Pardridge, Rapid transferrin efflux from brain to blood across the blood-brain barrier, Journal of Neurochemistry 76 (2001) 1597–1600.
[240] C. Zhu, Kinetics and mechanics of cell adhesion, Journal of Biomechanics 33 (2000) 23–33.

CHAPTER

5

Carrier-Mediated Transport Through Biomembranes

Ranjan K. Pradhan, Kalyan C. Vinnakota, Daniel A. Beard, and Ranjan K. Dash

Biotechnology and Bioengineering Center and Department of Physiology, Medical College of Wisconsin, Milwaukee, WI, USA

OUTLINE

5.1 Introduction 182

5.2 Physicochemical Principles and Kinetic Modeling of Carrier-Mediated Transport 183

5.2.1 Thermodynamics of Solute Transport 183

5.2.2 Kinetic Treatment of a Simple Carrier 183

5.2.2.1 Rapid Equilibrium Approximations 186

5.2.3 Kinetic Treatment of a Simple Pore 187

5.2.3.1 Rapid Equilibrium Approximations 188

5.2.4 Membrane Potential Dependency of Solute Transport 188

5.3 Experimentally Observable Features of Carrier-Mediated Transport Phenomena 190

5.4 Kinetic Modeling of Mitochondrial Ca^{2+} Uniporter 191

5.4.1 Historical Background 191

5.4.2 Kinetic Scheme for Mitochondrial Ca^{2+} Transport via the Ca^{2+} Uniporter 193

5.4.3 Derivation of Mitochondrial Ca^{2+} Uniporter Flux Expression 195

5.4.3.1 Full Cooperative Binding 196

5.4.3.2 Partial Cooperative Binding 196

5.4.3.3 No Cooperative Binding 196

5.4.4 $\Delta\Psi$ Dependency of Mitochondrial Ca^{2+} Uniporter Flux: Free Energy Barrier Formalism 197

5.4.4.1 Equilibrium Constant 198

5.4.4.2 Dissociation Constants 199

5.4.4.3 Rate Constants 200

5.4.5 Mitochondrial Ca^{2+} Uniporter Model Parameterization and Simulations 200

5.4.5.1 Limitations of the Ca^{2+} Uniporter Model 201

Transport in Biological Media

© 2013 Elsevier Inc. All rights reserved.

http://dx.doi.org/10.1016/B978-0-12-415824-5.00005-9

5.4.6 $\Delta\Psi$ Dependency of Mitochondrial Ca^{2+} Uniporter Flux: Alternate Formulation 201
5.4.7 Mg^{2+} Inhibition and Pi Regulation of Mitochondrial Ca^{2+} Uniporter Function 204
5.4.8 Kinetic Scheme for Mg^{2+} Inhibition of Mitochondrial Ca^{2+} Uniporter Function 204
5.4.9 Mitochondrial Ca^{2+} Uniporter Model with Mg^{2+} Inhibition 206
5.5 Other Modes of Carrier-Mediated Transport: Antiport and Cotransport 208
5.6 Summary and Conclusion 210
Acknowledgment 211
References 211

5.1 INTRODUCTION

Biological membranes are composed of phospholipid bilayers, which act as a selective barrier within or around a cell [1]. Transporters or carriers are specialized proteins spanning the phospholipid bilayer that facilitate the translocation of ions, metabolites and macromolecules across the membrane. Transport processes are usually classified as active (primary or secondary) or passive transport [2]. Primary active transport involves the movement of a solute against its electrochemical gradient facilitated by coupling to a process that provides the required free energy, e.g., Na^+-K^+ pump driven by ATP hydrolysis, whereas passive transport is always driven by the solute electrochemical gradient. Secondary active transport is also commonly known as ion-coupled solute transport, and is where the electrochemical gradient of an ion maintained by an active transport process is utilized to transport a different solute. This is another commonly observed mode of transport in biological membrane systems (e.g., Na^+-glucose cotransporter [3], H^+-monocarboxylate cotransporter [4] and Na^+-Ca^{2+} exchanger [5]).

Commonly encountered modes of passive carrier-mediated transport fall into three major categories:

i. The uniport, which involves transport of a single solute across the membrane,
ii. The antiport (exchange), which involves transport of a solute from one side in exchange for a second solute from the opposite side of the membrane, and
iii. The symport (cotransport), which involves transport of two solutes in the same direction across the membrane.

Unlike pores or channels, carrier proteins undergo enzyme-like binding and conformational changes to promote energetically downhill transport of their specific ions or metabolites. The thermodynamic driving force for solute translocation is the net electrochemical gradient of the solute across the membrane. Theoretical models based on hypothesized carrier catalytic cycles are employed to design and interpret experiments aimed at determining the mode and the kinetics of the carrier-mediated solute transport problem at hand. The present chapter aims to provide a self-contained introduction to the kinetic treatment of passive, carrier-mediated, solute transport (Sections 5.2 and 5.3), with details given for the kinetic treatment of mitochondrial Ca^{2+} uniporter (Section 5.4) and a brief introduction given for the kinetic treatment of other kinds of carriers (cotransporters and antiporters) (Section 5.5). For an extensive treatment of membrane transport phenomena, the reader may refer to *Stein*'s excellent monographs [6,7].

5.2 PHYSICOCHEMICAL PRINCIPLES AND KINETIC MODELING OF CARRIER-MEDIATED TRANSPORT

5.2.1 Thermodynamics of Solute Transport

The driving force for the passive translocation of a set of solutes across a membrane is given by the difference in electrochemical potential of those solutes across the membrane. The electrochemical potential of a solute with charge z in a dilute solution is given by:

$$\mu = \mu_0 + RT \log a + zF\psi, \tag{5.1}$$

where μ is the electrochemical potential, μ_0 is the reference electrochemical potential, a is the activity, ψ is the electrical potential, R is the ideal gas constant, T is the absolute temperature and F is Faraday's constant. Let us consider carrier-mediated translocation of a single solute molecule from aqueous phase 1 to aqueous phase 2, at electrical potentials ψ_1 and ψ_2, respectively, with respect to a reference potential. After approximating the activity of this solute molecule by its concentration $[S]$ in both aqueous phases and assuming the same reference potential for either aqueous phase, the driving force for the translocation of this solute is given by:

$$\mu_2 - \mu_1 = RT\left(\log[S_2] - \log[S_1]\right) + zF\left(\psi_2 - \psi_1\right), \tag{5.2}$$

where the subscripts 1 and 2 denote the aqueous phases separated by the biological membrane. For multiple solutes, one may sum their electrochemical potential differences across the membrane to obtain the net driving force for the overall translocation. From Eq. (5.2) one may observe that when the electrochemical potential difference is zero, i.e., at equilibrium, the concentration ratio for a charged solute is determined by the electrical potential difference:

$$\log[S_2]_{eq} - \log[S_1]_{eq} = -zF\left(\psi_2 - \psi_1\right)/RT, \tag{5.3}$$

$$[S_2]_{eq}/[S_1]_{eq} = \exp(-zF\left(\psi_2 - \psi_1\right)/RT), \tag{5.4}$$

where the subscript *eq* denotes equilibrium of solute between phase 1 and phase 2. In the absence of any electrical potential or for an electro-neutral solute, $[S_2]_{eq} = [S_1]_{eq}$.

5.2.2 Kinetic Treatment of a Simple Carrier

The process of translocation of a single solute by a simple carrier is illustrated in Fig. 5.1a and 5.1b as a catalytic cycle that involves four key steps: association of solute with the carrier, conformational change of solute-carrier complex, dissociation of solute from the solute-carrier complex and the conformational change of the unbound carrier. A defining characteristic of carrier-mediated transport is the existence of distinct conformational states of the unloaded carrier protein, which are exclusively accessible to the solute molecules on each side of the membrane. When there is only one state of the unloaded protein (Fig. 5.1c), it becomes a simple pore with properties that are distinct from those of the simple carrier. The kinetics of the individual steps in the catalytic cycle describing solute transport may follow the law of mass action, where the rate of reaction is given by the product of the rate constant for that step and the activities of participating reactants. For example, the rate of forward reaction of the solute

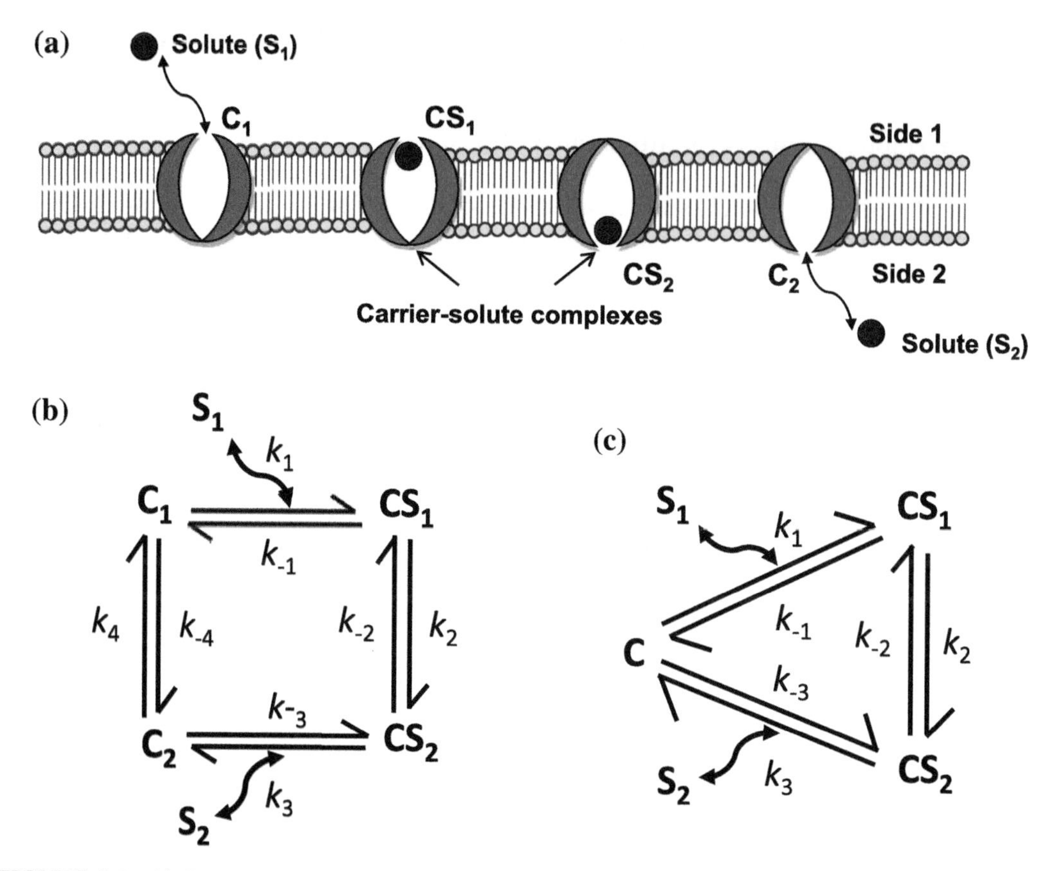

FIGURE 5.1 (a) Illustration of solute transport mediated by a simple carrier. (b) The catalytic cycle of a simple carrier. Solute S in aqueous phase 1 (S_1) is translocated to aqueous phase 2 (S_2) by first associating with the free carrier binding site accessible from side 1 (C_1) forming a solute-carrier complex CS_1. CS_1 undergoes a conformational change to CS_2, which dissociates to provide free carrier binding site C_2 accessible to the solute on side 2 and releases solute S to aqueous phase 2 (S_2). (c) The catalytic cycle of a simple pore differs from that of a simple carrier in having a single free carrier state. The rate constants for each elementary step i in panels B and C are denoted by k_i for the forward direction and k_{-i} for the reverse direction.

on side 1 binding to the carrier is given by $k_1[C_1][S_1]$ and that of dissociation of the complex CS_1 by $k_{-1}[CS_1]$, where k_1 is the rate for the forward reaction with units of $M^{-1}s^{-1}$ and k_{-1} is the rate constant of the reverse reaction with units of s^{-1}.

Using the law of mass action for the catalytic cycle described in Fig. 5.1b, and then using conservation of total carrier states, we arrive at the following set of equations describing the rates of change of bound and unbound states of the carrier protein:

$$\frac{d[C_1]}{dt} = k_4[C_2] - k_{-4}[C_1] - k_1[C_1][S_1] + k_{-1}[CS_1], \tag{5.5}$$

$$\frac{d[CS_1]}{dt} = k_1[C_1][S_1] - k_{-1}[CS_1] - k_2[CS_1] + k_{-2}[CS_2], \tag{5.6}$$

$$\frac{d[CS_2]}{dt} = k_2[CS_1] - k_{-2}[CS_2] - k_3[CS_2] + k_{-3}[C_2][S_2], \tag{5.7}$$

$$[C_2] = [C_{tot}] - [C_1] - [CS_1] - [CS_2]. \tag{5.8}$$

When the catalytic cycle described in Fig. 5.1b and in Eqs. (5.5–5.8) operates under a steady state turnover of the carrier states, the time derivatives of the carrier states become zero, which facilitates the calculation of their steady state distribution as the solution of a set of algebraic equations. After setting the time derivatives of the carrier states in Eqs. (5.5–5.8) to zero and then solving the resulting algebraic equations, we obtain the following steady state distribution of the carrier states:

$$[C_1] = [C_{tot}]\frac{k_3k_4(k_2 + k_{-1}) + k_{-2}k_{-1}(k_4 + k_{-3})[S_2]}{D}, \tag{5.9}$$

$$[CS_1] = [C_{tot}]\frac{k_1k_4(k_3 + k_{-2})[S_1] + k_{-3}k_{-2}(k_{-4} + k_1[S_1])[S_2]}{D}, \tag{5.10}$$

$$[CS_2] = [C_{tot}]\frac{k_1k_2(k_4 + k_{-3}[S_2])[S_1] + k_{-4}k_{-3}(k_{-1} + k_2)[S_2]}{D}, \tag{5.11}$$

$$[C_2] = [C_{tot}]\frac{k_{-4}k_{-1}(k_{-2} + k_3) + k_2k_3(k_{-4} + k_1[S_1])}{D}, \tag{5.12}$$

where $[C_{tot}]$ is the total concentration of the carrier protein; D is the denominator expression given by the sum of all numerator terms divided by $[C_{tot}]$:

$$\begin{aligned} D = {} & k_3k_4(k_2 + k_{-1}) + k_{-2}k_{-1}(k_4 + k_{-4}) + k_{-4}k_3(k_{-1} + k_2) \\ & + [k_1k_4(k_3 + k_{-2}) + k_1k_2(k_4 + k_3)][S_1] + [k_{-3}k_{-2}(k_{-1} + k_{-4}) \\ & + k_{-4}k_{-3}(k_{-1} + k_2)][S_2] + k_{-3}k_1(k_{-2} + k_2)[S_1][S_2]. \end{aligned} \tag{5.13}$$

When the carrier states achieve the distributions shown in Eqs. (5.9–5.12) much faster than the changes in $[S_1]$ and $[S_2]$, the net rate of translocation of the solute across the membrane is given by the net solute flux through the carrier:

$$J_{net} = k_2[CS_1] - k_{-2}[CS_2] = [C_{tot}]\frac{k_1k_2k_3k_4[S_1] - k_{-1}k_{-2}k_{-3}k_{-4}[S_2]}{D}. \tag{5.14}$$

The same flux expression will also be obtained using, for example, $J_{net} = k_{-4}[C_2] - k_4[C_1]$. At equilibrium, the forward and reverse reaction rates for each reaction step are equal in magnitude and the net solute flux through the carrier is zero, which results in the following thermodynamic constraint for the rate constants governing the 4-state carrier model:

$$\frac{[S_2]_{eq}}{[S_1]_{eq}} = \frac{k_1k_2k_3k_4}{k_{-1}k_{-2}k_{-3}k_{-4}}. \tag{5.15}$$

The right hand side of Eq. (5.15) is equal to the product of equilibrium constants of individual reaction steps of the catalytic cycle described in Fig. 5.1b.

5.2.2.1 *Rapid Equilibrium Approximations*

Further simplification of the steady state treatment is possible if the binding of the solute to the carrier is considered to be much faster than the conformational changes in the carrier. Assuming that the equilibrium constant for the dissociation of the solute-carrier complex is the same on both sides of the membrane, we arrive at the following algebraic equations governing the steady state distributions of $[CS_1]$ and $[CS_2]$ for the simple carrier Eqs. (Eqs. 5.5–5.8):

$$(k_{-4}[C_1] - k_4[C_2]) + (k_2[CS_1] - k_{-2}[CS_2]) = 0, \tag{5.16}$$

$$[C_1] + [C_2] + [CS_1] + [CS_2] = [C_{tot}], \tag{5.17}$$

$$K_S = \frac{k_{-1}}{k_1} = \frac{[C_1][S_1]}{[CS_1]}; \; K_S = \frac{k_3}{k_{-3}} = \frac{[C_2][S_2]}{[CS_2]}, \tag{5.18}$$

where K_S is the dissociation constant of the solute-carrier complex on either side of the membrane. The assumption of equal dissociation constants for the solute-carrier complex on either side may not hold when the physical conditions such as ionic strength and temperature are markedly different between the two phases. When the dissociation equilibrium constants (K_S) differ between the two phases on either side of the membrane, one may replace $[S_1]/K_S$ with $[S_1]/K_{S1}$ and $[S_2]/K_S$ with $[S_2]/K_{S2}$ in Eqs. (5.19–5.22); K_{S1} and K_{S2} are the dissociation equilibrium constants in aqueous phases 1 and 2.

Solving Eqs. (5.16–5.18), we obtain the following steady state distribution of $[CS_1]$ and $[CS_2]$ and the net solute flux through the carrier:

$$[CS_1] = \frac{[C_{tot}]}{D}\left(k_{-2}\frac{[S_2]}{K_S} + k_4\right)\frac{[S_1]}{K_S}, \tag{5.19}$$

$$[CS_2] = \frac{[C_{tot}]}{D}\left(k_2\frac{[S_1]}{K_S} + k_{-4}\right)\frac{[S_2]}{K_S}, \tag{5.20}$$

$$J_{net} = k_2[CS_1] - k_{-2}[CS_2] = \frac{[C_{tot}]}{D}\left(k_2k_4\frac{[S_1]}{K_S} - k_{-2}k_{-4}\frac{[S_2]}{K_S}\right), \tag{5.21}$$

where the denominator expression D is given by:

$$D = \left(1 + \frac{[S_1]}{K_S}\right)\left(k_{-2}\frac{[S_2]}{K_S} + k_4\right) + \left(1 + \frac{[S_2]}{K_S}\right)\left(k_2\frac{[S_1]}{K_S} + k_{-4}\right). \tag{5.22}$$

The equilibrium relationship and the thermodynamic constraint for the rate constants governing the simple carrier model (Eq. 5.15) is reduced to:

$$\frac{[S_2]_{eq}}{[S_1]_{eq}} = \frac{k_2 k_4}{k_{-2} k_{-4}}. \tag{5.23}$$

5.2.3 Kinetic Treatment of a Simple Pore

For a simple pore (Fig. 5.1c), the set of equations describing the rates of change of bound and unbound states of the protein are:

$$\frac{d[CS_1]}{dt} = k_1[C][S_1] - k_{-1}[CS_1] - k_2[CS_1] + k_{-2}[CS_2], \tag{5.24}$$

$$\frac{d[CS_2]}{dt} = k_2[CS_1] - k_{-2}[CS_2] - k_3[CS_2] + k_{-3}[C][S_2], \tag{5.25}$$

$$[C] = [C_{tot}] - [CS_1] - [CS_2]. \tag{5.26}$$

Under steady state operation of the catalytic cycle described in Fig. 5.1c and in Eqs. (5.24–5.26), the distribution of the pore states is given by:

$$[CS_1] = [C_{tot}]\frac{k_1(k_3 + k_{-2})[S_1] + k_{-3}k_{-2}[S_2]}{D}, \tag{5.27}$$

$$[CS_2] = [C_{tot}]\frac{k_1 k_2[S_1] + k_{-3}(k_{-1} + k_2)[S_2]}{D}, \tag{5.28}$$

$$[C] = [C_{tot}]\frac{k_{-1}(k_{-2} + k_3) + k_2 k_3}{D}, \tag{5.29}$$

where D is the denominator expression given by the sum of all numerator terms divided by $[C_{tot}]$:

$$D = k_{-1}(k_{-2} + k_3) + k_2 k_3 + k_1(k_3 + k_{-2} + k_2)[S_1] + k_{-3}(k_{-2} + k_{-1} + k_2)[S_2]. \tag{5.30}$$

Accordingly, the net flux of the solute across the membrane is given by the net solute flux through the simple pore:

$$J_{net} = k_2[CS_1] - k_{-2}[CS_2] = [C_{tot}]\frac{k_1 k_2 k_3[S_1] - k_{-1}k_{-2}k_{-3}[S_2]}{D}. \tag{5.31}$$

At equilibrium, the net solute flux through the carrier is zero, which results in the following thermodynamic constraint for the rate constants governing the simple pore model:

$$\frac{[S_2]_{eq}}{[S_1]_{eq}} = \frac{k_1 k_2 k_3}{k_{-1}k_{-2}k_{-3}}. \tag{5.32}$$

5.2.3.1 *Rapid Equilibrium Approximations*

Under the assumption of rapid equilibrium, it is straightforward to show that the steady state distribution of $[CS_1]$ and $[CS_2]$ and the net solute flux through the pore are given by:

$$[CS_1] = \frac{[C_{tot}]}{D}\frac{[S_1]}{K_S} \quad \text{and} \quad [CS_2] = \frac{[C_{tot}]}{D}\frac{[S_2]}{K_S}, \tag{5.33}$$

$$J_{net} = k_2[CS_1] - k_{-2}[CS_2] = \frac{[C_{tot}]}{D}\left(k_2\frac{[S_1]}{K_S} - k_{-2}\frac{[S_2]}{K_S}\right), \tag{5.34}$$

where the denominator expression D is given by

$$D = 1 + \frac{[S_1]}{K_S} + \frac{[S_2]}{K_S}. \tag{5.35}$$

The equilibrium relationship and the thermodynamic constraint for the rate constants governing the simple pore model (Eq. 5.32) is reduced to:

$$\frac{[S_2]_{eq}}{[S_1]_{eq}} = \frac{k_2}{k_{-2}}. \tag{5.36}$$

5.2.4 Membrane Potential Dependency of Solute Transport

To establish how the transport process depends on the electrical potential difference across the membrane, let us consider a charged solute with valence z. Combining Eq. (5.4) with Eqs. (5.15 and 5.23), we obtain the following thermodynamic constraints for the rate constants of the carrier catalytic cycles under two different assumptions (with and without rapid equilibrium assumption on the binding of solute to the carrier) for the simple carrier:

$$k_1k_2k_3k_4/k_{-1}k_{-2}k_{-3}k_{-4} = \exp\left(-zF\left(\psi_2 - \psi_1\right)/RT\right), \tag{5.37}$$

$$k_2k_4/k_{-2}k_{-4} = \exp\left(-zF\left(\psi_2 - \psi_1\right)/RT\right). \tag{5.38}$$

Here Eq. (5.37) corresponds to the simple carrier model and Eq. (5.38) corresponds to simple carrier model with rapid equilibrium assumption. When the electrical potential difference across the membrane is zero or when the solute is electro-neutral, the right hand sides of Eqs. (5.37–5.38) are reduced to unity.

The thermodynamic constraints stated in Eqs. (5.37 and 5.38) do not describe how the rate constants themselves are modified by the membrane potential. Erying's transition theory provides a representation for how an electrical potential difference across the membrane influences the individual transition rate constants [8–11]. Eyring's free energy is based on the existence of a free energy barrier associated with the transition state (*TS*) that the molecular system must pass through in moving from one state conformation to other. Rate constants for state transition depend on the transition state free energy according to:

$$k = \kappa\,(k_BT/h)\exp\left(-\Delta G^{TS}/RT\right), \tag{5.39}$$

where κ is the probability that the molecules in the transition state TS decay to form products, k_B is Boltzmann's constant, h is Planck's constant, ΔG^{TS} is the free energy change from the reactants to the transition state.

This thought is applied to the transition of $CS_1 \underset{k_{-2}}{\overset{k_2}{\rightleftarrows}} CS_2$ in the model of Fig. 5.1b. Here the states CS_1 and CS_2 are associated with minima in free energy along an abstract reaction coordinate representing progress of reaction between the states. The transition state energy ΔG_2^{TS} is associated with the transition state barrier, which is modified by the membrane potential. Let us define the fraction of trans-membrane potential difference span modifying the transition state barrier as β, which results in the addition of $\beta F(\psi_2 - \psi_1)$ to ΔG_2^{TS}. Therefore, the rate constant k_2 is given by:

$$k_2 = \kappa\left(\frac{k_BT}{h}\right)\exp\left(-\frac{\Delta G_2^{TS} + \beta zF(\psi_2 - \psi_1)}{RT}\right) = k_2^0\exp\left(-\frac{\beta zF(\psi_2 - \psi_1)}{RT}\right) \tag{5.40}$$

where $k_2^0 = \kappa(k_BT/h)\exp\left(-\Delta G_2^{TS}/RT\right)$ is the rate constant k_2 in the absence of a membrane potential. The reverse rate constant k_{-2} is given by the following expression:

$$k_{-2} = \kappa\left(\frac{k_BT}{h}\right)\exp\left(-\frac{\Delta G_{-2}^{TS} - (1-\beta)zF\Delta\psi}{RT}\right) = k_{-2}^0\exp\left(\frac{(1-\beta)zF(\psi_2 - \psi_1)}{RT}\right) \tag{5.41}$$

where $k_{-2}^0 = \kappa\,(k_BT/h)\exp\left(-\Delta G_{-2}^{TS}/RT\right)$ is the rate constant k_{-2} in the absence of a membrane potential, and ΔG_{-2}^{TS} is the free energy change from CS_2 to the transition state. Combining Eqs. (5.15, 5.37, 5.40 and 5.41), the carrier cycle equilibrium constant is given by:

$$\begin{aligned}\frac{k_1k_2k_3k_4}{k_{-1}k_{-2}k_{-3}k_{-4}} &= \frac{k_1k_2^0\exp\left(-\beta zF(\psi_2 - \psi_1)/RT\right)k_3k_4}{k_{-1}k_{-2}^0\exp\left((1-\beta)zF(\psi_2 - \psi_1)/RT\right)k_{-3}k_{-4}}\\ &= \frac{k_1k_2^0k_3k_4}{k_{-1}k_{-2}^0k_{-3}k_{-4}}\exp\left(-zF(\psi_2 - \psi_1)/RT\right)\\ &= \exp\left(-zF(\psi_2 - \psi_1)/RT\right)\end{aligned} \tag{5.42}$$

Eyring's theory for rate constants (see Fig. 5.2) is also applicable to kinetic models of pores in a similar manner (see [10]).

In the preceding analysis, we make the assumption that the binding of the solute is independent of the membrane potential since the solute binds to the carrier in a conducting medium. The free carrier is also assumed to be uncharged. While this may not necessarily be true, the preceding analysis could be easily extended to describe charge translocation across the membrane in the relevant carrier catalytic cycle steps (see Section 5.4 below).

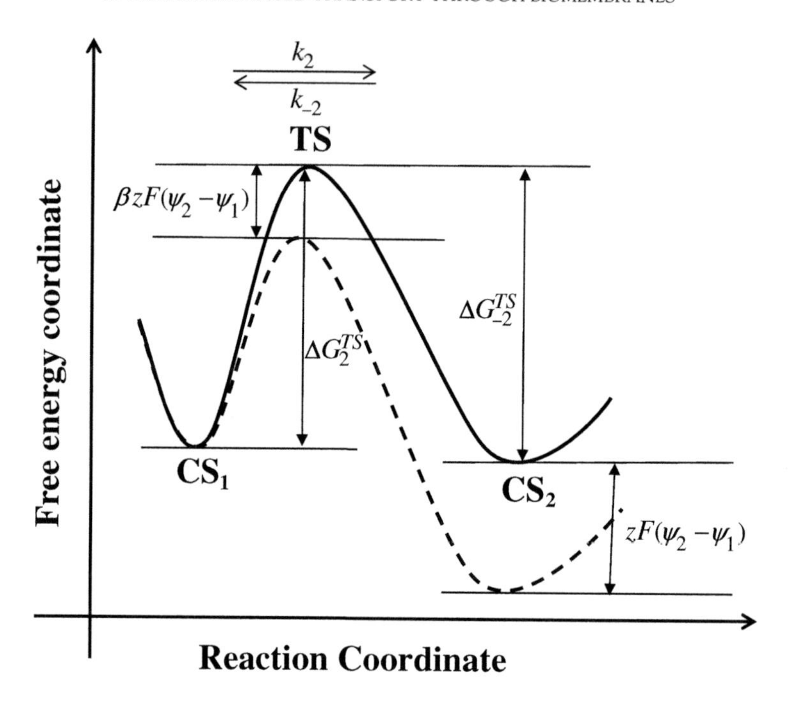

FIGURE 5.2 Free energy vs. reaction coordinate plot for the reaction $CS_1 \underset{k_{-2}}{\overset{k_2}{\rightleftarrows}} CS_2$ illustrating Eyring's free energy barrier theory. The solid line represents the free energy barrier in the absence a potential difference across the membrane, while the dashed line represents the modified free energy barrier when a potential difference exists across the membrane. CS_1 goes through a transition state (TS) with a free energy change ΔG_2^{TS} to form the product CS_2. ΔG_2^{TS} is modified by addition of $\beta_z F(\psi_2 - \psi_1)$ in the presence of potentials ψ_1 and ψ_2 on sides 1 and 2 of the membrane. Similarly, through the reverse reaction, CS_2 goes through the transition state TS with a free energy change ΔG_{-2}^{TS} to form the product CS_1. ΔG_{-2}^{TS} is modified by addition of $(\beta - 1)zF(\psi_2 - \psi_1)$ in the presence of potentials ψ_1 and ψ_2 on sides 1 and 2 of the membrane.

5.3 EXPERIMENTALLY OBSERVABLE FEATURES OF CARRIER-MEDIATED TRANSPORT PHENOMENA

An important role of the preceding theoretical treatment is to enable the design of experimental tests to determine the mode and potentially the molecular kinetic features of the solute transport phenomenon under investigation. For this purpose, it is useful to ask which properties of a carrier-mediated transport flux are helpful in excluding other modes of transport, such as simple diffusion through the membrane or transport through a pore (Fig. 5.1c).

A frequently employed experimental technique for flux measurement involves labeling the substance being transported with a radioactive tracer or a fluorescent dye on one side of the membrane and following the appearance of the tracer on the other side of the membrane. If one were to add a tracer label to S_1 and follow its movement across the membrane, the initial flux of the appearance of the tracer label on side 2 for a carrier is given by the following unidirectional flux, based on the 4-state carrier model (derived from Eqs. (5.19–5.22):

$$J_{1\to 2} = k_2[CS_1] = \frac{k_2[C_{tot}]\left(k_{-2}\frac{[S_2]}{K_S} + k_4\right)\frac{[S_1]}{K_S}}{\left(1+\frac{[S_1]}{K_S}\right)\left(k_{-2}\frac{[S_2]}{K_S}+k_4\right)+\left(1+\frac{[S_2]}{K_S}\right)\left(k_2\frac{[S_1]}{K_S}+k_{-4}\right)}, \tag{5.43}$$

and flux based on the catalytic cycle for a simple pore shown in Fig. 5.1c (derived from Eqs. 5.33–5.35):

$$J_{1\to 2} = k_2[CS_1] = k_2[C_{tot}]\frac{[S_1]}{K_S} \Big/ \left(1+\frac{[S_1]}{K_S}+\frac{[S_2]}{K_S}\right). \tag{5.44}$$

In a closed system, only the initial flux follows Eqs. (5.43–5.44), since the transport of the solute over time alters S_1 and S_2. To compute the concentrations of S_1 and S_2 over time, one must solve a set of ordinary differential equations that represent the transport experiment.

Terminology in experimental transport research defines a reference aqueous phase as *cis* and the phase across the membrane from the reference phase as *trans*. Let us define aqueous phase 1 as *cis* and phase 2 as *trans* to describe three common initial experimental conditions, viz., equilibrium exchange, zero-*trans* and infinite-*cis*. In an equilibrium exchange experiment, $[S_1]$ and $[S_2]$ across the membrane are equal for an electro-neutral solute or in the absence of a membrane potential for a charged solute, or are constrained by the thermodynamic constraint in Eq. (5.4) for a clamped membrane potential. $j_{1\to 2}$ is measured as a function of the equilibrium $[S_1]$ and $[S_2]$ concentrations, which is simpler for a membrane potential independent process when compared to a membrane potential dependent process. In a zero-*trans* experiment, $[S_2]$ is set to zero and $J_{1\to 2}$ is measured as a function of initial $[S_1]$. In an infinite *cis* experiment, $[S_1] \gg K_s$ and $J_{1\to 2}$ is measured as a function of initial $[S_2]$. Following *Lieb* [12], taking derivatives of Eqs. (5.43) and (5.44) with respect to $[S_2]$ shows that there can be no condition where $J_{1\to 2}$ through a pore may increase when $[S_2]$ is increased, whereas for a carrier $J_{1\to 2}$ under equilibrium exchange conditions can be greater than that under zero-trans conditions when $k_4 < k_{-2}$, a property known as trans-acceleration. A second property that is unique to carrier-mediated transport, even when the trans-acceleration is not observed, is the counterflow of tracer labels against an uphill gradient, which was first demonstrated by *Rosenberg and Wilbrandt* [13]. For a detailed exposition the reader may refer to *Lieb* [12].

5.4 KINETIC MODELING OF MITOCHONDRIAL Ca^{2+} UNIPORTER

5.4.1 Historical Background

Mitochondria are tubular organelles found in every eukaryotic cell and consist of two specialized membranes embedded with many transport proteins. While the outer membrane contains many channels formed by the protein *porin* to filter out large molecules, the inner membrane contains a group of transport proteins to allow the passage of different ions and metabolites from cytosol to the mitochondrial matrix. The mitochondrion possesses an intricate transport system for facilitating the transport of Ca^{2+} ions across its inner membrane to regulate cytosolic Ca^{2+}, to serve as a buffer during excess Ca^{2+} overload, and to modulate mitochondrial matrix Ca^{2+}, thereby controlling the activities of Ca^{2+}-sensitive enzymes (e.g., the dehydrogenases of the TCA cycle), for its bioenergetic function.

Ca^{2+} uniporter is a membrane protein located in the inner mitochondrial membrane (IMM) and serves as the primary pathway for Ca^{2+} transport into the energized mitochondria [14,15]. The mechanism of mitochondrial Ca^{2+} transport has been a subject of investigation from various standpoints for over five decades. Originally, mitochondrial Ca^{2+} uptake was described as an active transport mechanism [16]. However, with the evolution of chemiosmotic theory and measurement of a large negative IMM potential ($\Delta\Psi$) [17] led to our current understanding, which is that Ca^{2+} is transported into the energized mitochondria via the Ca^{2+} uniporter down the electrochemical gradient maintained across the IMM without utilizing any metabolic energy or directly coupling to other ion transport. Therefore, any kinetic study on mitochondrial Ca^{2+} uptake should concern both the effects of concentration gradient of Ca^{2+} and $\Delta\Psi$ across the IMM.

The earliest measurements of mitochondrial Ca^{2+} uptake in response to variations in extra-matrix $[Ca^{2+}]$ were performed in isolated respiring mitochondria from both rat livers and rat hearts [18–24]. These studies showed higher-order kinetics that are characteristic of cooperative binding, and a saturation mechanism associated with carrier-mediated transport. Most studies reported sigmoidicity in the plots of mitochondrial Ca^{2+} uptake vs. extra-matrix $[Ca^{2+}]$ with positive cooperativity and considerable variations in the Ca^{2+} binding affinity (K_m) and maximum uptake velocity (V_{max}). The Hill coefficient for the Ca^{2+} uptake has usually been reported to be around 2, indicating two Ca^{2+} ions bound in cooperative transport sites or one at a transport site and one at a separate activation site, such that binding of one Ca^{2+} ion at the activation site increases the affinity for the binding of another Ca^{2+} ion at the transport site. In addition, a large variation was reported in the apparent K_m value (K_m ranges from $\sim$1 to 90 μM; describing the extra-matrix $[Ca^{2+}]$ at which the transporter shows the half-maximal activity) under different experimental conditions with other divalent metal ions (e.g., Mg^{2+}).

The IMM $\Delta\Psi$ dependency of mitochondrial Ca^{2+} uptake has also been studied extensively using isolated mitochondrial preparations [15,22]. These studies suggest a non-linear Goldman-Hodgkin-Katz (GHK) type of dependency of Ca^{2+} uptake on $\Delta\Psi$ [25]. Until recently, the uniporter-mediated mitochondrial Ca^{2+} uptake has been described to be consistent with carrier, gated pore and with channel modes of transport [15]. Nevertheless, the mechanism of trans-acceleration has not been shown in any of the Ca^{2+} uniporter kinetic studies. In this regard, by comparing the turnover rates of Ca^{2+} per site, a large turnover rate for the uniporter was suggested, indicating the uniporter might work more like a gated pore [15]. Later, a patch clamp study in mitoplasts isolated from COS-7 cells demonstrated that the uniporter is a highly selective ion channel [26]. More recently, the structural identity of the uniporter was reported, which suggests that it is a 40 kDa protein that forms oligomers in the IMM, and resides within a high molecular weight complex. It is also thought to consist of two predicted trans-membrane helices [27,28]. An EF-hand-containing protein MICU1 has been suggested to interact with the uniporter and may serve as a putative site for binding of extra-matrix Ca^{2+} for uniporter operation [27,28]. However, identification of the actual structure and composition of the uniporter continues to be an active research topic in the field.

The kinetics of mitochondrial Ca^{2+} uptake is primarily determined by the catalytic properties of the Ca^{2+} uniporter, the electrochemical gradient of Ca^{2+} across the IMM and other regulatory factors (e.g., Mg^{2+} inhibition of the uniporter function, effects of cytosolic Pi and pH on the uniporter activity). Over the past five decades, the biophysical and catalytic properties of the Ca^{2+} uniporter in respiring mitochondria have been extensively studied using many initial velocity measurements [18–23] and mathematical models [25,29–32]. Mathematical modeling along with the experimental data has been shown to provide deeper insights into the kinetics and regulation of many membrane transport systems, including the Ca^{2+} uniporter. In the study of mitochondrial metabolism, the Ca^{2+} uniporter has been extensively modeled by many investigators, in attempts to understand its integrated functions in mitochondrial Ca^{2+} homeostasis and energy metabolism [33–35].

The initial mitochondrial Ca^{2+} uniporter model was developed by *Magnus-Keizer* [32], based on a simple binding scheme of Ca^{2+} for the uniporter (4-state model), which has been used by many researchers for integrated modeling studies of mitochondrial Ca^{2+} dynamics [33–35]. In this uniporter model, the $\Delta\Psi$ dependency of Ca^{2+} transport was described based on a linear GHK constant field approximation for electrodiffusion, with an offset potential $\Delta\Psi^* = 91$ mV to describe the $\Delta\Psi$ dependent data. However, this model fails to provide a unique, consistent explanation for the experimentally observed kinetics of uniporter-mediated Ca^{2+} transport, and hence was considered to have limited applications for studying integrated mitochondrial functions [30]. The major drawback of the Magnus-Keizer uniporter model is that it collapses for membrane potentials $\Delta\Psi \le \Delta\Psi^* = 91$ mV, and is not thermodynamically balanced. This model also fails to explain the extra-matrix $[Ca^{2+}]$ dependent data on Ca^{2+} uptake [20,21] measured in isolated rat liver and rat heart mitochondria. Furthermore, the Magnus-Keizer integrated model of mitochondrial Ca^{2+} handling [32] predicts a high steady state mitochondrial $[Ca^{2+}]$ ($\sim$15 μM) in response to a low cytoplasmic $[Ca^{2+}]$ ($\sim$1 μM) which is unusual for most cardiac cells.

Recently, our group has systematically developed a series of kinetic models [25,30,31] of the mitochondrial Ca^{2+} uniporter which mechanistically characterizes various driving forces that govern the uniporter function, based on a large body of experimental data [18–23] concerning the kinetics of mitochondrial Ca^{2+} uptake. Figure 5.3 shows the general schematics and proposed mechanisms (a 5 state carrier model and a reduced 3 state carrier model) for Ca^{2+} transport via the uniporter from the cytoplasmic side to the matrix side, driven by the electrochemical gradient of Ca^{2+} across the IMM, in the absence of any other effector interactions (e.g., Mg^{2+} and Pi) [30]. This preliminary Ca^{2+} uniporter model was developed on the basis of Michaelis-Menten kinetics for multi-state catalytic binding and an interconversion mechanism associated with carrier-mediated facilitated transport, combined with Eyring's free energy barrier theory for absolute reaction rates associated with interconversion or electrodiffusion (Ca^{2+} translocation). The model provides a biophysical basis for the catalytic cycle associated with Ca^{2+} transport via the uniporter and also depicts the mechanisms of cooperative binding of Ca^{2+} to the uniporter which are observed experimentally.

In the next subsections, we focus on details to illustrate how a step by step approach is applied to characterize mechanistically the kinetics of the mitochondrial Ca^{2+} uniporter and the regulation of its transport function by other cytosolic factors (e.g., Mg^{2+} and Pi).

5.4.2 Kinetic Scheme for Mitochondrial Ca^{2+} Transport via the Ca^{2+} Uniporter

The Ca^{2+} uniporter (Fig. 5.3a) is assumed to have two binding sites for Ca^{2+}, and these binding sites are assumed to be exposed to either the cytosolic (external) or matrix (internal) side of the IMM. In the 5-state carrier model (Fig. 5.3b), an ionized free Ca^{2+} from the external side of the IMM first binds to the unbound uniporter T (State 1) to form the intermediate complex TCa_e^{2+} (State 2), which then favors the binding of another ionized free Ca^{2+} (cooperative

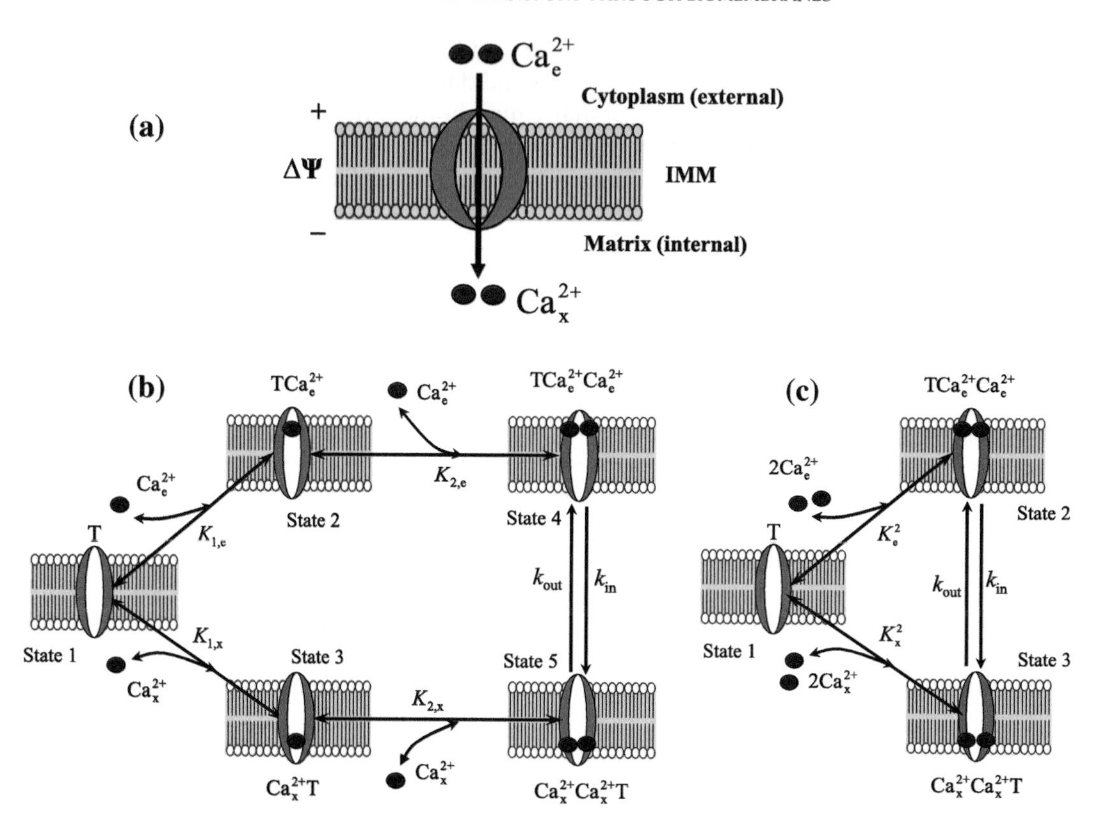

FIGURE 5.3 (a) A schematic representation of the mitochondrial Ca^{2+} uniporter embedded in the IMM and the transport of Ca^{2+} via the Ca^{2+} uniporter from the cytoplasmic side to matrix side driven by the electrochemical gradient of Ca^{2+} across the IMM. The arrow indicates the direction of Ca^{2+} movement. (b) A 5-state kinetic mechanism for Ca^{2+} transport into mitochondria via the Ca^{2+} uniporter. $K_{1,e}$, $K_{1,x}$, $K_{2,e}$ and $K_{2,x}$ are the dissociation constants associated with the two-step binding of external and internal Ca^{2+} to the uniporter. k_{in} and k_{out} are the rate constants involved in the translocation of Ca^{2+} bound uniporter complexes: $TCa_e^{2+}Ca_e^{2+}$ and $Ca_x^{2+}Ca_x^{2+}T$, which are influenced by $\Delta\Psi$. (c) A simplified 3-state kinetic mechanism for Ca^{2+} transport into mitochondria via the Ca^{2+}uniporter. In this case, the binding affinities of first external and internal Ca^{2+} for the uniporter are very large compared to the binding affinities of the second external and internal Ca^{2+} such that $K_{1,e} \gg 1, K_{1,x} \gg 1, K_{2,e} \ll 1$ and $K_{2,x} \ll 1$; $K_{1,e}.K_{2,e} = K_e^2$ and $K_{1,x}.K_{2,x} = K_x^2$.

binding) to form the ternary complex $TCa_e^{2+}Ca_e^{2+}$ (State 4). The complex $TCa_e^{2+}Ca_e^{2+}$ then undergoes conformal changes or flips upside down (Ca^{2+} translocation) to form the ternary complex $Ca_x^{2+}Ca_x^{2+}T$ (State 5). The complex $Ca_x^{2+}Ca_x^{2+}T$ in the internal side of the IMM goes through the reverse process, in which it dissociates in two steps to form the intermediate complex $Ca_x^{2+}T$ (State 3), unbound uniporter T (State 1), and two ionized free Ca^{2+}. The intermediate complexes TCa_e^{2+} and $Ca_x^{2+}T$ are assumed not to undergo any conformational changes, as they are likely to be present in negligible concentrations. Here $(K_{1,e}, K_{1,x})$ and $(K_{2,e}, K_{2,x})$ are the two pairs of dissociation constants associated with the two-step uniporter binding reactions with the external and internal Ca^{2+}; k_{in} and k_{out} are the forward and reverse rate constants in the interconversion of $TCa_e^{2+}Ca_e^{2+}$ and $Ca_x^{2+}Ca_x^{2+}T$, which limit the uniporter function. Since the conformational change or interconversion of $TCa_e^{2+}Ca_e^{2+}$ (State 4) and $Ca_x^{2+}Ca_x^{2+}T$ (State 5) involves translocation of positive charges (Ca^{2+}), the rate constants k_{in} and k_{out}

depend on the electrostatic potential difference $\Delta\Psi$ across the IMM. Furthermore, depending on the physical locations of the Ca^{2+} binding sites on the uniporter, the dissociation constants $K_{1,e}, K_{1,x}, K_{2,e}$ and $K_{2,x}$ can also be influenced by $\Delta\Psi$ (see Section 5.4.4 below). In Section 5.4.3, we show how the Ca^{2+} uniporter flux expressions are derived from the 5-state Ca^{2+} binding and translocation scheme of Fig. 5.3b.

5.4.3 Derivation of Mitochondrial Ca^{2+} Uniporter Flux Expression

Based on the 5-state Ca^{2+} binding and translocation scheme of Fig. 5.3b and with the assumptions of rapid equilibrium binding of the uniporter with the external and internal Ca^{2+}, the reactions for the uniporter-Ca^{2+} binding and the reaction for the conformational change of the ternary uniporter-2Ca^{2+} binding complex can be written as:

$$\begin{aligned}
&T + Ca_e^{2+} \underset{K_{1,e}}{\longleftrightarrow} TCa_e^{2+}, TCa_e^{2+} + Ca_e^{2+} \underset{K_{2,e}}{\longleftrightarrow} TCa_e^{2+}Ca_e^{2+}, \\
&T + Ca_x^{2+} \underset{K_{1,x}}{\longleftrightarrow} Ca_x^{2+}T, Ca_x^{2+}T + Ca_x^{2+} \underset{K_{2,x}}{\longleftrightarrow} Ca_x^{2+}Ca_x^{2+}T, \\
&TCa_e^{2+}Ca_e^{2+} \underset{k_{out}}{\overset{k_{in}}{\rightleftarrows}} Ca_x^{2+}Ca_x^{2+}T.
\end{aligned} \tag{5.45}$$

Then the relationship between various states of the uniporter can be written as:

$$\begin{aligned}
&[TCa_e^{2+}] = \frac{[Ca^{2+}]_e}{K_{1,e}}[T],\ [Ca_x^{2+}T] = \frac{[Ca^{2+}]_e}{K_{1,x}}[T], \\
&[TCa_e^{2+}Ca_e^{2+}] = \frac{[Ca^{2+}]_e^2}{K_{1,e}K_{2,e}}[T],\ [Ca_x^{2+}Ca_x^{2+}T] = \frac{[Ca^{2+}]_x^2}{K_{1,x}K_{2,x}}[T],
\end{aligned} \tag{5.46}$$

where the concentrations of free and Ca^{2+}-bound uniporter states are expressed with respect to the matrix volume; $[Ca^{2+}]_e$ and $[Ca^{2+}]_x$ denote the extra-matrix and matrix concentrations of Ca^{2+}; $K_{1,e}, K_{1,x}, K_{2,e}$ and $K_{2,x}$ are in the units of concentration (molar). Furthermore, since the total uniporter concentration $[T]_{tot}$ is constant, we have by mass conservation:

$$[T] + [TCa_e^{2+}] + [Ca_x^{2+}T] + [TCa_e^{2+}Ca_e^{2+}] + [Ca_x^{2+}Ca_x^{2+}T] = [T]_{tot}. \tag{5.47}$$

Upon substituting Eq. (5.46) into Eq. (5.47) and by rearranging, we can express the concentration of unbound uniporter [T] in terms of the total uniporter concentration $[T]_{tot}$ as $[T] = [T]_{tot}/D$, where D is the denominator expression given by:

$$D = 1 + \frac{[Ca^{2+}]_e}{K_{1,e}} + \frac{[Ca^{2+}]_x}{K_{1,x}} + \frac{[Ca^{2+}]_e^2}{K_{1,e}K_{2,e}} + \frac{[Ca^{2+}]_x^2}{K_{1,x}K_{2,x}}. \tag{5.48}$$

According to the kinetic scheme described in Fig. 5.3b, the Ca^{2+} transport flux through the Ca^{2+} uniporter can be expressed as:

$$J_{CU} = k_{in}[TCa_e^{2+}Ca_e^{2+}] - k_{out}[Ca_x^{2+}Ca_x^{2+}T] = \frac{[T]_{tot}}{D}\left(k_{in}\frac{[Ca^{2+}]_e^2}{K_{1,e}K_{2,e}} - k_{out}\frac{[Ca^{2+}]_x^2}{K_{1,x}K_{2,x}}\right). \tag{5.49}$$

The generalized flux expression (Eq. 5.49) contains four dissociation constants ($K_{1,e}, K_{1,x}, K_{2,e}$ and $K_{2,x}$) and two rate constants (k_{in} and k_{out}) – that is a total of six unknown kinetic parameters. Further assumptions on the cooperative binding of Ca^{2+} to the uniporter can lead to different forms of the uniporter model, and the flux expression (Eq. 5.49) can be modified as follows.

5.4.3.1 *Full Cooperative Binding*

The first dissociation constants $K_{1,e}$ and $K_{1,x}$ are assumed to be large compared to the second dissociation constants $K_{2,e}$ and $K_{2,x}$ with the constraints that $K_{1,e}.K_{2,e} = K_e^2$ and $K_{1,x}.K_{2,x} = K_x^2$ are finite. These approximations are valid under the conditions $K_{1,e} \gg 1, K_{1,x} \gg 1, K_{2,e} \ll 1$ and $K_{2,x} \ll 1$. In this case, the concentrations of TCa_e^{2+} and Ca_x^{2+} T can be considered negligible compared to the concentrations of the other binding states of the uniporter. In this case the 5-state kinetic scheme of Fig. 5.3b is reduced to a 3-state kinetic scheme of Fig. 5.3c. The Ca^{2+} uniporter flux expression (Eq. 5.49) is reduced to:

$$J_{CU} = \frac{[T]_{tot}\left(k_{in}\frac{[Ca^{2+}]_e^2}{K_e^2} - k_{out}\frac{[Ca^{2+}]_x^2}{K_x^2}\right)}{1 + \frac{[Ca^{2+}]_e^2}{K_e^2} + \frac{[Ca^{2+}]_x^2}{K_x^2}}. \tag{5.50}$$

5.4.3.2 *Partial Cooperative Binding*

The first dissociation constants $K_{1,e}$ and $K_{1,x}$ are assumed to be equal to the second dissociation constants $K_{2,e}$ and $K_{2,x}$, respectively; $K_{1,e} = K_{2,e} = K_e$ and $K_{1,x} = K_{2,x} = K_x$. In this case, the Ca^{2+} uniporter flux expression (Eq. 5.49) is reduced to:

$$J_{CU} = \frac{[T]_{tot}\left(k_{in}\frac{[Ca^{2+}]_e^2}{K_e^2} - k_{out}\frac{[Ca^{2+}]_x^2}{K_x^2}\right)}{1 + \frac{[Ca^{2+}]_e}{K_e} + \frac{[Ca^{2+}]_x}{K_x} + \frac{[Ca^{2+}]_e^2}{K_e^2} + \frac{[Ca^{2+}]_x^2}{K_x^2}}. \tag{5.51}$$

5.4.3.3 *No Cooperative Binding*

The dissociation constants associated with the binding of external and internal Ca^{2+} to the uniporter are assumed to satisfy the constraints: $K_{1,e} = K_e/2, K_{2,e} = 2K_e, K_{1,x} = K_x/2, K_{2,x} = 2K_x$. In this case, the Ca^{2+} uniporter flux expression (Eq. 5.49) is reduced to:

$$J_{CU} = \frac{[T]_{tot}\left(k_{in}\frac{[Ca^{2+}]_e^2}{K_e^2} - k_{out}\frac{[Ca^{2+}]_x^2}{K_x^2}\right)}{1 + 2\frac{[Ca^{2+}]_e}{K_e} + 2\frac{[Ca^{2+}]_x}{K_x} + \frac{[Ca^{2+}]_e^2}{K_e^2} + \frac{[Ca^{2+}]_x^2}{K_x^2}}. \tag{5.52}$$

The reduced Ca^{2+} uniporter flux expressions (Eqs. 5.50–5.52) contain only two dissociation constants (K_e and K_x) and two rate constants (k_{in} and k_{out}), a total of four unknown kinetic parameters. Furthermore, under equilibrium transport conditions, the flux of Ca^{2+} via the Ca^{2+} uniporter is zero (i.e., $J_{CU} = 0$). Therefore, the kinetic parameters K_e, K_x, k_{in} and k_{out} can be further constrained by the following equilibrium relationship (thermodynamic constraint):

$$K_{eq} = \frac{[Ca^{2+}]^2_{x,eq}}{[Ca^{2+}]^2_{e,eq}} = \frac{k_{in} \cdot K_x^2}{k_{out} \cdot K_e^2}, \tag{5.53}$$

where K_{eq} is the equilibrium constant for the uniporter-mediated Ca^{2+} transport across the IMM, which is a function of IMM $\Delta\Psi$.

5.4.4 $\Delta\Psi$ Dependency of Mitochondrial Ca^{2+} Uniporter Flux: Free Energy Barrier Formalism

The binding of Ca^{2+} to the uniporter and the translocation of Ca^{2+} via the uniporter are influenced by the electrostatic field of the charged IMM. Since intra-matrix is negatively charged and extra-matrix is positively charged, it is easier for the positively charged ions to enter into the matrix, but difficult for the positively charged ions to extrude from the matrix in the outward direction. To take this charge dependency into account, we assume that the kinetic parameters (the dissociation constants and the rate constants: K_e, K_x, k_{in} and k_{out}) depend on the electrostatic potential difference ($\Delta\Psi$) across the membrane. The $\Delta\Psi$-dependencies of the kinetic parameters are derived on the basis of biophysical principles and well-known laws of thermodynamics, electrostatics and superposition. In this approach, we assume that the total value of the membrane potential is the sum of local electric potentials, and each of these local electric potentials influences the corresponding stages of the Ca^{2+} binding and translocation processes.

Figure 5.4 illustrates various Ca^{2+} binding stages during Ca^{2+} translocation via the uniporter. In this case, the potential energy of Ca^{2+} passing through the uniporter is described by a potential energy profile (Fig. 5.4a). Every position of Ca^{2+} on the uniporter is assigned to an electrical potential value. We assume here that the difference in potentials between the adjacent positions of Ca^{2+} is proportional to the total potential difference across the membrane. Using the principle of superposition, the sum of potential differences between the consecutive positions of Ca^{2+} is equal to the total potential difference across the membrane. Thus, this approach divides the total drop in potential across the membrane into elementary stages. For simplicity, we assume here that the potential energy profile for Ca^{2+} translocation across the membrane is a single barrier (Fig. 5.4b). We define the reaction coordinate as the coordinate from Ca^{2+} bound at the external side to Ca^{2+} bound at the internal side of the membrane along the direction of Ca^{2+} translocation. The local maximum (peak) (State II) of the potential energy profile corresponds to the barrier (transition state *TS* described in Section 5.2.4) that impedes the Ca^{2+} translocation, while the local minima (States I and III) correspond to the uniporter-$2Ca^{2+}$ complex states on the either side of the membrane. The Ca^{2+} transport rate is determined by the probability of the uniporter to translocate Ca^{2+} from one binding site to the other, which depends on the height of the potential energy barrier, which in turn depends on $\Delta\Psi$.

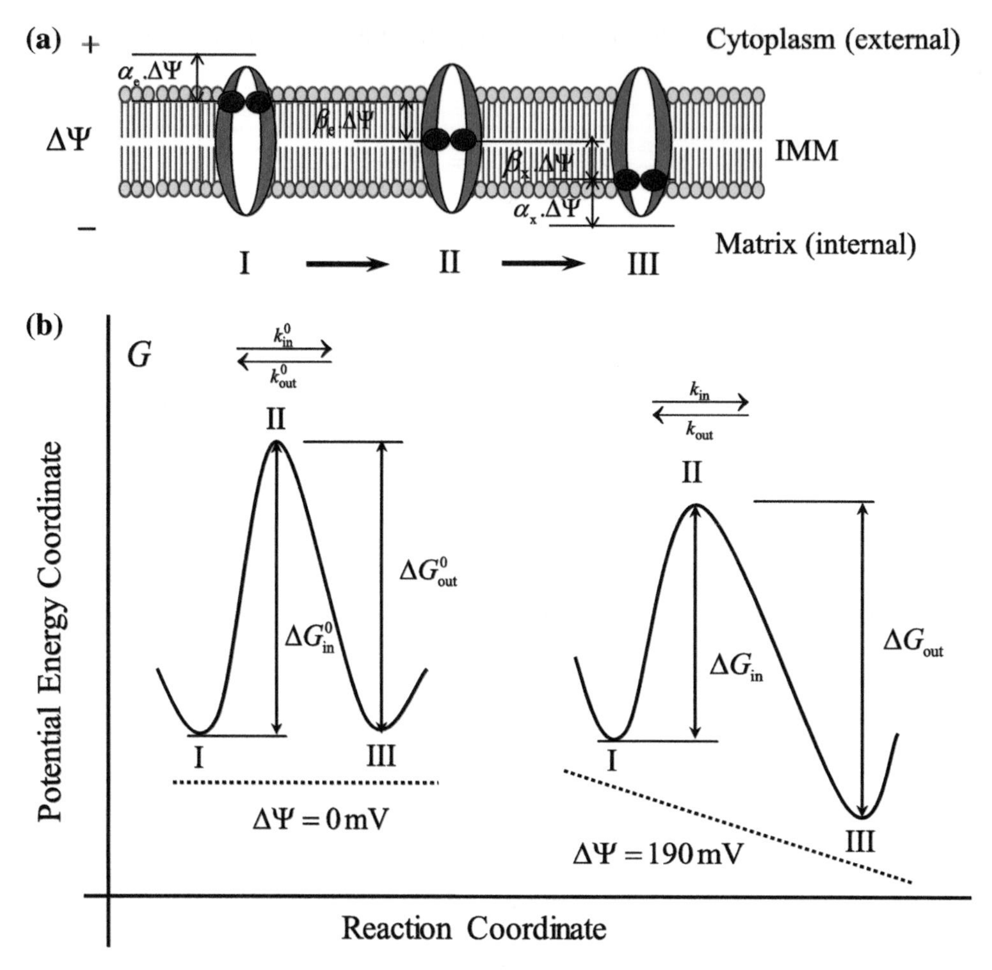

FIGURE 5.4 A schematic description of the potential energy barrier formalism for Ca^{2+} transport into mitochondria via the Ca^{2+} uniporter, reproduced from Dash et al. [30]. (a) (I–III) Consecutive states of the Ca^{2+}-bound uniporter functional unit in the process of Ca^{2+} translocation that is used to derive the dependence of the rate of Ca^{2+} transport on the electrostatic membrane potential $\Delta\Psi$. $\alpha_e(\alpha_x)$ is the ratio of the potential difference between Ca^{2+} bound at the site of uniporter facing the external (internal) side of the IMM and Ca^{2+} in the bulk phase to the total membrane potential; $\beta_e(\beta_x)$ is the displacement of external (internal) Ca^{2+} from the coordinate of maximum potential barrier. (b) The potential energy barrier profile along the reaction coordinate that is used to derive the dependence of the rate of Ca^{2+} transport on $\Delta\Psi$. The dashed line shows the profile of the potential created by the electric field of the charged membrane. The points I, II and III correspond to the Ca^{2+}-bound uniporter states depicted in the upper panel (A). The rate constants k_{in} and k_{out} are related to the changes in the Gibbs free energy ΔG_{in} and ΔG_{out}.

5.4.4.1 *Equilibrium Constant*

In a complete catalytic cycle of the uniporter operation, there is a net translocation of four elementary positive charges ($2Ca^{2+}$) across the IMM. Therefore, based on Eq. (5.4), the dependence of the equilibrium constant K_{eq} on $\Delta\Psi$ can be expressed as:

$$K_{eq} = \frac{[Ca^{2+}]^2_{x,eq}}{[Ca^{2+}]^2_{e,eq}} = \exp(2\Delta\Phi);\ \Delta\Phi = \frac{Z_{Ca}F\Delta\Psi}{RT}, \tag{5.54}$$

where $Z_{Ca} = 2$ is the valence of Ca^{2+}; F, R and T denote the Faraday's constant, ideal gas constant and absolute temperature, respectively; $\Delta\Phi$ is the non-dimensional potential difference across the IMM. In the absence of electric field ($\Delta\Psi = 0$), K_{eq} becomes 1.

5.4.4.2 Dissociation Constants

To derive the dependence of the dissociation constants of uniporter-Ca^{2+} complex on the membrane potential $\Delta\Psi$, let us consider the two-step binding of the external Ca^{2+} to the uniporter. The changes in Gibbs free energy for the two binding reactions are:

$$\begin{aligned}\Delta\mu_{1,e} &= \Delta\mu^0_{1,e} + Z_{Ca}F\alpha_e\Delta\Psi + RT\ln\left([T][Ca^{2+}]_e/[TCa^{2+}_e]\right),\\ \Delta\mu_{2,e} &= \Delta\mu^0_{2,e} + Z_{Ca}F\alpha_e\Delta\Psi + RT\ln\left([TCa^{2+}_e][Ca^{2+}]_e/[TCa^{2+}_eCa^{2+}_e]\right),\end{aligned} \tag{5.55}$$

where $\Delta\mu^0_{1,e}$ and $\Delta\mu^0_{2,e}$ are the standard changes in Gibbs free energy of the reactions; α_e is the ratio of the potential difference between Ca^{2+} bound at the site of uniporter facing the external side of the IMM and Ca^{2+} in the bulk phase to the total membrane potential $\Delta\Psi$ ($\Delta\Psi = \Psi_e - \Psi_x$, i.e., outside potential minus inside potential, so $\Delta\Psi$ is positive). An assumption inherent in this derivation is that both the Ca^{2+} binding sites on the uniporter are equidistant from the bulk medium. At equilibrium ($\Delta\mu_{1,e} = \Delta\mu_{2,e} = 0$), hence Eq. (5.55) gives:

$$\begin{aligned}K_{1,e} &= \left([T][Ca^{2+}]_e/[TCa^{2+}_e]\right)_{eq} = K^0_{1,e}\exp(-\alpha_e\Delta\Phi),\\ K_{2,e} &= \left([TCa^{2+}_e][Ca^{2+}]_e/[TCa^{2+}_eCa^{2+}_e]\right)_{eq} = K^0_{2,e}\exp(-\alpha_e\Delta\Phi),\end{aligned} \tag{5.56}$$

where $K^0_{1,e} = K_{1,e}(\Delta\Psi = 0) = \exp(-\Delta\mu^0_{1,e}/RT)$ and $K^0_{2,e} = K_{2,e}(\Delta\Psi = 0) = \exp(-\Delta\mu^0_{2,e}/RT)$. Eq. (5.56) suggests that the dissociation constants $K_{1,e}$ and $K_{2,e}$ associated with the binding of external Ca^{2+} to the uniporter are reduced (i.e., the association becomes easy) in the presence of an electric field, provided $\alpha_e > 0$. Similarly, for binding of the internal Ca^{2+} to the uniporter:

$$\begin{aligned}K_{1,x} &= \left([T][Ca^{2+}]_x/[Ca^{2+}_xT]\right)_{eq} = K^0_{1,x}\exp(+\alpha_x\Delta\Phi),\\ K_{2,x} &= \left([Ca^{2+}_xT][Ca^{2+}]_x/[Ca^{2+}_xCa^{2+}_xT]\right)_{eq} = K^0_{2,x}\exp(+\alpha_x\Delta\Phi),\end{aligned} \tag{5.57}$$

where α_x is the ratio of the potential difference between Ca^{2+} bound at the site of uniporter facing the internal side of the IMM and Ca^{2+} in the bulk phase to the total membrane potential $\Delta\Psi$. In contrast to $K_{1,e}$ and $K_{2,e}$, the dissociation constants $K_{1,x}$ and $K_{2,x}$ associated with the binding of internal Ca^{2+} to the uniporter are increased (i.e., the association becomes difficult) in the presence of electric field, provided $\alpha_x > 0$.

Thus, for any of the cooperative binding assumptions used earlier (Eqs. 5.50–5.52), the dissociation constants K_e and K_x can be obtained from Eqs. 5.56 and 5.57 as:

$$K_e = K^0_e\exp(-\alpha_e\Delta\Phi),\quad K_x = K^0_x\exp(+\alpha_x\Delta\Phi). \tag{5.58}$$

For generality, it is assumed here that K_e^0 and K_x^0 are distinct. Therefore, the dissociation constants K_e and K_x can be fully characterized by four unknown parameters: K_e^0, K_x^0, α_e and α_x. For positive α_e and α_x, the dissociation constant tends to decrease (easier for binding) on the outside and increase (more difficult for binding) on the inside of the IMM.

5.4.4.3 Rate Constants

According to Eyring's free energy barrier theory (introduced in Section 5.2.5), the rate at which an ion can jump from one binding site to the other is given by Eq. (5.9). In the present case, the heights of the free energy barrier (State II) from States I and III (Fig. 5.4b) can be defined by:

$$\Delta G_{in} = \Delta G_{in}^0 - 2Z_{Ca}F\beta_e\Delta\Psi, \quad \Delta G_{out} = \Delta G_{out}^0 + 2Z_{Ca}F\beta_x\Delta\Psi, \tag{5.59}$$

where ΔG_{in}^0 and ΔG_{out}^0 are the heights of the free energy barriers in the absence of an electric field ($\Delta\Psi = 0\,\text{mV}$), β_e is the displacement of external Ca^{2+} (State I) from the coordinate of maximum potential barrier (State II) and β_x is the displacement of internal Ca^{2+} (State III) from the coordinate of maximum potential barrier (State II). Note that $\Delta G_{in}^0 = \Delta G_{out}^0 = \Delta G^0$ subject to the condition $K_e^0 = K_x^0 = K^0$. For simplicity, we have assumed here that the uniporter has no net charge (neutral) and that the charge on the uniporter-$2Ca^{2+}$ complex is $2Z_{Ca}$. It is evident from Eq. (5.49) that the height of the free energy barrier in the inward direction is reduced, while the height in the outward direction is increased in the presence of an electric field (see Fig. 5.4b). This means it becomes easier for the Ca^{2+} ions to cross the barrier in the inward direction, but more difficult for the Ca^{2+} ions to exit the matrix in the presence of a positive membrane potential $\Delta\Psi$, measured from outside to inside. The rate constants of Ca^{2+} translocation can be written as:

$$k_{in} = k_{in}^0 \exp(2\beta_e\Delta\Phi), \quad k_{out} = k_{out}^0 \exp(-2\beta_x\Delta\Phi), \tag{5.60}$$

where $k_{in}^0 = \kappa(k_BT/h)\exp(-\Delta G_{in}^0/RT)$ and $k_{out}^0 = \kappa(k_BT/h)\exp(-\Delta G_{out}^0/RT)$ are the forward and reverse rate constants in the absence of an electric field ($\Delta\Psi = 0\,\text{mV}$). Thus, the rate constants k_{in} and k_{out} can be fully characterized by four unknown parameters: $k_{in}^0, k_{out}^0, \beta_e$ and β_x. Also note that $k_{in}^0 = k_{out}^0 = k^0$, subject to the condition $\Delta G_{in}^0 = \Delta G_{out}^0 = \Delta G^0$ or $K_e^0 = K_x^0 = K^0$. Now substituting Eq. (5.54) for K_{eq}, Eq. (5.58) for K_e and K_x, and Eq. (5.60) for k_{in} and k_{out} into Eq. (5.53), we obtain the following thermodynamic constraints between $k_{in}^0, k_{out}^0, k_x^0$ and k_e^0 and $\alpha_e, \alpha_x, \beta_e$ and β_x:

$$(k_{in}^0/k_{out}^0).\,(k_x^0/k_e^0)^2 = 1, \quad \alpha_e + \alpha_x + \beta_e + \beta_x = 1. \tag{5.61}$$

These thermodynamic constraints are useful in reducing the number of unknown parameters in the model, from a total of eight parameters to six parameters.

5.4.5 Mitochondrial Ca^{2+} Uniporter Model Parameterization and Simulations

The Ca^{2+} uniporter model presented in the preceding sections is characterized by six unknown parameters (reduced from eight through the two thermodynamic constraints of Eq. (5.61)), which were estimated using the available experimental data [20–22] on the extra-matrix $[Ca^{2+}]$ and $\Delta\Psi$ dependent mitochondrial Ca^{2+} uptake via the Ca^{2+} uniporter. Since in most of the experiments, matrix $[Ca^{2+}]$ was unknown and not perturbed, it was not possible to estimate all of the six unknown parameters uniquely and accurately. Hence, two feasible cases were considered for parameter estimation [30]. In one case (Case 1), it was assumed that $K_e^0 = K_x^0$, so that $\Delta G_{in}^0 = \Delta G_{out}^0$ and $k_{in}^0 = k_{out}^0$, and in another case (Case 2), K_e^0 and K_x^0 were assumed to be distinct so that ΔG_{in}^0 and ΔG_{out}^0, as well as k_{in}^0 and k_{out}^0 are distinct. For simplicity and better parameter identifiability, it was also assumed that $\alpha_e = 0$, which means that the Ca^{2+} binding sites on the external side of the uniporter are situated at a negligible distance from the bulk phase, so that the potential barrier that the external Ca^{2+} ions would have to overcome to bind to the uniporter would be negligible. With these assumptions, the number of unknown parameters in the model was further reduced to four in Case 1, and to five in Case 2.

This model characterized the possible mechanisms of both the extra-matrix [Ca^{2+}] and $\Delta\Psi$-dependencies of mitochondrial Ca^{2+} uptake via the uniporter. This model was found to be a great improvement over the earlier kinetic models of the uniporter found in the literature [32–35]. Most importantly, under the assumption of distinct binding constants for external and internal Ca^{2+} (Case 2), the model provides the best possible description of the observed $\Delta\Psi$ dependency of mitochondrial Ca^{2+} uptake in the entire $\Delta\Psi$ regime [30] and the parameter estimates followed the trend that $\alpha_x < 0, K_e^0 \gg K_x^0$ and $k_{in}^0 \ll k_{out}^0$.

5.4.5.1 *Limitations of the Ca^{2+} Uniporter Model*

The model was parameterized by using available experimental data [20–22] in which matrix [Ca^{2+}] was not known. Therefore, for parameter estimation, matrix [Ca^{2+}] was fixed at a physiologically realistic level of 250 nM [30]. However, with further model simulation analyses with varying matrix [Ca^{2+}], it was observed that the model predictions under the assumption of distinct binding constants for external and internal Ca^{2+} to the uniporter (Case 2) are highly sensitive to variations in matrix [Ca^{2+}], indicating limitations in the model in providing physiologically plausible description of the observed $\Delta\Psi$ dependency of the uniporter-mediated mitochondrial Ca^{2+} uptake [25]. This sensitivity was mainly attributed to the negative estimate of the $\Delta\Psi$-dependent biophysical parameter α_x under Case 2 that characterizes the binding of internal Ca^{2+} to the uniporter [30].

Reparameterization of the model with additional non-negativity constraints on the $\Delta\Psi$-dependent biophysical parameters $\alpha_e = \alpha_x = \alpha \geq 0$ showed that the two binding assumptions (Case 1 and Case 2) are indistinguishable from each other. This indicates that the external and internal Ca^{2+} binding constants for the uniporter may be equal (Case 1). The model predictions in this case are insensitive to variations in matrix [Ca^{2+}], but do not match the $\Delta\Psi$-dependent kinetic data in the domain $\Delta\Psi \leq 120$ mV (see [25]). This analysis led to further improvement of Ca^{2+} uniporter model, as described in Section 5.4.6. Since the effects of extra-matrix Mg^{2+} and Pi were not explicitly accounted for in the model, the estimates of K_e^0 and K_x^0 were different for different data sets depending on the amount of Mg^{2+} and Pi present in the incubation medium, which is addressed below in Sections 5.4.7–5.4.9.

5.4.6 $\Delta\Psi$ Dependency of Mitochondrial Ca^{2+} Uniporter Flux: Alternate Formulation

The limitation of the Ca^{2+} uniporter model described in the previous section can be overcome by applying a slightly different approach [25]. Here, we devise a generalized non-linear Goldman-Hodgkin-Katz (GHK) type of formulation for the rate constants k_{in} and k_{out}

associated with the translocation of Ca^{2+} via the Ca^{2+} uniporter. This formulation effectively redefines the biophysical parameters β_e and β_x associated with the free energy barrier for Ca^{2+} translocation in terms of the IMM $\Delta\Psi$. In this formulation, the rate constants k_{in} and k_{out} can be expressed as:

$$k_{in} = k_{in}^0 \frac{f(+\Delta\Phi)}{\exp(+2\alpha_e\Delta\Phi)} \text{ and } k_{out} = k_{out}^0 \frac{f(-\Delta\Phi)}{\exp(-2\alpha_x\Delta\Phi)}. \tag{5.62}$$

where $f(\Delta\Phi)$ is an unknown non-linear function to be determined. Substituting Eq. (5.62) for k_{in} and k_{out} and Eq. (5.58) for K_e and K_x into Eq. (5.50, 5.51 or 5.52), the Ca^{2+} uniporter flux expression is reduced to:

$$J_{CU} = \frac{[T]_{tot}}{D}\left(k_{in}^0 \frac{[Ca^{2+}]_e^2}{K_e^{02}} f(+\Delta\Phi) - k_{out}^0 \frac{[Ca^{2+}]_x^2}{K_x^{02}} f(-\Delta\Phi)\right), \tag{5.63}$$

where D is D_1 or D_2 or D_3. In order to derive the functional form of $f(\Delta\Phi)$ let us consider the equilibrium condition for trans-membrane Ca^{2+} transport via the Ca^{2+} uniporter ($J_{CU} = 0$), which in combination with Eqs. (5.53 and 5.62) gives:

$$\left(\frac{k_{in}^0}{k_{out}^0}\right)\left(\frac{K_x^0}{K_e^0}\right)^2 = 1 \text{ and } \frac{f(+\Delta\Phi)}{f(-\Delta\Phi)} = \exp(+2\Delta\Phi). \tag{5.64}$$

When $K_e^0 = K_x^0$ and $K_{in}^0 = K_{out}^0$, the kinetic constraint of Eq. (5.64) is automatically satisfied. However, the thermodynamic constraint of Eq. (5.64) provides multiple solutions for $f(\Delta\Phi)$. The general solution that satisfies the equilibrium condition for passive Ca^{2+} transport via the Ca^{2+} uniporter in the absence of $\Delta\Phi$ ($\lim_{\Delta\Phi\to 0} f(\Delta\Phi) = 1$) is given by:

$$f(\Delta\Phi) = \exp(\Delta\Phi)E(\Delta\Phi);\ E(\Delta\Phi) = \left(\frac{\Delta\Phi/nH}{\sinh(\Delta\Phi/nH)}\right)^{nH}, \tag{5.65}$$

where $E(\Delta\Phi)$ is an even function $E(+\Delta\Phi) = E(-\Delta\Phi)$; $nH \geq 0$ is an arbitrary number to be determined. Thus, the unknown function $f(\Delta\Phi)$ is fully characterized by only one unknown parameter nH, and hence the two rate constants k_{in} and k_{out} in Eq. (5.60) are fully characterized by only three unknown parameters k_{in}^0, k_{out}^0 and nH, in contrast to the four unknown parameters k_{in}^0, k_{out}^0, β_e and β_x in the previous formulation (Section 5.4.4). In standard linear GHK formulation (constant field approximation), the interconversion (electrodiffusion) of the uniporter-2Ca^{2+} complex is given by $nH = 1$, $E(\Delta\Phi) = 2\Delta\Phi/[\exp(+\Delta\Phi) - \exp(-\Delta\Phi)]$ and $f(\Delta\Phi) = 2\Delta\Phi/[1 - \exp(-2\Delta\Phi)]$. Substituting Eq. (5.65) into Eq. (5.63), the Ca^{2+} uniporter flux expression is reduced to:

$$J_{CU} = \frac{[T]_{tot}}{D}\left(k_{in}^0 \frac{[Ca^{2+}]_e^2}{K_e^{02}} \exp(+\Delta\Phi) - k_{out}^0 \frac{[Ca^{2+}]_x^2}{K_x^{02}} \exp(-\Delta\Phi)\right)\left(\frac{\Delta\Phi/nH}{\sinh(\Delta\Phi/nH)}\right)^{nH}. \tag{5.66}$$

Based on the previous formulation (Section 5.4.4), by substituting Eq. (5.58) for K_e and K_x and Eq. (5.60) for k_{in} and k_{out} into Eq. (5.50), and using the kinetic and thermodynamic constraints of Eq. (5.53), the Ca^{2+} uniporter flux expression is reduced to:

$$J_{CU} = \frac{[T]_{tot}}{D}\left(k_{in}^{0}\frac{[Ca^{2+}]_e^2}{K_e^{02}}\exp(+\Delta\Phi) - k_{out}^{0}\frac{[Ca^{2+}]_x^2}{K_x^{02}}\exp(-\Delta\Phi)\right)\exp\left(+(2\alpha_e + 2\beta_e - 1)\Delta\Phi\right). \tag{5.67}$$

Now, if we compare Eq. (5.66) with Eq. (5.67), we obtain the following functional relationship between the parameters $\alpha_e, \alpha_x, \beta_e, \beta_x$ and nH:

$$\exp\left(+(2\alpha_e + 2\beta_e - 1)\Delta\Phi\right) \approx \left(\frac{\Delta\Phi/nH}{\sinh(\Delta\Phi/nH)}\right)^{nH}. \tag{5.68}$$

Using Eq. (5.68) along with the thermodynamic constraint: $\alpha_e, \alpha_x, \beta_e, \beta_x = 1$, the biophysical parameters β_e and β_x can be expressed in terms of $\Delta\Phi$ as:

$$\beta_e = \frac{1}{2}\left[1 + \frac{nH}{\Delta\Phi}\ln\left\{\frac{(\Delta\Phi/nH)}{\sinh(\Delta\Phi/nH)}\right\}\right] - \alpha_e, \quad \beta_x = \frac{1}{2}\left[1 - \frac{nH}{\Delta\Phi}\ln\left\{\frac{(\Delta\Phi/nH)}{\sinh(\Delta\Phi/nH)}\right\}\right] - \alpha_x. \tag{5.69}$$

Therefore, in this formulation, the biophysical parameters β_e and β_x became continuous functions of $\Delta\Psi$ when $\alpha_e = \alpha_x = \alpha \geqslant 0$, compared to that seen in the previous formulation (Section 5.4.4, Eqs. (5.59–5.60)), in which β_e and β_x were constant with respect to $\Delta\Psi$. The main advantage of this formulation is that both the parameters β_e and β_x are characterized by only one unknown parameter nH, reducing further the number of unknown parameters in the model.

With this alternate $\Delta\Psi$ dependency formulation [25], the Ca^{2+} uniporter model satisfactorily described all the available experimental data [20–22] and provided a robust and unique set of kinetic parameters. It also retained all the characteristics of the previous Ca^{2+} uniporter model, described in Section 5.4.2, and provided more accurate descriptions of both the extra-matrix Ca^{2+} and $\Delta\Psi$ dependencies of mitochondrial Ca^{2+} uptake observed in the experimental studies [20–22]. An interesting observation from this model analysis of the available experimental data is that the Ca^{2+} binding sites on the uniporter were located at almost equal and negligible distances from the bulk phase on either side of the IMM (i.e., $\alpha_e = \alpha_x = \alpha \approx 0$), and both cytosolic and matrix Ca^{2+} have equal binding affinities for the uniporter (i.e., $K_e = K_x$). In addition, the model is also insensitive to variation in matrix $[Ca^{2+}]$, predicting a relatively stable physiological operation of the Ca^{2+} uniporter. However, additional kinetic data may be necessary to provide a more precise explanation of different binding assumptions used in the present formulation.

The Ca^{2+} uniporter model presented in the preceding sections characterizes the various driving forces governing the transport of Ca^{2+} via the Ca^{2+} uniporter, but does not explain how various cytosolic factors like Mg^{2+} and Pi influence the kinetics of the Ca^{2+} uniporter. This was addressed in a subsequent study [31], and is briefly described in the next sections.

5.4.7 Mg^{2+} Inhibition and Pi Regulation of Mitochondrial Ca^{2+} Uniporter Function

Mg^{2+} is a divalent cation abundantly found in most eukaryotic cells, and the intracellular concentration of free Mg^{2+} varies under various physiological and pathological conditions [36]. Alteration of cytosolic free Mg^{2+} significantly affects mitochondrial substrate and cation (e.g., Ca^{2+}) transport systems. Though the actual composition and locations of Mg^{2+} binding sites on the mitochondrial Ca^{2+} uniporter are largely unknown, a vast number of experimental studies on the kinetics of Ca^{2+} uptake in rat heart and rat liver mitochondria suggest that Mg^{2+} is a competitive inhibitor of Ca^{2+} uptake, which alters the sigmoidicity of the initial rate of Ca^{2+} uptake as well as the maximal activity of the uniporter [18–21]. Both competitive [19–21] and non-competitive [18] type of Mg^{2+} inhibition have been reported in these kinetic studies, giving contradictory conclusions. Furthermore, mitochondrial Ca^{2+} uptakes have been shown to be affected by cytosolic Pi, when measured both in the presence and absence of Mg^{2+} in the assay mediums. Such regulation aspects of the Ca^{2+} uniporter were not accounted for in the previous uniporter models described in earlier sections.

Recently, we extended our previous Ca^{2+} uniporter model to mechanistically characterize the regulation of the uniporter function by extra-matrix Mg^{2+} and Pi [31], observed in many isolated mitochondrial preparations [18–21] and studies in permealized cells [37]. A linear, mixed-type inhibition scheme is proposed and validated to explain the possible Ca^{2+} and Mg^{2+} interaction with the Ca^{2+} uniporter, seen in those experiments, both in the absence and presence of different amount of Pi. The model satisfactorily describes all the available experimental data obtained both in the presence and absence of Mg^{2+} in the assay media, and appropriately reproduces the experimentally observed Mg^{2+} inhibition of the uniporter function, in which the mitochondrial Ca^{2+} uptake profile is hyperbolic in the absence of Mg^{2+} and sigmoidal in the presence of Mg^{2+}. In the next sections, we briefly describe the basic principles used for building the kinetic model of Mg^{2+} inhibition and Pi regulation of the Ca^{2+} uniporter function.

5.4.8 Kinetic Scheme for Mg^{2+} Inhibition of Mitochondrial Ca^{2+} Uniporter Function

Figure 5.5 depicts the mechanism of Mg^{2+} inhibition of the uniporter-mediated mitochondrial Ca^{2+} transport. The binding scheme is similar to that shown in Fig. 5.3, except for the Mg^{2+} binding steps. Here, the uniporter functional unit is assumed to have two binding sites for Ca^{2+} as before (see Fig. 5.3b) and two distinct allosteric sites for Mg^{2+}, facing either the cytosolic side or the matrix side depending on the state of the uniporter.

In one process, two Ca^{2+} ions from the cytosolic side cooperatively bind to the uniporter T in two different steps to form the complexes TCa_e^{2+} and $T2Ca_e^{2+}$. The complex $T2Ca_e^{2+}$ then undergoes conformal changes to form the complex $2Ca_x^{2+}T$, complex $Ca_x^{2+}T$, unbound uniporter T, and two free Ca^{2+} ions in the matrix side. For $Ca^{2+} - Mg^{2+}$ interactions, two Mg^{2+} ions from the cytosolic side cooperatively bind to the uniporter states $E_1(T, T\,Ca_e^{2+}, T2Ca_e^{2+})$ in two different steps to form the complexes $E_1Mg_e^{2+}(TMg_e^{2+}, TMg_e^{2+}Ca_e^{2+}, TMg_e^{2+}2Ca_e^{2+})$ and $E_12Mg_e^{2+}(T2Mg_e^{2+}, T2Mg_e^{2+}Ca_e^{2+}, T2Mg_e^{2+}2Ca_e^{2+})$. In another process, two Ca^{2+} ions from the

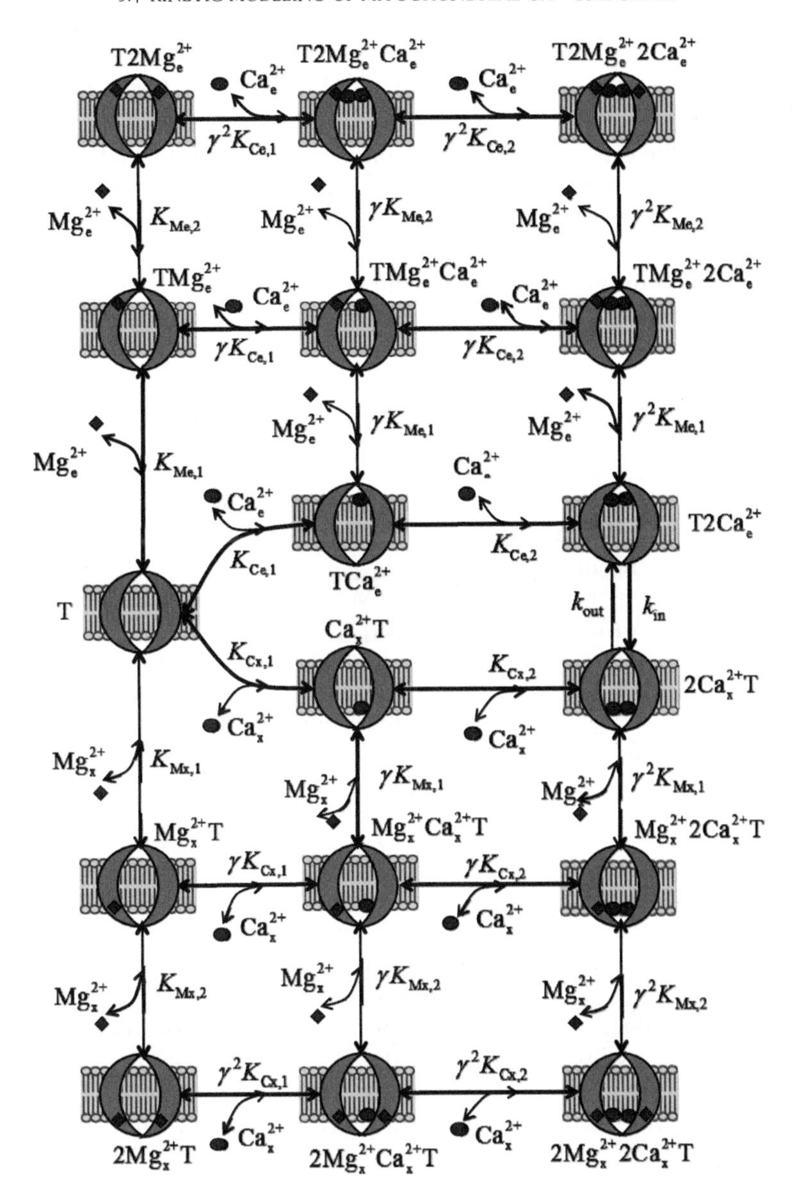

FIGURE 5.5 Mechanisms of Mg^{2+} inhibition of mitochondrial Ca^{2+} transport via the Ca^{2+} uniporter (redrawn from Pradhan et al. [31]). Shown are the schematics of various Ca^{2+} and Mg^{2+} bound uniporter states, both in the cytoplasmic and matrix sides of the IMM with a linear mixed-type inhibition scheme for Mg^{2+} inhibition of Ca^{2+} transport. $K_{Ce,1}$, $K_{Ce,2}$, $K_{Me,1}$ and $K_{Me,2}$ are the dissociation constants associated with the binding of first and second cytoplasmic and to the uniporter; $K_{Cx,1}$, $K_{Cx,2}$, $K_{Mx,1}$ and $K_{Mx,2}$ are the dissociation constants associated with the binding of first and second matrix Ca^{2+} and Mg^{2+} to the uniporter; $1 \leq \gamma \leq \infty$ is a parameter that controls the binding affinity of Ca^{2+} and Mg^{2+} to the uniporter. The translocation Ca^{2+} of across the IMM by the interconversion reaction $T2Ca_e^{2+} \leftrightarrow 2Ca_x^{2+}T$ is limited by two rate constants k_{in} and k_{out}, which are dependent on $\Delta\Psi$. Except the bound $2Ca^{2+}$ uniporter complexes and, the other uniporter complexes $T2Ca_e^{2+}$ are $2Ca_x^{2+}T$ are assumed not to undergo any conformational changes.

cytosolic side cooperatively bind to the uniporter states $E_2(T, TMg_e^{2+}, T2Mg_e^{2+}$ in two different steps to form the complexes $E_2Ca_e^{2+}(TCa_e^{2+}, TMg_e^{2+}Ca_e^{2+}, T2Mg_e^{2+}Ca_e^{2+})$ and $E_22Ca_e^{2+}(T2Ca_e^{2+}, TMg_e^{2+}2Ca_e^{2+}, T2Mg_e^{2+}2Ca_e^{2+})$.

In the matrix side, the Ca^{2+}–Mg^{2+} interaction is symmetrical to that in the cytoplasmic side. The parameter γ determines the type of Mg^{2+} inhibition suitable for the uniporter operation. For example, $\gamma = 1$ signifies a non-competitive inhibition, $\gamma = \infty$ signifies a competitive inhibition, and $1 < \gamma < \infty$ signifies a mixed-type inhibition. If Mg^{2+} does not bind to the Ca^{2+} uniporter (e.g., no competition), then Fig. 5.5 will be reduced to Fig. 5.3 with $K_{Ce,1} = K_{1,e}$ and $K_{Ce,2} = K_{2,e}$.

5.4.9 Mitochondrial Ca^{2+} Uniporter Model with Mg^{2+} Inhibition

Similar steps (Eqs.5.45–5.52) can be applied to derive the Ca^{2+} uniporter flux expressions for variant models with different cooperative binding assumptions, as described in Section 5.4.3. Also, similar binding assumptions can be made for different cooperative binding of Mg^{2+} to the uniporter: $K_{Me,1}K_{Me,2} = K_{Me}^2$ and $K_{Mx,1}K_{Mx,2} = K_{Mx}^2$ are finite, but $K_{Me,1} \gg 1$ mM, $K_{Me,2} \ll 1$ mM, $K_{Mx,1} \gg 1$ mM, and $K_{Mx,2} \ll 1$ mM (full cooperative binding); $K_{Me,1} = K_{Me,2} = K_{Me}$ and $K_{Mx,1} = K_{Mx,2} = K_{Mx}$ (partial cooperative binding); $K_{Me,1} = K_{Me}/2, K_{Me,2} = 2K_{Me}, K_{Mx,1} = K_{Mx}/2$, and $K_{Mx,2} = 2K_{Mx}$ (no cooperative binding). The detailed derivations of the Ca^{2+} uniporter flux expressions can be found in *Pradhan et al.* [25]. Based on the Mg^{2+} inhibition scheme described in Fig. 5.5, the net Ca^{2+} transport flux via the Ca^{2+} uniporter (J_{CU}) can be written as:

$$J_{CU} = k_{in}[T2Ca_e^{2+}] - k_{out}[2Ca_x^{2+}T] = \frac{[T]_{tot}}{D}\left(k_{in}\frac{[Ca^{2+}]_e^2}{K_{Ce}^2} - k_{out}\frac{[Ca^{2+}]_x^2}{K_{Cx}^2}\right), \tag{5.70}$$

where:

$$D = D_1 = 1+\frac{[Ca^{2+}]_e^2}{K_{Ce}^2}+\frac{[Ca^{2+}]_x^2}{K_{Cx}^2}+\frac{[Mg^{2+}]_e^2}{K_{Me}^2}+\frac{[Mg^{2+}]_x^2}{K_{Mx}^2}+\frac{[Ca^{2+}]_e^2[Mg^{2+}]_e^2}{\gamma^4 K_{Ce}^2 \cdot K_{Me}^2}+\frac{[Ca^{2+}]_x^2[Mg^{2+}]_x^2}{\gamma^4 K_{Cx}^2 \cdot K_{Mx}^2}, \tag{5.71}$$

$$\begin{aligned} D = D_2 = 1 &+ \frac{[Ca^{2+}]_e}{K_{Ce}} + \frac{[Ca^{2+}]_e^2}{K_{Ce}^2} + \frac{[Ca^{2+}]_x}{K_{Cx}} + \frac{[Ca^{2+}]_x^2}{K_{Cx}^2} + \frac{[Mg^{2+}]_e}{K_{Me}} + \frac{[Mg^{2+}]_e^2}{K_{Me}^2} + \frac{[Mg^{2+}]_x}{K_{Mx}} + \frac{[Mg^{2+}]_x^2}{K_{Mx}^2} \\ &+ \frac{[Ca^{2+}]_e[Mg^{2+}]_e}{\gamma \cdot K_{Ce} \cdot K_{Me}} + \frac{[Ca^{2+}]_e[Mg^{2+}]_e^2}{\gamma^2 \cdot K_{Ce} \cdot K_{Me}^2} + \frac{[Ca^{2+}]_e^2[Mg^{2+}]_e}{\gamma^2 \cdot K_{Ce}^2 \cdot K_{Me}} + \frac{[Ca^{2+}]_e^2[Mg^{2+}]_e^2}{\gamma^4 \cdot K_{Ce}^2 \cdot K_{Me}^2} \\ &+ \frac{[Ca^{2+}]_x[Mg^{2+}]_x}{\gamma \cdot K_{Cx} \cdot K_{Mx}} + \frac{[Ca^{2+}]_x[Mg^{2+}]_x^2}{\gamma^2 \cdot K_{Cx} \cdot K_{Mx}^2} + \frac{[Ca^{2+}]_x^2[Mg^{2+}]_x}{\gamma^2 \cdot K_{Cx}^2 \cdot K_{Mx}} + \frac{[Ca^{2+}]_x^2[Mg^{2+}]_x^2}{\gamma^4 \cdot K_{Cx}^2 \cdot K_{Mx}^2}, \end{aligned} \tag{5.72}$$

$$D = D_3 = 1 + 2\frac{[Ca^{2+}]_e}{K_{Ce}} + \frac{[Ca^{2+}]_e^2}{K_{Ce}^2} + 2\frac{[Ca^{2+}]_x}{K_{Cx}} + \frac{[Ca^{2+}]_x^2}{K_{Cx}^2} + 2\frac{[Mg^{2+}]_e}{K_{Me}} + \frac{[Mg^{2+}]_e^2}{K_{Me}^2} + 2\frac{[Mg^{2+}]_x}{K_{Mx}} + \frac{[Mg^{2+}]_x^2}{K_{Mx}^2}$$
$$+ 4\frac{[Ca^{2+}]_e[Mg^{2+}]_e}{\gamma \cdot K_{Ce} \cdot K_{Me}} + 2\frac{[Ca^{2+}]_e[Mg^{2+}]_e^2}{\gamma^2 \cdot K_{Ce}K_{Me}^2} + 2\frac{[Ca^{2+}]_e^2[Mg^{2+}]_e}{\gamma^2 \cdot K_{Ce}^2 \cdot K_{Me}} + \frac{[Ca^{2+}]_e^2[Mg^{2+}]_e^2}{\gamma^4 \cdot K_{Ce}^2 \cdot K_{Me}^2}$$
$$+ 4\frac{[Ca^{2+}]_x[Mg^{2+}]_x}{\gamma \cdot K_{Cx} \cdot K_{Mx}} + 2\frac{[Ca^{2+}]_x[Mg^{2+}]_x^2}{\gamma^2 \cdot K_{Cx} \cdot K_{Mx}^2} + 2\frac{[Ca^{2+}]_x^2[Mg^{2+}]_x}{\gamma^2 \cdot K_{Cx}^2 \cdot K_{Mx}} + \frac{[Ca^{2+}]_x^2[Mg^{2+}]_x^2}{\gamma^4 \cdot K_{Cx}^2 \cdot K_{Mx}^2}. \quad (5.73)$$

The effects of cytosolic and matrix Pi on the uniporter flux can be incorporated by exclusively modifying the binding constants Ca^{2+} and Mg^{2+} for the uniporter ($K_{Ce}, K_{Cx}, K_{Me}, K_{Mx}$):

$$K_{Ce}^0 = K_{Ce}^0 \cdot (1 + [Pi]_e/(K_{Pi} + [Pi]_e), \quad K_{Me}^0 = K_{Me}^0/(1 + [Pi]_e/(K_{Pi} + [Pi]_e),$$
$$K_{Cx}^0 = K_{Cx}^0 \cdot (1 + [Pi]_x/(K_{Pi} + [Pi]_x), \quad K_{Mx}^0 = K_{Mx}^0/(1 + [Pi]_x/(K_{Pi} + [Pi]_x), \quad (5.74)$$

where K_{Pi} is the Pi binding constant for the uniporter; K_{Ce}^0, K_{Cx}^0, K_{Me}^0 and K_{Mx}^0 on the right side of Eq. (5.74) are the true dissociation constants of Ca^{2+} and Mg^{2+} for the uniporter, independent of $\Delta\Psi$. Thus, Pi increases the values of K_{Ce}^0 and K_{Cx}^0, but decreases the values of K_{Me}^0 and K_{Mx}^0; or equivalently, Pi decreases the binding affinity of Ca^{2+} for the uniporter, but increases the binding affinity of Mg^{2+} for the uniporter. This formulation of the Pi dependency is purely based on the observations of Ca^{2+} uptake in isolated mitochondrial preparations with different amount of Pi present in the incubation medium [19], and may need additional structural/kinetic information on the Ca^{2+} uniporter for a more precise description of Pi regulation. [21]

The present Ca^{2+} uniporter model was used to reanalyze all the available experimental data on the kinetics of mitochondrial Ca^{2+} uptake, measured both in the presence and absence of Mg^{2+} and Pi in isolated mitochondria and permealized cell preparations [18–21,37], and to estimate the unknown model parameters. In our previous analyses [25] on the $\Delta\Psi$ dependency of mitochondrial Ca^{2+} uptake (Section 5.4.4), it was shown that the Ca^{2+}(Mg^{2+}) binding sites on the uniporter are located at equal and negligible distances from the bulk phase on either side of the IMM (i.e., $\alpha_e = \alpha_e = \alpha \approx 0$) and the binding affinities of the uniporter to both cytosolic and matrix Ca^{2+}(Mg^{2+}) are the same ($K_{Ce}^0 = K_{Cx}^0 = K_C^0$ and $K_{Me}^0 = K_{Mx}^0 = K_M^0$), implying $K_i^0 = K_o^0 = K^0$ (based on the kinetic constraint of Eq. (5.61)). Also the previous estimate of $nH = 2.65$ was shown to describe the observed $\Delta\Psi$ dependency of the Ca^{2+} uniporter flux quite well.

Based on these observations, the present Ca^{2+} uniporter model was utilized to reanalyze all the available experimental data [18–21,37], and to estimate the remaining four unknown kinetic parameters (K^0, K_C^0, K_M^0 and γ). The model was able to explain adequately the extra-matrix Pi, Mg^{2+}, Ca^{2+} and $\Delta\Psi$ dependent kinetic data on mitochondrial Ca^{2+} uptake in a wide variety of experimental preparations with a unique set of parameters [31]. Specifically, the model analyses of the diverse experimental data suggested that K_C^0 is of the order of only 4 μM (in contrast to the experimentally derived range 1–90 μM reported in the literature). It also clarified the earlier contradictory results on the type of Mg^{2+} inhibition of the uniporter function, and suggested that a mixed-type mechanism of Mg^{2+} inhibition of Ca^{2+} uptake is consistent with the available experimental data. Another important outcome of this modeling study was that it demonstrates how simpler binding schemes which assume a single Ca^{2+} and/or single Mg^{2+} binding to the uniporter are not able to explain the observed kinetics of this transporter. For more detailed analysis and model applications, the readers can refer to the original work [31]. However, understanding the regulation of mitochondrial Ca^{2+} transport system and its roles on mitochondrial bioenergetics and other cellular functions is still a subject of active research.

5.5 OTHER MODES OF CARRIER-MEDIATED TRANSPORT: ANTIPORT AND COTRANSPORT

Two additional, commonly observed modes of carrier-mediated transport are the antiport (exchange) and the cotransport (symport), which facilitate the simultaneous movement of two solutes across the membrane. An antiporter is a carrier protein that allows the obligatory exchange of one solute for another solute in the opposite direction, while a cotransporter is a carrier protein that allows the simultaneous transport of two solutes in the same direction. We will briefly describe below the kinetic treatments of these transporters.

The mitochondrial ATP/ADP antiporter [38,39] and the plasma membrane and mitochondrial inner membrane Na^+/Ca^{2+} exchanger [5,40], and Na^+/H^+ exchanger [41,42] are the most commonly studied antiporters, although there are many others. Figure 5.6 schematizes the frequently used kinetic models of the antiporters. In these carrier models (excluding ping-pong models), the interconversion between the two forms of the single solute-bound carrier states is so slow that it is usually ignored; only the two solute-bound carrier complexes undergo interconversion resulting in an obligatory equimolar exchange.

As shown in Fig. 5.6, in an antiporter, two solutes A and B share the same unbound carrier states C_1 or C_2. Either solute A or B can cross the membrane via the antiporter, but only in exchange for the second solute. The solute A moves from side 1 on the antiporter at a rate proportional to its concentration $[A]_1$, and B moves from side 2 at a rate proportional to $[B]_2$. Therefore, the exchange of A coming from side 1 with B coming from side 2 will occur at a rate proportional to the product of $[A]_1$ and $[B]_2$. At equilibrium, where the net transport of A from side 1 to side 2 is equal to the net transport of solute B from side 2 to side 1, we have:

$$[A]_1[B]_2 = [A]_2[B]_1. \tag{5.75}$$

Eq. (5.75) is the fundamental rule of solute transport through an antiporter. Similar to the simple carrier model discussed earlier, the antiporter can also be electrogenic in nature. To accurately account for the transfer of charges by the antiporter, Eq. (5.75) can be modified to include trans-membrane potential $\Delta\Psi$ and the charges Z_A and Z_B of the solutes A and B, which becomes:

$$\ln([A]_2/[A]_1) = \ln([B]_2/[B]_1) + (Z_A - Z_B)F(\Delta\Psi/RT). \tag{5.76}$$

When the net charge transferred across the membrane $(Z_A - Z_B)$ is zero, Eq. (5.76) reduces to Eq. (5.76), and the antiporter becomes electro-neutral.

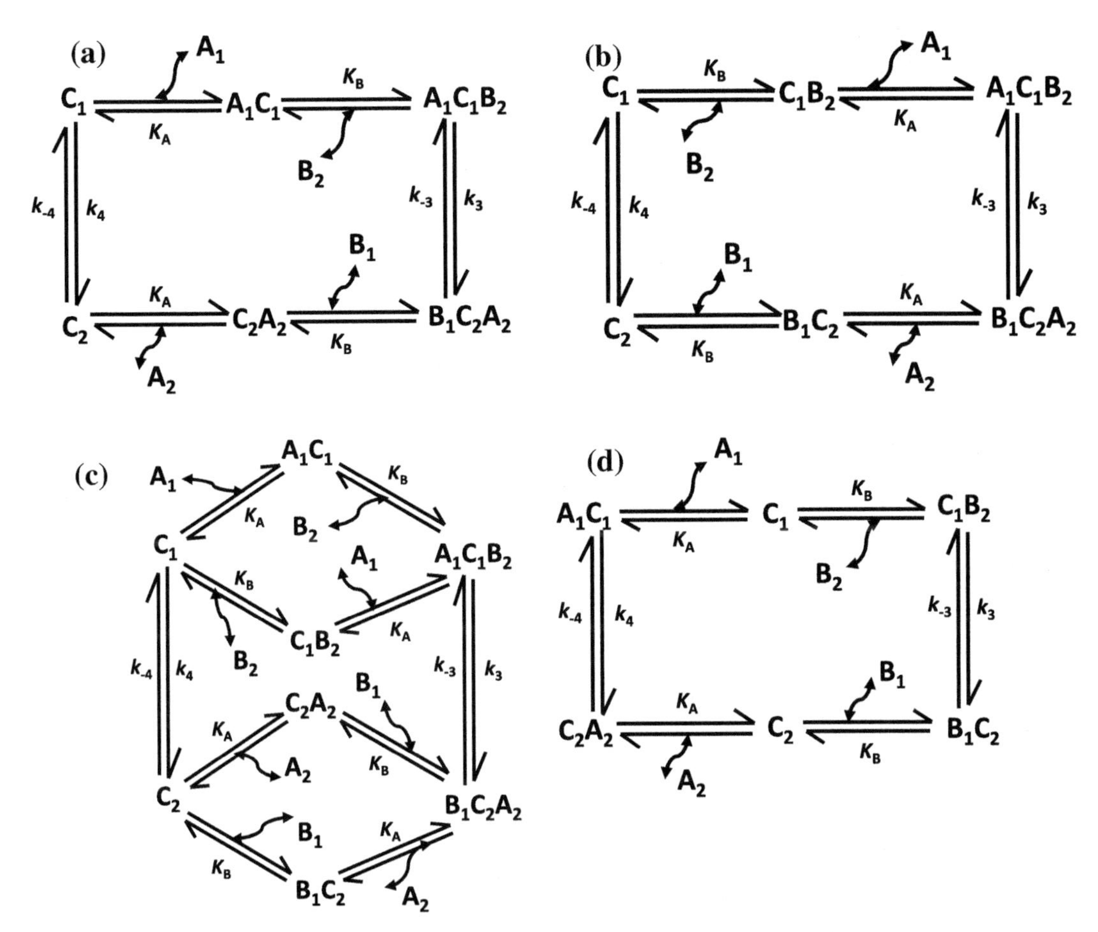

FIGURE 5.6 Schematic representation of kinetic models involving the movement of two solutes in the opposite direction across a membrane (antiporter). Kinetic models for a symporter can similarly be represented. (a) An ordered bi-bi model for two solutes in which the solute A from outside (side 1) binds first to the carrier and then the solute B binds from inside (side 2) before going for conformational change; (b) an ordered bi-bi model for two solutes in which the solute B from side 2 binds first to the carrier and then the solute A binds from the side 1 before going for conformational change; (c) a random bi-bi model involving two solutes in which both binding orders of the solutes to the carrier is considered; and (d) a ping-pong model in which the binding of one solute (A or B) can lead to conformational change before binding of the other solute to the carrier. C_1 and C_2 represent two distinct unbound carrier states, each binds to one external solute and one internal solute. K_A and K_B are the solute dissociation constants, while k_i and k_{-i} are the forward and reverse rate constants for carrier and solute-carrier complex conformational change.

Some antiporters bind more than one molecule of a particular solute. If m and n are the *stoichiometry* of the solutes A and B, Eq. (5.76) can be modified to:

$$[A]_1^m/[A]_2^m = [B]_1^n/[B]_2^n. \tag{5.77}$$

The stoichiometry of an antiporter is one of its most fundamental properties, which has been intensively studied for many exchangers in many tissues (e.g., Na^+/Ca^{2+} exchanger).

Recently, mathematical modeling of antiporter kinetics, based on the initial rate data, has significantly improved our understanding on the catalytic properties and regulation of these transporters in many species. Some recent examples include the biophysical modeling of mitochondrial ATP/ADP exchanger [39] and mitochondrial Na^+-Ca^{2+} exchanger [29,43] in cardiac cells.

A cotransporter (symporter) is a carrier protein that allows the transport of two different species (a solute and an ion) from one side of the membrane to the other at the same time [44,45]. Among many in existence, the most commonly studied cotransporters are the plasma membrane Na^+–glucose cotransporter [3] and H^+–lactate cotransporter [4]. In this case, the movement of two species is strictly coupled for active transport. Thus the gradient of one solute at steady state will be dependent on the gradient of other solute.

The kinetic models of cotransporters can be represented similarly to that of antiporters, as shown in Fig. 5.6. Depending on the order of binding of the solutes and/or ions to the carrier, the transport kinetics can take different forms such as:

i. An ordered binding scheme, in which solute A binds first and then solute B binds to the carrier, both from the external side, before going for conformational change,
ii. An ordered binding scheme in which solute B binds before solute A, both from the external side, before going for conformational change,
iii. A random binding scheme, in which there is a given probability of both solute A and solute B to binding to the carrier first, from the external side before going for conformation change, and
iv. A ping-pong binding scheme, in which the binding of one solute (A or B) can lead to conformational change before binding of the other solute to the carrier.

The rate of transport from side 1 to side 2 will be proportional to the concentrations of both A and B at side 1 – i.e., to the product $[A_1][B_1]$. Similarly, the rate of transport from side 2 to side 1 will be proportional to the concentrations of both A and B at side 2, i.e., to the product $[A]_2[B]_2$. At equilibrium, these two rates will be equal such that:

$$[A_1][B_1] = [A_2][B_2] \tag{5.78}$$

Eq. (5.78) is the fundamental rule of a cotransport system. Also at steady state, the ratios of the two solutes are inversely related to one another (i.e., $[B_1]/[B_2] = [A_2]/[A_1]$. The equation for transport flux can be easily derived using an approach similar to that used for the simple carrier models, discussed in earlier sections. Similarly, if the movement of solutes via the cotransporter is accompanied by the movement of electrical charges, then the cotransporter can be electrogenic in nature. Most extensively studied cotransport systems include the transport of pyruvate, lactate, and other monocarboxylate across the plasma membrane of skeletal and cardiac cells via passive diffusion and by monocarboxylate transporter (MCT). For details, readers are referred to the recent modeling studies on the MCT transporter isoforms 1 and 4 of *Vinnakota and Beard* [46].

5.6 SUMMARY AND CONCLUSION

In living cells, the uptake and extrusion of hydrophilic molecules are generally governed by specialized membrane proteins known as transporters or carriers. Unlike pores or channels, they undergo enzyme-like binding and conformational changes to promote energetically downhill transport of their specific molecules. Elucidation of the catalytic properties of carriers by a combination of experimental measurements of carrier fluxes and analysis of those data by kinetic models is a fundamental endeavor in this field. This chapter provides a systematic description of the formulation and analysis of kinetic models of carrier-mediated transport across biomembranes. Detailed analyses are provided for the transport of Ca^{2+} ions into the mitochondrial matrix via the Ca^{2+} uniporter located in the IMM. A brief introduction is also given to the kinetic treatment of cotransporters and antiporters. The analyses presented in this chapter can be easily extended to study the kinetics of carrier-mediated ion and metabolite transport in general.

Acknowledgment

This work was partially supported by the National Institute of Health grants R01-HL095122 (RKD) and P50-GM094503 (DAB).

References

[1] A.B. Harvey Lodish, Chris A. Kaiser, Monty Krieger, Matthew P. Scott, Molecular Cell Biology, sixth ed., W.H. Freeman, 2007.
[2] W. Stein, The Movement of Molecules Across Cell Membranes, Academic Press, New York, 1967.
[3] M. Mueckler, C. Caruso, S.A. Baldwin, M. Panico, I. Blench, H.R. Morris, W.J. Allard, G.E. Lienhard, H.F. Lodish, Sequence and structure of a human glucose transporter, Science 229 (1985) 941–945.
[4] R.C. Poole, A.P. Halestrap, Transport of lactate and other monocarboxylates across mammalian plasma membranes, The American Journal of Physiology 264 (1993) C761–782.
[5] L.J. Mullins, An electrogenic saga: consequences of sodium-calcium exchange in cardiac muscle, Society General of Physiologists Series 38 (1984) 161–179.
[6] W.D. Stein, Channels, Carriers, and Pumps: An Introduction to Membrane Transport, Academic Press, London, 1990.
[7] W.D. Stein, Transport and Diffusion Across Cell Membranes, Academic Press, San Diego CA, USA, 1986.
[8] P. Lauger, Ion transport through pores: a rate-theory analysis, Biochimica Biophysica Acta 311 (1973) 423–441.
[9] P. Lauger, B. Neumcke, Theoretical analysis of ion conductance in lipid bilayer membranes, Membranes 2 (1973) 1–59.
[10] J.P. Keener, J. Sneyd, Mathematical Physiology, Springer, New York, 1998.
[11] D.A. Beard, Biosimulation: Simulation of Living Systems, Cambridge University Press, Cambridge, UK, 2012.
[12] W.R. Lieb, A Kinetic Approach to Transport Studies. Red Cell Membranes a Methodological Approach, Academic Press, London, 1982.
[13] T. Rosenberg, W. Wilbrandt, Uphill transport induced by counterflow, The Journal of General Physiology 41 (1957) 289–296.
[14] P. Bernardi, Mitochondrial transport of cations: channels, exchangers, and permeability transition, Physiological Review 79 (1999) 1127–1155.
[15] T.E. Gunter, D.R. Pfeiffer, Mechanisms by which mitochondria transport calcium, American Journal of Physiology 258 (1990) C755–786.
[16] B. Chance, The energy-linked reaction of calcium with mitochondria, The Journal of Biological Chemistry 240 (1965) 2729–2748.
[17] H. Rottenberg, The measurement of membrane potential and delta pH in cells, organelles, and vesicles, Methods in enzymology 55 (1979) 547–569.
[18] M. Bragadin, T. Pozzan, G.F. Azzone, Kinetics of Ca^{2+} carrier in rat liver mitochondria, Biochemistry 18 (1979) 5972–5978.
[19] M. Crompton, E. Sigel, M. Salzmann, E. Carafoli, A kinetic study of the energy-linked influx of Ca^{2+} into heart mitochondria, European Journal of Biochemistry 69 (1976) 429–434.
[20] A. Scarpa, P. Graziotti, Mechanisms for intracellular calcium regulation in heart. I. Stopped-flow measurements of Ca^{2+} uptake by cardiac mitochondria, Journal of General Physiology 62 (1973) 756–772.

[21] A. Vinogradov, A. Scarpa, The initial velocities of calcium uptake by rat liver mitochondria, The Journal of Biological Chemistry 248 (1973) 5527–5531.

[22] D.E. Wingrove, J.M. Amatruda, T.E. Gunter, Glucagon effects on the membrane potential and calcium uptake rate of rat liver mitochondria, Journal of Biological Chemistry 259 (1984) 9390–9394.

[23] T.E. Gunter, D.E. Wingrove, S. Banerjee, K.K. Gunter, Mechanisms of mitochondrial calcium transport, Advances in Experimental Medicine and Biology 232 (1988) 1–14.

[24] F.L. Bygrave, K.C. Reed, T. Spencer, Cooperative interactions in energy-dependent accumulation of Ca^{2+} by isolated rat liver mitochondria, Nature New Biology 230 (1971) 89.

[25] R.K. Pradhan, F. Qi, D.A. Beard, R.K. Dash, Characterization of membrane potential dependency of mitochondrial Ca^{2+} uptake by an improved biophysical model of mitochondrial Ca^{2+} uniporter, PLoS One 5 (2010) e13278.

[26] Y. Kirichok, G. Krapivinsky, D.E. Clapham, The mitochondrial calcium uniporter is a highly selective ion channel, Nature 427 (2004) 360–364.

[27] J.M. Baughman, F. Perocchi, H.S. Girgis, M. Plovanich, C.A. Belcher-Timme, Y. Sancak, X.R. Bao, L. Strittmatter, O. Goldberger, R.L. Bogorad, V. Koteliansky, V.K. Mootha, Integrative genomics identifies MCU as an essential component of the mitochondrial calcium uniporter, Nature 476 (2011) 341–345.

[28] D. De Stefani, A. Raffaello, E. Teardo, I. Szabo, R. Rizzuto, A forty-kilodalton protein of the inner membrane is the mitochondrial calcium uniporter, Nature 476 (2011) 336–340.

[29] R.K. Dash, D.A. Beard, Analysis of cardiac mitochondrial Na^+/Ca^{2+} exchanger kinetics with a biophysical model of mitochondrial Ca^{2+} handling suggests a 3:1 stoichiometry, Journal of Physiology 586 (2008) 3267–3285.

[30] R.K. Dash, F. Qi, D.A. Beard, A biophysically-based mathematical model for the kinetics of mitochondrial calcium uniporter, Biophysical Journal 96 (2009) 1318–1332.

[31] R.K. Pradhan, F. Qi, D.A. Beard, R.K. Dash, Characterization of Mg2+ inhibition of mitochondrial Ca2+ uptake by a mechanistic model of mitochondrial Ca2+ uniporter, Biophysical Journal 101 (2011) 2071–2081.

[32] G. Magnus, J. Keizer, Minimal model of beta-cell mitochondrial Ca^{2+} handling, American Journal of Physiology 273 (1997) C717–733.

[33] S. Cortassa, M.A. Aon, E. Marban, R.L. Winslow, B. O'Rourke, An integrated model of cardiac mitochondrial energy metabolism and calcium dynamics, Biophysical Journal 84 (2003) 2734–2755.

[34] M.H. Nguyen, M.S. Jafri, Mitochondrial calcium signaling and energy metabolism, Annals of the New York Academy of Sciences 1047 (2005) 127–137.

[35] M.H. Nguyen, S.J. Dudycha, M.S. Jafri, The Effects of Ca^{2+} on cardiac mitochondrial energy production is modulated by Na^+ and H^+ dynamics, American Journal of Physiology and Cell Physiology 292 (2007) 2004–2020.

[36] A. Romani, A. Scarpa, Hormonal control of Mg^{2+} transport in the heart, Nature 346 (1990) 841–844.

[37] G. Szanda, A. Rajki, S. Gallego-Sandin, J. Garcia-Sancho, A. Spat, Effect of cytosolic Mg^{2+} on mitochondrial Ca^{2+} signaling, Pflugers Archiv 457 (2009) 941–954.

[38] M. Klingenberg, The ADP-ATP translocation in mitochondria, a membrane potential controlled transport, The Journal of Membrane Biology 56 (1980) 97–105.

[39] E. Metelkin, I. Goryanin, O. Demin, Mathematical modeling of mitochondrial adenine nucleotide translocase, Biophysical Journal 90 (2006) 423–432.

[40] K. Baysal, D.W. Jung, K.K. Gunter, T.E. Gunter, G.P. Brierley, Na^+-dependent Ca^{2+} efflux mechanism of heart mitochondria is not a passive $Ca^{2+}/2Na^+$ exchanger, American Journal of Physiology 266 (1994) C800–808.

[41] J.L. Seifter, P.S. Aronson, Properties and physiologic roles of the plasma membrane sodium-hydrogen exchanger, The Journal of Clinical Investigation 78 (1986) 859–864.

[42] A. Kapus, E. Ligeti, A. Fonyo, Na^+/H^+ exchange in mitochondria as monitored by BCECF fluorescence, FEBS Letters 251 (1989) 49–52.

[43] R.K. Pradhan, D.A. Beard, R.K. Dash, A biophysically based mathematical model for the kinetics of mitochondrial $Na^+ - Ca^{2+}$ antiporter, Biophysical Journal 98 (2010) 218–230.

[44] W.D. Stein, The cotransport systems, Current Opinion in Cell Biology 1 (1989) 739–745.

[45] W.D. Stein, Kinetics of transport: analyzing, testing, and characterizing models using kinetic approaches, Methods in Enzymology 171 (1989) 23–62.

[46] K.C. Vinnakota, D.A. Beard, Kinetic analysis and design of experiments to identify the catalytic mechanism of the monocarboxylate transporter isoforms 4 and 1, Biophysical Journal 100 (2011) 369–380.

CHAPTER

6

Blood Flow Through Capillary Networks

C. Pozrikidis and J.M. Davis

Department of Chemical Engineering, University of Massachusetts
Amherst, MA, USA

OUTLINE

6.1 Introduction 214

6.2 Equations of Steady Capillary Blood Flow 216

6.2.1 *Balances at a Bifurcation* 217

6.2.2 *Discharge Hematocrit* 217

6.2.3 *Tube Hematocrit and Hematocrit Ratio* 218

6.2.4 *Effective Viscosity* 218

6.2.5 *Cell Partitioning at a Bifurcation* 219

6.2.6 *Converging Bifurcations* 220

6.2.7 *Diverging Bifurcation* 220

6.2.8 *Klitzman and Johnson Cell Partitioning Law* 220

6.2.9 *General Expression for the Cell Partitioning Law* 221

6.2.10 *Empirical Cell Partitioning Law* 221

6.2.11 *Numerical Studies of Suspension Flow Through a Bifurcation* 221

6.3 Steady Flow Through Tree Networks 222

6.3.1 *Geometrical Construction* 222

6.3.2 *Numerical Method* 223

6.3.3 *Results and Discussion* 223

6.4 Steady Flow Through Homogeneous Networks 226

6.4.1 *Theoretical Model* 226

6.4.2 *Geometrical Construction* 228

6.4.3 *Numerical Method* 228

6.4.4 *Flow Through a Pristine Network* 228

6.4.5 *Dimensions and Parameters* 229

6.4.6 *Effective Hydraulic Permeability* 229

6.4.7 *Results and Discussion* 230

6.4.8 *Significance of the Bifurcation Law* 231

6.4.9 *Significance of the Viscosity Correlation* 233

6.4.10 *Discussion* 234

6.5 Equations of Unsteady Blood Flow 235

6.5.1 *Unsteady Flow Through a Straight Capillary* 235

6.5.2 *Circular Capillaries* 236

6.5.3 *Correlations* 237

6.5.4 *Balances at Bifurcations* 237

6.5.5 *Numerical Method* 238

6.5.6 *Single-Node Dynamics* 240

6.6 Unsteady Flow Through Tree Networks 243

6.6.1 *Steady Flow for Subcritical Exponents* 244

Transport in Biological Media

© 2013 Elsevier Inc. All rights reserved.

http://dx.doi.org/10.1016/B978-0-12-415824-5.00006-0

6.6.2 *Unsteady Flow for Supercritical Exponents* 246
6.6.3 *State Space* 248
6.6.4 *Summary and Discussion* 248
6.7 Summary and Outlook 249
References 250

6.1 INTRODUCTION

Theoretical models of blood flow through capillary networks are built on several assumptions. Most models adopt the continuum approximation, where blood is treated as a two-phase homogeneous medium consisting of plasma and suspended cells, predominantly red blood cells. The flow through a capillary tube is described by Poiseuille's law with an effective viscosity determining the flow rate for a given pressure drop. Theoretical, computational and laboratory studies have sought to establish relations or correlations between the effective viscosity and the capillary diameter, cell concentration and flow rate. The dependence of the effective viscosity on the flow rate is due to the flow-induced deformation and migration of red blood cells. The discharge hematocrit, defined as the ratio of the whole blood flow rate to the red cell flow rate, is deduced from a mass balance at converging bifurcations where two streams merge into one, or a partitioning law at diverging bifurcations where one stream splits into two.

The physiological relevance of each approximation involved in the continuum formulation can be improved, and additional effects can be incorporated. However, the basic implementation offers a convenient and instructive framework for obtaining insights, studying the significance of physiological and geometrical parameters of interest and detecting the possible onset of flow instability (e.g., [1–7]). Although cell-level models have been developed, high computational cost and storage capacity required for tracking the motion of the individual red blood cells using boundary-integral and domain-discretization methods are serious concerns (e.g., [8]). The main usefulness of particulate models hinges on their ability to furnish or confirm expressions for rheological properties in simple flow configurations, illustrate the significance of the cell membrane and interior physical properties and provide us with information on the geometry of the suspension microstructure.

Models of capillary networks in the microcirculation originate from an arterial entrance point or multiple arterial entrance points where the pressure and discharge hematocrit are specified. The model networks terminate at venular exit points where the pressure is specified and the discharge hematocrits are found as part of the solution. Tree-like branching networks and area- and space-filling networks, concisely called homogeneous networks, have been employed with reference to specific organs or tissue (e.g., [9–12]). In mathematics, branching networks with arbitrary coordination number are known as Bethe lattices or Cayley trees. Tree networks terminating at polygonal structures are hybrids of tree-like and homogeneous networks. Tree models are convenient for studying in a systematic fashion the effect of bifurcation parameters and network size. An important limitation is the absence of anastomosis and the rapid decay of the pressure across the vasculature. Homogeneous networks are more appropriate for modeling blood flow through normal and tumor tissue. Deterministic and stochastic geometrical parameters involved in tree and homogeneous networks have been deduced by analyzing physiological samples of different organs, healthy and malignant tissue (e.g., [13–16]). Several physiological networks have been documented by *Secomb* [17].

Significant variations in the red blood cell velocity and number density expressed by the tube hematocrit, defined precisely in Section 6.2, are observed in the microcirculation (e.g., [18]). The partitioning of the cells at a capillary bifurcation where one parental stream splits into two offspring is an important aspect of particulate capillary flow. In numerical simulations, the relationship between the discharge hematocrit of a daughter capillary tube to that of a parent tube is expressed by a theoretical or empirical bifurcation law for cell partitioning. In one extreme case, the fractional flow rate of the cells is equal to the fractional volumetric flow rate of whole blood. In the opposite extreme case, all cells are channeled into the daughter tube that receives the highest flow rate. The first scenario occurs in large vessels and the second scenario occurs in vessels whose diameter is comparable to the individual cell dimensions. Intermediate cases are parameterized by a bifurcation exponent, q, ranging from unity to infinity.

Numerical simulations using a discrete model of blood flow where the motion of the individual red cells is followed through the network have shown that the exponent, q, plays an important role in tree network dynamics (e.g., [4]). Further simulations using a continuum model confirmed this sensitivity and demonstrated that a supercritical Hopf bifurcation occurs at a critical value of q in a tree-like network with more than a few generations [6,7]. For higher values of q, the flow develops spontaneous self-sustained oscillations in the absence of external forcing. The occurrence of this instability has not yet been fully appreciated.

In this chapter, classical and recent formulations regarding the numerical simulation of blood flow through capillary networks are discussed. In Section 6.2, the equations governing steady blood flow through a capillary network with arbitrary structure are outlined. Two important components of the mathematical model are an equation providing us with the blood viscosity as a function of the discharge hematocrit and a partitioning law providing us with the fractional division of red blood cells at divergent bifurcations in terms of the fractional flow rates. The former is determined from available correlations, and the latter is determined either by available correlations or by a heuristic theoretical model involving an adjustable scalar parameter.

In Section 6.3, a geometrical model of a tree-like capillary network is discussed. Blood enters the network from a single arterial point and exits the network at multiple venular points after undergoing a cascade of bifurcations. One distinguishing feature of the tree model is that only divergent bifurcations where one capillary stream splits into two are present. Numerical simulations will be presented to demonstrate that the bifurcation law for cell partitioning has a significant effect on the spatial structure of blood flow through the network, whereas the viscosity law affects the nature of the predicted blood flow at steady state only in quantitative ways.

In Section 6.4, a new geometrical model of an area-filling, homogeneous capillary network relevant to flow in tissue is proposed. The idealized vasculature is a planar strip of a randomized honeycomb capillary lattice repeated periodically in one direction. Blood enters the network through inlet capillary segments on one side of the strip and exits the network through outlet capillary segments at the opposite side. The model is not meant to describe blood flow in any particular organ, but rather to serve as a prototype that is complementary to that of tree-like networks. Convergent and divergent bifurcations may be present in an area-filling network. The cell concentration at convergent bifurcations is computed based on a cell conservation law, whereas cell apportioning at divergent bifurcations is determined by a theoretical or empirical partitioning law. The simulations clarify the effect of random variations in the network geometry and capillary radius and elucidate the significance of the cell partitioning law with regard to the spatial distribution of the discharge hematocrit.

We find that the cell partitioning law plays an important role in determining the scattering of the capillary segment hematocrit in a homogeneous network.

In Section 6.5, the equations governing unsteady blood flow are derived and an algorithm for computing unsteady blood flow through a capillary network of arbitrary topology is presented. The evolution of the discharge hematocrit along each capillary segment is computed by integrating in time a one-dimensional convection equation using a finite-difference method that employs upwind differences. The convection velocity is determined by the pressure drop across the vessel and by the local and instantaneous effective capillary blood viscosity, which is obtained from available correlations. A mass balance at each bifurcation provides us with a linear system of equations for the pressure at each node in the network that is solved prior to each time step. In the case of a divergent bifurcation, the discharge hematocrit entering the descendent segments is determined from the partitioning law described for steady flow. The implementation of this method is demonstrated for a junction comprised of three segments and a central node. This small system exhibits a time sequence of convergent and divergent bifurcations as the pressure at the peripheral nodes is varied sinusoidally out of phase.

In Section 6.6, the numerical method for computing unsteady blood flow is applied to the tree-like capillary network discussed in Section 6.3. It is shown through numerical simulations that spontaneous, self-sustained oscillations can occur for sufficiently high values of the partitioning-law exponent, q, in the absence of external forcing. The transition to unsteady flow corresponds to a supercritical Hopf bifurcation. Using the *in vivo* viscosity correlation significantly decreases the critical value of the partitioning-law exponent relative to that based on the *in vitro* correlation. The reason is that both the effective viscosity and the variation of the effective viscosity with the discharge hematocrit are larger for the *in vivo* model. More generally, we find that the effect of self-sustained dynamics is more significant in systems with a strong dependence of the effective viscosity on the volume fraction of the suspended phase. A state map is presented to illustrate the critical value of the partitioning-law exponent in networks with different tree orders.

This chapter concludes in Section 6.7 where the new findings are critically summarized and suggestions for further research are outlined. The material included in this chapter is meant to be useful both as a learning and teaching resource and a state-of-the-art survey.

6.2 EQUATIONS OF STEADY CAPILLARY BLOOD FLOW

Blood flow through a capillary segment of length L_c and radius a_c is routinely described by Poiseuille's law with an effective fluid viscosity, μ_{eff}, that depends on the pressure drop across the capillary, number of suspended cells and capillary segment radius. The volumetric flow rate is given by

$$Q = \frac{\pi a_c^4}{8\mu_{\text{eff}}} \frac{\Delta p_c}{L_c} \tag{6.1}$$

where Δp_c is the pressure difference across the capillary end nodes, as shown in Fig. 6.1a (e.g., [19]). Due to entrance effects, imperfections and deviations from the circular shape, Poiseuille's law should be regarded as a scaling law involving effective geometrical and fluid parameters. In the case of a particulate medium, such as blood, the effective viscosity should be interpreted as a proportionality coefficient relating the flow rate, Q, to the pressure drop, Δp_c.

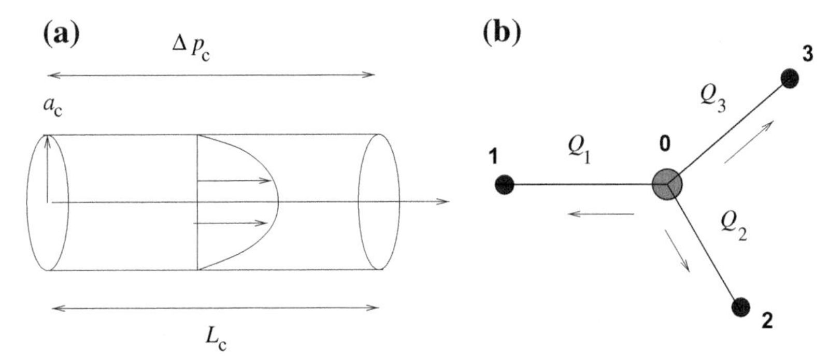

FIGURE 6.1 (a) Illustration of blood flow through a straight capillary tube of length L_c and radius a_c. (b) Illustration of a junction hosting three capillary segments numbered 1, 2 and 3. A mass balance provides us with an equation relating the nodal pressures at the four nodes.

6.2.1 Balances at a Bifurcation

Consider a bifurcation node of a capillary network where one parental capillary segment divides into two segments or two capillary segments converge into one segment, as shown in Fig. 6.1b. The local configuration involves four pressure nodes representing network junctions. A mass balance for incompressible fluids requires that:

$$Q_1 + Q_2 + Q_3 = 0 \tag{6.2}$$

or:

$$(p_0 - p_1)\frac{a_1^4}{\mu_{\mathrm{eff}_1} L_{c1}} + (p_0 - p_2)\frac{a_2^4}{\mu_{\mathrm{eff}_2} L_{c2}} + (p_0 - p_3)\frac{a_3^4}{\mu_{\mathrm{eff}_3} L_{c3}} = 0 \tag{6.3}$$

where a_1, a_2 and a_3 are the capillary radii and Q_1, Q_2 and Q_3 are the corresponding flow rates.

By convention, $Q_i > 0$ when blood is driven away from the junction along the ith capillary for $i = 1,2,3$. In the case of a converging bifurcation, two of the three flow rates, Q_1, Q_2, Q_3, are negative and the third flow rate is positive. In the case of a diverging bifurcation, two flow rates are positive and the third flow rate is negative.

6.2.2 Discharge Hematocrit

The discharge hematocrit, H_D, is defined as the ratio of the volumetric flow rate of the red blood cells, Q_{rbc}, to that of the whole blood, Q. By definition:

$$H_D \equiv \frac{Q_{\mathrm{rbc}}}{Q} \tag{6.4}$$

If blood is collected at the end of a tube inside a container, the discharge hematocrit will be precisely equal to the volume fraction of the cells inside the container. In network models of biphasic blood flow, a discharge hematocrit is assigned to each capillary segment at steady state. However, this is only an approximation due to entrance effects and possible phase separation in hemodynamics.

6.2.3 Tube Hematocrit and Hematocrit Ratio

The tube hematocrit, H_T, is defined as the volume fraction of the suspended red blood cells inside the tube, which is different than the volume fraction of the cells collected at the end of the tube expressed by the discharge hematocrit, H_D. For example, if the cells move with an exceedingly small velocity, the discharge hematocrit is nearly zero. The mean cell velocity, V_{rbc}, is related to the tube hematocrit and to the volumetric flow rate of the red blood cells, Q_{rbc}, by the equation:

$$V_{\mathrm{rbc}} = \frac{Q_{\mathrm{rbc}}}{SH_D} \tag{6.5}$$

where S is the tube cross-sectional area.

The ratio of the tube hematocrit, H_T, to the discharge hematocrit, H_D, is related to the mean cell velocity, V_{rbc}, by:

$$\chi \equiv \frac{H_T}{H_D} = \frac{U_m}{V_{\mathrm{rbc}}} \tag{6.6}$$

where $U_m \equiv Q/S$ is the mean blood velocity.

Pries et al. [20] proposed a correlation for the hematocrit ratio:

$$\chi = H_D + (1 - H_D)(1 + 1.7\mathrm{e}^{-0.7a_c} - 0.6\mathrm{e}^{-0.02a_c}) \tag{6.7}$$

where the capillary radius, a_c, is measured in μm. This correlation captures the Fåhraeus effect in which the discharge hematocrit exceeds the tube hematocrit. Physically, the blood cell or any other particle velocity is higher than the mean plasma velocity in capillary flow because the particles tend to concentrate near the centerline. The tube hematocrit deduced from the hematocrit ratio is useful only when the motion of the red blood cells needs to be followed.

6.2.4 Effective Viscosity

The effective viscosity of blood in a capillary segment, μ_{eff}, depends on the capillary radius, a_c, and either of the discharge hematocrit, H_D, or the tube hematocrit, H_T. *Pries et al.* [21] proposed a correlation based on *in vitro* data from 18 studies:

$$\frac{\mu_{\mathrm{eff}}}{\mu} = 1 + (\eta_{45} - 1)\frac{(1 - H_D)^{\delta} - 1}{(1 - 0.45)^{\delta} - 1} \tag{6.8}$$

where the capillary radius, a_c, is measured in μm, μ is the plasma viscosity:

$$\delta = (0.8 + \mathrm{e}^{-0.15a_c})\left(-1 + \frac{1}{f}\right) + \frac{1}{f} \tag{6.9}$$

is a dimensionless exponent, $f = 1 + 10(0.20a_c)^{12}$, and:

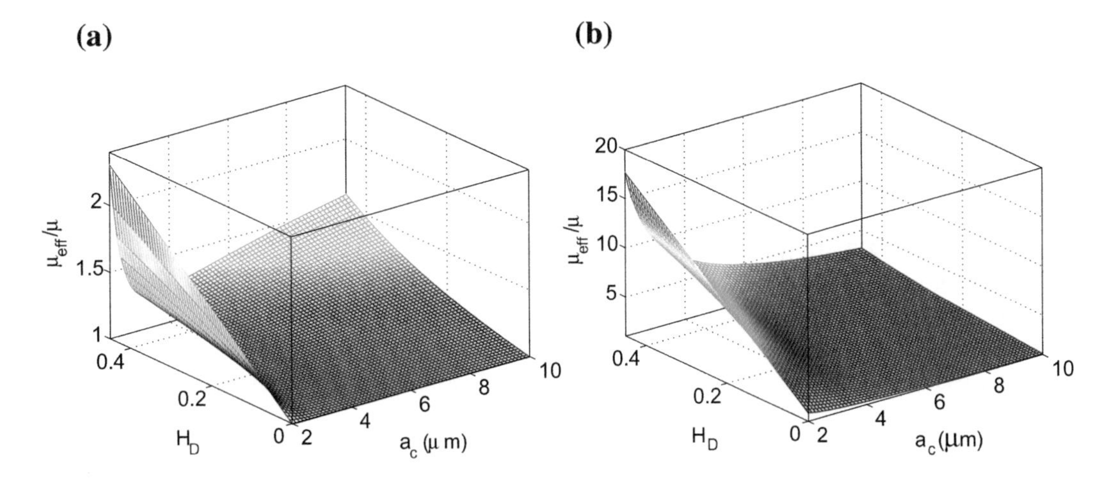

FIGURE 6.2 Effective blood viscosity as a function of the capillary radius, a_c, and discharge hematocrit, H_c, deduced from a correlation based on (a) *in vitro* and (b) *in vivo* experiments.

$$\eta_{45} = 220\,\mathrm{e}^{-2.6a_c} + 3.2 - 2.44\,\mathrm{e}^{-0.06(2a_c)^{0.645}} \tag{6.10}$$

is the relative blood viscosity at $H_D = 0.45$. A graph of the effective viscosity predicted by this correlation is shown in Fig. 6.2a. The correlation captures the Fåhraeus-Lindqvist effect in which the effective blood viscosity varies with the vessel diameter.

A similar correlation based on *in vivo* observations was proposed by *Pries et al.* [1,22]:

$$\frac{\mu_{\mathrm{eff}}}{\mu} = \beta\left[1 + \beta(\eta_{45} - 1)\frac{(1 - H_D)^C - 1}{(1 - 0.45)^C - 1}\right] \tag{6.11}$$

where:

$$\beta \equiv \left(\frac{a_c}{a_c - 0.55}\right)^2, \quad \eta_{45} = 6.0\,\mathrm{e}^{-0.17a_c} + 3.2 - 2.44\,\mathrm{e}^{-0.06(2a_c)^{0.645}} \tag{6.12}$$

and the capillary radius, a_c, is measured in μm. The correlation (6.11) seeks to patch the *in vivo* law for small vessels and the *in vitro* law for larger vessels. A graph of the effective viscosity predicted by this correlation is shown in Fig. 6.2b.

Comparing the two graphs in Fig. 6.2, we observe that the correlation based on the *in vivo* observations produces a significantly higher blood viscosity. These differences warrant an investigation of the effect of the constitutive law for the blood viscosity employed in the theoretical model.

6.2.5 Cell Partitioning at a Bifurcation

The empirical correlations for the effective viscosity require information on the discharge hematocrit of each segment of a capillary network. This information is obtained by performing a cell balance at each junction and introducing additional information from hydrodynamics.

Conservation of red blood cell volume at the junction, illustrated in Fig. 6.1b, requires that the volumetric flow rates of the cells balance to zero, that is:

$$Q_{\mathrm{rbc}_1} + Q_{\mathrm{rbc}_2} + Q_{\mathrm{rbc}_3} = 0 \tag{6.13}$$

or:

$$Q_1 H_{D_1} + Q_2 H_{D_2} + Q_3 H_{D_3} = 0 \tag{6.14}$$

where Q_i for $i=1,2,3$ are the whole blood flow rates and $Q_{\mathrm{rbc}_i} \equiv Q_i H_{D_i}$ are the corresponding red cell flow rates. We will assume that the three flow rates of the three capillary vessels involved, Q_i, and the hematocrits of one capillary vessel or two capillary vessels carrying fluid *toward* the junction are known. Our objective is to compute the hematocrit of the two capillary vessels or one capillary vessel carrying fluid *away* from the junction.

6.2.6 Converging Bifurcations

Consider a converging bifurcation where the first two streams merge into one stream, $Q_1 < 0, Q_2 < 0$ and $Q_3 > 0$. Solving Eq. (6.14) for H_{D_3} in terms of the known H_{D_1}, H_{D_2} and the three flow rates, we obtain:

$$H_{D_3} = \psi_1 H_{D_1} + \psi_2 H_{D_2} \tag{6.15}$$

where:

$$\psi_1 \equiv -Q_1/Q_3, \quad \psi_2 \equiv -Q_2/Q_3 \tag{6.16}$$

are known positive fractional flow rates determined by hydrodynamics.

6.2.7 Diverging Bifurcation

In the case of a diverging bifurcation where, for example, $Q_1 > 0, Q_2 > 0$ and $Q_3 < 0$, the third stream splits into two streams carrying fluid away from the bifurcation. It is convenient to introduce the positive cell fractional flow rates:

$$\phi_1 \equiv -\frac{Q_{\mathrm{rbc}_1}}{Q_{\mathrm{rbc}_3}} = \psi_1 \frac{H_{D_1}}{H_{D_3}}, \quad \phi_2 \equiv -\frac{Q_{\mathrm{rbc}_2}}{Q_{\mathrm{rbc}_3}} = \psi_2 \frac{H_{D_2}}{H_{D_3}} \tag{6.17}$$

where, by definition:

$$\psi_1 + \psi_2 = 1, \quad \phi_1 + \phi_2 = 1 \tag{6.18}$$

The outgoing discharge hematocrits are given by:

$$H_{D_1} = \frac{\phi_1}{\psi_1} H_{D_3}, \quad H_{D_2} = \frac{\phi_2}{\psi_2} H_{D_3} \tag{6.19}$$

6.2.8 Klitzman and Johnson Cell Partitioning Law

Klitzman and Johnson [23] proposed a simple partitioning law for capillaries with equal radii:

$$\phi_1 = \frac{\psi_1^q}{\psi_1^q + \psi_2^q}, \quad \phi_2 = \frac{\psi_2^q}{\psi_1^q + \psi_2^q} \tag{6.20}$$

where q is an adjustable dimensionless parameter. When $q=1$, all involved hematocrits are equal independent of the individual flow rates, $H_{D_1} = H_{D_2} = H_{D_3}$.

6.2.9 General Expression for the Cell Partitioning Law

A comprehensive expression for the cell partitioning law at a bifurcation can be written in terms of a partition function defined as [4]:

$$\frac{\phi_1}{\phi_2} \equiv \mathcal{F}_1\left(a_3, H_{D_3}; \psi_2, \frac{a_1}{a_2}\right) \tag{6.21}$$

Consequently:

$$\phi_1 = \frac{\mathcal{F}_1}{\mathcal{F}_1 + 1}, \quad \phi_2 = \frac{1}{\mathcal{F}_1 + 1} \tag{6.22}$$

To derive (6.20), we set $\mathcal{F}_1 = (\psi_1/\psi_2)^q$.

6.2.10 Empirical Cell Partitioning Law

Pries et al. [20,21] proposed the empirical relation:

$$\mathcal{F}_1 = \left(\frac{a_2}{a_1}\right)^{3.48/a_3} \cdot \left(\frac{\psi_1 - X_0}{1 - \psi_1 - X_0}\right)^q \tag{6.23}$$

where:

$$q = 1 + \frac{3.49}{a_3}(1 - H_{D_3}) \tag{6.24}$$

$X_0 = 0.2/a_3$ and the capillary radii, a_1, a_2 and a_3, are measured in μm. When $a_1 = a_2$ and $X_0 = 0$, we recover Eq. (6.20).

6.2.11 Numerical Studies of Suspension Flow Through a Bifurcation

Several authors have studied the motion of particles, drops, bubbles and biological cells through the bifurcating branches of a tube network (e.g., [24]). Of particular interest are (a) the ability of deformable particles to withstand the local extensional flow prevailing at the apex of a bifurcation and remain intact as they negotiate the changing environment, (b) the percentage of particles channeled through each branch and (c) the effect of the particles on the overall flow rates for given inlet and outlet conditions. Continuum and particulate models of flow through a bifurcation have been developed (e.g., [25]). Unfortunately, due to the inherent complexity, a theoretical partitioning function supported or motivated by the results of these simulations is not available at the present time.

6.3 STEADY FLOW THROUGH TREE NETWORKS

Tree models of capillary networks originate from a single arterial point and terminate at multiple venular points after undergoing a cascade of bifurcations, as shown in Fig. 6.3. Different trees can be constructed by employing deterministic and stochastic parameters. In this section, a model network used recently in numerical studies of particulate, continuum and unsteady blood flow is discussed [4,6,7].

One notable feature of the tree network is that all bifurcation nodes are diverging, as discussed in Section 6.2.6. In contrast, an area-filling network bifurcation may have converging and diverging bifurcations, as discussed in Section 6.4.

6.3.1 Geometrical Construction

The adopted capillary tree model originates from an arterial entrance point, which is the first end point of the first capillary segment in the xy plane with specified length, L_{inlet}, and radius, a_{inlet}. In the simulations presented in this section, we set $L_{\text{inlet}} = 100\ \mu\text{m}$ and $a_{\text{inlet}} = 6\ \mu\text{m}$. A cascade of bifurcations is then introduced to produce a branching network. The length of the projection of each generated capillary segment in the xy plane is:

$$L_c = \omega L_c' [1 + \epsilon_L (\varrho - 0.5)] \tag{6.25}$$

FIGURE 6.3 A capillary tree with six generations of bifurcating capillaries ($m = 6$) constructed using the procedure described in the text with mean bifurcation angle $\theta_0 = 0.15\pi$, tube length contraction ratio $\omega = 0.9$ and randomness parameters $\epsilon_L = 0.20$, $\epsilon_\theta = 0.10$, and $\epsilon_z = 0.50$. The radius of the inlet capillary tube is 6 μm and the radius of the outlet capillary tubes is 3 μm.

where L_c' is the length of the parental segment, ω is a specified contraction coefficient, ϵ_L is a specified dimensionless parameter and ϱ is a uniform deviate. In the simulations presented in this section, we set $\omega = 0.9$. The bifurcation angle in the xy plane is:

$$\theta = \pm\theta_0 + \epsilon_\theta(\varrho - 0.5)\pi \tag{6.26}$$

where θ_0 is a specified mean semi-angle and ϵ_θ is a specified dimensionless parameter. After a capillary bifurcation has been generated in the xy plane, the second end point of each newly created capillary segment is displaced normal to the xy plane by the distance:

$$z = L_{\text{inlet}}\,\epsilon_z(\varrho - 0.5) \tag{6.27}$$

where ϵ_z is a specified dimensionless parameter.

The radius of each generated capillary segment is reduced geometrically with respect to that of the parent segment, so that the last bifurcation yields capillary segments with a specified minimum radius, a_{exit}. The radial contraction ratio is:

$$r = \left(\frac{a_{\text{exit}}}{a_{\text{inlet}}}\right)^{1/m} \tag{6.28}$$

where m is the number of generations in the capillary tree. For the capillary tree shown in Fig. 6.3, $m = 6$. In the model network discussed in this section, the capillary tube radius decreases from 6.0 μm at the inlet segment to 3.0 μm at the exit segment.

6.3.2 Numerical Method

An iterative method is used to compute the nodal pressures according to the following steps:

1. The inlet discharge hematocrit is specified.
2. Guesses are made for the network capillary discharge hematocrits.
3. The effective blood viscosity in each capillary tube is calculated using the empirical relation (6.8) or (6.11).
4. The capillary pressures and flow rates are computed by Gauss-Seidel iterations (e.g., [26]).
5. The network capillary discharge hematocrits are updated using the empirical relations (6.19).
6. The process is repeated until convergence.

The tube hematocrit is computed *a posteriori* from the discharge hematocrit using the empirical relation (6.7).

6.3.3 Results and Discussion

The results of a simulation with a tree of order $m = 7$ conducted using the empirical relation (6.8) for the blood viscosity and the empirical relation (6.23) for cell partitioning at a bifurcation are shown in Fig. 6.4. The physical configuration of the network is shown in panel (a), and the distribution of the capillary pressure is shown in panel (b). The pressure declines

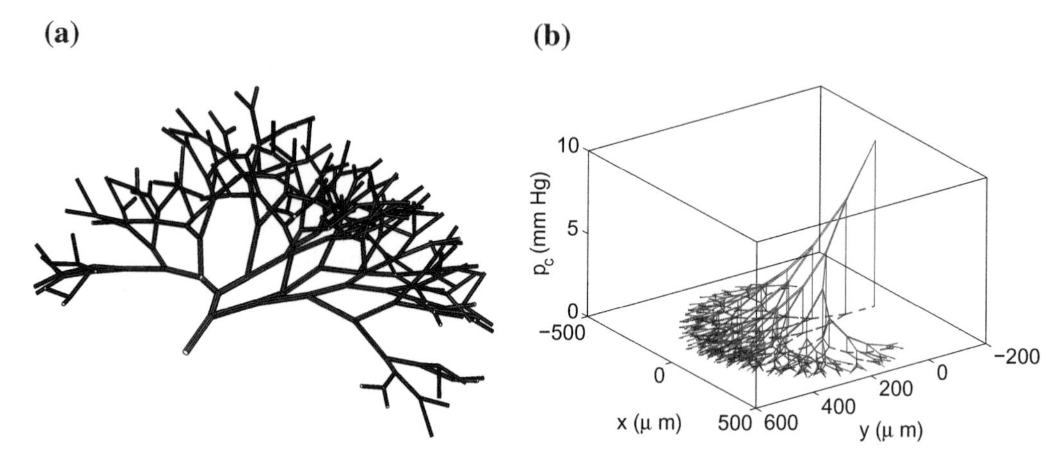

FIGURE 6.4 Blood flow through a capillary tree with seven generations of bifurcating capillaries ($m = 7$) constructed with $\theta_0 = 0.15\pi, \omega = 0.9, \epsilon_L = 0.20, \epsilon_\theta = 0.10$ and $\epsilon_z = 1.0$. The radius of the inlet capillary tube is 6 μm and the radius of the outlet capillary tubes is 3 μm. (a) Physical configuration, and (b) pressure distribution plotted on the projection of the network in the xy plane. The inlet hematocrit $H_D = 0.4375$.

rapidly from the inlet point to the outlets, with the largest pressure drop occurring in the first few branches of the capillary tree.

The distribution of the flow rate through the individual capillaries is shown in Fig. 6.5a. We note that the flow rate falls off rapidly to outlet values that are roughly two orders of magnitude less than the inlet value. The distribution of the discharge hematocrit, tube hematocrit, hematocrit ratio, red blood cell velocity and relative shifted blood viscosity, $\eta_{\text{eff}} = \mu_{\text{eff}}/\mu$, are shown in Fig. 6.5b. The solid or broken lines have been interpolated through the mean values of each generation. All variables plotted in Fig. 6.5b exhibit small or moderate variations across sibling branches of the network. The results of this calculation reveal that the exponent, q, varies in a narrow range, approximately $1.3 < q < 1.6$. Overall, blood flows uniformly through this network under the conditions considered.

Shown in Fig. 6.5c and d are the results of a corresponding simulation where the empirical relation (6.11) based on *in vivo* measurements of the blood viscosity is employed. The main difference is that the blood viscosity is predicted to be considerably higher, lowering by nearly one order of magnitude the capillary flow rates. The exponent, q, varies in a range that is comparable to that for the *in vitro* viscosity law. These comparisons indicate that the viscosity law employed affects the nature of the predicted blood flow at steady state in quantitative but not qualitative ways.

The results presented in Figs. 6.4 and 6.5 were obtained using the empirical relation (6.23) for cell partitioning at a bifurcation. To investigate the effect of the bifurcation law, now we adopt the *Klitzman and Johnson* [23] expressions (6.20) and vary the exponent, q. When $q = 1$, the discharge hematocrit is uniform through the entire network. However, the hematocrit ratio and thus the tube hematocrit vary across the network according to correlation (6.7).

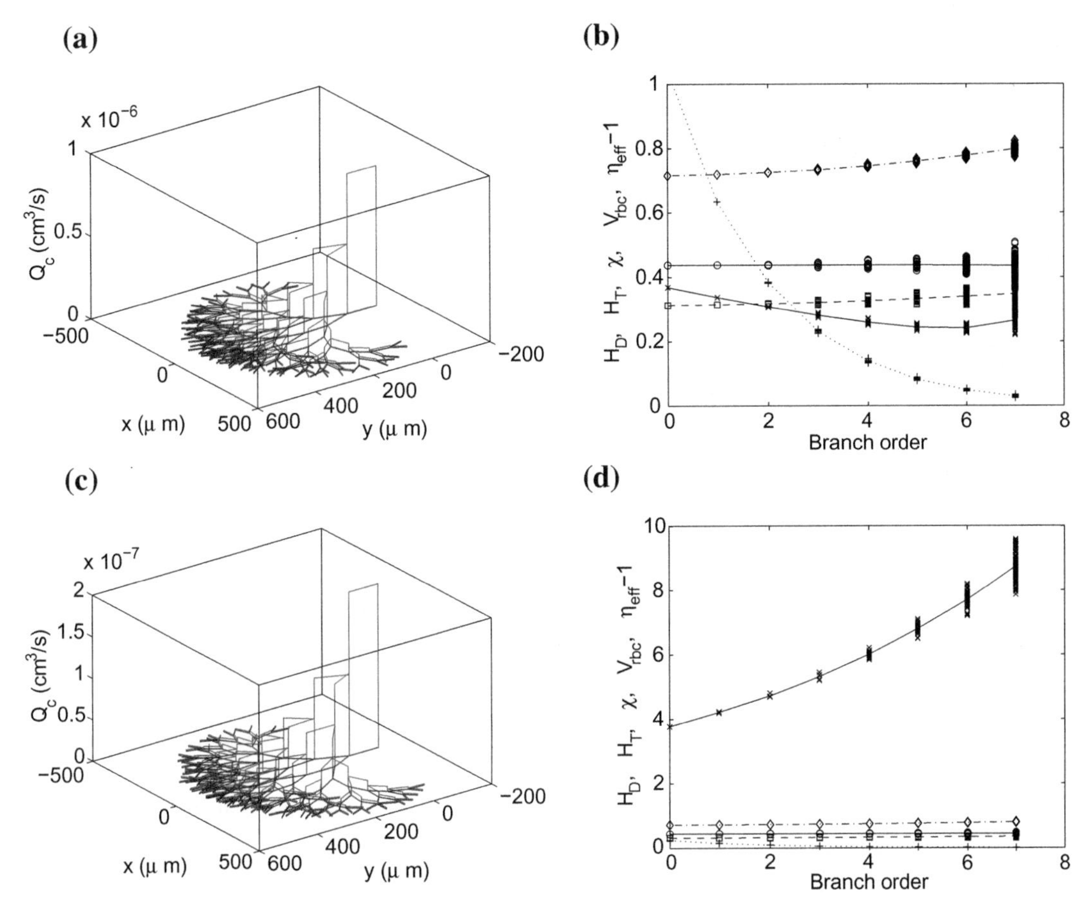

FIGURE 6.5 (a) Capillary flow rates displayed on the projection of a tree network in the xy plane and (b) distribution of H_D (circles), H_T (squares connected by the dashed line), hematocrit ratio, χ (diamonds connected by the dot-dashed line), red blood cell velocity in cm/s, $V_{\rm rbc}$ (crosses connected by the dotted line) and shifted relative viscosity, (×), plotted against the branch order from inlet to outlet, for the network shown in Fig. 6.4. (c, d) Results corresponding to those shown in (a, b) obtained using the *in vivo* viscosity correlation.

Results for exponent values $q = 1.5$ and 3.0 obtained using the network described in Fig. 6.4a are shown in Fig. 6.6. Not surprisingly, the results for $q = 1.5$ shown in Fig. 6.6a are similar to those shown in Fig. 6.5a. However, when $q = 3.0$, large variations of the tube hematocrit and effective viscosity are observed across sibling branches. Computations with higher values of q failed to converge using the Gauss-Seidel method with a uniform initial guess for the capillary tube hematocrits.

We conclude that the bifurcation law has a significant effect on the spatial structure of blood flow through a capillary network. Later in this chapter, we will see that the bifurcation exponent is a critical parameter for the stability of the network flow.

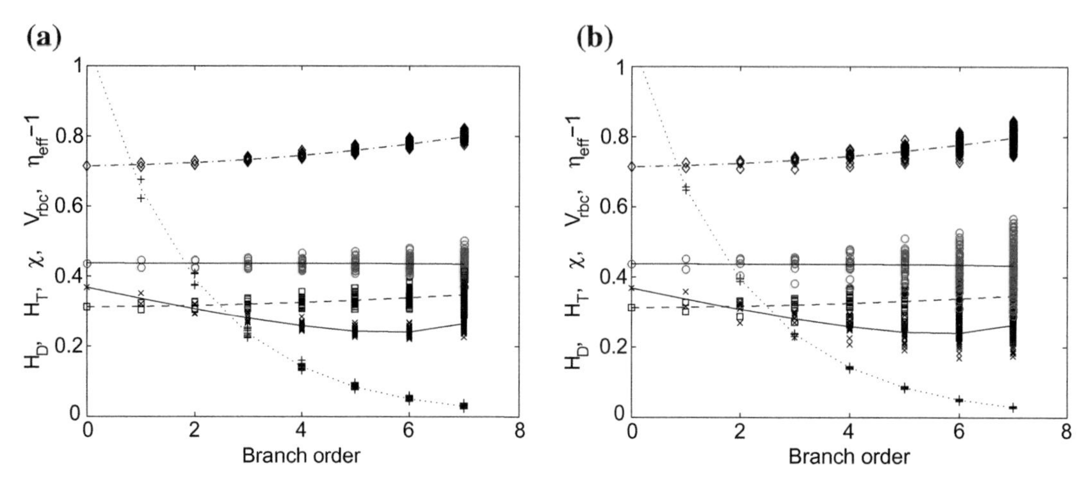

FIGURE 6.6 Same as Fig. 6.5b, but for the *Klitzman and Johnson* [23] bifurcation law with exponent (a) $q = 1.5$ and (b) 3.

6.4 STEADY FLOW THROUGH HOMOGENEOUS NETWORKS

Having discussed steady blood flow through tree networks in Section 6.3, we now turn to considering steady flow through homogeneous, area-filling networks originating from multiple entrance points and ending at multiple exit points. In contrast with tree networks, capillary bifurcations can be converging or diverging, depending on the local conditions.

6.4.1 Theoretical Model

The theoretical model employed in the numerical studies is a periodic honeycomb network of straight capillary tubes with circular cross-section forming an infinite two-dimensional strip. In the absence of geometrical imperfections, the capillary network is a pristine honeycomb lattice. One period of the strip enclosed by the vertical dashed lines is shown in Fig. 6.7a. The model was employed in a recent study of scalar transport across a defective or damaged network of conductive links [27]. The sides of each hexagonal unit cell arise from the Voronoi tessellation of a hexagonal (equilateral triangular) lattice whose nodes are located at the hexagon centers, with base vectors $\mathbf{a}_1$ and $\mathbf{a}_2$. A length scale is provided by the radius of each hexagonal unit, a. The radius of the inscribed circle is $b = \sqrt{3}/2a$.

The test section of the lattice shown in Fig. 6.7a consists of N whole hexagonal cells in the x direction, M whole hexagonal cells in the y direction and two horizontal rows of incomplete cells, where $N > 1$ is an arbitrary integer and M is an odd integer. In the configuration shown in Fig. 6.7a, $N = 6$ and $M = 5$. The length of each period of the test section is $L = 2Nb$, the exterior width of the strip is $W = \frac{1}{2}(3M + 5)a$ and the interior width of the strip is $D = \frac{1}{2}(3M + 1)a$. Each period of the test section contains N_l unique capillary segments consisting of N bottom segments N top segments and $N_l - 2N$ interior segments, where

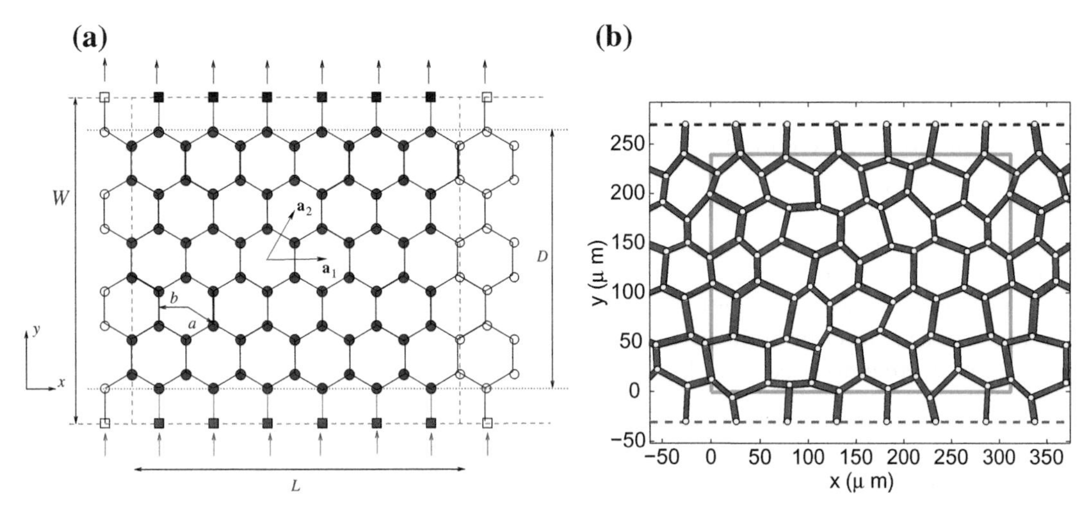

FIGURE 6.7 (a) Illustration of a 6 × 5 periodic test section of a honeycomb strip of a capillary network. Bottom (inlet) and top (outlet) nodes are denoted as filled squares, and interior nodes representing capillary junctions inside one period are denoted as filled circles. Blood enters the network from the bottom arterial segments with a specified discharge hematocrit, and exits the network from the top venous segments. (b) Network geometry after the interior nodes have been displaced randomly with amplitudes $\epsilon_x = 0.5$ and $\epsilon_y = 0.5$. Note that the inlet and outlet nodes are held fixed.

$N_l = N(3M + 4)$. Segment junctions identified as nodes are arranged on a honeycomb lattice consisting of two displaced hexagonal lattices.

High arterial pressure, p_a, is maintained at the bottom nodes, and a lower venous pressure, p_v, is maintained at the top nodes, so that blood flows upward through the test section. Blood enters the test section through the bottom links and exits the test section through the top links. Bottom and top nodes are marked as squares in Fig. 6.7a. The pressures and capillary flow rates are periodic along the infinite strip. It is convenient to express the driving pressure difference in terms of an effective pressure gradient along the y axis, γ, defined by the equation:

$$\Delta p \equiv p_a - p_v = \gamma W \tag{6.29}$$

where W is the exterior width of the strip shown in Fig. 6.7a.

To implement stochastic uncertainty, the interior nodes, marked as circles in Fig. 6.7a, are displaced randomly in the xy plane. The maximum displacement in the x direction is $\epsilon_x a$, and the maximum displacement in the y direction is $\epsilon_y a$, where ϵ_x and ϵ_y are specified dimensionless coefficients and a is the hexagonal cell radius. A typical randomized network with $\epsilon_x = 0.5$ and $\epsilon_y = 0.5$ is shown in Fig. 6.7b. The radius of each capillary tube is:

$$a_c = \bar{a}_c[1 + \epsilon_a(\varrho - 0.5)] \tag{6.30}$$

where $\bar{a}_c$ is the mean radius, ϵ_a is a specified dimensionless coefficient and ϱ is a random uniform deviate taking values in the range $[0, 1]$. When $\epsilon_a = 0$, all capillary segments have the same radius.

6.4.2 Geometrical Construction

The programmable construction of the network is discussed elsewhere [27]. Briefly, the six nodes of a master hexagonal unit are repeated in two directions with base vectors $\mathbf{a}_1 = (2b, 0)$ and $\mathbf{a}_2 = (b, \frac{3}{2}a)$, as shown in Fig. 6.7a. Only hexagons that fit inside the test section and one column to the right of the test section are retained. Next, unique capillary segments are identified by running around each hexagon and accepting new segments or rejecting existing segments. Possible junction images are also recorded and indices for the interior, bottom, left and right segments are introduced. Finally, unique nodes constituting the link end points are compiled and associated with the links through a connectivity matrix. The three closest neighbors of each interior node are identified and the interior nodes are randomly perturbed. Other indices are employed to indicate left, right, bottom or top nodes and periodic images, as needed.

6.4.3 Numerical Method

To develop an iterative solution scheme for computing the junction pressures, we write a mass balance at each interior node involving four nodal pressures multiplied by appropriate coefficients. We then solve for the pressure at the central node and iterate based on the derived expression. The algorithm involves guessing p_1, p_2, and p_3 and solving (6.3) for p_0 to obtain an improvement. The pressure is updated sequentially at all unique global nodes, and the iterations continue until the maximum correction falls below a specified threshold. The procedure is the counterpart of the point Gauss-Seidel (PGS) method for solving sparse systems of linear equations originating from Laplace's equation (e.g., [26]).

In the algorithm, initial guesses are made for the interior junction pressures and for the interior or outlet capillary discharge hematocrits. The discharge hematocrits of the inlet segments at the bottom of the test section are specified. After a preset number of inner Gauss-Seidel iterations for the nodal pressures, new capillary discharge hematocrits are calculated based on cell conservation at a converging bifurcation or an adopted partition law at a diverging bifurcation, and the segment blood viscosity is updated. Outer iterations are performed until the maximum correction falls below a specified threshold set to $10^{-6}\gamma W$. Typically, several hundreds inner iterations and less than 100 outer iterations are necessary at a computational cost that ranges between a few seconds to a few minutes of CPU time on a standard personal workstation.

6.4.4 Flow Through a Pristine Network

In the case of a pristine honeycomb lattice where the lengths of all capillary segments are equal and all capillary radii are the same, $L_c = a$ and $a_c = \bar{a}_c$, the capillary segment discharge hematocrits and blood viscosities are equal. Segments aligned with the y axis carry the same amount of fluid, while segments inclined with respect to the y axis carry half that amount of fluid. The balance equation (6.3) states that the pressure at the central node is the arithmetic mean of the pressure of the three neighboring nodes. This property is a discrete manifestation of the mean-value theorem for harmonic functions in a plane (e.g., [19]). In fact, the underlying difference equation is a numerical approximation to the Laplacian of a continuous function $\mathcal{P}$ in the xy plane, $\nabla^2\mathcal{P} = 0$ (graph Laplacian), with an error that is proportional to the third derivatives of $\mathcal{P}$ and is on the order of the square of the hexagon radius, a^2. Cursory inspection of the mass balance equations at the nodes reveals that the pressure distribution over the pristine test section is linear in y. This means that a three-dimensional plot of the pressure field along each capillary produces an inclined plane of hexagonal cells.

6.4.5 Dimensions and Parameters

We will present and discuss the results of numerical simulations in physical units corresponding to capillary blood flow. A reference capillary blood flow rate, Q_{ref}, and associated blood velocity, v_{ref}, can be defined based on Poiseuille's law:

$$Q_{\text{ref}} = \pi\bar{a}_c^2 v_{\text{ref}} = \gamma\frac{\pi\bar{a}_c^4}{8\mu} \tag{6.31}$$

where μ is the plasma viscosity, $\bar{a}_c$ is the mean capillary radius and γ is the effective pressure gradient defined in (6.29). Rearranging, we obtain:

$$\gamma = \frac{8\mu}{\bar{a}_c^2}v_{\text{ref}} \tag{6.32}$$

This equation is used to compute the pressure gradient γ from a specified blood flow velocity. For example, for $v_{\text{ref}} = 1$ mm/s, and $\bar{a}_c = 2.5$ μm, we find that $\gamma = 0.0044$ mm Hg/μm.

The standard parameters used in the simulations presented in Section 4.6 are radius of a hexagonal cell $a = 30$ μm; mean capillary radius $\bar{a}_c = 3.5$ μm; plasma viscosity $\mu = 1.85 \times 10^{-6}$ g/(μm s); reference blood velocity $v_{\text{ref}} = 1.0$ mm/s; and pressure gradient $\gamma = 0.0044$ mm Hg/μm. The pressure gradient, γ, derives from the reference blood velocity, v_{ref}. Similar results were obtained for other parameters in the physiological range of the microcirculation. With regard to the chosen standard radius of each hexagonal unit cell, 30 μm, we note that the diffusion distance of oxygen in healthy tissue is typically less than 100 μm.

6.4.6 Effective Hydraulic Permeability

A relative overall effective hydraulic permeability of the capillary network can be defined in terms of the capillary blood flow rate as:

$$\kappa_{\text{eff}} \equiv \frac{Q^{\text{tot}}}{NQ_{\text{ref}}} \tag{6.33}$$

where Q^{tot} is the flow rate of whole blood entering one period of the network through the N inlet segments. The total volumetric rate of red blood cells entering the test section is

$$Q_{\text{rbc}}^{\text{tot}} = (H_D)_{\text{in}}Q^{\text{tot}} = \kappa(H_D)_{\text{in}}NQ^{\text{ref}} \tag{6.34}$$

where $(H_D)_{\text{in}}$ is the inlet hematocrit.

6.4.7 Results and Discussion

Results of a simulation for a network with dimensions $N = 16$ and $M = 9$ are shown in Fig. 6.8. The standard physical parameters stated in Section 6.4.4, the geometrical parameters $\epsilon_x = \epsilon_y = 0.5$ and $\epsilon_a = 0.20$, and the inlet hematocrit $(H_D)_{\text{in}} = 0.4375$ are employed. Correlation (6.8) was used to determine the blood viscosity and correlation (6.23) was used to determine the partitioning of cells at divergent bifurcations. The relative effective hydraulic permeability of the network is found to be $\kappa_{\text{eff}} = 0.779$. As a reference, we note that the corresponding relative effective hydraulic permeability of a pristine network, $\epsilon_x = \epsilon_y = \epsilon_a = 0$, is somewhat, but not considerably, higher, $\kappa_{\text{eff}} = 0.806$. The nearly linear pressure distribution shown in Fig. 6.8b can be contrasted with the fast decaying distribution in branching tree networks discussed in Section 6.3. This feature clearly distinguishes homogeneous (area- or space-filling) from tree networks from the perspective of hemodynamics.

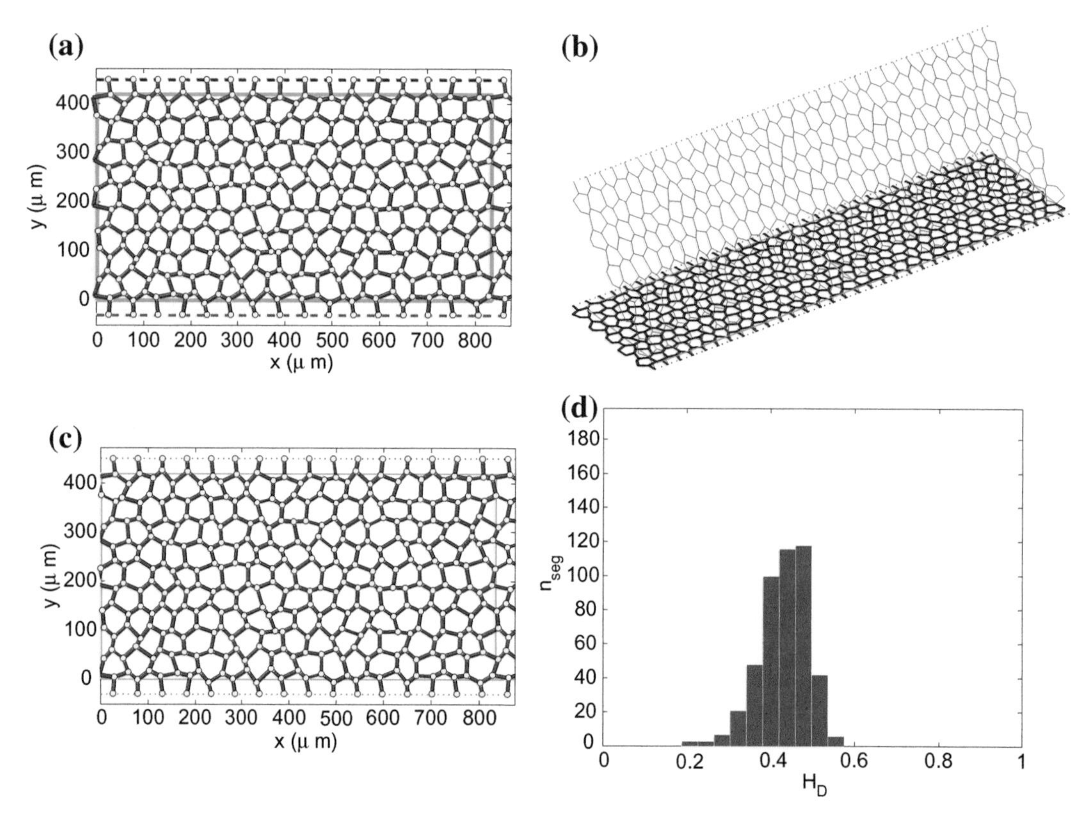

FIGURE 6.8 (a) A typical 16 × 9 network with node randomness parameters $\epsilon_x = \epsilon_y = 0.5$ and capillary radius randomness parameter $\epsilon_a = 0.20$. The thickness of the capillary segments shown scales with the capillary radius. (b) Pressure distribution over the network. (c) The thickness of the capillary segments shown scales with the capillary discharge hematocrit. (d) Histogram of the discharge hematocrit showing the number of segments inside a specified hematocrit window.

The thickness of the capillary segments plotted in Fig. 6.8c scales with the capillary discharge hematocrit. Close inspection reveals several thin segments carrying a low volume of red blood cells, and thicker segments carrying a higher volume of red blood cells. The histogram of the capillary discharge hematocrit shown in Fig. 6.8d reveals significant scattering around the inlet value, $(H_D)_{in} = 0.4375$. In fact, the discharge hematocrit of some capillary segments is almost half the inlet discharge hematocrit. Further simulations revealed that the broad spreading of the discharge hematocrit is due to the network deformation from the pristine honeycomb configuration and to random variations in the capillary tube radii. We conclude that moderate spatial variations of the hydraulic conductivity of the individual capillary cause a significant spread in the hematocrit.

Simulations were performed using test sections with different dimensions to assess the effect of the network size. For the conditions corresponding to Fig. 6.8, the relative effective hydraulic permeability of a 32×17 network was found to be $\kappa_{eff} = 0.773$, and that of a 64×32 network was found to be $\kappa_{eff} = 0.785$. Both are very close to the relative effective hydraulic permeability of the smaller 16×9 network described in Fig. 6.8. Histograms of the discharge hematocrit shown in Fig. 6.9a and c indicate that the larger networks exhibit similar behavior even though the number of capillary segments differs by a factor of 4. Histograms of the blood effective viscosity shown in Fig. 6.9b and d also reveal similar behavior. Under the conditions considered, the cells cause a moderate increase in the blood viscosity with respect to the cell-free plasma viscosity. We conclude that perturbed honeycomb networks of small size are able to capture the salient features of capillary blood flow.

6.4.8 Significance of the Bifurcation Law

An important objective of the simulations is to clarify the significance of the cell partitioning law at divergent bifurcations. For this purpose, simulations were performed on a 32×17 network whose nodes are perturbed with amplitudes $\epsilon_x = \epsilon_y = 0.5$, for uniform capillary radii, $\epsilon_a = 0$. The Klitzman and Johnson law (6.20) with a specified value of the exponent, q, was used at diverging bifurcations. Results for the capillary discharge hematocrit obtained using the standard parameters stated in Section 6.4.4 for inlet hematocrit $(H_D)_{in} = 0.4375$ are represented by the plus (+) symbol in Fig. 6.10a.

When $q = 1$, all segments have the same discharge hematocrit, equal to the inlet hematocrit, even though the capillary flow rates are not the same. As q increases, the range of variation of the hematocrit considerably widens inside a nearly linear envelop. The iterative solution procedure fails to converge approximately when $q > 4.5$ due to the exceedingly high discharge hematocrit of certain segments. The discharge hematocrits obtained using the empirical correlation (6.23) are represented by the $\times$ symbol near the right vertical frame in Fig. 6.10a. The data vary in a range corresponding to $q \simeq 1.5$. This value of q may be regarded as typical of homogeneous physiological networks.

The simulations confirm that the bifurcation exponent, q, has a profound effect on the distribution of the discharge hematocrit. The effect becomes more pronounced when the capillary tubes are assigned different radii, as shown in Fig. 6.10b for $\epsilon_a = 0.10$. In this case, the critical value of q below which the calculations fail to converge decreases to approximately 4.0. Simulations for lower inlet hematocrits reveal similar behavior, as shown in Fig. 6.10c and d

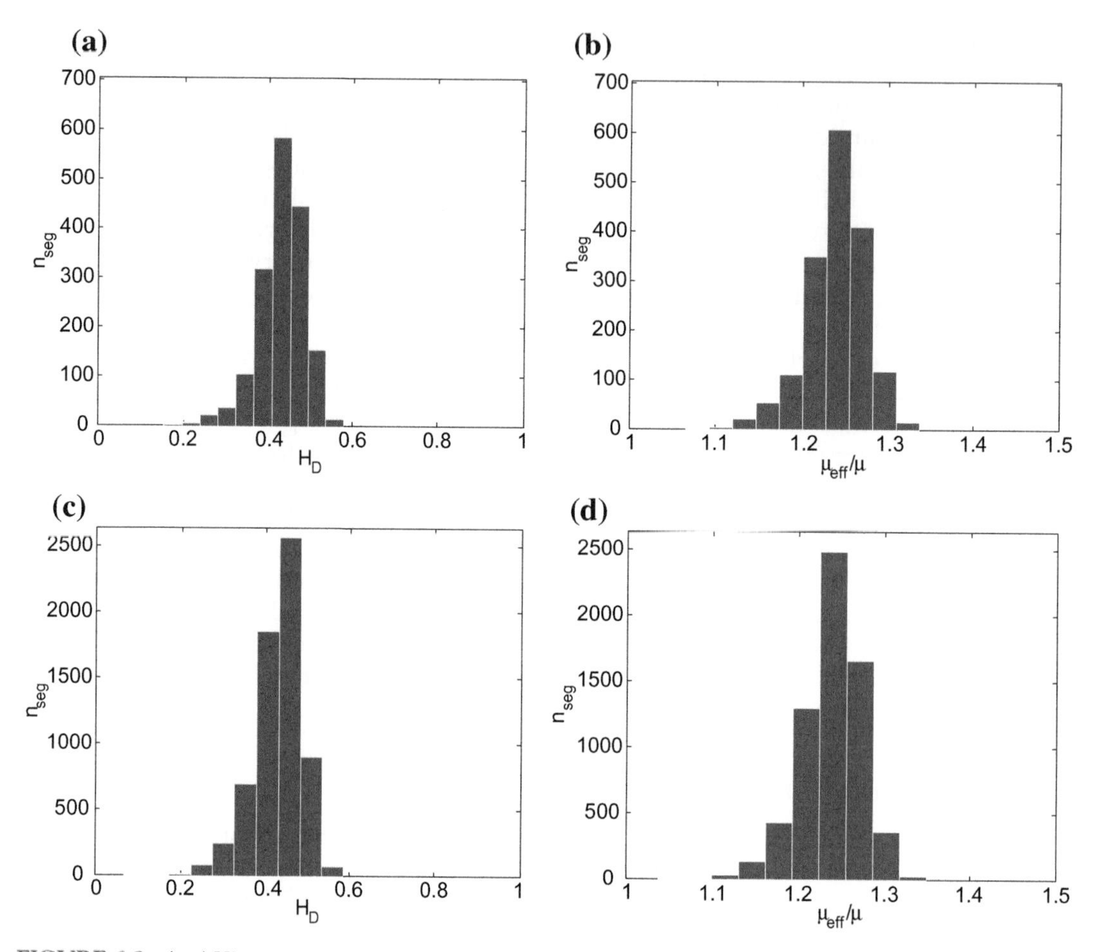

FIGURE 6.9 (a, c) Histograms of the capillary discharge hematocrit corresponding to the conditions described of Fig. 6.9 for a larger (a) 32×17 or (c) 64×33 network. (b, d) Corresponding histograms of the effective viscosity reduced by the plasma viscosity.

for inlet hematocrit $(H_D)_{\text{in}} = 0.2188$. When $q = 4.5$, some capillary segments are nearly entirely devoid of cells.

A direct visual impression of the broad range of the discharge hematocrit at high values of q can be obtained from Fig. 6.11a for a 16×9 randomized network. The thickness of the capillary segments shown in this figure scales with the capillary discharge hematocrit. For example, the thicknesses of all inlet segments at the bottom of the test section are all the same. The Klitzman and Johnson partitioning law (6.20) with $q = 4.5$ is employed at divergent bifurcations. Close inspection reveals capillary pathways connecting the inlet to the outlet. These pathways are reminiscent of percolation paths in damaged networks (e.g., [27]). Corresponding results are shown in Fig. 6.11b where the Pries et al. correlation (6.23) is employed. Close inspection reveals that the discharge hematocrit is distributed more uniformly across the capillary network.

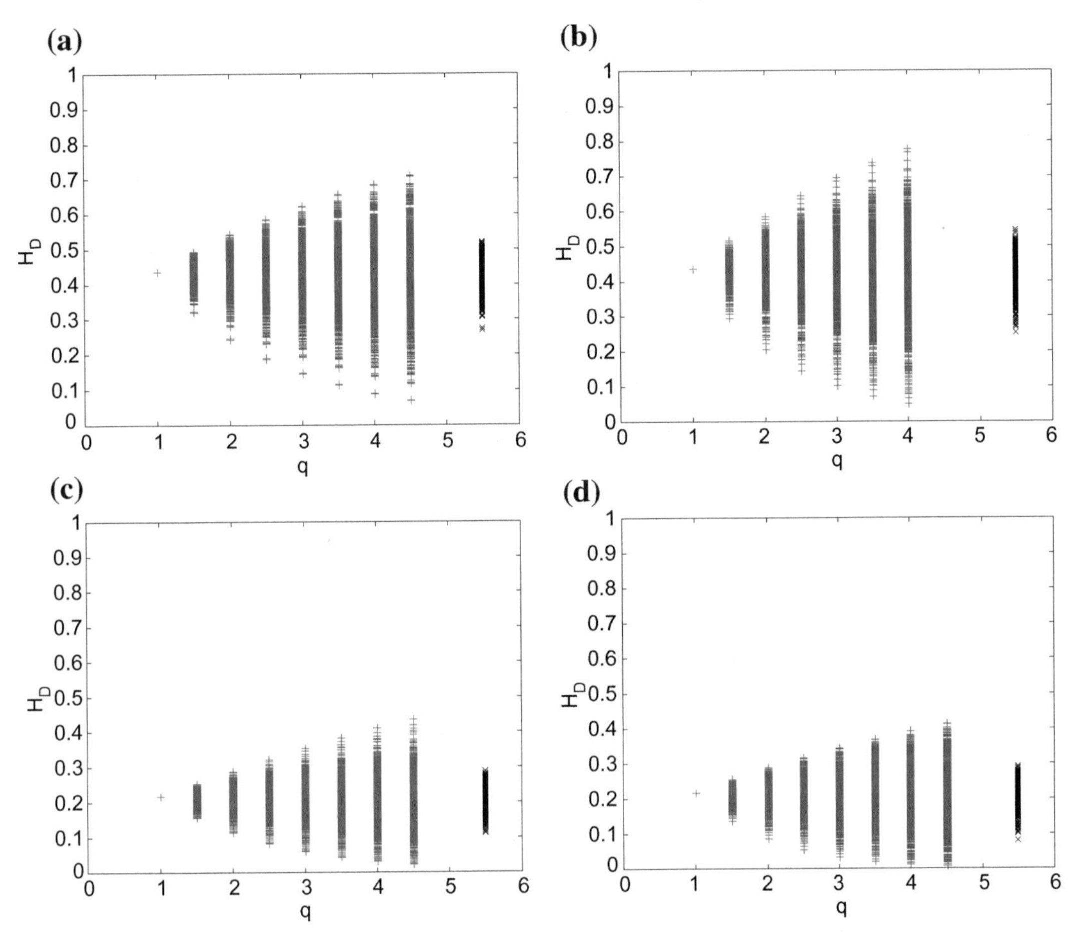

FIGURE 6.10 (a, b) Effect of the cell partitioning-law exponent, q, on the capillary discharge hematocrit of a 32 × 17 lattice for inlet hematocrit $(H_D)_{\rm in} = 0.4375$, node perturbation amplitude $\epsilon_x = \epsilon_y = 0.5$ and capillary radius perturbation amplitude (a) $\epsilon_a = 0$ or (b) 0.10. (c, d) Same as (a, b) but for inlet hematocrit $(H_D)_{\rm in} = 0.2188$.

Similar results were obtained for conditions other than the standard conditions discussed in Section 6.4.4. This includes different reference capillary blood velocities and different mean capillary radii. The results presented in this section are typical of flow through the model network under physiological conditions.

6.4.9 Significance of the Viscosity Correlation

All results presented previously in this section were obtained using correlation (6.8) for the blood viscosity. The results shown in Fig. 6.12a and b were obtained using the alternative correlation (6.11) corresponding to those shown in Fig. 6.9a and b. The comparison indicates that the viscosity law has a minor effect on the distribution of the capillary discharge hematocrit.

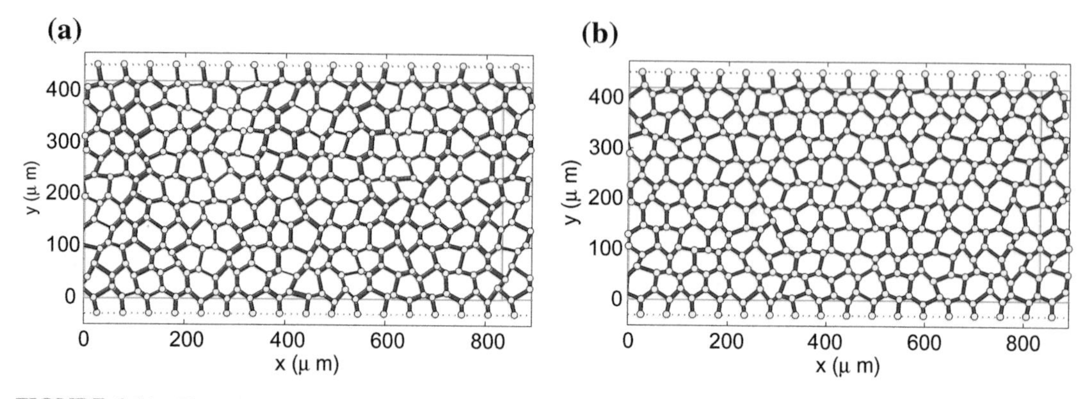

FIGURE 6.11 Flow in a 16 × 9 network with $\epsilon_x = 0.5, \epsilon_y = 0.5, \epsilon_a = 0.10$, for inlet hematocrit $(H_D)_{in} = 0.2188$. The Klitzman and Johnson law (6.20) with $q = 4.5$ is employed in (a) and the Pries et al. law (6.23) is employed in (b). The thickness of the capillary segments in the illustrations scales with the capillary discharge hematocrit.

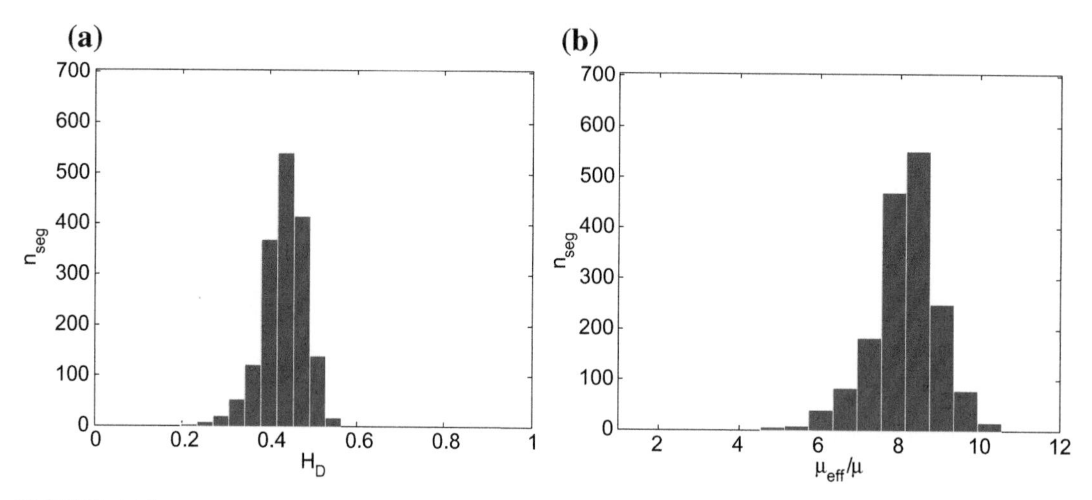

FIGURE 6.12 Histograms of (a) the capillary discharge hematocrit and (b) effective viscosity on a 32 × 17 network obtained using correlation (6.8) for the blood viscosity.

The main effect of the viscosity law is to shift the blood viscosity to higher values by a significant margin.

6.4.10 Discussion

We have presented numerical simulations of steady blood flow through an homogeneous, area-filling network of capillary tubes with multiple entrance points generated from a randomized honeycomb lattice. The simulations complement those for tree-like networks originating from a single arterial entrance point discussed in Section 6.3. Space-filling and tree networks are physiological relevant in different parts of the microcirculation. We have found that the discharge hematocrit varies over a broad range due to nonuniformities in the capillary length and radius distributions. The cell partitioning law at a bifurcation plays an important role in determining the distribution of blood cells through the network. The honeycomb network considered in the simulations is ideal for supplying oxygen and carrying nutrients into a growing or existing tissue. However, care must be taken that geometrical tolerances in the network construction are sufficiently small, otherwise areas of cell depletion and flow instability may arise.

6.5 EQUATIONS OF UNSTEADY BLOOD FLOW

While the vast majority of mathematical models and numerical simulations of blood flow through capillary networks are restricted to steady flow, the assumption of time independence requires careful examination and justification in view of the geometrical complexity of the capillary networks and the nonlinear dependence of the effective viscosity and nodal pressures on the discharge hematocrit in each segment. In fact, temporal fluctuations are often observed in the microcirculation. For example, *Krogh* [28] observed variations in the rate and direction of blood flow through the capillaries of frogs. Oscillations in the red blood cell velocity [29] and pressure [30] have been observed in animal and human tissue [31]. While most oscillations in the velocity of blood flowing through capillary networks are attributed to vasomotion or variations in the diameter of blood vessels, such unsteady behavior can also be caused by an inherent hydrodynamic instability.

Kiani et al. [32] developed a mathematical model of blood flow in a microvascular network based on the tracking of 'slugs' containing all red blood cells entering a vessel in a computational time increment. In fact, their formulation is an implementation of the method of characteristics for hyperbolic partial differential equations. The results for rat mesenteric networks revealed periodic fluctuations in the red blood cell velocity and discharge hematocrit. These predictions were in agreement with *in vivo* observations of temporal oscillations in hamster cremaster microvessels. *Carr and Lacoin* [33] and *Carr et al.* [34] subsequently derived the equations governing unsteady flow through capillaries and carried out numerical simulations using the method of characteristics under simplifying assumptions. Their results confirmed the occurrence of spontaneous, self-sustained oscillations under restricted conditions for simple networks with arcade topology. Recently, *Davis and Pozrikidis* [6] implemented a numerical method for computing unsteady blood flow through a branching, tree-like capillary network of arbitrary size. The results revealed the onset of spontaneous, self-sustained oscillations in the absence of external forcing. The transition to unsteady flow occurs by a supercritical Hopf bifurcation.

In this section, the equations governing unsteady blood flow are derived, and the implementation of a numerical method for computing flow through capillary networks with arbitrary topology is discussed.

6.5.1 Unsteady Flow Through a Straight Capillary

To develop the mathematical model, we consider the flow of a suspension of red blood cells through a straight tube with arbitrary cross-section, as shown in Fig. 6.13. As an approximation, cell-free zones near the vessel walls are ignored and the particles are assumed to be

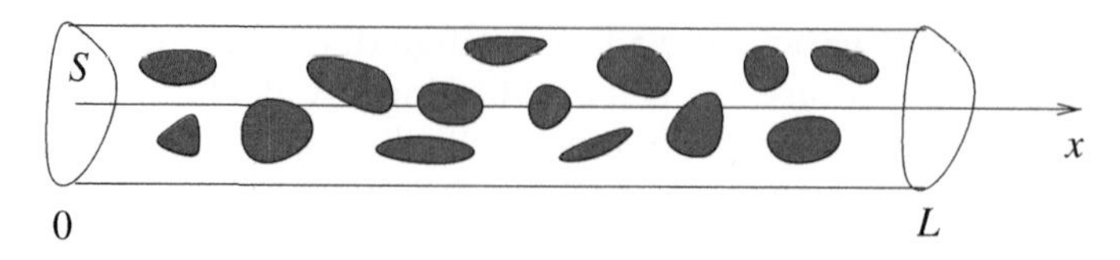

FIGURE 6.13 Illustration of unsteady blood flow through a straight capillary tube with arbitrary cross-section.

distributed uniformly over the entire cross-section. A cell population balance over a differential control volume confined between two parallel plates normal to the x axis requires that:

$$\frac{\partial H_T}{\partial t} + \frac{1}{S}\frac{\partial Q_{\rm rbc}}{\partial x} = 0 \tag{6.35}$$

where $H_T(x, t)$ is the tube hematocrit and $Q_{\rm rbc}(x, t)$ is the volumetric flow rate of the red blood cells. In terms of the mean cell velocity, $V_{\rm rbc} = Q_{\rm rbc}/(SH_T)$, we obtain:

$$\frac{\partial H_T}{\partial t} + \frac{1}{S}\frac{\partial (V_{\rm rbc} H_T S)}{\partial x} = 0 \tag{6.36}$$

where S is the available cross-sectional area.

It is useful to introduce the hematocrit ratio, $\chi \equiv H_T/H_D = U_m/V_{\rm rbc}$, where $U_m = Q/S$ is the mean suspension velocity. The evolution equation (6.36) takes the form:

$$\frac{\partial H_T}{\partial t} + \frac{1}{S}\frac{\partial (U_m H_D S)}{\partial x} = 0 \tag{6.37}$$

If the cross-sectional area, S, is constant, U_m is independent of x, yielding the simpler form:

$$\frac{\partial H_T}{\partial t} + U_m \frac{\partial H_D}{\partial x} = 0 \tag{6.38}$$

In terms of the discharge hematocrit, H_D, and the hematocrit ratio, $\chi(x, t)$, Eq. (6.38) becomes:

$$\frac{\partial (\chi H_D)}{\partial t} + U_m \frac{\partial H_D}{\partial x} = 0 \tag{6.39}$$

Expanding the first derivative in (6.39), we obtain:

$$\left(\chi + H_D \frac{\mathrm{d}\chi}{\mathrm{d}H_D}\right)\frac{\partial H_D}{\partial t} + U_m \frac{\partial H_D}{\partial x} = 0 \tag{6.40}$$

6.5.2 Circular Capillaries

The flow of a suspension of cells through a circular capillary of length L and radius a is described by Poiseuille's law with an effective fluid viscosity, $\mu_{\rm eff}$, determined by the volume fraction and velocity of the suspended cells and by the radius of the capillary. A pressure difference across the segment end nodes, Δp, produces a flow rate:

$$Q = -\frac{\mathrm{d}p}{\mathrm{d}x}\frac{\pi a^4}{8\mu_{\rm eff}} = \frac{\Delta p}{L}\frac{\pi a^4}{8\mu_{\rm eff}} \tag{6.41}$$

As for any particulate medium, the effective viscosity, μ_{eff}, is a proportionality coefficient relating the flow rate to the pressure gradient. Deviations of the segment from a circular cross-section, imperfections and entrance effects are incorporated into this coefficient. The mean velocity of the suspension is:

$$U_m = \frac{Q}{\pi a^2} = -\frac{\mathrm{d}p}{\mathrm{d}x}\frac{a^2}{8\mu_{\text{eff}}} \tag{6.42}$$

Conversely, the pressure gradient is given by:

$$\frac{\mathrm{d}p}{\mathrm{d}x} = -\frac{8\mu_{\text{eff}}}{\pi a^4}Q = -\frac{8\mu_{\text{eff}}}{a^2}U_m \tag{6.43}$$

where Q and U_m are constant in x. Integrating across the length of the capillary, L, we find that:

$$U_m = \frac{a^2}{8\tilde{\mu}}\frac{p_0 - p_L}{L} \tag{6.44}$$

where:

$$\tilde{\mu} \equiv \frac{1}{L}\int_0^L \mu_{\text{eff}}\,\mathrm{d}x \tag{6.45}$$

is an averaged viscosity.

6.5.3 Correlations

The correlations for the hematocrit ratio and effective viscosity discussed in Section 6.2 for steady flow also apply for unsteady flow. However, in the case of unsteady flow, the discharge hematocrit depends on time and position along each segment, $H_{D_i} = H_{D_i}(x,t)$, where the subscript refers to segment i in a network of N_{seg} segments. Consequently, correlation (6.7) yields the local and instantaneous hematocrit ratio $\chi(x,t)$ that appears in the conservation equation (6.40). The derivative $\mathrm{d}\chi/\mathrm{d}H_D$ in Eq. (6.40) depends only on a and is thus constant for each capillary segment. Similarly, correlations (6.8) and (6.11) express the local and instantaneous effective viscosity, $\mu_{\text{eff}}(x,t)$. The instantaneous average viscosity in each network segment is computed from Eq. (6.45).

6.5.4 Balances at Bifurcations

As in the case of steady blood flow, the dynamics at a bifurcation is governed by mass and population balances. Applied to the junction illustrated in Fig. 6.1b, the overall balance requires that $\sum_{i=1}^{3} Q_i = 0$, which can be expressed as:

$$\sum_{i=1}^{3}(P_0 - P_i)\frac{a_i^4}{\tilde{\mu}_i L_i} = 0 \tag{6.46}$$

where P_0 is the pressure at the central node. Conservation of the volume of the red blood cells requires that:

$$\sum_{i=1}^{3} Q_i^{\text{rbc}} = 0 \tag{6.47}$$

yielding:

$$Q_1 H_{D_1} + Q_2 H_{D_2} + Q_3 H_{D_3} = 0 \tag{6.48}$$

where the discharge hematocrit is evaluated at the end of the segment at the bifurcation. In the case of a converging bifurcation, Eq. (6.48) can be solved for the discharge hematocrit at the entrance of the descendent segment. This is identical to Eq. (6.15) for steady blood flow. In the case of unsteady blood flow, H_{D_1} and H_{D_2} are interpreted as the discharge hematocrit values for blood entering the junction and exiting segments 1 and 2, respectively.

In the case of a diverging bifurcation, the red blood cell concentration in each descendent segment depends on the corresponding fractional flow rates. This dependence is incorporated into the model as an effective boundary condition for H_D at the entrance of the children segments at the bifurcation. The correlations described in Sections 6.2.7–6.2.10 also apply for unsteady blood flow. Because $Q_1 > 0, Q_2 > 0$ and $Q_3 < 0$ when stream 3 splits into streams 1 and 2, H_{D_3} is identified with the discharge hematocrit of blood exiting segment 3 and entering the bifurcation, while H_{D_1} and H_{D_2} are identified with the discharge hematocrit exiting the bifurcation and entering segments 1 and 2, respectively.

6.5.5 Numerical Method

The numerical method used to determine the network dynamics involves the following steps:

- *Step 1:* A network with N_{seg} segments is constructed, the connectivity among the network segments is identified and each segment in the network is discretized into $N_{\text{div},n}$ divisions, where $n \in \{1, 2, \ldots, N_{\text{seg}}\}$. For example, for a tree network, the geometrical construction proceeds as in Section 6.3.1. The first segment of the network is then discretized into N_{div} divisions. To ensure that the grid spacing, Δx, is approximately constant for this branching network, the number of points used for the nth successive segment is given by $\texttt{floor}(L_n/L_1)N_{\text{pts}}$, where $N_{\text{pts}} = N_{\text{div}} + 1$.
- *Step 2:* The initial discharge hematocrit of the nth segment, $H_D(x_n, t = 0) \equiv H_{D,n}^{\text{init}}$, is specified, for $n \in \{1, 2, \ldots, N_{\text{seg}}\}$.
- *Step 3:* The pressure at each of the terminal nodes, p_i, is specified. Since $U_m \propto \Delta p$, the pressure difference across a segment has no effect in Stokes flow except to alter the dimensional time, which is inversely proportional to Δp.
- *Step 4:* The inlet discharge hematocrit for each terminal segment, $H_{D,i}^{(0)}(t)$, is specified to describe the flow into the network from a peripheral node.
- *Step 5:* The effective viscosity in each segment is calculated using Eq. (6.8) and then is integrated over each segment using the trapezoidal rule to yield the average effective viscosity according to Eq. (6.45).

- *Step 6:* The pressure at the central node, p_0, is computed from Eq. (6.46). The direction of flow in each segment is determined from the sign of $\Delta p_i \equiv p_i - p_0$. If $\Delta p_i > 0$ for only one segment, then that segment is the parent segment for a diverging bifurcation. If $\Delta p_i < 0$ for only one segment, then that segment is the descendent segment for a converging bifurcation. For a network with more than one interior node, the capillary pressures are found by Gauss-Seidel iterations.
- *Step 7:* The mean velocity, $U_{m,n}$, and flow rate, Q_n, through the nth segment are computed using Eqs. (6.41) and (6.44). This enables the computation of the fractional flow rates, ψ_1 and ψ_2, using (6.16).
- *Step 8:* The inlet discharge hematocrit of the nth segment, $H_{D,n}(x_n = 0, t)$ or $H_{D,n}(x_n = L_n, t)$ depending on the direction of flow, is calculated using Eq. (6.19) for a diverging bifurcation or from Eq. (6.15) for a converging bifurcation. If the node of segment n with higher pressure corresponds to a peripheral node, the inlet discharge hematocrit for that segment is determined by the appropriate boundary condition in *Step 4*.
- *Step 9:* The hematocrit ratio $\chi(x_n, t)$ is calculated for each segment using Eq. (6.7).
- *Step 10:* The discharge hematocrit at each grid point in each network segment is advanced by one time step using first-order upwind differences. Applying the evolution Eq. (6.40) at the point $x_{n,i}$ of the nth capillary segment at time instant t_j, and approximating $\partial H_D/\partial t$ with a forward difference and $\partial H_D/\partial x$ with a backward difference, we obtain a two-level explicit method expressed by:

$$H^{j+1}_{D,n,i} = cH^{j}_{D,n,i-1} + (1 - c)H^{j}_{D,n,i} \tag{6.49}$$

where:

$$c = \frac{|U^{j}_{m,n,i}|}{\chi^{j}_{n,i} + H^{j}_{D,n,i}(d\chi/dH_D)_n} \frac{\Delta t}{\Delta x_n} \tag{6.50}$$

This equation is implemented if $x_{n,1}$ corresponds to the end of segment n with higher pressure, such that $U^{j}_{m,n,i} > 0$. The direction of flow in a network segment can vary for a network of arbitrary topology due, for example, to a variation in the pressure at the peripheral nodes or to the onset of instability. If $x_{n,1}$ corresponds to the end of segment n with a lower pressure, such that $U^{j}_{m,n,i} < 0$, the two-level explicit method becomes:

$$H^{j+1}_{D,n,i} = cH^{j}_{D,n,i+1} + (1 - c)H^{j}_{D,n,i} \tag{6.51}$$

We note that $c > 0$ regardless of the direction of motion in each segment. The time step, Δt, is held constant in each simulation so that the Courant-Friedrichs-Lewy (CFL) stability criterion is met, $\max_{i,j,n}(c) \leqslant 1$. Values of c are monitored at each time step of every simulation to ensure numerical stability. For a typical simulation, $\max(c) < 0.3$.
- *Step 11:* Time is incremented, and values of the relevant variables are stored for post-processing.

The procedure is repeated from Step 3 onward for another time step.

For peripheral segments through which blood enters the network, the inlet discharge hematocrit is determined from the boundary condition at the network inlet given in *Step 4*. At the inlet to other segments, $H^{j+1}_{D,n>1,1}$ or $H^{j+1}_{D,n>1,N_{\mathrm{pts}}(n)}$ is determined from Eq. (6.15) or (6.19), with H_D at the end of the parent segment(s) and the flow rates Q_n determined at time t_j. Since the time step Δt is small and U_m depends weakly on H_D, the mean velocity in each segment is treated as a constant during each time step.

6.5.6 Single-Node Dynamics

The numerical method was applied to a simple network consisting of three capillary segments that meet at a central node, as illustrated in Fig. 6.1b. The length and radius of each segment are denoted by L_i and a_i, where $i = 1, 2, 3$. In the simulations presented in this section, we take $L_1 = L_2 = 80\ \mu\text{m}$, $L_3 = 100\ \mu\text{m}$ and $a_i = 6.0\ \mu\text{m}$. The pressure at each of the three terminal nodes measured in mm Hg is assumed to vary sinusoidally in time with period $T = 0.4$ s and angular frequency $\omega = 2\pi/T$:

$$p_3 = 8\left[1 + b\sin(\omega t)\right] \quad p_2 = 8\left[1 + b\sin\left(\omega t + \frac{\pi}{2}\right)\right]$$
$$p_1 = 8\left[1 + b\sin\left(\omega t + \frac{\pi}{4}\right)\right] \tag{6.52}$$

where $b = 0.4$ is the dimensionless amplitude of the pressure variations. Consequently, the nature of the junction flow changes periodically in time from diverging to converging. If blood flows into the network from a peripheral node, the inlet discharge hematocrit entering the corresponding network segment is:

$$H^{(0)}_{D,1} \equiv H_{D,1}(x_1 = L_1) = 0.35 \quad H^{(0)}_{D,2} \equiv H_{D,2}(x_2 = L_2) = 0.30$$
$$H^{(0)}_{D,3} \equiv H_{D,3}(x_3 = 0) = 0.40 \tag{6.53}$$

The temporal evolution of the nodal pressures and mean velocity in each segment is illustrated in Fig. 6.14. Because the pressure at each one of the three exterior nodes varies harmonically in time, the pressure at the central node, p_0, exhibits a sinusoidal response. The mean velocity in each segment also varies sinusoidally in time because the flow is driven by the pressure difference between the central and exterior nodes. The effective viscosity of each segment is affected by the discharge hematocrit according to Eq. (6.8) or (6.11). However, the variation of the effective viscosity does not induce noticeable deviations of the mean velocity of each segment from the sinusoidal shape.

The evolution of the discharge hematocrit entering or exiting each segment at the junction, $H^{\mathrm{bif}}_{D_n}$, exhibits a more complicated behavior, as illustrated in Fig. 6.15. The discharge hematocrit at the junction varies in the range $0 \leqslant H^{\mathrm{bif}}_{D_n} \leqslant 0.37$. The sharp variations are due to changes in the direction of the flow in each segment as the nodal pressures are varied. The reversal in the flow direction can be accompanied by a transition from converging to diverging flow. The discharge hematocrit entering a segment at the junction is determined either from conservation of cell volume at the junction, expressed by Eq. (6.15), or by a cell partitioning law, expressed by Eq. (6.19).

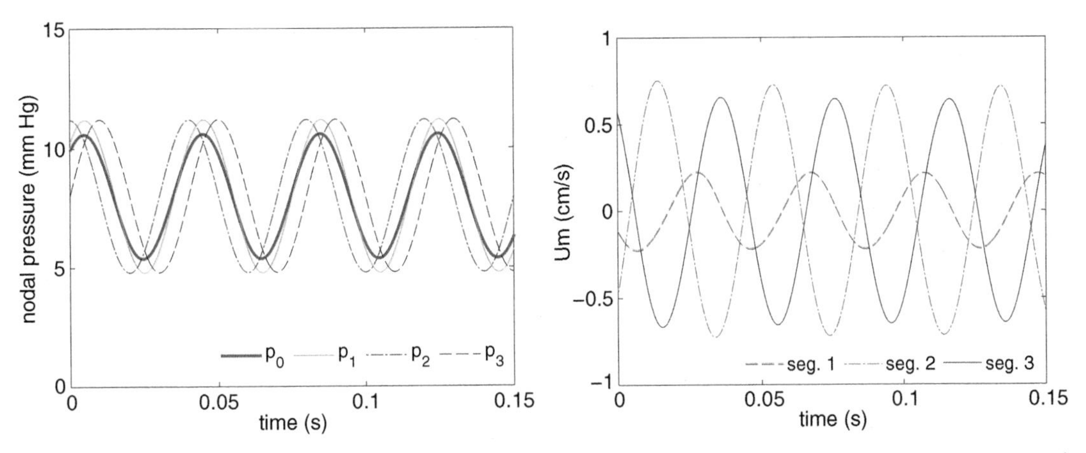

FIGURE 6.14 Temporal evolution of the (a) pressure at each node and (b) the mean velocity in each network segment.

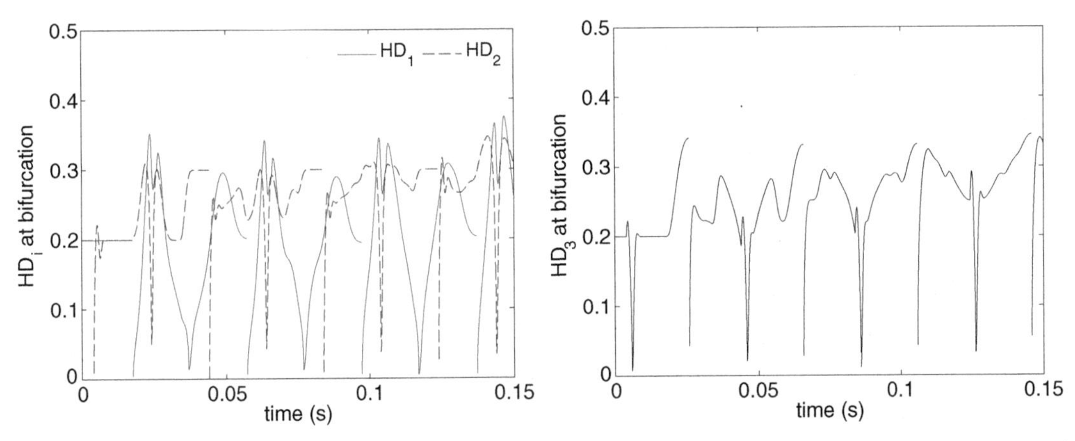

FIGURE 6.15 The evolution in the discharge hematocrit at the bifurcation for (a) segments 1 and 2 and (b) segment 3.

Further insight into the variation of the discharge hematocrit at the junction is provided by the instantaneous distribution of the discharge hematocrit along the network segments. The spatial variation in H_{D_1} in segment 1 is illustrated in Fig. 6.16 at several times. Initially, the flows in segments 1 and 2 converge into segment 3, and blood is supplied into segment 1 from the boundary node (right of the plot) at $H_{D,1}(x = L_1) = 0.35$, as determined by *Step 4* of the numerical procedure. A high-H_D front moves from right to left in the plot over the first two time intervals shown (curves 1 and 2). By $t = 0.0225$ s (curve 3), blood flows into segment 1 from the central node and the hematocrit wave moves to the right. Variations in $H_{D,1}$ also occur near $x_1 = 0$ for curve 3 due to the flow into the segment from the central node with a hematocrit value that differs from the initial hematocrit for segment 1. A similar behavior is observed at different times. The direction of propagation of the hematocrit front varies as the direction of flow changes, with a corresponding variation in $H_{D,1}$ near $x_1 = 0$ due to the different partitioning of red blood cells as the flow changes from converging to diverging.

The distribution of the discharge hematocrit along segment 3 at several times is illustrated in Fig. 6.17. Curves 1–3 describe fronts propagating toward the junction (right of the figure), corresponding to flow into segment 3 from the peripheral node with inlet hematocrit $H_{D,3}^{(0)} = 0.40$. For curves 4–6, the flow in this segment is reversed and blood enters the segment from the junction. The discharge hematocrit entering segment 3 from the junction is determined by a cell partitioning law and is considerably less than $H_{D,3}^{(0)}$. Significant variations in

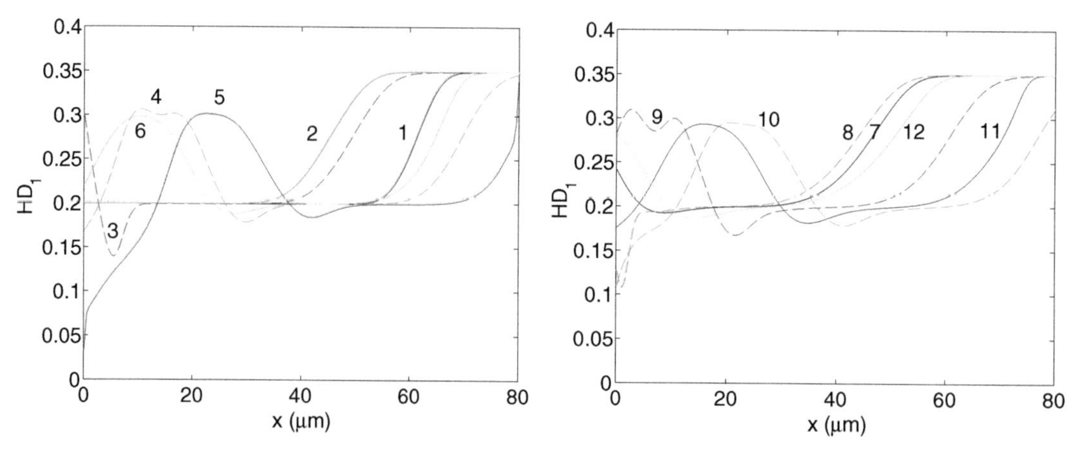

FIGURE 6.16 Spatial distribution of the discharge hematocrit in segment 1 at time intervals of $\Delta t = 0.0075$ s. The numbers order the curves with respect to time. The solid curves correspond to flow toward the junction (from right to left), while the dashed curves correspond to flow away from the junction (from left to right).

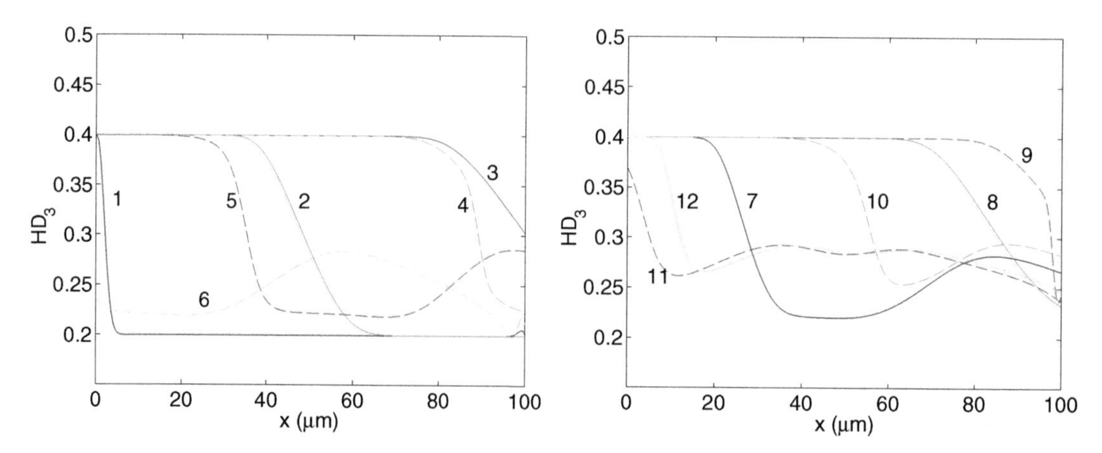

FIGURE 6.17 Distribution of the discharge hematocrit in segment 3 at time intervals of $\Delta t = 0.0075$ s. The numbers order the curves with respect to time. The solid curves correspond to flow toward the junction (from left to right), while the dashed curves correspond to flow away from the junction (from right to left).

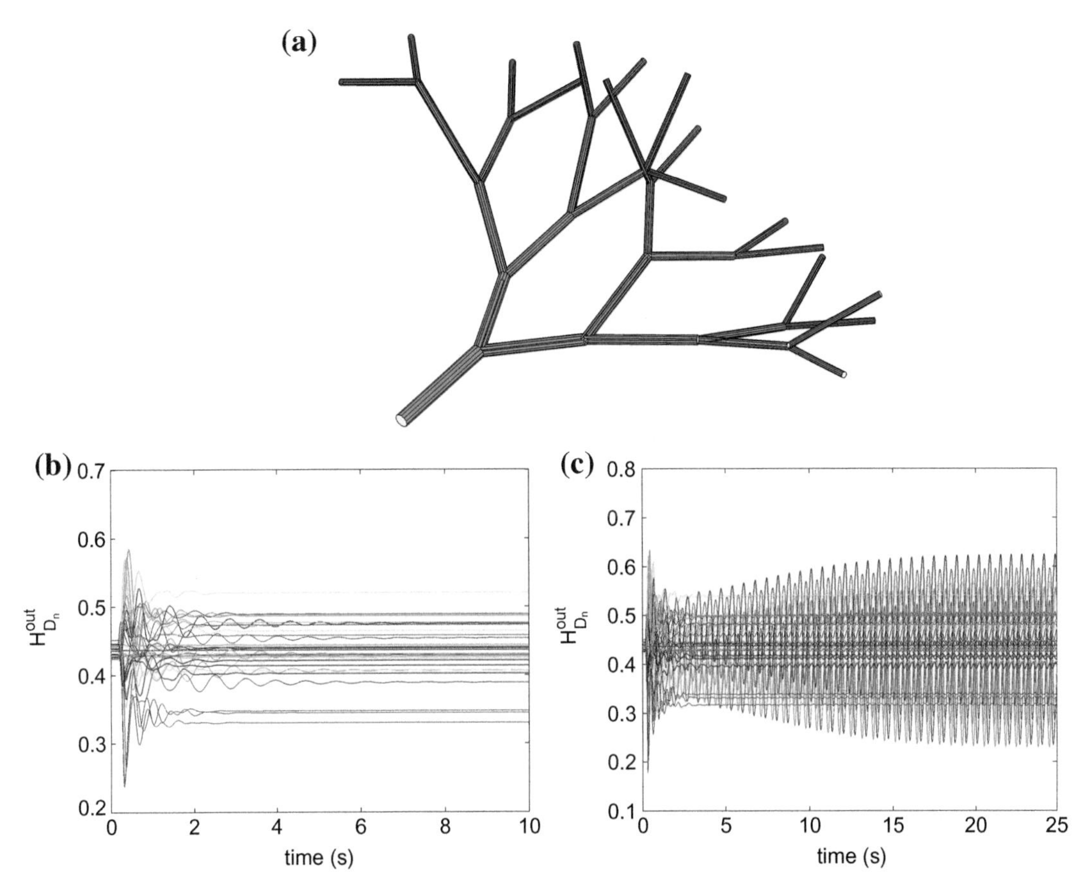

FIGURE 6.18 (a) Capillary tree with four generations of bifurcating capillaries ($m = 4$) constructed with $\theta_0 = 0.15\pi, \omega = 0.9, \epsilon_L = 0.20, \epsilon_\theta = 0.10, \epsilon_z = 1.0$ and $\epsilon_a = 0.10$. The radius of the inlet capillary tube is 6 μm and the radius of the outlet capillary tubes is 2.7 μm. The temporal evolution of the discharge hematocrit exiting each segment, $H^{out}_{D_n}$, is shown for a partitioning-law exponent of (b) $q = 4.0$, and (c) $q = 7.0$. The inlet hematocrit, $H_D = 0.4375$, and the *in vivo* viscosity correlation (6.11) is used.

H_{D_3} are therefore observed as the flow direction changes in response to the sinusoidal pressure variation at the peripheral nodes.

6.6 UNSTEADY FLOW THROUGH TREE NETWORKS

The numerical method for computing unsteady blood flow discussed in Section 6.5 was applied to the tree model of a capillary network discussed in Section 6.3 and is shown in Fig. 6.3. The geometrical construction of the network was discussed in Section 6.3.1. The *Klitzman and Johnson* [23] expression (6.20) was adopted for investigating the effect of the cell partitioning law.

Consider a network with four generations, as shown in Fig. 6.18a. The temporal evolution of the discharge hematocrit exiting each network segment, $H_{D_n}^{\text{out}}$, is shown in Fig. 6.18b and c for two values of the exponent in the cell partitioning law, q. Beginning with a uniform initial distribution of the discharge hematocrit, equal to the discharge hematocrit entering the first segment, $H_D(x_n, t = 0) \equiv H_{D,n}^{\text{init}} = H_{D_1}^{(0)} = 0.4375$, the network is allowed to evolve in time until a steady state is attained for $q = 1.0$. This value of the partitioning-law exponent corresponds to a distribution of the discharge hematocrit in proportion to the flow rate through the descendent segments at a bifurcation. The resulting hematocrit distribution is the starting point for further investigations of the network dynamics.

There is a critical value of the partitioning-law exponent, $q = q_0$, that delineates the regime of steady flow from the regime of self-sustained dynamics. For $q \leqslant q_0$, the network evolves to a steady state for any arbitrary initial condition. An example is shown in Fig. 6.18b for $q = 4.0$. Under these circumstances, there is a transient period where the discharge hematocrit entering each segment changes from the initial value to the value specified by the partitioning law in Eq. (6.20), and a steady state is subsequently attained. When $q > q_0$, the network exhibits spontaneous, self-sustained oscillations, as illustrated in Fig. 6.18c. The temporal variation in the discharge hematocrit in each segment alters the effective viscosity and thus the resistance of each segment to flow, thereby causing oscillations in the segment flow rates and nodal pressures. These spontaneous oscillations develop independent of the initial condition for the discharge hematocrit in each segment. The critical exponent q_0 decreases with a decrease in the diameter of the terminal segments or with a decrease in the diameter ratio for capillary segments in successive generations [6].

6.6.1 Steady Flow for Subcritical Exponents

If the pressures at the inlet and outlet nodes are held fixed, the flow ultimately attains a steady state for sufficiently small value of the partitioning-law exponent, $q \leqslant q_0$. Beginning with a uniform initial distribution of the discharge hematocrit, $H_{D_n}(x, t = 0) = H_{D_1}^{(0)} = 0.4375$, the system is allowed to evolve in time according to Eq. (6.40) until a steady state is attained. To facilitate comparison with computations for steady blood flow in Section 6.3, the numerical method is applied to a capillary tree of order $m = 7$. The physical configuration of the network is identical to that shown in Fig. 6.4a.

The distribution of the discharge hematocrit, tube hematocrit, hematocrit ratio, red blood cell velocity and relative shifted blood viscosity, $\eta_{\text{eff}} = \mu_{\text{eff}}/\mu$, are shown in Fig. 6.19 for the *in vitro* viscosity correlation. The solid and broken lines interpolate through the mean values of each generation. As discussed in Section 6.3.3 for steady blood flow, the variations in H_D, H_T, χ and η_{eff} within each generation become more pronounced as the partitioning-law exponent q becomes higher. While the iterative approach to steady flow is prone to divergence for sufficiently large q, a solution is found using the unsteady equations for any value of the partitioning-law exponent, $q \geqslant 1.0$.

The response curves plotted in Fig. 6.19 are repeated in Fig. 6.20 for the same network but with the *in vivo* viscosity correlation. Because the effective viscosity predicted by the *in vivo* correlation is significantly higher than that predicted by the *in vitro* correlation, the red-blood-cell velocity is significantly lower for the computations based on the *in vivo* correlation.

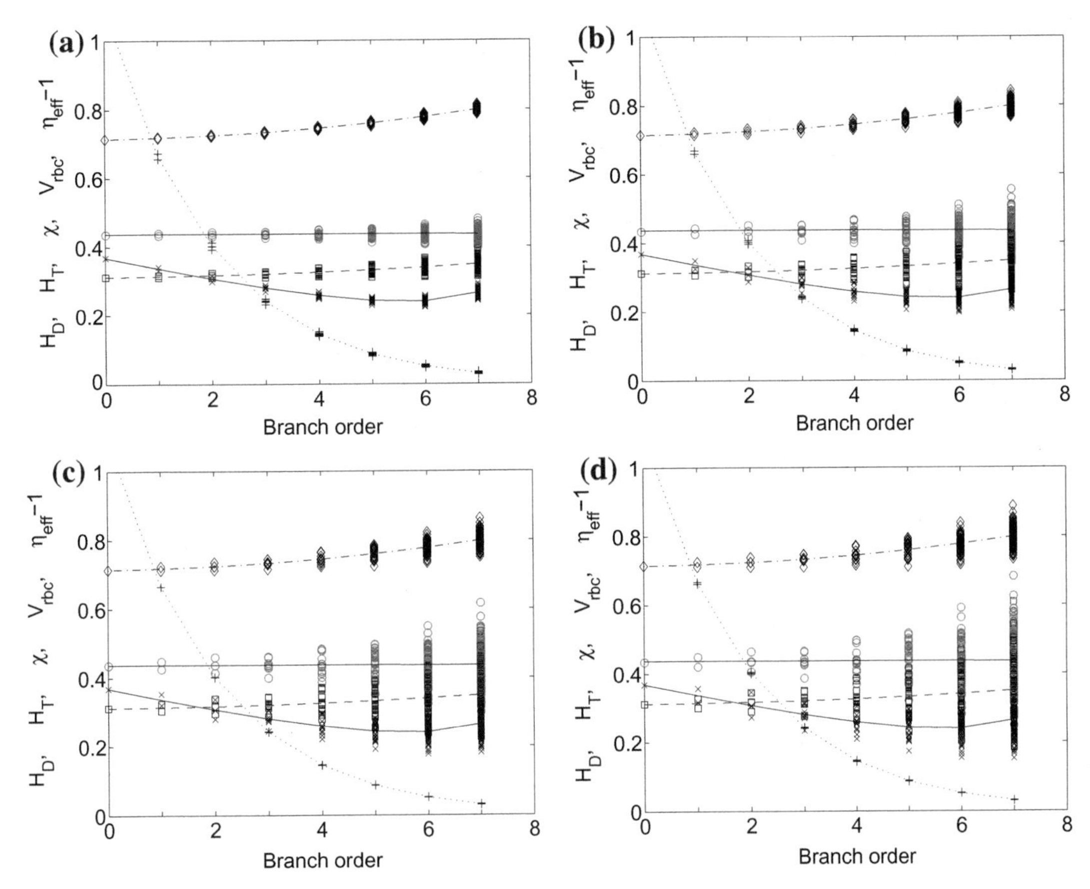

FIGURE 6.19 Unsteady blood flow through a capillary tree with seven generations of bifurcating capillaries ($m = 7$) constructed with $\theta_0 = 0.15\pi, \omega = 0.9, \epsilon_L = 0.20, \epsilon_\theta = 0.10$ and $\epsilon_z = 1.0$. The radius of the inlet capillary tube is 6 μm and the radius of the outlet capillary tubes is 3 μm. The partitioning-law exponent is (a) $q = 1.5$, (b) $q = 3.0$, (c) $q = 5.0$ and (d) $q = 10.0$. The inlet hematocrit $H_D = 0.4375$, and the *in vitro* viscosity correlation (6.8) is used. The distributions of H_D (circles), H_T (squares connected by the dashed line), hematocrit ratio, χ (diamonds connected by the dot-dashed line), red blood cell velocity in cm/s, V_{rbc} (crosses connected by the dotted line) and shifted relative viscosity, (×), are plotted against the branch order from inlet to outlet.

Moreover, the qualitative trends of the effective viscosity are different. For the *in vitro* model, the mean effective viscosity decreases monotonically from the first to sixth generation but increases slightly for the seventh generation. For the *in vivo* model, the mean effective viscosity increases monotonically from the first to last generation. The same trends are evident in Fig. 6.5 for computations based on the steady model. In addition, the variation of H_D, H_T and χ within each generation is less pronounced for the *in vitro* viscosity correlation, as is evident by comparing Figs. 6.19 and 6.20. Qualitatively similar results were found for many different networks that were constructed with random perturbations to the segment lengths and/or radii.

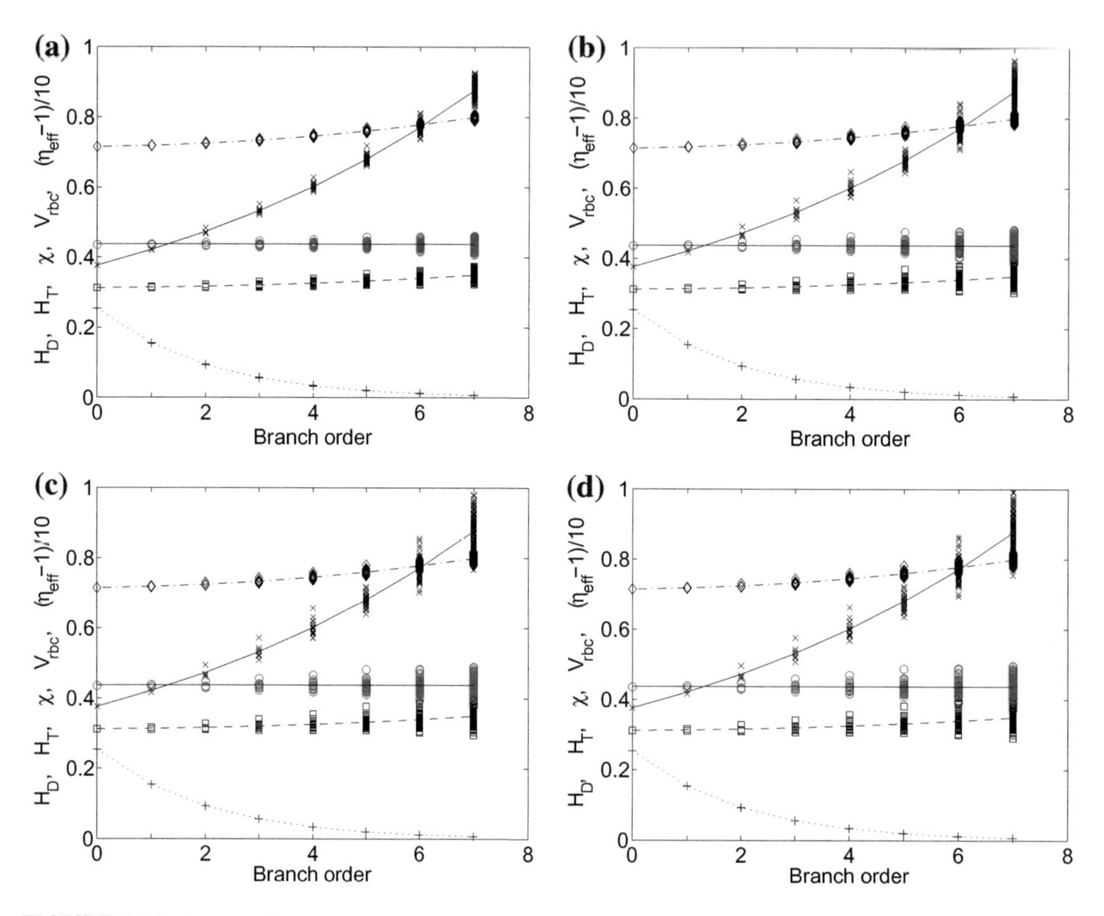

FIGURE 6.20 Same as for Fig. 6.19 but with the *in vivo* viscosity correlation (6.11).

6.6.2 Unsteady Flow for Supercritical Exponents

As the parameter q increases, a transition to time-periodic behavior occurs through a Hopf bifurcation [6]. For the cases considered in this work, a steady or time-periodic state is established at long times depending on the value of q. For $q < q_0$, self-sustained dynamics could not be forced even with a significantly nonuniform hematocrit in each segment in the initial condition.

An illustration of the Hopf bifurcation and unsteady dynamics is presented in Fig. 6.21 for a simple network with two generations subject to random perturbations in the segment lengths and radii. Beginning with a uniform initial condition, $H_{D,i}^{\text{init}} = H_D^{(0)} = 0.4375$, the network dynamics is followed until a stable steady or time-periodic state is found. Once time-periodic behavior is established, the amplitude of the oscillations:

$$\delta = 0.5\left[\max\left(H_{D,N_{\text{seg}}}^{\text{out}}\right) - \min\left(H_{D,N_{\text{seg}}}^{\text{out}}\right)\right] \tag{6.54}$$

is determined as a function of the exponent q, where the discharge hematocrit is evaluated at the exit to segment $n = N_{seg} = 7$.

Regardless of the correlation used for the effective viscosity, stable, time-periodic states bifurcate from stable, steady states. The transition from steady to self-sustained dynamics occurs at $q = q_0 \simeq 7.596$ if the *in vitro* viscosity correlation is employed, as seen in Fig. 6.21a. As q is increased further, the oscillation amplitude becomes increasingly pronounced. If the *in vivo* viscosity correlation is used for the same network, the transition to self-sustained

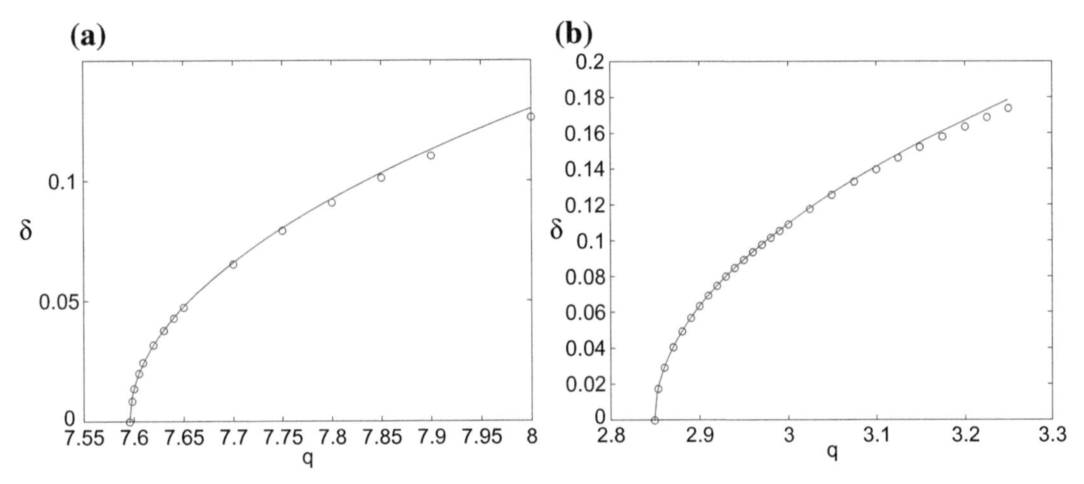

FIGURE 6.21 Amplitude of the time-periodic state, δ, plotted against the partitioning-law exponent, q, for a network with $m = 2, \epsilon_L = 0.2$ and $\epsilon_a = 0.1$. The parametric dependence of the time-periodic response in the vicinity of criticality, $q \to q_0^+$, is successfully fit to a quadratic polynomial, $q = q_0 + q_2\delta^2$. (a) *In vitro* viscosity correlation: $q_0 \approx 7.5955, q_2 \approx 23.795$ and (b) *in vivo* viscosity correlation: $q_0 \approx 2.849, q_2 \approx 12.530$.

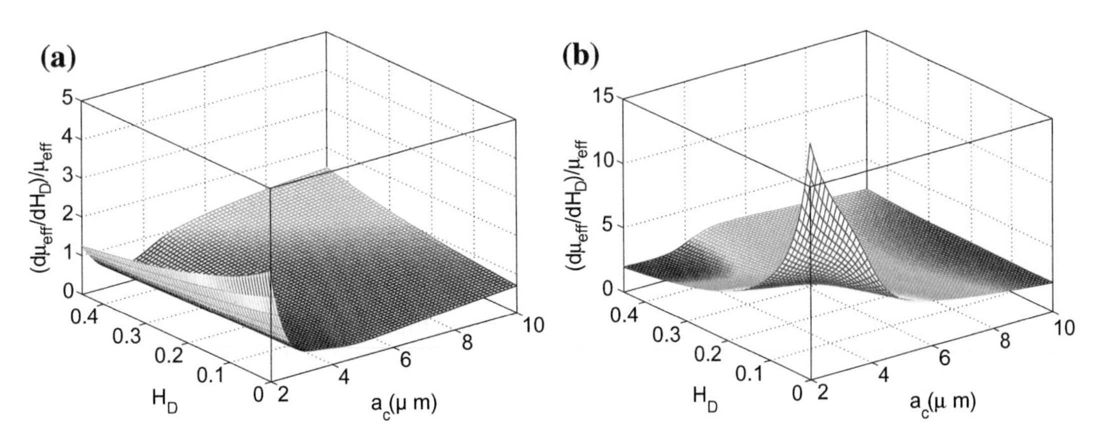

FIGURE 6.22 Relative change in the effective viscosity with the discharge hematocrit for correlations based on (a) *in vitro* measurements and (b) *in vivo* observations.

dynamics occurs at $q = q_0 \simeq 2.85$, as seen in Fig. 6.21b. The critical partitioning-law exponent is significantly smaller when the *in vivo* viscosity correlation is used because both the effective viscosity and the sensitivity of the effective viscosity to variations in the discharge hematocrit are larger than for the *in vitro* viscosity correlation. As seen in Fig. 6.21, the correlation based on *in vivo* observations produces a significantly higher blood viscosity. As seen from the plot of $(\partial \mu_{\text{eff}}/\partial H_D)/\mu_{\text{eff}}$ in Fig. 6.22, the relative change in the effective viscosity with the hematocrit is also significantly higher for the *in vivo* correlation.

According to the Hopf bifurcation theory, only the even-numbered coefficients are nonzero in the asymptotic expansion of the bifurcation parameter, q, with respect to the amplitude, δ, of the time-periodic state:

$$q = q_0 + q_2\delta^2 + O(\delta^4) \tag{6.55}$$

where q_0 is the bifurcation exponent at the Hopf point (e.g., [35]). As determined from the successful quadratic polynomial fit to the parametric dependence of the stable, time-periodic response in the vicinity of criticality, $q \to q_0^+$, the transition to unsteady behavior illustrated in Fig. 6.21 corresponds to a supercritical Hopf bifurcation. We have found through direct computation that the qualitative features of the dynamics, including the supercritical Hopf bifurcation corresponding to a transition from steady flow to time-periodic behavior, are insensitive to the particular choice of parameters governing the network, including the geometric parameters and correlations used for the hematocrit ratio and effective viscosity.

6.6.3 State Space

A state map was constructed to document the critical value of the bifurcation-law exponent, $q = q_0$, for the onset of self-sustained oscillations in networks with different tree orders, m. Five randomly perturbed networks were generated with $\epsilon_L = 0.2$ and $\epsilon_a = 0.1$ for each value of m, and the mean value of q_0 is reported in Fig. 6.23. For $m = 1$, corresponding to one bifurcation and three segments, time-periodic response is not observed for any value of q. The regime of self-sustained, unsteady behavior is shown for $2 \leqslant m \leqslant 6$ in Fig. 6.23. The results demonstrate that q_0 increases with m. In all cases, the oscillations are most pronounced in the penultimate generation because the segment radii are smaller, and thus the influence of the discharge hematocrit on the effective viscosity for the segment is stronger. Good agreement with the predictions of the discrete cell model has been confirmed for the onset of oscillations in computer animations and comparisons of the segment flow rates [6].

6.6.4 Summary and Discussion

The numerical method discussed in Section 6.5 for computing unsteady blood flow was applied to the tree-like capillary network discussed in Section 6.3. It was shown through numerical simulations that spontaneous, self-sustained oscillations can occur for sufficiently high values of the bifurcation-law exponent, q, in the absence of external forcing. Decreasing the radius of the outlet capillary segments increases the effective viscosity and decreases the critical exponent for the onset of instability, q_0. Similarly, the use of the *in vivo* viscosity

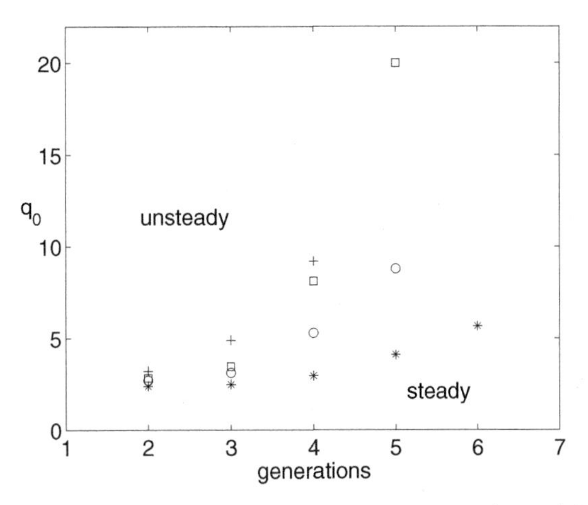

FIGURE 6.23 State-space map for a tree network with *in vivo* viscosity correlation delineates the regions of steady flow and self-sustained dynamics. The symbols correspond to networks with $a_{\text{inlet}} = 6.0\ \mu\text{m}$, $a_{\text{exit}} = 2.8\ \mu\text{m}$ (crosses); $a_{\text{inlet}} = 6.0\ \mu\text{m}$, $a_{\text{exit}} = 2.6\ \mu\text{m}$ (squares); $a_{\text{inlet}} = 6.0\ \mu\text{m}$, $a_{\text{exit}} = 2.4\ \mu\text{m}$ (circles); and $a_{\text{inlet}} = 8.0\ \mu\text{m}$, $a_{\text{exit}} = 2.4\ \mu\text{m}$ (asterisks).

correlation significantly decreases q_0 relative to that based on the *in vitro* viscosity correlation because both the effective viscosity and the variation of the effective viscosity with the discharge hematocrit are larger for the *in vivo* model. With all other network parameters fixed, increasing the diameter of the fist segment also lowers q_0. If the radius of the first segment is increased with m to maintain the same diameter ratio for subsequent segments regardless of the number of generations, q_0 is found to be essentially constant and independent of m. More generally, it has been found that the self-sustained dynamics is more significant in systems with a stronger dependence of the effective viscosity on the volume fraction of the suspended phase, such as in suspensions of rigid particles [7].

When the cell size is smaller than the capillary radius, the particulate nature of the fluid is not apparent, and q is near unity. As the cell size becomes comparable to the capillary radius, the cells are channeled into the daughter branches with the highest flow rate where 'winner takes all' and q tends to infinity. Accordingly, instability is expected to occur at the very last branches of a capillary network where the value of q is high. For a given network tree order, m, oscillations can be induced for a sufficiently high value of q by increasing the effective viscosity, by decreasing the ratio of the vessel diameter from one generation to the next, or by decreasing the diameter of the terminal vessels. With all other parameters fixed, oscillations are inhibited or suppressed by increasing m.

6.7 SUMMARY AND OUTLOOK

We have discussed the equations governing steady and unsteady blood flow through tree-like and area-filling capillary networks under the continuum approximation. The mathematical formulation culminates in a coupled set of nonlinear algebraic equations in the case of steady flow or partial differential equations in the case of unsteady flow. The nonlinearity is due to the dependence of the effective viscosity on the cell volume fraction expressed by the hematocrit. A linear system is obtained for fluids with constant viscosity.

In Section 6.3, we discussed steady flow through a tree-like network, and in Section 6.4, we discussed steady flow through a honeycomb network driven by a specified effective pressure gradient. Simulations of unsteady flow indicate the possibility of flow instability in sufficiently long tree networks under certain conditions (Section 6.6 and Refs. [6,7]). The instability first appears as a Hopf bifurcation of a steady solution branch where a steady flow becomes oscillatory. Recent simulations by the present authors have shown that a similar but stronger instability occurs in the case of a periodic honeycomb network discussed in Section 6.4.

The planar, area-filling network discussed in Section 6.4 can be extended to three-dimensional space-filling networks in several ways. In the simplest extension, a stack of two-dimensional networks are connected through lateral segments in an infinite or periodic configurations. In another extension, the hexagonal planar cells are replaced by one of the five known space-filling regular polyhedra. The effect of the cell partitioning exponent, q, and geometrical irregularities in these three-dimensional networks is expected to be similar to that in two-dimensional networks presently considered.

An algorithm was presented in Section 6.5 for computing unsteady blood flow through a capillary network with arbitrary topology. The evolution of the discharge hematocrit along each capillary segment is computed by integrating in time a one-dimensional convection equation using an upwind finite-difference method. The convection velocity is determined by the local and instantaneous effective capillary blood viscosity and the pressure drop across the vessel. A mass balance at each bifurcation provides us with a linear system of equations for the pressure at each node in the network. In the case of a converging bifurcation, the discharge hematocrit entering a descendant segment is determined by conservation of red blood cell volume. A partitioning law is required at a diverging bifurcation to relate the discharge hematocrit entering each descendant segment to the fractional flow rate. The effective blood viscosity, the ratio of the tube to the discharge hematocrit and the partitioning law are deduced from available correlations. The implementation of this method was demonstrated for a junction comprised of three segments and a central node.

An important assumption of the continuum model is that the red blood cells are distributed uniformly over the capillary cross-section. We have found that the results of the continuum formulation agree remarkably well with those of a discrete model where the motion of individual cells is tracked from inlet to outlet [6]. This agreement is excellent even when severe oscillations appear and the hematocrit approaches zero over most of a capillary segment. This agreement suggests that the continuum model furnishes accurate predictions even when the system parameters are such that the underlying assumptions are called into question.

References

[1] A.R. Pries, T.W. Secomb, T. Geßner, M.B. Sperandio, J.F. Gross, P. Gaehtgens, Resistance to blood flow in microvessels in vivo, Circulation Research 75 (1994) 904–915.

[2] A.R. Pries, T.W. Secomb, P. Gaehtgens, Biophysical aspects of blood flow in the microvasculature, Cardiovascular Research 32 (1996) 654–667.

[3] A.R. Pries, T.W. Secomb, Blood flow in microvascular networks, in: R.F. Tuma, W.N. Duran, K. Ley (Eds.), Handbook of Physiology: Microcirculation, second ed., Academic Press, San Diego, 2008, pp. 3–36.

[4] C. Pozrikidis, Numerical simulation of blood flow through microvascular capillary networks, Bulletin of Mathematical Biology 71 (2009) 1520–1541.
[5] D. Obrist, B. Weber, A. Buck, P. Jenny, Red blood cell distribution in simplified capillary networks, Philosophical Transactions of the Royal Society A 368 (2010) 2897–2918.
[6] J.M. Davis, C. Pozrikidis, Numerical simulation of unsteady blood flow through capillary networks, Bulletin of Mathematical Biology 73 (2011) 1857–1880.
[7] J.M. Davis, C. Pozrikidis, Hydrodynamic instability of a suspension of spherical particles through a branching network of circular tubes, Acta Mechanica (2012). doi:10.1007/s00707-011-0575-y.
[8] C. Pozrikidis, Computational Hydrodynamics of Capsules and Biological Cells, Taylor & Francis, 2010.
[9] R. Karshafian, P.N. Burns, M.R. Henkelman, Transit time kinetics in ordered and disordered vascular trees, Physics in Medicine and Biology 48 (2003) 3225–3237.
[10] N. Tsafnat, G. Tsafnat, T.D. Lambert, A three-dimensional fractal model of tumour vasculature, in: Proc. 26th Ann. Int. Conf. IEEE EMBS, 2004, pp. 683–686.
[11] A. Bui, I.D. Šutalo, R. Manasseh, K. Liffman, Dynamics of pulsatile flow in fractal models of vascular branching networks, Medical and Biological Engineering and Computing 47 (2009) 763–772.
[12] N. Safaeian, M. Sellier, T. David, A computational model of hemodynamic parameters in cortical capillary networks, Journal of Theoretical Biology 271 (2011) 145–156.
[13] J.W. Baish, R.K. Jain, Fractals and cancer, Cancer Research 60 (2000) 3683–3688.
[14] B.R. Masters, Fractal analysis of the vascular tree in the human retina, Annual Review of Biomedical Engineering 6 (2004) 427–452.
[15] L. Risser, F. Plouraboué, A. Steyer, P. Cloetens, G. Le Duc, C. Fonta, From homogeneous to fractal normal and tumorous microvascular networks in the brain, Journal of Cerebral Blood Flow and Metabolism 27 (2007) 293–303.
[16] D.J. Gould, T.J. Vadakkan, R.A. Poché, M.E. Dickinson, Multifractal and lacunarity analysis of microvascular morphology and remodeling, Microcirculation 18 (2010) 136–151.
[17] T.W. Secomb, Microvascular networks: 3D structural information, 2005. <http://www.physiology.arizona.edu/people/secomb/network.html>.
[18] A.S. Popel, P.C. Johnson, Microcirculation and hemorheology, Annual Review of Fluid Mechanics 37 (2005) 43–49.
[19] C. Pozrikidis, Introduction to Theoretical and Computational Fluid Dynamics, second ed., Oxford University Press, New York, 2011.
[20] A.R. Pries, T.W. Secomb, P. Gaehtgens, J.F. Gross, Blood flow in microvascular networks. Experiments and simulation, Circulation Research 67 (1990) 826–834.
[21] A.R. Pries, D. Neuhaus, P. Gaetgens, Blood viscosity in tube flow: dependence on diameter and hematocrit, American Journal of Physiology 263 (1992) H1770–H1778.
[22] A.R. Pries, T.W. Secomb, P. Gaehtgens, Biophysical aspects of blood flow in the microvasculature, Cardiovascular Research 32 (1996) 654–667.
[23] B. Klitzman, P.C. Johnson, Capillary network geometry and red cell distribution in the hamster cremaster muscle, American Journal of Physiology – Heart and Circulatory Physiology 242 (2) (1982) H211–H219.
[24] C. Pozrikidis, Passage of a liquid drop through a bifurcation, Engineering Analysis with Boundary Elements 36 (2012) 93–103.
[25] J.K.W. Chesnutt, J.S. Marshall, Effect of particle collisions and aggregation on red blood cell passage through a bifurcation, Microvascular Research 78 (2009) 301–313.
[26] C. Pozrikidis, Numerical Computation in Science and Engineering, second ed., Oxford University Press, New York, 2008.
[27] C. Pozrikidis, A.I. Hill, Conduction through a damaged honeycomb lattice, International Journal of Heat Mass Transfer 55 (2012) 2052–2061.
[28] A. Krogh, The Anatomy and Physiology of Capillaries, Yale University Press, New Haven, 1922.
[29] P.C. Johnson, H. Wayland, Regulation of blood flow in single capillaries, American Journal of Physiology 212 (1967) 1405–1415.
[30] C. Wiederhielm, J.W. Woodbury, S. Kirk, R.F. Rushmer, Pulsatile pressures in the microcirculation of frog mesentery, American Journal of Physiology 207 (1964) 173–176.
[31] G.P. Rogers, A.N. Schechter, C.T. Noguchi, H.G. Klein, A.W. Hiehuis, R.F. Bonner, Periodic microcirculatory flow in patients with sickle-cell disease, New England Journal of Medicine 311 (1984) 1534–1538.

[32] M.F. Kiani, A.R. Pries, L.L. Hsu, I.H. Sarelius, G.R. Cokelet, Fluctuations in microvascular blood flow parameters caused by hemodynamic mechanisms, American Journal of Physiology 266 (1994) H1822–H1828.
[33] R.T. Carr, M. Lacoin, Nonlinear dynamics of microvascular blood flow, Annals of Biomedical Engineering 28 (2000) 641–652.
[34] R.T. Carr, J.B. Geddes, F. Wu, Oscillations in a simple microvascular network, Annals of Biomedical Engineering 33 (2005) 764–771.
[35] G. Iooss, D.D. Joseph, Elementary Stability and Bifurcation Theory, second ed., Springer-Verlag, New York, 1990.

CHAPTER

7

Models of Cerebrovascular Perfusion

T. David and R.G. Brown

Blue Fern High Performance Computing Centre, University of Canterbury, Christchurch, New Zealand

OUTLINE

7.1 Introduction 254

7.2 From Arteries to Cells and Back Again (Cerebral Anatomy and Physiology) 254

7.2.1 Cerebral Arterial Structure 254

7.2.2 Circle of Willis 256

7.2.2.1 Variability of the CoW 256

7.2.3 Penetrating Arteries and the Cortex 256

7.3 Structure of Arterial Blood Vessels 257

7.4 A Simple Description of Cerebral Autoregulation 258

7.4.1 Organ Autoregulation 258

7.4.2 Local Autoregulation: 'Functional Hyperemia' 258

7.5 Vascular Trees and Their Numerical Simulation 259

7.6 Simple Models of Autoregulated Cerebral Perfusion 259

7.6.1 Blood Flow Through the Circle of Willis 259

7.6.2 Simple Models of Autoregulated Cerebral Perfusion 261

7.7 More Complex Models 264

7.7.1 Arteriolar Models of Autoregulation 264

7.7.2 Perfusion via the Capillary Bed 267

7.7.2.1 Oxygen Transport into the Cerebral Tissue 267

7.8 Conclusions 272

Acknowledgments 272

References 272

Transport in Biological Media

© 2013 Elsevier Inc. All rights reserved.

http://dx.doi.org/10.1016/B978-0-12-415824-5.00007-2

7.1 INTRODUCTION

As human beings, we need food to function on a day-to-day basis. Our organs are more particular about how often they receive nutrients: none more so than the brain. Muscular tissue in the legs, for example, can forego blood and the consequential flowing nutrients for up to an hour before succumbing to at least some dysfunction. Whereas the brain can last only a few minutes at most before cells die. For this purpose the brain has evolved a number of ways in which the cerebral cells are continually fed. Deep in the cerebrovascular tree one important way is to continually alter the diameters of the small arteries that feed the tissue. This is called 'functional hyperemia'. Over the past 5 years our research group has developed a numerical model that simulates functional hyperemia but also a number of other regulatory mechanisms such as the myogenic (or pressure induced) regulation that allow oxygen to constantly flow all the way from the major arteries of the brain including the circle of Willis down to the capillary beds deep inside the cortical structure of the cerebral tissue.

The chemical pathways that allow blood flow regulation to take place are complex and incompletely understood, yet in a world where vascular disease is one of the major contributors to non-traumatic death it is becoming increasingly important to not only understand how our vasculature functions normally but also pathologically. In order to understand the models that have so far been developed, we present a description of the major cerebral arteries and the structure of these arteries, especially that of the muscular wall. We then present mathematical and numerical models which simulate the way in which the arterial network allows blood to flow at a constant rate providing nutrients to the tissue.

The full simulation of blood flow regulation from major arteries to capillary bed is computationally intense, especially when the complex cellular mechanisms mapped onto arterial stress whose elements number into the million are taken into account. This requires considerable computational capacity and an algorithm which allows parallelism to enable results to come forward in a timely manner. The final section gives an indication of the computational challenges that cerebrovascular autoregulation demands and provides possible answers.

7.2 FROM ARTERIES TO CELLS AND BACK AGAIN (CEREBRAL ANATOMY AND PHYSIOLOGY)

7.2.1 Cerebral Arterial Structure

It is important to realize that the cerebral vasculature is a highly complex and variable structure and while the larger arteries between the heart and the proximal brain tend to follow similar patterns between individuals, the further into the cardiovascular system one delves, the more variable and difficult to classify the arteries become. This outline will therefore consider the more important cerebral arteries, providing a generalized structure compiled from a number of anatomical studies. Distal to the terminations of the arteries considered, are arteriole and capillary beds perfusing a particular structure in the brain and we shall look at these later on. The brain is fed essentially by four, then anastomosing into three arteries. They are the left and right internal carotids and the left and right vertebrals which anastomose into the basilar artery.

The major arteries of the brain are depicted in Fig. 7.1 [29,30]. To aid the reinforcement of the various arteries of the cerebral vasculature, Fig. 7.1c provides a schematic of the complex

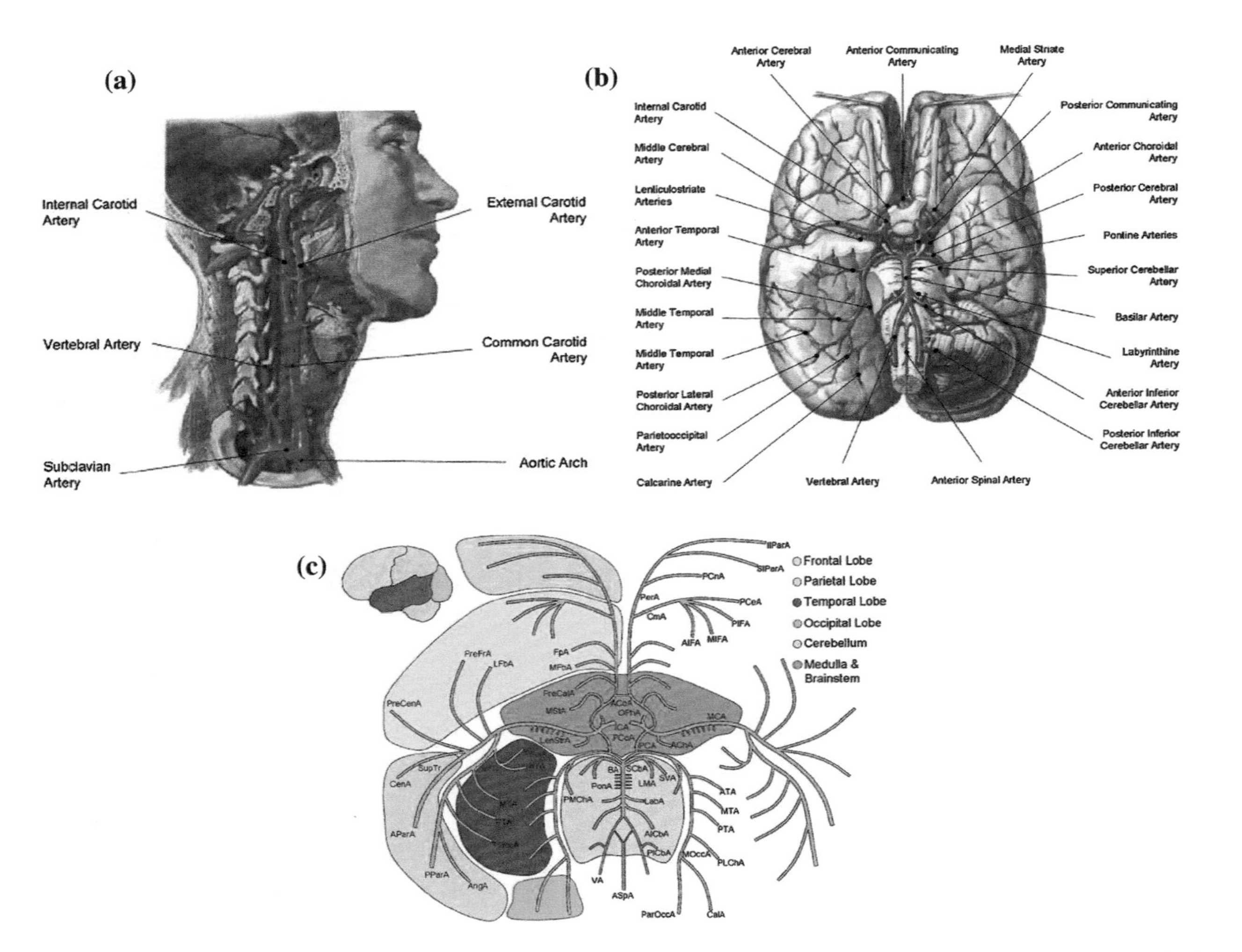

FIGURE 7.1 Topology of the major arteries in the cerebral vasculature. (a) A sagittal view of the major arteries supplying blood into the brain [29]. (b) An inferior transverse view of the brain, illustrating the afferent and efferent arteries visible from this viewpoint [29]. (c) A schematic of the cerebral arteries, labeled by their abbreviations and illustrating the general regions of the brain which they supply with blood [30].

3-D anatomical structure projected onto a 2-D plane. Also shown are the general cerebral territories supplied by each artery. At the level of the midbrain, just below the hypothalamus, is the major anastomosis in the brain, known as the circle of Willis (CoW).

7.2.2 Circle of Willis

The circle of Willis is an arterial ring sited just at the base of the brain (around eye level) and is completed by the anterior communicating artery (ACoA) and two posterior communicating arteries (PCoAs). The ACoA connects the two ACAs where the segment of the ACA, proximal to the ACoA, is known as the A1 segment and the segment distal to the ACoA is known as the A2 segment. The PCoA connects the ICA to the PCA on the left and right sides of the brain, respectively, where the segment of the PCA proximal to the PCoA is known as the P1 segment the segment distal to the PCoA is known as the P2 segment. Collateral Circulation and Collateral pathways for blood flow exist as anastomotic connections between supply systems and provide a means for blood to be rerouted if the primary blood supply to a given region of the brain is reduced. They can become important in situations when a larger afferent artery, or even one of the smaller cerebral arteries, becomes blocked. The effectiveness of collateral circulation depends upon a number of factors including the size of the anastomotic vessels, the length of time involved in the development of an occlusion and the size and location of the stenotic or occluded vessel.

7.2.2.1 Variability of the CoW

There have been a number of studies performed on the anatomy of the CoW, investigating the particular anatomical configuration, dimensions of the blood vessels, and the relationship between CoW geometries and neurological dysfunction among the general population. Details of the experiments and their related publications can be found in [1].

Variations among the human population in the anatomic configurations of the circle of Willis have been found in a number of studies. All of the studies, however, indicate that the presence of a complete circle of Willis aids in the ability to resist neurologic impairment. In the study performed by *Alpers et al.* [2], 837 brains were examined, of which 350 were tabulated in terms of anatomical configurations. The findings of the study included the presence of variations as either unilateral or bilateral (where appropriate) and recorded the prevalence of a variation in combination with others. The important results of the study are that only approximately 50% of the brains examined possessed a normal complete CoW. Similar results are reported in [3].

7.2.3 Penetrating Arteries and the Cortex

As in the case of the heart, the brain is essentially perfused from the 'outside in'. The pial arteries lie in the folds of the surface of the brain (called sulci) and upon reaching a certain diameter, change direction by 90° and penetrate the cortical tissue. These penetrating arteries lying perpendicular to the surface of the cortex then bifurcate into a complex network of small arteries. Work by *Lauwers et al.* [4,5] provides a significant amount of both visual and statistical data concerning the makeup of the arterial network and the capillary bed. This set of data is useful in providing a framework for developing numerical approximations to both the topology and physical size of the vascular network.

7.3 STRUCTURE OF ARTERIAL BLOOD VESSELS

As blood is transported from the heart to the tissue, the blood vessels undergo repeated branching to progressively smaller sizes and increasing number. Although the blood vessels form a continuum, they can be generally classified into arteries, arterioles, capillaries, venules and veins. For the present study, the focus will be on the delivery of blood to the cerebral tissue and not its transport back to the heart via the venules and veins, and as a result only the structure of the first three categories will be outlined (see Fig. 7.2).

The tunica intima consist of a layer of endothelial cells, a delicate connective tissue basement membrane, a thin layer of connective tissue called the lamina propria and a layer of fenestrated (having openings) fibers called the internal elastic membrane. The internal elastic membrane separates the tunica intima from the next layer, the tunica media. The tunica media consists of smooth muscle cells arranged in a circular fashion around the blood vessel. The tunica media also contains variable amounts of elastic and collagen fibers, depending on the size of the vessel.

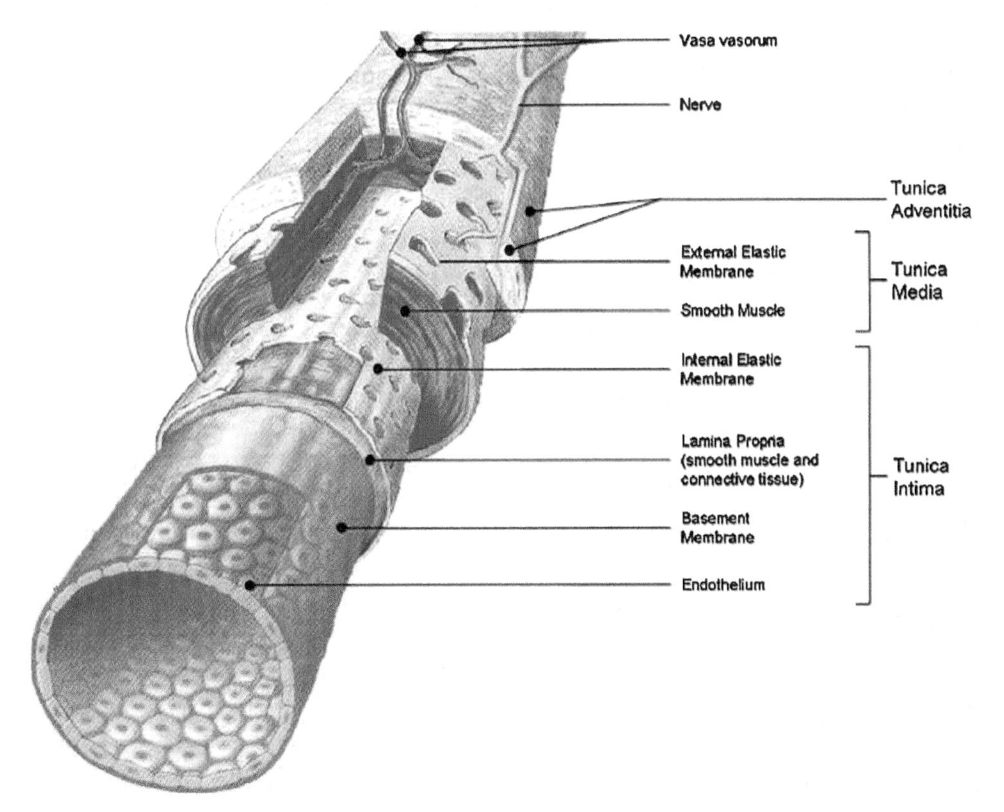

FIGURE 7.2 The generalized structure of the layers of the blood vessel wall, observable in most vessels except for the capillaries and venules [31].

Small arteries and arterioles are generally innervated to a greater extent than other blood vessel types. The afferent and efferent arteries outlined previously, supplying blood to the brain and distributing it to the various cerebral territories, would generally fall under the classification of elastic arteries. These arteries have the largest diameters and the highest blood pressures occurring within them, ranging between systolic and diastolic values. The tunica intima is relatively thick owing to the large amount of elastic tissue contained within it, although the fibers of the internal and external elastic membranes are typically not recognizable as distinct layers. The muscular arteries include most of the smaller unnamed arteries and their walls are relatively thick compared to their diameter, mainly because the tunica media contains 25–40 layers of smooth muscle. The presence of the large amount of smooth muscle allows these vessels to significantly regulate their diameter and thereby the resistance to blood flow and for this reason, the muscular arteries are often termed resistance arteries. Muscular arteries can range between 40 and 300 μm where the amount of smooth muscle decreases with the diameter such that those arteries which are in the 40 μm diameter range have only approximately three or four layers of smooth muscle in their tunica media. Arterioles have thick muscular walls and are in fact the primary site of vascular resistance. They range from approximately 40 μm to as small as 9 μm in diameter. The capillaries are the smallest of a body's blood vessels and most closely interact with the tissue. They consist primarily of a single layer of endothelial cells resting on a basement membrane. Outside the basement membrane is a delicate layer of loose connective tissue that merges with the connective tissue surrounding the capillary. Most capillaries range between 7 and 9 μm in diameter and they branch without a change in their diameter. Capillaries are variable in length, but in general they are approximately 1 mm in length. Substances cross capillary walls by diffusing through the endothelial cells, through fenestrae (small pores in endothelial cells), or the junction gaps between the endothelial cells.

7.4 A SIMPLE DESCRIPTION OF CEREBRAL AUTOREGULATION

7.4.1 Organ Autoregulation

The human body is subjected to a variety of environments during the day which can alter the systemic blood pressure in a 'global' way. Placing your hand in a bucket of ice is an easy way to increase blood pressure, or try counting aloud backwards from 100 subtracting 7 each time in front of your friends; unfortunately continued stress at work is another! These pressure changes require that body organs regulate their total vascular resistance in order to maintain a constancy of blood supply for nutrients (mostly oxygen).

7.4.2 Local Autoregulation: 'Functional Hyperemia'

Neurological functions such as visual acuity and speech (or even memory) are implemented by small neighborhoods of neurons interacting with each other. Hence in contrast to whole organ regulation the brain requires *local* blood flow changes to take place. The spatially varying metabolic rate from neuronal activity requires replenishment of both oxygen and glucose in the cerebral tissue. Variations in the local vascular resistance to changing requirements of nutrient flow are termed *functional hyperemia*.

7.5 VASCULAR TREES AND THEIR NUMERICAL SIMULATION

In order to investigate models of autoregulation at both organ and local levels, it is necessary to simulate in some sense the dynamics of a changing vascular resistance. We have modeled autoregulation at a number of levels: we have developed a model in which the vascular tree is treated a single lumped resistance with dynamic radius, and a model where each individual vessel throughout the vascular tree is modeled. To do this we have utilized previously published tree algorithms and added a dynamic component to model the changing resistance.

We adopted the tree branching algorithm developed for the abdominal fractal vascular network by *Steele et al.* [6]. This model is based on two variables: a power exponent k (describing the relationship between parent vessel radius r_p and daughter vessels radii d_{r1} and d_{r2}), and an asymmetry ratio γ (describing the relative ratio between two daughter vessels). For this particular study a binary tree emulates the vascular system stemming from a major artery, say the middle cerebral artery (MCA), for example. The values of k and γ change depending on the location of the vessel in the tree – i.e., values are different for arteries, arterioles and capillaries. The length and radius of a vessel are related by a length-to-radius ratio (denoted by rr_l and chosen as 20 for this model). Here we take values of k and γ as those given in [6].

Figure 7.3a shows a sketch of the basic network indicating an area of variable metabolic rate. The model has the ability to emulate actual vascular networks by using the statistical data of *Lauwers et al.* [5] (data of radii and length of all vessels in a small section of the cortex). The mean and the standard deviation from this data population is implemented into the code so that each arterial segment in the tree has a probability density function that emulates this physiological data. Additionally we are able to define trees in 3-D by finding the coordinates of a daughter vessel when it is born. With volume minimizing considerations, *Kamiya and Togawa* [7] showed in their studies that the angles between a parent and its daughters are a function of their radius. Details of how the tree is formulated in 3-D are available in [8]. Randomization of the plane in which the daughter arteries lie is done in a similar manner to that of *Schreiner and Buxbaum* [9]. A sample tree generated by the algorithm is depicted in Fig. 7.3b.

7.6 SIMPLE MODELS OF AUTOREGULATED CEREBRAL PERFUSION

7.6.1 Blood Flow Through the Circle of Willis

Numerical models of flow in the major arteries (notably the circle of Willis) have been completed using 3-D geometric data extracted from MR images. These models have provided patient-specific results and a detailed view of the complex flow fields induced by the curvature and bifurcations found in the major cerebral arteries. However from a clinical viewpoint, they do not provide information useful in either diagnosis or treatment due to the significant time required to solve the nonlinear equations of time-dependent motion of the blood. Particular efforts over the past 5 years have been made to utilize the 1-D equations of flow where the elasticity of the arteries is taken into account. *Sherwin et al.* [10] exemplified this and work has been initiated whereby the peripheral resistance is incorporated into the end boundary condition

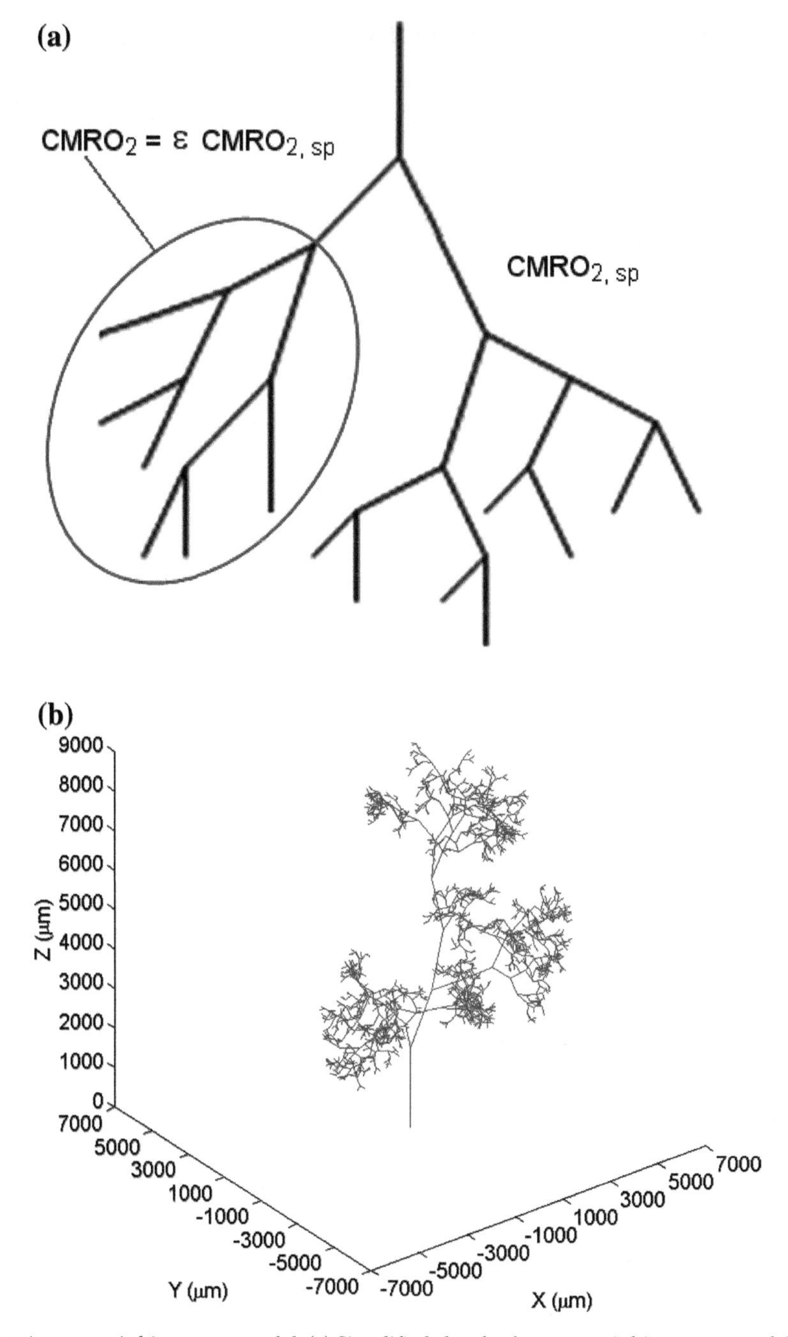

FIGURE 7.3 Asymmetric binary tree model. (a) Simplified sketch of asymmetric binary tree model, indicating an area of variable metabolic rate. (b) 3-D image of a numerically generated vascular tree.

along with a resistance and compliance capacitor to ensure no reflections back into the vascular network tree [11]. The time taken to solve the equations for even a full body of up to 55 arteries is insignificant compared to that for the Navier-Stokes equations in 3-D. Mass flow rates are provided as part of the solution and this is adequate enough to allow autoregulation to be simulated.

We provide below descriptions of autoregulated models for cerebral perfusion in increasing complexity.

7.6.2 Simple Models of Autoregulated Cerebral Perfusion

For arteries downstream of the circle of Willis, the diameters are such as to produce laminar axially dominated flow. Under the assumption therefore of Poiseuille flow, the resistance through a single artery flow is inversely proportional to the fourth power of the vessel radius as given in (7.1):

$$Q = \frac{\pi r_v^4}{8\mu l_v}\Delta P \tag{7.1}$$

Here r_v is the radius of the vessel, l_v its length, μ the blood viscosity and ΔP the pressure difference across the vessel. Simple linear models can be formed where the vessel is considered as a resistance and the arterial vasculature is a linked set of resistances. Pressures (which may be time-dependent) known *a priori* at both the inlets and outlets are sufficient boundary conditions to solve for the flows in the arterial network. Figure 7.4 shows a linked set of resistances which model the circle of Willis. Here the outer (efferent) arteries can be varied to account for autoregulation. Conservation of mass at each branch point in the network and the use of Poiseuille flow for each vessel (resistance) provide sufficient equations to solve both pressure at the branch points and flows in the vessels in the circle of Willis.

The autoregulation algorithm can be implemented in a variety of ways from a purely phenomenological viewpoint to one where highly complex cellular chemistry is modeled. In the initial work by *Alzaidi* [12] the autoregulation is modeled using control theory where the vasoconstriction and vasodilation phenomena are modeled using a PI controller governed by the following equation:

$$U(t) = K_p \mathrm{err}(t) + K_i \int_0^T \mathrm{err}(t)dt$$

The PI controller is a well-known control loop feedback mechanism, which attempts to correct the error between the current flowrate and a desired set point flowrate by calculating and then outputting a corrective action $U(t)$. In this case the desired set points are set to the reference average values. These reference conditions do not change with time. The proportional term has physiological significance as it represents the response when the vessel experiences fluctuations from the normal flow condition. The integral term acts as a low-pass filter so that the controller will not respond to high-frequency changes in the flow encountered as a result of the cardiac cycle. K_i and K_p are the integral and proportional gains, respectively, and err is the error representing the difference between the current flow and reference set point flow. The integral term of the controller equation forces the

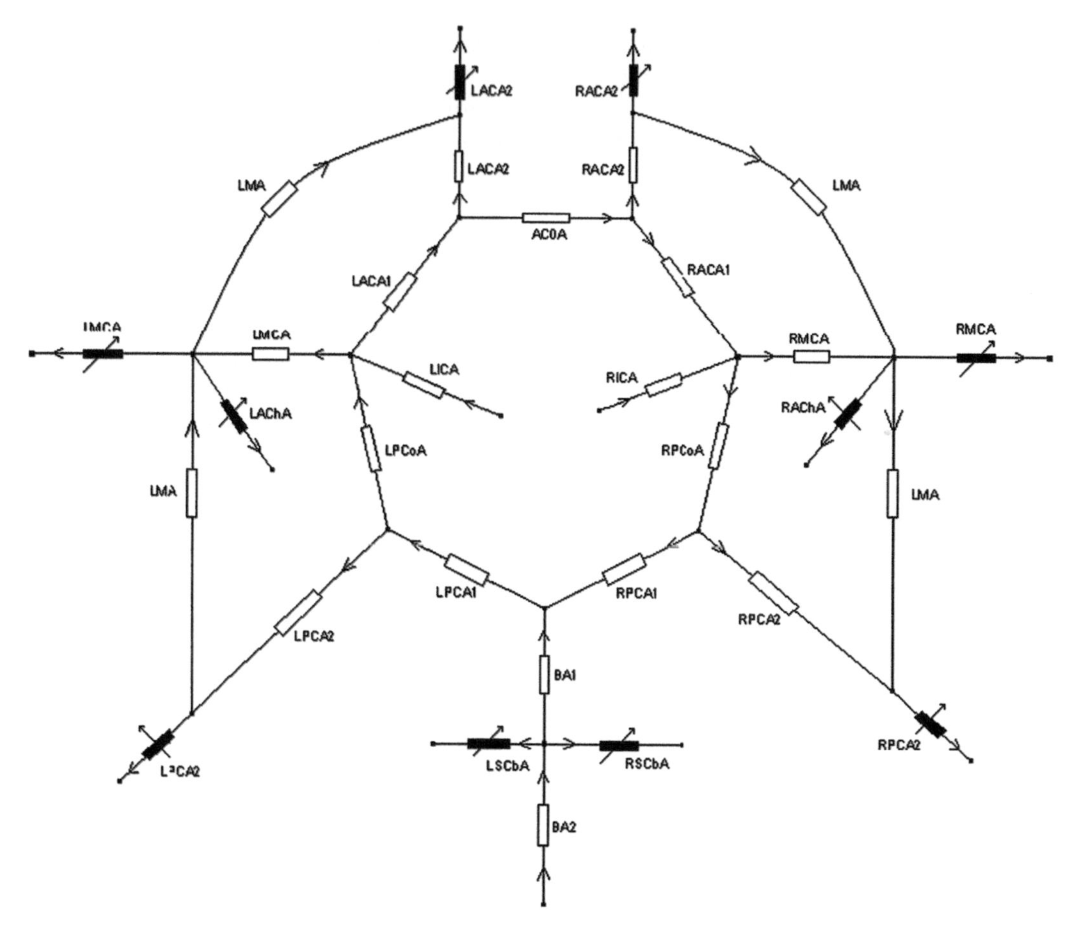

FIGURE 7.4 Linear model of CoW.

error to zero. Figure 7.5 shows the time-dependent flow response in the anterior cerebral artery to a 20 mmHg pressure drop. Included is the experimental data of *Newell et al.* [13].

These phenomenological models, although replicating physiological results, do not tend to provide any insight into the root cause and function of autoregulation. Nor can they shed any light on pathological conditions related to cerebral blood flow such as embolic stroke. Work by *David et al.* [14] has utilized a relatively simple autoregulatory model and implemented it into a vascular tree as described above in Section 7.5. The autoregulation mechanism is based on the phenomena that the metabolic consumption of oxygen produces carbon dioxide within the tissue. Variations in pH dilate the perfusing vessel and allow the CO_2 to be convected away with an increase in blood flow locally. The phenomenon can be modeled by a conservation equation for tissue CO_2 given by:

$$\frac{d}{dt}\mathrm{CO}_{2,\mathrm{tissue}} = \mathrm{CMRO}_2 + \mathrm{CBF}(\mathrm{CO}_{2,\mathrm{artery}} - \mathrm{CO}_{2,\mathrm{tissue}}) \tag{7.2}$$

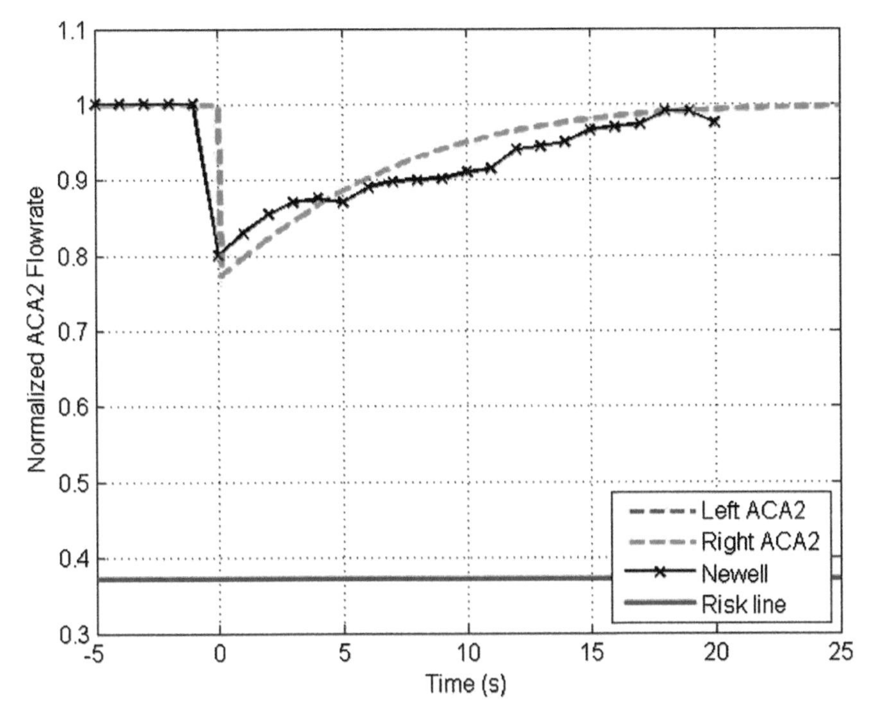

FIGURE 7.5 Flowrate response of the ACA2 arteries as a result of MAP drop of 20 mmHg.

where $CMRO_2$ is the cerebral metabolic rate of oxygen consumption, CBF the cerebral blood flow, $CO_{2,artery}$ is the carbon dioxide concentration of the incoming arterial blood and $CO_{2,tissue}$ is the carbon dioxide concentration in the tissue. The radius of a vessel in the cerebrovascular tree network is then modeled by a mean-reverting differential equation, given as:

$$\frac{dr}{dt} = \frac{1}{\tau}(CO_{2,sp} - CO_{2,tissue}) \tag{7.3}$$

where $CO_{2,sp}$ is the solution to the homogeneous state of the conservation equation (7.2) and τ is the time-constant governing the reaction speed of the vascular smooth muscle. This model can be implemented onto every small arteriole in the vascular tree. By altering the rate of oxygen consumption ($CMRO_2$) in a small neighborhood of the vascular tree, the arterioles can dilate (or contract) to ensure sufficient nutrients are perfused to the tissue as and when needed. Figure 7.6a shows part of the vascular tree where a steady state is reached after an increase of the metabolic rate by 50%. Figure 7.6b shows the relationship between the mean of the radii (where the metabolic rate is increased/varied) of the vascular tree and the metabolic rate required by the cerebral cortex.

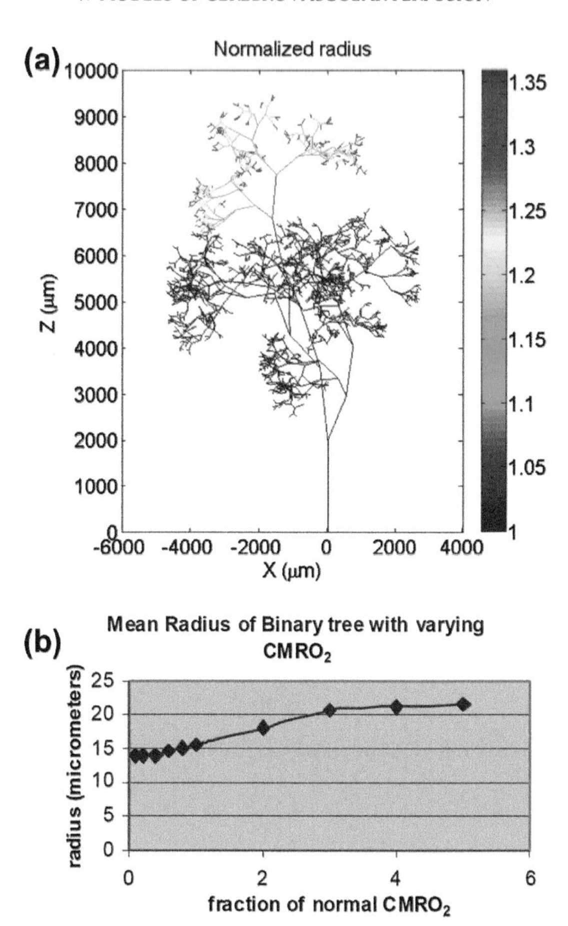

FIGURE 7.6 Results from autoregulating tree model. (a) A visualization of a tree after a steady state is reached after an increase of the metabolic rate by 50%. The colors indicate the normalized radius (note: before the increase, the vessels were at equilibrium and hence all dark blue). (b) Mean radius of the vessels in the tree at different levels of metabolic rate, relative to normal.

7.7 MORE COMPLEX MODELS

7.7.1 Arteriolar Models of Autoregulation

Motivated by the desire for a physiologically correct model that can incorporate the complex chemical pathways underlying metabolic autoregulation and hence give insights into the causes and function of autoregulation, we have been investigating the specific ion transports and cell apparatus involved. Of particular importance is a phenomenon known as functional hyperemia, a mechanism by which increased neuronal activity is matched by a rapid and regional increase in blood supply. This mechanism is facilitated by a process known as *neurovascular coupling* – the orchestrated intercellular communication between neurons, astrocytes and micro-vessels. An important step in this process is the release of potassium into the extracellular space by two potassium ion channels, BK and KIR. Previous models of neurovascular coupling have not included the mechanisms involving these channels. Here we provide such a model, which successfully accounts for the arteriolar dilation caused by the release of glutamate into the synaptic space between neurons. This model can achieve an approximate 20% vasodilation due to the rise in perivascular potassium concentration from 3 to 6 mM. It also successfully emulates the experimental finding that further increase in perivascular potassium can cause subsequent vasoconstriction. Our results suggest that the interaction of the changing smooth muscle cell membrane potential and the changing potassium-dependent resting potential of the KIR channel are responsible for this seemingly paradoxical effect.

This model includes all major mechanisms that occur in the neurovascular coupling process as proposed by *Filosa and Blanco* [15], starting with the release of glutamate at the synapse due to neural activity and ending with the dilation of an arteriolar segment. The major processes modeled are as follows: (1) Neural activity causes the release of glutamate into the synaptic space. (2) Glutamate then binds to metabotropic receptors on the top arm of the astrocyte leading to the release of IP3 into the cell. (3) IP3 stimulates the release of calcium from intracellular stores, causing a rise in cytosolic calcium that travels in a wave to the endfoot of the astrocyte. AA-derived EETs are also produced due to this rise in calcium. (4) Both EETs and increases in cytosolic calcium gate the BK potassium channels that are present in the endfeet of the astrocyte causing a release of potassium into the perivascular space between the smooth muscle cells surrounding the arteriole, and the endfoot. (5) The rise in potassium in the perivascular space further gates the KIR potassium channels in the smooth muscle cells (SMCs), causing them to open, extruding further potassium into this perivascular space, hyperpolarizing the SMC membrane in doing so. (6) This hyperpolarization closes voltage-operated calcium channels in the SMCs, preventing the influx of Ca^{2+}. (7) The decreased SMC cytosolic calcium then mediates the dilation of the arteriole through the consequent detachment of myosin-actin crossbridges and relaxation of smooth muscle. See Fig. 7.7 for a graphical depiction of this process.

The foundation for this model of the smooth muscle and endothelial cell processes follows that of *Koenigsberger et al.* [16,17], with equations added to include a KIR channel flux in the SMCs. We use a Hodgkin-Huxley formation and an equilibrium distribution of the open channel states from [18], a conductance value from [19] and a relationship for the resting potential of this channel from [20].

The cellular calcium concentration and the subsequent development of myosin-actin crossbridges in SMCs are related by the crossbridge phosphorylation and latch-state model of *Hai and Murphy* [21]. In their model, an elevation of calcium induces contraction through the formation of crossbridges between myosin and actin. Myosin can be in four possible states: free non-phosphorylated crossbridges [M], free phosphorylated crossbridges [Mp], attached phosphorylated crossbridges [Amp] and attached dephosphorylated crossbridges [AM]. It is the fraction of myosin in the two attached states that determines the active stress, and hence contraction that can be effected in the arteriolar wall. Equations describing the motion of the

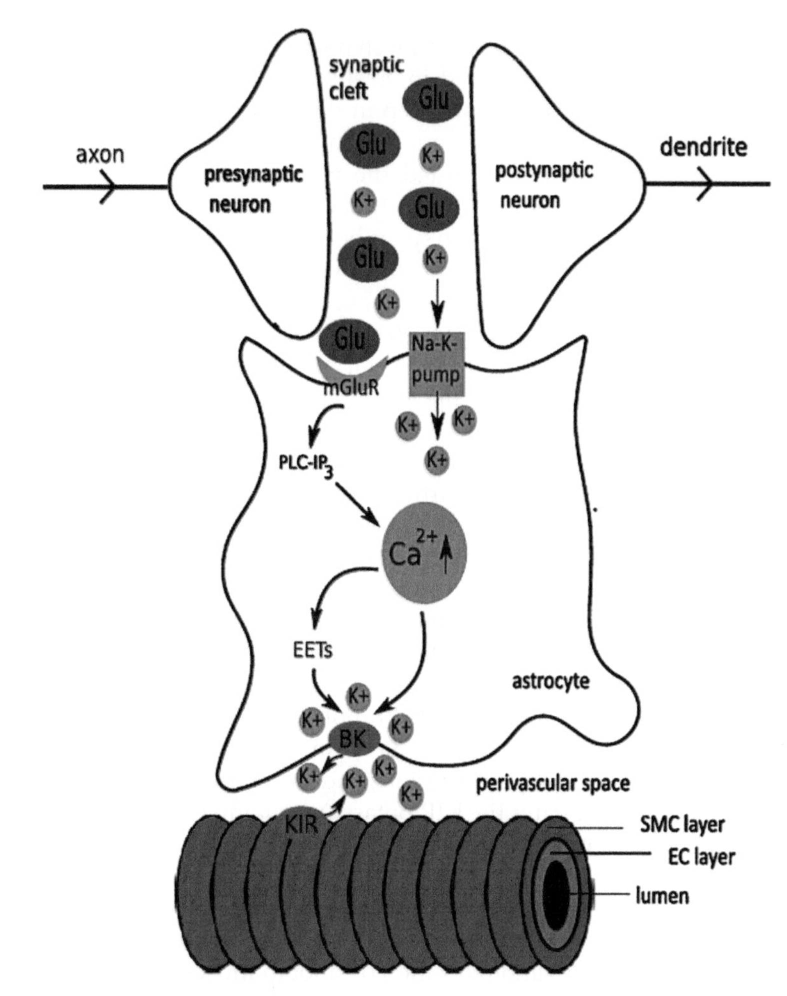

FIGURE 7.7 Neurovascular coupling model. See main text for description.

arterial wall are taken from the model of *Kudryashov and Chernyavskii* [22] where they define the active force on smooth muscle due to muscular tonus as k_2F (where k_2 is a constant and F is the concentration of contracting actin-myosin filaments).

To simulate the first processes of neurovascular coupling in the astrocyte, we assumed a square pulse (amplitude 0.8) for the fraction of bound glutamate receptors for a number of seconds in the synaptic space. We then varied perivascular potassium concentration within the physiological range of values, and solved the set of arteriole equations at each of these values. According to experiment, a modest increase of potassium should induce vasodilation [23–25] but paradoxically, a further increase in perivascular potassium should then cause vasoconstriction [25,26].

The results successfully support the hypothesis of a potassium-mediated exchange of information between arteriole and astrocyte. We show that a glutamate-induced calcium increase can gate BK channels in the astrocyte such that potassium is released into the perivascular space. We then show that we can achieve an approximate 20% vasodilation due to the increase of perivascular potassium concentration from 3 to 6 mM. This increase is sufficient to change the resistance of the asymmetric binary tree of *David et al.* [8] to the same degree that our previous phenomenological metabolic model did, making this model a promising substitute. We also successfully simulate the paradoxical effect where increased potassium causes first an increase and then a decrease in radius. This phenomenon has not been previously explained, but our investigations into the cause of this effect find that this is due to the interactions between the voltage changes of the hyperpolarizing SMC and the changing value of the resting potential of the KIR channel, which shifts according to the potassium concentration dependency described by *Oonuma et al.* [20].

7.7.2 Perfusion via the Capillary Bed

Cerebral capillary beds have a crucial role to control and transfer adequate oxygen and required nutrients to the neuron cells and the removal of metabolic substances from the cortical tissue. The behavior of cerebral capillaries and blood flow in the downstream areas of cortical networks are not well known because of their tortuous, complex architecture and high density of cortical capillary networks in the subsurface of the cortex. Despite many studies on the regulation of cerebral arterial blood flow, little work has been done on the capillary system and measuring the hemodynamic parameters of the cerebral capillary network. Most of these works were based on physiological observations. Using confocal laser-scanning microscopy techniques, the velocity of red blood cells in the individual capillaries was reported to be in the range of 0.3–3.2 mm/s.

The analysis of hemodynamic parameters and functional reactivity of cerebral capillaries is still controversial. To assess the hemodynamic parameters in the cortical capillary network, a generic model was created [27] using 2-D Voronoi tessellations, in which each edge represents a capillary segment [28]. This method is capable of creating an appropriate generic model of cerebral capillary network relating to each part of the brain cortex, because the geometric model is able to vary the capillary density. The modeling presented here is based on morphometric parameters extracted from physiological data of the human cortex [5]. The pertinent hemodynamic parameters were obtained by numerical simulation based on effective blood viscosity as a function of hematocrit and microvessel diameter, phase separation and plasma skimming effects. The hemodynamic parameters of capillary networks with two different densities (consistent with the variation of the morphometric data in the human cortical capillary network) were analyzed. The results show pertinent hemodynamic parameters for each model. The heterogeneity (coefficient variation) and the mean value of hematocrits, flow rates and velocities of both the network models were specified.

7.7.2.1 Oxygen Transport into the Cerebral Tissue

Having obtained both hematocrit and blood flow data within the model of the cerebral tissue, oxygen concentrations can be resolved throughout the tissue using diffusion principles. The analysis of oxygen transport throughout the 3-D model is based on the diffusion law in the steady state. The consumption rate of oxygen in the model is represented as a dynamic state by a Michaelis-Menten equation. Due to the existence of complex and unparallel structure of segments in the cortical capillary networks, the Krogh method cannot be considered a proper solution to the oxygen transport in the model. The pertinent Poisson equation is numerically solved using a Green's function method to estimate the oxygen partial pressure field in the tissue and space such that the conservation law of oxygen is satisfied. The difference in the intravascular oxygen contents at the inlets and outlets is equal to the diffusive oxygen fluxes to the tissue which is in balance with the oxygen consumption rate by the tissue in the model. Figure 7.8 depicts the modeled processes.

We model the tissue as an infinite three-dimensional domain, with an (finite-sized) embedded one-dimensional branched network of capillary vessels representing oxygen sources. We label this computational domain $\mathcal{C}$. In addition to the tissue oxygen distribution $P(x), x \in \Omega$, the oxygen tension in the capillary network $P_b(x),\ x \in \mathcal{C}$ is tracked using conservation equations, taking into account diffusive mass exchange through the capillary walls. Zero boundary conditions are imposed at infinity, and the total oxygen efflux from the capillaries is constrained to be matched with the total metabolic oxygen consumption.

We assume that oxygen in the tissue follows Henry's law, i.e., that the relationship between oxygen concentration C and oxygen partial pressure P can be written as $P = \alpha C$, where α is the solubility constant. We can therefore consider partial pressure and concentration interchangeably. Assuming Fickian diffusion, the Poisson equation for oxygen partial pressure is:

$$D\alpha \nabla^2 P(x) = f(P(x)) \tag{7.4}$$

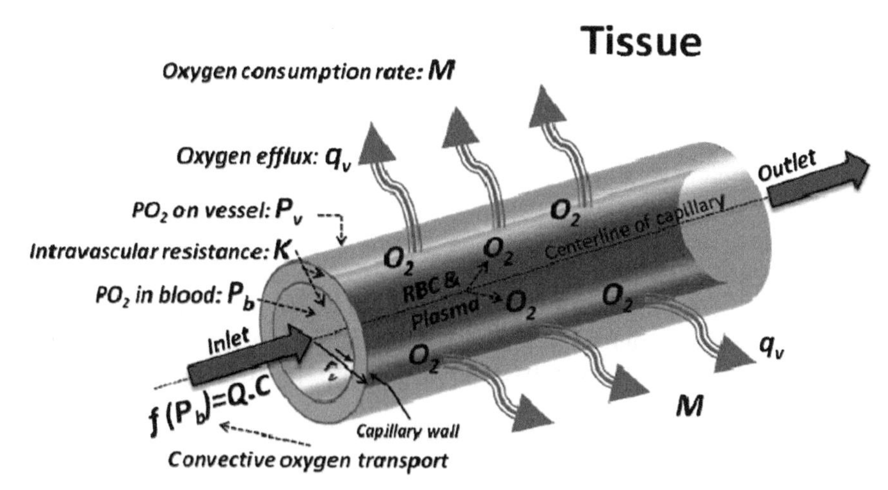

FIGURE 7.8 Schematic representation of oxygen transport from a part of a capillary segment to the relevant tissue.

where $f(P(x))$ encapsulates the reaction and source terms. In particular:

$$f(P(x)) = \begin{cases} M(P) - g(P - P_b), \ x \in C \end{cases} \tag{7.5}$$

$M(P)$ represents the metabolic consumption of oxygen which is assumed to have a Michaelis-Menten dependence on partial pressure as follows:

$$f(P(x)) = -\frac{M_0 P}{P(x) + P_{cr}}$$

where M_0 is the maximum cerebral metabolic rate, assumed to be uniform throughout the tissue, and the Michaelis constant P_{cr} is the pressure at half-maximal consumption. If the tissue oxygen pressure in a region is high enough, the rate of consumption $M(P)$ matches the maximum oxygen demand M_0 whereas there is a reduction of $M(P)$ in the regions with the low tissue pressure. The physiological reason comes from the dependency of mitochondrial respiration in the tissue cells and the oxygen supply. $g(P - P_b)$ represents the diffusive mass transport across the capillary wall, a function of the pressure difference between the blood and the tissue pressures.

The Green's function $G(x, s)$ for $x, s \in \mathbb{R}^3$ is given by the solution to the PDE:

$$D\alpha\nabla^2 G(x, s) = \delta(x - s) \tag{7.6}$$

where $\delta(\cdot)$ is the (three-dimensional) Dirac delta function. Multiplying through by $f(P(s))$ and integrating over s we then have:

$$D\alpha\nabla^2 \int_\Omega G(x, s) f(P(s)) ds = f(P(x)) = D\alpha\nabla^2 P(x) \tag{7.7}$$

This gives the pressure distribution $P(x)$ as the solution to an integral equation:

$$P(x) = \int_\Omega G(x, s) f(P(s)) ds \tag{7.8}$$

By discretizing the tissue into small hex elements and discretizing the capillary vessels into segments, the approximate solution is computed numerically by iteration, where in the discretized formulation the integral operator is replaced by a large matrix multiplication (the resulting Green's function matrix has approximately $n^3 \times n^3$ entries, where n is the number of spatial points in each coordinate direction). At each iteration, it is also necessary to iteratively recompute the blood oxygen pressure and transfer rates.

Figure 7.9 shows the spatial heterogeneity that results from these simulations and Fig. 7.10 shows two examples of oxygen concentration for differing metabolic rates viewed in three cut planes in the capillary tissue volume perpendicular to the z-axis. Here it is assumed that the metabolic rate is constant throughout the tissue domain. It is interesting to note that even for the 'normal' metabolic rate of 10 cm^3 O_2/100 cm^3/min the oxygen concentration is not homogeneous. As the metabolic rate decreases the tissue oxygen concentration becomes more homogeneous.

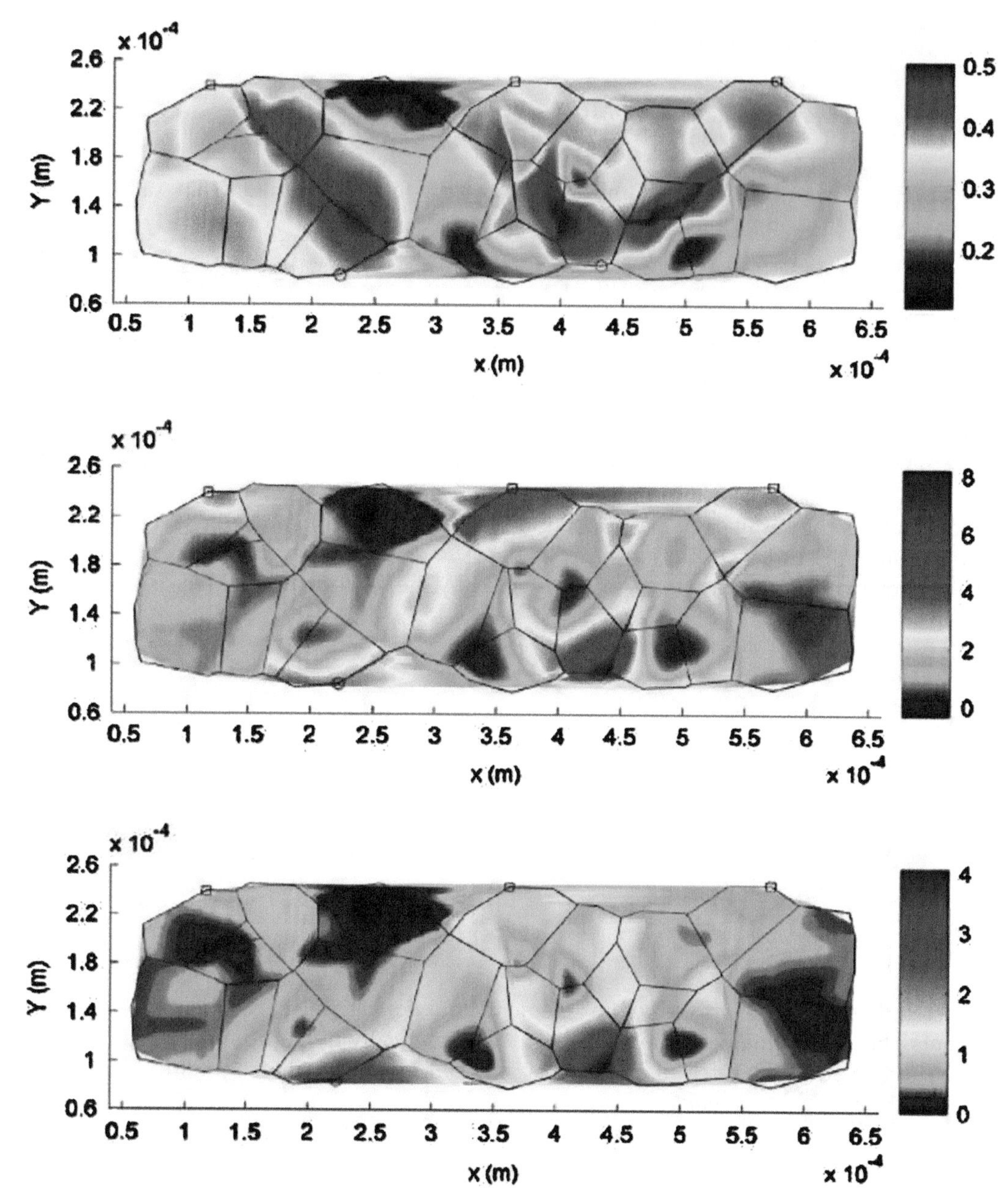

FIGURE 7.9 2-D spatial distribution of hematocrit (top), blood flow rate (middle), and blood velocity (bottom). The squares and the circles represent three nodes of high pressure (terminal arterioles) and two nodes of low pressure (collecting venules), respectively.

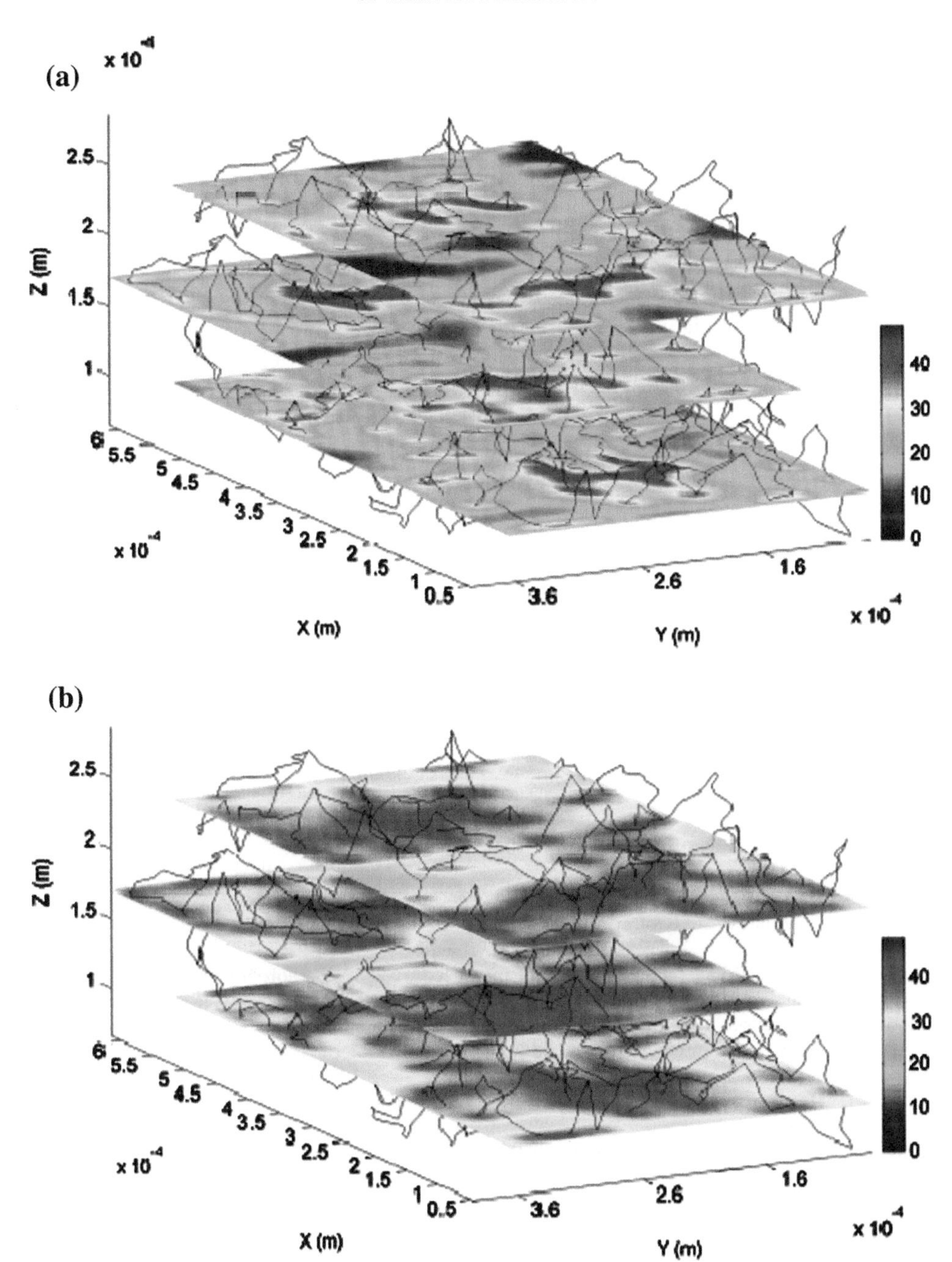

FIGURE 7.10 The distribution of partial oxygen pressure (oxygen tension) of the tissue presented in three slides perpendicular to the *z*-axis with respect to the two values of maximum oxygen consumption rate by the tissue, 10 cm^3 O_2/100 cm^3/min (a) and 5 cm^3 O_2/100 cm^3/min (b).

7.8 CONCLUSIONS

We have described in the above sections a number of models that aim toward understanding the regulation of the brain's blood supply and the subsequent perfusion into the cerebral tissue. These models have moved from the relatively simple (phenomenological) to the complex where cellular functions, hematocrit concentration and oxygen diffusion are modeled in detail.

As we move onwards in discovering the intricacies of the brain, its vasculature and function, we move even closer to a watershed; a computational watershed. Do we continue to develop more complex models of the neuron/astrocyte/perfusing arteriole unit and multiply up to forming a full discrete model (requiring compute resources that may very well be unavailable) or do we take results from the single unit complex model and develop a phenomenological model which takes orders of magnitude less compute resource and multiply up? If we take the latter pathway then there is an implied assumption that in doing so we find the same result as if we had taken the former. This is a nontrivial assumption and one that may cause model developers to follow a path that leads to incorrect science.

Acknowledgments

We would like to acknowledge the help from Hannah Farr, Navid Safaeian, Samara Alzaidi, Steve Moore, and IBM.

References

[1] T. David, S. Moore, Modeling perfusion in the cerebral vasculature, Medical Engineering and Physics 30 (10) (2008) 1227–1245.

[2] B.J. Alpers, R.G. Berry, R.M. Paddison, Anatomical studies of the circle of Willis in normal brain, AMA Archives of Neurology and Psychiatry 81 (4) (1959) 409–418.

[3] M.J. Krabbe-Hartkamp, J. Van Der Grond, F. De Leeuw, J. De Groot, A. Algra, B. Hillen, M. Breteler, W. Mali, Circle of Willis: morphologic variation on three-dimensional time-of-flight MR angiograms, Radiology 207 (1) (1998) 798–805.

[4] F. Cassot, F. Lauwers, C. Fouard, S. Prohaska, V. Lauwers-Cances, A novel three-dimensional computer-assisted method for a quantitative study of microvascular networks of the human cerebral cortex, Microcirculation 13 (1) (2006) 1–18.

[5] F. Lauwers, F. Cassot, V. Lauwers-Cances, P. Puwanarajah, H. Duvernoy, Morphometry of the human cerebral cortex microcirculation: general characteristics and space-related profiles, NeuroImage 39 (3) (2008) 936–948.

[6] B.N. Steele, M.S. Olufsen, C.A. Taylor, Fractal network model for simulating abdominal and lower extremity blood flow during resting and exercise conditions, Computer Methods in Biomechanics and Biomedical Engineering 10 (1) (2007) 39–51.

[7] A. Kamiya, T. Togawa, Optimal branching structure of the vascular tree, Bulletin of Mathematical Biophysics 34 (1) (1972) 431–438.

[8] T. David, S. Alzaidi, H. Farr, Coupled autoregulation models in the cerebro-vasculature, Journal of Engineering Mathematics 64 (4) (2009) 403–415.

[9] W. Schreiner, B.F. Buxbaum, Computer optimization of vascular trees, IEEE Transactions on Biomedical Engineering (1993) 482–491.

[10] S. Sherwin, V. Franke, J. Peiró, K. Parker, One-dimensional modelling of a vascular network in space-time variables, Journal of Engineering Mathematics 47 (3–4) (2003) 217–250.

[11] J. Alastruey, S.M. Moore, K.H. Parker, T. David, J. Peir, S.J. Sherwin, Reduced modelling of blood flow in the cerebral circulation: coupling 1-D, 0-D and cerebral auto-regulation models, International Journal for Numerical Methods in Fluids 56 (2008) 1061–1067.

[12] S.S. Alzaidi, Computational Models of Cerebral Hemodynamics, PhD, University of Canterbury, 2009.

[13] D.W. Newell, R. Aaslid, A. Lam, T.S. Mayberg, H.R. Winn, Comparison of flow and velocity during dynamic autoregulation testing in humans, Stroke: A Journal of Cerebral Circulation 25 (4) (1994) 793–797.
[14] T. David, S. Alzaidi, H. Farr, Computational models of functional hyperaemia in the cerebro-vasculature, in: 18th World IMACS/MODSIM Congress, Cairns, Australia, 2009, pp. 656–663.
[15] J.A. Filosa, V.M. Blanco, Neurovascular coupling in the mammalian brain, Experimental Physiology 92 (4) (2007) 641–646.
[16] M. Koenigsberger, R. Sauser, J.-L. Bény, J.-J. Meister, Role of the endothelium on arterial vasomotion, Biophysical Journal 88 (6) (2005) 3845–3854.
[17] M. Koenigsberger, R. Sauser, J.-L. Bény, J.-J. Meister, Effects of arterial wall stress on vasomotion, Biophysical Journal 91 (5) (2006) 1663–1674.
[18] Y. Kurachi, Voltage-dependent activation of the inward-rectifier potassium channel in the ventricular cell membrane of guinea-pig heart, Journal of Physiology 366 (1985) 365–385.
[19] J.M. Quayle, C. Dart, N.B. Standen, The properties and distribution of inward rectifier potassium currents in pig coronary arterial smooth muscle, Pflugers Archiv European Journal of Physiology 416 (2) (1996) 715–726.
[20] H. Oonuma, K. Iwasawa, H. Iida, T. Nagata, H. Imuta, Y. Morita, K. Yamamoto, R. Nagai, M. Omata, T. Nakajima, Inward rectifier K(+) current in human bronchial smooth muscle cells: inhibition with antisense oligonucleotides targeted to Kir2.1 mRNA, American Journal of Respiratory Cell and Molecular Biology 26 (3) (2002) 371–379.
[21] C.M. Hai, R.A. Murphy, Ca^{2+}, crossbridge phosphorylation, and contraction, Annual Review of Physiology 51 (60) (1989) 285–298.
[22] N.A. Kudryashov, I.L. Chernyavskii, Numerical simulation of the process of autoregulation of the arterial blood flow, Fluid Dynamics 43 (1) (2008) 32–48.
[23] W. Kuschinsky, M. Wahl, O. Bosse, K. Thurau, Perivascular potassium and pH as determinants of local pial arterial diameter in cats. A microapplication study, Circulation Research 31 (2) (1972) 240–247.
[24] J.G. McCarron, W. Halpern, Potassium dilates rat cerebral arteries by two independent mechanisms, American Journal of Physiology 259 (3 Pt 2) (1990) H902–H908.
[25] H.J. Knot, P.A. Zimmermann, M.T. Nelson, Extracellular K(+)-induced hyperpolarizations and dilatations of rat coronary and cerebral arteries involve inward rectifier K(+) channels, Journal of Physiology 492 (Pt 2) (1996) 419–430.
[26] T. Horiuchi, H.H. Dietrich, K. Hongo, R.G. Dacey, Mechanism of extracellular K+-induced local and conducted responses in cerebral penetrating arterioles, Stroke: A Journal of Cerebral Circulation 33 (11) (2002) 2692–2699.
[27] N. Safaeian, M. Sellier, T. David, A computational model of hemodynamic parameters in cortical capillary networks, Journal of Theoretical Biology 271 (2011) 145–156.
[28] N. Safaeian, T. David, M. Sellier, General model for cortical capillary networks and an investigation on pertinent functional reactivity to the different blood inflows, in: 6th World Congress of Biomechanics (WCB2010), Springer, Singapore, 2010, pp. 450–453.
[29] F.H. Netter, Atlas of Human Anatomy, fourth ed., Saunders Elsevier, Philadelphia, PA, 2006.
[30] S. Moore, Computational 3D Modelling of Hemodynamics in the Circle of Willis, PhD, University of Canterbury, 2007.
[31] R. Seeley, T. Stephes, P. Tate, Anatomy and Physiology, McGraw-Hill, New York, 2003.

CHAPTER

8

Mechanobiology of the Arterial Wall

Anne M. Robertson[a] *and Paul N. Watton*[b]

[a]Department of Mechanical Engineering and Materials Science, University of Pittsburgh, Pittsburgh, PA, USA

[b]Institute of Biomedical Engineering, Department of Engineering Science, University of Oxford, UK

OUTLINE

8.1 Introduction 276

8.2 Overview of the Arterial Wall 278
- *8.2.1 Brief Overview of the Architecture of the Arterial Wall* 279
- *8.2.2 Design Requirements for the Arterial Wall* 281
  - 8.2.2.1 Role of Compliance of the Arterial Wall 281
  - 8.2.2.2 Constraints on the Compliance of the Arterial Wall 282
  - 8.2.2.3 Other Considerations 283

8.3 The Extracellular Matrix 283
- *8.3.1 Collagen and the Arterial Wall* 283
  - 8.3.1.1 Collagen Molecules and Supramolecular Assemblies 284
- *8.3.2 Elastin in the Arterial Wall* 291

8.4 Vascular Cells 292
- *8.4.1 Endothelial Cells* 293
- *8.4.2 Vascular Smooth Muscle Cells* 295
- *8.4.3 Fibroblasts* 297
- *8.4.4 Matrix Assembly by Vascular Cells* 297
  - 8.4.4.1 Attachment of Collagen Fibers to the Extracellular Matrix 300

8.5 Architecture of the Arterial Wall 301
- *8.5.1 Tunica Intima* 301
  - 8.5.1.1 Glycocalyx 301
  - 8.5.1.2 Basement Membrane 302
  - 8.5.1.3 Subendothelium 303
  - 8.5.1.4 Internal Elastic Lamina (IEL) 304
- *8.5.2 Tunica Media* 306
  - 8.5.2.1 Tunica Media of Elastic Arteries 306

http://dx.doi.org/10.1016/B978-0-12-415824-5.00008-4

© 2013 Elsevier Inc. All rights reserved.

8.5.2.2 Tunica Media of Muscular Arteries 308
8.5.3 *Tunica Adventitia* 308

8.6 Constitutive Models for the Arterial Wall 309
8.6.1 *Multiple Mechanism Models* 311
8.6.2 *Isotropic Mechanism* 312
8.6.3 *Anisotropic Mechanisms: Kinematics of Fiber Recruitment* 313
8.6.4 *N-Fiber Anisotropic Models* 314
8.6.5 *Anisotropic Models with a Distribution of Fiber Orientations* 316
8.6.6 *Distributions of Fiber Recruitment Stretch* 317
8.6.7 *Multi-Mechanism Models: Growth, Remodeling and Damage (GR&D)* 319
8.6.7.1 Positive and Negative Growth of a Constituent 319
8.6.7.2 Remodeling of the Constituents 320
8.6.7.3 Damage Models 320
8.6.7.4 Growth, Remodeling and Damage for the Elastin Mechanism 321
8.6.7.5 Growth, Remodeling and Damage of the Collagen Mechanism 323

8.7 Modeling Vascular Disease: Intracranial Aneurysms 324
8.7.1 *Background* 324
8.7.2 *Computational Modeling of Intracranial Aneurysms* 326
8.7.3 *Example: Fluid-Solid-Growth Model of Aneurysm Evolution* 329
8.7.4 *Discussion* 331

References 333

8.1 INTRODUCTION

'... circulatory systems are among our greatest corporeal glories. ... that splendor of design, its overall coherence, its exquisite detail, and its unobtrusive operation'. – Steve Vogel, 1992 [303]

The proper functioning of the arterial wall is vital to the health of the individual, as evidenced by the numerous debilitating medical problems which can arise when the wall falls prey to the effects of disease, aging or deleterious genetic variations. Conversely, the arterial wall's remarkable capabilities for growth, repair and continual renewal under diverse physiological loads have yet to be reproduced in tissue engineered vessels, though rapid advances are being made [220,225,329].

The mechanical integrity of the arterial wall is dependent on its central passive load bearing components including collagen fibers, elastin fibers and elastin lamellae, as well as on the proper functioning of its cellular components: endothelial cells, vascular smooth muscle cells and fibroblasts. The intramural cellular content is responsible for the collagen production, repair and degradation that is essential for maintenance of wall integrity during the human lifespan as well as for recovery from external damage. However, the demands on our vascular system are not static over time. These same cells must also be able to sense changes in mechanical loads and chemical stimuli and orchestrate a collective response to adapt to these changes.

Acting in concert, the cellular material in our arteries can alter the quantity distribution, orientation and mechanical properties of the collagen fabric for pre- and postnatal vascular growth as well as for effective remodeling in response to changing stimuli. As it will become clear in this chapter, there are many fundamental unanswered questions regarding these processes that impact how we treat nearly all cardiovascular diseases.

Biomechanists have developed mathematical models of the arterial wall (constitutive equations) in an attempt to better understand the complex processes of arterial growth, remodeling and damage. These models can be used to explore hypotheses about the coupled role of biology, chemistry and mechanics and, in some cases, to develop predictive models of disease. For example, mathematical models have been developed and used to great effect to study abdominal aortic aneurysms [137,312,315], intracranial aneurysms [16,319,320], hypertension [98] and vasospasm [134]. Damage models have been developed to study the process of wall degradation under supraphysiological loading such as found during cerebral angioplasty as well as the enzymatic damage that can arise in response to abnormal flow conditions [22,23,39,173,174].

It is becoming increasingly well understood that the most idealized aspects of these modeling efforts relate to the role of biology. Some of these idealizations reflect a need for further research, while others reflect a gap between knowledge within engineering/mathematics and biology communities. Mounting evidence demonstrates that purely mechanical models are insufficient for many vascular diseases [91,136,137,242]. Rather, there is a need for increased sophistication by the biomechanics community to integrate the available biological information into models of vascular disease and to work with vascular biologists to acquire additional data through experiments motivated by these modeling studies. In anticipation of growing efforts in this direction, we devote several sections of this chapter to a detailed review of the components and architecture of the arterial wall.

TABLE 8.1 Definitions of Abbreviations

AFI	Aneurysm formation index	IEL	Internal elastic lamina
CFD	Computational fluid dynamics	GAGS	Glycosaminoglycans
EC	Endothelial cell	OSI	Oscillatory shear index
ECM	Extracellular matrix	MMP	Matrix metalloproteinase
EEL	External elastic lamina	MPM	Multi-photon microscope
EGL	Endothelial glycocalyx	RVE	Representative volume element
FACIT	Fibril-associated collagens with interrupted triple helices	SAH	Subarachnoid hemorrhage
FSG	Fluid solid growth	TIMP	Tissue inhibitors of metalloproteinase
GON	Gradient oscillatory number	VSMC	Vascular smooth muscle cell
G&R	Growth and remodeling	WSS	Magnitude of the wall shear stress vector
GR&D	Growth, remodeling, and damage	WSSG	Wall shear stress gradient
IA	Intracranial aneurysm	$\overline{\text{WSSG}}$	Magnitude of the wall shear stress gradient

We begin this chapter with a brief overview of the arterial wall in Section 8.2, including discussion of its primary functional requirements and constraints. In Section 8.3, we turn attention to the extracellular matrix (*ECM*) and its core components, i.e., collagen fibers, elastin fibers and elastin lamellae. The vascular cells within this structural scaffold are the focus of Section 8.4 with particular emphasis on their function as transducers of mechanical stimuli and their vital role in maintenance and adaption of the ECM. In Section 8.5, we return to the architecture of the arterial wall in more detail, with a discussion of how the cellular and extracellular building blocks are integrated into the arterial wall structure. In Section 8.6, we briefly discuss current mathematical models (constitutive equations) for the arterial wall, including some illustrative examples. The application of the earlier material to vascular disease is then considered in Section 8.7, through the particular example of cerebral aneurysms. The abbreviations used here and elsewhere in the text are defined in Table 8.1.

8.2 OVERVIEW OF THE ARTERIAL WALL

In 1846, Wertheim presented the results of his force elongation measurements on human tissues including arteries, muscles, tendon, veins and bone to the l'Académie des Sciences [323]. He recognized, for what appears to be the first time, the fundamental difference in mechanical response of soft tissues such as arteries compared with bone, wood and inorganic materials:

> Bone tissue extends by significantly following the law of proportionality to the loads, that is to say, in the same way as the inorganic body and wood. So, if we take the loads as abscissae and the corresponding elongations as ordinates, the line that represents elongations is walking a straight line. It is not the same for the soft body parts in their natural state of humidity, where the law of their elongation is represented by a curve which closely approximates a hyperbola whose top is fixed at the origin of coordinates. ***Translated from page 395 of [323].***

Wertheim did not present his experimental data, but rather coefficients for the curves he fit to the data, (e.g., Fig. 8.1). We now know that the loading curves he reported for these soft tissues are typical for healthy vascular tissue, displaying high flexibility at low loads (the toe region) followed by a relatively rapid transition to high and nearly constant stiffness with increasing loads, Fig. 8.1.[1] In arteries, this transition region has been found to coincide with the strain regime experienced during normal functioning. In 1909, *Osborne* proposed an explanation as to *why* a loading curve of this shape would be beneficial for biological structures such as the bladder [324], a subject we return to in Section 8.2.2. It was not until 1957 that *Roach and Burton* proposed an explanation as to *how* the shape of this mechanical loading curve is produced by the arterial wall. As discussed in Section 8.5, their explanation has been supported by their own work as well as more recent studies. Reproducing the mechanical properties of the arterial wall is one of the central challenges in tissue engineering of vascular grafts [220]. Efforts are also being made to design fiber reinforced materials that reproduce this response, with the objective of creating more realistic *in vitro* test systems for studies of vascular disease [21].

[1] Like others at that time, *Wertheim* believed bone and most inorganic materials behaved linearly. As correctly noted by *Roy* in 1880, when a larger range of loads is considered, it becomes clear that these metals are also nonlinear, curving downward in the opposite sense to that of soft tissue [245].

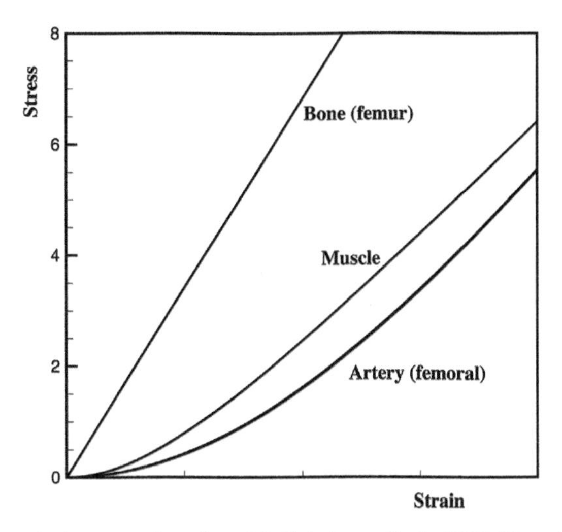

FIGURE 8.1 Stress versus strain curves based on *Wertheim*'s results in [323]. Here, the range for artery, muscle and bone is 20%, 2.5% and 0.75%, respectively. Stress for artery and muscle is given in gm/mm^2, while that for bone is given in kg/mm^2.

In the remainder of this chapter, we address these subjects as well as the more general questions of: (i) What design requirements must be met by the arterial wall? (ii) What are the components and architecture of the arterial wall? (iii) How does this structure function to satisfy these design requirements? While complete coverage of these questions is beyond the scope of a single chapter, we use them to motivate the material covered in this chapter and to point out open questions in the field of vascular mechanobiology. Answers to these questions are important for improving treatments of acquired and inherited diseases as well as for designing tissue engineered blood vessels.

8.2.1 Brief Overview of the Architecture of the Arterial Wall

The arterial wall contains vascular cells (e.g., endothelial cells, vascular smooth muscle cells and fibroblasts) housed within an organized network of macromolecules, namely, the extracellular matrix. The ECM is largely composed of two types of macromolecules: proteins (such as collagen, elastin, fibronectin and laminin) as well as a class of unbranched, polysaccharide chains called glycosaminoglycans (*GAGs*). Most GAGs are found covalently attached to protein cores, forming proteoglycans. Due to the stiffness of the GAGs as well as their highly anionic nature, they draw water into the ECM to form a gel that provides the tissue with compressive resistance and enables diffusive processes within the arterial wall [11]. While the ECM was previously viewed as an inert scaffold, it is now understood to play a vital dynamic role in regulation of cells inside the ECM – influencing their structure, migration, proliferation and function through cell-matrix interactions [11].

In a tubular segment of arterial wall, the ECM and cellular content are found to be organized in a system of concentric cylindrical layers, see Fig. 8.2a. These layers are generally viewed within the context of three specific regions which are separated by elastin lamellae (that lessen cell migration between layers) and generally distinguished by cell type. In a prototypical healthy artery,

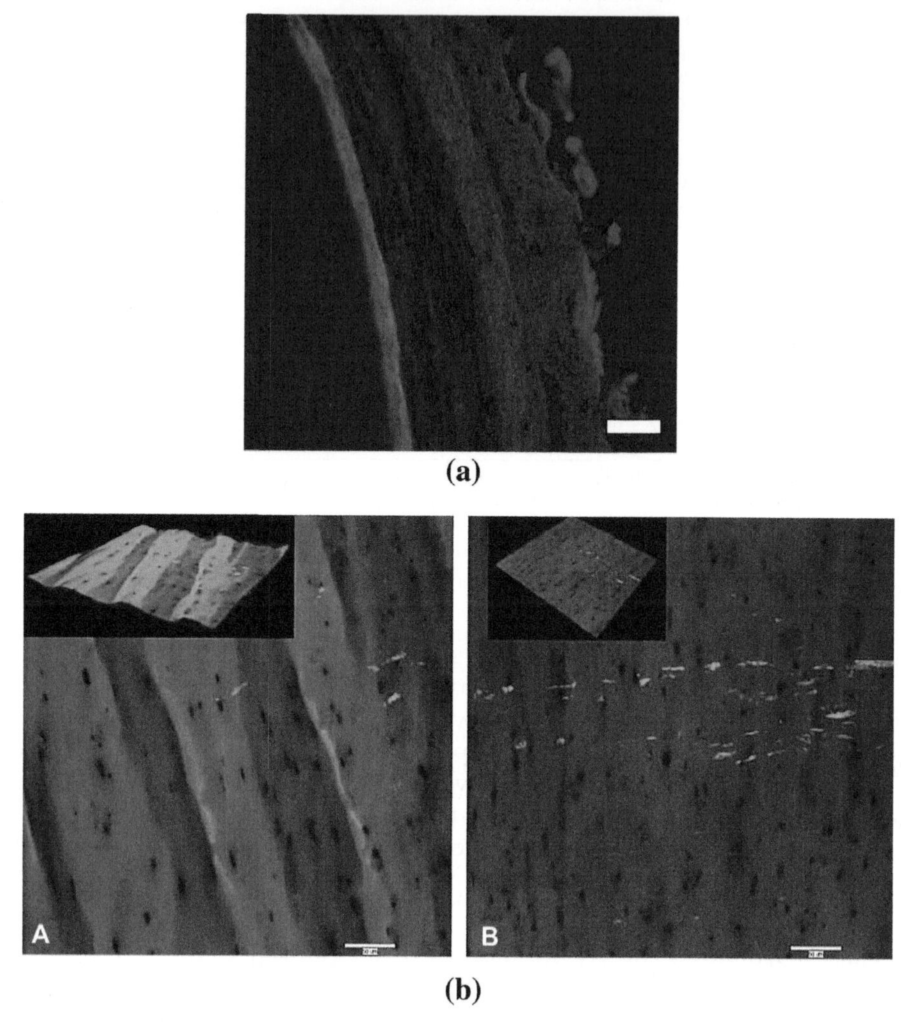

FIGURE 8.2 (a) Fluorescence microscopy images of cross-sectional preparation of the human left vertebral artery (cerebral), fixed at 30% stretch. Immunohistochemical staining of the arterial wall reveals elastin (green) localized in the internal elastic lamina, cell nuclei (blue, DAPI stain), collagen fibers (red). (b) 3-D reconstruction of confocal microscopy image slices taken of a human anterior cerebral artery (ACA) revealing the autofluorescent internal elastic lamina, under (A) zero stretch and (B) 30% strain. All three scale bars = 50 μm. (Reprinted from [243] with permission from Springer.)

the innermost region, or *tunica intima*, is lined with endothelial cells (*ECs*), attached to a collagen network called the basement membrane. The middle layer, or *tunica media*, is largely composed of concentric layers of vascular smooth muscle cells (*VSMCs*) within an ECM composed of collagen and elastin fibers. The outer layer or *tunica adventitia* contains fibroblasts within a collagenous ECM. In most arteries, the medial layer is separated from the intima and adventitia by concentric elastin sheets referred to as the internal elastin lamina (*IEL*) and external elastin lamina (*EEL*), that line these respective vascular regions. Stehbens provides the succinct and unequivocal definition that the intima 'rightfully includes the endothelium, internal elastin lamina (IEL) and all intervening tissue' [282] and we use this definition here. It should be noted that in some works, the IEL is considered part of the medial layer (e.g., [281]). The fenestrae (windows) in the IEL enhance transport between the lumen and the media, Fig. 8.2b. In thicker walled arteries, the adventitia and outer media are also supplied by small blood vessels on the adventitial side, the *vasa vasorum*. As will be elaborated on throughout this chapter, the distribution and volumetric fraction of these components vary across the vascular system and in response to stimuli through the process of growth and remodeling. The wall components and structure are also modified by disease and aging.

8.2.2 Design Requirements for the Arterial Wall

The human heart can be conceptualized as a pair of periodic, positive displacement pumps and therefore blood leaving the heart is pulsatile in nature. The left pump collects oxygenated blood from the lungs and drives it through the systemic circulation, while the right pump collects de-oxygenated blood from the systemic circulation and drives it through the pulmonary circulation. Both pumps have two chambers: the smaller upper chamber, the *atrium*, elevates the pressure of the incoming blood and drives it into the larger, lower chamber, the *ventricle* during atrial systole. Both ventricles of the heart contract together during a phase called ventricular systole, ejecting close to 140 mL of nearly incompressible blood from the adult human heart. This volume must be taken up by an expansion of the remainder of the circulatory system. Therefore, by necessity, the vasculature must be compliant to avoid rupture [303]. In the remainder of this subsection, we will discuss functional requirements and physical conditions that have driven qualitative aspects of the arterial elastic response as well as specific quantitative features. In fact, as discussed below, there are other benefits from a flexible vascular system.

8.2.2.1 Role of Compliance of the Arterial Wall

It is clearly desirable that the material and geometric properties of the vasculature contribute toward the reduction of mechanical load on the heart while still satisfying physiological requirements of the body regarding distribution of nutrients and collection of waste products. It is easy to see analytically for the idealized case of laminar flow in a straight, rigid pipe that less energy is required to pump fluid at a steady flow rate than to pump flow with an oscillatory component superposed on top of this same steady component (e.g., [339]). There is simply less energy wasted accelerating the fluid and driving it back and forth against viscous drag forces. Therefore, one means to reduce the energy expenditures of the heart is to decrease the magnitude of the flow pulsations. *Borelli* (1608–1679) appears to be the first to propose the role of the elastic artery wall in flattening the velocity waveform [37] (p. 243 of English Translation [186]),

> … the blood does not leave the heart in a continuous flow as do streams and rivulets from springs. It flows by jerks but regularly. … Although the heart does not pour blood into the arteries during its diastoles, the blood does not stop and remain completely immobile and stagnant in the arteries, viscera, flesh and veins when the heart is as rest. This results from the fact that the arteries themselves are constricted by contraction of their circular fibers.

Hales observed this decrease in pulsatility in the horse circulatory system and clarified the capacitance role of the elastic arteries, noting the similarity of this role to the pulse dampeners used in fire engines (pp. 22–23 [108]). *Hales* is often given credit for discovering this phenomenon, later named the 'Windkessel effect' by *Frank*, who placed *Hales'* observations in a mathematical context [86] (English Translation [87,250]). *Frank*'s zero-dimensional analysis is based on several idealizations including the neglect of reflected waves, which have been included in more recent studies (see [269] and references cited, therein). Nonetheless, *Frank*'s model, now termed a two-element Windkessel model, provides a simple way to approximate the drop in pressure in the aorta during diastole.

While the capacitance effect of the artery will tend to smooth out the flow waveform and reduce the energetic demands on the heart, there are other factors at play. In particular, reflections of the pressure pulse at arterial bifurcations and narrowed regions can lead to proximal regions of increased amplitude of the pressure pulse [200,213]. Hence, the geometry of the vascular anatomy as well as the properties of the arterial wall determine the efficiency of the circulation (see, e.g., [93,200,339]).

8.2.2.2 Constraints on the Compliance of the Arterial Wall

Although it would seem energetically useful for the amplitude of the flow waveform to be dampened close to the heart, there are constraints on how flexible the wall can be in the physiological pressure range without adverse consequences. One constraint arises from a limit of the strain that arterial cells can be exposed to without imparting damage to their intercellular junctions or attachments to the ECM. Another restriction for the arterial elastic response is the need to avoid localized bulges in the wall arising from multiple solutions for radii at a particular transmural pressure. Such limit point instabilities can be avoided in materials for which the inflation pressure-radius relationship is monotonic. The shape of this curve depends on the constitutive equation for the material and, for tubular geometries, on the boundary conditions at the ends of the tubes. In a paper communicated by *Lord Rayleigh* in 1891, *Mallock* reported experimental results demonstrating this limit point instability during inflation of a cylindrical tube, composed of India-rubber and capped at the ends [184], Fig. 8.3. As will be discussed in Sections 8.4 and 8.7, the complex flows associated with a bulge of the kind shown are undesirable in the vascular system.

Mallock also reported a similar loss of uniqueness in the case of pressure inflation of spherical membranes composed of India-rubber. About 10 years later, *Osborne* demonstrated the remarkable result that this limit point instability is avoided in feline and monkey bladders

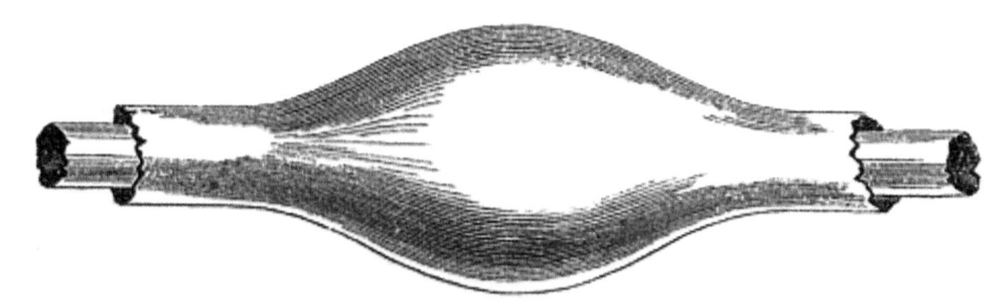

FIGURE 8.3 Schematic of uneven inflation of a cylindrical tube composed of India-rubber as seen by Mallock in his experimental studies of limit point instabilities in cylindrical tubes and hollow spheres. (Reprinted from [184] with permission from Proceedings of the Royal Society of London.)

[324]. *Osborne* attributed this qualitative difference with rubber membranes to the fibrous structure within these biological membranes as well as the 'initial rigidity' of the rubber membranes after which the stiffness diminishes with increasing strain. Over 75 years later, his qualitative explanation was made rigorous and generalized using results from finite elasticity theory [27,47].

8.2.2.3 Other Considerations

There are a number of other important design requirements for the arterial wall that we will only briefly mention here. For example, it is essential that arteries are sufficiently reinforced to withstand periods of elevated blood pressure, such as might occur due to emotional stress or physical exertion involving the Valsalva maneuver [255]. As will be discussed in Section 8.3, the collagen fibers in the wall serve this protective purpose. This is a stringent demand since blood pressure can rise well above physiological levels. For example, the Valsalva maneuver during heavy weight lifting has been reported to increase arterial pressure to 480/350 mm Hg [116]. The association between rupture of cerebral aneurysms (see Section 8.7) and defecation, coitus and weight lifting demonstrates the catastrophic outcomes that are possible in pathologies with compromised wall strength [80,116,175,236].

On average, human blood vessels will undergo on the order of 10^9 loading cycles during a 75-year lifetime and therefore arteries must also be resistant to fatigue. The arterial wall regularly replaces (turns over) its central load bearing component (collagen) as a means of withstanding this continual cyclic loading. However, this is not the case for the elastin components of the wall (see Sections 8.3 and 8.4). Furthermore, from an energetic point of view, it is desirable for the elastic energy stored in the arterial wall during systole to be nearly completely recovered during diastole (*high resilience*), though some level of viscous losses are valuable for damping traveling pressure waves in the wall. This loss of energy arises from the viscoelastic nature of the components of the arterial wall and their coupled interactions. In Sections 8.3–8.5, we return to these requirements in the context of the building blocks and structure of the arterial wall.

8.3 THE EXTRACELLULAR MATRIX

We now turn attention to the collagen and elastin components of the extracellular matrix including collagen fibers and networks as well as elastin fibers and lamellae. The physical composition and morphology of these components play a vital role in ensuring the structural integrity of the vessel wall and determining its mechanical properties. In this section, we consider the ultrastructure as well as the superstructure of these components in preparation for a discussion of their role and distribution within the layered architecture of the arterial wall, Section 8.5.

8.3.1 Collagen and the Arterial Wall

Collagen is the most prevalent protein in mammals (25–30% of total protein mass) [11,161,239] and is one of the central load bearing components in the body. It is found abundantly in tissues such as tendon, skin, cornea, cartilage, bone and vascular tissue and is estimated to contribute between 20 and 50% of the dry weight of arteries [161]. Its importance is not solely due to its role as a supporting scaffold; as will be elaborated on later in this chapter, it has a profound influence on cell function.

8.3.1.1 Collagen Molecules and Supramolecular Assemblies

It is now understood that collagen is not a single molecular type but rather a heterogeneous family of molecules, all including at least one collagenous domain in which three polypeptide chains (alpha chains), are wrapped into a triple-stranded right-handed superhelix [33]. Each of these α chains are turned in a left-handed helix, with three amino acids per turn and glycine spaced as every third amino acid, Fig. 8.4. When the three α chains are wrapped together, glycine is situated in the interior of all three chains. Since glycine is the smallest

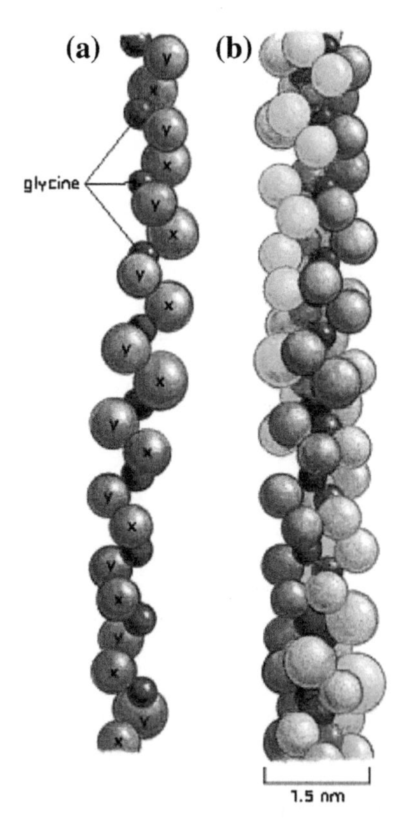

FIGURE 8.4 (a) Schematic of a section of a single α chain forming a left-handed helix with three amino acids per turn. Each sphere represents an amino acid. The chain is formed of Gly-X-Y sequences in which Gly is glycine and X, Y are typically proline and hydroxyproline, respectively. (b) Schematic of a segment of collagen molecule formed by three polypeptide α chains wrapped into a triple-stranded right helix. As the smallest amino acid, glycine is uniquely capable of fitting in the interior of the helix. A typical fibrillar collagen molecule is about 300 nm long. (Reprinted from [11] with permission from Garland Science.)

amino acid, this configuration enables the closest packing of the α chains [115]. The amino acid chain is thus formed of a series of Gly-X-Y sequences in which X and Y are other amino acids; X is often proline and Y hydroxyproline.

Although 25 different α chains exist, only 28 distinct combinations of the chains have been identified to date [33,239]. Each combination has been assigned a Roman numeral (I–XXVIII) and the value generally indicates the order of their discovery [239]. While the term *collagen* is used to refer to the family of proteins with the general structural features just described, designation of specific collagen types is used to identify proteins with similar supramolecular organization. Differences in size and location of the triple-helical domains are the primary factor responsible for these distinct structures and influence their physiochemical properties.

Collagen molecules can be assembled into supramolecular structures including fibrils, microfibrils, filaments and network-like structures. These in turn can be assembled into higher-order structures. For example, fibrils can assemble into fibers and lamellae while network-like collagens can assemble into basement membranes and anchoring fibrils [33]. The fraction of the collagen molecule occupied by the triple-helical domain can vary markedly between collagen molecules influencing its ultrastructure. For example, type I collagens have a single triple helix section, occupying more than 95% of the molecule resulting in the stiff, rod-like shape of these fibrils. In other collagen molecules, such as type IV collagens, the triple-helical domain is split between numerous regions, lending flexibility to the ultrastructure [239].

The structural integrity of the collagen molecule and its higher-order assemblies is of vital importance for the proper functioning of the arterial wall and its ability to withstand mechanical loading. One source of this integrity is the hydrogen bonds between the α chains that act to stabilize the helical conformation. Thermal or chemical degradation of these bonds causes the helix to unfold. The peptide bonds joining amino acids are located within the triple helix, sheltering them from proteolytic attack by enzymes, with the exception of collagenases [115]. This feature of the collagen molecule is essential for collagen turnover during healthy growth and remodeling (see Section 8.4). Lysine, another amino acid found in collagen, plays a role in cross-linking between the α chains and between collagen molecules. The stability of the triple helix is further increased by a high proline and hydroxyproline content in the chains [239]. The mechanical stiffness of the arterial collagen can increase with age due to glycation, a process which leads to increased cross-linking between collagen molecules [15]. This increased cross-linking is believed to be an important source of diminished vascular compliance, and is accelerated in diabetics [9].

8.3.1.1.1 COLLAGEN TYPES AND SUBFAMILIES

Collagen molecules can be divided into subfamilies based on their supramolecular assemblies. The subfamilies and collagen molecules most relevant to the arterial wall are fibril-forming or fibrillar collagens (types I, III, V), basement membrane network collagens (type IV), beaded filament forming collagen (type VI), network forming collagens (type VIII), transmembrane collagens (type XIII), and fibril-associated collagens with interrupted triple helices (*FACIT*) (type IX, XII, XIV) [11,33,148]. Additional collagen types have been identified in the arterial wall but will not be discussed here (see, e.g., [148,222]).

8.3.1.1.2 FIBRIL-FORMING COLLAGENS (TYPES I, III, V)

Fibril-forming collagens play the most important role in load bearing within the arterial wall and therefore their proper manufacture and assembly are particularly vital for the health and proper functioning of the vascular system. The collagen molecules of fibril-forming collagens assemble to form mature cylindrical-shaped collagen fibrils that are 10–500 nm in diameter in mammals [295], Fig. 8.5. These fibrils can in turn be combined into fibers on the order of 0.1–100 μm thick and these in turn can form fiber bundles [11,33]. We return to this assembly

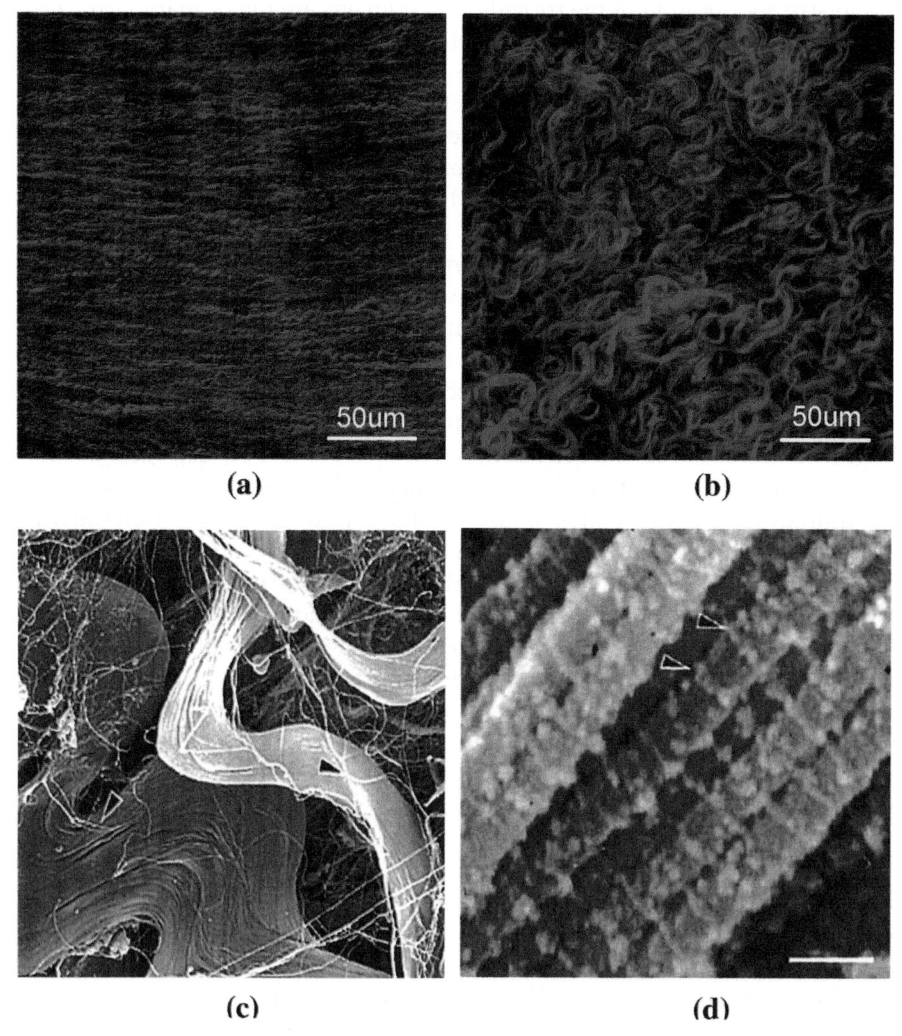

FIGURE 8.5 Arterial collagen fibers. Multi-photon images of collagen from (a) media and (b) adventitia of an unloaded left common carotid rabbit artery. SEM images of rat aortic adventitia showing (c) flat collagen fibers with much smaller fibrils and fibril bundles departing from the collagen fibers (at arrowheads) (×2200) and (d) bundles of collagen fibrils (155,000×, bar = 100 nm). (Figures (a) and (b) published with permission from Dr. M.R. Hill, Dr. A.M. Robertson and X. Duan. Figures (c) and (d) reproduced from [295] with permission from Dr. T. Ushiki.)

process in Section 8.4.4 and concentrate here on a description of the morphology and functional role of fibrillar collagens in the arterial wall.

Due to their non-centrosymmetric nature, fibers formed from fibrillar collagen have the capacity for second harmonic generation and thus can be imaged using multi-photon microscopy (*MPM*) without the use of exogenous stains or fixation, see Fig. 8.5a and b [120]. This enables the architecture of these fibers to be visualized and quantified in the context of a 3-D segment of the arterial wall [118,120]. The fibers as well as the fibrils are also visible using scanning electron microscopy [295], see Fig. 8.5c and d.

The collagen fiber morphology varies across the arterial wall and throughout the vasculature, including features such as fiber diameter, orientation and tortuosity as well as fibril diameter. For example, in the rabbit common carotid artery shown in Fig. 8.5a and b, the average fiber diameter is approximately fourfold less in the media compared with the adventitia and the medial fibers are more highly aligned than the adventitial fibers. Figures 8.5a–c illustrate that collagen fibers have a crimped and wavy nature in unloaded arterial tissue in both the media and adventitia. It is this feature of the collagen architecture that gives rise to the strong nonlinearity in the mechanical response curve of the arterial wall (see Fig. 8.1), i.e., the fibers are gradually *recruited* to load bearing as they straighten under deformation [40,120,241]. In these figures of rabbit common carotid artery, the waviness of the adventitial fibers can be seen to display a greater amplitude and wavelength than the medial collagen fibers, with bands of adventitial fibers following the same curve (see Fig. 8.5b and c). For example, average diameter, distribution of diameters and shapes of areal cross-sections of collagen fibrils are observed to vary across arterial layers of the human thoracic aorta [70], see Fig. 8.6. We return to a discussion of the collagen architecture within the arterial wall in Section 8.5.

Collagen Types I and III contribute nearly all the passive mechanical resistance of the arterial wall to circumferential and axial loading [189] and represent roughly 60% and 30% of the *total* arterial collagen, respectively, with the remaining 10% coming from type V and other collagens [36]. However, these ratios vary with location in the vascular tree, age and disease state as well as within the layers of the arterial wall [161]. In fact, this heterogeneity, coupled with the fact that earlier works were affected by experimental artifacts associated with pepsin digestion [189], could explain (apparently) conflicting published results for mass fractions of collagen types in the arterial wall.

While the α chains making up collagen I and III differ, both these molecules have uninterrupted triple-helical regions occupying about 95% of their length. This gives rise to their rod-like nature. However, collagen III fibrils are reported to be thinner, of more uniform diameter and more loosely packed compared with type I fibrils [204]. More specifically, type III fibrils are reported to be approximately 45 nm in diameter and loosely packed into $0.5-1.5\ \mu\text{m}$ diameter fibers while type I fibrils are approximately 75 nm in diameter, densely packed to form $2-10\ \mu\text{m}$ diameter fibers and bound together in fiber bundles [204]. Note that while the diameters differ, collagen typing based on diameter may not always be appropriate, e.g., in wound healing and some diseases [204].

Collagen V is found in the basal lamina underlying endothelial cells as well as that surrounding vascular smooth muscle cells [268]. It is diffusely distributed in the intercellular space of the intima [268] and is reported in all layers of cerebral vessels [201]. When the vessel wall is compromised, intimal collagen can be exposed to blood products and contribute to platelet adhesion and activation [77,217]. Whereas collagen I, III, and IV support platelet

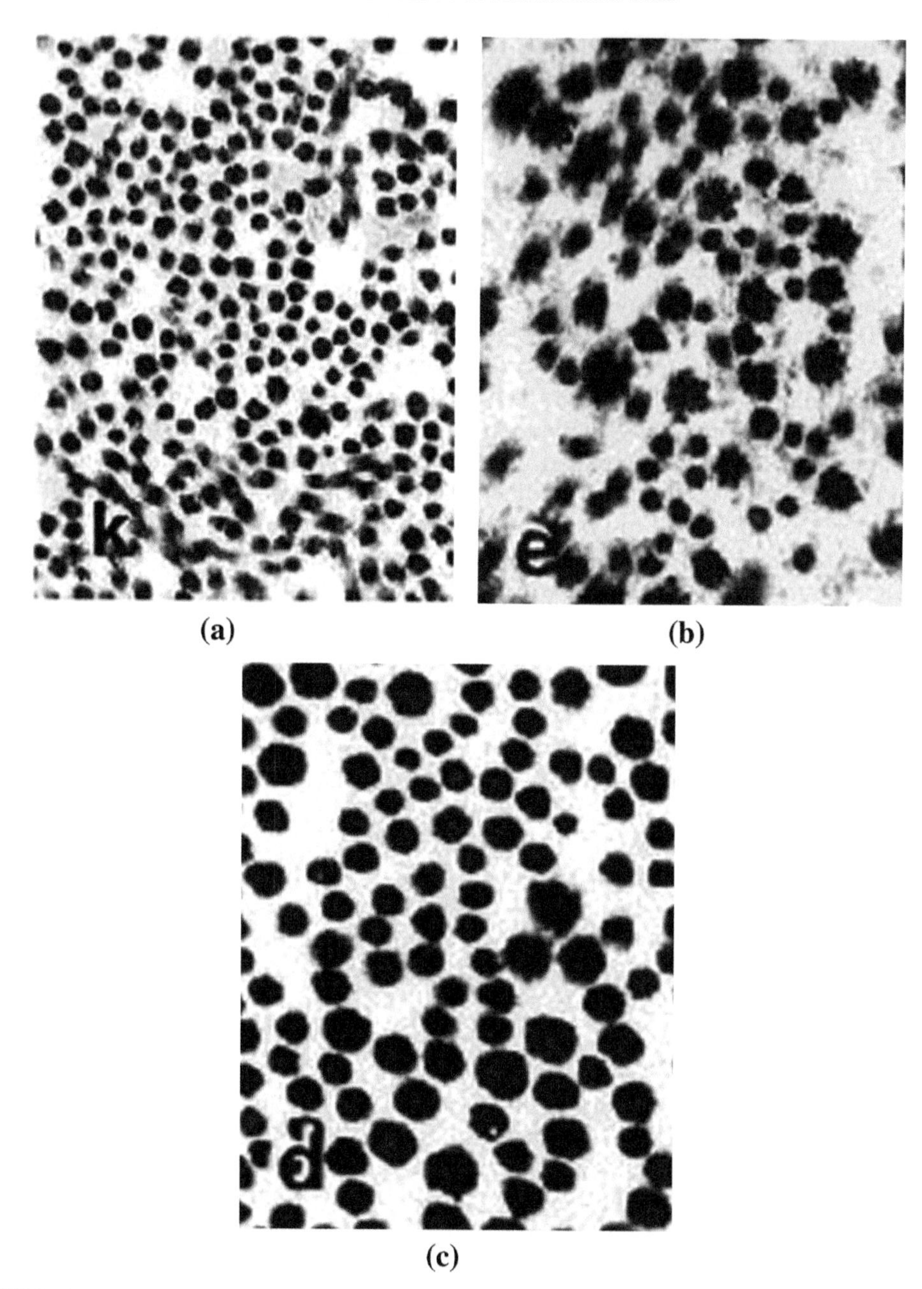

FIGURE 8.6 Cross-section of collagen fibrils from the (a) intima, (b) media, and (c) adventitia of human thoracic aorta. The distribution of fibril diameter is more uniform in the intima and adventitia compared with the media, where the cross-sections can be irregular, even appearing as 'collagen flowers' (×35,000). (Reprinted from [70] with permission from John Wiley & Sons.)

adhesion over a wide range of shear rates, type V collagen requires static conditions for platelet adhesion [249]. Hence it is suggested that collagen V can serve an anti-thrombogenic role in the vasculature, diminishing thrombosis in vessels that have only minor damage to the endothelium. *In vitro* studies demonstrate adhesion and proliferation of endothelial cells [112], monocytes and VSMCs [252] are also diminished on collagen V substrates relative to collagen I or III substrates.

While collagen V only represents a small mass fraction of the arterial wall, it plays an essential regulatory role in collagen fibril formation [33,34,81]. Much of what we know about this role has been gleaned from studies in the corneal stroma where collagen V constitutes 10–20% of the total collagen compared with 2–5% in most other tissues [34,285]. For instance, cell culture and *in vivo* mouse studies of the corneal stroma, both demonstrated that reducing the ratio of type V to type I collagen increased fibril diameter and decreased the quantity of fibrils [32,34,285].

8.3.1.1.3 COLLAGEN IV (BASEMENT MEMBRANE FORMING COLLAGEN)

Collagen IV is exclusively found in *basal laminae*. These are thin, flexible sheets that serve to compartmentalize tissues, provide structural support and serve as a reservoir for enzymes and cytokines that influence cell-cell activities, see, e.g., [11,149,170]. The endothelial cells lining the interior of the vessel wall are structurally supported by a basement membrane and which includes a basal lamina as the central component [11] and hence collagen IV is found in the basement membrane. In addition, it is also found in the basal lamina surrounding VSMCs, where it is believed to influence VSMC activities such as migration, proliferation, as well as their phenotype [1,24].

In type IV collagen molecules, the Gly-X-Y repeat units are frequently interrupted. This creates locations of structural flexibility within these molecules, to the collagen networks they create and the basement laminae scaffold formed from these networks. In fact, this flexibility enables type IV collagen molecules to assemble into a chicken-wire like network [33]. Furthermore, these interruptions also serve as sites for cell binding and interchain cross-linking [149].

8.3.1.1.4 COLLAGEN VI (BEADED FILAMENT FORMING, NETWORK FORMING)

Type VI collagen was initially isolated from the intimal layer of the human aorta [61]. It is now understood to be distributed throughout all layers of large arteries [25,189] as well as through many other connective tissues. It can be found assembled in various supramolecular forms including beaded microfibrils and hexagonal networks. While its precise function is still not understood, it is believed to play a role in cell migration, differentiation and apoptosis/proliferation [33]. For example, in cell culture, collagen VI microfibrils influence adhesion and mobility of human VSMCs [151]. It is of particular importance due to its assumed role in anchoring cells, including platelets and VSMCs, to other cells and neighboring ECM [150]. It is found in the subendothelium where it influences platelet activation and thrombogenicity [24,330]. Platelet adhesion and aggregation onto collagen VI is highly sensitive to shear rate, being most reactive at low shear rates (e.g., on the order of 100 s^{-1}) [244,249].

8.3.1.1.5 COLLAGEN VIII (NETWORK FORMING)

Type VIII collagen is a short-chain non-fibrillar network forming collagen. It was first identified in cell culture ECs [251]; however, it is now known to also be expressed by contractile VSMCs of healthy artery as well as by synthetic intimal VSMCs in atherosclerotic lesions [179,222]. Type VIII collagen influences a broad range of vascular processes in both health and disease. It has demonstrated particular importance in mechanical stability of the vascular wall, vascular repair and atherogenesis. A comprehensive review of its role in atherogenesis can be found in [222].

Collagen VIII is believed to serve as a bridge between ECM components and thereby influence the mechanical properties and mechanobiology of the vascular wall. For example, in the intima of healthy arteries, it is found in association with the endothelial basement membrane and the microfibrils of the IEL [155]; in the medial layer, it is found in association with the basal laminae of VSMCs and elastin fibers [222]; in the adventitia, it is found to link elastin fibers.

Platelet adhesion is diminished on substrates of type VIII collagen compared with I and III, suggesting a possible anti-thrombogenic role for endothelial and subendothelial type VIII collagen [222]. Its expression in the adventitia is an early marker of vessel injury and it is closely associated with cell proliferation and migration following vessel injury by angioplasty [28,274,277].

8.3.1.1.6 COLLAGENS XII (FACIT)

Type XII collagen is a member of the fibril-associated collagens with interrupted triple helices. They are so named because their triple helix structure is interrupted by one or two non-helical domains. Similar to type IV collagen, the interruptions add to the flexibility of the molecule. However, rather than forming fibrils themselves, FACIT collagens bind in regular intervals to the surface of fibril-forming collagens (see Fig. 8.7); e.g., FACIT collagen XII is found attached to the surface of collagen I fibrils. They alter the surface properties and assembly of fibrils [11,33] and are believed to influence fibril-fibril interactions as well as the interaction of fibrils with other macromolecules [11,115]. Fibroblasts produce collagen XII and this production is known to increase when the fibroblast substrate is stretched [293]. The production rate can also be altered by disease (e.g., it is increased in abdominal aortic aneurysm tissue compared with control arteries [69]); however, the significance of this upregulation is not known. In fact, little is known about the role of collagen type XII in the arterial wall.

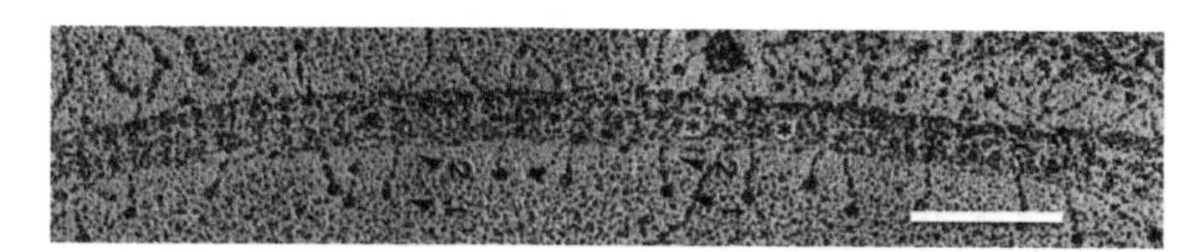

FIGURE 8.7 Rotary-shadowed image of type II collagen containing fibril with FACIT type IX collagen molecules seen attached to the fibril surface. Bar = 100 nm. © 1988. Rockefeller University Press. (Originally published in the Journal of Cell Biology 106:991–997 [300].)

8.3.2 Elastin in the Arterial Wall

Elastin molecules are the dominant protein within the elastin fibers of the arterial wall. These fibers (approximately 0.2–1.5 microns in diameter [295]) can fuse to form lamellae (e.g., IEL, EEL and medial lamellae), Figs. 8.2 and 8.8, Section 8.5. While the term elastin is often used to denote the entire structural element such as the fiber or lamella, strictly speaking, elastin is simply the molecule within the fiber. These structures are also sometimes referred to as *elastic* fibers or *elastic* lamellae. We prefer the former terms since both elastin fibers and collagen fibers are often idealized as mechanically elastic.

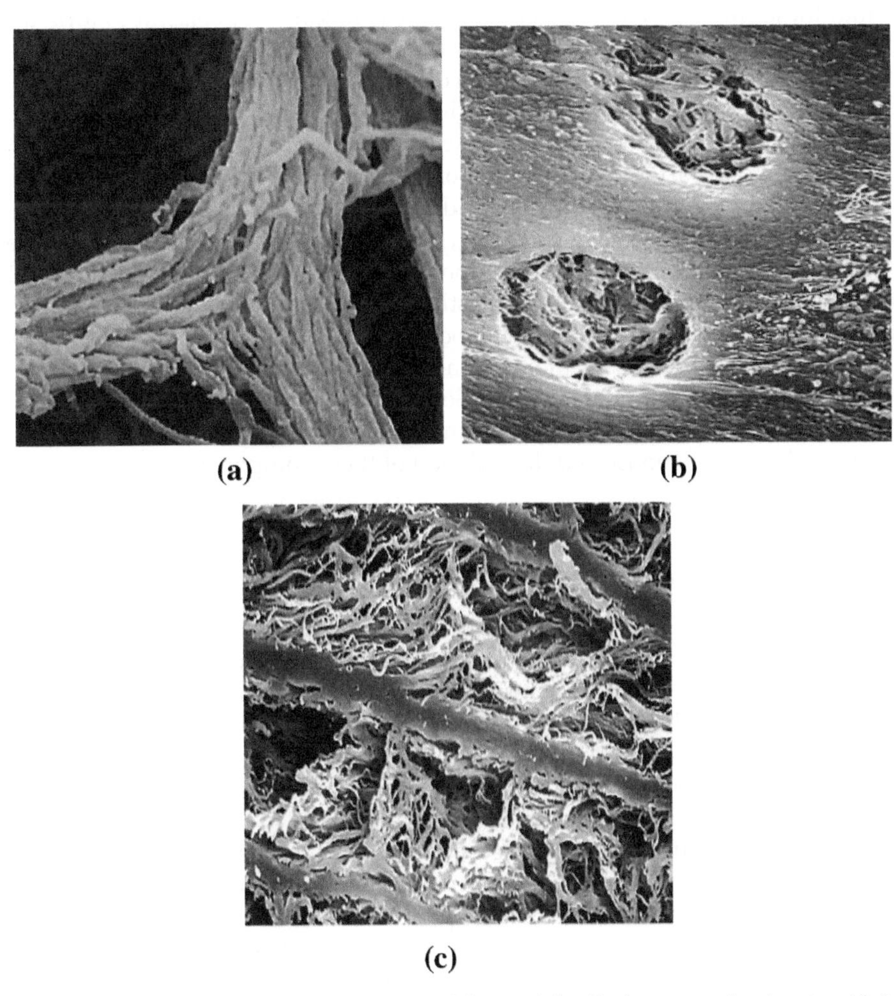

FIGURE 8.8 SEM images of the elastic components of the arterial wall of rat aorta after formic acid digestion to remove non-elastin components. (a) Elastin fiber in the adventitia with visible bundles of fibrils (0.1–0.2 μm), (2500×). (b) Internal elastic lamina (IEL) seen as smooth elastin sheet formed of fibrils with two fenestrae. Elastic fibrils can also be seen through the fenestrae, where they form an underlying meshwork or 'wire fence' like structure (3000×). (c) Layered laminae and interlaminar elastic fibers in transverse sections of the medial layer (1500×). (Images reproduced from [295] with permission from Dr. T. Ushiki.)

Elastin is a chemically inert and extremely hydrophobic protein that is encoded by a single gene. In contrast to the highly structured collagen molecule, the elastin molecule is formed of relatively loose, unstructured polypeptide chains. While it is generally agreed that the elastin molecule is responsible for storing the majority of the elastic energy, providing high flexibility of the arterial wall at low loads as well as the wall's resilience [192], controversy remains about the physical mechanism responsible for the high compliance [11]. Elastin fiber assembly is a complex process involving production of a monomeric form of elastin (tropoelastin) that is secreted into the extracellular space onto 10–15 nm microfibril scaffolds, composed largely of fibrillins. While many aspects of elastin fiber assembly are still not understood, it is believed that these microfibrils are essential for cross-linking tropoelastin to form an amorphous elastin polymer [128,307]. We do not elaborate further on the elastin fiber assembly process and refer interested readers to references such as [192,307].

The majority of arterial elastin is laid down during the perinatal period and the arterial wall has limited capacity for forming or repairing elastin components after puberty [67,307]. Therefore, damage to the arterial elastin components from causes such as supraphysiological loading during angioplasty or fatigue damage during aging cannot be properly repaired. This is in sharp contrast to arterial collagen fibers, which are regularly replaced (turned over). The absence of elastin turnover could have an important role in stabilizing the wall morphology, in that the unloaded configuration of the elastin components remains as a reference for the artery during the process of growth and remodeling [297]. Elastin is also believed to be the primary component responsible for the residual stresses and axial prestretches in the arterial wall (see, e.g., [297] and cited references). In Section 8.5, we return to a discussion of elastin fibers and their distribution in the artery wall and in Section 8.6 to constitutive models used to describe the mechanical response and degradation of these components.

8.4 VASCULAR CELLS

The artery is not simply an inert elastic tube. It is a living structure whose functionality is continually maintained by vascular cells, i.e., endothelial cells, vascular smooth muscle cells and adventitial fibroblasts. The proper formation, degradation and repair of collagen within the ECM is at the heart of the ability of our vascular system to maintain its integrity over time (via collagen turnover), adapt to changing mechanical and chemical stimuli (via growth and remodeling) and recover from external damage. Moreover, the morphology, structure and functionality of vascular cells that maintain it are intimately linked to their local extracellular environment. Environmental cues that are sensed by these cells can be classified into biochemical or biomechanical cues. Biochemical cues are numerous and include pH, oxygenation, growth factors, cytokines, chemokines, hormones and lipoproteins [3]. Biomechanical stimuli arise due to the blood flow within the artery which gives rise to cyclic deformation of the arterial wall, frictional drag forces on the inner wall (shear stress), transmural pressure and interstitial fluid forces due to the movement of fluid through the ECM. Mechanosensors on the cells convert the mechanical stimuli into chemical signals; this is referred to as *mechanotransduction*. Activation of secondary messengers, i.e., the molecules that transduce the signals from the mechanoreceptors to the nucleus, follows. This leads to an increase in the activity of transcription factors which can bind to the DNA to activate genes that regulate cell functionality, e.g., cell proliferation, apoptosis, differentiation, morphology, alignment, migration and synthesis [142].

Malfunctions in the formation of structural proteins, e.g., collagen molecules arising from genetic mutations, can result in vastly altered mechanical properties of the ECM and even lead to failure of the arterial wall. Yet, fundamental questions remain unanswered in regard to these processes. For example, as elaborated in Section 8.4.4, the precise process by which the three-dimensional collagen fiber architecture is created, such as that shown in Fig. 8.5a–c, and the role of the vascular cells in this process is still a subject of investigation. We now discuss the mechanobiology of ECs, VSMCs and adventitial fibroblasts in more detail.

8.4.1 Endothelial Cells

> Endothelial cells derived from human umbilical veins were first successfully cultured in vitro in 1973. It is these landmark studies which helped initiate the growth of modern vascular biology [209].

A monolayer of endothelial cells, supported by the underlying basement membrane (see Section 8.5.1), lines the entire vascular system. This layer of cells, the endothelium, forms the cellular interface between circulating blood and underlying tissue and, in this location, the ECs are able to play a vital role in controlling transport into the wall and responding to changes in flow within the vessel. ECs are highly active and participate in many physiological functions such as control of vasomotor tone through the release of vasodilators and vasoconstrictors that regulate contractility of VSMCs; hemostatic balance, regulated transfer of water, nutrients and leukocytes across the vascular wall and angiogenesis [2,232]. Dimensions of ECs vary with the particular vascular bed; however, when they are exposed to unidirectional shear flow, they are typically long, narrow and flat with dimensions on the order of $50\ \mu m \times 10\ \mu m \times 1\ \mu m$ (see [4]). In the vasculature of an adult, there are approximately 10^{13} ECs; if lined end-to-end, it has been calculated that they would wrap more than four times around the circumference of the earth [2].

As blood flows through the arteries, it applies a pressure and a viscous drag force to the endothelium. While the viscous drag force per unit area of artery (the wall shear stress vector) is of fundamental interest, typically only scalar functions of this vector field, such as the magnitude (WSS), are considered. For example, deviations of the time-averaged magnitude of WSS from physiological levels have been of great interest in studies of atherosclerosis and cerebral aneurysms (e.g., [182,263]). However, it is well recognized that ECs are influenced by more than just the time-averaged WSS; they are also influenced by the temporal variation of both the direction and magnitude of the wall shear stress vector. For example, flows with large ratios of peak to mean WSS are associated with proatherogenic patterns of gene expression [54,55]. Furthermore, ECs are also able to sense spatial gradients in the wall shear stress (*WSSGs*) [166]. High values of spatial WSSGs typically occur in regions of flow detachment and at stagnation points. Interestingly, recent *in vivo* animal investigations suggest that a combination of high WSS and a high positive spatial WSSG induces protease production by ECs and VSMCs which leads to the type of destructive remodeling that is observed in cerebral aneurysms [193]. We return to this subject in Sections 8.6 and 8.7.

Both the intramural stress state of the arterial wall (e.g., arising from cyclic stretch) as well as the wall shear stress vector influence the morphology of the ECs. In relatively straight arterial segments, where the flow is nearly one-dimensional, the ECs are aligned and elongated in the direction of flow and perpendicular to the direction of cyclic stretch [310]. However, in locations of the arterial tree where the geometry is more complex, e.g., bifurcations, the flow can oscillate in direction during the cardiac cycle and the cyclic deformation of ECs can be biaxial; in these locations the ECs have a polygonal morphology [310]. It is observed that elongated ECs have well-organized, parallel actin stress fibers whereas the polygonal ECs have short, randomly oriented actin filaments, mainly localized at the cell periphery [58]. It is conjectured that remodeling of the EC cytoskeletal structure acts to minimize alterations in EC intracellular stress/strain during the cardiac cycle [58].

The question of how the mechanical stimuli are sensed by the ECs remains an area of active research [203]. Several proteins and cell structures have been proposed as mechanosensors. These include membranal structures (ion channels, tyrosine kinase receptors, caveolae, G proteins, primary cilia and the endothelial cytoskeleton), cell-matrix molecules (integrins) and cell-cell junction molecules [203,235]. The endothelial glycocalyx (see Section 8.5.1) lining the lumenal surface of the endothelium also plays a role in mechanotransduction [311]. Rather than these mechanosensors acting independently, it is thought that the cytoskeleton of the EC communicates the forces that arise due to deformation of the endothelium (by shear stress) to mutliple sites of mechanotransduction (see Fig. 8.9); this is referred to as a model of decentralized mechanotransduction [66].

The endothelium regulates the transfer of water, nutrients, leukocytes and other materials across the vascular wall. This regulation of permeability is important for vascular homeostasis and wound healing [63]. The edges of endothelial cells in general overlap and the degree of adhesion between these surfaces (or intercellular junctions) is a central factor in determining permeability of the endothelium. Large macromolecules such as low density lipoproteins are generally unable to pass through the endothelium of straight arterial segments where the intercellular junctions are intact (see [58] and references therein). However, in regions of disturbed flow, increased permeability to macromolecules occurs, a consequence of discontinuous intercellular junctional distributions that can arise from factors such as accelerated EC turnover rate and disrupted junctional proteins [58].

Local hemodynamic conditions can determine whether ECs are in a 'quiescent' or 'activated' state. Quiescent ECs are of vasodilatory phenotype and inhibit leukocyte adhesion, platelet aggregation and exhibit anti-inflammatory, anti-coagulant, anti-adhesive, anti-proliferative and anti-oxidant characteristics [109]. Aird [6] summarizes that 'the activated phenotype generally consists of some combination of increased cell adhesiveness, shift in hemostatic balance to the procoagulant side, secretion of inflammatory mediators and change in cell survival/proliferation'.

Computational fluid dynamic analysis of the vasculature illustrates that shear stress is non-uniform, i.e., spatially heterogeneous, within the vasculature tree. Hence, it is perhaps not so surprising that EC phenotypes display heterogeneity in both structure and function in the vascular tree [6,66]. Several recent articles by *Aird* (see [3,4,5,6,232]) suggest that this heterogeneity is the result of both *nature* and *nurture*, i.e., a consequence of the environment and epigenetics. It represents a balance between stability and plasticity in gene expression and phenotype [3] and enables the ECs to adapt and function in different environments within the vasculature [232].

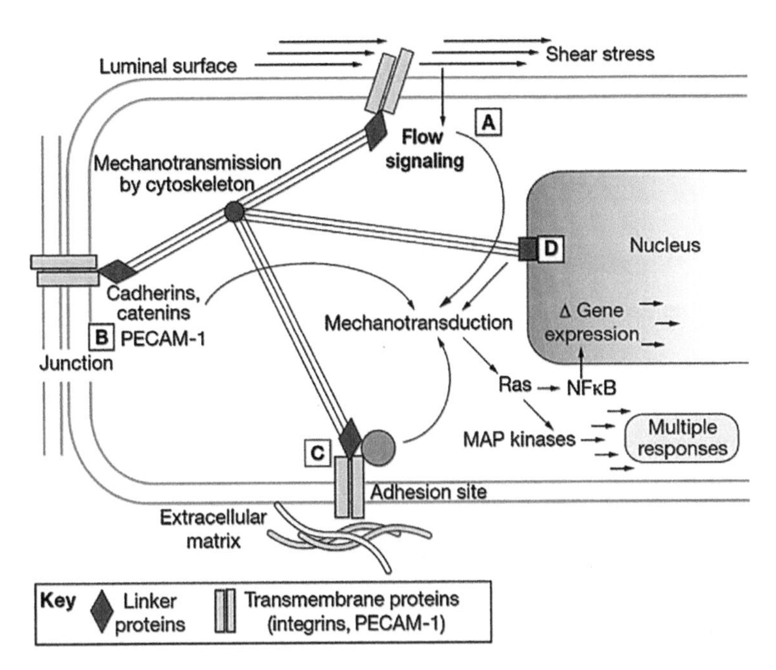

FIGURE 8.9 The decentralized model of endothelial mechanotransduction by shear stress. The cytoskeleton has a central role in the transmission of tension changes throughout the cell. (a) Direct signaling can occur through deformation of the luminal surface, possibly via the glycocalyx. (b) Mechanotransduction is also mediated via junctional signaling, that is, the transmission of forces to intercellular junction protein complexes via the cortical and/or filamentous cytoskeleton. (c) Cytoskeletal forces are also transmitted to adhesion sites. Transmembrane integrins bound to the extracellular matrix serve as a focus for deformation. (d) Nuclear deformation is also likely to result in mechanically induced signaling, possibly via laminins in the nuclear membrane. (Reprinted from [66] with permission from Nature Publishing Group.)

On this note, it is often stated that high WSS is atheroprotective whereas low WSS is associated with atherogenesis. However, given the plasticity in gene expression and phenotype, it is perhaps preferable to think in terms of increases or decreases in mechanical stimuli from homeostatic levels, rather than absolute magnitudes of shear stress. Futhermore, given the plasticity of ECs, we believe it may be beneficial to keep in mind that homeostatic values could potentially be both spatially heterogeneous and temporally adaptive.

8.4.2 Vascular Smooth Muscle Cells

In their contractile state, VSMCs are spindle-shaped cells, approximately 100 μm in length and 5 μm in width. The majority of these cells are located in the medial layer where they are approximately circumferentially aligned and layered (see Fig. 8.10). In vascular homeostasis, they are partially contracted, i.e., they have a *basal tone*. This enables them to relax or contract to respond to changes in their environment to regulate the diameter of the blood vessel. In fact, their principal function is regulation of blood vessel diameter, blood pressure and blood flow distribution [13]. Their contractile tone allows the redistribution

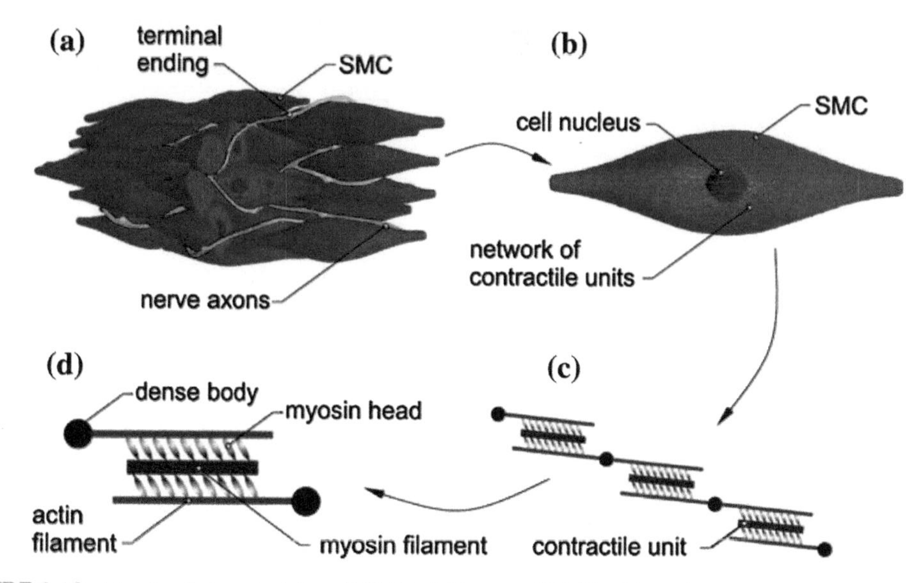

FIGURE 8.10 Structural characteristics of VSMCs: (a) layers of VSMCs, (b) single VSMC with network of contractile units around the centrally positioned nucleus, (c) contractile units connected in series via dense bodies, and (d) isolated, basic contractile unit mainly composed of an actin and myosin filament connected by cross bridges. (Reprinted from [258] with permission from Elsevier.)

of local flow in relation to organ-specific metabolic demand [164] and they control local blood flow at the arteriolar level [188]. They are cyclically stretched by 10% with a 25–50% mean strain in the healthy arterial wall [188]. Peak force development typically occurs at 90% of the distended passive diameter at 100 mmHg [8]. Their contractile state can be activated by electrical, pharmacological and mechanical stimuli [117]. Contractile VSMCs in blood vessels of adult animals proliferate at extremely low rates and exhibit low synthetic activity.

A VSMC can display a range of phenotypes from the quiescent (contractile) state just discussed to a proliferative (synthetic) phenotype, capable of synthesizing large quantities of ECM. The synthetic VSMCs are less elongated and show a more cobblestone shape compared with the contractile cells (see, e.g., [13,234]). Generally, synthetic VSMCs proliferate and migrate at much higher rates than the contractile cells. The VSMCs of even adult arteries have a remarkable capacity for switching between phenotypes, and can therefore induce rapid changes in vessel caliber when in their contractile state and switch over to a synthetic state when vessel remodeling or repair is needed [234]. Interestingly, it is this plasticity that predisposes the VSMC to adverse phenotypic switching and contribution to development and/or progression of vascular diseases such as atherosclerosis [164].

As for ECs, the location of a VSMC along the phenotype spectrum is determined by an interplay between epigenetic and environmental factors. Two such environmental factors are the structure and content of the ECM [234]. For example, the contractile phenotype is activated by the presence of the fibrillar form of collagen I, while VSMC proliferation is promoted by the presence of the monomeric form of collagen type I. The ECM architecture can also influence the VSMC phenotype: cells located in 3-D collagen matrix are less proliferative compared with those in a 2-D matrix [234].

8.4.3 Fibroblasts

Arterial fibroblasts are largely localized in the adventitial layer where they regulate the ECM content through both the production and degradation of its components [195]. For example, fibroblasts have the capacity to synthesize approximately 3.5 million procollagen molecules (a precursor for the collagen molecule) per day. However, only a small fraction of the procollagen molecules are released from the cell, rather between 10% and 90% are degraded intracellularly prior to secretion. The availability of procollagen within the cell provides a mechanism for the arterial wall to rapidly adapt to changing structural needs [190]. Fibroblasts can also secrete a variety of other ECM components including proteoglycans, fibronectin, tenascin, laminin and fibronectin. They control the balance between degradation and creation of the ECM through the production of matrix metalloproteinases (MMP) and their inhibitors, tissue inhibitors of metalloproteinase (TIMP) and thereby have an important role in both normal maintenance and in disease. It is now understood that fibroblasts also play important roles in wound healing and are involved in the initiation, modulation and maintenance of the inflammatory response [74].

Fibroblasts display heterogeneous phenotypes ranging from limiting states denoted as *inactive* and *active* fibroblasts. The active phenotype is associated with proliferation, differentiation and upregulation of ECM proteins as well as release of factors that influence vascular function and structure [336]. Under pathological conditions, fibroblasts can undergo phenotypic changes between these states. For instance, during inflammation, fibroblasts are activated and differentiate toward a migratory and contractile myofibroblast phenotype, Fig.8.11. During vessel injury, fibroblasts differentiate into myofibroblasts and migrate into the wound bed, proliferate and synthesize a new collagen-rich matrix. While in most wounds, myofibroblasts do not persist, when they do remain, excessive ECM deposition is observed, leading to altered tissue structure and pathological wound healing [195].

In culture, fibroblasts do not adhere to other cells or form sheets of cells. However, when they are grown on 3-D substrate, they can transmit forces to other cells by applying tension to the substrate. The internal skeleton in fibroblasts is minimal and so adhesions to these substrates are a major factor in determining their shape and function. Fibroblasts securely attach to the adventitial ECM using matrix adhesion contacts on their cell surface, in particular integrins [309]. These adhesions are mechanosensitive [110]. They serve both as strain gauges [56,253] and as sensors that probe the mechanical properties of local ECM [30,59]. They enable fibroblasts to transduce mechanical into chemical information, and they integrate these signals with growth factor derived stimuli to achieve specific changes in gene expression [57] and synthesis of ECM molecules.

8.4.4 Matrix Assembly by Vascular Cells

There appear to be two central theories for the collagen matrix assembly [31,35]. The first theory involves the fibropositor model in which cells create the ECM architecture by directing

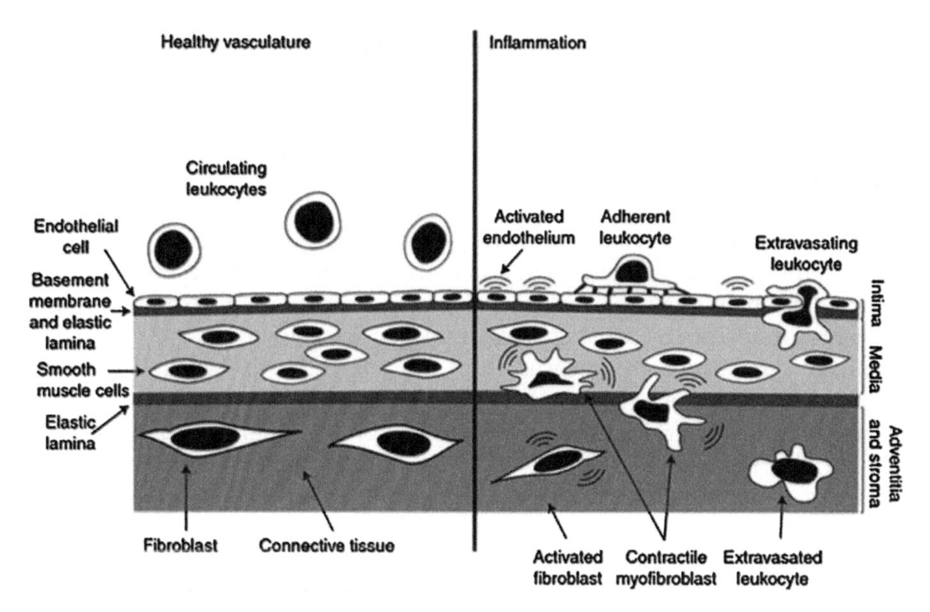

FIGURE 8.11 Fibroblast activation in vascular inflammation. Schematic illustration of the healthy and inflamed vessel wall, and the influence of intramural fibroblasts on endothelial cells and leukocytes. During inflammation, fibroblasts become activated, secrete chemokines and cytokines and differentiate toward a migratory and contractile myofibroblast phenotype. (Reprinted from [74] with permission from John Wiley & Sons.)

the production and placement of the collagen fibrils during early embryonic development when collagen concentration is low [33,45,145]. Fibril assembly starts within the cell (e.g., fibroblast) where three pro-α chains (α chains with propeptide end attachments) are self-assembled into procollagen molecules composed of a single triple-stranded helix domain with propeptides at both ends. Once procollagen is secreted from the cell, the propeptides are removed thereby forming collagen molecules that self-assemble within infoldings in the plasma membrane (see Fig. 8.12).

First, relatively short protofibrils (diameter 20 nm and length of 4–12 nm) with tapered ends are assembled at the cell surface (nucleation). The length of the mature fibril is attained through fusion of numerous protofibrils that are placed end-to-end with overlapping tapered regions. The diameter is believed to be increased by lateral merging of the protofibrils [33]. In a study of embryonic tendon fibroblasts, *Canty et al.* found, that during a brief period of embryonic development, protofibrils were found to be assembled within the cell, inside projected extrusions of the cell membrane that are parallel to the tendon axis [45]. The tip of these extrusions, called 'fibropositors', is the site of protofibril deposition into narrow channels between cells [45]. It is conjectured that fibril alignment is established during this brief period after which fibrils are assembled extracellularly as just described. However, fibropositors have not been identified after embryonic development and hence cannot be used to explain changes in collagen during remodeling, later in life. *In vitro*, collagen fiber alignment and position have been shown to be altered through the mechanical action of cells physically pulling on fibers [196], see Fig. 8.13.

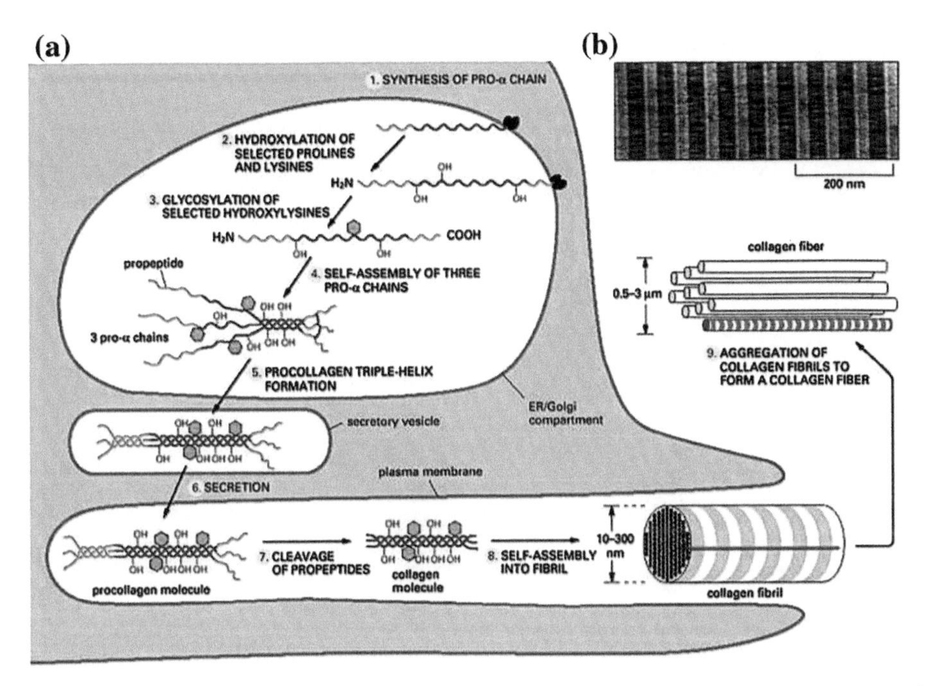

FIGURE 8.12 Schematic of collagen fiber formation. (a) Collagen fibril formation starts within the cell and is completed in an infolding in the plasma membrane. Collagen fibrils are then bundled to form collagen fibers which are visible under light microscope. (b) Electron micrograph of a collagen fibril that has been stained to show the regularly spaced striations. (Reprinted from [11] with permission from Garland Science.)

In the alternative theory, the fiber alignment is the outcome of the differential response of strained and unstrained collagen to enzymatic attack [31,83,106]. The collagen matrix forms as a result of fibril condensation from a 'liquid crystal like' solution of collagen monomers [31,97]. The matrix is then enzymatically sculpted into a load adapted structure. In this theory, the role of the cell in remodeling is to produce the collagen monomer and control vulnerability to enzymatic attack by modifying the force on the fibers, rather than directing the deposition and orientation of collagen fibrils. Fiber growth is conjectured to result from preferential uptake of collagen monomers by strained fibrils (strain-induced polymerization) [31].

The vital role of collagen in providing proper functioning of the arterial wall is glaringly apparent in diseases for which collagen manufacture is flawed. To illustrate this importance, we discuss just a few of these diseases. As already noted, the glycine amino acid in the basement membrane collagens is naturally replaced at some locations by another amino acid, interrupting the superhelix structure and lending flexibility to the collagen molecule. However, this replacement can also arise from mutations in collagen genes, leading to undesirable flexibility in other collagen molecules and causing systemic connective tissue diseases [33]. Such a mutation in the COL3A1 gene is established for vascular (type IV) Ehlers-Danlos syndrome (EDS), leading to errors in type III collagen synthesis [183]. Vascular fragility of both small and large blood vessels are outcomes of the structural flaws in type III collagen. In

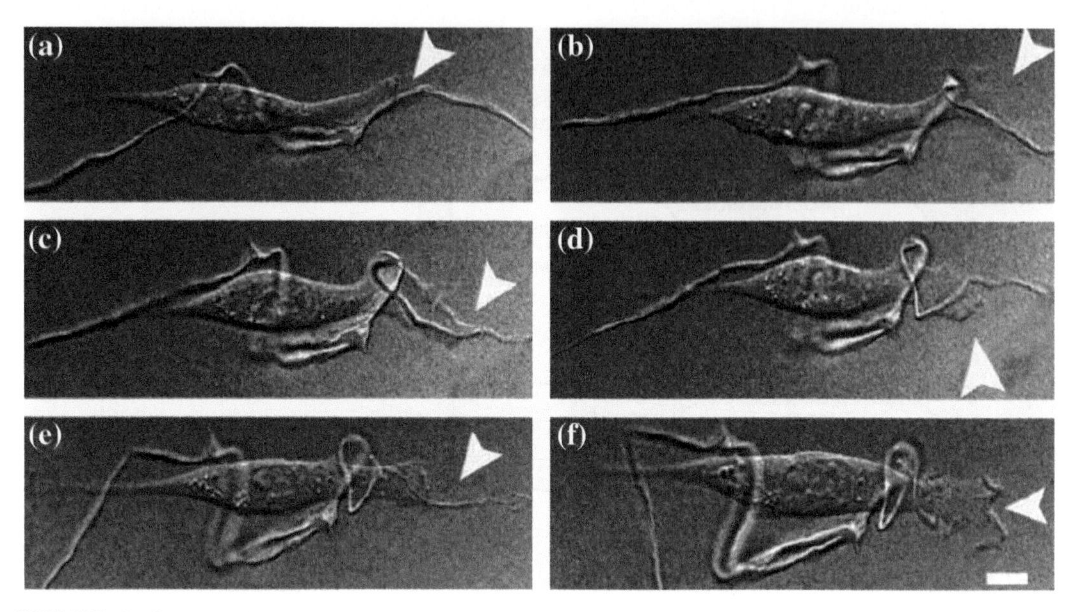

FIGURE 8.13 Fibroblast moving an individual collagen fiber in 'hand-over-hand' process of cell extension, retraction and release. Shown in (a)–(f) are time lapse images of the motion of labeled type I collagen fibers on the upper surface of fibroblasts that had been cultured on a coverslip. The cell orients itself parallel to the fiber and then extends the front of the lamellipodia (white arrow) along the fiber during a 20–40 s period (a). The subsequent contraction of the cell moves the fiber relative to the cell (during a period of about 85 s) (b). The extension and contraction process repeats as shown in (c)–(f) until the end of the fiber is reached. The displacement during a single retraction was approximately 3.5 μm. Scale bar is 5 μm. (Reprinted from [196] with permission from Nature Publishing Group.)

addition to causing excessive bruising and bleeding, these defects can be life threatening when they result in arterial rupture.

Six distinct genes encode for the α chains of type IV collagen. Mutations in the COL4A1 gene have been found to inhibit the normal assembly and secretion of type IV collagen that in turn alters the structure of the basement membrane supporting the endothelial cells and surrounding the VSMC [101,167]. The range of vascular problems associated with COL4A1 mutations underscores the influential role of collagen IV on vascular structure and function. For example, a COL4A1 mutation affecting a lysine residue at the Y position of collagen IV has been shown to cause defects in the basement membrane and focal detachment of the endothelium that are associated with altered vascular tone, endothelial cell function and blood pressure regulation [299]. Mice with a mutation affecting the Gly position of the Gly-X-Y repeat unit of type IV collagen display fragile brain arterioles and increased tortuosity of retinal vessels [101]. In humans, mutations of the COL4A1 gene are associated with a spectrum of diseases of small cerebral vessels leading, for example, to aneurysms of the carotid siphon [305].

8.4.4.1 Attachment of Collagen Fibers to the Extracellular Matrix

Maintenance of the ECM requires a continual turnover of the collagen fibers within the arterial wall. For example, in [215] the collagen half-life of the aorta and mesenteric arteries of a rat was found to be 60–70 days in normotensive animals and reduced to 17 days in hypertensive conditions. Vascular cells work on the collagen and configure the fibers in a state of stretch [12]. For the purposes of mathematical models, it is often hypothesized that the stretch that the fibers are configured to the matrix is invariant of the current configuration of the artery and constant in time; this approach has served the basis for many subsequent models of arterial growth and remodeling. Exact definitions can differ between authors; however, the stretch that the fibers are attached to the extracellular matrix is commonly referred to as the fiber *deposition* stretch [99], *attachment* stretch [315] or *pre*-stretch [159]. We return to this subject in Section 8.6.7.

8.5 ARCHITECTURE OF THE ARTERIAL WALL

The overall mechanical properties and biological functioning of the arterial wall are determined by the collective contributions of the wall components. The combined response is in turn influenced by the supramolecular structure and spatial distribution of these components, their volumetric ratios as well as their physical and chemical interactions. The importance of these morphological features, the *wall architecture*, is particularly evident when comparing the structure/function relationship of vessels close to the heart *elastic arteries* with those located more peripherally, the *muscular arteries* (e.g., [325]).

As discussed in Section 8.2.2.1, there are energetic advantages to diminishing the amplitude of the pulsatile waveform close to the heart. This reduction in pulsatility is achieved by the highly elastic character of the elastic arteries such as the aorta, main pulmonary artery, common carotids and common iliacs, that enables them to store blood during systole and release it during diastole. This distension is largest in the pulmonary artery, where it reaches nearly 8% of the mean radius [200]. The effectiveness of the capacitance role in elastic arteries is evidenced by the drop in ratio of peak to mean flow from approximately six in the aortic arch to less than two in the femoral arteries [199]. Downstream of the elastic arteries, the functional role of the blood vessels gradually shifts to controlling pressure in the muscular arteries and eventually to shunting flow to areas of metabolic need in the smallest arteries and arterioles. Muscular arteries include vessels such as the coronary, cerebral, femoral and brachial arteries. While the structure of elastic arteries gradually transitions to that of the muscular arteries, the prototypical structures of these vessels are quite different. In this section, we discuss the detailed architecture of the arterial wall with emphasis on these distinctions.

8.5.1 Tunica Intima

8.5.1.1 Glycocalyx

A polymeric network called the endothelial glycocalyx layer (EGL) or simply glycocalyx coats the luminal side of the endothelial monolayer, Fig. 8.14. The EGL is composed of proteoglycans, glycosaminoglycans (GAGs), glycoproteins as well as adhering plasma proteins. It is found in both the macro and micro vasculature and plays an important role in regulation

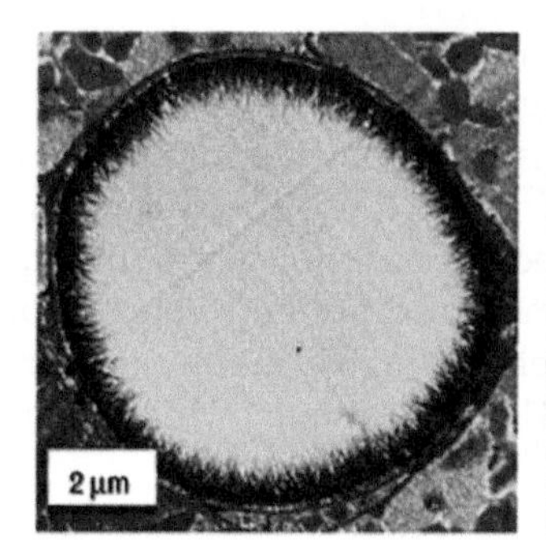

FIGURE 8.14 Electron microscopy images of the endothelial glycocalyx in a coronary capillary. (Reproduced from [214] with permission from Wolters Kluwer Health.)

of vascular permeability for macromolecules, modulation of leukocyte and platelet adhesion as well as transduction of shear forces to the endothelial intracellular cytoskeleton [224,291,321]. As early as the 1980s an association between atherosclerosis and EGL thickness had been identified [171] and it is now known to be associated with other diseases including diabetes and ischemia. The glycocalyx thickness is a consequence of the balance between renewal and degradation that can be effected by biochemical as well as mechanical influences such as shear stress. For example, exposure to a high fat, high cholesterol diet has been shown to reduce the EGL thickness in murine carotid artery bifurcations [29]. The EGL is now being explored as a potential diagnostic and therapeutic target in cardiovascular diseases. While there remain technological challenges for *in vivo* assessment of the EGL, recent advances in imaging technology have been employed to obtain promising results [337]. For example, using two-photon laser scanning microscopy, *Reitsma et al.* were able to assess EGL thickness in mounted intact, viable carotid arteries of mice [233].

8.5.1.2 Basement Membrane

The intimal monolayer of ECs is attached to a supporting basement membrane, a composite structure consisting of a thin basal lamina as well as a layer of underlying anchoring fibers [11,202], Fig. 8.15.[2] Type IV collagen and laminin are central components of the basal lamina, which also contains collagen type VIII, the proteoglycans entactin (nidogen) and sulfated proteoglycans [202]. Type IV collagen and laminin individually self-assemble into suprastructural networks, providing structure to the basement membrane [170]. Collagen types XV and XVIII are also found in association with the basement membrane [170,281,305]. Type XV is believed to be involved in anchoring cells to the membrane, while type XVIII is an inhibitor of angiogenesis and EC migration [148].

[2] Frequently, the terms *basement membrane* and *basal lamina* are used interchangeably, though strictly speaking the basal lamina is a component of the basement membrane [202]. *Kefalides et al.* defined the basement membrane as the composite of the basal laminae and two distinct layers on either side of the basal lamina: the *lamina lucinda* lying between the cells and the *basal lamina* and the *reticular lamina* composed of the anchoring fibrils [147]. The lamina lucinda is now believed to be a tissue processing artifact [202].

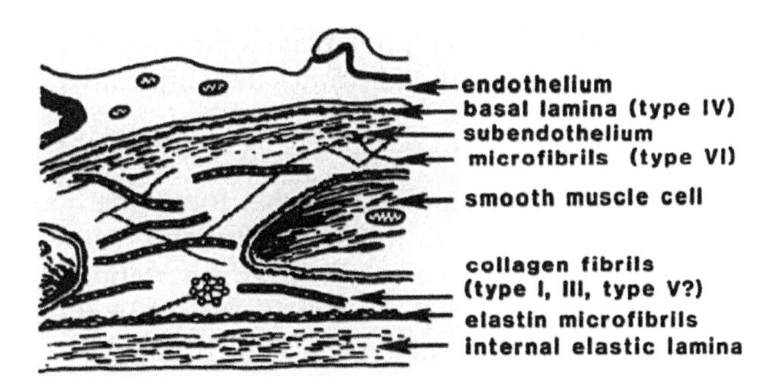

FIGURE 8.15 Schematic of central components and collagen distribution within the intimal layer, reproduced from [189] with permission from Wolters Kluwer Health. Collagen types VI, XV, and XVIII have also been found associated with the basement membrane and are not shown here.

In addition to providing structural support, the basement membrane serves as a reservoir for enzymes, growth factors and other cytokines [170,202,305,338]. In this role, the basement membrane influences cellular events such as proliferation, adhesion, migration and differentiation. The basement membrane also serves as a semipermeable selective barrier and substrate for cell adhesion and migration during vascular wall wound healing [305].

8.5.1.3 *Subendothelium*

The intima is similar in healthy elastic and muscular arteries with the exception that in some elastic arteries (e.g., aorta and coronary vessels) a layer of diffusely distributed, thin fibrillar components of collagen types I, III, V and isolated VSMCs can be found between the basement membrane and the IEL [244,268]. The term subendothelium is sometimes used to define this region, though the definition varies and is often not precisely stated [283]. Type VI collagen, which generally plays a role in cell adhesion, is also found in this area [24,330] as is collagen VIII, which is seen as a bridge between the endothelial basement membrane and the microfibrils of the IEL [155].

When intimal collagens are exposed to blood products due to vessel injury, they have a great influence on the thromobogenicity of the arterial wall. For example, collagens I, III, VI can recruit and then activate circulating platelets, leading to platelet plug formation and occlusion at the site of vessel damage [77,217]. In contrast, collagen types V and VIII are believed to play anti-thrombogenic roles [222,249].

The thickness of the region between the basal lamina and IEL increases with age, demonstrating an increase in both VSMCs and collagen fibers [268,281]. In particular, the ratio of collagen type I to III in the intima increases with maturation [281]. This region may be quite small or nonexistent in healthy muscular arteries [268]. Previous biomechanical models of the arterial wall have largely neglected the contributions of the intima. More recently, it has been recognized that in the aged vessel wall, this layer can have a significant influence on the wall properties [125,322]. For example, in nonstenotic human left anterior descending coronary arteries (age 71.5 ± 7.3 years), *Holzapfel et al.* found the contribution of the intimal layer to the load bearing capacity and mechanical strength of the artery to be significant relative to that of the media and adventitia [125]. A recent analysis of collagen fiber orientation in the intima of human aorta and common iliac arteries using polarized light microscopy provides extensive data on fiber orientation in this layer and demonstrates its complexity [259].

8.5.1.4 Internal Elastic Lamina (IEL)

The internal elastic lamina is a fenestrated sheet that forms the boundary between the intimal and medial layers, influencing both its mechanical and mass transport properties. The size and number of these fenestrae vary in the arterial system and change with maturation [100,118]. The main function of the fenestrae (or pores) in the IEL, clearly seen in Figs. 8.2b,c and 8.8b, appears to be the enhancement of passage of water, nutrients and electrolytes across the wall. Using a two-dimensional model for macromolecular transport, *Tada and Tarbell* concluded transport of ATP but not low density lipoproteins is sensitive to IEL pores distribution [288]. Furthermore, three-dimensional models of transmural flow of water through these pores suggest shear stresses on VSMCs adjacent to the IEL could be large enough to influence cell proliferation and migration [286,287]. The density and area fraction of these pores have also indirectly been shown to have a modest influence on the mechanical properties of the IEL [42,43]. Note that the IEL is also conjectured to play a role in preventing direct contact between precursor VSMC and ECs [152]; however, physical contact between ECs protruding through these pores and establishing contact with VSMCs have been reported [238].

As will be discussed in Section 8.7, the loss of the IEL is a defining step in cerebral aneurysm initiation. IEL degeneration has been recreated in animal models of aneurysm initiation solely as the result of supra-physiological hemodynamic loading at arterial bifurcations [94,194]. This damage is conjectured to arise from a process of enzymatic degradation rather than mechanical overloading. Hemodynamic derived changes to the aneurysm wall after damage to the IEL (initiation) are the subject of intense experimental research [140,194] and biomechanical modeling [131,318,320], including the development of damage models that include enzymatic damage to the arterial wall [173].

The IEL in cerebral arteries of adults also suffers from mechanical damage in the form of cracks, Fig. 8.16. As early as the 1920s, Reuterwall reported ‘tears’ in human IELs. These breaks in the IEL have been termed ‘Reuterwall’s tears’ by *Hassler* [113]. Such tears, visible as 700–3000 micron long gaps in the IEL, were nearly always found oriented in the circumferential direction (Chapter 5 of [113]), e.g., Fig. 8.16a. They were most common in larger cerebral vessels such as the basilar and vertebral arteries and generally located away from bifurcations formed by two larger vessels. These tears were uncommon in individuals less than 30 years of age (Chapter 5 of [113]). Cracks have also been seen in experimental arteriovenous fistulas created between the common carotid artery and jugular vein [39]. In this latter case, no evidence of elastolytic activity was found, so the cause was hypothesized to be due to direct overstressing (acute rupture) or from fatigue-type wear. Histological examination of the IEL from common carotid arteries subjected to longitudinal [39] and circumferential uniaxial loading [118,243] has also shown damage to the IEL in the form of mechanically induced tears, Fig. 8.16b and c.

In vivo, as the IEL is progressively damaged and possibly fails, the mechanical loads will be transferred to the stiffer collagen fibers [104,241,326,333], leading to loss of the toe region in the vessel, Fig. 8.1. *Scott et al.* hypothesized overload of this kind was responsible for the loss of toe region after supraphysiological loading of cerebral vessels [260]. *Robertson et al.*

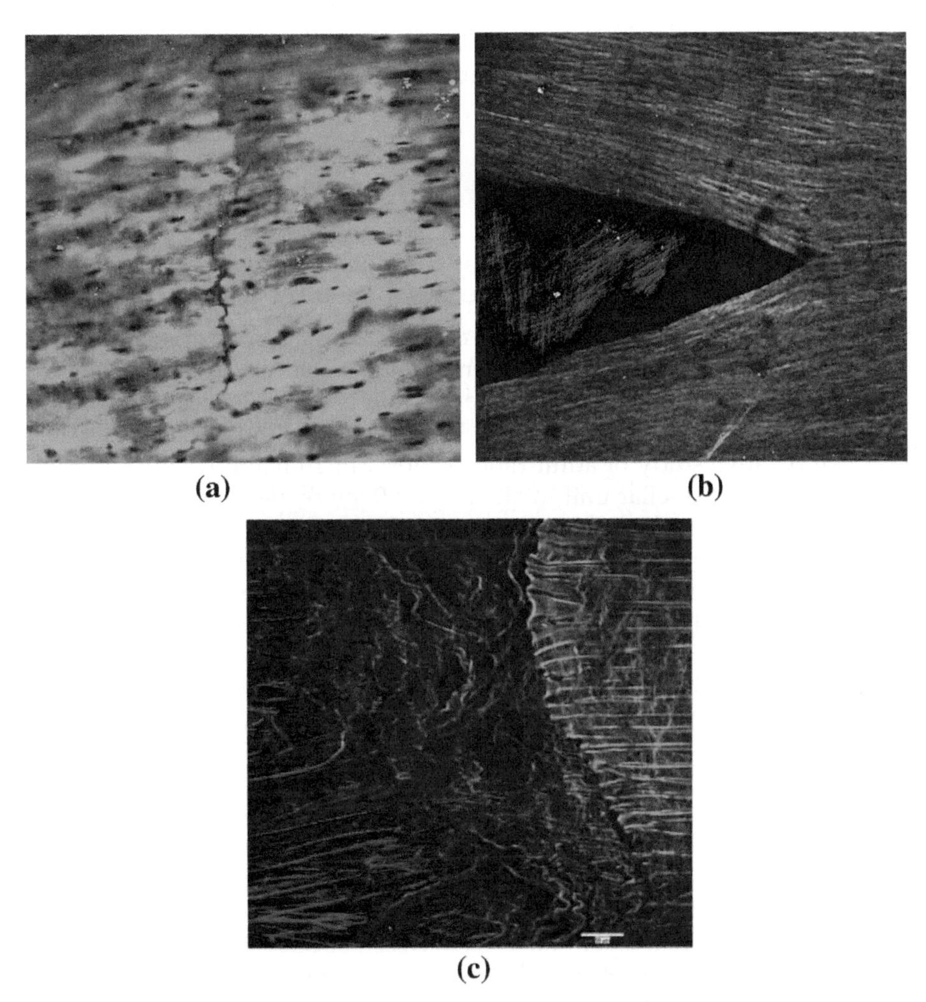

FIGURE 8.16 Cracks in the internal elastic lamina of human cerebral vessels. Multi-photon microscopy images from en face preparations of the human basilar artery displaying autofluorescent elastin, by utilizing two-photon emission (2PE) spectroscopy. The fenestrated internal elastic lamina can be seen. (a) Sample fixed at 30% strain, with a crack oriented in the direction of applied load (circumferential direction, vertical in image). (b) Sample loaded in uniaxial tension in circumferential direction (vertical in image) until macro scale crack propagation is seen. Crack opened perpendicular to the direction of applied load. The IEL (green) can be seen to be retracted and underlying medial collagen fibers (red) are visible using the signal from second harmonic generation [243]. (c) Zoomed image of lateral side of an induced crack. Bars $=50$ μm. (Figures (a) and (b) reprinted from [243] with permission from Springer. Figure (c) unpublished with permission from Drs. M.J. Hill and A.M. Robertson.)

introduced a constitutive model that can capture this loss of toe region using a non-zero recruitment stretch for collagen [172,331,333], i.e., the stretch (>1) relative to the unloaded reference configuration that collagen fibers begin to bear load. Subsequently, damage theories built on this concept were developed and used to model acute overloading during cerebral angioplasty [173,174]. The long-term biological response to mechanical damage of this kind has important clinical implications and is not well understood. Mechanical damage theories for the arterial wall are gaining increasing attention and are considered in Section 8.6.7.3 [20,243,304,322].

8.5.2 Tunica Media

8.5.2.1 Tunica Media of Elastic Arteries

In 1964, *Wolinsky and Glagov* published their classic paper on rabbit aorta, demonstrating that under physiological pressures, the tunica media of the elastic artery can clearly be seen to be composed of nearly equally spaced layers of fenestrated elastin lamellae surrounding nearly circumferentially aligned VSMCs, collagen fibers and fine elastin fibers. They later termed this structural unit the 'lamellar unit' [326], Figs. 8.17a and 8.8c. *Wolinsky and Glagov* undertook a comparative study of adult thoracic aorta in 10 mammalian species including humans and identified a lamellar unit in all species. Remarkably, the tension (force/width) supported by the arterial wall ranged from 7.82 N/m in a mouse to 203 N/m in a sow, the tension per lamellar unit only ranged from 1.1 N/m to 3.1 N/m. They concluded that these mammalian species adapted to the increased wall tension in vessels with larger lumen by simply adding more lamellar units, rather than increasing the thickness or the ECM content of these layers. It was observed that sufficient layers were added to maintain the tension per lamellae approximately constant across species and they conjectured that the structure and number of the lamellar unit is determined during fetal and postnatal periods as an adaptive response (now termed growth and remodeling) to rapidly changing intramural loads during this period.

The number of lamellar units decreases with peripheral distance from the heart, until there are very few remaining lamellae in the peripheral muscular arteries. Close to the heart, the number can be considerable, with 78 lamellar units in the media of ascending thoracic aorta [70]. At a fixed peripheral distance, *Wolinsky and Glagov* found little variation in width of the lamellar unit across the medial layer and even between species. At a reference point in the descending thoracic aorta, they observed a range in average thickness of only 6–18 μm between mice and humans [327]. In studies of the media of human thoracic aorta (45–74 years), *Dingemans et al.* found the lamellae unit to be 13. 9 μm thick, varying less than 1% across the medial layer [70]. Not surprisingly, in larger arteries, the outer media requires a separate blood supply. Small blood vessels (vasa vasorum) are found in the adventitia for vessels with greater than 29 lamellar units and are not found in smaller mammals with fewer than 29 layers [328].

Further investigation of the components of the lamellar unit have revealed the complexity of architecture within this structure and underscore the many unanswered questions regarding the association between cellular function and extracellular matrix during normal remodeling, disease and maturation. The VSMCs in the lamellar unit are almost completely covered by a thin, basal lamina that also bridges the gap between the cells, Fig. 8.17. These cells are connected to the fenestrated lamellar sheets by fibers containing fibrillin and type IV collagen (see Fig. 8.17) [70,76] as well as by thin elastin protrusions from longitudinal ridges on the cells. The interlamellar elastin fibers are clearly seen in Fig. 8.8c. Type VIII collagen has also been identified in association with the basal laminae of VSMCs and elastin fibers [222].

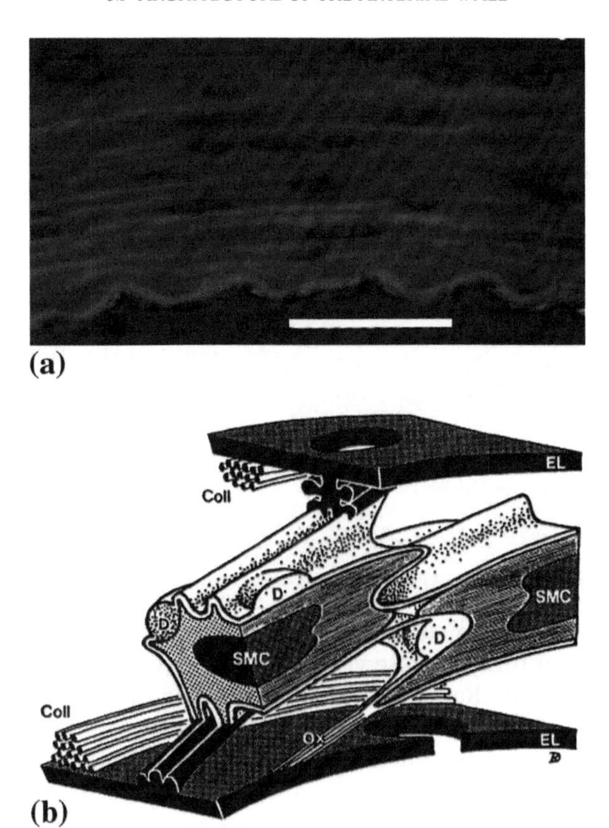

FIGURE 8.17 Components of the medial layer in elastic arteries. (a) Layering of elastin lamellae and smooth muscle cells seen in a cross-sectional view of a normal rabbit left common carotid artery. Lumen side is down in figure with nuclei in endothelial monolayer (red, CD31), IEL (green), elastic lamella in media (green, autofluorescence), cell nuclei (blue DAPI). Imaged under confocal microscope, bar = 50 mm, unpublished, from Dr. M. R. Hill, *with permission.* (b) Schematic of a lamellar unit containing two smooth muscle cells (SMCs) enveloped in a thin, basal lamina, positioned between two lamellar sheets (EL) with round fenestrations. The VSMCs are connected to the lamellae by an oxytalan fiber (Ox) containing fibrillin and type VI collagen as well as by thin elastin protrusions that attach to longitudinal ridges outside the main cell body (smoothed and filled in for illustrative purposes). Collagen fibers (Coll) are located close to both elastic lamellae. (Reprinted from [70] with permission from John Wiley & Sons.)

The relative amounts of type I, III and V collagen vary with location and age. *Dingemans et al.* identified all three fibrillar collagens within the media of human thoracic aorta (45–74 years) [70]. *Menashi et al.* reported a 2:1 ratio of collagen types I and III in the media of both normal and stenotic aorta (mean age approximately 67 years) [191]. A detailed study of the layer-dependent collagen fiber orientation within non-atherosclerotic, human elastic arteries found differences in orientation between the aortas and common iliac arteries [259]. Two families of medial collagen fibers were identified in the media of the aorta, while only a single preferred direction was found in the common iliac arteries, demonstrating the need for vessel-dependent structural information.

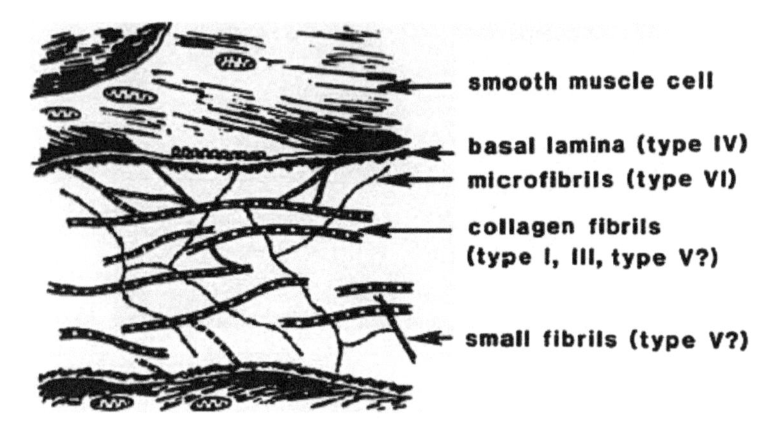

FIGURE 8.18 Schematic of representative region of medial layer of a muscular artery. (Reproduced from [189] with permission from Wolters Kluwer Health.)

8.5.2.2 *Tunica Media of Muscular Arteries*

The architecture of the tunica media in elastic and muscular arteries is substantially different, reflecting the different roles of these vessels [238]. As the muscular arteries are reached, there are few elastin laminae. Rather, the media is dominated by nearly circumferentially aligned layers of smooth muscle cells with interspersed collagen fibers and small amounts of elastin fibers. As a consequence the internal elastin laminae is much more prominent. Active contraction of VSMCs endows muscular arteries with the capacity to actively change their arterial cross-section (see Section 8.4.2). In addition to providing a means to maintain blood pressure, this feature also provides a means to elicit more permanent changes in vessel caliber in conjunction with vessel remodeling. This capability is also evident in the smaller muscular arteries and the arterioles which are the primary regulator of peripheral resistance.

8.5.3 Tunica Adventitia

In both elastic and muscular arteries, the adventitial layer is composed of collagen fibers, some elastin fibers (Figs. 8.8a and 8.19) as well as a sparse population of vascular cells (mainly fibroblasts, Section 8.4.3). This layer is a larger percentage of the wall thickness in muscular arteries, where it can be as thick as the medial layer (e.g., Fig. 8.2a). Collagen fibers can be found in bands in the aorta as well as in bundles or individual fibrils interconnected to the bundles [295], Fig. 8.5b and c. Orientation of the collagen fibers in the adventitia shows a great variability among vessels and, as for collagen in the medial layer, this orientation depends on the loading conditions of the vessel. *Schriefl et al.* found two fiber families with close to axial alignment in the adventitia of human thoracic and abdominal aortas as well as in common iliac arteries [259]. *Canham et al.* report a nearly circumferential alignment in adventitia of coronary vessels fixed at distending pressure [44]. A much wider range of orientations has been found in cerebral vessels at physiological loading conditions [82]. The inner surface of the adventitia is lined with the external elastin lamina, which is generally less pronounced than the IEL. It is absent in some arteries, for example, the cerebral arteries.

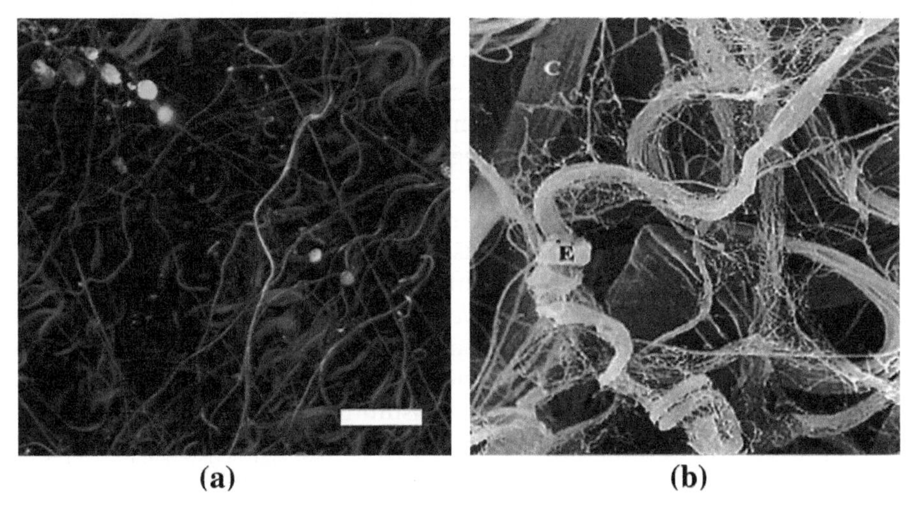

FIGURE 8.19 Adventitial collagen and elastin. (a) Adventitia of a rabbit common carotid artery imaged using multi-photon microscopy (2PE for elastin and second harmonic generation for collagen) revealing elastin fibers (green) and tortuous collagen fibers (red), bar = 50 μm, and (b) SEM image of elastin (E) and collagen (C) fibers in adventitia of a mouse aorta. (Figure (a) reprinted from [243] with permission from Springer. Figure (b) reprinted from [295] with permission from Dr. T. Ushiki.)

The importance of the vascular adventitia not only in vascular disease but also in normal maintenance and homeostasis of vessels is increasingly being recognized [114]. For years it was thought that the adventitia merely provided a passive structural support for the blood vessel to prevent overstretch of the arterial wall under acute loading conditions [68]. However, recent studies suggest the adventitia plays an important role in vascular inflammation, Fig. 8.11. Myofibroblasts and progenitor cells can also be found in the adventitia. The latter cells are capable of differentiating into VSMCs and migrating into the medial and intimal layers. The adventitia also enables the arterial wall to interact with the surrounding tissue, i.e., exchanges signals and cells.

8.6 CONSTITUTIVE MODELS FOR THE ARTERIAL WALL

Given the complexity of the arterial wall, it is important to start this section with a reminder that no material – either organic or inorganic – has a uniquely defined constitutive model. Rather, we seek to define a material model that includes what are perceived as the most salient elements for the application of interest. It is therefore helpful to first recall some of these complexities in the structure and function of the arterial wall as well as commonly employed idealizations.

As discussed in the previous sections, the arterial wall is a multi-layered composite material made up of passive structural material (e.g., collagen fibers, elastin fibers and other structural proteins such as glycoproteins), actively contracting VSMCs, and other vascular cells. The ECM components vary across the thickness of the arterial wall as well as along its length, lending a strong heterogeneity to the wall material properties. Numerous biomechanical studies idealize the three-dimensional wall as a membrane (or two-dimensional surface) (e.g., [160,315,333,340]). Each material point on the surface represents the through thickness wall properties as well as the lateral boundary conditions [210]. This can be appropriate when bending and transmural shear stresses are negligible such as in pressure inflation of some materials. However, in cases where the load distribution across the wall is of interest such as in layer-specific growth and remodeling or damage, necessarily a more complex, computationally intensive multi-layer model must be used [124,174,243,297].

Other idealizations depend on the degree to which individual fibers and matrix components are included in the model. *Structural models* explicitly include information on the tissue composition, structure and load carrying mechanisms of individual components, such as collagen and elastin fibers. In doing so, they provide insight into the function and mechanics of tissue components, at the expense of requiring constitutive data for each of the wall components and adding complexity to the governing equations. *Phenomenological models* describe the bulk mechanical response of the arterial wall without consideration of the role of specific components. Between these two extremes are *structurally motivated models*, that include some structural information, such as the fiber orientation, without directly incorporating the mechanical properties of individual components. For example, the mechanical response for individual fibers is not prescribed.

The vessel wall is subject to large deformations during normal physiological operation and hence a linearized elasticity theory is inappropriate. While in some applications it can be suitable to consider small on large theories of elasticity, (e.g., [18,60,103]), for the most part, all arterial models are based on nonlinear elasticity. Linear and nonlinear viscoelastic models have also been introduced to account for energy dissipation during tissue deformation, though these are less commonly used [48,219,346].

Constitutive equations for the arterial wall can further be categorized by the types of biological processes they are capable of modeling. As discussed in detail in Section 8.4, the intramural cells alter the ECM in response to mechanical and chemical cues. Changes in the quantity, distribution, orientation and mechanical properties of the ECM are known to occur as part of a healthy response to changing stimuli (e.g., growth and remodeling) as well as during pathological and damage processes in disease and aging. Further, this response is of a multi-scale nature, involving multiple spatial and temporal scales. For example, remodeling of arterial collagen in response to altered hemodynamic conditions involves physical scales ranging from the molecular scale (collagen molecules) up through at least the centimeter scale (local vascular region). This remodeling process is believed to depend on hemodynamic cues on the order of seconds, while the remodeling process occurs over a time scale of weeks.

Classic models of nonlinear elasticity [294] are used to characterize the elastic response of the arterial wall without accounting for changes to its structure. In contrast, growth and remodeling (G&R) theories have been introduced to model the deposition and degradation of components, such as collagen fibers, during normal maintenance and growth, as well as the remodeling that occurs in response to altered loads. Damage theories are receiving increasing attention as a means to account for mechanically and enzymatic derived changes in mechanical properties of the ECM [22,23,173,174,218,219,243]. Theoretical frameworks for growth, remodeling and damage are introduced in Section 8.6.7. Applications of these theories to cerebral aneurysms are then considered in Section 8.7.

There is insufficient room in this chapter to comprehensively cover the vast subject of constitutive modeling of the arterial wall. We therefore refer the reader to classic texts such as [294] as well as [53,122] for background material in continuum mechanics. The recent review paper by *Holzapfel and Ogden* [126] provides a thorough discussion of fiber-based constitutive models that are motivated by the earlier work of *Spencer* [280]. The remainder of this section is focused on the passive response of the arterial wall, and in particular will not cover the active response due to vascular smooth muscle cells. A review of models for active contributions from smooth muscle cells can be found in [126] including early phenomenological modeling by *Rachev and Hayashi* [226] and by *Zulliger et al.* [347]. See, for example, [296,325] for more recent approaches for modeling active and passive contributions of VSMC.

8.6.1 Multiple Mechanism Models

Indirect and direct experimental evidence supports *Roach and Burton*'s conjecture that the nonlinear nature of the loading curve for the arterial wall, Fig. 8.1, arises from the highly flexible elastin components acting in parallel with stiffer collagenous fibers that are recruited to load bearing beyond a critical level of stretch [84,120,241,246,260]. *Scott et al.* later evaluated the mechanical response of cerebral arteries loaded beyond their elastic limit, obtaining data that suggested the elastin mechanism could fail prior to the collagen component, resulting in an increase in unloaded vessel radius [260]. Motivated by this earlier work, *Wulandana and Robertson* introduced a dual mechanism constitutive model that included kinematic criteria for collagen recruitment at finite strain as well as failure of the elastin mechanism separately from that of collagen [331–333]. *Watton et al.* introduced a more structurally motivated fiber model for the collagen mechanism, describing collagen recruitment in terms of critical conditions for collagen fiber stretch and have used this approach in many subsequent studies of abdominal aortic and cerebral aneurysms [315] (Section 8.7). *Li and Robertson* [172] considered both anisotropic collagen fiber recruitment and collagen fiber dispersion and used it in the context of cerebral angioplasty [174]. Gradual fiber recruitment is implicit in the integral G&R theories discussed later in this section.

Most current constitutive models for the arterial wall now consider the mechanical response to arise from additive contributions of distinct components which we term as *mechanisms*, as opposed to individual material components of the wall. These mechanisms are most often introduced as collective isotropic and anisotropic mechanisms. For example, the isotropic component might be viewed as representing the IEL as well as other ECM contributions of a nearly isotropic nature. The anisotropic contribution is often assumed to arise from one or more families of collagen fibers and possibly VSMC, defined by their orientation (angle) relative to a reference direction in a chosen configuration. Some studies have also considered the anisotropic nature of the elastin fibers [107,237]. It is also frequently assumed that all mechanisms can be modeled as nonlinearly elastic, and in particular as hyperelastic. In this case, a strain energy function $W(\mathbf{C})$ defined with respect to unit volume in a reference configuration κ_0 can be introduced:

$$W = W_{iso} + W_{aniso} \tag{8.1}$$

where $\mathbf{C}$ is the right Cauchy Green tensor $\mathbf{C} = \mathbf{F}^T\mathbf{F}$ and $\mathbf{F}$ is the deformation gradient tensor relative to κ_0. The arterial wall is also generally idealized as incompressible (determinant of

$\mathbf{F} = 1$), though slightly compressible formulations are frequently introduced for numerical purposes (e.g., [122]). For incompressible, hyperelastic materials, the strain energy function can be shown to be related to the Cauchy stress tensor through (e.g., [279]):

$$\mathbf{T} = -p\mathbf{I} + 2\mathbf{F}\frac{\partial W_{iso}}{\partial \mathbf{C}}\mathbf{F}^T + 2\mathbf{F}\frac{\partial W_{aniso}}{\partial \mathbf{C}}\mathbf{F}^T \tag{8.2}$$

where p is the Lagrange multiplier arising from the incompressibility constraint and $\mathbf{I}$ is the identity matrix. These mechanisms will, in general, have different constitutive responses and different unloaded configurations. For example, the arterial collagen is in a wavy or crimped state in the unloaded artery and is gradually recruited to load bearing as the vessel is strained. In such cases, multiple reference configurations are needed to identify the kinematic state for recruitment. Further, damage to the isotropic and anisotropic mechanisms (e.g., elastin and collagen fibers) can occur independently and prior to failure of the arterial wall. The explicit treatment of these multiple mechanisms makes it possible to separately model damage and failure of the components [331,333].

8.6.2 Isotropic Mechanism

After imposing invariance requirements [102,294], the incompressibility condition and material isotropy, the strain energy function for the isotropic mechanism can be expressed, without loss in generality, as a function of the first and second principal invariants of $\mathbf{C}$ [280,294], which are the same as the invariants of $\mathbf{B} = \mathbf{F}\mathbf{F}^T$, i.e:

$$W_{iso} = W_{iso}(I_I, I_{II}) \tag{8.3}$$

where $I_I = \text{tr}(\mathbf{B}), I_{II} = 1/2[(\text{tr } \mathbf{B})^2 - \text{tr}(\mathbf{B}^2)]$. Most strain energy functions for the isotropic mechanism in the arterial wall depend only on the first invariant. The two most commonly used models of this form are:

Neo-Hookean Model:

$$W_{iso} = \frac{\alpha}{2}(I_1 - 3), \quad \mathbf{T}_{iso} = -p\mathbf{I} + \alpha\mathbf{B} \tag{8.4}$$

Exponential Model:

$$W_{iso} = \frac{\alpha}{2\gamma}(e^{\gamma(I_1-3)^n} - 1), \quad \mathbf{T}_{iso} = -p\mathbf{I} + \alpha n(I_1 - 3)^{n-1}e^{\gamma(I_1-3)^n}\mathbf{B} \tag{8.5}$$

where α and γ are material constants and the constant n used in the exponential model is often chosen as one or two. Alternatively, strain energy functions that depend on both the first and second invariants of B are considered, with the Mooney-Rivlin being a frequently used model,

Mooney-Rivlin Model:

$$W_{iso} = \frac{\alpha}{2}(I_1 - 3) + \frac{\beta}{2}(I_2 - 3), \quad \mathbf{T}_{iso} = -p\mathbf{I} + \alpha\mathbf{B} - \beta\mathbf{B}^{-1} \tag{8.6}$$

where α and β are material constants. Equation (8.2) has been used with the strain energy functions in Eqs. (8.4)–(8.6) to obtain the corresponding Cauchy stress tensor, $\mathbf{T}$.

8.6.3 Anisotropic Mechanisms: Kinematics of Fiber Recruitment

The non-random orientation of the collagen fibers in the arterial wall is largely responsible for its anisotropic material response. Though in some instances it can be appropriate to consider a single fiber type with a single orientation, more generally the fibers display multiple orientations (or angles) and unloaded reference configurations, Section 8.5. In preparation for a discussion of constitutive modeling of such diverse families of fibers, we first discuss the kinematics needed to describe the deformation of an infinitesimal segment of a single fiber and then generalize this formulation in the following sections.

In the ensuing discussion, we consider an infinitesimal material segment of an arbitrary collagen fiber that is buckled (or crimped) in the unloaded configuration of the body, κ_0, Fig. 8.20. Under suitable stretch, this fiber region will straighten and begin to resist further stretching. We denote the configuration where the fiber is ready to bear load (recruited) as κ_R and the current configuration of the material at some arbitrary time t as κ, Fig. 8.20. At the current time t, the contribution of the infinitesimal fiber region to the strain energy function depends on the actual fiber stretch λ_f, namely its stretch in κ relative to κ_R.

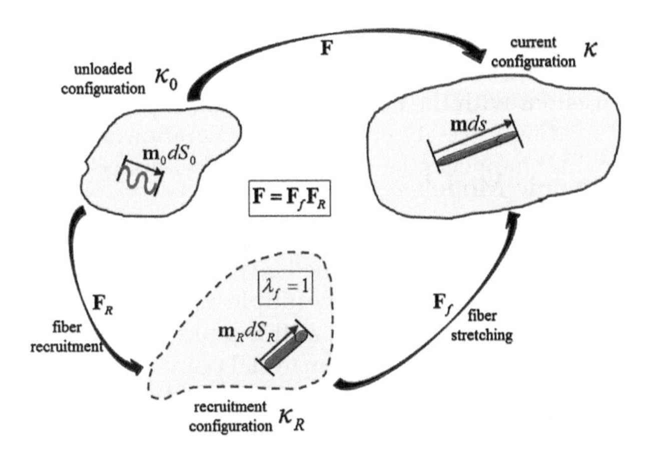

FIGURE 8.20 Schematic of reference configurations and notation used for collagen fiber recruitment kinematics. In the unloaded configuration of the body κ_0, the infinitesimal fiber segment is crimped (buckled). Under stretch, this fiber segment gradually uncrimps. The configuration where the fiber is ready to bear load (recruited) is denoted as κ_R and the current configuration as κ. At the current time t, the contribution of the infinitesimal fiber region to the strain energy function depends on the *real stretch* of the collagen fiber segment, $\lambda_f = ds/dS_R$, which is the stretch in κ relative to its unloaded state in κ_R. For convenience, the fiber segment can be mapped back to κ_0 under an affine deformation, with stretch $\lambda_R = dS_R/dS_0$ between κ_0 and κ_R. It follows that, $\lambda_f = \lambda/\lambda_R$ where $\lambda = \sqrt{(\mathbf{C} : \mathbf{m}_0 \otimes \mathbf{m}_0)}$ and $\mathbf{C} = \mathbf{F}^T\mathbf{F}$.

We can put this in mathematical terms by noting that an infinitesimal material line element (fiber segment) $\mathbf{m}_R dS_R$ in κ_R will be mapped to $\mathbf{m}ds = \mathbf{F}_f \mathbf{m}_R dS_R$ in κ, where $\mathbf{m}_R$ and $\mathbf{m}$ are unit vectors and $\mathbf{F}_f$ is the deformation gradient tensor for this mapping. It follows that:

$$\lambda_f = ds/dS_R = \sqrt{\mathbf{C}_\mathbf{f} : (\mathbf{m}_R \otimes \mathbf{m}_R)} \tag{8.7}$$

where $\mathbf{C}_f = \mathbf{F}_f^T \mathbf{F}_f$. It is often useful to describe this deformation relative to the unloaded configuration of the body κ_0. For this reason, we note that the infinitesimal fiber segment in κ_R can be mapped back to κ_0 under an affine deformation through $\mathbf{m}_0 dS_0 = \mathbf{F}_R^{-1} \mathbf{m}_R dS_R$, where $\mathbf{m}_0$ is a unit vector and $\mathbf{F}_R$ is the deformation gradient for the mapping from κ_0 to κ_R. It then follows that:

$$\lambda_R = dS_R/dS_0 = \sqrt{\mathbf{C}_R : (\mathbf{m}_0 \otimes \mathbf{m}_0)} \tag{8.8}$$

and therefore:

$$\lambda_f = \lambda/\lambda_R, \quad \text{where} \quad \lambda = ds/dS_0 = \sqrt{\mathbf{C} : (\mathbf{m}_0 \otimes \mathbf{m}_0)} \tag{8.9}$$

The recruitment stretch λ_R is a (spatially and temporally heterogeneous) material property of the fiber-reinforced material. We point out that recently, direct methods have been developed for measuring both the collagen fiber orientations and the collagen fiber recruitment stretches λ_R [118,120]. Consequently, the geometric configuration of the crimped fiber in κ_0 is not inherently relevant for the fiber modeling (and remodeling). Rather, the process of mapping the fiber vector $\mathbf{m}_R$ back to κ_0 is merely done for mathematical convenience so that a single reference configuration can be used to define the mechanical response of all constituents. In fact, the affine mapping from κ_R back to κ_0 describes compression of the fiber, rather than the actual crimping (buckling) process. As discussed below, the contribution of the collagen fiber to the strain energy function is explicitly defined to be zero under compression, to be consistent with the negligible loading expected during buckling.

8.6.4 N-Fiber Anisotropic Models

In formulating structurally motivated models, we consider a representative volume element (*RVE*) at an arbitrary material point in the body, rather than tracking the behavior of individual fibers. Within this region, fibers of multiple orientations and levels of tortuosity can exist and hence, the constitutive equation at this point reflects the combined contribution of all these fibers. In *N*-fiber models, each material point in the body is assumed to contain *N* families of fibers, with each family distinguished by distinct orientations, mechanical properties and/or recruitment stretch. While *N*-fiber models were previously developed [278,280], *Holzapfel et al.* appear to be the first to apply these theories to the arterial wall [124].

Following earlier work [172,243,315], here, we consider a generalization of the model used by *Holzapfel et al.*, in which fiber recruitment can initiate at non-zero strains of the underlying material, such as would be expected to occur for fibers that are crimped (or tortuous) in the unloaded arterial wall, Fig. 8.6.

The strain energy function for each of the fiber families is assumed to depend solely on the real stretch of this fiber λ_f (stretch beyond uncrimping):

$$W_{aniso} = W_{aniso}(\lambda_f), \quad \mathbf{T}_{aniso} = \lambda_f \frac{dW_{aniso}(\lambda_f)}{d\lambda_f} \mathbf{m} \otimes \mathbf{m} \tag{8.10}$$

Holzapfel et al. [124] introduced an *N*-fiber model using an exponential strain energy function that has found widespread application. Each fiber is modeled as an exponential function of the fiber stretch. When this model is generalized to include fiber recruitment at finite (non-zero) strain, it can be written as:

$$W_{aniso}(\lambda_f) = \mathsf{H}(\lambda_f - 1)\frac{\alpha}{4\gamma}\left(e^{\gamma(\lambda_f^2-1)^2} - 1\right) \tag{8.11}$$

where the symbol H is the unit step function. Namely, the anisotropic contribution is zero when the fiber is buckled ($\lambda_f < 1$). Equation (8.10) can be used to obtain the Cauchy stress corresponding to the strain energy function in Eq. (8.11):

$$\mathbf{T}_{aniso} = \mathsf{H}(\lambda_f - 1)\alpha\lambda_f^2(\lambda_f^2 - 1)e^{\gamma(\lambda_f^2-1)^2}\mathbf{m} \otimes \mathbf{m} \tag{8.12}$$

The classical model without fiber recruitment is recovered when λ_R is set equal to one, so that λ_f is equivalent to λ in Eqs. 8.11 and 8.12. It is straightforward to see that Eqs. 8.11 and 8.12 can be used to define the relationship for an *N*-fiber model:

$$\begin{aligned} W_{aniso} &= \sum_{i=1}^{N} \mathsf{H}(\lambda_f^{(i)} - 1)\frac{\alpha^{(i)}}{4\gamma^{(i)}}\left(e^{\gamma^{(i)}(\lambda_f^{(i)^2}-1)^2} - 1\right) \\ \mathbf{T}_{aniso} &= \sum_{i=1}^{N} \mathsf{H}(\lambda_f^{(i)} - 1)\alpha^{(i)}\lambda_f^{(i)^2}(\lambda_f^{(i)^2} - 1)e^{\gamma^{(i)}(\lambda_f^{(i)^2}-1)^2}\mathbf{m}^{(i)} \otimes \mathbf{m}^{(i)} \end{aligned} \tag{8.13}$$

The difference in orientation of the *N*-fibers enters directly through the definition of $\mathbf{m}_0^{(i)}$ which will in turn be reflected in $\mathbf{m}^{(i)}$. Fibers of different orientation will in general experience different stretches, and hence will have different values for true fiber stretch $\lambda_f^{(i)}$, even in cases when the recruitment stretch is the same. Equation (8.13) also includes the possibility that the fibers have distinct material properties, defined through $\alpha^{(i)}$ and $\gamma^{(i)}$. In writing Eq. (8.13), we have neglected any contributions arising from coupled effects between the fibers. See [280] for a formulation which includes coupled effects. The *N*-fiber exponential constitutive model in Eq. (8.13) has been used to model the contributions from collagen fibers of different orientations (e.g., circumferential, axial [124,172]) as well as the passive response of vascular smooth muscle [296,325].

As noted in [124], the constants in the exponential model can be chosen to ensure collagen fibers have little contribution at low pressures. However, where a microstructural model is of interest, it is valuable to recognize that collagen has been shown to remain crimped until finite strain levels are reached [120]. In fact, as will be elaborated on below, the single material constant neo-Hooken strain energy function was found to be well suited for collagen fibers when gradual collagen recruitment was included in the model, commencing at finite strain [120]. It should also be emphasized that, while collagen is known to undergo irreversible changes at strains larger than 4%, under physiological loads, the arterial wall is subjected to strains that are much greater than this relative to the unloaded state (e.g., [308]). Hence, if collagen is immediately recruited to load bearing, it will be beyond its elastic limit in physiological settings. Therefore, while an exponential model can be used to model the phenomenological data, it does not realistically model the recruitment role of collagen fibers, nor account for its limited extensibility.

8.6.5 Anisotropic Models with a Distribution of Fiber Orientations

We now turn to a generalization of the *N*-fiber theory that allows for a distribution of fiber orientations in the regional volume element, rather than idealizing the fibers as oriented in a few primary directions. For convenience, we define this distribution of fiber orientations, after the fibers have been mapped back to κ_0. In this configuration, the orientation of an arbitrary fiber direction is denoted as $\mathbf{m}_0$ and, hence, the orientation density function can be written as $\rho = \rho(\mathbf{m}_0)$ (e.g., [95,168]). Using local rectilinear coordinates with unit base vectors $(\mathbf{e}_1, \mathbf{e}_2, \mathbf{e}_3)$, Fig. 8.21a, the vector $\mathbf{m}_0$ can be written with respect to spherical coordinates θ and ϕ, through:

$$\mathbf{m}_0(\theta,\phi) = \cos\theta\,\mathbf{e}_1 + \sin\theta\cos\phi\,\mathbf{e}_2 + \sin\theta\sin\phi\,\mathbf{e}_3 \tag{8.14}$$

Hence, the orientation density function can be written as a function of $\theta \in [-\pi/2, \pi/2]$ and $\phi \in [0, \pi]$, Fig. 8.21a. Note that each fiber is only counted once, since for an arbitrary $\mathbf{m}_0$, the fiber $-\mathbf{m}_0$ is not included. The distribution function is non-negative and defined such that $\rho(\theta,\phi)\cos\theta\, d\theta\, d\phi$ represents the proportion of fibers with angles in the range $[\theta, \theta + d\theta]$ and $[\phi, \phi + d\phi]$ where, for definiteness, $\rho(\theta,\phi)$ is taken to be normalized on a unit semi-sphere:

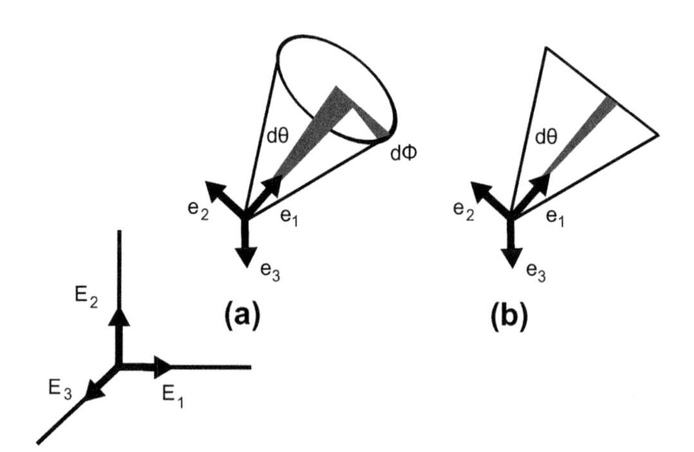

FIGURE 8.21 Schematic displaying the geometric variables used in modeling the distribution of collagen fiber orientation for (a) general 3-D distribution and (b) planar or 'fan splay' distribution. (Reproduced from [120] with permission from Elsevier.)

$$1 = \frac{1}{2\pi} \int_0^{\pi} \int_{-\pi/2}^{\pi/2} \rho(\theta, \phi) \cos\theta \, d\theta \, d\phi \tag{8.15}$$

Several symmetries in the orientation distribution have been found to be useful for studies of soft tissue. At material points where a coordinate axis can be chosen such that orientation distribution function is independent of ϕ, we can simply write $\rho = \rho(\theta)$. This symmetry is referred to as *conical splay*, and the materials are said to be transversely isotropic. For *planar splay*, the distribution of fibers in an RVE lies within a single plane (e.g., [88, 89]). In this case, a local rectilinear basis can be chosen such that all fibers lie within the $\mathbf{e}_1 \otimes \mathbf{e}_2$ plane, Fig. 8.21b. A 2-D orientation distribution function $\rho_{2D}(\theta)$ can then be defined on $\theta \in [-\pi/2, \pi/2]$ with normalization condition:

$$1 = \frac{1}{\pi} \int_{-\pi/2}^{\pi/2} \rho_{2D}(\theta) d\theta \tag{8.16}$$

The strain energy function for the material is then the result of the integrated response of fibers over all angles. For deformations that do not not break these respective symmetries, we can write the strain energy function in a simpler form:

$$\begin{aligned} &\text{Conical splay} \quad W_{aniso} = \frac{1}{2} \int_{-\pi/2}^{\pi/2} w_f(\lambda_f) \rho(\theta) \cos\theta \, d\theta \\ &\text{Planar splay} \quad W_{aniso} = \frac{1}{\pi} \int_{\pi/2}^{-\pi/2} w_f(\lambda_f) \rho_{2D}(\theta) d\theta \end{aligned} \tag{8.17}$$

A variety of orientation distribution functions have been used including a π-periodic von Mises distribution [95], Gamma distribution [248] and a Bingham distribution [7,96]. Experimental studies of collagen fiber orientation in the media of rabbit carotid arteries [120] and in all layers of the human thoracic aorta, abdominal aorta and common iliac arteries [259] support this two-dimensional idealization for the orientation distribution function. Recently, a methodology was developed to measure $\rho_{2D}(\theta)$ from stacks of projected images obtained using multi-photon microscopy, Fig. 8.5 [120,243]. In this case, the orientation can be imposed directly without the need to select a functional form.

8.6.6 Distributions of Fiber Recruitment Stretch

In the unloaded state, the medial collagen fibers are found in a crimped state and are gradually recruited, beginning at finite strain, as the vessel is loaded, Fig. 8.22. The transition region from the highly flexible toe region to the stiff regions of the loading curve has been shown to arise from the gradual recruitment of the medial collagen fibers, rather than a strong nonlinearity of the fibers themselves [120]. The collagen undulation level therefore determines the onset of the stiffer region of the loading curve and can have a profound influence on the mechanical behavior of the vessel wall [96] as well as the growth and remodeling process [187]. *Hill et al.* appear to be the first to measure 3-D collagen fiber tortuosity in

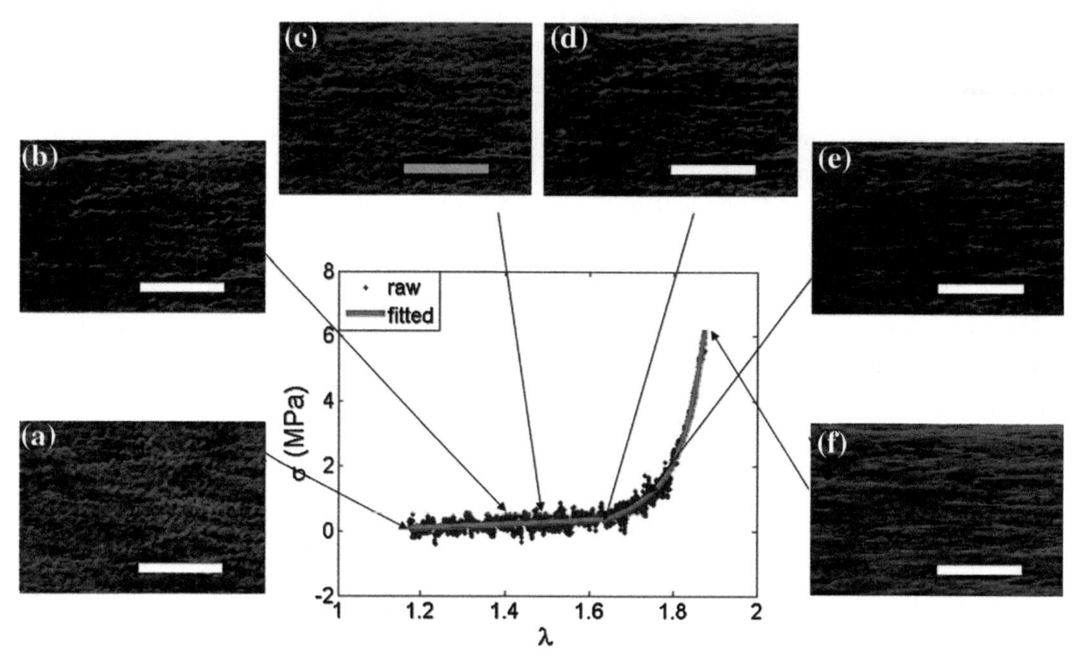

FIGURE 8.22 Medial collagen recruitment in a single sample from a rabbit common carotid artery during uniaxial loading. Images shown are projected stacks of the sample under increasing unaxial stretch λ. Samples were loaded in circumferential direction – shown left to right in images. Corresponding loading state for each image is shown on the applied Cauchy stress versus stretch curve. Data obtained using protocol described in [120]. (Unpublished image, produced with permission from Drs. M.R. Hill and A.M. Robertson.)

arteries [120] in their study of collagen fabric in the media of rabbit common carotid arteries. Fiber tortuosity had previously been measured in fixed arterial segments [246]. Following earlier work of *Lanir* [169], *Hill et al.* described the distribution of collagen fiber recruitment stretches using a recruitment probability distribution function, denoted by $\rho_R(\lambda_R)$. In this case, $\rho_R(\lambda_R)d\lambda_R$ represents the fraction of fibers with recruitment stretch in the range $[\lambda_R, \lambda_R + d\lambda_R]$. The contribution to the strain energy from this fiber ensemble is $\rho_R(\lambda_R)w_f^*(\lambda_f)d\lambda_R$. Therefore, the strain energy potential of the whole fiber ensemble in the RVE for the ensemble and the corresponding normalization condition is:

$$w_f(\lambda) = \int_1^{\lambda} \rho_R(\lambda_R) w_f^* \left(\frac{\lambda}{\lambda_R}\right) d\lambda_R, \quad 1 = \int_1^{\infty} \rho_R(\lambda_R) d\lambda_R \tag{8.18}$$

The collagen fibers in an RVE are assumed to bear no load when the local tissue stretch is less than the minimum recruitment stretch, $\lambda_R^{\min}$. Here, this condition is directly imposed by defining $\rho_R(\lambda_R)$ to be zero for $\lambda_R < \lambda_R^{\min}$. It can also be imposed using a Heaviside function [120,243]. Other approaches have been introduced for modeling the recruitment process such as that in [41] where a random distribution of fiber crimp is used and unfurling of the fiber as well as fiber failure are considered.

8.6.7 Multi-Mechanism Models: Growth, Remodeling and Damage (GR&D)

As discussed in Section 8.4, the constituents in the arterial wall can change in time due to orchestrated production and removal of ECM. For example, the mass of individual components in the wall can increase due to the production of ECM components by the cells (e.g., fibroblasts and VSMCs) and decrease due to their breakdown by MMPs. Following [130], we use the terms *positive/negative growth* to describe the net increase/decrease in mass of a wall component (also called growth/atrophy). For example, positive growth can result from cell proliferation or migration as well as from manufacture of ECM at a faster rate than removal. Further, new constituents can be produced with different material properties, orientations, and deposition stretches from the prior material. This process, which endows the wall with an altered morphology and mechanical behavior, is termed *remodeling*. Finally, the components of the wall can be mechanically or chemically damaged. In particular, their mechanical properties can change without a change in mass. For example, the number of cross-links in the collagen fibers can be enzymatically altered, resulting in a change in stiffness without changing the mass of these fibers. Or, the fiber stiffness and strength can be diminished due to repetitive loading (fatigue damage), or from supraphysiological mechanical loading [173]. Following [173,243], we make a distinction between negative growth and damage. In particular, we use the term *damage* to describe the process of breaking down the mechanical structure of a component of the ECM and thereby altering its mechanical properties with or without changes in the volume fraction of the component.[3] Below we provide a brief background on formulations for *GR&D* followed by discussion specific to the extracellular matrix in the arterial wall.

8.6.7.1 Positive and Negative Growth of a Constituent

To simulate the local influence of growth/atrophy of a constituent on the mechanical response of the composite in the RVE, the strain energy function of a constituent 'k' is often multiplied by a scalar function representing the normalized mass density of this constituent, $\hat{m}^{(k)}$, i.e.:

$$W = \hat{m}^{(k)} \widehat{W}^{(k)} \tag{8.19}$$

where $\widehat{W}^{(k)}$ is the strain energy function of the constituent at time $t = 0$ and:

$$\hat{m}^{(k)} = \frac{m^{(k)}(t)}{m^{(k)}(t=0)} \tag{8.20}$$

denotes the normalized mass density of a constituent, i.e., ratio of mass density of constituent k at time t to its mass density at time $t = 0$. As noted below, $m^{(k)}(t)$ can be a prescribed function of time, or a more complex function of the biological response of the wall to mechanical or chemical stimuli [254,319].

[3] In some works, negative growth is used as a way of accounting for enzymatic damage. However, in structurally motivated models, where the volume fraction of constituents enters as a measurable quantity, the distinction between damage and negative growth can be important.

8.6.7.2 *Remodeling of the Constituents*

Simulating the remodeling of arterial wall requires accounting for how the natural reference configurations of the individual constituents evolve during adaption in response to altered environmental conditions. In general, two mathematical approaches can be adopted: the formulations can be integral based or rate based. We do not discuss the integral-based formulations here; rather we refer the interested reader to the seminal work of *Humphrey* which applied an integral-based formulation to address the remodeling of a 1-D collagenous tissue held at (altered) fixed length, see [133]. This model has provided the basis for many subsequent studies and developments, e.g., see [17,130,264,298,341] and the interested reader is referred to [137] for discussion and comparison of the two approaches for modeling abdominal aortic aneurysm evolution. Below we outline (primarily) a rate-based formulation to address the remodeling of the natural reference configurations of the collagenous constituents as this is utilized for the example of intracranial aneurysm evolution in the next section (see Section 8.7). We also point out that while integral-based formulations can have the potential for the inclusion of more meaningful remodeling parameters, e.g., they can incorporate parameters that explicitly account for the lifetimes of individual constituents whereas parameters in rate-based formulations are generally phenomenological in nature, they often have greater theoretical complexity and computational expense. An advantage of a rate-based remodeling formulation over an integral-based one is that it is not as computationally expensive, i.e., the number of reference configurations to keep track of is fixed and does not increase with the number of computational time-steps of a simulation.

8.6.7.3 *Damage Models*

Damage to the components within the arterial wall has been modeled using continuum damage theories [173,198,243] and we briefly outline this approach here. Following earlier work (e.g., [275]), we introduce a scalar damage variable (d) and take a strain space based approach, by assuming the dependence of the strain energy on d can be explicitly written as:

$$W = (1 - d)W^o(\mathbf{C}) \tag{8.21}$$

where W^o is the effective strain energy of the hypothetical undamaged material, and the internal variable d is defined to be in the range $[0, 1]$ with zero corresponding to no damage and one to total damage. After imposing the Clausius-Planck inequality, it can be shown that (e.g., [122,243,275]):

$$\mathbf{T} = (1 - d)\mathbf{T}^o, \quad \mathbf{T}^o = 2\left(\mathbf{F}\frac{\partial W^o}{\partial \mathbf{C}}\mathbf{F}^\mathrm{T}\right) \tag{8.22}$$

with the additional requirement that the internal dissipation be non-negative for all times and material points in the body:

$$\mathscr{D}_{in} = W^o\dot{d} \geqslant 0 \tag{8.23}$$

where $\mathbf{T}^o$ is the effective Cauchy stress tensor for the hypothetical undamaged material. As in [173], we consider three possible damage modes, with damage variables denoted as d_j

with $j = 1, 2, 3$. Damage accumulation in time will be defined through the dependence of the damage variable on an accumulation variable α_j. While details can be found elsewhere (e.g., [198,243]), we briefly define the functional form of α_j for these three modes.

8.6.7.3.1 DISCONTINUOUS DAMAGE

When a rubber sample is cyclically loaded and unloaded in uniaxial tension, the applied stress needed to reach a given level of strain decreases with increased loading cycles. This phenomenon is termed *stress softening* or the *Mullins effect*, so named due to the early studies by *Mullins* on stress softening in rubber materials with imbedded particles [206,207]. Damage of this kind has been modeled by setting the current value of the accumulation variable α_1 equal to the maximum effective strain energy the material has experienced [275]:

$$\alpha_1(t) = \max_{s \in [0,t]} W^o(s) \tag{8.24}$$

In this damage mode, termed *discontinuous damage*, α_1 only increases (damage accumulates) when the effective strain energy increases beyond the previous maximum.

8.6.7.3.2 CONTINUOUS DAMAGE

Mechanical damage can increase during cyclic loading even if the effective strain energy remains below the maximum strain experienced in prior cycles. To address this phenomenon, *Miehe* [198] introduced a contribution to damage evolution depending on the arc length of the effective strain energy. In this case, the mechanical damage variable d_2 depends on an accumulation variable $\alpha_2(t)$ that is a function of accumulated equivalent strain:

$$\alpha_2(t) = \int_0^t \left| \frac{dW^o}{ds} \right| ds \tag{8.25}$$

In this damage mode, termed *continuous damage*, α_2 increases continuously, regardless of whether the prior maximum has been surpassed.

8.6.7.3.3 ENZYMATIC DAMAGE

The scalar damage formation has also been used to prescribe damage accumulation due to the exposure to enzymes that are a consequence of the walls response to abnormal hemodynamics [173,243]. In this damage mode:

$$\alpha_3 = \frac{1}{T} \int_0^t f(\text{WSS}, \text{WSSG}) ds \tag{8.26}$$

Specific examples will be considered below where the functional form of $f(\text{WSS}, \text{WSSG})$ is chosen based on *in vivo* experimental data.

8.6.7.4 Growth, Remodeling and Damage for the Elastin Mechanism

As discussed in the beginning of this chapter, the elastic nature of the artery wall contributes to the overall efficiency of the circulatory system. However, the production of structurally sound elastin components in adult arteries is believed to be negligible and, therefore, we do not consider positive growth and remodeling of the elastin mechanism. In contrast, degradation and damage to the elastin mechanism in the adult artery are of particular importance, because of minimal opportunity for repair or replacement. In this section, we briefly outline some of the constitutive equations which have been used to model the chemically and mechanically induced changes to the isotropic mechanism that were discussed in Section 8.6.2.

8.6.7.4.1 DEGRADATION OF THE ELASTIN MECHANISM

To simulate the effects of the change in mass associated with enzymatic degradation of elastin, the normalized density of elastin, say $\hat{m}_E$, can be prescribed to evolve as a function of space and time, i.e., $\hat{m}_E = \hat{m}_E(\mathbf{X}, t)$ where $\hat{m}_E \in [0,1]$ for degradation. The simplest manner to achieve this is to prescribe the spatial and temporal evolution of the function $\hat{m}_E(\mathbf{X}, t)$, e.g., see [17,315].

For example, in the simulation of intracranial aneurysm (IA) evolution presented in the next section, the inception stage of IA formation is modeled by prescribing a localized degradation of elastin to create a small outpouching of the arterial domain. This perturbs the hemodynamic environment and leads to a perturbation in the wall shear stress distribution $\mathrm{WSS}(\mathbf{X}, t)$. Subsequent elastin degradation can then be explicitly linked to deviations of the wall shear stress from homeostatic levels, e.g.:

$$\frac{\partial \hat{m}_E}{\partial t} = -f_D(\mathrm{WSS}(\mathbf{X}, t))\, a_D \hat{m}_E \tag{8.27}$$

where $f_D(\mathrm{WSS}) \in [0,1]$. The parameter a_D thus relates to the maximum rate of degradation of elastin. As an illustrative example linking elastin degradation solely to low WSS, one can consider elastin degradation to be maximum below a threshold WSS_X and nonexistent for WSS greater than a larger critical value WSS_{crit} with a gradual transition between these values. In particular, for $0 \leqslant \mathrm{WSS}_X < \mathrm{WSS}_{crit}$ if: $\mathrm{WSS} \geqslant \mathrm{WSS}_{crit}, f_D(\mathrm{WSS}) = 0$; $\mathrm{WSS} \leqslant \mathrm{WSS}_X, f_D(\mathrm{WSS}) = 1$ and $f_D(\mathrm{WSS})$ is a monotonically decreasing function of WSS in the interval $\mathrm{WSS}_X \leqslant \mathrm{WSS} \leqslant \mathrm{WSS}_{crit}$ (for further details see [316,319]).

8.6.7.4.2 DAMAGE TO THE ELASTIN MECHANISM

Scalar damage theories such as those defined in Section 8.6.7.3 were proposed to model both mechanical and enzymatic damage to the elastin mechanism [173,243]. The discontinuous damage theory in Eqs. 8.21–8.24 has been applied in a three-layer heterogenous wall model to predict damage during angioplasty (to both the elastin and collagen mechanisms) [174,243].

Animal studies have found that damage to the IEL in native and non-native cerebral bifurcations is associated with a combination of elevation of WSS above a threshold level and positive elevated WSSG [193,197]. This degradation has been found to be progressive in that damage increased with exposure time [194,205] and dose dependent in that the damage level increased with increasing magnitude of WSS [197]. *Li and Robertson* proposed an enzymatic damage model [173] that captured all these features, Eq. (8.26), and can also be used to model elastin damage resulting from pathologically low and high WSS:

$$\alpha_3 = \frac{1}{T}\int_0^t \mathrm{H}(\zeta)\mathrm{H}(\eta)(\zeta + b\eta)ds \quad \text{where } \zeta = \frac{\mathrm{WSS} - \mathrm{WSS}_T}{\mathrm{WSS}_T}, \quad \eta = \frac{\mathrm{WSSG} - \mathrm{WSSG}_T}{\mathrm{WSSG}_T} \tag{8.28}$$

where H() denotes the unit step function and b, WSS_T and $WSSG_T$ are material constants. Equation (8.28) satisfies the criterion that damage does not increase unless WSS is sufficiently elevated above a threshold value and the WSSG is elevated above a positive threshold value (see, [xx], for the definition of WWG). Further, the evolution Eq. (8.28) satisfies the condition that damage increases with exposure time and dosage. More data are needed to determine whether the rate of accumulation should increase with increased η (amount by which WSSG exceeds the threshold value) or whether it is only necessary that η be positive, in which case b can be set to zero.

Both the negative growth model in Eq. (8.27) and the enzymatic damage model in Eq. (8.28) similarly use a mutliplicative reduction type variable to diminish the magnitude of the strain energy function. The change in mass that can occur from enzymatic damage is not explicitly included in Eq. (8.28). Further experimental investigation would be valuable to determine whether the diminished mechanical properties do, in fact, scale with the remaining mass of elastin, as in Eq. (8.27).

8.6.7.5 Growth, Remodeling and Damage of the Collagen Mechanism

8.6.7.5.1 THE COLLAGEN FIBER ATTACHMENT STRETCH λ_{f_a}

Collagen fibers are in a continual state of deposition and degradation in the current configuration κ. Vascular cells (fibroblasts in the adventitia and vascular smooth muscle cells in the media) work on the collagen fibers to attach them to the matrix in a state of stretch in this configuration. Consequently, the recruitment configuration κ_R is inferred from the stretch at which fibers are configured and attached to the extracellular matrix in the current configuration κ. On this note, it is important to recognize that the matrix is pulsating and thus it is desirable for the definition of the attachment stretch to explicitly take this into account. *Watton et al.* [315] hypothesized that the fibers are configured to the matrix to achieve a maximum stretch during the cardiac cycle and introduced the terminology *attachment stretch* where $\lambda_{f_a} \geq 1$.

8.6.7.5.2 COLLAGEN REMODELING

Collagen remodeling relates to changes in the reference configurations κ_R of the fibers with no net change in mass of the fibers. This can be simulated by adapting the fiber orientations defined by the vectors $\mathbf{m}_0$ or by altering the magnitudes of the fiber recruitment stretches λ_R. Evolving the recruitment stretches simulates the mechanical consequences of: (i) fiber deposition and degradation in altered configurations; (ii) fibroblasts configuring the collagen to achieve a maximum stretch during the cardiac cycle, i.e., the *fiber attachment stetch* λ_{f_a}. These hypotheses imply that the reference configuration κ_R of the collagen fibers evolves such that the maximum stretch of the fiber during the cardiac cycle remodels toward λ_{f_a}. A simple numerical algorithm to simulate this is to adopt linear differential equations for the remodeling of the recruitment stretches, i.e.:

$$\frac{\partial \lambda_R}{\partial t} = \xi_1 \left(\frac{\lambda_f |_{\max} - \lambda_{f_a}}{\lambda_{f_a}} \right) \tag{8.29}$$

where $\lambda_f |_{\max}$ denotes the maximum stretch experienced by the collagen fiber during the cardiac cycle and $\xi_1 > 0$ is a remodeling rate parameter. Note numerical schemes that address remodeling of fiber orientations have been simulated, e.g., see [72] and references therein.

8.6.7.5.3 GROWTH/ATROPHY OF THE COLLAGEN FABRIC

In vascular homeostasis the mass of the collagenous constituents is constant even though the fibers are in a continual state of deposition and degradation. However, in response to perturbations to the mechanical environment vascular cells can respond by up- (down)-regulating synthesis and down- (up)-regulating degradation leading to a net increase (decrease) in mass. We outline one algorithm proposed to simulate this [313], the key assumptions of which are:

- the reference configuration of the cells is equal to the reference configuration of the constituents that they are maintaining;
- the number of cells is proportional to the mass of constituents they are maintaining;
- in vascular homeostasis, the mass of the constituents is constant.

From these assumptions, the simplest (linear) equation for adapting the normalized mass-density of the collagenous constituents can be derived to be (see [316] for further details):

$$\frac{\partial \hat{m}_f}{\partial t} = \hat{m}_f \xi_2 \left(\frac{\lambda_f \left|_{\max} - \lambda_{f_a}\right.}{\lambda_{f_a}} \right) \tag{8.30}$$

where $\xi_2 \geqslant 0$ is a growth rate parameter. Note that this equation is derived by considering perturbations of the stretches of the vascular cells from homeostatic levels. It is the assumption that the reference configuration of the cells is equal to that of the constituent that they are maintaining leads to them being expressed as a function of the stretches of the collagen fiber.

8.7 MODELING VASCULAR DISEASE: INTRACRANIAL ANEURYSMS

Modeling both the biology and mechanics of the arterial wall is fundamental to furthering our understanding many arterial diseases. To illustrate this approach, we now consider a specific vascular pathology: intracranial aneurysms.

8.7.1 Background

The term *aneurysm* commonly refers to a pathologic dilatation of an artery. It comes from the Greek word $\alpha\nu\varepsilon\upsilon\rho\iota\sigma\mu\alpha$, a juxtaposition of $\alpha\nu\alpha$ meaning across and $\epsilon\upsilon\rho\iota\sigma$ meaning broad. Aneurysms have been recognized since ancient Egyptian times. In fact, the first description is attributed to the Egyptian polymath Imhotep (2655–2600 BC) and can be found in the Ebers Papyrus [223]. They most commonly occur in the abdominal aorta that supplies blood to the legs, and in the cerebral arteries that supply blood to the brain. The majority of intracranial aneurysms (*IAs*) are categorized as saccular aneurysms as they appear as sac-like outpouchings of the arteries inflated by the pressure of the blood. They are sometimes referred to as berry aneurysms; however, this terminology is perhaps misleading as the morphology can be complex, e.g., notice the irregular geometries of the IAs depicted in Fig. 8.23.

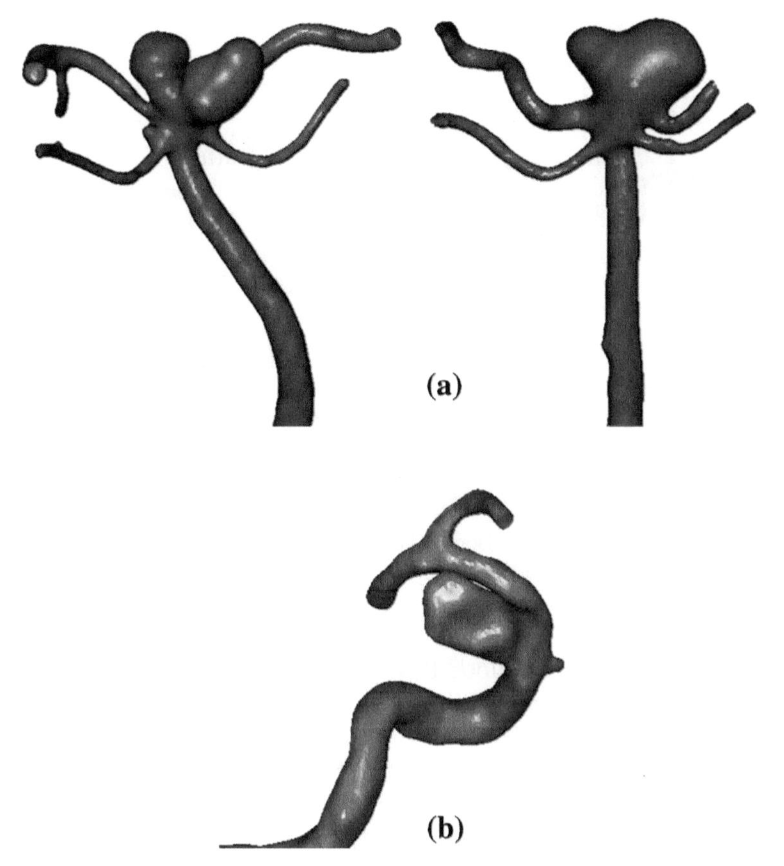

FIGURE 8.23 Examples of (a) basilar tip IAs and (b) a sidewall IA of the right internal carotid artery.

Prevalance rates of IAs in populations without comorbidity are estimated to be 3.2% [302]. Most remain asymptomatic; however, there is a small but inherent risk of rupture: 0.1–1% of detected IAs rupture every year [139]. Subarachnoid hemorrhage (*SAH*) due to rupture is associated with a 50% chance of fatality [105] and of those that survive, nearly half have long-term physical and mental sequelae [129]. Pre-emptive treatment can prevent aneurysm SAH and thus reduce the associated (large) financial burden, e.g., the total annual economic cost of aneurysm SAH is £510M in the UK [240]. However, management of unruptured IAs by interventional procedures, i.e., minimally invasive endovascular approaches or surgical-clipping, is highly controversial and not without risk [156]. Moreover, treatment is expensive: recent developments in imaging technology have led to a dramatic increase in coincidentally detected asymptomatic IAs. Indeed, endovascular intervention is now the major driving force behind the increase of costs in the US healthcare system [129]. Given the very low risk of IA rupture, there is both a clinical and an economic need to identify those IAs which would benefit from intervention.

IAs preferentially develop at or close to the apices of cerebral artery bifurcations at specific locations around the circle of Willis, a circle of arteries located at the base of the brain

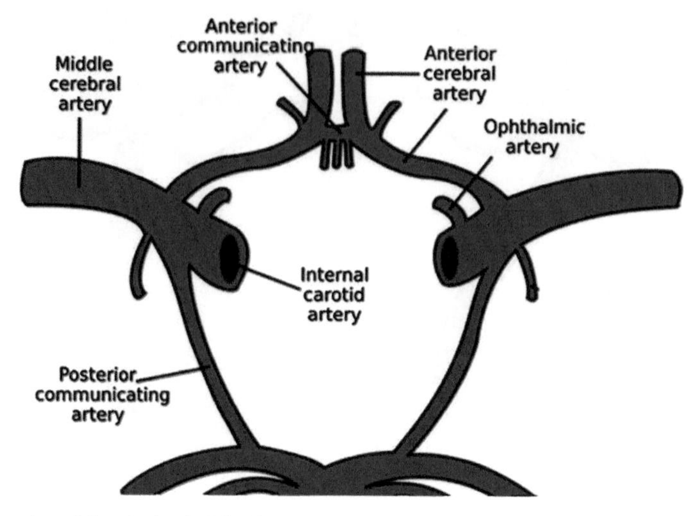

FIGURE 8.24 Arteries of the circle of Willis (source: Wikipedia).

(see Fig. 8.24). In such locations the blood flow dynamics are more complex, e.g., stagnation points, regions of accelerating/decelerating flow and regions where the flow oscillates in direction. Such disturbed flow can affect the functionality of ECs and hence it is hypothesized that the hemodynamic environment plays a role in the pathophysiological processes that give rise to IA formation. Development is associated with apoptosis of VSMCs (animal model) [157], disrupted internal elastin laminae, the breakage and elimination of elastin fibers [90] and a thinned medial layer. A large heterogeneity in wall structure and cellular content was found in ruptured aneurysms, though in most unruptured IAs, the inner surface of the aneurysm sac was found to be completely covered with normally shaped arterial ECs [90,146]. Although the endothelium may still be able to sense local hemodynamic stimuli, their response to the abnormal intrasaccular flow and ultimate influence on the biology of the aneurysm wall (and its stability or rupture) remain poorly understood.

8.7.2 Computational Modeling of Intracranial Aneurysms

Modeling of vascular disease holds the potential to yield insight into the disease process and lead to the development of computational tools to assist clinical diagnosis and treatment. Consequently, such research has grown extensively in recent years; for recent review articles see, e.g., [131,265]. Broadly speaking, computational modeling has four themes: morphological characterization, computational fluid dynamic analysis, structural analysis and aneurysm evolution models. The latter have gained increasingly in sophistication over the past decade, i.e., aneurysm evolution models combine CFD analyses, structural analyses and constitutive models to reflect the mechanobiology of the arterial wall.

Morphological analysis has emerged as a possible means to assess rupture risk [228,301]. Quantification of the topology of aneurysm sacs on large databases using simple 2-D measures and more complex 3-D measures provide data that can be used in the search for correlations

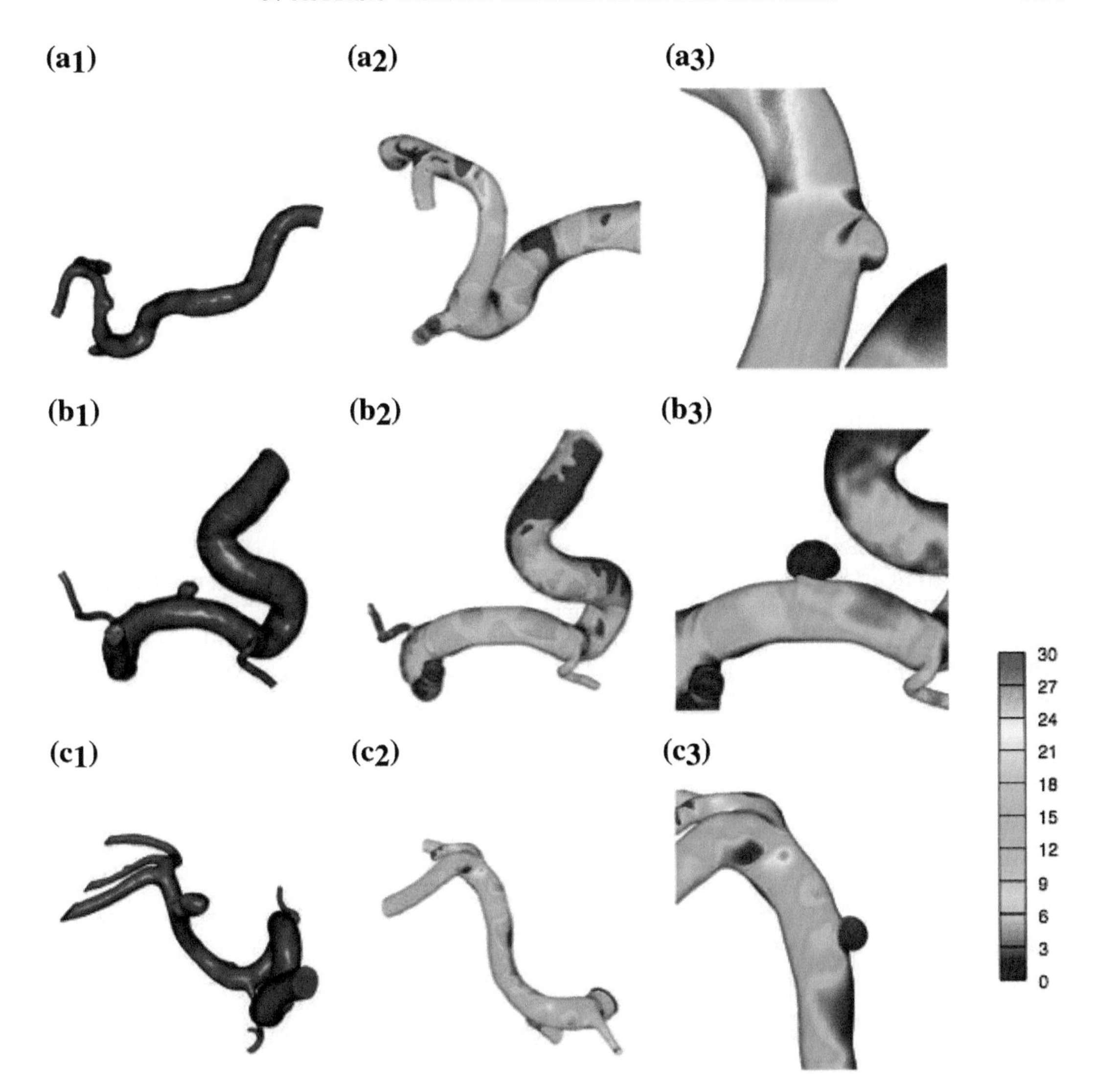

FIGURE 8.25 Segmented clinical imaging data depicting IAs (a_1–c_1). IAs are removed and the geometry of the healthy artery reconstructed (a_2–c_2). Computational models of IAs on patient-specific geometries with degradation of elastin linked to low WSS (a_3–c_3). The color map depicts WSS (Pa).

between geometric measures and rupture risk [221]. Such analyses can be performed in real time and are straightforward to interpret and thus have great clinical potential. However, to date, the results of purely morphological approaches have been inconclusive.

CFD research on IAs has provided us with tremendous insights regarding the variability of flow within the aneurysm dome and illustrated some of the challenges and complexities we face in attempting to further our understanding of the relationship between flow and IA inception [19,71,85,162,185,271,276]; IA enlargement [38,229,273,290]; IA rupture [50–52,272,334]; thrombus formation [230,231]; interventional treatment, e.g., stents ([14,92,138,153,154]) or coils [141]. To date, the majority of studies assume rigid boundaries for the CFD analysis; however, more recently flexible boundaries [26,289,292,335] are modeled, though these studies are limited by the dearth of data on the mechanical properties of the aneurysm wall.

CFD studies have the potential to provide a hemodynamic identifier for an aspect of an aneurysms pathology, e.g., why it formed in a particular location, the reason for its continued enlargement, the likelihood of its rupture. Many indices are considered, e.g., wall shear stress (WSS), spatial wall shear stress gradients ($\overline{\text{WSSG}}$), oscillatory shear index (OSI $\in [0, .5]$), aneurysm formation index (*AFI*), gradient oscillatory number (*GON* $\in [0, 1]$). The *AFI* $\in [-1, 1]$, see [185], is defined to be the cosine of the angle made between the instantaneous wall shear stress vector and its time-averaged (mean) value; evaluating the minimum of the AFI over the cardiac cycle, say min(AFI), characterizes the maximum deviation of the wall shear stress vector from its mean direction. The GON, see [271], is a measure of the average deviation of the spatial gradient of the wall shear stress from its mean direction over the cardiac cycle.

Figure 8.26 illustrates examples of the spatial distributions of the aforementioned indices for a sidewall aneurysm on the internal carotid artery. It can be seen that a jet enters the aneurysm (a) and this results in a local elevation in WSS at the apex of the dome. Notice though that the overall WSS distribution within the aneurysm sac is relatively low compared to values in the parent artery. The spatial $\overline{\text{WSSG}}s$ are elevated around the aneurysm neck and at the bifurcation. Regions of oscillatory flow, characterized by high values of OSI and low values of min(AFI), are observed in regions of high curvature of the parent and a complex distribution is observed within the aneurysm sac. Note that while the OSI provides a time-averaged measure of the oscillation of the WSS vector about its mean direction during the cardiac cycle,

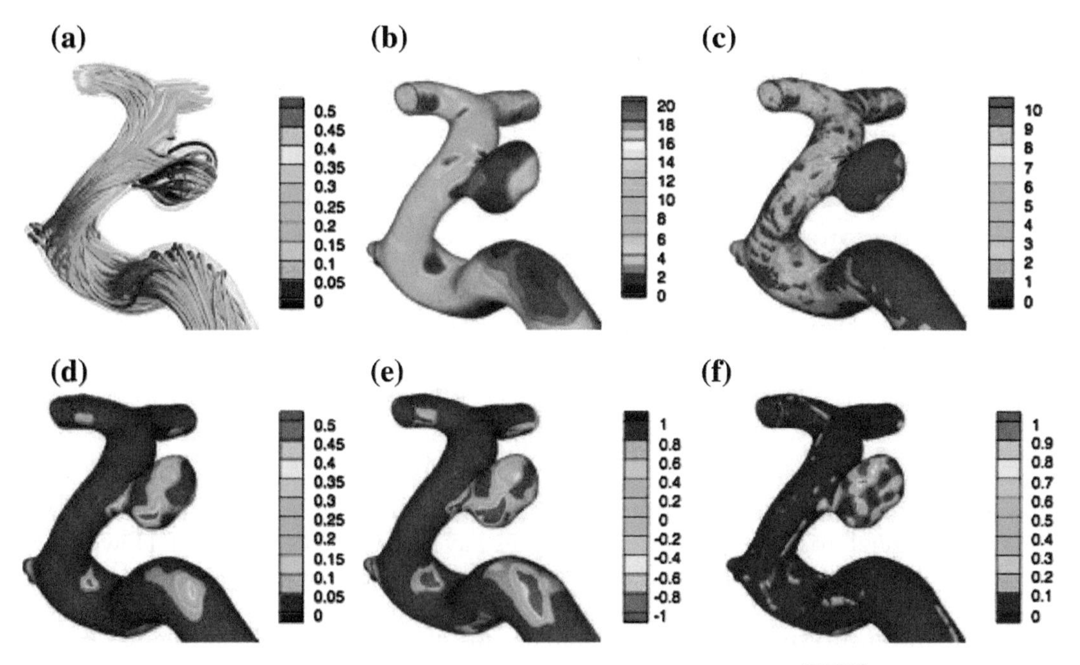

FIGURE 8.26 (a) Streamlines (m/s) and spatial distributions of (b) WSS (Pa), (c) $\overline{\text{WSSG}}$ (kPa/m), (d) OSI, (e) min(AFI), and (f) GON.

min(AFI) quantifies the cosine of the angle between maximum deviation of the WSS vector from its mean direction. Hence each value of min(AFI) has a unique geometrical interpretation whereas there is ambiguity in the interpretation of values of OSI. Lastly, (f) illustrates the distribution of the GON index. It can be seen that a complex distribution occurs within the aneurysm sac. It has been suggested that this index is linked to regions of aneurysm inception [271] although others have questioned its significance [85]. However, despite the plethora of CFD studies, or perhaps as a consequence of such studies, the significance of CFD findings is currently the focus of lively debate among the clinical and modeling communities, e.g., see [144,49,242,284].

Aneurysm rupture occurs when tissue stress exceeds tissue strength. In theory, patient-specific stress analyses have the greatest to offer with regard to the potential for improved predictive measures. Unfortunately, limited data exist in the literature for the mechanical properties of aneurysm tissue [62] and the spatially heterogeneous thickness distribution cannot currently be obtained from imaging data. Moreover, there is little data on the strength of the tissue which can also be spatially heterogeneous. To date, structural analyses have often focused on conceptual geometrical models of IAs which use idealized geometries [64,65,132,163,247,262,267] while relatively few utilize patient-specific geometries [177,178,227,301,343–345]. Lastly, in recent years there has been a focus on modeling the evolution of IAs, e.g., see [10,16,17,75,78,79,159,160, 172,173,180,256,257,270,313,316,317,319,320,333]. We will proceed to discuss this aspect of research in more detail.

8.7.3 Example: Fluid-Solid-Growth Model of Aneurysm Evolution

The physiological mechanisms that give rise to the development of an aneurysm involve the complex interplay between the local mechanical forces acting on the arterial wall and the biological processes occurring at the cellular level. Consequently, models of aneurysm evolution must take into consideration: (i) the biomechanics of the arterial wall; (ii) the biology of the arterial wall; and (iii) the complex interplay between (i) and (ii), i.e., the mechanobiology of the arterial wall. *Humphrey and Taylor* [135] recently emphasized the need for a new class of *Fluid-Solid-Growth* models to study aneurysm evolution and proposed the terminology *FSG* models. These combine fluid and solid mechanics analyses of the vascular wall with descriptions of the kinetics of biological growth, remodeling and damage, *GR&D*. In this section, we briefly overview a novel fluid–solid-growth FSG computational framework for modeling aneurysm evolution. It utilizes and extends the abdominal aortic aneurysm evolution model developed by *Watton et al.* [312,315] which was later adapted to model IA evolution [313,317] and extended to consider transmural variations in G&R [256,257]. The aneurysm evolution model incorporates microstructural G&R variables into a realistic structural model of the arterial wall [123]. These describe the *normalized mass-density* and *natural reference configurations* of the load bearing constituents and enable the G&R of the tissue to be simulated as an aneurysm evolves. More specifically, the natural reference configurations that collagen fibers are recruited to load bearing remodels to simulate the mechanical consequences of: (i) fiber deposition and degradation in altered configurations as the aneurysm enlarges; (ii) fibroblasts configuring the collagen to achieve a maximum strain during the cardiac cycle, denoted the *attachment strain*. The normalized mass-density evolves to simulate growth/atrophy of the constituents (elastin and collagen). The aneurysm evolution model has been integrated into a novel FSG framework [316] so that G&R can be explicitly linked to hemodynamic stimuli. More recently, the G&R framework has been extended to link both *growth* and *remodeling* to cyclic deformation of vascular cells (see [314]).

Figure 8.27 depicts the *FSG* methodology. The computational modeling cycle begins with a structural analysis to solve the systolic and diastolic equilibrium deformation fields (of the artery/aneurysm) for given pressure and boundary conditions. The structural analysis quantifies the stress, stretch and the cyclic deformation of the constituents and vascular cells (each of which can have different natural reference configurations). The geometry of the aneurysm is subsequently exported to be prepared for hemodynamic analysis: first the geometry is integrated into a physiological geometrical domain; the domain is automatically meshed; physiological flow rate and pressure boundary conditions are applied; the flow is solved assuming rigid boundaries for the hemodynamic domain. The hemodynamic quantities of interest, e.g., WSS, $\overline{\text{WSSG}}$, are then exported and interpolated onto the nodes of the structural mesh: each node of the structural mesh contains information regarding the mechanical stimuli obtained from the hemodynamic and structural analyses. G&R algorithms simulate cells responding to the mechanical stimuli and adapting the tissue: the constitutive model of the aneurysmal tissue is updated. The structural analysis is re-executed to calculate the new equilibrium deformation fields. The updated geometry is exported for hemodynamic analysis. The cycle continues and as the tissue adapts an aneurysm evolves. To simulate IA inception, a localized loss of elastin is prescribed within a small circular patch of the arterial domain (the elastin is modeled with a neo-Hookean constitutive model [318]). The collagen fabric adapts to restore homeostasis and a small perturbation to the geometry alters the spatial distribution of hemodynamic stimuli that act on the lumenal layer of the artery. This enables subsequent degradation of elastin to be linked to deviations of hemodynamic stimuli from homeostatic levels via evolution equations. As the elastin degrades and the collagen fabric adapts (via G&R) an IA evolves. This approach was adopted to investigate the evolution of IAs assuming degradation of elastin was linked to high WSS or high $\overline{\text{WSSG}}$ [316]. Given that a region of elevated WSS occurs downstream of the distal neck of the model IA and elevated spatial $\overline{\text{WSSG}}$*s* occur in the proximal/distal neck regions, this approach led to IAs that enlarged axially along the arterial domain, i.e., it did not yield IA with characteristic 'berry' topologies. Consequently,

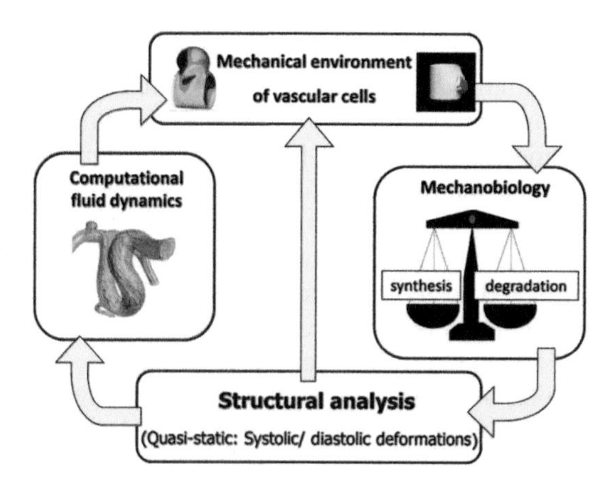

FIGURE 8.27 Fluid-solid-growth computational framework for modeling aneurysm evolution.

Watton et al. linked elastin degradation to low WSS and restricted the degradation of elastin to a localized region of the arterial domain [319]: this yielded IAs of a characteristic saccular shape that enlarged and stabilized in size. Although interesting insights were obtained in both studies, an inherent limitation was that the IAs evolved on a cylindrical section of artery and consequently the spatial distribution of hemodynamic stimuli is non-physiological. This motivated the application of the FSG modeling framework to patient-specific vascular geometries. For a detailed description of the model methodology, we refer the interested reader to [314,320]. Here we briefly illustrate the application of the *FSG* modeling framework to three clinical cases.

Figure 8.25 (left column) illustrates three clinical cases depicting IAs. The IA is removed and replaced with a short cylindrical nonlinear-elastic (two-layered) membrane model of a healthy arterial wall, e.g., see [315]. It is on this section that IA evolution is simulated. The cylindrical section is smoothly reconnected to the upstream and downstream sections of the parent artery (middle column; see [261] for methodology). In all three cases, IA inception is prescribed, i.e., an initial degradation of elastin is prescribed in a localized region of the domain, the collagen fabric adapts to restore homeostasis and a small localized outpouching of the artery develops. This perturbs the hemodynamic environment: subsequent degradation of elastin is linked to low levels of WSS. It can be seen that the modeling framework gives rise to IAs with different morphologies, i.e., IAs with: asymmetries in geometries (a_3); well-defined necks (b_3); no neck (c_3). Interestingly, for case a_3, which depicts an IA at (perhaps) a relatively early stage of formation (crudely inferred from its small size), the qualitative asymmetries of the simulated IA (see (a_3)) are in agreement with the patient aneurysm (a_1) and thus (tentatively) support the modeling hypotheses for elastin degradation (low WSS drives degradation) and collagen adaption.

Figure 8.28 illustrates the evolution of aneurysm depicted in Fig. 8.25a in more detail. Due to the asymmetry in the distribution of the WSS, the region of elastin degradation evolves asymmetrically (see Fig. 8.28(a)), i.e., the elastin degrades at a greater rate in the proximal region of the aneurysm. The asymmetry in the evolution of the elastin degradation creates a substantial asymmetry in the evolution of the aneurysm geometry: the proximal side of the dome develops a well-defined aneurysm neck whereas the distal region of the aneurysm flattens to connect with the downstream section of the artery smoothly. Figures 8.28c and d illustrates the evolution of the Green-Lagrange strain of the elastin. The strains increase to the greatest extent at the upper proximal region of the dome. As the aneurysm enlarges, the asymmetry in the strain distribution increases. In contrast to the evolution of the elastin strain, the collagen fiber Green-Lagrange strains increase negligibly even though the deformations are large, see Fig. 8.28d, i.e., maximum values of 0.14 for collagen as opposed to 10 for elastin. This is due to the evolution of the natural reference configurations that the fibers are recruited to load bearing. For further details on this particular example, the interested reader is referred to [320].

8.7.4 Discussion

It is envisaged that models of aneurysm evolution can ultimately lead to predictive models that have diagnostic relevance on a patient-specific basis. Given that this will yield substantial healthcare and economic benefits, there is significant growth of research in this area. However, while models of aneurysm evolution have gained increasing sophistication over the

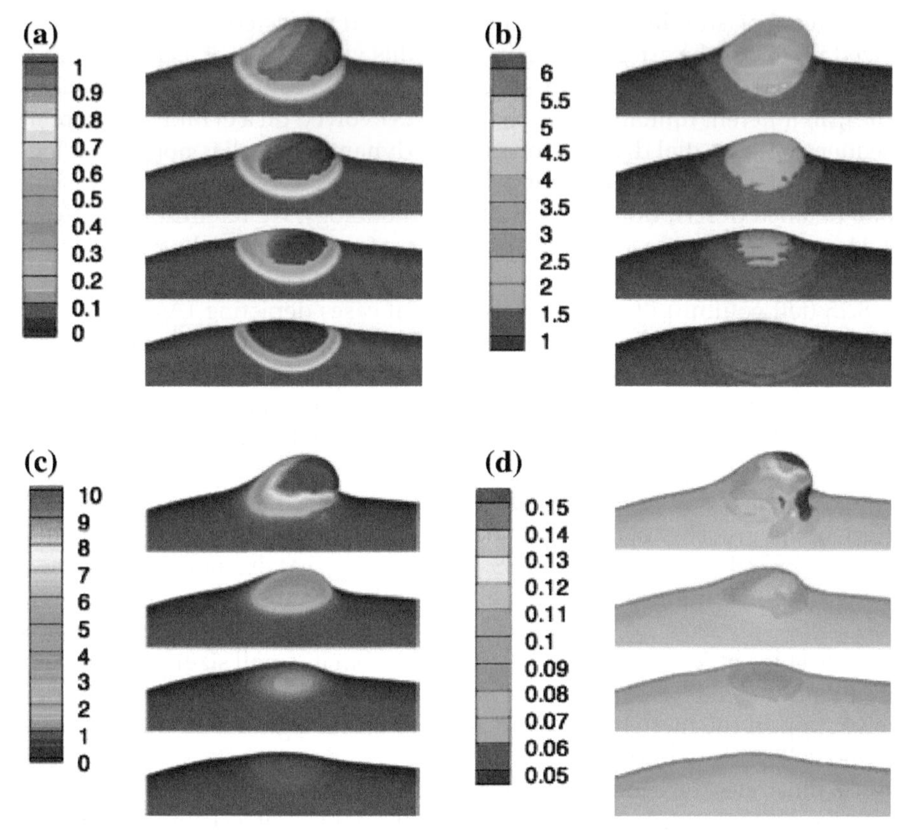

FIGURE 8.28 Evolution of (a) the elastin (normalized) mass density $\hat{m}_E$, (b) the average collagen fiber (normalized) mass density $\hat{m}_C$, (c) the elastin strain (E_{22}), and (d) the Green-Lagrange strain of the medial collagen fibers as the aneurysm depicted in Figure 8.25c enlarges in size. Note that the collagen strains increase negligibly due to the evolution of the natural reference configurations that they are recruited to load bearing.

past decade, many further improvements are still needed to reach clinical value. For instance, there is a need to incorporate explicit representations of vascular cells (endothelial cells, fibroblast cells and vascular smooth muscle cells) as well as their interactions, and the signaling networks [91,121] that link the stimuli acting on them to their functionality in physiological, supraphysiological and pathological conditions. There is also a need for implementation of more sophisticated constitutive models to represent, e.g., the collagen fiber recruitment and orientation distributions [120] and, with respect to initiation, the active and passive response of vascular smooth muscle cells [208]. As already mentioned, there is a tremendous need for mechanical and ECM structural data. Lastly, improved understanding and modeling of how this complex microstructure adapts in pathological conditions are needed: the modeling framework needs to be validated and/or calibrated against physiological data; animal models undoubtedly have a role to play in this respect (e.g., [140,193,342]). Such enhancements will offer the potential for patient-specific predictive models of vascular disease evolution and intervention. They will benefit patients immensely because the decision on whether to/how to intervene will be founded upon a robust concentration of knowledge with respect to patient-specific vascular physiology, biology and biomechanics. Of course, the challenging and multi-disciplinary nature of such research implies collaborations are essential.

References

[1] C.M. Aguilera, S.J. George, J.L. Johnson, A.C. Newby, Relationship between type IV collagen degradation, metalloproteinase activity and smooth muscle cell migration and proliferation in cultured human saphenous vein, Cardiovascular Research 58 (3) (2003) 679–688.

[2] W.C. Aird, Spatial and temporal dynamics of the endothelium, Journal of Thrombosis and Haemostasis 3 (7) (2005) 1392–1406, http://dx.doi.org/10.1111/j.1538-7836.2005.01328.x.

[3] W.C. Aird, Mechanisms of endothelial cell heterogeneity in health and disease, Circulation Research 98 (2) (2006) 159–162, http://dx.doi.org/10.1161/01.RES.0000204553.32549.a7.

[4] W.C. Aird, Phenotypic heterogeneity of the endothelium: I. Structure, function, and mechanisms, Circulation Research 100 (2) (2007) 158–173, http://dx.doi.org/10.1161/01.RES.0000255691.76142.4a.

[5] W.C. Aird, Phenotypic heterogeneity of the endothelium: II. Representative vascular beds, Circulation Research 100 (2) (2007) 174–190, http://dx.doi.org/10.1161/01.RES.0000255690.03436.ae.

[6] W.C. Aird, Endothelium in health and disease, Pharmacological Reports 60 (1) (2008) 139–143.

[7] V. Alastrué, M.A. Martínez, M. Doblaré, A. Menzel, Anisotropic micro-sphere-based finite elasticity applied to blood vessel modelling, Journal of the Mechanics and Physics of Solids 57 (1) (2009) 178–203.

[8] J. van den Akker, B.G. Tuna, A. Pistea, A.J.J. Sleutel, E.N.T.P. Bakker, E. van Bavel, Vascular smooth muscle cells remodel collagen matrices by long-distance action and anisotropic interaction, Medical and Biological Engineering and Computing 50 (7) (2012) 701–715, http://dx.doi.org/10.1007/s11517-012-0916-6.

[9] D. Aronson, Cross-linking of glycated collagen in the pathogenesis of arterial and myocardial stiffening of aging and diabetes, Journal of Hypertension 21 (2003) 3–12.

[10] C.A. Figueroa, S. Baek, C.A. Taylor, J.D. Humphrey, A computational framework for fluidsolid-growth modeling in cardiovascular simulations, Computer Methods in Applied Mechanics and Engineering 198 (45–46) (2009) 3583–3602, http://dx.doi.org/10.1016/j.cma.2008.09.013.

[11] B. Alberts, A. Johnson, J. Lewis, M. Raff, K. Roberts, P. Walter, The extracellular matrix of animals, in: Molecular Biology of the Cell, fourth ed., Garland Science, New York, 2002.

[12] B. Alberts, A. Johnson, J. Lewis, M. Raff, K. Roberts, P. Walter, Molecular Biology of the Cell, fifth ed., Garland Science, New York, 2008.

[13] M.R. Alexander, G.K. Owens, Epigenetic control of smooth muscle cell differentiation and phenotypic switching in vascular development and disease, Annual Review of Physiology 74 (2012) 13–40, http://dx.doi.org/10.1146/annurev-physiol-012110-142315.

[14] L. Augsburger, P. Reymond, D. Rufenacht, N. Stergiopulos, Intracranial stents being modeled as a porous medium: flow simulation in stented cerebral aneurysms, Annals of Biomedical Engineering 39 (2) (2011) 850–863, http://dx.doi.org/10.1007/s10439-010-0200-6.

[15] N.C. Avery, A.J. Bailey, The effects of the Maillard reaction on the physical properties and cell interactions of collagen, Pathologie-Biologie (Paris) 54 (7) (2006) 387–395.

[16] S. Baek, K.R. Rajagopal, J.D. Humphrey, Competition between radial expansion and thickening in the enlargement of an intracranial saccular aneurysm, Journal of Elasticity 80 (1–3) (2005) 13–31, http://dx.doi.org/10.1007/s10659-005-9004-6.

[17] S. Baek, K.R. Rajagopal, J.D. Humphrey, A theoretical model of enlarging intracranial fusiform aneurysms, Journal of Biomechanical Engineering 128 (1) (2006) 142, http://dx.doi.org/10.1115/1.2132374.

[18] S. Baek, R.L. Gleason, K.R. Rajagopal, J.D. Humphrey, Theory of small on large: potential utility in computations of fluid-solid interactions in arteries, Computer Methods in Applied Mechanics and Engineering 196 (31–32) (2007) 3070–3078, http://dx.doi.org/10.1016/j.cma.2006.06.018.

[19] H. Baek, M.V. Jayaraman, G.E. Karniadakis, Wall shear stress and pressure distribution on aneurysms and infundibulae in the posterior communicating artery bifurcation, Annals of Biomedical Engineering 37 (12) (2009) 2469–2487, http://dx.doi.org/10.1007/s10439-009-9794-y.

[20] S. Baek, T.J. Pence, On mechanically induced degradation of fiber-reinforced hyperelastic materials, Mathematics and Mechanics of Solids 16 (4) (2011) 406–434.
[21] L. Bailly, C. Geindreau, L. Orgeas, V. Deplano, Towards a biomimetism of abdominal healthy and aneurysmal arterial tissues, Journal of the Mechanical Behavior of Biomedical Materials 10 (2012) 151–165.
[22] D. Balzani, J. Schroeder, D. Gross, Modeling of eigenstresses and damage in arterial walls, Proceedings in Applied Mathematics and Mechanics 6 (1) (2006) 127–128.
[23] D. Balzani, J. Schroeder, D. Gross, Simulation of discontinuous damage incorporating residual stresses in circumferentially overstretched atherosclerotic arteries, Acta Biomaterialia 2 (6) (2006) 609–618.
[24] M.J. Barnes, R.W. Farndale, Collagens and atherosclerosis, Experimental Gerontology 34 (4) (1999) 513–525.
[25] R.I. Bashey, A. Martinez-Hernandez, S.A. Jimenez, Isolation, characterization, and localization of cardiac collagen type VI. Associations with other extracellular matrix components, Circulation Research 70 (5) (1992) 1006–1017.
[26] Y. Bazilevs, M.C. Hsu, Y. Zhang, W. Wang, T. Kvamsdal, S. Hentschel, J.G. Isaksen, Computational vascular fluid-structure interaction: methodology and application to cerebral aneurysms, Biomechanics and Modeling in Mechanobiology 9 (4) (2010) 481–498, http://dx.doi.org/10.1007/s10237-010-0189-7.
[27] M.F. Beatty, Topics in finite elasticity: hyperelasticity of rubber, elastomers, and biological tissues with examples, Applied Mechanics Reviews 40 (12) (1987) 1699–1734.
[28] M.P. Bendeck, S. Regenass, W.D. Tom, C.M. Giachelli, S.M. Schwartz, C. Hart, M.A. Reidy, Differential expression of alpha 1 type VIII collagen in injured platelet-derived growth factor-BB-stimulated rat carotid arteries, Circulation Research 79 (3) (1996) 524–531.
[29] B.M. van den Berg, J.A.E. Spaan, T.M. Rolf, H. Vink, Atherogenic region and diet diminish glycocalyx dimension and increase intima-to-media ratios at murine carotid artery bifurcation, American Journal of Physiology. Heart and Circulatory Physiology 290 (2) (2006) H915–H920.
[30] A. Bershadsky, M. Kozlov, B. Geiger, Adhesion-mediated mechanosensitivity: a time to experiment, and a time to theorize, Current Opinion in Cell Biology 18 (5) (2006) 472–481, http://dx.doi.org/10.1016/j.ceb.2006.08.012.
[31] A.P. Bhole, B.P. Flynn, M. Liles, N. Saeidi, C.A. Dimarzio, J.W. Ruberti, Mechanical strain enhances survivability of collagen micronetworks in the presence of collagenase: implications for load-bearing matrix growth and stability, Philosophical Transactions Series A, Mathematical, Physical, and Engineering Sciences 367 (1902) (2009) 3339–3362, http://dx.doi.org/10.1098/rsta.2009.0093.
[32] D.E. Birk, Type V collagen: heterotypic type I/V collagen interactions in the regulation of fibril assembly, Micron 32 (3) (2001) 223–237.
[33] D.E. Birk, P. Bruckner, Collagens, suprastructures, and collagen fibril assembly, in: R.P. Mecham (Ed.), The Extracellular Matrix: An Overview, Biology of Extracellular Matrix, Spring-Verlag, New York, 2011, pp. 77–115.
[34] D.E. Birk, J.M. Fitch, J.P. Babiarz, K.J. Doane, T.F. Linsenmayer, Collagen fibrillogenesis in vitro: interaction of types I and V collagen regulates fibril diameter, Journal of Cell Science 95 (Pt. 4) (1990) 649–657.
[35] D.E. Birk, F.H. Silver, R.L. Trelstad, Matrix assembly, in: E.D. Hay (Ed.), Cell Biology of Extracellular Matrix, Plenum Press, New York, 1997, pp. 221–254.
[36] J.P. Borel, G. Bellon, Vascular collagens. General review, Pathologie-Biologie 33 (4) (1985) 254–260.
[37] G.A. Borelli, De Motu Animalium, vol. II, Petrum Van der Aa, Lugdunum Batavorum, 1710.
[38] L. Boussel, V. Rayz, C. McCulloch, A. Martin, G. Acevedo-Bolton, M. Lawton, R. Higashida, W.S. Smith, W.L. Young, D. Saloner, Aneurysm growth occurs at region of low wall shear stress: patient-specific correlation of hemodynamics and growth in a longitudinal study, Stroke 39 (11) (2008) 2997–3002, http://dx.doi.org/10.1161/STROKEAHA.108.521617.
[39] N.D. Broom, G. Ramsey, R. Mackie, B.J. Martins, W.W. Stehbens, A new biomechanical approach to assessing the fragility of the internal elastic lamina of the arterial wall, Connective Tissue Research 30 (2) (1993) 143–155.
[40] A.C. Burton, Relation of structure to function of the tissues of the wall of blood vessels, Physiological Reviews 34 (1954) 619–642.
[41] F. Cacho, P.J. Elbischger, J.F. Rodrguez, M. Doblar, G.A. Holzapfel, A constitutive model for fibrous tissues considering collagen fiber crimp, International Journal of Non-Linear Mechanics 42 (2007) 391–402.
[42] G.J. Cambell, M.R. Roach, G.J. Campbell, Fenestrations in the internal elastic lamina at bifurcations of human cerebral arteries, Stroke 12 (4) (1981) 489–496.
[43] G.J. Campbell, M.R. Roach, The use of ligament efficiency to model fenestrations in the internal elastic lamina of cerebral arteries. II – Analysis of the spatial geometry, Journal of Biomechanics 16 (10) (1983) 883–891.

[44] P.B. Canham, H.M. Finlay, J.G. Dixon, D.R. Boughner, A. Chen, Measurements from light and polarised light microscopy of human coronary arteries fixed at distending pressure, Cardiovascular Research 23 (11) (1989) 973–982.

[45] E.G. Canty, Y. Lu, R.S. Meadows, M.K. Shaw, D.F. Holmes, K.E. Kadler, Coalignment of plasma membrane channels and protrusions (fibripositors) specifies the parallelism of tendon, The Journal of Cell Biology 165 (4) (2004) 553–563.

[47] M.M. Carroll, Pressure maximum behavior in inflation of incompressible elastic hollow spheres and cylinders, Quarterly of Applied Mathematics 45 (1) (1987) 141–154.

[48] S. Canič, C.J. Hartley, D. Rosenstrauch, J. Tambaca, G. Guidoboni, A. Mikelič, Blood flow in compliant arteries: an effective viscoelastic reduced model, numerics, and experimental validation, Annals of Biomedical Engineering 34 (4) (2006) 575–592.

[49] J.R. Cebral, H. Meng, Counterpoint: realizing the clinical utility of computational fluid dynamics – closing the gap, AJNR. American Journal of Neuroradiology 33 (3) (2012) 396–398, http://dx.doi.org/10.3174/ajnr.A2994.

[50] J.R. Cebral, M.A. Castro, J.E. Burgess, R.S.P.M.J. Sheridan, C.M. Putman, Characterization of cerebral aneurysm for assessing risk of rupture using patient-specific computational hemodynamics models, AJNR. American Journal of Neuroradiology 26 (10) (2005) 2550–2559.

[51] J.R. Cebral, F. Mut, D. Sforza, R. Löhner, E. Scrivano, P. Lylyk, C. Putman, Clinical application of image-based CFD for cerebral aneurysms, International Journal for Numerical Methods in Biomedical Engineering 27 (7) (2011) 977–992, http://dx.doi.org/10.1002/cnm.

[52] J.R. Cebral, F. Mut, J. Weir, C. Putman, Quantitative characterization of the hemodynamic environment in ruptured and unruptured brain aneurysms, AJNR. American Journal of Neuroradiology 32 (1) (2011) 145–151, http://dx.doi.org/10.3174/ajnr.A2419.

[53] P. Chadwick, Continuum Mechanics: Concise Theory and Problems, Dover, Mineola, NY, 1999.

[54] I. Chatziprodromou, D. Poulikakos, Y. Ventikos, On the influence of variation in haemodynamic conditions on the generation and growth of cerebral aneurysms and atherogenesis: a computational model, Journal of Biomechanics 40 (16) (2007) 3626–3640, http://dx.doi.org/10.1016/j.jbiomech.2007.06.013.

[55] S. Chien, Effects of disturbed flow on endothelial cells, Annals of Biomedical Engineering 36 (4) (2007) 554–562.

[56] M. Chiquet, A.S. Renedo, F. Huber, M. Flück, How do fibroblasts translate mechanical signals into changes in extracellular matrix production? Matrix Biology 22 (1) (2003) 73–80, http://dx.doi.org/10.1016/S0945-053X(03)00004-0.

[57] M. Chiquet, L. Gelman, R. Lutz, S. Maier, From mechanotransduction to extracellular matrix gene expression in fibroblasts, Biochimica et Biophysica Acta 1793 (5) (2009) 911–920, http://dx.doi.org/10.1016/j.bbamcr.2009.01.012.

[58] J.J. Chiu, S. Chien, Effects of disturbed flow on vascular endothelium: pathophysiological basis and clinical perspectives, Physiological Reviews 91 (1) (2011) 327–387, http://dx.doi.org/10.1152/physrev.00047.2009.

[59] D. Choquet, D.P. Felsenfeld, M.P. Sheetz, Extracellular matrix rigidity causes strengthening of integrin-cytoskeleton linkages, Cell 88 (1) (1997) 39–48.

[60] B.J. Chung, The Study of Blood Flow in Arterial Bifurcations: The Influence of Hemodynamics on Endothelial Cell Response to Vessel Wall Mechanics, Ph.D. Thesis, University of Pittsburgh, 2004.

[61] E. Chung, K. Rhodes, E.J. Miller, Isolation of three collagenous components of probable basement membrane origin from several tissues, Biochemical and Biophysical Research Communications 71 (4) (1976) 1167–1174.

[62] V. Costalat, M. Sanchez, D. Ambard, L. Thines, N. Lonjon, F. Nicoud, H. Brunel, J.P. Lejeune, H. Dufour, P. Bouillot, J.P. Lhaldky, K. Kouri, F. Segnarbieux, C.A. Maurage, K. Lobotesis, M.C. Villa-Uriol, C. Zhang, A.F. Frangi, G. Mercier, A. Bonafé, L. Sarry, F. Jourdan, Biomechanical wall properties of human intracranial aneurysms resected following surgical clipping (IRRAs project), Journal of Biomechanics 44 (15) (2011) 2685–2691, http://dx.doi.org/10.1016/j.jbiomech.2011.07.026.

[63] P.M. Cummins, N. von Offenberg Sweeney, M.T. Killeen, Y.A. Birney, E.M. Redmond, P.A. Cahill, N. von Offenberg Sweeney, Cyclic strain-mediated matrix metalloproteinase regulation within the vascular endothelium: a force to be reckoned with, American Journal of Physiology Heart and Circulatory Physiology 292 (1) (2007) H28–H42, http://dx.doi.org/10.1152/ajpheart.00304.2006.

[64] J.C. Daniel, A. Tongen, D.P.A. Warne, P.G. Warne, A 3D nonlinear anisotropic spherical inflation model for intracranial saccular aneurysm elastodynamics, Mathematics and Mechanics of Solids 15 (3) (2010) 279–307, http://dx.doi.org/10.1177/1081286508100498.

[65] G. David, J.D. Humphrey, Further evidence for the dynamic stability of intracranial saccular aneurysms, Journal of Biomechanics 36 (2003) 1143–1150.

[66] P.F. Davies, Hemodynamic shear stress and the endothelium in cardiovascular pathophysiology, Nature Clinical Practice Cardiovascular Medicine 6 (1) (2009) 16–26, http://dx.doi.org/10.1038/ncpcardio1397.

[67] E.C. Davis, Elastic lamina growth in the developing mouse aorta, The Journal of Histochemistry and Cytochemistry 43 (1995) 1115–1123.

[68] H. Di Wang, M.T. Rätsep, A. Chapman, R. Boyd, Adventitial fibroblasts in vascular structure and function: the role of oxidative stress and beyond, Canadian Journal of Physiology and Pharmacology 88 (3) (2010) 177–186, http://dx.doi.org/10.1139/Y10-015.

[69] A. Didangelos, X. Yin, K. Mandal, A. Saje, A. Smith, Q. Xu, M. Jahangiri, M. Mayr, Extracellular matrix composition and remodeling in human abdominal aortic aneurysms: a proteomics approach, Molecular & Cellular Proteomics 10 (8) (2011) M111 008,128.

[70] K.P. Dingemans, P. Teeling, J.H. Lagendijk, A.E. Becker, Extracellular matrix of the human aortic media: an ultrastructural histochemical and immunohistochemical study of the adult aortic media, The Anatomical Record 258 (1) (2000) 1–14.

[71] C. Doenitz, K.M.M. Schebesch, R. Zoephel, A. Brawanski, A mechanism for rapid development of intracranial aneurysms: a case study, Neurosurgery 67 (5) (2010) 1213–1221, http://dx.doi.org/10.1227/NEU.0b013e3181f34def.

[72] N.J.B. Driessen, Cox MaJ, C.V.C. Bouten, F.P.T. Baaijens, Remodelling of the angular collagen fiber distribution in cardiovascular tissues, Biomechanics and Modeling in Mechanobiology 7 (2) (2008) 93–103, http://dx.doi.org/10.1007/s10237-007-0078-x.

[74] A. Enzerink, A. Vaheri, Fibroblast activation in vascular inflammation, Journal of Thrombosis and Haemostasis 9 (4) (2011) 619–626, http://dx.doi.org/10.1111/j.1538-7836.2011.04209.x.

[75] T. Eriksson, M. Kroon, G.A. Holzapfel, Influence of medial collagen organization and axial in situ stretch on saccular cerebral aneurysm growth, ASME Journal of Biomechanical Engineering 131 (2009) 101,010.

[76] V. Everts, A. Niehof, D. Jansen, W. Beertsen, Type VI collagen is associated with microfibrils and oxytalan fibers in the extracellular matrix of periodontium, mesenterium and periosteum, Journal of Periodontal Research 33 (2) (1998) 118–125.

[77] R.W. Farndale, J.J. Sixma, M.J. Barnes, P.G. de Groot, The role of collagen in thrombosis and hemostasis, Journal of Thrombosis and Haemostasis 2 (4) (2004) 561–573.

[78] Y. Feng, S. Wada, K.I. Tsubota, T. Yamaguchi, Growth of intracranial aneurysms arised from curved vessels under the influence of elevated wall shear stress – a computer simulation study, JSME International Journal Series C 47 (4) (2004) 1035–1042, http://dx.doi.org/10.1299/jsmec.47.1035.

[79] Y. Feng, S. Wada, T. Ishikawa, K.I. Tsubota, T. Yamaguchi, A rule-based computational study on the early progression of intracranial aneurysms using fluid-structure interaction: comparison between straight model and curved model, Journal of Biomechanical Science and Engineering 3 (2) (2008) 124–137, http://dx.doi.org/10.1299/jbse.3.124.

[80] G.G. Ferguson, Physical factors in the initiation, growth, and rupture of human intracranial saccular aneurysms, Journal of Neurosurgery 37 (6) (1972) 666–677.

[81] J.H. Fessler, N. Shigaki, L.I. Fessler, Biosynthesis and properties of procollagens V, Annals of the New York Academy of Sciences 460 (1985) 181–186.

[82] H.M. Finlay, L. McCullough, P.B. Canham, Three-dimensional collagen organization of human brain arteries at different transmural pressures, Journal of Vascular Research 32 (1995) 301–312.

[83] B.P. Flynn, G.E. Tilburey, J.W. Ruberti, Highly sensitive single-fibril erosion assay demonstrates mechanochemical switch in native collagen fibrils, Biomechanics and Modeling in Mechanobiology, 12 (2) (2013) 291–300.

[84] E. Fonck, G. Prod'hom, S. Roy, L. Augsburger, D.A. Rüfenacht, N. Stergiopulos, Effect of elastin degradation on carotid wall mechanics as assessed by a constituent-based biomechanical model, American Journal of Physiology. Heart and Circulatory Physiology 292 (2007) H2754–H2763.

[85] M.D. Ford, Y. Hoi, M. Piccinelli, L. Antiga, D. Steinman, An objective approach to digital removal of saccular aneurysms: technique and applications, The British Journal of Radiology 82 (2009) S55–S61, http://dx.doi.org/10.1259/bjr/67593727.

[86] O. Frank, Die Grundform des Arteriellen Pulses, Zeitschrift für Biologie 37 (1899) 483–526.

[87] O. Frank, The basic shape of the arterial pulse. First treatise: mathematical analysis. 1899, Journal of Molecular and Cellular Cardiology 22 (3) (1990) 255–277.

[88] A.D. Freed, Anisotropy in hypoelastic soft-tissue mechanics. I. Theory, Journal of Mechanics of Materials and Structures 3 (2008) 911–928.

[89] A.D. Freed, T.C. Doehring, Elastic model for crimped collagen fibrils, Journal of Biomechanical Engineering 127 (4) (2005) 587, http://dx.doi.org/10.1115/1.1934145.

[90] J. Frosen, A. Piippo, A. Paetau, M. Kangasniemi, M. Niemela, J. Hernesniemi, J. Jaaskelelainen, Remodelling of saccular cerebral artery aneurysm wall is associated with rupture. Histological analysis of 24 unruptured and 42 ruptured cases, Stroke 35 (2004) 2287–2293.

[91] J. Frosen, R. Tulamo, A. Paetau, E. Laaksamo, M. Korja, et al., Saccular intracranial aneurysm: pathology and mechanisms, Acta Neuropathologica 123 (2012) 773–786.

[92] W. Fu, Z. Gu, X. Meng, B. Chu, A. Qiao, Numerical simulation of hemodynamics in stented internal carotid aneurysm based on patient-specific model, Journal of Biomechanics 43 (7) (2010) 1337–1342, http://dx.doi.org/10.1016/j.jbiomech.2010.01.009.

[93] Y.C. Fung, Biomechanics: Mechanical Properties of Living Tissues, Springer-Verlag, New York, 1993.

[94] L. Gao, Y. Hoi, D.D. Swartz, J. Kolega, A. Siddiqui, H. Meng, Nascent aneurysm formation at the basilar terminus induced by hemodynamics, Stroke 39 (7) (2008) 2085–2090, http://dx.doi.org/10.1161/STROKEAHA.107.509422.

[95] T.C. Gasser, R.W. Ogden, G.A. Holzapfel, Hyperelastic modelling of arterial layers with distributed collagen fibre orientations, Journal of the Royal Society, Interface 3 (6) (2006) 15–35, http://dx.doi.org/10.1098/rsif.2005.0073.

[96] T.C. Gasser, S. Gallinetti, X. Xing, C. Forsell, J. Swedenborg, J. Roy, Spatial orientation of collagen fibers in the abdominal aortic aneurysm's wall and its relation to wall mechanics, Acta Biomaterialia 8 (2012) 3091–3103.

[97] M.M. Giraud-Guille, Twisted liquid crystalline supramolecular arrangements in morphogenesis, International Review of Cytology 166 (1996) 59–101.

[98] R.L. Gleason, J.D. Humphrey, A 2D constrained mixture model for arterial adaptations to large changes in flow, pressure and axial stretch, Mathematical Medicine and Biology 22 (4) (2005) 347–369, http://dx.doi.org/10.1093/imammb/dqi014.

[99] R.L. Gleason, L.A. Taber, J.D. Humphrey, A 2-D model of flow-induced alterations in the geometry, structure, and properties of carotid arteries, Journal of Biomechanical Engineering 126 (3) (2004) 371–381, http://dx.doi.org/10.1115/1.1762899.

[100] J.M. González, A.M. Briones, B. Starcher, M.V. Conde, B. Somoza, C. Daly, E. Vila, I. Mcgrath, M.C. González, S.M. Arribas, M.C. Gonz, C. González, Influence of elastin on rat small artery mechanical properties, Experimental Physiology 90 (4) (2005) 463–468, http://dx.doi.org/10.1113/expphysiol.2005.030056.

[101] D.B. Gould, F.C. Phalan, G.J. Breedveld, S.E. van Mil, R.S. Smith, J.C. Schimenti, U. Aguglia, M.S. van der Knaap, P. Heutink, S.W. John, Mutations in Col4a1 cause perinatal cerebral hemorrhage and porencephaly, Science 308 (5725) (2005) 1167–1171.

[102] A.E. Green, P.M. Naghdi, A note on invariance under superposed rigid body motions, Journal of Elasticity 9 (1979) 1–8.

[103] A.E. Green, R.S. Rivlin, R.T. Shields, General theory of small elastic deformations superposed on finite elastic deformations, Proceedings of the Royal Society A: Mathematical, Physical and Engineering Science 211 (1952) 128–154.

[104] S.E. Greenwald, Ageing of the conduit arteries, The Journal of Pathology 211 (2) (2007) 157–172.

[105] J.P. Greving, G.J.E. Rinkel, E. Buskens, A. Algra, Cost-effectiveness of preventive treatment of intracranial aneurysms: new data and uncertainties, Neurology 73 (4) (2009) 258–265, http://dx.doi.org/10.1212/01.wnl.0b013e3181a2a4ea.

[106] R. Grytz, I.A. Sigal, J.W. Ruberti, G. Meschke, J.C. Downs, Lamina cribrosa thickening in early glaucoma predicted by a microstructure motivated growth and remodeling approach, Mechanics of Materials: An International Journal 44 (2012) 99–109.

[107] N. Gundiah, M.B. Ratcliffe, L.A. Pruitt, The biomechanics of arterial elastin, Journal of the Mechanical Behavior of Biomedical Materials 2 (2009) 288–296.

[108] S. Hales, Statical Essays: Containing Haemastatics; or an Account of Some Hydraulic and Hydrostatical Experiments made on the Blood and Blood-Vessels of Animals, second ed., vol. II, Innes, Manby, Woodward, London, 1740.

[109] I. Halliday, M.A. Atherton, C.M. Care, M.W. Collins, D.E. Evans, P.C. Evans, D.R. Hose, A.W. Khir, C. Koenig, R. Krams, P.V. Lawford, S.V. Lishchuk, G. Pontrelli, V. Ridger, T.J. Spencer, Y. Ventikos, D.C. Walker, P.N. Watton, C.S. König, Multi-scale interaction of particulate flow and the artery wall, Medical Engineering & Physics 33 (7) (2011) 840–848, http://dx.doi.org/10.1016/j.medengphy.2010.09.007.

[110] D. Hanein, A.R. Horwitz, The structure of cell-matrix adhesions: the new frontier, Current Opinion in Cell Biology 24 (1) (2012) 134–140, http://dx.doi.org/10.1016/j.ceb.2011.12.001.

[112] K. Hashimoto, M. Hatai, Y. Yaoi, Inhibition of cell adhesion by type V collagen, Cell Structure and Function 16 (5) (1991) 391–397.

[113] O. Hassler, Morphological studies on the large cerebral arteries, with reference to the aetiology of subarachnoid haemorrhage, Acta Psychiatrica Scandinavica 36 (Suppl. 154) (1961) 1–145.

[114] M.J. Haurani, P.J. Pagano, Adventitial fibroblast reactive oxygen species as autocrine and paracrine mediators of remodeling: bellwether for vascular disease? Cardiovascular Research 75 (4) (2007) 679–689, http://dx.doi.org/10.1016/j.cardiores.2007.06.016.

[115] E.D. Hay, Matrix Assembly. Cell Biology of Extracellular Matrix, 20th ed., Plenum, New York, 1991.

[116] M.J. Haykowsky, J.M. Findlay, A.P. Ignaszewski, Aneurysmal subarachnoid hemorrhage associated with weight training: three case reports, Clinical Journal of Sport Medicine 6 (1) (1996) 52–55.

[117] M.A. Hill, G.A. Meininger, Arteriolar vascular smooth muscle cells: mechanotransducers in a complex environment, The International Journal of Biochemistry & Cell Biology 44 (9) (2012) 1505–1510, http://dx.doi.org/10.1016/j.biocel.2012.05.021.

[118] M.R. Hill, A Novel Approach for Combing Biomechanical and Micro-structural Analyses to Assess the Mechanical and Damage Properties of the Artery Wall, Ph.D. Thesis, 2011.

[120] M.R. Hill, X. Duan, G.A. Gibson, S. Watkins, A.M. Robertson, A theoretical and non-destructive experimental approach for direct inclusion of measured collagen orientation and recruitment into mechanical models of the artery wall, Journal of Biomechanics 45 (5) (2012) 762–771, http://dx.doi.org/10.1016/j.jbiomech.2011.11.016.

[121] H. Ho, V. Suresh, W. Kang, M.T. Cooling, P.N. Watton, P.J. Hunter, Multiscale modelling of intracranial aneurysms: cell signalling, hemodynamics and remodelling, IEEE Transactions on Biomedical Engineering Letters 58 (2011) 2974–2977.

[122] G.A. Holzapfel, Nonlinear Solid Mechanics: A Continuum Approach for Engineering, John Wiley & Sons, New York, 2000.

[123] G.A. Holzapfel, T.C. Gasser, R. Ogden, A new constitutive framework for arterial wall mechanics and a comparative study of material models, Journal of Elasticity 61 (1–3) (2000) 1–48.

[124] G.A. Holzapfel, T.C. Gasser, A viscoelastic model for fiber-reinforced composites at finite strains: continuum basis, computational aspects and applications, Computer Methods in Applied Mechanics and Engineering 61 (2000) 4379–4403.

[125] G.A. Holzapfel, G. Sommer, C.T. Gasser, P. Regitnig, Determination of the layer-specific mechanical properties of human coronary arteries with non-atherosclerotic intimal thickening, and related constitutive modelling, American Journal of Physiology. Heart and Circulatory Physiology 289 (5) (2005) H2048–H2058, http://dx.doi.org/10.1152/ajpheart.00934.2004.

[126] G.A. Holzapfel, R.W. Ogden, Constitutive modelling of arteries, Proceedings of the Royal Society A: Mathematical, Physical and Engineering Sciences 466 (2118) (2010) 1551–1597.

[128] A. Houghton, M. Mouded, S. Shapiro, Consequences of elastolysis, in: W. Parks, R. Mecham (Eds.), Extracellular Matrix Degradation, Biology of Extracellular Matrix, Springer-Verlag, Berlin, 2011.

[129] M.C. Huang, A.A. Baaj, K. Downes, A.S. Youssef, E. Sauvageau, H.R. van Loveren, S. Agazzi, Paradoxical trends in the management of unruptured cerebral aneurysms in the United States: analysis of nationwide database over a 10-year period, Stroke 42 (6) (2011) 1730–1735, http://dx.doi.org/10.1161/STROKEAHA.110.603803.

[130] J.D. Humphrey, K.R. Rajagopal, A constrained mixture model for growth and remodeling of soft tissues, Mathematical Models and Methods in Applied Sciences 12 (3) (2002) 407–430.

[131] J.D. Humphrey, Vascular mechanics, mechanobiology and remodelling, Journal of Mechanics in Medicine and Biology 9 (2009) 243–257.

[132] J.D. Humphrey, S.K. Kyriacou, The Use of Laplace's Equation in Aneurysm Mechanics, Neurological Research 18 (1996) 204–208.

[133] J.D. Humphrey, Remodelling of a collagenous tissue at fixed lengths, Journal of Biomechanical Engineering 121 (1999) 591–597.

[134] J.D. Humphrey, S. Baek, L.E. Niklason, Biochemomechanics of cerebral vasospasm and its resolution: I. A new hypothesis and theoretical framework, Annals of Biomedical Engineering 35 (9) (2007) 1485–1497.

[135] J.D. Humphrey, C.A. Taylor, Intracranial and abdominal aortic aneurysms: similarities, differences, and need for a new class of computational models, Annual Review of Biomedical Engineering 10 (March) (2008) 221–246, http://dx.doi.org/10.1146/annurev.bioeng.10.061807.160439.

[136] J.D. Humphrey, Vascular adaptation and mechanical homeostasis at tissue, cellular, and sub-cellular levels, Cell Biochemistry and Biophysics 50 (2) (2008) 53–78.

[137] J.D. Humphrey, G.A. Holzapfel, Mechanics, mechanobiology, and modeling of human abdominal aorta and aneurysms, Journal of Biomechanics 45 (5) (2008) 805–814.

[138] Y. Imai, K. Sato, T. Ishikawa, T. Yamaguchi, Inflow into saccular cerebral aneurysms at arterial bends, Annals of Biomedical Engineering 36 (9) (2008) 1489–1495, http://dx.doi.org/10.1007/s10439-008-9522-z.

[139] S. Juvela, Treatment options of unruptured intracranial aneurysms, Stroke 35 (2) (2004) 372–374, http://dx.doi.org/10.1161/01.STR.0000115299.02909.68.

[140] R. Kadirvel, Y.H. Ding, D. Dai, H. Zakaria, A.M. Robertson, M.A. Danielson, D.A. Lewis, H.J. Cloft, D.F. Kallmes, The influence of hemodynamic forces on biomarkers in the walls of elastase-induced aneurysms in rabbits, Neuroradiology 49 (12) (2007) 1041–1053, http://dx.doi.org/10.1007/s00234-007-0295-0.

[141] N.M.P. Kakalis, A.P. Mitsos, J.V. Byrne, Y. Ventikos, The haemodynamics of endovascular aneurysm treatment: a computational modelling approach for estimating the influence of multiple coil deployment, IEEE Transactions on Medical Imaging 27 (6) (2008) 814–824, http://dx.doi.org/10.1109/TMI.2008.915549.

[142] J.D. Kakisis, C.D. Liapis, B.E. Sumpio, Effects of cyclic strain on vascular cells, Endothelium: Journal of Endothelial Cell Research 11 (2004) 17–28.

[144] D.F. Kallmes, Point: CFD – computational fluid dynamics or confounding factor dissemination, AJNR. American Journal of Neuroradiology 33 (3) (2012) 395–396, http://dx.doi.org/10.3174/ajnr.A2993.

[145] Z. Kapacee, S.H. Richardson, Y. Lu, T. Starborg, D.F. Holmes, K. Baar, K.E. Kadler, Tension is required for fibripositor formation, Matrix Biology: Journal of the International Society for Matrix Biology 27 (4) (2008) 371–375.

[146] K. Kataoka, M. Taneda, T. Asai, A. Kinoshita, M. Ito, R. Kuroda, K. Kataoka, Structural fragility and inflammatory response of ruptured cerebral aneurysms: a comparative study between ruptured and unruptured cerebral aneurysms, Stroke 30 (7) (1999) 1396–1401, http://dx.doi.org/10.1161/01.STR.30.7.1396.

[147] N.A. Kefalides, R. Alper, C.C. Clark, Biochemistry and metabolism of basement membranes, International Review of Cytology 61 (1979) 167–228.

[148] C.M. Kelleher, S.E. McLean, R.P. Mecham, Vascular extracellular matrix and aortic development, Current Topics in Developmental Biology 62 (2004) 153–188.

[149] J. Khoshnoodi, V. Pedchenko, B.G. Hudson, Mammalian collagen IV, Microscopy Research and Technique 71 (5) (2008) 357–370.

[150] C.M. Kielty, S.P. Whittaker, M.E. Grant, C.A. Shuttleworth, Attachment of human vascular smooth muscles cells to intact microfibrillar assemblies of collagen VI and fibrillin, Journal of Cell Science 103 (Pt. 2) (1992) 445–451.

[151] C.M. Kielty, M.J. Sherratt, C.A. Shuttleworth, Elastic fibres, Journal of Cell Science 115 (Pt. 14) (2002) 2817–2828.

[152] C.M. Kielty, S. Stephan, M.J. Sherratt, M. Williamson, C.A. Shuttleworth, Applying elastic fibre biology in vascular tissue engineering, Philosophical Transactions of the Royal Society of London. Series B, Biological Science 362 (1484) (2007) 1293–1312.

[153] M. Kim, D.B. Taulbee, M. Tremmel, H. Meng, Comparison of two stents in modifying cerebral aneurysm hemodynamics, Annals of Biomedical Engineering 36 (2008) 726–741, http://dx.doi.org/10.1007/s10439-008-9449-4.

[154] Y.H. Kim, X. Xu, J.S. Lee, The effect of stent porosity and strut shape on saccular aneurysm and its numerical analysis with lattice Boltzmann method, Annals of Biomedical Engineering 38 (7) (2010) 2274–2292, http://dx.doi.org/10.1007/s10439-010-9994-5.

[155] R. Kittelberger, P.F. Davis, D.W. Flynn, N.S. Greenhill, Distribution of type VIII collagen in tissues: an immunohistochemical study, Connective Tissue Research 24 (3–4) (1990) 303–318.

[156] R.J. Komotar, J. Mocco, R.A. Solomon, Guidelines for the surgical treatment of unruptured intracranial aneurysms: the first annual J. Lawrence pool memorial research symposium – controversies in the management of cerebral aneurysms, Neurosurgery 62 (1) (2008) 183–194, http://dx.doi.org/10.1227/01.NEU.0000296982.54288.12.

[157] S. Kondo, N. Hashimoto, H. Kikuchi, F. Hazama, I. Nagata, H. Kataoka, N. Hashoimotot, Apoptosis of medial smooth muscle cells in the development of saccular cerebral aneurysms in rats, Stroke 29 (1) (1998) 181–188, discussion 189.

[159] M. Kroon, G.A. Holzapfel, Modeling of saccular aneurysm growth in a human middle cerebral artery, ASME Journal of Biomechanical Engineering 130 (2008) 51,012.

[160] M. Kroon, G.A. Holzapfel, A theoretical model for fibroblast-controlled growth of saccular cerebral aneurysms, Journal of Theoretical Biology 257 (1) (2009) 73–83.

[161] E.J. Kucharz, The Collagens: Biochemistry and Pathophysiology, Springer-Verlag, New York, 1992.

[162] Z. Kulcsár, A. Ugron, M. Marosfoi, Z. Berentei, G. Paál, I. Szikora, Hemodynamics of cerebral aneurysm initiation: the role of wall shear stress and spatial wall shear stress gradient, AJNR. American Journal of Neuroradiology 32 (3) (2011) 587594, http://dx.doi.org/10.3174/ajnr.A2339.

[163] S.K. Kyriacou, J.D. Humphrey, Influence of size, shape and properties on the mechanics of axisymmetric saccular aneurysms, Journal of Biomechanics 29 (1996) 1015–1022.

[164] P. Lacolley, V. Regnault, A. Nicoletti, Z. Li, J.B. Michel, The vascular smooth muscle cell in arterial pathology: a cell that can take on multiple roles, Cardiovascular Research 95 (2) (2012) 194–204, http://dx.doi.org/10.1093/cvr/cvs135.

[166] J.A. LaMack, M.H. Friedman, Individual and combined effects of shear stress magnitude and spatial gradient on endothelial cell gene expression, American Journal of Physiology. Heart and Circulatory Physiology 293 (5) (2007) H2853–H2859, http://dx.doi.org/10.1152/ajpheart.00244.2007.

[167] S. Lanfranconi, H.S. Markus, COL4A1 mutations as a monogenic cause of cerebral small vessel disease, Stroke 41 (8) (2010) e513–e518.

[168] Y. Lanir, Constitutive equations for fibrous connective tissues, Journal of Biomechanics 16 (1) (1983) 1–12.

[169] Y. Lanir, Plausibility of structural constitutive equations for isotropic soft tissues in finite static deformations, Journal of Applied Mechanics 61 (1994) 695–702.

[170] V.S. LeBleu, B. Macdonald, R. Kalluri, Structure and function of basement membranes, Experimental Biology and Medicine (Maywood) 232 (9) (2007) 1121–1129.

[171] J.C. Lewis, R.G. Taylor, N.D. Jones, R.W. Stclair, J.F. Cornhill, Endothelial surface characteristics in pigeon coronary-artery atherosclerosis. 1. Cellular alterations during the initial-stages of dietary-cholesterol challenge, Laboratory Investigation 46 (2) (1982) 123–138.

[172] D. Li, A.M. Robertson, A structural multi-mechanism constitutive equation for cerebral arterial tissue, International Journal of Solids and Structures 46 (14–15) (2009) 2920–2928, http://dx.doi.org/10.1016/j.ijsolstr.2009.03.017.

[173] D. Li, A.M. Robertson, A structural multi-mechanism damage model for cerebral arterial tissue, Journal of Biomechanical Engineering 131 (10) (2009) 101,013, http://dx.doi.org/10.1115/1.3202559.

[174] D. Li, A.M. Robertson, L. Guoyu, M. Lovell, Finite element modeling of cerebral angioplasty using a structural multi-mechanism anisotropic damage model, International Journal for Numerical Methods in Engineering 92 (5): 457–474, http://dx.doi.org/10.1002/nme.4342.

[175] W.A. Littler, A.J. Honour, P. Sleight, Direct arterial pressure, pulse rate, and electrocardiogram during micturition and defecation in unrestricted man, American Heart Journal 88 (2) (1974) 205–210.

[177] J. Lu, X. Zhou, M.L. Raghavan, Inverse method of stress analysis for cerebral aneurysms, Biomechanics and Modeling in Mechanobiology 7 (6) (2008) 477–486, http://dx.doi.org/10.1007/s10237-007-0110-1.

[178] B. Ma, J. Lu, R.E. Harbaugh, M.L. Raghavan, Nonlinear anisotropic stress analysis of anatomically realistic cerebral aneurysms, ASME Journal of Biomechanical Engineering 129 (1) (2007) 88–96, http://dx.doi.org/10.1115/1.2401187.

[179] J.R. MacBeath, C.M. Kielty, C.A. Shuttleworth, Type VIII collagen is a product of vascular smooth-muscle cells in development and disease, The Biochemical Journal 319 (3) (1996) 993–998.

[180] I.M. Machyshyn, P.H.M. Bovendeerd, A.A.F. van de Ven, P.M.J. Rongen, F.N. van de Vosse, A model for arterial adaptation combining microstructural collagen remodeling and 3D tissue growth, Biomechanics and Modeling in Mechanobiology 9 (2010) 671–687.

[182] A.M. Malek, S.L. Alper, S. Izumo, Hemodynamic shear stress and its role in atherosclerosis, JAMA 282 (21) (1999) 2035–2042.

[183] F. Malfait, A. De Paepe, Bleeding in the heritable connective tissue disorders: mechanisms, diagnosis and treatment, Blood Reviews 23 (5) (2009) 191–197.

[184] A. Mallock, Note on instability of India-rubber tubes and bellows when distended by fluid pressure, Proceedings of the Royal Society 49 (1891) 458–463.

[185] A. Mantha, C. Karmonik, G. Benndorf, C. Strother, R. Metcalfe, Hemodynamics in a cerebral artery before and after the formation of an aneurysm, AJNR. American Journal of Neuroradiology 27 (2006) 1113–1118.

[186] P. Maquet, On the Movement of Animals, Springer-Verlag, New York, 1989.

[187] G. Martufi, T.C. Gasser, Turnover of fibrillar collagen in soft biological tissue with application to the expansion of abdominal aortic aneurysms, Journal of the Royal Society, Interface (2012).

[188] T. Matsumoto, K. Nagayama, Tensile properties of vascular smooth muscle cells: bridging vascular and cellular biomechanics, Journal of Biomechanics 45 (5) (2012) 745–755, http://dx.doi.org/10.1016/j.jbiomech.2011.11.014.

[189] R. Mayne, Collagenous proteins of blood vessels, Arteriosclerosis 6 (6) (1986) 585–593, http://dx.doi.org/10.1161/01.ATV.6.6.585.

[190] R.J. McAnulty, Fibroblasts and myofibroblasts: their source, function and role in disease, The International Journal of Biochemistry & Cell Biology 39 (4) (2007) 666–671, http://dx.doi.org/10.1016/j.biocel.2006.11.005.

[191] S. Menashi, J.S. Campa, R.M. Greenhalgh, J.T. Powell, Collagen in abdominal aortic aneurysm: typing, content, and degradation, Journal of Vascular Surgery: Official Publication, the Society for Vascular Surgery [and] International Society for Cardiovascular Surgery, North American Chapter 6 (6) (1987) 578–582.

[192] J.P. Mecham, J.E. Heuser, The elastic fiber, in: Elizabeth D. Hay (Ed.), Cell Biology of the Extracellular Matrix, Plenum Press, New York, 1991.

[193] H. Meng, Z. Wang, Y. Hoi, L. Gao, E. Metaxa, D.D. Swartz, J. Kolega, D.D. Swart, Complex hemodynamics at the apex of an arterial bifurcation induces vascular remodelling resembling cerebral aneurysm initiation, Stroke 38 (6) (2007) 1924–1931, http://dx.doi.org/10.1161/STROKEAHA.106.481234.

[194] H. Meng, E. Metaxa, L. Gao, N. Liaw, S.K. Natarajan, D.D. Swartz, A.H. Siddiqui, J. Kolega, J. Mocco, Progressive aneurysm development following hemodynamic insult, Journal of Neurosurgery 114 (4) (2011) 1095–1103, http://dx.doi.org/10.3171/2010.9.JNS10368.

[195] S. Meran, R. Steadman, Fibroblasts and myofibroblasts in renal fibrosis, International Journal of Experimental Pathology 92 (3) (2011) 158–167, http://dx.doi.org/10.1111/j.1365-2613.2011.00764.x.

[196] A.S. Meshel, Q. Wei, R.S. Adelstein, M.P. Sheetz, Basic mechanism of three-dimensional collagen fibre transport by fibroblasts, Nature Cell Biology 7 (2) (2005) 157–164.

[197] E. Metaxa, M. Tremmel, S.K. Natarajan, J. Xiang, R.A. Paluch, M. Mandelbaum, A.H. Siddiqui, J. Kolega, J. Mocco, H. Meng, Characterization of critical hemodynamics contributing to aneurysmal remodeling at the basilar terminus in a rabbit model, Stroke 41 (2010) 1774–1782.

[198] C. Miehe, Discontinuous and continuous damage evolution in Ogden-type large-strain elastic materials, European Journal of Mechanics – A/Solids 14 (5) (1995) 697–720.

[199] W.R. Milnor, Hemodynamics, second ed., Williams & Wilkins, Baltimore, 1989.

[200] W.R. Milnor, Cardiovascular Physiology, Oxford University Press, New York, 1990.

[201] C. Mimata, M. Kitaoka, S. Nagahiro, K. Iyama, H. Hori, H. Yoshioka, Y. Ushio, Differential distribution and expressions of collagens in the cerebral aneurysmal wall, Acta Neuropathologica 94 (3) (1997) 197–206.

[202] J.H. Miner, Basement membranes, The Extracellular Matrix: An Overview, Springer-Verlag, 2011., pp. 117–145.

[203] M.R.K. Mofrad, R.D. Kamm (Eds.), Cellular Mechanotransduction – Diverse Perspectives from Molecules to Tissues, Cambridge University Press, New York, 2010.

[204] G.S. Montes, Structural biology of the fibres of the collagenous and elastic systems, Cell Biology International 20 (1) (1996) 15–27.

[205] M. Morimoto, S. Miyamoto, A. Mizoguchi, N. Kume, T. Kita, N. Hashimoto, Mouse model of cerebral aneurysm: experimental induction by renal hypertension and local hemodynamic changes, Stroke 33 (2002) 1911–1915.

[206] L. Mullins, Effect of stretching on the properties of rubber, Rubber Chemistry and Technology 21 (1948) 281–300.

[207] L. Mullins, Softening of rubber by deformation, Rubber Chemistry and Technology 42 (1969) 339–362.

[208] S. Murtada, M. Kroon, G.A. Holzapfel, A calcium-driven mechanochemical model for prediction of force generation in smooth muscle, Biomechanics and Modeling in Mechanobiology 9 (2010) 749–762.

[209] R.L. Nachman, E.A. Jaffe, Endothelial cell culture: beginnings of modern vascular biology, The Journal of Clinical Investigation 114 (8) (2004) 19–22, http://dx.doi.org/10.1172/JCI200423284.endothelial.

[210] P.M. Naghdi, Mechanics of solids, in: C. Truesdell (Ed.), Handbuch der Physik, in: The Theory, vol. VIa/2, Springer-Verlag, 1972, pp. 425–640.

[213] W.W. Nichols, S.J. Denardo, I.B. Wilkinson, C.M. McEniery, J. Cockcroft, M.F. O'Rourke, Effects of arterial stiffness, pulse wave velocity, and wave reflections on the central aortic pressure waveform, Journal of Clinical Hypertension 10 (4) (2008) 295–303.

[214] M. Nieuwdorp, M.C. Meuwese, H. Vink, J.B.L. Hoekstra, J.J.P. Kastelein, E.S.G. Stroes, The endothelial glycocalyx: a potential barrier between health and vascular disease, Current Opinion in Lipidology 16 (5) (2005) 507–511.

[215] R. Nissen, G.J. Cardinale, S. Udenfriend, Increased turnover of arterial collagen in hypertensive rats, Proceedings of the National Academy of Sciences of the United States of America 75 (1) (1978) 451–453, http://dx.doi.org/10.1073/pnas.75.1.451.

[217] B.P. Nuyttens, T. Thijs, H. Deckmyn, K. Broos, Platelet adhesion to collagen, Thrombosis Research 127 (Suppl.) (2011) S26–S29.

[218] H.S. Oktay, Continuum Damage Mechanics of Balloon Angioplasty, Ph.D. Thesis, University of Maryland, Baltimore County, Baltimore, 1993.
[219] E. Peña, V. Alastrué, A. Laborda, M.A. Martínez, M. Doblaré, A constitutive formulation of vascular tissue mechanics including viscoelasticity and softening behaviour, Journal of Biomechanics 43 (5) (2010) 984–989.
[220] M. Peck, D. Gebhart, N. Dusserre, T.N. McAllister, N. L'Heureux, The evolution of vascular tissue engineering and current state of the art, Cells Tissues Organs 195 (1–2) (2012) 144–158.
[221] M. Piccinelli, Steinman Da, Y. Hoi, F. Tong, A. Veneziani, L. Antiga, Automatic neck plane detection and 3D geometric characterization of aneurysmal sacs, Annals of Biomedical Engineering (2012), http://dx.doi.org/10.1007/s10439-012-0577-5.
[222] G.A. Plenz, M.C. Deng, H. Robenek, W. Volker, Vascular collagens: spotlight on the role of type VIII collagen in atherogenesis, Atherosclerosis 166 (1) (2003) 1–11.
[223] N.V. Polevaya, M.Y.S. Kalani, G.K. Steinberg, V.C.K. Tse, The transition from hunterian ligation to intracranial aneurysm clips: a historical perspective, Neurosurgery 20 (6) (2006) 1–7.
[224] A.R. Pries, T.W. Secomb, P. Gaehtgens, The endothelial surface layer, Pflugers Archiv 440 (5) (2000) 653–666.
[225] C. Quint, Y. Kondo, R.J. Manson, J.H. Lawson, A. Dardik, L.E. Niklason, Decellularized tissue-engineered blood vessel as an arterial conduit, Proceedings of the National Academy of Sciences of the United States of America 108 (22) (2011) 9214–9219.
[226] A. Rachev, K. Hayashi, Theoretical study of the effects of vascular smooth muscle contraction on strain and stress distributions in arteries, Annals of Biomedical Engineering 27 (4) (1999) 459–468.
[227] M.L. Raghavan, B. Ma, M.F. Fillinger, Non-invasive determination of zero-pressure geometry of arterial aneurysms, Annals of Biomedical Engineering 34 (9) (2006) 1414–1419., http://dx.doi.org/10.1007/s10439-006-9115-7.
[228] M.L. Raghavan, B. Ma, R.E. Harbaugh, Quantified aneurysm shape and rupture risk, Journal of Neurosurgery 102 (2005) 355–362.
[229] M. Raschi, F. Mut, G. Byrne, C.M. Putman, S. Tateshima, F. Viñuela, T. Tanoue, K. Tanishita, J.R. Cebral, CFD and PIV analysis of hemodynamics in a growing intracranial aneurysm, International Journal for Numerical Methods in Biomedical Engineering (2011), http://dx.doi.org/10.1002/cnm.
[230] V.L. Rayz, L. Boussel, G. Acevedo-Bolton, A.J. Martin, W.L. Young, M.T. Lawton, R. Higashida, D. Saloner, Numerical simulations of flow in cerebral aneurysms: comparison of CFD results and in vivo MRI measurements, Journal of Biomechanical Engineering 130 (5) (2008) 051,011, http://dx.doi.org/10.1115/1.2970056.
[231] V.L. Rayz, L. Boussel, J.R. Leach, A.J. Martin, M.T. Lawton, C. McCulloch, D. Saloner, L. Ge, Flow residence time and regions of intraluminal thrombus deposition in intracranial aneurysms, Annals of Biomedical Engineering 38 (10) (2010) 3058–3069, http://dx.doi.org/10.1007/s10439-010-0065-8.
[232] E.R. Regan, W.C. Aird, Dynamical systems approach to endothelial heterogeneity, Circulation Research 111 (1) (2012) 110–130, http://dx.doi.org/10.1161/CIRCRESAHA.111.261701.
[233] S. Reitsma, M.G.A.O. Egbrink, H. Vink, B.M. van den Berg, V.L. Passos, W. Engels, D.W. Slaaf, M.A.M.J. van Zandvoort, Endothelial glycocalyx structure in the intact carotid artery: a two-photon laser scanning microscopy study, Journal of Vascular Research 48 (4) (2011) 297–306.
[234] S.S.M. Rensen, P.A.F.M. Doevendans, G.J.J.M. van Eys, Regulation and characteristics of vascular smooth muscle cell phenotypic diversity, Netherlands Heart Journal 15 (2007) 100–108.
[235] N. Resnick, H. Yahav, A. Shay-Salit, M. Shushy, S. Schubert, L.C.M. Zilberman, E. Wofovitz, Fluid shear stress and the vascular endothelium: for better and for worse, Progress in Biophysics 81 (3) (2003) 177–199.
[236] M.R. Reynolds, J.T. Willie, G.J. Zipfel, R.G. Dacey, Sexual intercourse and cerebral aneurysmal rupture: potential mechanisms and precipitants, Journal of Neurosurgery 114 (4) (2011) 969–977, http://dx.doi.org/10.3171/2010.4.JNS09975.
[237] R. Rezakhaniha, E. Fonck, C. Genoud, N. Stergiopulos, Role of elastin anisotropy in structural strain energy functions of arterial tissue, Biomechanics and Modeling in Mechanobiology 10 (2011) 599–611.
[238] J.A.G. Rhodin, Architecture of the vessel wall, in: R.M. Berne, N. Sperelakis (Eds.), Vascular Smooth Muscle, in: The Cardiovascular System, vol. 2, APS, Baltimore, 1979, pp. 1–31.
[239] S. Ricard-Blum, The collagen family, in: R.O. Hynes, K.M. Yamada (Eds.), Extracellular Matrix Biology, Cold Spring Harbor Perspectives in Biology, Cold Spring Harbor Laboratory Press, 2011.
[240] O. Rivero-Arias, A. Gray, J. Wolstenholme, Burden of disease and costs of aneurysmal subarachnoid haemorrhage (aSAH) in the United Kingdom, Cost Effective Resource Allocation 8 (2) (2010) 6, http://dx.doi.org/10.1186/1478-7547-8-6.

[241] M.R. Roach, A.C. Burton, The reason for the shape of the distensibility curves of arteries, Canadian Journal of Biochemistry and Physiology 35 (8) (1957) 681–690.
[242] A.M. Robertson, P.N. Watton, Computational fluid dynamics in aneurysm research: critical reflections, future directions, AJNR. American Journal of Neuroradiology 33 (6) (2012) 992–995, http://dx.doi.org/10.3174/ajnr.A3192.
[243] A.M. Robertson, M.R. Hill, D. Li, Structurally motivated damage models for arterial walls – theory and application, in: D. Ambrosi, A. Quarteroni, G. Rozza (Eds.), in: Modelling of Physiological Flows, Modeling, Simulation and Applications, vol. 5, Springer-Verlag, 2011.
[244] J.M. Ross, L.V. McIntire, J.L. Moake, J.H. Rand, Platelet adhesion and aggregation on human type VI collagen surfaces under physiological flow conditions, Blood 85 (7) (1995) 1826–1835.
[245] C.S. Roy, The elastic properties of arterial wall, The Journal of Physiology 3 (1880–1882) 125–159.
[246] S. Roy, C. Boss, R. Rezakhaniha, N. Stergiopulos, Experimental characterization of the distribution of collagen fiber recruitment, Journal of Biorheology 24 (2011) 84–93.
[247] J.M. Ryan, J.D. Humphrey, Finite element based predictions of preferred material symmetries in saccular aneurysms, Annals of Biomedical Engineering 27 (1999) 641–647.
[248] M.S. Sacks, Incorporation of experimentally-derived fiber orientation into a structural constitutive model for planar collagenous tissues, ASME Journal of Biomechanical Engineering 125 (2) (2003) 280–287.
[249] E.U. Saelman, H.K. Nieuwenhuis, K.M. Hese, P.G. de Groot, H.F. Heijnen, E.H. Sage, S. Williams, L. McKeown, H.R. Gralnick, J.J. Sixma, Platelet adhesion to collagen types I through VIII under conditions of stasis and flow is mediated by GPIa/IIa (alpha 2 beta 1-integrin), Blood 83 (5) (1994) 1244–1250.
[250] K. Sagawa, R.K. Lie, Schaefer J (1990) Translation of Otto Frank's paper? Die Grundform des arteriellen Pulses zeitschrift fur biologie 37: 483–526, Journal of Molecular and Cellular Cardiology 22 (3) (1899) 253–254.
[251] H. Sage, W.R. Gray, Studies on the evolution of elastin? II. Histology, Comparative Biochemistry and Physiology Part B: Comparative Biochemistry 66 (1) (1980) 13–22.
[252] N. Sakata, S. Jimi, S. Takebayashi, M.A. Marques, Type V collagen represses the attachment, spread, and growth of porcine vascular smooth muscle cells in vitro, Experimental and Molecular Pathology 56 (1) (1992) 20–36.
[253] A. Sarasa-Renedo, M. Chiquet, Mechanical signals regulating extracellular matrix gene expression in fibroblasts, Scandinavian Journal of Medicine & Science in Sports 15 (4) (2005) 223–230, http://dx.doi.org/10.1111/j.1600-0838.2005.00461.x.
[254] A. Sheidaei, S.C. Hunley, S. Zeinali-Davarani, L.G. Raguin, S. Baek, Simulation of abdominal aortic aneurysm growth with updating hemodynamic loads using a realistic geometry, Medical Engineering and Physics 33 (2011) 80–88.
[255] W.I. Schievink, J.M. Karemaker, L.M. Hageman, D.J. van der Werf, Circumstances surrounding aneurysmal subarachnoid hemorrhage, Surgical Neurology 32 (4) (1989) 266–272.
[256] H. Schmid, P.N. Watton, M.M. Maurer, J. Wimmer, P. Winkler, Y.K. Wang, O. Roehrle, M. Itskov, O. Röhrle, Impact of transmural heterogeneities on arterial adaptation: application to aneurysm formation, Biomechanics and Modeling in Mechanobiology 9 (3) (2010) 295–315, http://dx.doi.org/10.1007/s10237-009-0177-y. <http://www.ncbi.nlm.nih.gov/pubmed/19943177>.
[257] H. Schmid, A. Grytsan, E. Poshtan, P.N. Watton, M. Itskov, Influence of differing material properties in media and adventitia on arterial adaptation – application to aneurysm formation and rupture, Computer Methods in Biomechanics and Biomedical Engineering (July) (2011) 1–21, http://dx.doi.org/10.1080/10255842.2011.603309.
[258] A. Schmitz, M. Böl, On a phenomenological model for active smooth muscle contraction, Journal of Biomechanics 44 (11) (2011) 2090–2095, http://dx.doi.org/10.1016/j.jbiomech.2011.05.020.
[259] A.J. Schriefl, G. Zeindlinger, D.M. Pierce, P. Regitnig, G.A. Holzapfel, Determination of the layer-specific distributed collagen fibre orientations in human thoracic and abdominal aortas and common iliac arteries, Journal of the Royal Society, Interface 9 (71) (2012) 1275–1286, http://dx.doi.org/10.1098/rsif.2011.0727.
[260] S. Scott, G.G. Ferguson, M.R. Roach, Comparison of the elastic properties of human intracranial arteries and aneurysms, Canadian Journal of Physiology and Pharmacology 50 (4) (1972) 328–332.
[261] A. Selimovic, M.C. Villa-Uriol, G.A. Holzapfel, Y. Ventikos, P.N. Watton, A computational framework to explore the role of the pulsatile haemodynamic environment on the development of cerebral aneurysms for patient-specific arterial geometries, in: C.T. Lim, J.C.H. Goh (Eds.), Sixth World Congress of Biomechanics (WCB 2010) IFMBE Proceedings, vol. 31, Springer, 2010., pp. 759–762.
[262] P. Seshaiyer, J.D. Humphrey, On the potentially protective role of contact constraints on saccular aneurysms, Journal of Biomechanics 34 (2001) 607–612.

[263] D.M. Sforza, C.M. Putman, J.R. Cebral, Hemodynamics of cerebral aneurysms, Annual Review of Fluid Mechanics 41 (2009) 91–107.
[264] A. Sheidaei, S.C. Hunley, S. Zeinali-Davarani, L.G. Raguin, S. Baek, Simulation of abdominal aortic aneurysm growth with updating hemodynamic loads using a realistic geometry, Medical Engineering and Physics 33 (2011) 80–88.
[265] D.M. Sforza, C.M. Putman, J.R. Cebral, Computational fluid dynamics in brain aneurysms, International Journal for Numerical Methods in Biomedical Engineering (2011), http://dx.doi.org/10.1002/cnm.
[266] R.E. Shadwick, Mechanical design in arteries, The Journal of Experimental Biology 202 (1999) 3305–3313.
[267] A.D. Shah, J.D. Humphrey, Finite strain elastodynamics of intracranial saccular aneurysms, Journal of Biomechanics 32 (1999) 593–599.
[268] B.V. Shekhonin, S.P. Domogatsky, V.R. Muzykantov, G.L. Idelson, V.S. Rukosuev, Distribution of type I, III, IV and V collagen in normal and atherosclerotic human arterial wall: immunomorphological characteristics, Collagen and Related Research 5 (4) (1985) 355–368.
[269] Y. Shi, P. Lawford, R. Hose, Review of zero-D and 1-D models of blood flow in the cardiovascular system, Biomedical Engineering Online 10 (33) (2011).
[270] Y. Shimogonya, T. Ishikawa, Y. Imai, N. Matsuki, T. Yamaguchi, A realistic simulation of saccular cerebral aneurysm formation: focussing on a novel haemodynamic index, the gradient oscillatory number, International Journal of Computational Fluid Dynamics 23 (8) (2009) 583–589, http://dx.doi.org/10.1080/10618560902953575.
[271] Y. Shimogonya, T. Ishikawa, Y. Imai, N. Matsuki, T. Yamaguchi, Can temporal fluctuation in spatial wall shear stress gradient initiate a cerebral aneurysm? A proposed novel hemodynamic index, the gradient oscillatory number (GON), Journal of Biomechanics 42 (4) (2009) 550–554, http://dx.doi.org/10.1016/j.jbiomech.2008.10.006.
[272] M. Shojima, M. Oshima, K. Takagi, R. Torii, M. Hayakawa, K. Katada, A. Morita, T. Kirino, Magnitude and role of wall shear stress on cerebral aneurysm: computational fluid dynamic study of 20 middle cerebral artery aneurysms, Stroke 35 (11) (2004) 2500–2505, http://dx.doi.org/10.1161/01.STR.0000144648.89172.0f.
[273] M. Shojima, S. Nemotot, A. Morita, M. Oshima, E. Wantanabe, N. Saito, S. Nemoto, E. Watanabe, Role of shear stress in the blister formation of cerebral aneurysms, Neurosurgery 67 (5) (2010) 1268–1275, http://dx.doi.org/10.1227/NEU.0b013e3181f2f442.
[274] N.E. Sibinga, L.C. Foster, C.M. Hsieh, M.A. Perrella, W.S. Lee, W.O. Endege, E.H. Sage, M.E. Lee, E. Haber, Collagen VIII is expressed by vascular smooth muscle cells in response to vascular injury, Circulation Research 80 (4) (1997) 532–541.
[275] J.C. Simo, J.W. Ju, Strain and stress-based continuum damage models – I. Formulation, International Journal of Solids and Structures 23 (1987) 821–840.
[276] P.K. Singh, A. Marzo, B. Howard, D.A. Rufenacht, P. Bijlenga, A.F. Frangi, P.V. Lawford, S.C. Coley, D.R. Hose, U.J. Patel, Effects of smoking and hypertension on wall shear stress and oscillatory shear index at the site of intracranial aneurysm formation, Clinical Neurology and Neurosurgery 112 (2010) 306–313.
[277] S. Sinha, C.M. Kielty, A.M. Heagerty, A.E. Canfield, C.A. Shuttleworth, Upregulation of collagen VIII following porcine coronary artery angioplasty is related to smooth muscle cell migration not angiogenesis, International Journal of Experimental Pathology 82 (5) (2001) 295–302.
[278] A.J.M. Spencer, Theory of invariants, in: A.C. Eringen (Ed.), in: Continuum Physics, vol. I, Academic Press, New York, 1971, pp. 239–253.
[279] A.J.M. Spencer, Continuum Mechanics, Dover Publications, Inc., 1980.
[280] A.J.M. Spencer, Constitutive theory for strongly anisotropic solids, in: A.J.M. Spencer (Ed.), Continuum Theory of the Mechanics of Fibre-Reinforced Composites, in: CISM Courses and Lectures, vol. 282, Springer, 1984.
[281] H.C. Stary, D.H. Blankenhorn, A.B. Chandler, S. Glagov, W.J. Insull, M. Richardson, M.E. Rosenfeld, S.A. Schaffer, C.J. Schwartz, W.D. Wagner, et al., A definition of the intima of human arteries and of its atherosclerosis-prone regions. A report from the Committee on Vascular Lesions of the Council on Arteriosclerosis, American Heart Association, Arteriosclerosis, Thrombosis 12 (1) (1992) 120–134.
[282] G.R. Stehbens, Definition of the Arterial Intima, Cardiovascular Pathology 5 (1996) 177.
[283] W.E. Stehbens, Analysis of definitions and word misusage in vascular pathology, Cardiovascular Pathology 10 (5) (2001) 251–257.
[284] C.M. Strother, J. Jiang, Intracranial aneurysms, cancer, X-rays and CFD, AJNR. American Journal of Neuroradiology 33 (6) (2012) 991–992.

[285] M. Sun, S. Chen, S.M. Adams, J.B. Florer, H. Liu, W.W.Y. Kao, R.J. Wenstrup, D.E. Birk, Collagen V is a dominant regulator of collagen fibrillogenesis: dysfunctional regulation of structure and function in a corneal-stroma-specific Col5a1-null mouse model, Journal of Cell Science 124 (23) (2011) 4096–4105.

[286] S. Tada, J.M. Tarbell, Interstitial flow through the internal elastic lamina affects shear stress on arterial smooth muscle cells, American Journal of Physiology. Heart and Circulatory Physiology 278 (5) (2000) H1589–H1597.

[287] S. Tada, J.M. Tarbell, Flow through internal elastic lamina affects shear stress on smooth muscle cells (3D simulations), American Journal of Physiology. Heart and Circulatory Physiology 282 (2) (2002) H576–H584.

[288] S. Tada, J.M. Tarbell, Internal elastic lamina affects the distribution of macromolecules in the arterial wall: a computational study, American Journal of Physiology. Heart and Circulatory Physiology 287 (2) (2004) H905–H913.

[289] K. Takizawa, T. Brummer, T.E. Tezduyar, P.R. Chen, A comparative study based on patient-specific fluid-structure interaction modeling of cerebral aneurysms, Journal of Applied Mechanics 79 (1) (2012) 1665–1710, http://dx.doi.org/10.1115/1.4005071.

[290] T. Tanoue, S. Tateshima, J.P. Villablanca, F. Vinuela, K. Tanishita, F. Viñuela, Wall shear stress distribution inside growing cerebral aneurysm, AJNR. American Journal of Neuroradiology 32 (9) (2011) 1732–1737, http://dx.doi.org/10.3174/ajnr.A2607.

[291] J.M. Tarbell, Z.D. Shi, Effect of the glycocalyx layer on transmission of interstitial flow shear stress to embedded cells, Biomechanics and Modeling in Mechanobiology (2012), http://dx.doi.org/10.1007/s10237-012-0385-8.

[292] T.E. Tezduyar, K. Takizawa, T. Brummer, P.R. Chen, Space time fluid structure interaction modeling of patient-specific cerebral aneurysms, International Journal for Numerical Methods in Biomedical Engineering 27 (February) (2011) 1665–1710, http://dx.doi.org/10.1002/cnm.

[293] J. Trachslin, M. Koch, M. Chiquet, Rapid and reversible regulation of collagen XII expression by changes in tensile stress, Experimental Cell Research 247 (2) (1999) 320–328.

[294] C. Truesdell, R.A. Toupin, Handbuch der Physik, Springer-Verlag, 1960.

[295] T. Ushiki, Collagen fibers, reticular fibers and elastic fibers. A comprehensive understanding from a morphological viewpoint, Archives of Histology and Cytology 65 (2) (2002) 109–126.

[296] A. Valentin, L. Cardamone, S. Baek, J.D. Humphrey, A. Valentín, A. Valentín, Complementary vasoactivity and matrix remodelling in arterial adaptations to altered flow and pressure, Journal of the Royal Society Interface 6 (32) (2009) 293–306, http://dx.doi.org/10.1098/rsif.2008.0254.

[297] A. Valentin, J.D. Humphrey, G.A. Holzapfel, A multi-layered computational model of coupled elastin degradation, vasoactive dysfunction, and collagenous stiffening in aortic aging, Annals of Biomedical Engineering 39 (2011) 2027–2045.

[298] A. Valentín, G.A. Holzapfel, Constrained mixture models as tools for testing competing hypotheses in arterial biomechanics: a brief survey, Mechanics Research Communications 42 (2012) 126–133.

[299] T. Van Agtmael, M.A. Bailey, U. Schlotzer-Schrehardt, E. Craigie, I.J. Jackson, D.G. Brownstein, I.L. Megson, J.J. Mullins, Col4a1 mutation in mice causes defects in vascular function and low blood pressure associated with reduced red blood cell volume, Human Molecular Genetics 19 (6) (2010) 1119–1128.

[300] L. Vaughan, M. Mendler, S. Huber, P. Bruckner, K.H. Winterhalter, M.I. Irwin, R. Mayne, D-periodic distribution of collagen type IX along cartilage fibrils, The Journal of Cell Biology 106 (3) (1988) 991–997.

[301] M.C. Villa-Uriol, G. Berti, D.R. Hose, A. Marzo, A. Chiarini, J. Penrose, J. Pozo, J.G. Schmidt, P. Singh, R. Lycett, I. Larrabide, A.F. Frangi, @neurIST complex information processing toolchain for the integrated management of cerebral aneurysms, Interface Focus 1 (3) (2011) 308–319, http://dx.doi.org/10.1098/rsfs.2010.0033.

[302] M.H.M. Vlak, A. Algra, R. Brandenburg, G.J.E. Rinkel, Prevalence of unruptured intracranial aneurysms, with emphasis on sex, age, comorbidity, country, and time period: a systematic review and meta-analysis, The Lancet Neurology 10 (7) (2011) 626–636, http://dx.doi.org/10.1016/S1474-4422(11)70109-0.

[303] S. Vogel, Vital Circuits, on Pumps, Pipes and the Workings of Circulatory Systems, Oxford University Press, New York, 1992.

[304] K.Y. Volokh, Modeling failure of soft anisotropic materials with application to arteries, Journal of the Mechanical Behavior of Biomedical Materials 4 (8) (2011) 1582–1594.

[305] I. Volonghi, A. Pezzini, E. Del Zotto, A. Giossi, P. Costa, D. Ferrari, A. Padovani, Role of COL4A1 in basement-membrane integrity and cerebral small-vessel disease. The COL4A1 stroke syndrome, Current Medicinal Chemistry 17 (13) (2010) 1317–1324.

[307] J.E. Wagenseil, R.P. Mecham, New insights into elastic fiber assembly, Birth Defects Research. Part C, Embryo Today: Reviews 81 (2007) 229–240.

[308] S.A. Wainwright, W.D. Biggs, J.D. Currey, J.M. Gosline, Mechanical Design in Organisms, John Wiley & Sons, New York, 1976.

[309] J.H.C. Wang, B.P. Thampatty, An introductory review of cell mechanobiology, Biomechanics and Modeling in Mechanobiology 5 (1) (2006) 1–16, http://dx.doi.org/10.1007/s10237-005-0012-z.

[310] J.H.C. Wang, P. Goldschmidt-Clermont, F.C.P. Yin, Contractility affects stress fiber remodeling and reorientation of endothelial cells subjected to cyclic mechanical stretching, Annals of Biomedical Engineering 28 (10) (2000) 1165–1171.

[311] W. Wang, Change in properties of the glycocalyx affects the shear rate and stress distribution on endothelial cells, Journal of Biomechanical Engineering 129 (3) (2007) 324–329, http://dx.doi.org/10.1115/1.2720909.

[312] P.N. Watton, N. Hill, Evolving mechanical properties of a model of abdominal aortic aneurysm, Biomechanics and Modeling in Mechanobiology 8 (1) (2009) 25–42, http://dx.doi.org/10.1007/s10237-007-0115-9.

[313] P. Watton, Y. Ventikos, Modelling evolution of saccular cerebral aneurysms, The Journal of Strain Analysis for Engineering Design 44 (November 2008) (2009) 375–389, http://dx.doi.org/10.1243/03093247JSA492.

[314] P.N. Watton, H. Huang, Y. Ventikos, Multi-scale modelling of vascular disease: abdominal aortic aneurysm evolution, in: L. Geris (Ed.), Computational Modeling in Tissue Engineering, Springer, Berlin/Heidelberg, 2012.

[315] P.N. Watton, N.A. Hill, M. Heil, A mathematical model for the growth of the abdominal aortic aneurysm, Biomechanics and Modeling in Mechanobiology 3 (2) (2004) 98–113, http://dx.doi.org/10.1007/s10237-004-0052-9.

[316] P.N. Watton, N.B. Raberger, G.A. Holzapfel, Y. Ventikos, Coupling the hemodynamic environment to the evolution of cerebral aneurysms: computational framework and numerical examples, Journal of Biomechanical Engineering 131 (10) (2009) 101,003, http://dx.doi.org/10.1115/1.3192141.

[317] P.N. Watton, Y. Ventikos, G.A. Holzapfel, Modelling the growth and stabilization of cerebral aneurysms, Mathematical Medicine and Biology: A Journal of the IMA 26 (2) (2009) 133–164, http://dx.doi.org/10.1093/imammb/dqp001.

[318] P.N. Watton, Y. Ventikos, G.A. Holzapfel, Modelling the mechanical response of elastin for arterial tissue, Journal of Biomechanics 42 (9) (2009) 1320–1325, http://dx.doi.org/10.1016/j.jbiomech.2009.03.012.

[319] P.N. Watton, A. Selimovic, N.B. Raberger, P. Huang, G.A. Holzapfel, Y. Ventikos, Modelling evolution and the evolving mechanical environment of saccular cerebral aneurysms, Biomechanics and Modeling in Mechanobiology 10 (1) (2011) 109–132, http://dx.doi.org/10.1007/s10237-010-0221-y.

[320] P.N. Watton, Y. Ventikos, G.A. Holzapfel, Modelling cerebral aneurysm evolution, in: T. McGloughlin (Ed.), Biomechanics and Mechanobiology of Aneurysms, in: Studies in Mechanobiology, Tissue Engineering and Biomaterials, vol. 7, Springer-Verlag, Heidelberg, 2011, pp. 307–322, http://dx.doi.org/10.1007/841. (Chapter 12).

[321] S. Weinbaum, J.M. Tarbell, E.R. Damiano, The structure and function of the endothelial glycocalyx layer, Annual Review of Biomedical Engineering 9 (2007) 121–167, http://dx.doi.org/10.1146/annurev.bioeng.9.060906.151959.

[322] H. Weisbecker, D.M. Pierce, P. Regitnig, G.A. Holzapfel, Layer-specific damage experiments and modeling of human thoracic and abdominal aortas with non-atherosclerotic intimal thickening, Journal of the Mechanical Behavior of Biomedical Materials 12C (2012) 93–106.

[323] G. Wertheim, Mémorie sur l'elasticité et al cohésion des principaux tissues du corps humain (3rd Ser), Annals de chemie et de physique 21 (1847) 385–414.

[324] W.A. Osborne, with a note by W. Sutherland, The elasticity of rubber balloons and hollow viscera, Proceedings of the Royal Society of London Series B. 81 (551) (1909) 485–499.

[325] H.P. Wagner, J.D. Humphrey, Differential passive and active biaxial mechanical behaviors of muscular and elastic arteries: basilar versus common carotid, Journal of Biomechanical Engineering 133 (5) (2011) 051009.

[326] H. Wolinsky, S. Glagov, Structural basis for the static mechanical properties of the aortic media, Circulation Research 14 (1964) 400–413.

[327] H. Wolinsky, S. Glagov, A lamellar unit of aortic medial structure and function in mammals, Circulation Research 20 (1967) 99–111.

[328] H. Wolinsky, S. Glagov, Nature of species differences in the medial distribution of aortic vasa vasorum in mammals, Circulation Research 20 (1967) 409–421.

[329] W. Wu, R.A. Allen, Y. Wang, Fast-degrading elastomer enables rapid remodeling of a cell-free synthetic graft into a neoartery, Nature Medicine (2012).

[330] X.X. Wu, R.E. Gordon, R.W. Glanville, H.J. Kuo, R.R. Uson, J.H. Rand, Morphological relationships of von Willebrand factor, type VI collagen, and fibrillin in human vascular subendothelium, The American Journal of Pathology 149 (1) (1996) 283–291.

[331] R. Wulandana, A.M. Robertson, Use of a multi-mechanism constitutive model for inflation of cerebral arteries, in: First Joint BMES/EMBS Conference, Atlanta, GA, vol. 1, 1999, p. 235.
[332] R. Wulandana, A Nonlinear and Inelastic Constitutive Equation for Human Cerebral Arterial and Aneurysm Walls, Ph.D. Dissertation, University of Pittsburgh, Pittsburgh, PA, 2003.
[333] R. Wulandana, A.M. Robertson, An inelastic multi-mechanism constitutive equation for cerebral arterial tissue, Biomechanics and Modeling in Mechanobiology 4 (4) (2005) 235–248, http://dx.doi.org/10.1007/s10237-005-0004-z.
[334] J. Xiang, S.K. Natarajan, M. Tremmel, D. Ma, J. Mocco, L.N. Hopkins, A.H. Siddiqui, E.I. Levy, H. Meng, Hemodynamic-morphologic discriminants for intracranial aneurysm rupture, Stroke 42 (1) (2011) 144–152, http://dx.doi.org/10.1161/STROKEAHA.110.592923.
[335] G. Xiong, C.A. Figueroa, N. Xiao, C.A. Taylor, Simulation of blood flow in deformable vessels using subject-specific geometry and spatially varying wall properties, International Journal for Numerical Methods in Biomedical Engineering 27 (7) (2011) 1000–1016, http://dx.doi.org/10.1002/cnm.
[336] F. Xu, J. Ji, L. Li, R. Chen, W. Hu, Activation of adventitial fibroblasts contributes to the early development of atherosclerosis: a novel hypothesis that complements the response-to-injury hypothesis and the inflammation hypothesis, Medical Hypotheses 69 (4) (2007) 908–912, http://dx.doi.org/10.1016/j.mehy.2007.01.062.
[337] W.Y. Yen, B. Cai, M. Zeng, J.M. Tarbell, B.M. Fu, Quantification of the endothelial surface glycocalyx on rat and mouse blood vessels, Microvascular Research 83 (3) (2012) 337–346.
[338] P.D. Yurchenco, Basement membranes: cell scaffoldings and signaling platforms, Cold Spring Harbor Perspectives in Biology 3 (2) (2011).
[339] M. Zamir, The Physics of Pulsatile Flow, Springer-Verlag, New York, 2000.
[340] J.A. Zasadzinski, B. Wong, N. Forbes, G. Braun, G. Wu, Novel methods of enhanced retention in and rapid, targeted release from liposomes, Current Opinion in Colloid and Interface Science 16 (3) (2011) 203–214.
[341] S. Zeinali-Davarani, A. Sheidaei, S. Baek, A finite element model of stress-mediated vascular adaptation: application to abdominal aortic aneurysms, Computer Methods in Biomechanics and Biomedical Engineering 14 (2011) 803–817.
[342] Z. Zeng, D.F. Kallmes, M.J. Durka, Y. Ding, D. Lewis, R. Kadirvel, A.M. Robertson, Hemodynamics and anatomy of elastase-induced rabbit aneurysm models: similarity to human cerebral aneurysms? AJNR. American Journal of Neuroradiology 32 (2011) 595–601.
[343] X. Zhao, M.L. Raghavan, J. Lu, Characterizing heterogeneous properties of cerebral aneurysms with unknown stress-free geometry: a precursor to in vivo identification, Journal of Biomechanical Engineering 133 (5) (2011) 051,008, http://dx.doi.org/10.1115/1.4003872.
[344] X. Zhao, M.L. Raghavan, J. Lu, Identifying heterogeneous anisotropic properties in cerebral aneurysms: a pointwise approach, Biomechanics and Modeling in Mechanobiology 10 (2) (2011) 177–189, http://dx.doi.org/10.1007/s10237-010-0225-7.
[345] X. Zhou, M.L. Raghavan, R.E. Harbaugh, J. Lu, Patient-specific wall stress analysis in cerebral aneurysms using inverse shell model, Annals of Biomedical Engineering 38 (2) (2010) 478–489, http://dx.doi.org/10.1007/s10439-009-9839-2.
[346] Y. Zou, Y. Zhang, The orthotropic viscoelastic behavior of aortic elastin, Biomechanics and Modeling in Mechanobiology 10 (5) (2011) 613–625, http://dx.doi.org/10.1007/s10237-010-0260-4.
[347] M.A. Zulliger, A. Rachev, N. Stergiopulos, A constitutive formulation of arterial mechanics including vascular smooth muscle tone, American Journal of Physiology. Heart and Circulatory Physiology 287 (3) (2004) 1335–1343, http://dx.doi.org/10.1152/ajpheart.00094.2004.

CHAPTER

9

Shear Stress Variation and Plasma Viscosity Effect in Microcirculation

Xuewen Yin and Junfeng Zhang

Bharti School of Engineering, Laurentian University, Sudbury, Ontario, Canada

OUTLINE

9.1 Introduction 350

9.2 Models and Methods 351

9.2.1 The Lattice-Boltzmann Method (LBM) for Fluid Dynamics 351

9.2.2 Red Blood Cell (RBC) Model and Membrane Mechanics 352

9.2.3 Intercellular Aggregation 353

9.2.4 The Immersed-Boundary Method (IBM) for Fluid-Membrane Interaction 354

9.2.5 Viscosity Update Algorithm 355

9.3 Algorithm Validations 356

9.3.1 The Laplace Relationship for Stationary Bubbles 356

9.3.2 The Dispersion Relationship for Capillary Waves 357

9.3.3 The Drag Coefficient for Circular Cylinders 358

9.3.4 RBC Deformation and Rotation in Shear Flows 360

9.4 WSS Variation Induced by Blood Flows in Microvessels 360

9.4.1 Single-File RBC Flows 361

9.4.1.1 Cell Deformation 362

9.4.1.2 Shear Stress Variation 364

9.4.1.3 Effects of Cell and Flow Properties 365

9.4.1.4 Hematocrit Effect 366

9.4.2 Cell-Free Layer (CFL) and Wall Shear Stress (WSS) Variation in Multiple RBC Flows 368

9.4.2.1 Cell-Free Layer and WSS Correlation 368

9.4.2.2 Effects of Cell Deformability, Flow Rate and Aggregation on WSS Variation 371

© 2013 Elsevier Inc. All rights reserved.

http://dx.doi.org/10.1016/B978-0-12-415824-5.00009-6

9.5 Suspending Viscosity Effect 372
9.5.1 Single RBC Rotation in Shear Flows 372
9.5.2 Single RBC Migration in Channel Flows 374
9.5.3 Multiple RBC Flows in Microvessels 376
9.5.4 RBC Motion and Deformation in Bifurcated Microvessels 380
9.6 Summary 386
Acknowledgments 387
References 387

9.1 INTRODUCTION

In its continuous circulation through bodies, blood performs many crucial biological functions, including supplying oxygen and nutrients, removing carbon dioxide and metabolic wastes, circulating white blood cells, antibodies, and platelets for immunization and self-repair and regulating body temperature and pH. Such functions are mainly accomplished within the microvascular network of microvessels and capillaries, and the microcirculation has been the subject of extensive studies [40]. At a microscopic level, blood can be considered as a concentrated suspension of several cellular components, including red blood cells (RBCs) or erythrocytes, white blood cells or leukocytes and platelets, all of which are suspended in an aqueous solution: the so-called plasma. Among these cellular components, RBCs are the most important due to their large number density, crucial role in oxygen transport and direct influence on blood flow behaviors. A normal human RBC exhibits a biconcave shape with a diameter of roughly 8 μm and a thickness of about 2 μm [40], with a highly deformable membrane and viscous interior fluid called cytoplasm. RBC flow is bounded within the interior surfaces of blood vessels and capillaries, which are lined with a monolayer of endothelial cells (ECs). These ECs can sense and respond to mechanical stimuli from the blood flow. *In vivo* observations indicate that wall shear stress (WSS) on the microvessel wall is a key factor in determining the shape and orientation of ECs [29,36]. It has also been observed that WSS can induce a series of biochemical reactions in ECs, and these reactions can further result in phenomena such as ion channel activation, vessel dilation [2] and changes in various gene expressions [2,31,38].

In addition to the cell properties and microvessel structure, another important factor related to blood flow behaviors is the plasma viscosity. For example, *Drochon et al.* [12] had utilized an 8-cP Dextran solution in filtration experiments to measure changes in cell deformability due to membrane modifications. *Secomb and Hsu* [49] had also studied the RBC transit time through micropores with different suspending viscosities using a theoretical model. Moreover, experimental studies in resuscitation with plasma expanders by *Intaglietta and coworkers* [10,60,62] have demonstrated that a higher suspending viscosity could be beneficial in maintaining adequate functional capillary density to ensure oxygen delivery. *Waschke et al.* [65] varied the suspending viscosity from 1.4 to 7.7 cP and noticed no significant change in cerebral perfusion. These observations appear contradictory to the intuitive anticipation that a high suspending viscosity will increase blood flow resistance and therefore impair the normal blood functions in the microcirculation. In spite of these interesting *in vivo* observations, fundamental investigations on how the plasma viscosity affects the RBC deformation and motion, the RBC flow behaviors in microvessels and the flow-vessel interactions are relatively inadequate, and the underlying mechanisms are not clear yet.

In this chapter, we summarize our recent computational research on the shear stress variation induced by blood flows and the plasma viscosity effect on hemodynamics in microvessels. Firstly, the immersed-boundary lattice-Boltzmann method (IB-LBM) is summarized. We then employ this method to study the WSS variation induced by blood flows in microvessels. The effects of cell deformability, flow rate, aggregation, hematocrit and cell-free layer structure on the WSS variation characteristics are also carefully examined. In addition, we investigate the motion and deformation of a single RBC and multiple RBCs in straight and bifurcated microvessels with different suspending viscosities. The results show that a higher viscosity can facilitate the cell-free layer development and enhance the hematocrit phase separation at a bifurcation. Such information observed in these simulations can advance our understanding of the blood-vessel interaction and the beneficial effect of a high plasma viscosity in hemodilution and resuscitation experiments, and could be valuable for fundamental studies and biomedical applications in microcirculation.

9.2 MODELS AND METHODS

Blood flows in large vessels are typically treated as non-Newtonian flows with empirical constitutive relations. Due to the transverse migration of the flexible RBCs occurring in microvessels, the RBCs are depleted from the vessel surface, and this results in a cell-free layer (CFL) near the vessel surface. To incorporate the CFL effect in microvessels, multiple layer models have been proposed [23,52]. However, this continuum approach provides inherent difficulties in including the effects of cell properties on hemodynamics and it is not suitable for investigations of the microscopic flow structures. Consider that in small microvessels and capillaries, the blood flow must be modeled as a discrete particulate flow of individual RBCs.

In this section, we describe in detail the immersed-boundary lattice-Boltzmann method (IB-LBM), which had been developed to simulate microscopic blood flows [70]. In this model, the lattice-Boltzmann method (LBM) is adopted to solve the flow field for its advantages in dealing with complex boundaries and potential efficiency in parallel computation [69], and the immersed-boundary method (IBM) is utilized to incorporate the flow-cell interaction. Our model considers important RBC features, such as the biconcave shape, neo-Hookean elastic membrane mechanics, high-viscosity cytoplasm and intercellular aggregation. An efficient algorithm has also been proposed to dynamically update the domain viscosity values according to simultaneous RBC configurations.

9.2.1 The Lattice-Boltzmann Method (LBM) for Fluid Dynamics

In LBM, a fluid is modeled as pseudo-particles moving over a lattice domain at discrete time steps. The major variable in LBM is the density distribution $f_i(\mathbf{x}, t)$, indicating the particle amount moving with the ith lattice velocity $\mathbf{c}_i$ at position $\mathbf{x}$ and time t. The time evolution of density distributions is governed by the so-called lattice-Boltzmann equation, which can be considered as a discrete version of the Boltzmann equation in classical statistical physics [55]:

$$f_i(\mathbf{x} + \mathbf{c}_i \Delta t, t + \Delta t) = f_i(\mathbf{x}, t) + \Gamma_i(f) \tag{9.1}$$

where Δt is the time step and Γ_i is the collision operator incorporating the change in f_i due to the particle collisions. The collision operator is typically simplified by the single-time-relaxation approximation [8]:

$$\Gamma_i(f) = -\frac{f_i(\mathbf{x},t) - f_i^{eq}(\mathbf{x},t)}{\tau} \tag{9.2}$$

where τ is a relaxation parameter and the equilibrium distribution f_i^{eq} can be expressed as [72]:

$$f_i^{eq} = \rho w_i \left[1 + \frac{\mathbf{u}\cdot\mathbf{c}_i}{c_s^2} + \frac{1}{2}\left(\frac{\mathbf{u}\cdot\mathbf{c}_i}{c_s^2}\right)^2 - \frac{u^2}{2c_s^2}\right] \tag{9.3}$$

Here $\rho = \sum_i f_i$ is the fluid density and $\mathbf{u} = \sum_i f_i \mathbf{c}_i/\rho$ is the fluid velocity. Other parameters, including the lattice sound speed, c_s, and weight factors, w_i, are lattice structure dependent. Through the Chapman-Enskog expansion [55], one can recover the macroscopic continuity and momentum (Navier-Stokes) equations from the above-defined LBM microdynamics:

$$\begin{aligned} &\frac{\partial \rho}{\partial t} + \nabla\cdot(\rho\mathbf{u}) = 0, \\ &\frac{\partial \mathbf{u}}{\partial t} + (\mathbf{u}\cdot\nabla)\mathbf{u} = -\frac{1}{\rho}\nabla P + \frac{\mu}{\rho}\nabla^2\mathbf{u} \end{aligned} \tag{9.4}$$

where $\mu = (2\tau - 1)c_s^2\Delta t/2\rho$ is the fluid viscosity and $P = c_s^2\rho$ is the fluid pressure. For the D2Q9 (2-D and nine-lattice velocity) lattice model utilized in this work, the nine-lattice velocities can be expressed as:

$$\mathbf{c}_0 = (0,0) \tag{9.5}$$

$$\mathbf{c}_i = [\cos(i-1)\pi/2,\ \sin(i-1)\pi/2]\Delta x/\Delta t, \quad i = 1-4 \tag{9.6}$$

$$\mathbf{c}_i = [\cos(2i-9)\pi/4,\ \sin(2i-9)\pi/4]\Delta x/\Delta t, \quad i = 5-8 \tag{9.7}$$

where Δx is the lattice grid. The lattice weight factors for the D2Q9 model are $w_0 = 4/9, w_{1-4} = 1/9$ and $w_{5-8} = 1/36$; and the lattice sound speed is $c_s = (1/\sqrt{3})\Delta x/\Delta t$. It can be seen from the above description that the microscopic LBM dynamics is local (i.e., only the very neighboring lattice nodes are required to update the density distribution f_i), and hence a LBM algorithm is advantageous for parallel computations. Also its particulate nature makes it relatively easily for systems with complex boundaries [69]. Such merits are advantageous for RBC flow simulations in complicated microvascular geometries.

9.2.2 Red Blood Cell (RBC) Model and Membrane Mechanics

In this chapter, unless otherwise specified, an individual RBC is represented by a two-dimensional (2-D) biconcave capsule with an elastic membrane. The RBC shape at the stress-free state is described by the following empirical equation proposed by *Evans and Fung* [18]:

$$\bar{y} = \pm 0.5(1 - \bar{x}^2)^{1/2}(c_0 + c_1\bar{x}^2 + c_2\bar{x}^4), \quad -1 \leqslant \bar{x} \leqslant 1 \tag{9.8}$$

where $c_0 = 0.207, c_1 = 2.002$ and $c_2 = 1.122$. The non-dimensional coordinates $(\bar{x}, \bar{y})$ are normalized by the radius of a human RBC $a = 3.91$ μm.

The RBC membrane is highly deformable with a finite bending resistance, which becomes profound in regions with large curvature [17,24]. In our present two-dimensional (2-D) model, following *Bagchi et al.* [4], the neo-Hookean elastic component of membrane stress is expressed as:

$$S_e = E_s \frac{\epsilon^3 - 1}{\epsilon^{3/2}} \tag{9.9}$$

where E_s is the membrane elastic modulus and ϵ is the stretch ratio. In addition, the bending resistance can also be incorporated by relating the membrane curvature change to the membrane stress with [4,41]:

$$S_b = E_b \frac{d}{dl}(\kappa - \kappa_0) \tag{9.10}$$

where E_b is the bending modulus, and κ and κ_0 are, respectively, the instantaneous and initial stress-free membrane curvatures. The geometric parameter, l, is a measure of the arc length along the membrane surface. Therefore, the total membrane stress **S** induced as a result of the cell deformation is a sum of the two terms discussed above:

$$\mathbf{S} = S_e\mathbf{t} + S_b\mathbf{n} \tag{9.11}$$

Here **n** and **t** are, respectively, the local normal and tangential directions at a position on the membrane.

9.2.3 Intercellular Aggregation

RBCs can also aggregate and form one-dimensional rouleaux (like stacks of coins) or 3-D aggregates [7,40,54]. This process is reversible, and the 1-D rouleaux and the 3-D aggregates can be broken, for example, by increasing the flow shear rate. This phenomenon is particularly important in the microcirculation, since such rouleaux or aggregates can dramatically influence the blood flow within the microvessels. However, the mechanisms underlying this aggregation are still the subject of investigation. At present, there are two theoretical descriptions of this process: the bridging model and the depletion model [7,40]. The bridging model assumes that macromolecules, such as fibrinogen or dextran, can adhere to the adjacent RBC surfaces and bridge them [9,33]. The depletion model attributes the RBC aggregation to a polymer depletion layer between RBC surfaces, which is accompanied by a decrease in the osmotic pressure [6,16]. Both the bridging and depletion models can describe certain aggregation phenomena; however, they appear to be incapable of explaining some specific observations [7]. Detailed discussions of these two models can be found elsewhere (see, for example, the review by *Baumler et al.* [7]). In general, it can be assumed that the attractive interaction between RBC surfaces will occur when the surfaces are close, and a repulsive interaction would occur when the separation distance is sufficiently small. The repulsive interaction includes the steric forces from the RBC surfaces [30]. In this chapter, we follow the approach proposed by *Liu et al.* [30] to model the intercellular interaction energy φ as a Morse potential:

$$\varphi(r) = D_e\left[e^{2\beta(r_0-r)} - 2e^{\beta(r_0-r)}\right] \tag{9.12}$$

where r is the surface separation, r_0 and D_e are, respectively, the zero force separation and surface energy and β is a scaling factor controlling the interaction decay behavior. The interaction force from such a potential is its negative derivative, i.e., $f(r) = -\partial\varphi/\partial r$. This Morse potential was selected for its simplicity and realistic physical description of the intercellular aggregation; however, it does not imply any particular underlying mechanism. In our simulations, r_0 was chosen as 2.5 lattice units (0.49 μm) because of the limitation of the grid size and the immersed-boundary treatment employed in the computational scheme. This distance is relatively large when compared to experimental data [37], and the scheme could be improved by adopting a smaller lattice grid and/or the adaptive mesh refinement technology [57]. To improve the computational efficiency, as is typical in molecular dynamics simulations, we have adopted a cut-off distance $r_c = 3.52$ μm. This value has been selected to generate an effective aggregation among RBCs with affordable computational time. The scaling factor β was adjusted so that the attractive force beyond the cut-off distance r_c was negligible.

9.2.4 The Immersed-Boundary Method (IBM) for Fluid-Membrane Interaction

In the 1970s, *Peskin* [39] developed the immersed-boundary method (IBM) to simulate flexible membranes in fluid flows. The membrane-fluid interaction is accomplished by distributing membrane forces as local fluid forces and updating membrane configuration according to local flow velocity. Since then, the IBM has been widely employed to study various situations, including cell deformation in micropipettes, leukocyte adhesion and movement, multiphase flows, RBC deformation and aggregation in shear flows and the behavior of biofilms. Figure 9.1 displays a segment of membrane and the nearby fluid domain, where filled circles represent membrane markers and open circles represent fluid nodes. In IBM, the membrane force $\mathbf{f}(\mathbf{x}_m)$ at a membrane marker $\mathbf{x}_m$ induced by membrane deformation is distributed to the nearby fluid grid points $\mathbf{x}_f$ by:

$$\mathbf{F}(\mathbf{x}_f) = \sum_{\mathbf{x}_m} D(\mathbf{x}_f - \mathbf{x}_m)\mathbf{f}(\mathbf{x}_m) \tag{9.13}$$

through a discrete delta function $D(\mathbf{x})$, which is chosen to approximate the properties of the Dirac delta function [39]. In our 2-D system, $D(\mathbf{x})$ is given as:

$$D(\mathbf{x}) = \frac{1}{4(\Delta x)^2}\left(1+\cos\frac{\pi x}{2\Delta x}\right)\left(1+\cos\frac{\pi y}{2\Delta x}\right), \quad |x| \leqslant 2\Delta x \text{ and } |y| \leqslant 2\Delta x \tag{9.14}$$

$$D(\mathbf{x}) = 0, \quad \text{otherwise} \tag{9.15}$$

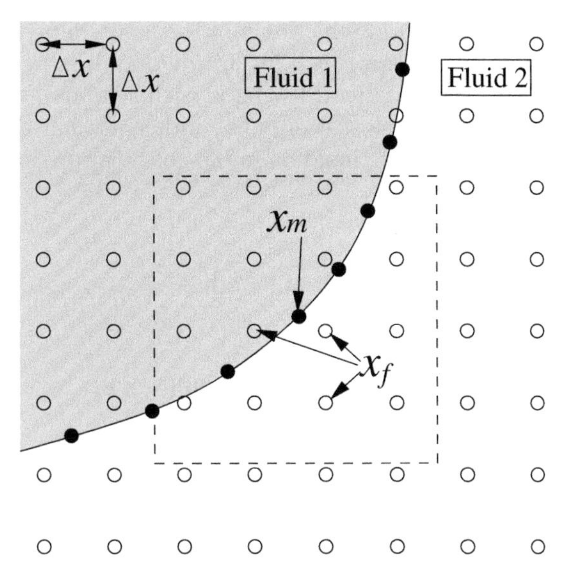

FIGURE 9.1 A schematic of the immersed-boundary method. The open circles are fluid nodes and the filled circles represent the membrane nodes. The membrane force calculated at node $\mathbf{x}_m$ is distributed to the fluid nodes $\mathbf{x}_f$ in the $2\Delta x \times 2\Delta x$ square (dashed lines) through Eq. (9.13); and the position $\mathbf{x}_m$ is updated according to $\mathbf{x}_f$ velocities through Eq. (9.16).

The membrane velocity $\mathbf{u}(\mathbf{x}_m)$ can be obtained in a similar way according to the local flow field:

$$\mathbf{u}(\mathbf{x}_m) = \sum_{\mathbf{x}_f} D(\mathbf{x}_f - \mathbf{x}_m)\mathbf{u}(\mathbf{x}_f) \tag{9.16}$$

The no-slip requirement between the fluid and the membrane is automatically satisfied, since the membrane points are moving exactly at the same velocity (within negligible numerical errors) as the local flow field. Here both the force distribution (Eq. (9.13)) and velocity interpolation (Eq. (9.16)) should be carried out in a $4\Delta x \times 4\Delta x$ region (dashed square in Fig. 9.1) [39].

9.2.5 Viscosity Update Algorithm

The fluids separated by the membrane can have different properties, such as density and viscosity. For normal RBCs, the viscosity of cytoplasm is five times that of the suspending plasma. As the membrane moves with the fluid flow, the fluid properties (here the viscosity) have to be updated accordingly. While this process seems straightforward, it is not trivial. Typically, an index field is introduced in order to identify the relative position of a fluid node with respect to the membrane. In previous IBM studies, *Tryggvason et al.* [59] solved the index field through a Poisson equation over the entire domain at each time step, but the available information from the explicitly tracked membrane was not fully employed [61]. On the other hand, *Shyy and coworkers* [35,61] suggested to update the fluid properties directly according to the normal distance to the membrane surface. We prefer the latter approach for its simple algorithm and computational efficiency. However, this approach originally proposed formulations of the Heaviside function which were coordinate dependent [70]. To overcome this drawback, here we modify this approach by simply defining the index d of a fluid node as the shortest distance to the membrane. If the node is close to two or more membrane segments, the index taken is that to the closest one. The index also has a sign, which is positive outside of the membrane and negative inside. Similar to the approach of *Shyy and coworkers* [35], a Heaviside function is defined as:

$$H(d) = \begin{cases} 0, & d < -2\Delta x \\ \frac{1}{2}\left(1 + \frac{d}{2\Delta x} + \frac{1}{\pi}\sin\frac{\pi d}{2\Delta x}\right), & -2\Delta x \leqslant d \leqslant 2\Delta x \\ 1, & d > 2\Delta x \end{cases} \tag{9.17}$$

A varying fluid property α can then be related to the index d by:

$$\alpha(\mathbf{x}) = \alpha_{in} + (\alpha_{out} - \alpha_{in})H[d(\mathbf{x})] \tag{9.18}$$

where the subscripts *in* and *out* indicate, respectively, the bulk values inside and outside of the membrane. It can be seen from Eq. (9.17) that, as the membrane moves, only the nodes close to the membrane ($|d| \leqslant 2\Delta x$) will be affected. Therefore, in the computer program we just need to calculate the index and update the fluid properties in the vicinity of the membrane. Such a process should be more efficient than solving a Poisson equation over the entire fluid domain [59].

9.3 ALGORITHM VALIDATIONS

Before addressing the results of numerical simulations of wall shear stress variation induced by the blood flows and the viscosity effects on RBC dynamics in capillaries, we have conducted simulations to verify the algorithm, the program and the physical models employed in this chapter. The three cases: stationary bubbles, dynamic capillary waves and flow passing over cylinders, have analytical solutions, which allow for a direct comparison between the LBM results and the theoretical predictions. The results of these tests will be presented as non-dimensional values (i.e., normalized by the simulation unit length, mass and time step). Additionally, we have simulated RBC deformation and rotation in a 13-cP medium under different shear rates, and our results of RBC deformation and rotation frequencies have been compared with previous experimental and numerical data [20,51].

9.3.1 The Laplace Relationship for Stationary Bubbles

The bubble test is usually employed in numerical studies of multiphase flows to verify the Laplace relationship for a circular bubble or droplet with a constant interfacial tension. To mimic the interfacial tension effect, the membrane force is simply given as:

$$\mathbf{T} = \sigma \mathbf{t} \tag{9.19}$$

where σ is the interfacial tension and $\mathbf{t}$ is the tangential direction of the membrane surface. According to the Laplace relationship, the pressure difference across the interface $\Delta\mathbf{P} = P_{in} - P_{out}$ (P_{in} and P_{out} are, respectively, the pressure inside and outside of the bubble) is inversely proportional to the bubble radius R:

$$\Delta\mathbf{P} = \sigma / R \tag{9.20}$$

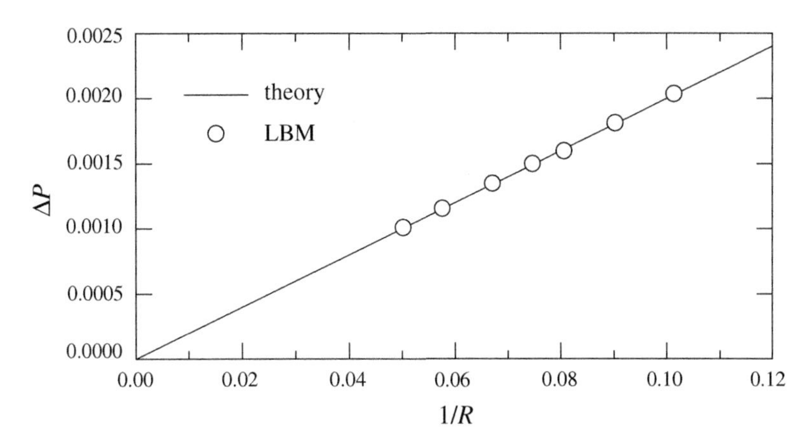

FIGURE 9.2 Pressure difference across the interface for different bubble sizes from LBM simulations (symbols) compared with the theoretical prediction from the Laplace relationship (Eq. (9.20)). (Reprinted from J. Zhang, P.C. Johnson and A.S. Popel, *Physical Biology*, 4:285–295, 2007. With kind permission from Institute of Physics.)

In this test, we set $\sigma = 0.02$ and varied the radius R, and the pressure difference was measured when a stable state was reached. The LBM results are plotted in Fig. 9.2 (symbols) with those from Eq. (9.20) (solid line) in excellent agreement.

9.3.2 The Dispersion Relationship for Capillary Waves

While the bubble test allows us to verify the equilibrium state, simulations of capillary waves provide an effective method to test the interfacial dynamics. In such simulations, a rectangular domain is initially separated by a sinusoidal interface in the middle. The wavelength λ is equal to the domain width. Due to the unbalanced interfacial tension, the interface undergoes a damped oscillation in the viscous fluid. The normal mode analysis shows that the wave amplitude $h(t)$ can be expressed as [53]:

$$h(t) = h_0 \cos[Re(\omega)t]e^{-I_m(\omega)t} \tag{9.21}$$

where $h_0 = h(t = 0)$ and the complex angular frequency ω is related to the wavenumber $k = 2\pi/\lambda$ through the dispersion relation [53]:

$$\omega^2 = \frac{\sigma k^3(1 - k/q)}{2\rho} \tag{9.22}$$

where $q = (k^2 - i\omega/\upsilon)^{1/2}$.

Several wavenumbers were tested, and the interface was followed to obtain its oscillation frequency. In Fig. 9.3a, the frequencies from these LBM simulations (symbols) are compared with the numerical solutions of Eq. (9.22) (solid line). Again, excellent agreement was observed, especially for small wavenumbers. For a large wavenumber k, the wavelength, λ, was small, and therefore the interface was less accurately represented in the system because of the finite

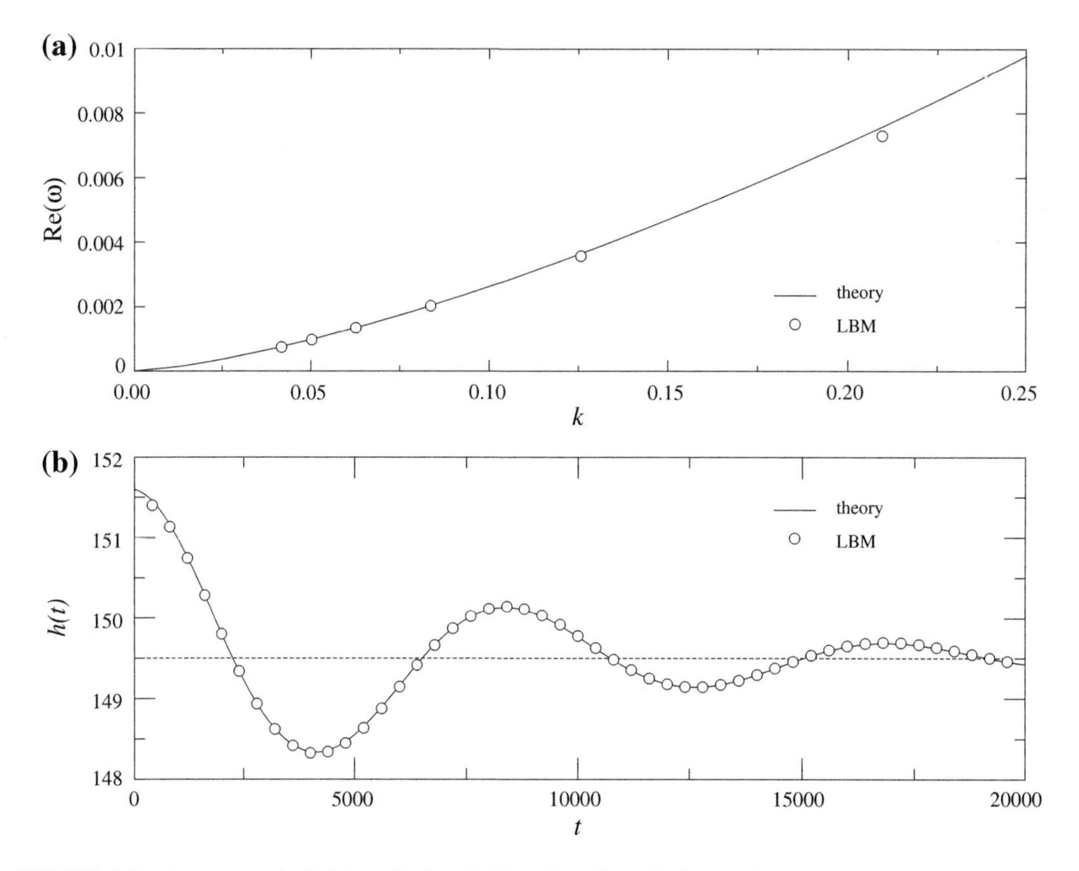

FIGURE 9.3 Comparing the LBM results (symbols) with analytical solutions (solid lines) of capillary waves: (a) the dispersion relation and (b) the wave amplitude evolution. The dashed line in (b) indicates the equilibrium planar interface position. (Reprinted from J. Zhang, P.C. Johnson and A.S. Popel, *Physical Biology*, 4:285–295, 2007. With kind permission from Institute of Physics.)

grid length h. Similar difficulties have also been reported in other LBM simulations of capillary waves [53]. As an example, we also present the time evolution of the wave amplitude for $\lambda = 150$ in Fig. 9.3b. The obtained LBM simulation results (symbols) were in good agreement with the analytical description (solid line) for both the oscillation and decay.

9.3.3 The Drag Coefficient for Circular Cylinders

Next, we extended our method to another extreme case to simulate solid objects. This simulation can help to examine the no-slip boundary condition and flow-structure interaction, which are both critical in our RBC simulation. Here we placed a circular ring in the fluid and treated the membrane as Hookean elastic, with a relatively large elastic modulus. To hold the ring in place, each membrane node was also linked to a fixed virtual node, which was not involved in the force distribution and position updating in IBM. Periodic boundary conditions were applied in both directions of the square domain, and hence the system simulated was actually a square array of cylinders. To generate the fluid flow, a body force was employed. According to *Sangani and Acrivos* [48], the drag coefficient C_D can be accurately approximated by:

$$C_d = \frac{F}{\rho \upsilon U} = \frac{4\pi}{\ln c^{-1/2} - 0.738 + c - 0.887c^2 + 2.038c^3 + O(c^4)} \tag{9.23}$$

for creeping flows and dilute arrays. Here F is the drag force experienced by the cylinder and U is the mean flow velocity. The parameter c is the fractional area occupied by the cylinder in the domain, so that $c = \pi R^2/L^2$, where R is the cylinder radius and L is the domain size. In our simulation, we set $R = 10$, $L = 100$, and $F = 9.69 \times 10^{-5}$. The resulting mean flow velocity $U = 1.581 \times 10^{-5}$. The Reynolds number was $R_e = 2U\,R/\upsilon = 6.32 \times 10^{-4}$, thus satisfying the creeping flow requirement. The drag coefficient from LBM simulation was $C_{LBM}\,D = F/\rho\upsilon U = 12.254$, while the analytical value from Eq. (9.23) by inserting $c = 0.0314$ was $C_D = 12.325$; the relative error was only 0.58%. Figure 9.4 also shows the flow field passing around the cylinder, demonstrating that the cylinder deformation was negligible and the no-slip boundary condition over the cylinder surface was well satisfied.

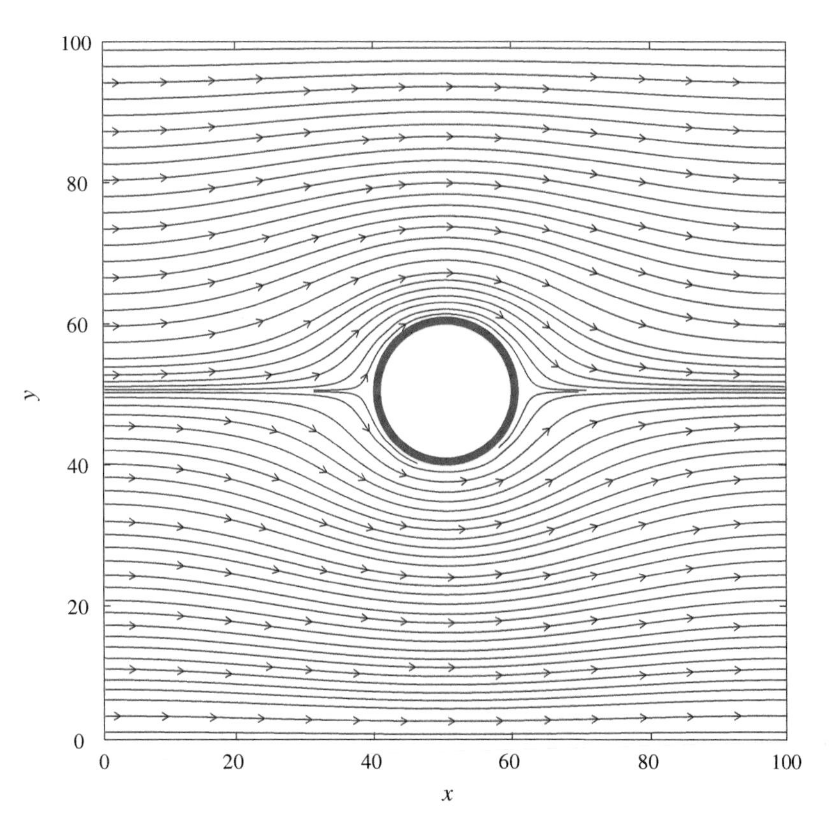

FIGURE 9.4 The flow field passing over a circular cylinder. (Reprinted from J. Zhang, P.C. Johnson and A.S. Popel, *Physical Biology*, 4:285–295, 2007. With kind permission from Institute of Physics.)

9.3.4 RBC Deformation and Rotation in Shear Flows

Although pure shear flows do not exist *in vivo*, the RBC behaviors in such flows have been extensively investigated both experimentally and numerically, for the practical convenience and analytical simplicity. Shear flows have also been frequently utilized to retrieve RBC membrane properties, such as the shear and bending moduli and membrane viscosity [58]. If the shear stress is weak, the RBC will not only deform, but also rotate and flip in the flow. This is the so-called tumbling motion [21], which resembles that of a soft solid particle in such a flow. As the shear strength increases, the elongated cell is observed, instead of rotating continuously, to swing back and forth about a certain orientation. Further increasing the shear stress will decrease the swinging angle amplitude and increase the elongation. When the shear stress is strong enough, the swinging motion becomes unobservable and the cell exhibits a stable shape and orientation [19]. Moreover, in all these situations, the cell membrane is always rotating around the cell, even when the cell is in a stable configuration at strong shear flows. This is usually called the tank-treading motion. The detailed behaviors of these tumbling, swinging or tank-treading motions (including the cell rotating/swinging frequency, deformed shape, elongation and orientation angle) depend on the particular flow condition and cell properties.

In order to further verify the IB-LBM algorithm and RBC parameters used, we simulate the RBC deformation and rotating frequency suspended under different shear rates in a 13-cP medium [51]. The deformed cell length measured in the flow direction (i.e., the horizontal direction) is normalized with the initial cell length of our RBC model $2a_0 = 7.82\ \mu\text{m}$. The results of the normalized cell length and tank-tread frequency are displayed in Fig. 9.5 along with the data reported in previous experimental and numerical studies [20,51]. A reasonable agreement is observed there, although our model predicts a larger cell deformation in the low shear rate regime ($<100\ \text{s}^{-1}$) when compared to those reported in the previous studies. This may, in part, be due to the relatively low circularity index defined as $2\pi A^{1/2}/l_0$, where A is the enclosed area and l_0 is the membrane circumference length of our 2D RBC model [51]. The circularity index calculated from Eq. (9.8) is 0.679, while for normal human RBCs it is in a range of $0.72-0.86$ [51]. This lower circularity index indicates that our RBC model has a relatively long membrane circumference and, when flattened in shear flows, the cell would have a longer length. Nevertheless, we accept such a discrepancy as a reasonable consequence of the 2-D simplification for real 3-D systems.

9.4 WSS VARIATION INDUCED BY BLOOD FLOWS IN MICROVESSELS

Wall shear stress (WSS) has attracted extensive interests for its influence on EC morphology and functions. Previous studies typically assume a steady WSS and the dynamic feature in microcirculation has not been well addressed. In this section, we study the WSS behavior associated with microscopic blood flows using the IB-LBM model we have developed. Simulations were carried out over rectangular D2Q9 lattice grids, and the flow is induced by applying a body force equivalent to a desired pressure gradient. The WSS on the channel surfaces can be readily obtained from the velocity field. Two sets of simulations, single-file RBC flows and multiple RBC suspension flows, have been conducted in order to investigate the

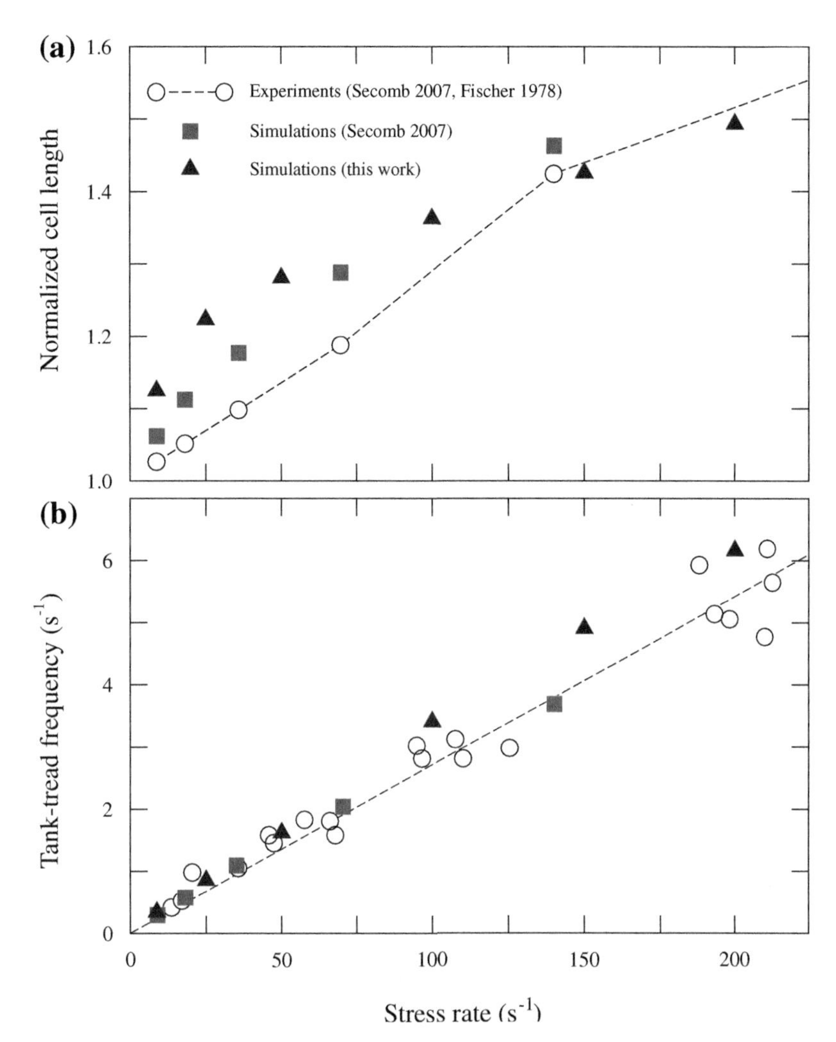

FIGURE 9.5 Comparison of simulated cell deformation (a) and rotating frequency (b) with previous experimental and numerical results in shear flows with a 13-cP suspending viscosity. The dashed line in (a) connects the experimental data points (open circles), while the dashed line in (b) is a linear fitting of the experimental data. (Reprinted from J. Zhang, *Microcirculation*, 18:562–573, 2011. With kind permission from John Wiley and Sons.)

relationship between the RBC motion and the induced wall shear stress. Parallel computation has been employed for multiple RBC simulations, and a typical run takes about 10 h for 8 Intel Xeon L5420 nodes at 2.5 GHz.

9.4.1 Single-File RBC Flows

For simplicity and to avoid any possible bias of the cell deformation from the initial RBC inclinations, in this section, a RBC was modeled as a circular 2-D capsule [45] with the

TABLE 9.1 Simulated Cases for Single RBC Flows with Different Cell Deformability, Channel Size and Applied Pressure Gradient

Case	Cell Deformability	Channel Size H (μm)	Pressure Gradient P_x(kPa/m)
SF-C	Normal	7.80	157.8
SF-R	Rigid	7.80	157.8
SF-h	Normal	5.85	280.5
SF-H	Normal	11.70	70.1
SF-V2	Normal	7.80	315.6
SF-V3	Normal	7.80	473.4

membrane mechanics described in Section 9.2.1. The radius was selected as 2.08 μm, such that the circular RBC model here has a same enclosed area as the radial cross-sectional cut of a typical biconcave humane RBC. The cell membrane was discretized into 100 segments of equal lengths. The lattice mesh size in the current study is $\Delta x = 0.195$ μm. Mesh resolution independence has also been examined in a previous study [68], showing that the grid spacing used in our simulations is fine enough to produce accurate results. Table 9.1 lists the six cases we have examined, where different channel size, cell deformability and pressure gradient were employed. The situation with a normal deformable cell (i.e., $E_s = 6 \times 10^{-6}$ N/m and $E_b = 2 \times 10^{-19}$ Nm [66]) flowing in a 40 lattice (7.8 μm) wide channel under a pressure gradient of $P_x = 157.8$; kPa/m is considered as the control case (Case SF-C in Table 9.1). Each other case differentiates from the control case SF-C with one aspect: Case SF-R with a rigid (less deformable) cell, Case SF-h with a smaller (30 lattice, 5.85 μm, wide) channel, Case SF-H with a larger (60 lattice, 11.7 μm, wide) channel, and Cases SF-V2 and SF-V3 with larger pressure gradients (315.6 and 473.4 kPa/m, respectively). To model the reduced deformability in Case SF-R, we have increased both the elastic and bending modulus values of cell membrane to 20 times of their normal ones, i.e., $E_s = 1.2 \times 10^{-4}$ N/m and $E_b = 4 \times 10^{-18}$ Nm [70,71]. Measurements of the cell shape, velocity and wall shear stress were taken after the flow and cell deformation have reached their steady states.

9.4.1.1 Cell Deformation

To further investigate the relationship between RBC motion and induced wall shear stress variation, we have plotted the stable cell shapes of the six simulated situations in Fig. 9.6. Clearly the originally circular cells have deformed greatly as they are traveling through the channel. In general, the cell deformation results in a bullet-like front to reduce hydrodynamic resistance, and a concave end due to the high pressure built up behind it [50]. This shape is similar to those observed in *in vivo* and *in vitro* experiments [45,47]. In Case SF-R, the deformation is much less evident due to its high membrane rigidity. Effects of the channel size (Cases SF-h and SF-H) and flow velocity (Cases SF-V2 and SF-V3) on the cell deformation have also been examined. In a smaller channel (Case SF-h), the cell has been squeezed to a narrower width and stretched to a longer length. Conversely, the cell in a wider channel (Case SF-H) can deform into a broader profile. Meanwhile, the concave bending-in at the rear of the cell in Case SF-H is less severe, since a smaller pressure gradient is applied to generate a same flow

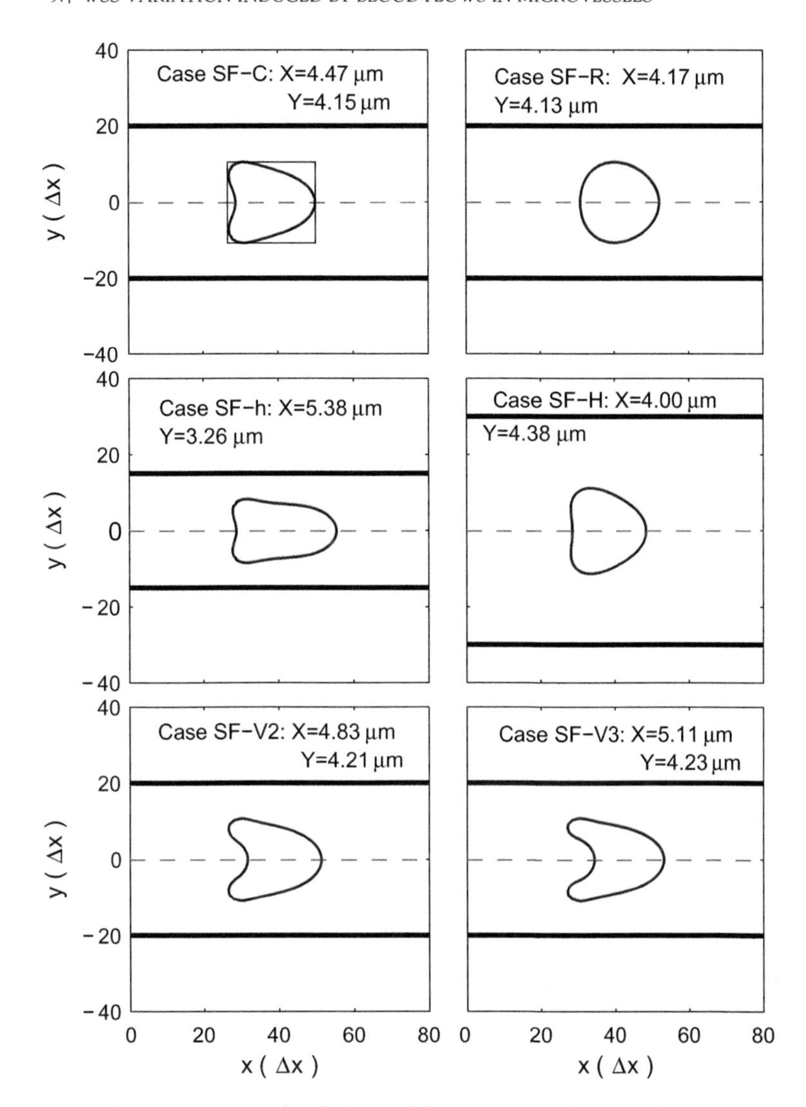

FIGURE 9.6 Deformation of single RBCs flowing (left to right) in microvessels. The deformed cell length X and width Y are also displayed for individual cases. An $X \times Y$ rectangular is plotted for Case SF-C to illustrate how X and Y are measured. The dashed lines are the channel centerlines, and the thick lines represent the channel walls. (Reprinted from W. Xiong and J. Zhang, *Annals of Biomedical Engineering*, 38:2649–2659, 2010. With kind permission from Springer Science and Business Media.)

velocity in a wider channel. On the other hand, the response to increases in flow velocity (see Cases SF-V2 and SF-V3 in Fig. 9.6) is less dramatic. Only slight increases in cell lengths are observed when the flow velocity has been doubled (Case SF-V2) or even tripled (Case SF-V3). These increases in flow velocity result in even smaller increases in cell widths. This might be explained by the double (Case SF-V2) or triple (Case SF-V3) increases in applied pressure gradient, which is also indicated by the larger concave deformation at the cell end (Fig. 9.6).

9.4.1.2 Shear Stress Variation

We then examine the shear stress variation on the channel wall while a RBC flows by. As an example, Fig. 9.7 shows the stable flow situation (a) and the shear stress distribution on the top channel wall (b). It can be seen in Fig. 9.7a that, for a position far away from the cell, the velocity profile across the channel is nearly parabolic as of typical Poiseuille flows. However, the velocity is uniform inside the cell, since at the stable state the deformed cell is actually moving as a rigid body. Also a high pressure region can be observed behind the cell, which is responsible for the concave cell end as discussed before. The disturbed flow field around the cell will introduce a corresponding change in shear stress on the channel walls. A normalized wall shear stress τ_w^* is defined as:

$$\tau_w^* = \frac{\tau_w}{P_x H/2} \tag{9.24}$$

where τ_w is the wall shear stress. Note that the denominator is the shear stress resulting from a uniform Poiseuille flow with no cells. Figure 9.7b plots the distribution of this normalized shear stress along the top channel wall. Due to the system symmetry, the shear stress on the bottom wall is identical to that on the top wall, and thus it is not shown here. Consistent to the nearly parabolic velocity profile of a Poiseuille flow in the region away from the cell, the normalized shear stress in those locations is close to unity. Both the mean velocity and shear stress in this region are slightly (~2%) below those of a Poiseuille flow with no cells, since the existence of a RBC in the channel increases the system hydrodynamic resistance. However, the cell occupies only 6% of the flow domain, and therefore its effect on the far region flow can be neglected.

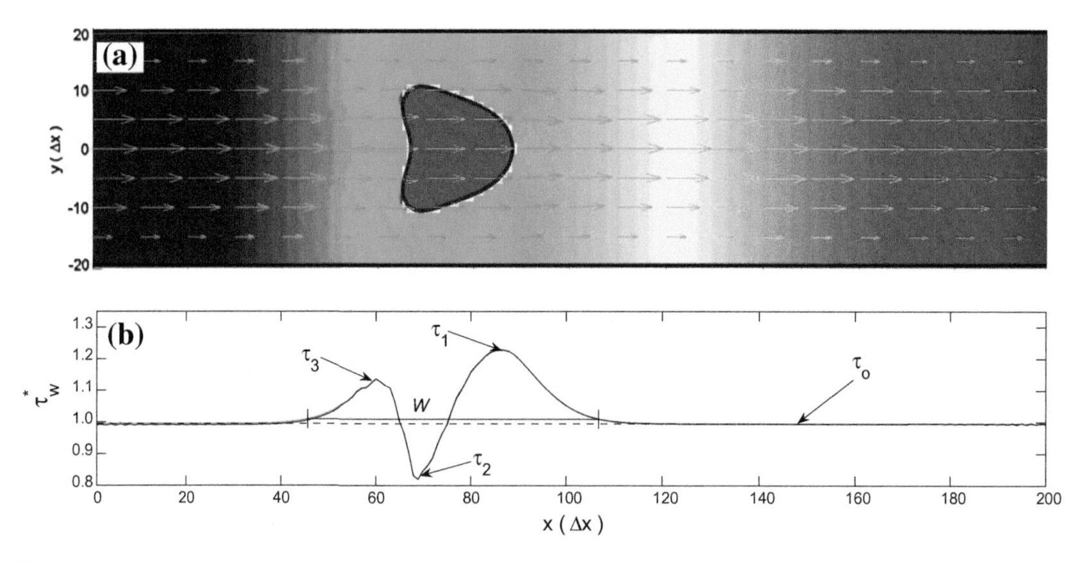

FIGURE 9.7 (a) Stable flow situation of Case SF-C with the pressure distribution (colorful background) and flow field (arrows) and (b) the shear stress variation along the top channel wall. Also illustrated in (b) are the extreme shear stress values τ_1, τ_2 and τ_3 and the influencing range W of the variation. (Reprinted from W. Xiong and J. Zhang, *Annals of Biomedical Engineering*, 38:2649–2659, 2010. With kind permission from Springer Science and Business Media.)

From the perspective of a point located on the channel wall, as the front of a RBC approaches, the shear stress increases first to a maximum value, τ_1, due to a larger velocity gradient across the cell-wall gap. When the cell front passes this location, the shear stress rapidly decreases, and reaches a minimum point, τ_2, at the end of the cell, when the local flow becomes the wake flow region behind the moving cell. During the recovery process from this low shear stress back to that of normal channel flow, another peak, τ_3, is observed; however, its magnitude is slightly lower than the previous maximum value τ_1.

9.4.1.3 *Effects of Cell and Flow Properties*

This interesting peak-valley-peak structure has been commonly observed in all other simulated cases, and the three extreme shear stress values are summarized in Table 9.2. Also provided in this table are the influential range W and the time duration $\widetilde{T}$ for a fixed wall point of the above-discussed shear stress variation. Here W is defined as the distance between the front and back points of $\tau_w^* = \tau_0 + 0.1(\tau_1 - \tau_0)$ (see Fig. 9.6), and τ_0 is the shear stress in the region far away from the cell. For a fixed point on the wall, the time duration of the shear stress variation $\widetilde{T}$ as a RBC flows by can be calculated as $\widetilde{T} = W/V$, where V is the cell velocity.

It should be noted that the channel size is the major factor influencing the shear stress variation amplitude. In a small channel (Case SF-h), due to the narrow gap between the channel wall and cell membrane, the shear stress variation (0.66–1.28) is most profound among all simulated cases. On the other hand, the shear stress in Case SF-H with a wide channel exhibits its smallest variation within a range of 0.97–1.06. Recall this analysis is for the normalized shear stress τ_w^*, and the nominal shear stress in the small channel is higher than that in a wider channel with the same mean flow velocity. Therefore, the actual shear stress variation in a smaller channel will be even more profound than that listed in Table 9.2. The effects of cell rigidity and flow velocity are less dramatic, although the variation magnitude indeed increases with both of them.

The spatial influential range of the shear stress variation, W, can be related to the cell length when the channel size is the same, as observed in Fig. 9.6 and Table 9.2. For the same channel size $H = 40$ lattice unit (7.8 μm), Case SF-R (rigid cell) has the smallest cell length $X = 4.17$ μm and variation range $W = 10.83$ μm, while Case SF-V3 (three times velocity) has the largest cell length $X = 5.11$ μm and variation range $W = 11.57$ μm. This is understandable since the variation

TABLE 9.2 Characters of Shear Stress Variation. See Fig. 9.7 for the Definitions of τ_1, τ_2, τ_3 and W. The Time Duration $\widetilde{T}$ Is Calculated from the Influencing Range W and Cell Velocity.

Case	τ_1	τ_2	τ_3	W (μm)	$\widetilde{T}$ (ms)
SF-C	1.23	0.82	1.14	11.15	13.19
SF-R	1.26	0.92	1.18	10.83	13.05
SF-h	1.28	0.66	1.24	10.06	12.25
SF-H	1.06	0.97	1.03	15.34	16.73
SF-V2	1.24	0.76	1.15	11.53	6.92
SF-V3	1.25	0.77	1.14	11.57	4.63

is initiated when the cell front approaches, and it can only recover back to the undisturbed state when the cell end has passed away. However, when the channel size is involved (e.g., comparing Cases SF-h, SF-C and SF-H), the above analysis does not hold anymore. Case SF-h has the longest cell length $X = 5.38\ \mu m$ but the smallest variation range $W = 10.06\ \mu m$, and Case SF-H is observed with the smallest cell length $X = 4.00\ \mu m$ and the largest variation range $W = 15.34\ \mu m$. This implies that the effect of channel size is much stronger and has suppressed that of the cell length. A possible explanation is that, as the cell approaches (or leaves) a particular location, a larger region is required in a wider channel for the flow to adjust from (or restore back to) the normal Poiseuille flow. This observation is similar to that of longer entrance lengths for pipe flows in wider pipes [11]. Although the effect of flow velocity on shear stress variation range is limited, for a fixed position on the channel wall, the time duration $\widetilde{T}$ is more strongly dependent on the flow velocity, as shown in Table 9.2. At a higher velocity, it takes a shorter time for the cell to pass a particular point and therefore the shear stress disturbance lasts a short time period as well.

9.4.1.4 Hematocrit Effect

In actual microcirculation situations, RBCs are usually flowing through a small microvessel or capillary in a single-file fashion with hematocrit up to ~30%. To study the influence of hematocrit, we have modified Case SF-C by adding more cells in the channel, while other cell and flow parameters remain the same. The corresponding hematocrit values are 4.54% for Case SF-C (one cell), 8.91% for Case SF-N2 (two cells), 17.82% for Case SF-N4 (four cells), and 30.72% for Case SF-N7 (seven cells) in a 39 μm long and 7.8 μm wide channel. Figure 9.8 displays the deformed cells, the axial velocity distributions and the induced wall shear stress profiles for these three cases (SF-N2, SF-N4 and SF-N7) with multiple cells. The measured cell velocities are, respectively, 0.845 (Case SF-C), 0.833 (Case SF-N2), 0.809 (Case SF-N4) and 0.764 mm/s (Case SF-N7). Clearly, under the same applied pressure gradient ($P_x = 157.8$ kPa/m), the cell moving velocity decreases with hematocrit because of the increasing flow resistance. Correspondingly, the cell deformation becomes less pronounced at higher hematocrit (Fig. 9.8a, b, c and e).

As for the shear stress variation, the particular peak-valley-peak structure is still clearly observed in Case N2 (Fig. 9.8b), since the separation between the two cells (~15 μm) is long enough when compared to the variation range induced by individual cells ($W = 11.15\ \mu m$, see Table 9.2). When the hematocrit increases to 17.82% with decreases in cell separation up to ~7 μm in Case N4, the rear peak from one cell begins to merge to the front peak generated by the following cell. The shear stress variation profile then cannot be completely separated for individual cells (Fig. 9.8d). At an even higher hematocrit of 30.72% in Case N7, the original two peaks, one from the front cell and the other from the following cell, totally combine together, resulting in a single peak between the two close cells (Fig. 9.8f). However, the valleys in the shear stress profiles appear not be affected by the increasing hematocrit very much.

We have also examined the shear stress variation magnitude in these three cases with higher hematocrit (N2, N4 and N7). The peak/valley values of the normalized wall shear stress are, respectively, 1.23/0.82/1.14 for Case N2 (which is the same as those in Case SF-C, see Table 9.2), 1.20/0.81/1.7 for Case N4 and 1.12/0.83 for Case N7 (Fig. 9.8b, d and f). It appears that the variation magnitude decreases slightly with the cell hematocrit, and this is probably due to the slower cell motion at higher hematocrit. The shear stress variation waves

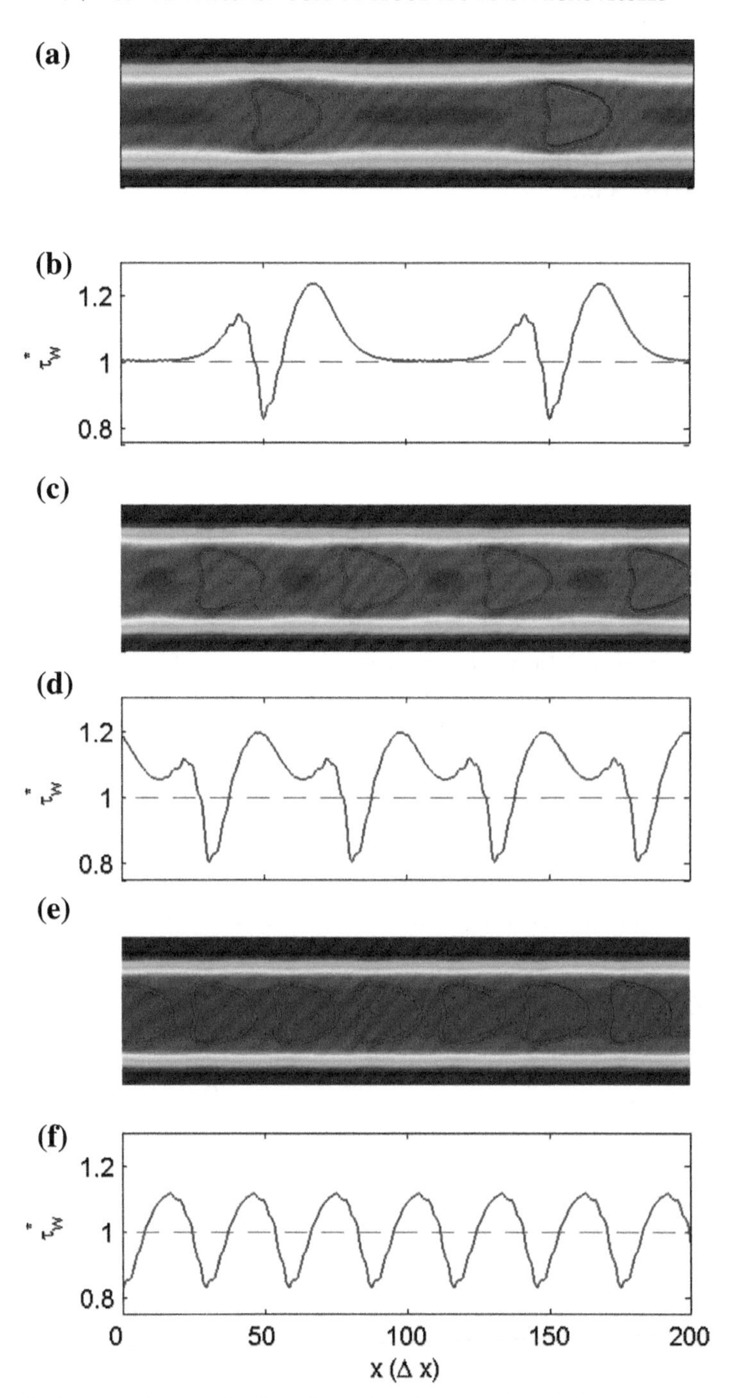

FIGURE 9.8 Stable flow situations and cell configurations (a, c and e) and shear stress variations (b, d and f) of Cases SF-N2, SF-N4 and SF-N7 with respective hematocrit values of 8.91% (SF-N2), 17.82% (SF-N4) and 30.72% (SF-N7). (Reprinted from W. Xiong and J. Zhang, *Annals of Biomedical Engineering*, 38:2649–2659, 2010. With kind permission from Springer Science and Business Media.)

induced by sequential cells cannot cancel out each other but are still profound even at a hematocrit of 30.72%.

9.4.2 Cell-Free Layer (CFL) and Wall Shear Stress (WSS) Variation in Multiple RBC Flows

When multiple RBCs are passing through the vessel, they tend to migrate toward the center of mainstream flow and keep away from the wall of the vessel. This is how the cell-free layer (CFL) is formed. The relationship between WSS variation and cell-free layer is of great interest in microcirculatory research. Recently, *Namgung et al.* [34] have calculated WSS variation based on their CFL measurements. The local instantaneous WSS τ was obtained from the measured CFL thickness δ by the relationship:

$$\tau = \mu_p \frac{V_{RBC}}{\delta} \tag{9.25}$$

by assuming a linear shear flow across the CFL. Here μ_P is the plasma viscosity and V_{RBC} is the velocity of the edge RBC on the inner side of the CFL, and V_{RBC} has been considered as a constant for its relatively small variation [27,34]. According to this relationship, a low (or high) WSS would be resulted in at a location with a wide (or narrow) cell-wall gap (see Fig.9.1 in Ref. [34]). However, simulations on WSS variation in single-file RBC flows in the previous section have showed that the minimum WSS (valley) is actually observed at the cell back end, which is the location of the minimum cell-wall gap distance.

The simulations in the previous Section 9.4.1 were for single-file RBC flows in small microvessels, while the experiments and analysis in Ref. [34] were for concentrated RBC flows in large microvessels. In this section, we will extend our simulations to multiple RBC flows in a larger straight microvessel. Here, we consider 27 RBCs flowing in a 19.8 μm wide channel with normal membrane elasticity [71]. The flow is generated by applying a pressure gradient of 24.55 kPa/m, so that the maximum velocity of a pure plasma (with a viscosity $\mu_p = 1.2$ cP) under this pressure gradient will be $U_0 = 1$ mm/s.

9.4.2.1 Cell-Free Layer and WSS Correlation

We first examine the linear shear assumption (Eq. (9.25)) with our simulated flow field. A part of the simulation domain is shown in the inset of Fig. 9.9, with the deformed cells in gray, and the flow field by arrows. We consider the cell (dark gray in Fig. 9.9 inset) with its center at about (70, 25), and plot the streamwise velocity u across the cell-wall gap at locations $x=60$, 70 and 80 (thick lines, Fig. 9.9b). The y-locations of the cell membrane at these three x-locations are indicated by the small circles and straight lines are obtained according to the linear shear assumption (Eq. (9.25)). We see the membrane velocity is approximately constant, indicating that the cell is mainly moving as a rigid body, and its deformation and rotation are relatively less pronounced at this quasi-steady state. The velocity profiles across the cell-wall gap exhibit strong nonlinearity, and the straight lines according to Eq. (9.25) clearly cannot describe the velocity change. The WSS τ then should be evaluated from the velocity gradient near the wall:

$$\tau = \mu_p \left. \frac{\partial u}{\partial y} \right|_{y=0} \tag{9.26}$$

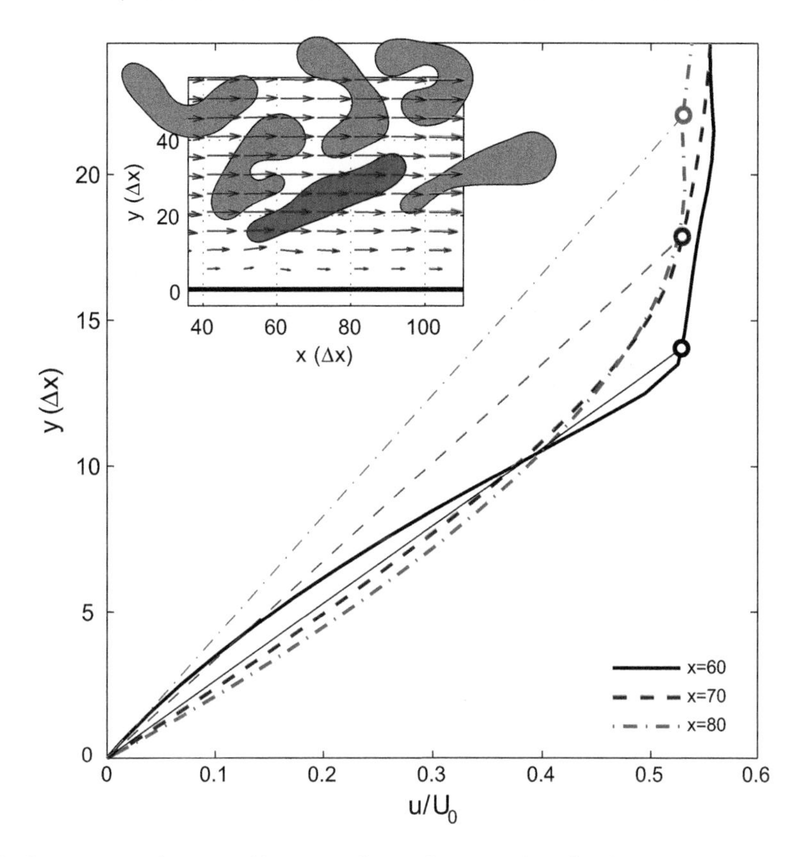

FIGURE 9.9 Streamwise velocity profiles across the gap between the cell at (75, 20) and wall (inset) at different locations $x=60$, 70 and 80 (thick lines). The circles display the vertical positions of the cell membrane at those locations, and the thin lines are the linear velocity distributions according to Eq. (9.25).

In addition to the inaccuracy from the linear velocity assumption across the cell-wall gap, it is also clear from Fig. 9.9 that the linear shear assumption predicts an incorrect WSS variation trend as observed in the simulated flow field. When moving along the wall surface from $x=60$, 70, to 80, the cell-wall gap expands, and the WSS from Eq. (9.25) will decrease, as indicated by the increasing slopes of dashed lines in Fig. 9.9. However, the local velocity gradient $(\partial u/\partial y)_{y=0}$, as well as the WSS, from our simulation is actually increasing from $x=60$ to $x=80$ (Fig. 9.9). In spite of the model simplicity, this comparison shows that the linear shear assumption in Eq. (9.25) is oversimplified and it cannot predict either the magnitude or the variation trend of the dynamic WSS on microvessel walls.

For an overall picture of the flow structure of the 27 cells flowing in the microvessel, Fig. 9.10b displays the deformed cell configurations (gray patches), flow field (arrows) and pressure distribution (background color, light for high and dark for low pressures). The flow field shown by arrows demonstrates a blunt velocity profile across the channel, and this has been well observed in previous studies [40,71]. The WSS distributions on the top (Fig. 9.10a) and bottom (Fig. 9.10c) walls can be readily obtained from the simulated flow field using

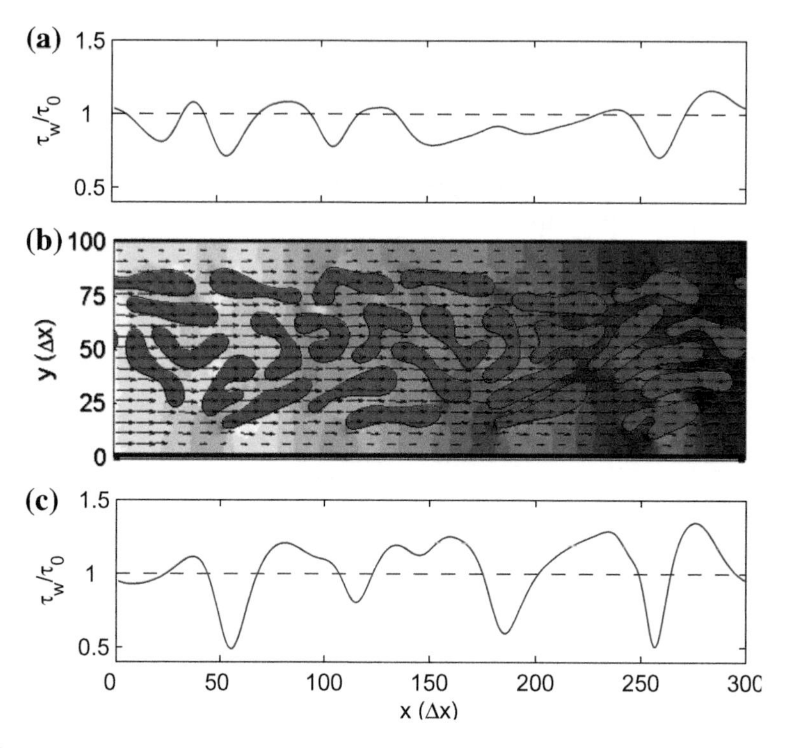

FIGURE 9.10 The flow (arrows) and pressure (gray background, light for high and dark for low pressures) fields (b) and WSS distributions along the top (a) and bottom (c) walls.

Eq. (9.26). Because of the irregular edge of the RBC-core, it is not surprising to see the WSS distributions exhibit considerable variations. The average WSS:

$$\tau_{avg} = \frac{1}{2L}\left(\int_{top} \tau dx + \int_{bottom} \tau dx\right) \tag{9.27}$$

agrees well with the nominal shear stress $\tau_0 = \frac{H}{2}\frac{dP}{dx}$ within a ~1.5% difference, and we attribute this small difference to the numerical errors in differentiation (Eq. (9.26)) and integration (Eq. (9.27)) as well as the unsteady effect in the flow [71]. Here $L=300\ \Delta x$ and $H=100\ \Delta x$ are, respectively, the length and width of our simulation domain.

Observing the WSS distributions and the edge shape of the RBC-core together in Fig. 9.10, a clear correlation can be found. In general, a local minimum WSS is usually found at positions where the cell-wall gap is narrow (e.g., $x=110$ at the top wall), while a large cell-wall distance typically corresponds to a high WSS (e.g., $x=230$ at the bottom wall). Again this is consistent with the above analysis of velocity profiles and the findings of the previous section of single-file RBC flows. However, this is contradictory to the linear shear assumption of Eq. (9.25). In addition, we also notice that there is usually a high pressure region in front of and a low pressure region behind a flowing cell near the wall. See the cells with centers at (70, 25) near the top wall and at (260, 70) near the bottom wall as an example. Such a pressure difference is less evident for a cell flowing relatively farther away from the walls (compare the cell at (210, 75) vs. the cell at (160, 70) near the top wall), since it experiences less resistance from the wall. Similar reverse pressure differences have also been observed in a previous three-dimensional simulation of RBCs in a square channel [13]. An intuitive hypothesis about the relationship between such reverse pressure differences and the WSS variation could be that this reverse pressure gradient will slow down the streamwise flow through the narrow cell-wall gap and therefore reduce the WSS in this area.

9.4.2.2 Effects of Cell Deformability, Flow Rate and Aggregation on WSS Variation

Next, we consider how the WSS is influenced by the RBC deformability, the flow rate through the channel and the aggregation among RBCs. We consider the case above with normal cell deformability, a nominal maximum velocity $U_0 = 1$ mm/s and no aggregation considered as the control case (Case MF-C in Table 9.3). Starting from Case MF-C, we then, respectively, reduce the cell deformability by fivefold (Case MF-R), increase the nominal maximum velocity by twice (Case MF-F) and introduce a moderate aggregation (Case MF-A) [71]. Detailed simulation parameters and results for these cases are listed in Table 9.3. The correlation between WSS variation and cell-wall gap distance observed in Fig. 9.10b has been confirmed in all these simulations. The overall CFL thickness δ is determined as the averaged value of the closest cell-wall distances along the top and bottom walls. Unlike the treatment in Ref. [34], we do not use the local instantaneous cell-wall gap distance as a measure of the CFL variation, since it is not continuous when jumping from one edge cell to another, and even may be not well defined at some locations where there are no cells near the wall (see x=~230 at the top and bottom walls in Fig. 9.10b). Here we consider the CFL as an overall, global description of the RBC flow configuration in microvessels, instead of a local, distributed parameter for the edge cell positions. The WSS variation in each case is characterized by the average standard deviations of the WSS distributions σ_τ on the two walls, normalized with the corresponding nominal WSS τ_0. The mean WSS values are not reported here since they are always close to the nominal WSS τ_0 with relative errors <2%.

As observed in previous experimental and numerical studies [46,56,71], we see that the CFL thickness increases when the aggregation is introduced (Case MF-A), or when the flow rate is increased (Case MF-F). Accompanying with the CFL increase, the WSS variation becomes less profound. On the other side, higher cell rigidity (Case MF-R) inhibits the cell deformation and migration, and the developed CFL is thinner, associated with an increase in WSS variation magnitude. Such a relationship between the overall CFL thickness and the

TABLE 9.3 Simulation Parameters and Results of Average CFL Thickness and WSS Variation. Definitions and Implementation of these Model Parameters in Simulations Can Be Found in Refs. [70,71].

Case	E_s (N/m)	E_b (Nm)	D_e (μJ/μm^2)	dP/dx (kPa/m)	δ (μm)	σ_τ/τ_0
MF-C	6×10^{-6}	2×10^{-19}	–	24.55	2.13	0.574
MF-R	3×10^{-5}	1×10^{-18}	–	24.55	1.01	0.601
MF-F	6×10^{-6}	2×10^{-19}	–	49.10	2.56	0.533
MF-A	6×10^{-6}	2×10^{-19}	5.2×10^{-7}	24.55	2.89	0.532

WSS variation can be understood by considering the following two aspects: for a wider CFL, (1) the reverse pressure difference between the front and back sides of an edge cell is less significant (see Fig. 9.10b), and (2) its effect on WSS is even weaker because of the larger cell-wall distance. We also would like to clarify that the less significant WSS variation associated with a wider global CFL observed here is different than the lower WSS magnitude predicted by Eq. (9.25) for a larger local cell-wall gap.

9.5 SUSPENDING VISCOSITY EFFECT

In addition to the various cellular properties, another important factor related to blood flow behaviors is the plasma viscosity. As summarized in Section 9.1, though there are several interesting *in vivo* observations reported in previous studies, fundamental investigations on the plasma viscosity effects on the RBC deformation and motion, the RBC flow behaviors in microvessels and the flow-vessel interactions are relatively inadequate. In this section, we simulate the suspending viscosity effect on RBC dynamics and microscopic hemorheology. Four flow types are considered: single RBC rotation in shear flows, single RBC flowing in a direct channel, multiple RBCs in channel flows and single RBC in a bifurcated microvessel. The results and analysis are presented below.

9.5.1 Single RBC Rotation in Shear Flows

We first simulate a single RBC in shear flows with the suspending viscosity varying from $\mu_p = 1.2$ (the normal value of human plasma), 2.0–3.0 cP. To generate a shear flow, the modified bounce-back method [28] is applied on the top and bottom boundaries with identical magnitudes but opposite directions, U and $-U$, respectively. The general periodic boundary condition [69] is imposed at the left and right boundaries. The nominal shear rate $\dot{\gamma} = \frac{2U}{H}$ and shear stress $\tau = \mu_p \dot{\gamma}$ then can be easily adjusted by tuning the magnitude of the imposed top/bottom boundary velocity U. Here H is the gap distance between the top and bottom walls and μ_p is the suspending viscosity. Simulation snapshots with a constant shear rate $\dot{\gamma} = 100\,\text{s}^{-1}$ and a constant shear stress $\tau = 0.12$ Pa are displayed in Fig. 9.11. These shear rate and shear stress values are similar to those employed in previous simulations of RBCs in shear flows [41,42,51]. For $\mu_p = 1.2$ cP, the RBC performs a tumbling and flipping motion (Fig. 9.11), similar to those observed in previous studies [42]. When suspended in a more viscous medium ($\mu_p = 2.0$ or 3.0 cP), the RBC exhibits tank-treading rotation with a flattened, elongated shape, with holding either the shear rate constant $\dot{\gamma} = 100\ \text{s}^{-1}$ (Fig. 9.11 b and c) or the shear stress constant $\tau = 0.12$ Pa (Fig. 9.11 d and e). This indicates that a RBC appears more flexible in a more viscous fluid, even the actual cell properties are the same (i.e., same E_s and E_b values).

To further elucidate this fact, in addition to the viscosity ratio $\lambda = \mu_c/\mu_p$, we now introduce another non-dimensional parameter: the membrane rigidity $E_s^* = E_s/(\mu_p \dot{\gamma} a)$ [4]. The former measures the relative resistance due to the cytoplasm flow, while the latter represents the relative resistance from the membrane elasticity, compared to the hydrodynamic force. The values calculated for the five cases in Fig. 9.11 are listed in Table 9.4. We see that the relative cytoplasm resistance decreases with the suspending viscosity, and the relative membrane

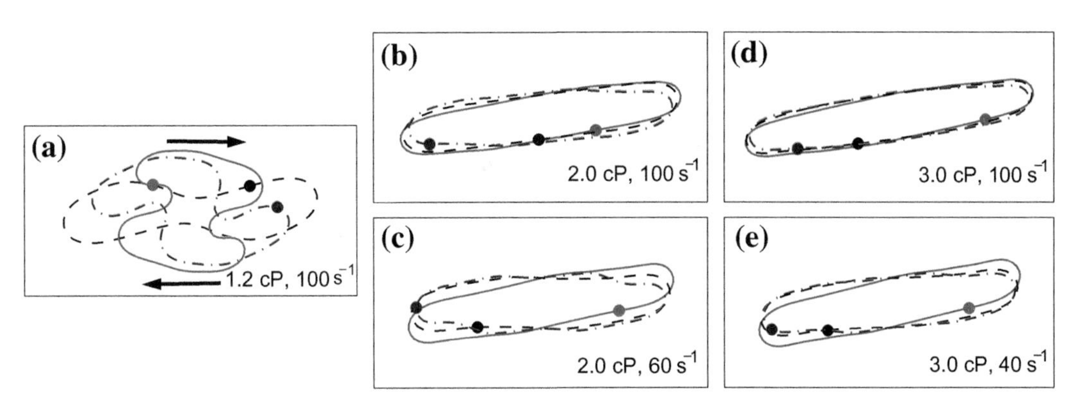

FIGURE 9.11 Single RBC deformation and rotation in shear flows with different suspending viscosities and shear rates. For each case, three consecutive cell configurations are displayed, respectively, in solid, dashed and dot-dashed lines. A membrane marker is also shown to indicate the membrane rotation. (Reprinted from J. Zhang, *Microcirculation*, 18:562–573, 2011. With permission from John Wiley and Sons.)

resistance remains constant (for a constant shear stress $\tau = 0.12$ Pa) or decreases with μ_p (for a constant shear rate $\dot{\gamma} = 100\ \mathrm{s}^{-1}$). As a result of the decrease in the total relative resistance, the RBC motion changes from the tumbling-flipping type for $\mu_p = 1.2$ cP to the tank-treading type for $\mu_p = 2.0$ or 3.0 cP.

We next examine the cell inclination angle θ and membrane rotation angle ϕ during the tank-treading processes with higher suspending viscosities. Figure 9.12 displays the θ and ϕ variations with time. It can be seen that, after the initial transition period, both angles vary periodically. The periods were measured as $T_\phi = 7.3 \times 10^6 \Delta t$ for the membrane rotation and $T_\theta = 3.7 \times 10^6 \Delta t$ for the inclination variation. The former is approximately twice of the latter due to the initial symmetric cell shape, which agrees qualitatively with previous experiments.

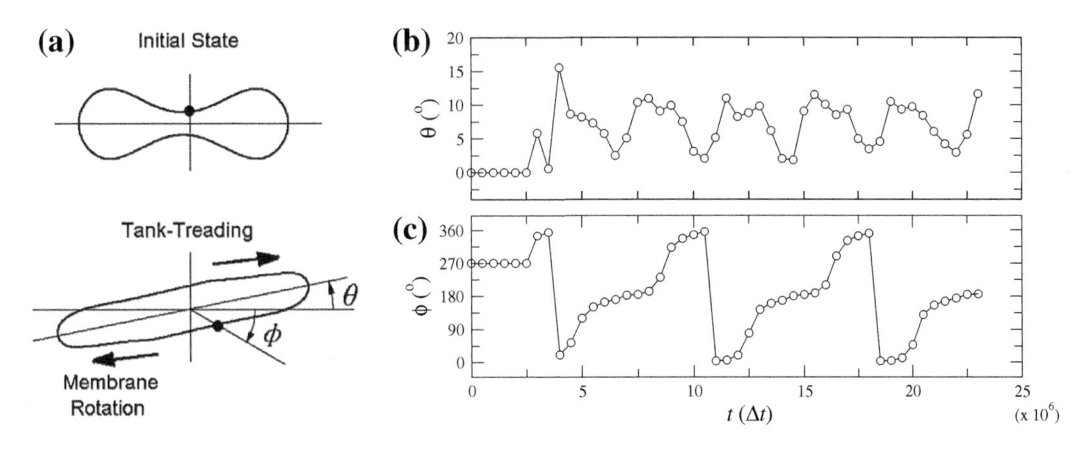

FIGURE 9.12 Variation of (b) the cell inclination angle θ and (c) the membrane rotation angle ϕ for the RBC tank-treading motion in the shear flow with $\mu_p = 2.0$ cP and $\dot{\gamma} = 142.0\ \mathrm{s}^{-1}$ (Case d in Table 9.4 and Fig. 9.11). The definition of θ and ϕ is illustrated in (a).

TABLE 9.4 Non-Dimensional Relative Resistances and Cell Inclination Variations of Single RBCs in Shear Flows

Case	μ_p (cP)	$\dot{\gamma}(s^{-1})$	τ (Pa)	λ	E_s^*	$\bar{\theta}$(°)	$\Delta\theta$(°)
(a)	1.2	100	0.12	5.0	12.79	–	–
(b)	2.0	100	0.20	3.0	7.67	7.3	4.4
(c)	3.0	100	0.30	2.0	5.12	9.8	5.3
(d)	2.0	60	0.12	3.0	12.79	6.7	9.7
(e)	3.0	40	0.12	2.0	12.79	9.5	7.4

The average inclination angle $\bar{\theta} = (\theta_{max} + \theta_{min})/2$ and the variation $\Delta\theta = \theta_{max} - \theta_{min}$ are also listed in Table 9.4. Here θ_{max} and θ_{min} are, respectively, the maximum and minimum inclination angles. Again our results show similar trends to the previous work by *Abkarian et al.* [1], including the increase in θ and the decrease in $\delta\theta$ with the shear rate for a same suspending viscosity (Cases d vs. b and Cases e vs. c), and the increase in $\Delta\theta$ with the suspending viscosity for a same shear rate (Case b vs. c). The comparison further demonstrates that our model has correctly incorporated the cell and flow dynamics and interaction between them (at least qualitatively). However, a quantitative comparison with these experimental data is difficult for the inherent limitations of this 2-D model and the extremely high suspending viscosity of the experiments (22–47 cP). Such a high suspending viscosity will cause a severe numerical instability in simulations, and it is not relevant to actual microcirculation situations, where the suspending viscosity is always lower than the RBC cytoplasm viscosity (1.2 cP). For these reasons, existing simulations of RBC dynamics usually have lower suspending viscosity than the cytoplasm viscosity.

9.5.2 Single RBC Migration in Channel Flows

One interesting phenomenon of RBC dynamics is that, when flowing in a tube or microvessel, a RBC tends to migrate toward the central region, resulting in a cell-free layer (CFL) near the wall surface and a concentrated core in the center [32,40]. This migration is attributed to the cell rotation (tank-treading motion) under a velocity gradient as in shear flows [32]. This particular flow structure is important to both the blood hemorheology and mass transfer in the microcirculation [27,64]. The cell migration phenomenon has been simulated in several previous studies [42]; however, as with the single cells in shear flows, the suspending viscosity effect has not been addressed.

To simulate the suspending viscosity effect on cell migration, a RBC was considered in a $H = 100\ \Delta x$ (19.55 μm) wide channel and a body force g was applied to induce a channel flow, and a nominal flow velocity V is defined as:

$$V = \frac{gH^2}{8\mu_p} \tag{9.28}$$

which is the maximum velocity at the channel center according to the Poiseuille law, if there are no cells within the channel. With RBCs suspending in the fluid, the resulting velocity will

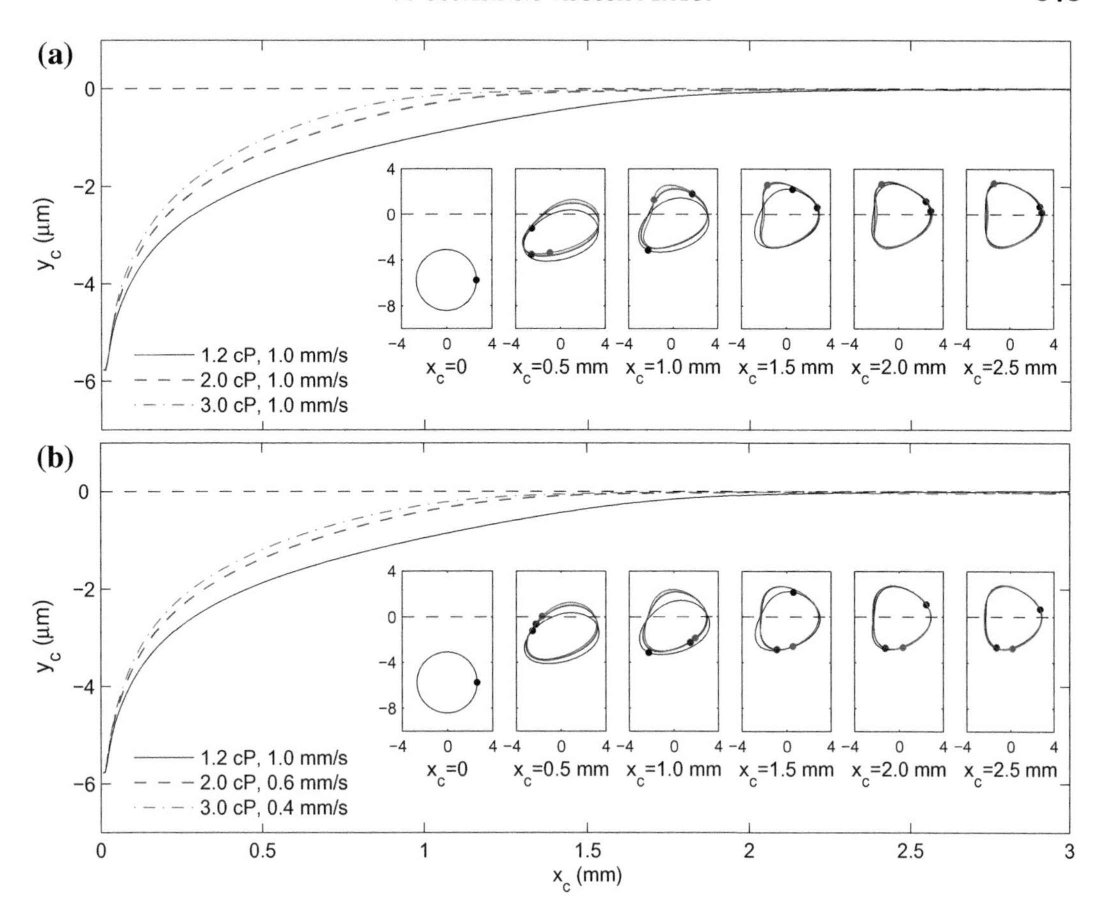

FIGURE 9.13 Trajectories of the cell center in channel flows under a same nominal velocity $V = 1.0$ mm/s (a) and a same driving force $g = 25.18$ kN/m^3 (b) with different suspending viscosities. Also displayed in the insets are cell shapes at several representative axial locations. The dashed lines represent the channel centerline. A membrane marker is employed for each cell to show its rotation during the migration process, and the values of the suspending viscosity and nominal velocity are provided in the legends. (Reprinted from J. Zhang, *Microcirculation*, 18:562–573, 2011. With permission from John Wiley and Sons.)

be less than this nominal value because of the increased flow resistance. Nevertheless, this nominal velocity serves as a characteristic representation of the flow situation, since the actual flow velocity cannot be calculated in advance and it is hard to control during a simulation [71]. Again, to avoid any possible bias of the cell trajectory induced by initial RBC inclinations and therefore to make the analysis simpler and more rigorous, we adopted a circular shape with a radius of 2.66 μm instead of a biconcave cell [5]. The RBC was initially positioned at 5.77 μm below the channel centerline. Again five cases were studied with $\mu_p = 1.2$, 2.0, and 3.0 cP by holding a constant nominal velocity V=1.0 mm/s or a constant driving force $g = 25.18$ kN/m^3. Figure 9.13 displays the cell center trajectories of these cases as well as the cell shape deformations. Clearly for all simulated cases the cell migrates away from the lower wall toward the channel center. The membrane markers (small filled circles in Fig. 9.13) also indicate the cell rotation during the migration. At the end of the migration process, the cell has moved to the channel center and a parachute-like shape is observed, which is similar to findings of previous studies [47,67]. At this stage, the flow structure is approximately symmetric and the cell rotation also becomes negligible. The driving force for the cell migration therefore gradually disappears, and no further evident changes in cell shape and orientation are observed.

When comparing the constant nominal velocity (Fig. 9.13a) and the constant driving force (Fig. 9.13b) situations, we see that the cell migration is more evident in a more viscous fluid. For a transverse migration displacement of 5.77 μm, it takes a greater axial travel distance for the cell in a $\mu_p = 1.2$ cP medium (Fig. 9.13) than in a higher viscosity medium (Fig. 9.13). This confirms our observation in Section 9.3.4 that RBCs behave more like liquid drops in a more viscous medium. Additionally, these results are in qualitative agreement with previous experiments on cell deformation and alignment in tubes [47] and simulations of RBC motion through bifurcated microvessels [68].

9.5.3 Multiple RBC Flows in Microvessels

For a more realistic representation of blood flows in microvessels, we included multiple RBCs in straight channels with varying tube hematocrit, H_T (the volume fraction occupied by RBCs in the tube), and varying channel width, H (Table 9.5 and Fig. 9.14a). The shape of each cell is described by Eq. (9.8) with necessary translation and rotation and, therefore, all cells have the same initial shape and size. The flow is induced by applying a body force g to the fluid in the axial direction. Once the force is applied, the blood suspension begins to flow, and this is accompanied by large cell deformations due to local flow conditions. Meanwhile, during the flow development, RBCs near the boundaries migrate toward the channel center. As a result, cell-free layers (CFLs) are formed near the channel boundaries and the central region has a higher RBC concentration (Fig. 9.14b). Detailed descriptions of this flow development and these CFL formations can be found in [71]. Due to the high RBC concentration in the central region, the velocity profile across the channel exhibits a blunt shape (Fig. 9.14c), as observed in previous studies [40]. During the development process, the cross-sectional flow rate:

$$Q = \int_0^H \bar{u}(y)dy \tag{9.29}$$

TABLE 9.5 Setup Parameters for the Three Multiple RBC Flow Systems Simulated

System	RBC Number N	Channel Width H (μm)	Domain Length L (μm)	Tube Hematocrit H_T (%)
I	15	11.73	58.65	29.6
II	12	11.73	58.65	23.9
III	27	19.55	63.3	29.6

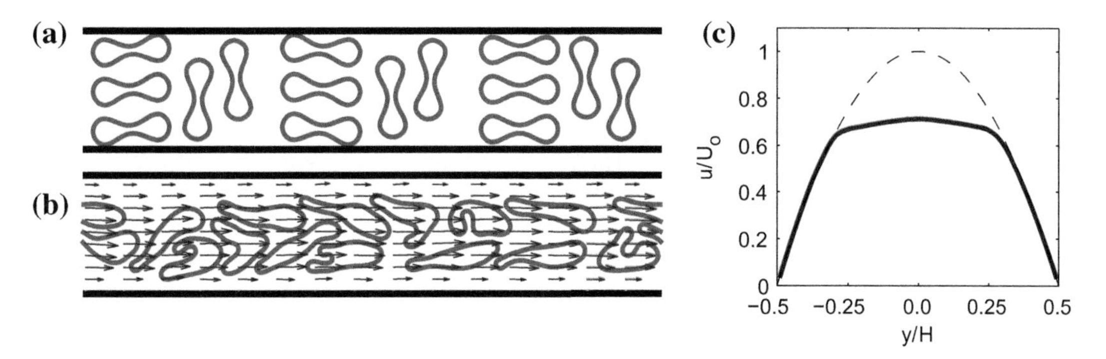

FIGURE 9.14 The RBC configurations at the initial (a) and quasi-steady (b) states in the channel flow simulation with $\mu_p = 1.2$ cP (Case a in Table 9.6). The blunt velocity profile at the quasi-steady state (thick line) is also compared with the parabolic profile of a pure plasma flow with no RBCs (dashed line) in panel (c). (Reprinted from J. Zhang, *Microcirculation*, 18:562–573, 2011. With permission from John Wiley and Sons.)

increases gradually. Here $\bar{u}(y)$ is the axial fluid velocity averaged over the channel length. Unlike a pure fluid, the flow rate Q exhibits random fluctuations and only a quasi-steady state can be established. This is because the RBC configuration is constantly changing [71]. The RBC flux can also be calculated from the flow field and RBC configurations via:

$$Q_{RBC} = \int_0^H \bar{u}(y)\bar{\beta}(y)dy \tag{9.30}$$

where $\bar{\beta}(y)$ is the channel-length averaged phase index that identifies the fluid domains (1 for cytoplasm and 0 for plasma, see Eq. (9.17)). The discharge hematocrit (the volume fraction occupied by RBCs in a discharge reservoir), H_D, and the relative apparent viscosity, μ_{rel}, can then be obtained from the flow rate and the RBC flux at the quasi-steady state as:

$$H_D = \frac{Q_{RBC}}{Q}, \quad \mu_{rel} = \frac{Q_0}{Q} \tag{9.31}$$

where $Q_0 = gH^3/12\mu_p$ is the volumetric flow rate of a pure plasma flow under the same body force g.

For each of the three systems listed in Table 9.5, five cases with different suspending viscosity μ_p, nominal velocity V, and driving force g have been considered. Table 9.6 lists these input parameters as well as the CFL thickness δ, the discharge hematocrit H_D and the relative apparent viscosity μ_{rel} obtained at the quasi-steady states for the five cases of System I. Here the CFL thickness δ is defined as the nearest distance between cell membranes and channel walls, and the values presented in Table 9.6 are the averaged ones from those measured near the top and bottom walls at a same time instant.

For all these five cases, we have $H_D > H_T = 29.6\%$, which is the well-known Fahraeus effect [40,54]. The relative apparent viscosity μ_{rel} is always larger than 1, indicating an increased flow resistance due to the existence of RBCs. However, no strong correlation is observed between the flow behaviors (including δ, H_D and μ_{rel}) and the suspending viscosity,

TABLE 9.6 Simulation Parameters and Flow Behaviors of Multiple RBC Flows for the Five Cases of System I

Case	μ_p (cP)	V (mm/s)	g (kN/m^3)	δ (μm)	H_D (%)	μ_{rel}
(a)	1.2	1.0	69.77	1.603	37.5	1.29
(b)	2.0	1.0	116.29	1.661	38.0	1.27
(c)	3.0	1.0	174.43	1.723	38.3	1.24
(d)	2.0	0.6	69.77	1.628	38.2	1.28
(e)	3.0	0.4	69.77	1.654	38.6	1.27

TABLE 9.7 Cell-Free Layer Thickness δ and Development Length $L_{1/2}$ for Multiple RBC Flows Under Different Situations

System	I		II		III	
Case	δ (μm)	$L_{1/2}$(μm)	δ (μm)	$L_{1/2}$ (μm)	δ (μm)	$L_{1/2}$ (μm)
(a)	1.603	83.0	2.069	104.1	2.182	308.5
(b)	1.661	93.0	1.994	77.0	2.350	277.5
(c)	1.723	76.5	2.186	78.1	2.452	236.3
(d)	1.628	84.0	2.044	91.6	2.226	262.5
(e)	1.654	67.5	1.941	71.1	2.281	223.9

either under the constant flow (Cases a, b and c) or constant forcing (Cases a, d and e) conditions. This finding is also confirmed with the results for Systems II and III (see the CFL thickness data in Table 9.7). Unlike these previous 2-D numerical studies, we have not attempted to compare our simulation results to the empirical relationships proposed by *Pries et al.* [44]. We believe that 2-D studies, although valuable, can only provide qualitative information. More discussions on this point can be found in Ref. [71].

With no evident effect of suspending viscosity on flow characteristics at the quasi-steady state, we then examine the process of flow development by looking at the CFL thickness change with time. Figure 9.15 plots the CFL thickness near the lower wall of System I under the flow situation of Case (a) (Table 9.6). We define a characteristic development time $T_{1/2}$, at which instant the CFL thickness increases from its initial value, δ_0, to $(\delta_0 + \delta)/2$, where δ is the mean CFL thickness at the quasi-steady state. Furthermore, a development length $L_{1/2} = VT_{1/2}$ is introduced as a characteristic measure of the distance traveled by RBCs along the channel during the developing process. With the periodic boundary condition applied along our channel model, the systems we are simulating actually are infinitely long with repeated units such as those shown in Fig. 9.14b, and therefore the *developmental length* concept is not applicable to our channel model. Nevertheless, the developmental length $L_{1/2}$ defined above provides a characteristic representation of the entrance length, over which the quasi-steady state is established.

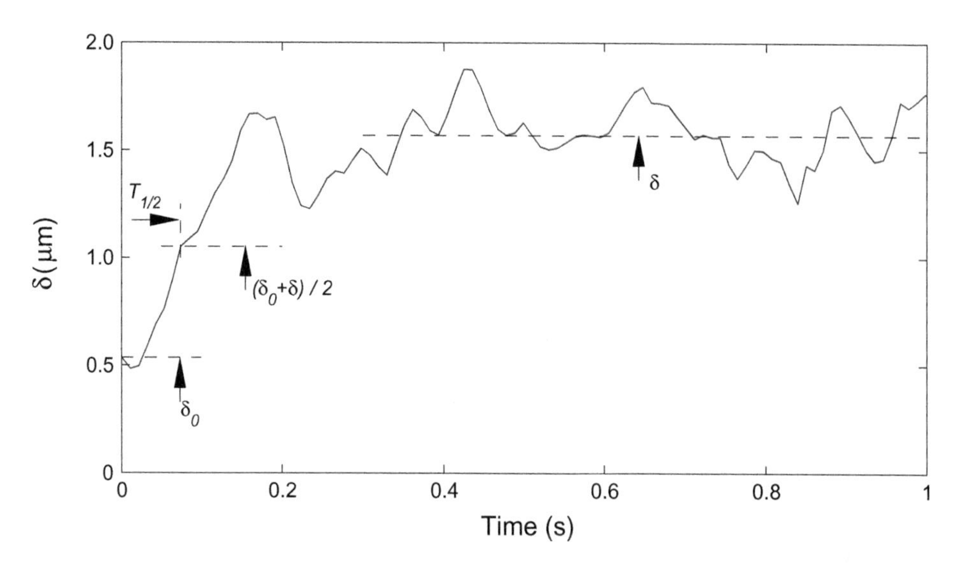

FIGURE 9.15 The CFL thickness change near the lower wall for System I with a suspending viscosity of $\mu_p = 1.2$ cP and a nominal velocity $V = 1$ mm/s (Case a in Table 9.6). (Reprinted from J. Zhang, *Microcirculation*, 18:562–573, 2011. With permission from John Wiley and Sons.)

Table 9.7 shows the CFL thickness, δ, and the developmental length, $L_{1/2}$, for all cases and systems simulated. Comparing System I with Systems II and III, it is evident that the CFL thickness is larger both in System II (due to the lower hematocrit) and in System III (due to the larger channel width). This trend is consistent with previous experimental and numerical investigations [46,52]. Additionally, the CFL thickness decreases slightly at a lower flow velocity for the same suspending viscosity (e.g., Cases c vs. e for System I). This is reasonable since the CFL development is driven by flow hydrodynamic forces.

On the other hand, we notice that, in general, the developmental length $L_{1/2}$ decreases with the suspending viscosity for both constant flow and forcing conditions, and the change is more profound with thicker CFLs, either due to a lower hematocrit (System II) or a wider channel size (System III). The shorter developmental length in a more viscous medium can be associated with the faster migration process discussed in Section 9.5.2 due to the increases in apparent cell deformability. Several exceptional cases, for which $L_{1/2}$ increases with μ_p, exist (e.g., Cases a vs. b and Cases a vs. d for System I, and Cases b vs. c for System II). We attribute this to the native uncertainty of such complex systems. Also, we note that the CFL thickness increases from ~1.7 μm in System I, ~2.0 μm in System II and to ~2.3 μm in System III; therefore, the statistical noise in δ is relatively strong in System I and relatively weak in System III. For this reason, we see two exceptional cases in System I, one in System II and none in System III. A similar exceptional case for the CFL thickness change with flow velocity can be found when comparing Cases b vs. d for System II. Here δ increases slightly from 1.994 to 2.044 μm as the nominal flow velocity V slows down from 1.0 to 0.6 mm/s.

The change in developmental length may have little influence for *in vitro* experiments of blood flows in long tubes, since the region being observed usually greatly exceeds this short developmental length. However, in the actual microcirculation environment, arterioles are bifurcated in consecutive levels till capillaries. According to the morphometric analysis by *Kassab et al.* [26], for the diameter range of 10–20 μm considered in this study, the length of a arteriole segment between two bifurcations is typically between 50 and 90 μm. Such an arteriole length is of the same order as the RBC flow development length. This change in developmental length will directly affect the flow structure before the downstream bifurcation, and hence the phase separation of RBC flux into its two daughter branches at the bifurcation [5,14,43].

The above analysis may imply a possible mechanism of the beneficial effect from a higher suspending viscosity in resuscitation from hemodilution [62]. In a hemodilution situation, the hematocrit decreases and a longer distance is required to establish a thicker CFL (from Systems I to II). The increase in developmental length will shift the flow structure in front of the downstream bifurcation to less developed conditions, and further affect the regular plasma skimming and phase separation processes at the bifurcation [14,43]. The abnormal RBC flux in the daughter arteriole will certainly impair the microcirculation functions, with the functional capillary density as an indicator among others, in the downstream tissue. When a more viscous plasma expander is introduced, the developmental length will be reduced (from Cases a to b–e for System II), and the blood flow structure in front of the bifurcation is restored to a more developed condition, which is closer to the original more developed state.

This above analysis might be considered as a preliminary hypothesis of a possible mechanism for the beneficial effect from a more viscous plasma expander, which has been observed in resuscitation experiments. More detailed investigations are necessary to confirm this preliminary idea and to explore any other underlying mechanism involved in this complex process. We acknowledge the inherent limitations of the numerical model utilized in this study. First of all, it should not be expected to obtain quantitative data from a 2-D model for the real three-dimensional systems, as addressed in Section 9.3.4 and existing literature [5,70,71]. Also, our current model does not consider the cell membrane viscosity either, which could be important in determining the cell motion and deformation. In addition, the increase in plasma viscosity is often caused by a higher concentration of macromolecules, and these macromolecules would also enhance the aggregation among RBCs. This present work has not considered the aggregation for simplicity, and this issue could be examined in the future.

9.5.4 RBC Motion and Deformation in Bifurcated Microvessels

When blood flows through bifurcated microvessels, RBCs are distributed unequally into different branches (the so-called hematocrit phase separation), depending on the flow and cell properties. In this section, we examine the effects of the flow-rate ratio, cell deformability and suspending viscosity. Similar to those employed in previous simulations [5,25], a symmetric microvascular bifurcation model is constructed as shown in Fig. 9.16. The widths of flow inlet (w_0) and outlets (w_1 and w_2, respectively, for the upper and lower branches) are set as $w_0 = w_1 = w_2 = 8$ μm. Straight segments of the vessel walls are connected with circular curves of a 3-μm radius to avoid sharp corners.

Several different types of boundary conditions have been applied to the bifurcation model. The no-slip boundary requirement over the straight and curved microvessel walls is

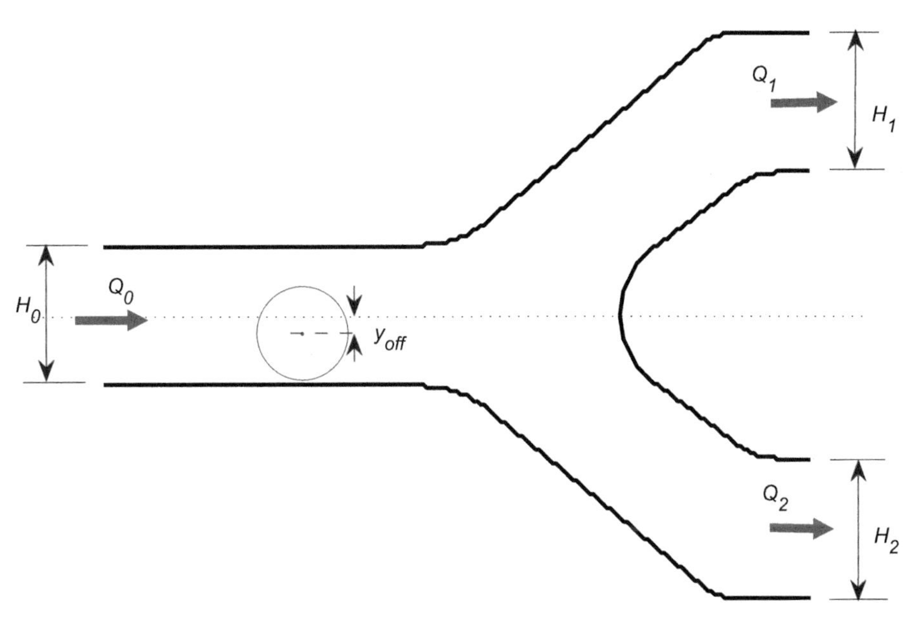

FIGURE 9.16 The symmetric bifurcation model used in this study. See text for detailed descriptions of the boundary conditions imposed. (Reprinted from W. Xiong and J. Zhang, *Biomechanics and Modeling in Mechanobiology*, 11:575–583, 2012. With kind permission from Springer Science and Business Media.)

implemented with the extrapolation scheme proposed by *Guo et al.* [22]. To introduce a flow through the bifurcation, parabolic velocity profiles are imposed on the two outlets, with central maximum velocity values of, respectively, U_1 and U_2. Meanwhile, a linear extrapolation treatment is applied at the mother vessel inlet [69]. The flow rates through the two daughter branches (Q_1 and Q_2) are related to the imposed velocity magnitudes (U_1 and U_2) by $Q_i = 2U_i w_i/3$ ($i = 1$ or 2). The flow rate through the mother vessel is then $Q_0 = Q_1 + Q_2$, according to the principle of mass conservation for incompressible flows. The flow-rate ratio between the upper branch and the mother vessel, $\Psi_1 = Q_1/Q_0$, can then be adjusted by tuning the imposed maximum velocities U_1 and U_2 at the outlets, with the inlet maximum velocity, U_0, maintained constant at 1 mm/s, similar to those in previous studies [5,25]. Also in this part, a RBC is represented as a circular capsule with a 5.32-μm diameter [5]; and its membrane is discretized into 100 equal segments.

To examine the cell motion and trajectory as it flows through the bifurcation, an undeformed cell is initially placed in the mother vessel about 10 μm upstream from the bifurcation, and the simulation is terminated once the cell has been transported into one branch. The suspending viscosity is varied from 1.2 cP (normal plasma viscosity) to 2.0 cP in order to investigate the effect of plasma viscosity on RBC motion and deformation, while the interior cytoplasm viscosity is maintained constant at 6.0 cP (normal cytoplasm viscosity). In addition, a repulsion force is introduced when the distance between a membrane node and the vessel wall is less than $2\Delta x$, to maintain the validity of the immersed-boundary treatment [68].

Before simulating RBC motion and deformation in the bifurcation model, we first studied the pure plasma flows through the bifurcated microvessel with no RBCs. A separating streamline, above which the plasma will flow into the upper branch, can be found in the mother vessel. The vertical distance between this separating streamline and the mother vessel centerline is termed as the critical offset y_s. Since the velocity profile at the straight segment of the mother vessel can be approximated as a Poiseuille flow, the flow rate to the upper branch Q_1 can be analytically obtained by integrating the parabolic velocity profile from the critical offset location to the inlet upper wall. An analytical relationship between the critical offset y_s and flow-rate ratio Ψ_1 can then be obtained. For $\Psi_1 = 0.62$, the critical offset calculated from this analytical relationship was $y_s = 0.645$ μm, while the value obtained from our simulation of a pure plasma flow with no cell was $y_s = 0.650$ μm. This agreement indicates that the LBM model and boundary conditions are correctly implemented in our program.

To examine the interaction between the plasma flow and the cell motion, we have simulated the trajectories of normal RBCs with different values of critical offset downward from the mother vessel centerline. Figure 9.17 shows the consecutive cell motion and deformation at $\Psi_1 = 0.6$ with two initial offsets $y_{off} = 0.82$ μm (a) and $y_{off} = 0.92$ μm (b), both below the critical offset for the pure plasma flow ($y_s = 0.54$ μm) (dashed lines in Fig. 9.17). However, the RBC placed at $y_{off} = 0.82$ μm in Fig. 9.17a moves to the upper branch, while the RBC at $y_{off} = 0.92$ μm in Fig. 9.17b enters the lower branch. Similar simulations have been conducted at various Ψ_1 values, and the critical offsets of cell destination are plotted in Fig. 9.19. In general, for the same flow-rate ratio, we notice that the critical offset y_s for a RBC has shifted significantly toward the low flow-rate branch compared with y_s for pure plasma flow; that is,

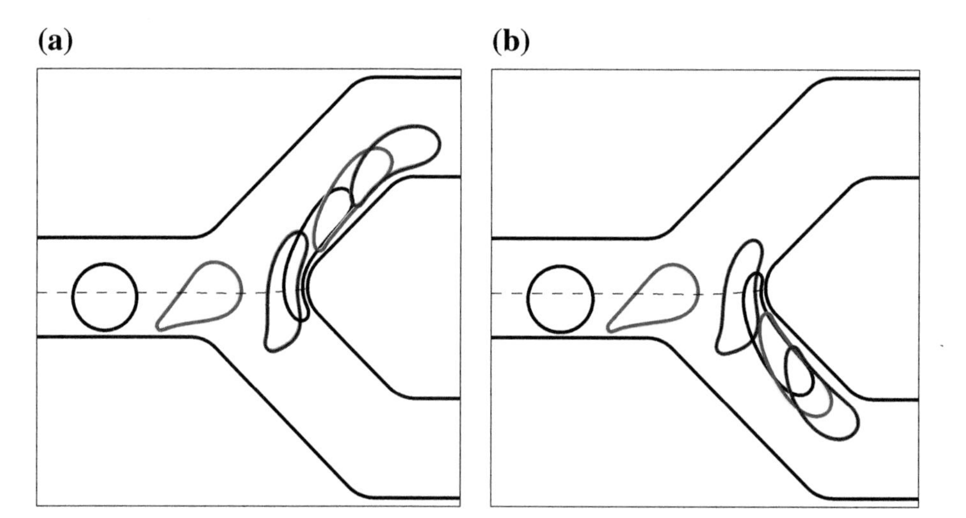

FIGURE 9.17 Trajectories of normal RBCs with offsets $y_{off} = 0.82$ μm (a) and $y_{off} = 0.92$ μm (b) at the same flow-rate ratio $\Psi_1 = 0.6$. Both cells are located below the separating streamline for pure plasma flow at the same flow-rate ratio. The cell shapes are plotted at time $t = 0, 10.2, 25.4, 61.0, 76.3$ and 86.4 ms, respectively. The dashed lines indicate the separating streamlines for a pure plasma flow at the same flow-rate ratio. (Reprinted from W. Xiong and J. Zhang, *Biomechanics and Modeling in Mechanobiology*, 11:575–583, 2012. With kind permission from Springer Science and Business Media.)

RBCs can enter the high flow-rate branch, even when its initial offset is below the separating streamline for pure plasma flows. These simulations again demonstrate that, in microvessels whose diameter is similar to that of the blood cell, simply assuming that the cell motion will follow the streamlines from a pure plasma flow is not adequate for the complex flow field in the bifurcation region.

We have further studied the effect of RBC deformability on the cell trajectory in the bifurcated microvessel. Here a less deformable cell is modeled by increasing the membrane elastic and bending moduli by 10 times. Figure 9.18 shows the motion of a normal RBC (a) and a less flexible RBC (b) in the bifurcated microvessels. Flow-rate ratio here is $\Psi_1 = 0.6$, and the initial cell location is below the vessel centerline with the same offset of $y_{off} = 0.82$ μm. The normal cell quickly deforms into an asymmetric shape as the flow starts. When it reaches the bifurcation, the cell is pressed against the vessel wall by the incoming flow, moves slowly along the wall and then enters the high flow-rate branch (Fig. 9.18a). The less flexible cell under the same fluid condition, however, is less deformed during traveling and relatively easily enters the low flow-rate branch (Fig. 9.18b). Such a difference in cell trajectory indicates that RBC deformability can have a significant influence on cell movement in bifurcated microvessels. The critical offset, y_s, separating the cell destination is then examined under different flow-rate ratios. For comparison, the critical offset for normal and less flexible RBCs at different flow-rate ratios is plotted in Fig. 9.19a. Significant deviation between the critical offsets for cells and that for pure plasma flows can be observed in this figure. Due to cell migration when traveling along the mother vessel and the comparable sizes of the cell and the bifurcation, the critical offset for a cell is much larger than that for the pure plasma flow prediction at the same flow rate. Also, an obvious difference between the curves representing the normal and less flexible cells can be observed: the critical offset for a less flexible cell is in general smaller than that for a normal cell. This may be associated with the faster migration

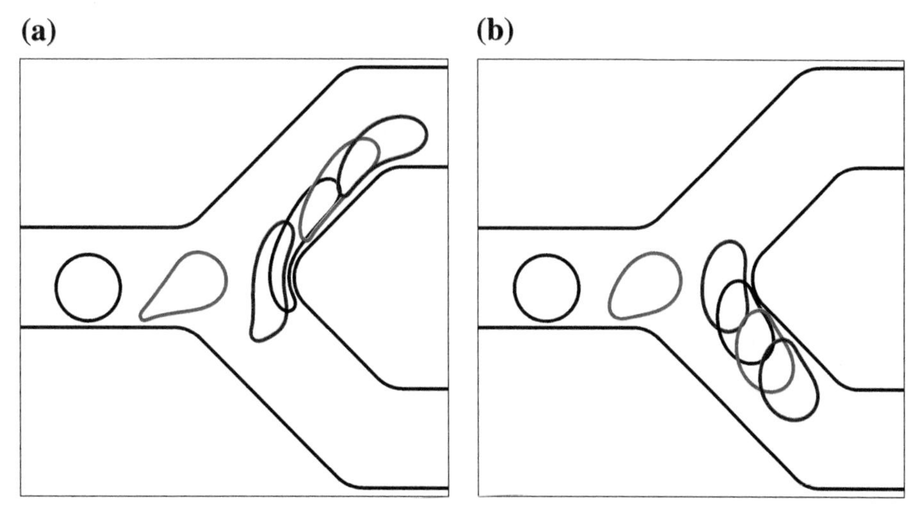

FIGURE 9.18 Trajectories of a normal RBC (a) and a less flexible RBC (b) with the same offset $y_{off} = 0.82$ μm and the same flow-rate ratio $\Psi_1 = 0.6$. The cell shapes are plotted at time $t =$ 0, 10.2, 25.4, 61.0, 76.3 and 86.4 ms, respectively. (Reprinted from W. Xiong and J. Zhang, *Biomechanics and Modeling in Mechanobiology*, 11:575–583, 2012. With kind permission from Springer Science and Business Media.)

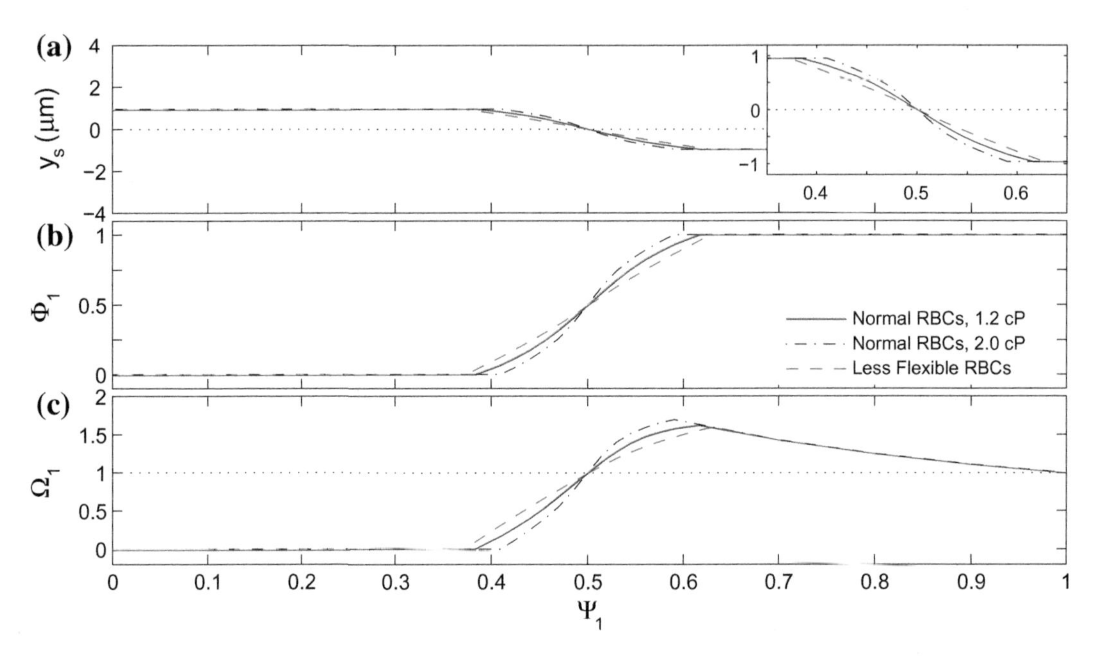

FIGURE 9.19 Critical offsets y_s (a), RBC flux ratio Φ_1 (b), and discharge hematocrit ratio Ω_1 (c) for less flexible RBCs (dashed) and normal RBCs with suspending viscosities of 1.2 cP (solid) and 2.0 cP (dot-dashed) under different flow-rate ratios Ψ_1. The inset in (a) shows the critical offsets for flow-rate ratios in the range of $0.35 < \Psi_1 < 0.65$. (Reprinted from W. Xiong and J. Zhang, *Biomechanics and Modeling in Mechanobiology*, 11:575–583, 2012. With kind permission from Springer Science and Business Media.)

of normal cells toward the centerline compared with less flexible cells, as observed in previous studies [3].

To further illustrate the effect of the critical offset on the cell partitioning at the bifurcation, we have calculated the RBC flux ratio Φ_1 and the discharge hematocrit ratio Ω_1 between the upper branch and the mother vessel [5,15]:

$$\Phi_1(\Psi_1) = \frac{Q_{RBC,1}}{Q_{RBC,0}} = \frac{\int_{y_s(\Psi_1)}^{y_t} s(y)u_c(y)dy}{\int_{y_b}^{y_t} s(y)u_c(y)dy}, \quad \Omega_1(\Psi_1) = \frac{H_{D,1}}{H_{D,0}} = \frac{\Phi_1(\Psi_1)}{\Psi_1} \tag{9.32}$$

where Q_{RBC} is the volumetric RBC flux and H_D is the discharge hematocrit, with subscripts 0 and 1 indicating the mother and the upper daughter vessels, respectively. In the above equation, $s(y)$ is the density distribution of cell centers as RBCs pass the location of 10 μm in front of the bifurcation center in the mother vessel, and this distribution is assumed to be uniform as in previous studies [5,15]. The horizontal velocity of cell centers $u_c(y)$ is obtained by separate simulations of cell movement in an 8-μm straight channel at different transverse offsets; the results were then approximated by a quadratic function [5], with a coefficient of determination $R^2 > 0.995$. The integral limits y_t and y_b are the top and bottom extreme positions of cell center, which are ±0.96 μm due to the cell size and a 0.38 μm gap between the cell and vessel wall for the immersed-boundary implementation. This gap can also be physically interpreted as the cell-free layer in microvessels. Results of Φ_1 and Ω_1 calculated for RBCs with different deformability are displayed in Fig. 9.19b and c. The figures show that RBC flux from the mother vessel is separated at the bifurcation, disproportionally to the flow-rate ratio. The high flow-rate branch receives even more of the incoming RBC flux ($\Phi_1 > \Psi_1$ for $\Psi_1 > 0.5$ in Fig. 9.19b), and a higher hematocrit is observed ($\Omega_1 > 1$ for $\Psi_1 > 0.5$ in Fig. 9.19c). This demonstrates the well-known phase separation and plasma skimming phenomena. For a flow-rate ratio Ψ_1 beyond a certain interval around 0.5 ([0.383, 0.617] for normal cells and [0.372, 0.628] for less flexible cells in this study), all cells enter the high flow-rate branch, and no cells go to the low flow-rate branch. This is illustrated in Fig. 9.19b and c by the $\Phi_1 = 0$ and $\Omega_1 = 0$ curves for the low Ψ_1 regime and the $\Phi_1 = 1$ and $\Omega_1 = 1/\Psi_1$ curves for the high Ψ_1 regime. The relative lower Φ_1 and Ω_1 values at $\Psi_1 > 0.5$ for less flexible cells (dashed lines in Fig. 9.19b and c) indicate that a higher cell rigidity can reduce the phase separation and plasma skimming processes in microcirculation. The higher hematocrit ratio in the high flow-rate branch has also been reported in previous experimental and numerical studies [15,43]. The observation of smaller critical offsets and reduced phase separation and plasma skimming for less flexible RBCs here is also consistent with the previous study of a similar 2-D system by *Barber et al.* [5]. However, a quantitative agreement cannot be found. The cell critical offsets from our simulations are in general larger in magnitude than those presented in *Barber et al.* [5]. There the critical offsets for rigid cells over the entire flow-rate ratio range, and for flexible cells in certain flow-rate ratio ranges, were found to be less than those from the pure plasma flow prediction, while our simulations show that the critical offsets for RBCs are always greater than or equal to those for pure plasma flows. Increased cell migration toward the centerline during the travel in the mother vessel leads to a larger critical offset at the initial location. However, the complex flow situation and the cell-flow interaction at the bifurcation region are hard to predict, and they also have a direct influence in determining the cell destination.

Simulations have also been carried out with different suspending viscosities to examine the corresponding effect on cell motion and trajectory. Figure 9.19 shows the results of the critical offset y_s (a), the RBC flux ratio Φ_1 (b), and the hematocrit ratio Ω_1 (c) for less deformable RBCs (dashed), normal RBCs with a normal suspending viscosity of 1.2 cP (solid) and normal RBCs with an elevated suspending viscosity of 2.0 cP (dot-dashed). Unlike cell rigidity, an increased suspending viscosity results in larger critical offsets, and consequently larger cell flux and hematocrit ratios in the high flow-rate branch (see the dot-dashed curves in Fig. 9.19). To understand this phenomenon, here we introduce two-dimensionless parameters: the dimensionless shear rate $G \equiv \mu_p U_0 a / w_0 E_s$, and the viscosity ratio between the interior cytoplasm (μ_c) and suspending plasma (μ_p) $\lambda \equiv \mu_c/\mu_p$ [41]. Here $a = 2.66$ µm is the cell radius. As the suspending plasma viscosity, μ_p, increases and all other parameters remain the same, the dimensionless shear rate, G, increases, and the viscosity ratio, λ, decreases. Both the G increase and λ decrease will enhance the apparent cell deformability [41,70], and therefore enhance the cell migration toward the centerline as the cell travels toward the bifurcation. This indicates that a larger plasma viscosity can enhance the phase separation and plasma skimming phenomena. According to our simulations, the magnitudes in critical offset change induced by increasing cell rigidity (10 times) and by increasing suspending viscosity (1.2–2.0 cP) are approximately of the same order (Fig. 9.19). As such, nonuniform RBC partitioning at bifurcations seems to be much more sensitive to the viscosity of the surrounding plasma than to the cell rigidity. The elevated plasma viscosity observed in some pathological situations [54] therefore could be beneficial in reducing or compensating for the negative effect on phase separation caused by an increase in RBC rigidity. These simulations here may also be helpful in understanding the higher functional capillary density in tissues when the suspending viscosity of plasma is increased, as observed by *Intaglietta and coworkers* [10,62,63].

9.6 SUMMARY

We have employed the immersed-boundary lattice-Boltzmann method to simulate the shear stress variation induced by blood flows and the viscosity effect on RBC dynamics in microvessels. When a single RBC flows in microvessels, a typical peak-valley-peak structure is observed, and it is analyzed in terms of its magnitude, spatial influencing range and temporal elapsed duration. Effects of red cell deformability, microvessel size and flow velocity have been investigated. The corresponding variation characters have also been related to cell deformation and flow field. Simulation results show that the variation magnitude is mainly determined by the gap size between cell and vessel wall, while the spatial range of the shear stress variation depends on the cell length as well as the microvessel size. For a certain point on the vessel wall, the shear stress variation lasts a short time at a higher flow velocity, and vice versa. As the cell concentration in the microvessel increases, the shear stress variation structure changes accordingly with the two peaks from two close cells merging together, and eventually only one peak is observed at a hematocrit of 30.72%. However, the effect of hematocrit on the variation magnitude of shear stress is less obvious, and the dynamic nature of shear stress is still significant.

We have also found that the linear shear stress assumption is oversimplified in [34] regarding the relation between cell-free layer and wall shear stress (WSS). For multiple RBC flows, the typical peak-valley-peak WSS mode has also been observed, and the influences of cell deformation, flow properties were also discussed. When we summarized the WSS characteristics in single-file RBC flows and in multiple RBC flows, together with the analysis of shear stress distribution in microvessels, we may deduce that the underlying mechanism behind the peak-valley-peak WSS mode is attributed to the reverse pressure difference between the front and back side of the passing RBC.

We also investigate the motion and deformation of a single RBC or multiple RBCs in straight or bifurcated microvessels with different suspending viscosity. For a single RBC in simple shear or channel flows, RBC appears more flexible as indicated by the tank-treading motion in shear flows and the strong transverse migration in channel flows. For the multiple RBC flows in a channel, our results show that the flow velocity (flow rate) is the major parameter in determining the flow structure. A thinner cell-free layer is found with a lower flow rate, even with the driving force (pressure gradient) held constant. The decrease in cell-free layer thickness will inhibit the plasma skimming process at a capillary entrance so that RBCs may have opportunity to enter. This may imply the possible mechanism of the high functional capillary density associated with a high suspending viscosity observed in experiments. Also this work represents the first systematic study on the suspending viscosity effect on RBC behaviors in various flow situations.

For the symmetric bifurcation model employed, the critical offset position in the mother branch, which separates the RBC flux toward the two branches, has been calculated. The RBC flux and the hematocrit partitioning between the two daughter branches have also been studied. Effects of the flow-rate ratio, cell deformability and suspending viscosity have been examined. Simulation results indicate that increased cell rigidity and suspending viscosity have counter-effects on cell trajectory through a bifurcation: the cell trajectory shifts toward the low flow-rate branch for less deformable cells, and toward the high flow-rate branch for more viscous plasma. These results imply that a higher cell rigidity would reduce the regular phase separation of hematocrit and plasma skimming processes in microcirculation, while an increased viscosity has the opposite effect. This has implications for relevant studies in fundamental biology and biomedical applications.

Acknowledgments

This work was supported by the Natural Science and Engineering Research Council of Canada (NSERC) and the Laurentian University Research Fund (LURF). This work was made possible by the facilities at WestGrid (www.westgrid.ca) and SHARCNet (www.sharcnet.ca).

References

[1] M. Abkarian, M. Faivre, A. Viallat, Swinging of red blood cells under shear flow, Physical Review Letters 98 (2007) 188302.

[2] J. Ando, K. Yamamoto, Vascular mechanobiology: endothelial cell responses of fluid shear stress, Circulation Journal 73 (2009) 1983–1992.

[3] P. Bagchi, Mesoscale simulation of blood flow in small vessels, Biophysical Journal 92 (6) (2007) 1858–1877.

[4] P. Bagchi, P.C. Johnson, A.S. Popel, Computational fluid dynamic simulation of aggregation of deformable cells in a shear flow, Journal of Biomechanical Engineering 127 (2005) 1070–1080.

[5] J.O. Barber, J.P. Alberding, J.M. Restrepo, T.W. Secomb, Simulated two-dimensional red blood cell motion, deformation, and partitioning in microvessel bifurcations, Annals of Biomedical Engineering 36 (2008) 1690–1698.

[6] H. Baumler, E. Donath, Does dextran really significantly increase the surface potential of human red blood cells? Studia Biophysica 120 (1987) 113–122.

[7] H. Baumler, B. Neu, E. Donath, H. Kiesewetter, Basic phenomena of red blood cell rouleaux formation, Biorheology 36 (1999) 439–442.

[8] P. Bhatnagar, E. Gross, K. Krook, A model for collisional processes in gases I: Small amplitude processes in charged and neutral one-component system, Physical Review B 94 (3) (1954) 511–525.

[9] D.E. Brooks, Electrostatic effects in dextran mediated cellular interactions, Journal of Colloid and Interface Science 43 (1973) 714–726.

[10] P. Cabrales, A.G. Tsai, M. Intaglietta, Increased plasma viscosity prolongs microhemodyanamic conditions during small volume resuscitation from hemorrhagic shock, Resuscitation 77 (2008) 379–386.

[11] Y.A. Cengel, J.M. Cimbala, Fluid Mechanics: Fundamentals and Applications. second ed., McGraw Hill, New York, NY, 2010.

[12] A. Drochon, D. Barthes-Biesel, C. Bucherer, C. Lacombe, J.C. Lelievre, Viscous filtration of red blood cell suspensions, Biorheology 30 (1993) 1–8.

[13] M.M. Dupin, I. Halliday, C.M. Care, L. Alboul, L.L. Munn, Modeling the flow of dense suspensions of deformable particles in three dimensions, Physical Review E 75 (2007) 066707.

[14] G. Enden, A.S. Popel, A numerical study of the shape of the surface separating flow into branches in microvascular bifurcations, Journal of Biomechanical Engineering 114 (1992) 398–405.

[15] G. Enden, A.S. Popel, A numerical study of plasma skimming in small vascular bifurcations, Journal of Biomechanical Engineering 116 (1994) 79–88.

[16] E. Evans, D. Needham, Attraction between lipid bilayer membranes in concentrated solutions of nonadsorbing polymers: comparison of mean-field theory with measurements of adhesion energy, Macromolecules 21 (1988) 1822–1831.
[17] E.A. Evans, Bending elastic modulus of red blood cell membrane derived from buckling instability in micropipette aspiration tests, Biophysical Journal 43 (1983) 27–30.
[18] E.A. Evans, Y.C. Fung, Improved measurements of the erythrocyte geometry, Microvascular Research 4 (1972) 335–347.
[19] T.M. Fischer, H. Schmid-Schonbein, Tank tread motion of red cell membranes in viscometric flow: behavior of intracellular and extracellular markers, Blood Cells 3 (1977) 351–365.
[20] T.M. Fischer, M. Stohr-Liesen, H. Schmid-Schonbein, Red-cell as a fluid droplet–tank tread-like motion of human erythrocyte-membrane in shear-flow, Science 202 (1978) 894–896.
[21] H.L. Goldsmith, J. Marlow, Flow behavior of erythrocytes. I. rotation and deformation in dilute suspensions, Proceedings of the Royal Society of London Series B-Biological Sciences 182 (1972) 351–384.
[22] Z. Guo, C. Zheng, B. Shi, An extrapolation method for boundary conditions in lattice Boltzmann method, Physics of Fluids 14 (2002) 2007–2010.
[23] B.B. Gupta, K.M. Nigam, M.Y. Jaffrin, A 3-layer semi-empirical model for flow of blood and other particulate suspensions through narrow tubes, Journal of Biomechanical Engineering–Transactions of the ASME 104 (1982) 129–135.
[24] R.M. Hochmuth, R.E. Waugh, Erythrocyte membrane elasticity and viscosity, Annual Review of Psychology 49 (1987) 209–219.
[25] T. Hyakutake, S. Tominaga, T. Matsumoto, S. Yanase, Numerical study on flows of red blood cells with liposome-encapsulated hemoglobin at microvascular bifurcation, Journal of Biomechanical Engineering 130 (2008) 011014.
[26] G.S. Kassab, C.A. Rider, N.J. Tang, Y.C.B. Fung, Morphometry of pig coronary arterial trees, American Journal of Physiology: Heart and Circulatory Physiology 265 (1993) H350–H365.
[27] S. Kim, R.L. Kong, A.S. Popel, M. Intaglietta, P.C. Johnson, Temporal and spatial variations of cell-free layer width in arterioles, American Journal of Physiology 293 (2007) H1526–H1535.
[28] A.J.C. Ladd, Numerical simulations of particulate suspensions via a discretized Boltzmann equation. Part I. Theoretical foundation, Journal of Fluid Mechanics 271 (1994) 285–309.
[29] B.L. Langille, S.L. Adamson, Relationship between blood flow direction and endothelial cell orientation at arterial branch sites in rabbits and mice, Circulation Research 48 (1981) 481–488.
[30] Y. Liu, L. Zhang, X. Wang, W.K. Liu, Coupling of Navier-Stokes equations with protein molecular dynamics and its application to hemodynamics, International Journal for Numerical Methods in Fluids 46 (2004) 1237–1252.
[31] A.M. Malek, G.H. Gibbons, V.J. Dzau, S. Izum, Fluid shear stress differentially modulates expression of genes encoding basic fibroblast growth factor and platelet-derived growth factor b chain in vascular endothelium, Journal of Clinical Investigation 92 (1993) 2013–2021.
[32] G. Mchedlishvili, N. Maeda, Blood flow structure related to red cell flow: a determinant of blood fluidity in narrow microvessels, Japanese Journal of Physiology 51 (2001) 19–30.
[33] E.W. Merill, E.R. Gilliland, T.S. Lee, E.W. Salzman, Blood rheology: effect of fibrinogen deduced by addition, Circulation Research 18 (1966) 437–446.
[34] B. Namgung, P.K. Ong, P.C. Johnson, S. Kim, Effect of cell-free layer variation on arteriolar wall shear stress, Annals of Biomedical Engineering 39 (2011) 359–366.
[35] N.A. N'Dri, W. Shyy, R. Tran-Son-Tay, Computational modeling of cell adhesion and movement using a continuum-kinetics approach, Biophysical Journal 85 (2003) 2273–2286.
[36] R.M. Nerem, M.J. Levesque, Vascular endothelial mophology as an indicator of the pattern of blood flow, Journal of Biomechanical Engineering 103 (1981) 172–177.
[37] B. Neu, H.J. Meiselman, Depletion-mediated red blood cell aggregation in polymer solutions, Biophysical Journal 83 (2002) 2482–2490.
[38] K. Okahara, J. Kambayashi, T. Ohnishi, Y. Fujiwara, T. Kawasaki, M. Monden, Shear stress induces expression of cnp gene in human endothelial cells, FEBS Letters 373 (1995) 108–110.
[39] C.S. Peskin, Numerical analysis of blood flow in the heart, Journal of Computational Physics 25 (3) (1977) 220–252.

[40] A.S. Popel, P.C. Johnson, Microcirculation and hemorheology, Annual Review of Fluid Mechanics 37 (2005) 43–69.
[41] C. Pozrikidis, Effect of membrane bending stiffness on the deformation of capusles in simple shear flow, Journal of Fluid Mechanics 440 (2001) 269–291.
[42] C. Pozrikidis, Computational Hydrodynamics of Capsules and Biological Cells, Chapman & Hall/CRC Press, New York, NY, 2010.
[43] A.R. Pries, K. Ley, M. Claassen, P. Gaehtgens, Red-cell distribution at microvascular bifurcations, Microvascular Research 38 (1989) 81–101.
[44] A.R. Pries, D. Neuhaus, P. Gaehtgens, Blood viscosity in tube flow: dependence on diameter and hematocrit, American Journal of Physiology 263 (1992) H1770–H1778.
[45] A.R. Pries, T.W. Secomb, Microvascular blood viscosity in vivo and the endothelial surface layer, American Journal of Physiology Heart and Circulatory Physiology 289 (2005) H2657–H2664.
[46] W. Reinke, P. Gaehtgens, P.C. Johnson, Blood viscosity in small tubes–effect of shear rate, aggregation, and sedimentation, American Journal of Physiology 253 (1987) H540–H547.
[47] H. Sakai, A. Sato, N. Okuda, S. Takeoka, N. Maeda, E. Tsuchida, Peculiar flow patterns of RBCs suspended in viscous fluids and perfused through a narrow tube 25 μm, American Journal of Physiology Heart and Circulatory Physiology 297 (2009) H583–H589.
[48] A.S. Sangani, A Acrivos, Slow flow past periodic arrays of cylinders with application to heat transfer, International Journal of Multiphase Flow 8 (1982) 193–206.
[49] T.W. Secomb, R. Hsu, Analysis of red blood cell motion through cylindrical micropores: effects of cell properties, Biophysical Journal 71 (1996) 1095–1101.
[50] T.W. Secomb, R. Hsu, A.P. Pries, Blood flow and red blood cell deformation in nonuniform capillaries: effects of the endothelial surface layer, Microcirculation 9 (2002) 189.
[51] T.W. Secomb, B. Styp-Rekowska, A.R. Pries, Two-dimensional simulation of red blood cell deformation and lateral migration in microvessels, Annals of Biomedical Engineering 35 (2007) 755–765.
[52] M. Sharan, A.S. Popel, A two-phase model for blood flow in narrow tubes with increased viscosity near the wall, Biorheology 38 (2001) 415–428.
[53] D. Stelitano, D.H. Rothman, Fluctuations of elastic interfaces in fluids: theory, lattice Boltzmann model, and simulation, Physical Review E 62 (2000) 6667–6680.
[54] J.F. Stoltz, M. Singh, P. Riha, Hemorheology in Practice, IOS Press, Amsterdam, Netherlands, 1999.
[55] S. Succi, The Lattice Boltzmann Equation, Oxford University Press, Oxford, 2001.
[56] N. Tateishi, Y. Suzuki, M. Soutani, N. Maeda, Flow dynamics of erythrocytes in microvessels of isolated rabbit mesentery: cell-free layer and flow resistance, Journal of Biomechanics 27 (1994) 1119–1125.
[57] J. Toolke, S. Freudiger, M. Krafczyk, An adaptive scheme using hierarchical grids for lattice Boltzmann multiphase flow simulations, Computational Fluids 35 (2006) 820–830.
[58] R. Tran-Son-Tay, S.P. Sutera, P.R. Rao, Determination of red blood cell membrane viscosity from rheoscopic observations of tank-treading motion, Biophysical Journal 46 (1984) 65–72.
[59] G. Tryggvason, B. Bunner, A. Esmaeeli, D. Juric, N. Al-Rawahi, W. Tauber, J. Han, S. Nas, Y.-J. Jan, A front-tracking method for the computations of multiphase flow, Journal of Computational Physics 169 (2001) 708–759.
[60] A.G. Tsai, B. Friesenecker, M. McCarthy, H. Sakai, M. Intaglietta, Plasma viscosity regulates capillary perfusion during extreme hemodilution in hamster skinfold model, American Journal of Physiology: Heart and Circulatory Physiology 275 (1998) H2170–H2180.
[61] H.S. Udaykumar, H.-C. Kan, W. Shyy, R. Tran-Son-Tay, Multiphase dynamics in arbitrary geometries on fixed Cartesian grids, Journal of Computational Physics 137 (1997) 366–405.
[62] B.Y.S. Vázquez, J. Martini, A.C. Negrete, P. Cabrales, A.G. Tsai, M. Intaglietta, Microvascular benefits of increasing plasma viscosity and maintaining blood viscosity: counterintuitive experimental findings, Biorheology 46 (2009) 167–179.
[63] B.Y.S. Vázquez, R. Wettstein, P. Cabrales, A.G. Tsai, M. Intaglietta, Microvascular experimental evidence on the relative significance of restoring oxygen carrying capacity vs. blood viscosity in shock resuscitation, Biochimica et Biophysica Acta 1784 (2008) 1421–1427.
[64] C.H. Wang, A.S. Popel, Effect of red-blood-cell shape on oxygen-transport in capillaries, Mathematical Biosciences 116 (1993) 89–110.

[65] K.F. Waschke, H. Krieter, G. Hagen, D.M. Albrecht, K. Van Ackern, W. Kuschinsky, Lack of dependence of cerebral blood flow on blood viscosity after blood exchange with a Newtonian O_2 carrier, Journal of Cerebral Blood Flow and Metabolism 14 (1994) 871–876.
[66] R.E. Waugh, R.M. Hochmuth, Chapter 32: Mechanics and deformability of hematocytes, in: J.D. Bronzino (Ed.), The Biomedical Engineering Handbook. second ed., CRC Press LLC, Boca Raton, 2000.
[67] W. Xiong, J. Zhang, Shear stress variation induced by red blood cell motion in microvessel, Annals of Biomedical Engineering 38 (2009) 2649–2659.
[68] W. Xiong, J. Zhang, Two-dimensional lattice Boltzmann study of red blood cell motion through microvascular bifurcation: cell deformability and suspending viscosity effects, Biomechanics and Modeling in Mechanobiology 11 (2012) 575–583.
[69] J. Zhang, Lattice Boltzmann method for microfluidics: models and applications, Microfluidics and Nanofluidics 10 (2011) 1–28.
[70] J. Zhang, P.C. Johnson, A.S. Popel, An immersed boundary lattice Boltzmann approach to simulate deformable liquid capsules and its application to microscopic blood flows, Physical Biology 4 (2007) 285–295.
[71] J. Zhang, P.C. Johnson, A.S. Popel, Effects of erythrocyte deformability and aggregation on the cell free layer and apparent viscosity of microscopic blood flows, Microvascular Research 77 (2009) 265–272.
[72] J. Zhang, D.Y. Kwok, Contact line and contact angle dynamics in superhydrophobic channels, Langmuir 22 (2006) 4998–5004.

CHAPTER

10

Targeted Drug Delivery: Multifunctional Nanoparticles and Direct Micro-Drug Delivery to Tumors

Clement Kleinstreuer[a], *Emily Childress*[a], *and Andrew Kennedy*[b]

[a]NC State University, Raleigh, NC, USA, [b]Sarah Cannon Research Institute, Nashville, TN, USA

OUTLINE

10.1 Introduction 392

10.2 Diagnostic Imaging and Image-Guided Drug Delivery 392

10.2.1 Imaging Techniques 392

10.2.2 Image-Guided Drug Delivery 394

10.3 Free Transport 394

10.3.1 Passive and Active Targeting 395

10.3.2 Nanodrug Carriers 395

10.3.2.1 Lipids 396

10.3.2.2 Polymers 397

10.3.2.3 Carbons 397

10.3.2.4 Metals 398

10.3.2.5 Ceramics 398

10.3.2.6 Viruses 398

10.3.2.7 Multifunctional Nanoparticles 398

10.3.2.8 Immunotherapy 399

10.3.3 Summary 400

10.4 Forced Transport 400

10.5 Direct Transport 401

10.5.1 Implementation of Optimal Targeted Drug Delivery 401

10.5.2 Applications of Optimal Micro-Drug Delivery 404

10.5.2.1 Microsphere Delivery to Liver Tumors by a Smart Microcatheter System 404

10.5.2.2 Drug-Aerosol Delivery with a Smart Inhaler System to Lung Sites 408

10.6 Conclusions 413

Acknowledgments 413

References 414

Transport in Biological Media

http://dx.doi.org/10.1016/B978-0-12-415824-5.00010-2

© 2013 Elsevier Inc. All rights reserved.

10.1 INTRODUCTION

One of the main goals of targeted drug delivery (TDD) is the development of 'smart drugs' which, once injected and circulating in the body, detect, image, diagnose and subsequently treat (and hopefully cure) a disease. Towards achieving this objective, the techniques for diagnostic imaging and drug targeting are continuously being revised and improved. Over the last two decades, two basic modes of targeting have emerged through the use of either drug-loaded nanoparticles (NPs) or multifunctional nanodrugs (NDs). In the first approach, called *passive targeting*, NDs somehow migrate towards cancerous or inflamed tissue by taking advantage of the unique features and hemodynamics of the diseased site. In the second technique, traditionally labeled *active targeting*, ligands are incorporated on the NDs' surface to selectively bind to overexpressed antigens or receptors on malignant cells. While both have demonstrated great promise in disease treatment, the major challenges that remain include controlled drug release, cellular uptake, and the unknown toxicities of nanoparticles. More importantly, both techniques rely on the drugs coming near enough to the tumor for their targeting to be effective. Hence, forcing or directly propelling the drug particles towards diseased sites would be a major improvement. As an example, *magnetic drug targeting*, which guides drugs to the diseased site via an externally applied magnetic field, was found to improve site-specific accumulation. However, this method may be unfeasible in some cases, due to the location of the tumor or locally high blood perfusion rates. Consequently, current methodologies do not offer a wide range of targeting capabilities. Thus, a newer technique called *optimal targeted drug delivery* (OTDD), discussed in Section 10.5, uses knowledge of the local fluid flow and best upstream drug injection position to achieve site-specific targeting. This chapter explores three major transport mechanisms for drug-particle targeting: free forced and direct particle-transport (see Fig. 10.1), and also discusses the current state of diagnostic imaging.

10.2 DIAGNOSTIC IMAGING AND IMAGE-GUIDED DRUG DELIVERY

A prerequisite to successful drug delivery is the identification of the target site. This not only includes the location and extent of a solid tumor or an inflamed tissue area but also the identification of the network of supporting blood vessels. Furthermore, the *in vivo* detection of drugs is also of interest because it facilitates image-guided drug delivery (IGDD) which promises improved therapeutic response and minimal systemic toxicity. This section summarizes current diagnostic imaging techniques as well as the goals of IGDD.

10.2.1 Imaging Techniques

The main diagnostic imaging techniques are ultrasound, magnetic resonance imaging (MRI), X-ray computed tomography (CT), positron emission tomography (PET) and single photon emission computed tomography (SPECT). With the exception of ultrasound (which utilizes mechanical excitation), the remaining techniques generate images by recording the physical properties of tissue when it is exposed to a certain type of electromagnetic radiation.

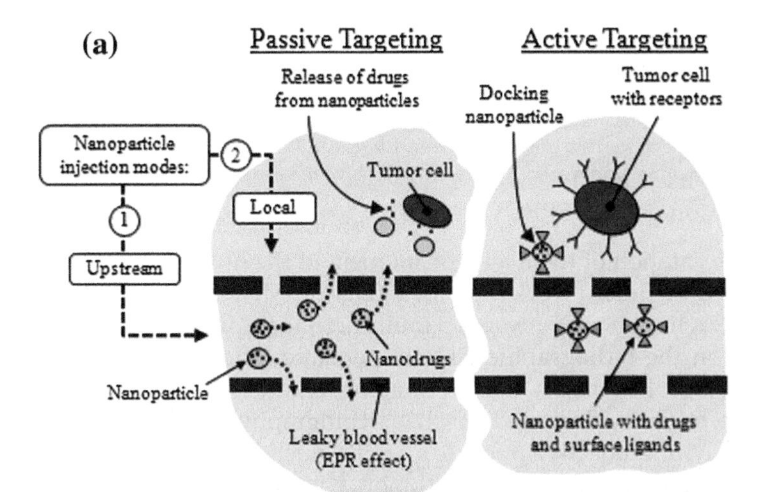

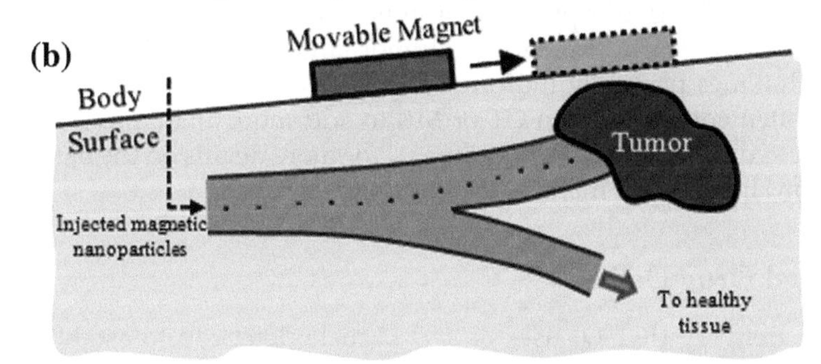

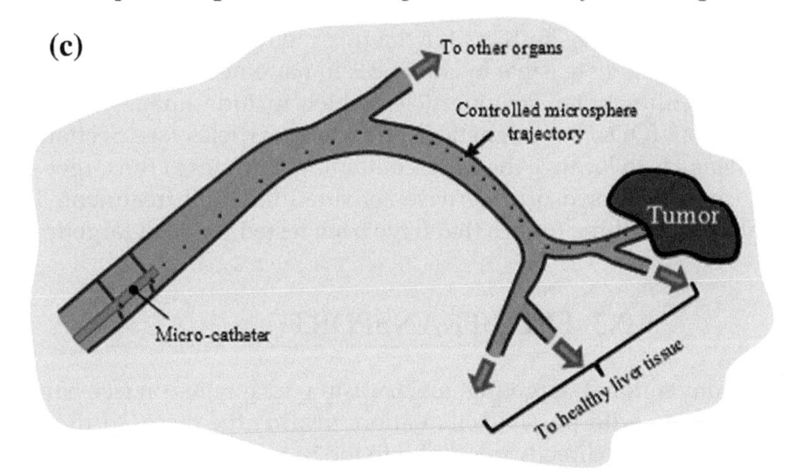

FIGURE 10.1 Schematics of drug-particle transport and targeting.

Briefly, tissue atoms can absorb or emit photons, making tissue structure visible. The frequency ranges of different types of electromagnetic radiation allow for the classification of different imaging techniques/machines [1]:

- $10^4 - 10^8$ Hz for MRI
- $10^{18} - 10^{20}$ Hz in X-rays also used in CT
- $>10^{20}$ Hz in gamma-radiation used in SPECT and PET

Standard MRI relies on the excitation of the magnetic moment of the hydrogen nuclei present in tissue, water and fat molecules; thus, it works best for soft tissue. Modern spiral CT-scanning, which is best for hard tissue, relies on an X-ray and a multi-detector assembly rotating around the patient who translates on the radiographic table to generate (after data-reconstruction) slice-volumes, 1 mm in thickness. In contrast to X-ray imaging, where radiation is applied from the outside, for imaging to work in nuclear medicine (i.e., scintigraphy) the subject has to emit ionizing radiation. Specifically, a radioactive tracer (e.g., radio-nuclide technetium) is injected, which preferentially accumulates in the patient's tumor, emitting gamma-quanta during nuclide decay. The data is recorded by gamma-cameras as image intensities in SPECT or PET. Specifically, PET-imaging can be used to provide real-time, non-invasive, quantitative monitoring of the tissue distribution, pharmacokinetics and tumor-targeting efficacy of NPs. These two molecular imaging techniques primarily indicate the physiological function of the tumor or organ, and hence are often combined with CT or MR to add more anatomical features. The interested reader is referred to the book by *Birkfellner* [1] for more details on the basic methods, machines and algorithms involved in medical image processing.

10.2.2 Image-Guided Drug Delivery

Image-guided drug delivery (IGDD) uses *in vivo* particle detection to establish the best drug delivery route and dosage, to determine when to trigger drug release, and to assess treatment effectiveness. Thus, one of its main goals is to develop platforms for multifunctional and multiplexed drug delivery systems for the treatment of cancer (see *Luo et al.* [2], among others). This has been made feasible by advances in nanomedicine which have made it possible to fabricate multifunctional nanoparticles which include image-contrast agents such as dyes, quantum dots (QDs) or magnetic/radioactive particles (see Section 10.3.2.7) [3]. The remaining challenge is to localize the drug-containing particles at the target site, after which the nanodrugs can be released or otherwise activated for local treatment. Thus, the following sections highlight the three modes that have been tested for such targeting.

10.3 FREE TRANSPORT

Free transport is the conventional 'targeting' mechanism which relies on free nanoparticle convection in the bloodstream with possible circulation towards the diseased tissue. In this mode, nanodrugs (NDs) are either directly injected into the tumor or into the bloodstream. In the latter case, the particles are advected toward the tumor where they preferentially attach via passive and active 'targeting'. To improve the success of this transport, NDs are being designed to have an increased circulation time, improved targeting specificity and decreased toxicity to healthy cells. This transport mode along with the current nanodrug carriers are summarized in this section.

10.3.1 Passive and Active Targeting

To support the rapid growth of tumors, a complex array of tumor vessels is created via angiogenesis. In contrast to their healthy neighbors, these vessels are often highly variable, having increased lengths and diameters, poor lymphatic drainage and leaky walls [4–7]. Due to these characteristics, drugs of even large molecular size (typically larger than 40–50 kDa) can more readily enter the tumor interstitium and linger for extended periods. This 'passive targeting' was first observed and documented by *Maeda* and colleagues who called it the enhanced permeability and retention (EPR) effect [7,8]. To enhance this effect, current research has focused on lengthening the circulation time of NDs so that they have a better chance of accumulating in the tumor. To accomplish this, they must evade clearance by non-target organs or by the mononuclear phagocyte system (MPS), the body's primary mechanism of clearing foreign particles [6,9–11].

Once in the vicinity of the tumor, accumulation and cellular uptake can be enhanced by 'active targeting', which utilizes connective ligands incorporated on the drug's surface [9,12]. These ligands, which can be antibodies, aptamers, peptides, vitamins or carbohydrates, allow the drug to selectively target tumors due to the overexpressed antigens or receptors on their cells [12,13]. Aptamers, which are DNA or RNA oligonucleotide sequences that selectively bind to targeted cells with very high affinity and selectivity, are attractive possibilities, because they can be adapted to recognize cancer cells and thus avoid the need to predetermine the cancer's target antigens [6,14]. After binding to the tumor cell via these ligands, the drug can then be released and will diffuse across the cell membrane or be digested through receptor-mediated endocytosis where the drug is then internally released [6,12,13]. The success of active targeting depends on the number of ligand-receptor sites, their level of affinity and their binding strength [12,15].

10.3.2 Nanodrug Carriers

The success of free transport greatly depends on the design of the drug carrier. Given their small nano-size (at least in one dimension), nanoparticles (NPs) make ideal drug carriers because they can diffuse across a variety of barriers, such as capillary walls or membrane pores, between cells forming protective layers, or directly into diseased tissue, leading to local drug accumulation [16]. The latter ensures cancer-cell killing and reduces toxic side-effects.

The characteristics of these nanodrug carriers, including their size, shape and surface properties, influence both their circulation time and targeting ability. For instance, to avoid clearance by the spleen, liver, and kidneys and to extravasate through the tumor vasculature, carriers should be between 10 and 100 nm in size [6,9,12,17]. However, when in the tumor interstitium, larger particles are more desirable because they can be retained longer due to higher hydraulic resistance [17]. Additionally, elliptical or tubular particles take longer to endocytose than spheres, exhibit distinct biodistribution due to the unique fluid forces they experience as a result of their shape and are able to contain more drugs and/or ligands due to their increased surface areas [12,17,18]. Furthermore, the hydrophobicity of a particle's

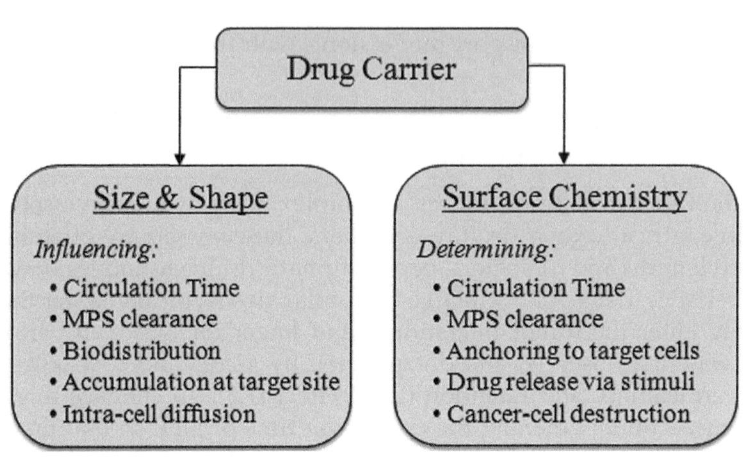

FIGURE 10.2 Characteristics of drug carriers.

surface can also affect its clearance from the body. For example, hydrophobic particles more readily attract proteins that trigger immunogenic responses, and they are more susceptible to agglomeration [9]. Thus, NDs have often been coated with polyethylene glycol (PEG) to make them hydrophilic [9–11]. While this coating can extend the drug's circulation time by preventing phagocytosis, it may also prevent tumor endocytosis [12]. Additionally, the particle's surface charge can also affect cellular uptake, with positively charged particles being more readily endocytosed [12]. Given that such conflicting characteristics are desirable in the different stages of targeting, it is suggested that ideal drug carriers should be able to dynamically alter their size, shape or surface properties *in vivo* [17]. Figure 10.2 summarizes the influence of these characteristics, and the remainder of this section reviews some of the current nanodrug carrier designs.

10.3.2.1 Lipids

Liposomes, which consist of a lipid bilayer surface enclosing an aqueous core, were first described by *Bangham* and colleagues in the 1960s and were the first drug carriers used for targeted drug delivery [19–21]. Due to their structure of hydrophilic heads stabilized by surfactants and multiple hydrophobic tails, they are able to encapsulate both hydrophilic and hydrophobic drugs which could then be released via diffusion or cell internalization [11,22–24]. Their small nano-size, easy surface manipulation, biocompatibility, biodegradability and their multi-route options for injection make them attractive carriers [13,25,26]. Thus, they have been approved for multiple clinical trials, and a liposome formulation of doxorubicin, called Doxil®, is already on the market [11,22,24,26].

Despite their advantages, the major limitation of liposomes as drug carriers has been premature and uncontrolled drug release due to their leaky structure [12,13,23,26]. To circumvent this, PEG, ammonium sulfate gradient or cholesterol can be incorporated to improve their *in vivo* stability [13,15,26]. Alternatively, lipid nanocapsules, described as a cross between polymer nanocapsules and liposomes, can instead be used for improved stability [23].

10.3.2.2 Polymers

Polymer drug carriers have been used in the form of nanoparticles, micelles, dendrimers or hydrogels, and are composed of either natural or synthetic polymers, such as PLGA (poly-D, L-lactide co-glycolide) or PLA (polylactide) [6,12,13,25]. Nano-size micelles are usually spherical in shape and can either be solid or hollow for housing the drug. Dendrimers, on the other hand, are macromolecules which branch out into tree-like forms from a central core [6,11,25]. They are attractive because they are monodisperse, highly symmetric, water-soluble and are able to encapsulate hydrophobic nanodrugs [6,13]. Alternatively, hydrogels are made of a network of synthetic or natural polymer chains, are hydrophilic and are composed almost entirely of water [27,28].

To load drugs onto these carriers, covalent attachment, adsorption to the surface or entrapment in the polymer matrix can be employed [12]. Entrapment, which often involves simply mixing the drug with the polymer, is the most popular technique. However, this method does not easily lend itself to more than one drug being loaded. Thus, covalent attachment, which more readily allows for multiple drugs to be delivered, is an effective alternative.

For controlled drug release, the polymer's properties can be adjusted such that it erodes, disintegrates or swells in the presence of an acidic environment or increased temperatures, which are common around tumors [12,17,27]. These environmental changes can also be used to alter the shape or surface characteristics of NDs for improved targeting. Further, enhanced targeting can also be accomplished through the incorporation of multiple connective ligands on the NPs' surface. An example of how to fabricate such NPs can be found in [29].

These characteristics, along with their biocompatibility, biodegradability and water solubility make polymers one of the most popular drug carriers studied [6,13,25]. Some issues that remain to be addressed, however, include the heterogeneity of manufactured polymer nanoparticles and the possible cytotoxic effects resulting from non-targeted cellular uptake [13,25]. Polymeric nanostructures, their characteristics and their functionalities are discussed in *Patil et al.* [29], *Kateb et al.* [16], *Parveen et al.* [30] and *Venkataraman et al.* [18].

10.3.2.3 Carbons

Carbon-based particles are usually in the form of carbon nanotubes (CNTs), which are made up of thin sheets (e.g., just 0.3 nm thick) of carbon, present as benzene rings, which has been rolled into single or multi-wall tubes [16]. This novel structure belongs to the family of fullerenes, and is distinctly different from the carbon forms of graphite and diamonds. CNTs are ultralight with a high aspect ratio (i.e., $d_p = O$ (10 nm) and $L = O$ (100 μm)), and exhibit high mechanical strength as well as high electrical/thermal conductivities [31,32]. Due to their ultrahigh surface area relative to their volume, nanodrugs, peptides and/or DNA can be readily attached to the CNT walls and tips [6,25,31,33]. Drugs can also be loaded inside CNTs [31]. As with liposomes and polymers, ligands can also be incorporated on the surface of CNTs for enhanced targeting. Further, controlled drug release can be made possible by attaching drugs to the CNTs' surface via a linker that is responsive to changes in pH or temperature [31]. Despite these advantages, their major limitation is their possible cytotoxicity [6,10,16,31,32]. This is, in part, due to the metal catalyst and amorphous carbon contaminants which are often present and difficult to remove [31].

10.3.2.4 Metals

Current metal-based nanoparticles include nano-shells (metallic covered silica particles), gold nano-cages (hollow, porous metallic particles) and gold nanoparticles [13]. They are attractive carriers because they are very small ($d_p < 50$ nm), have stable shapes, can be electrically charged, have high surface reactivity and are biocompatible [13]. These NPs are often used for imaging or hypothermia-based therapies in which the particles absorb light and, consequently, emit heat to destroy tumor cells [13]. This form of targeting can be advantageous since only the lighted areas are affected. Magnetic nanoparticles (MNPs) are also useful metal-based NPs because they can be employed to enhance contrast agents in MRI, or they can function as drug carriers which can be directed towards tumors via a locally applied magnetic field [13]. Often, magnetic iron oxide particles (Fe_3O_4 or Fe_2O_3), with a synthetic or a natural polymer surface coating to help avoid particle agglomeration, are used in this application [27,34]. As with the other drug carriers, metal-based NPs can also be modified to include targeting ligands and drugs, which can be released from the drug carriers by local changes in the physiological conditions (e.g., temperature and/or pH) near/at the target site [13,27]. The major limitation of these particles, however, is that they are not biodegradable; thus, it is still unclear what adverse effects their buildup may cause [13,16,27,34]. In fact, *in vitro* studies have recently demonstrated the toxicity of gold NPs [35,36], which makes further investigations into their toxicity a necessity.

10.3.2.5 Ceramics

Ceramic NPs, often in the form of mesoporous silica NPs, are novel nanodrug carriers that are porous and less than 50 nm in mean diameter. They are desirable for transporting drugs which are susceptible to changes in pH since they are unaffected by this condition. Furthermore, with their multitude of pores they have an increased surface area for incorporating drugs, and simultaneously their surfaces can be modified to include connective ligands. However, like the metal-based carriers, they are not biodegradable, which may cause undesirable effects [10,25,36].

10.3.2.6 Viruses

Lastly, viruses have also been proposed as potential drug carrier candidates due to the innate attraction of certain viruses to specific overexpressed receptors on tumor cells and due to their ability to kill these cells [6,37]. Through genetic alteration, infection is not only avoided but viruses can also be designed to target tumor cells [37]. Like the other carriers, targeting ligands can also be conjugated to the virus' surface. Several viruses that have been identified as promising drug carriers include cowpea mosaic virus, canine parvovirus, herpesvirus, adenovirus, retrovirus, and bacteriophages. For example, hepatitis B core particles have been demonstrated as potential delivery agents to liver cells [38]. As expected, a major limitation of using viruses as drug carriers is their potential for developing severe complications [37].

10.3.2.7 Multifunctional Nanoparticles

To further improve the effectiveness of free transport, multifunctional nanoparticles can be designed with components that detect diseases at the cellular level, enhance medical images, anchor the particles to diseased sites and release therapeutic compounds. This can be accomplished through the inclusion of protective coatings, the incorporation of different targeting ligands and the encapsulation of image-contrast agents, therapeutic nanodrugs and ferric crystals for magnetic steering (see Fig. 10.3). Examples of such nanostructures include: polymer uni-molecular micelles loaded with PET isotope chelator, tumor targeting ligands, and doxorubicin for drug detection, targeting and treatment of tumors [39], fluorescent quantum dots (QDs) loaded with aptamers and doxorubicin for imaging, targeting and therapeutics [40], carbon nanotubes for detecting cancerous cells and delivering nanodrugs to these cells [33] and luminescent mesoporous NPs that combine simultaneous cancer-cell imaging plus drug delivery [41]. Figures 10.3c and 10.3e depict semi-porous silica (or polymer) particles, covered with ligands as well as a hydrophilic coating and loaded with nanodrugs plus ferric nano-crystals, in case of magnetic steering towards the diseased site. In another example, ultrasound in combination with temperature and/or pH-responsive properties of NPs, loaded with nanodrugs and ferric crystals, are currently being tested for both local drug delivery to tumors and hyperthermia treatment (see [27]).

10.3.2.8 Immunotherapy

An alternative approach to nanodrug delivery for cancer treatment is immunotherapy, which involves the use of vaccines that actively engage the patient's immune system in

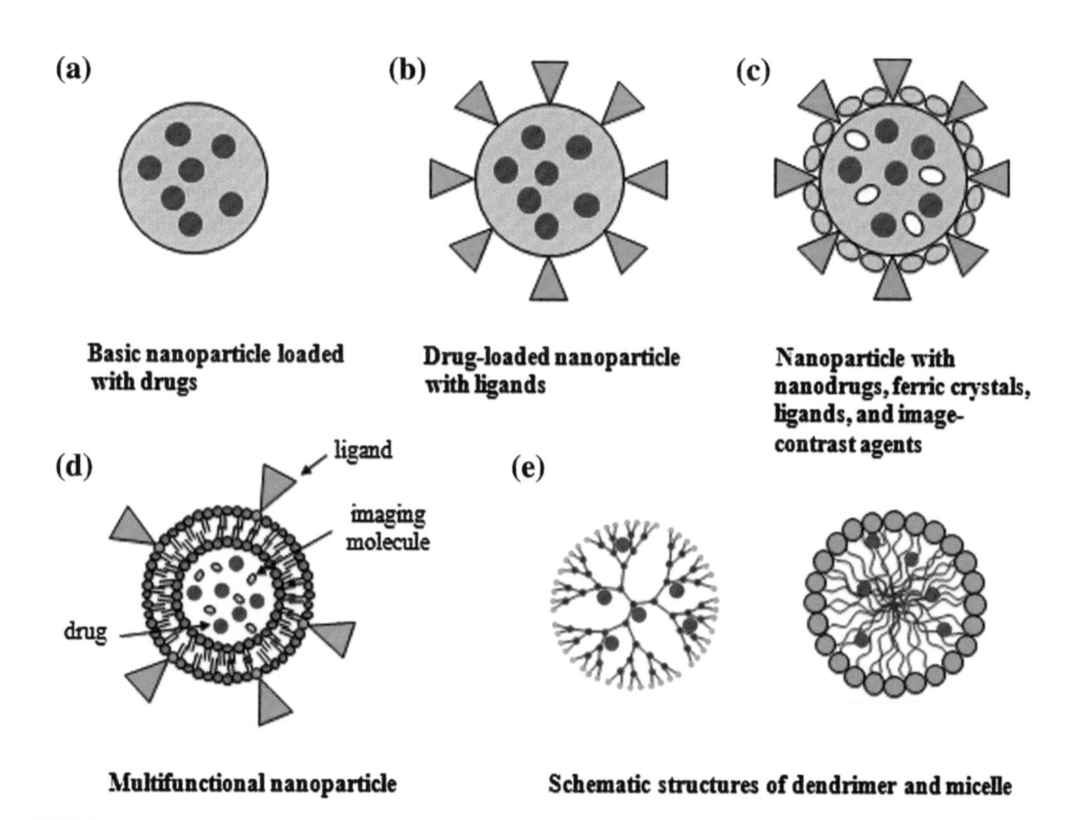

FIGURE 10.3 Schematics of multifunctional nanoparticles.

killing cancer cells or even entire tumors. This type of vaccine contains snippets of malignant proteins as well as fragments of cancer cells from the patient. When injected, the vaccine molecules perform dual functions: they train the body's immune cells (i.e., the T-cells) to recognize the cancer cells and then they activate the T-cells to kill the diseased tissue. Additionally, T-cells can be genetically modified to recognize cancer proteins and then destroy them. Immunotherapy is a precarious approach because tumor cells are not exactly a pathogen, such as bacteria and viruses, which the body can recognize as a foreign invader. Hence, this vaccine-based strategy is currently being applied in special cases after surgery and/or chemotherapy to detect the first signs of recurrent growth before any scans can visualize it [42,43].

10.3.3 Summary

In conclusion, the advantages of employing NPs in medicine as imaging, diagnostic and therapeutic agents can be summarized as follows:

- Drug-loaded NPs can be effectively formulated using LoC (lab-on-a-chip) devices [12].
- NPs reach tumor sites even via capillaries, i.e., no vessel embolization.
- Once near a predetermined disease site, NPs make targeted drug delivery possible.
- NPs protect the encapsulated drug from degradation, and surfactant-layers avoid particle flocculation.
- Upon release from NPs, nanodrugs exhibit high aqueous solubility.
- Being of molecular size, nanodrugs can readily cross cell membranes and other barriers.
- Via successful targeting, NPs decrease the toxic side-effects of aggressive drugs.

Advances in new drug formulations and multifunctional nanoparticles as drug carriers have measurably improved the pharmacokinetics and biodistribution of drugs, subject to poor water solubility, low stability and unwanted toxicity. Even so, for such nanoparticles to avoid clearance from circulation or destruction by the immune system, achieving high cancer-cell selectivity and sufficient target concentrations are necessary, and these areas form the current research challenges [17]. Figure 10.3 summarizes the characteristics of NPs which come close to achieving these two goals. As demonstrated in this section, all parameters are dependent on the 'circulation time', i.e., the probability of finding the target cells, and the associated biodistribution. Additionally, while these nanostructures must remain stable until successful drug delivery, they also should be biodegradable and biocompatible. However, as pointed out by *Fadeel & Garcia-Bennett* [36], the factors influencing the biocompatibility and the toxicity of inorganic nanoparticles have to be understood as well.

10.4 FORCED TRANSPORT

To improve the accumulation of drugs in tumors, forced transport uses an external magnetic field to steer nanodrugs towards specific sites. This is accomplished by incorporating drugs on ferric nanoparticles and injecting them near the tumor. An external magnet ensures high NP concentrations at the tumor site and hence rapid nanodrug-diffusion into the cancer cells [34,44]. The factors influencing the effectiveness of magnetic drug targeting include the magnetic properties of the nanodrug carrier, the strength of the magnetic field, the concentration of the magnetic NPs, the location of the tumor and the blood flow rate [34,44,45]. Because of the limited magnetic field penetration of the external magnet, however, the diseased site has to be located near the body surface and in a region of relatively low blood perfusion [46]. As a result, magnetic implants, such as spherical seeds, stents or wires, have been proposed to increase the local magnetic field gradient around the target site, allowing the use of magnetic drug targeting in deeper tissue [44]. A limitation of this technique may be the difficulty in implanting these devices.

10.5 DIRECT TRANSPORT

Direct drug-particle transport through optimal targeted drug delivery (TDD) has the potential to overcome the major problems associated with the free and forced transport mechanisms discussed so far. At present, the delivery of drugs or drug carriers to predetermined target sites is a *non-directional* event; i.e., injected particles are subjected to the local fluid flow conditions, possible clearance and/or migration towards healthy tissue. For example, the clinical ^{90}Y microsphere deposition efficiency on liver tumors is estimated to be about 40% to 60% [47]. Delivery of inhaled drug-aerosols to a predetermined lung site, say, via a pressurized metered dose inhaler is less than 25%, as discussed by *Kleinstreuer et al.* [48]. Optimal TDD is a new methodology to bring drug particles *directly* from the release point, i.e., the exit of an optimally located catheter or inhaler nozzle, to the (solid) tumor or predetermined lung site [49].

Focusing on optimal liver tumor targeting, the new methodology works as follows. Using patient-specific blood flow geometries, pressure waveforms and flow rates, tens-of-thousands of microspheres (representing spherical chemo-drugs, radioactive spheres, etc.) are randomly released from a suitable cross-sectional inlet plane – say, the hepatic artery in the case of liver-tumor targeting. Then, the departure locations of those depositing on selected targets (i.e., tumor surfaces) are determined via backtracking. This allows the construction of a particle release map, where color-coded particle-departure areas of the map (i.e., microsphere-release zones) are directly linked to the individual branch exits or target surfaces. Taking yttrium-90 microsphere delivery to a hepatic tumor as an example, Fig. 10.4 contrasts traditional ^{90}Y-microsphere delivery with controlled delivery using an optimally positioned catheter. *In order to have deterministic particle trajectories connecting optimal injection/release points and targets, the only requirements are that the fluid flow is laminar and the drug-particles are spherical and of at least micron-size to avoid Brownian-motion effects.*

10.5.1 Implementation of Optimal Targeted Drug Delivery

In order to achieve the goal of optimal TDD, a computational *medical management program* (MMP) is devised and implemented (Fig. 10.5). Briefly, an MMP is a step-by-step guide for disease diagnostics and treatment on a patient-specific basis. The predictive capabilities of the computational MMP are currently being demonstrated via targeting of liver tumors with radioactive microspheres, i.e., radioembolization (RE), as discussed in Section 10.5.2.1. Similar applications include optimal targeting with drug-eluting beads (chemotherapy), or a combination of both (chemo-radioembolization) onto solid tumors. In Section 10.5.2.2,

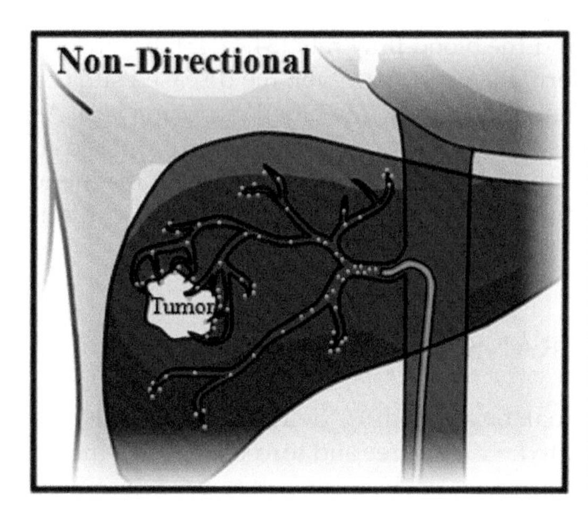

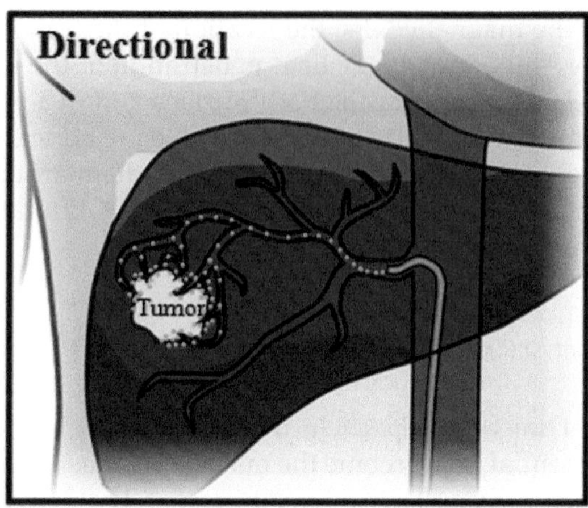

FIGURE 10.4 Non-directional vs. directional ^{90}Y-microsphere delivery. (Adaptation of an original by Andrew L. Richards [54]. Adapted with permission.)

computational and experimental results of this direct targeting methodology are shown when applied to the delivery of inhaled drug-aerosols onto diseased lung sites or into the deeper lung regions for mass transfer to the systemic system. In general, computational medical management programs are time and cost savers. That has been discussed by *Wan et al.* [50], who focused on a quasi–1-D model of the entire cardiovascular system with the goal of best recommendations for bypass surgery.

The major implementation steps for the proposed MMP, taking liver-tumor targeting as the key application, are:

(i) Geometric file conversion of the patient-specific tumor and vasculature images;
(ii) Local arterial flow/pressure measurements to deduce patient-specific blood flow waveforms for model input/output conditions;
(iii) Generation of a suitable mesh for computational particle-hemodynamics analysis on a patient-by-patient basis; and
(iv) Component integration to create the computational MMP.

The associated requirements for optimal drug targeting include the determination of:

- Best range of physical microparticle (drug) characteristics, i.e., diameter and density;
- Optimal drug release position, i.e., best axial and radial catheter location;
- Best drug injection interval during the local cardiac cycle; and
- Sufficient drug dosage to assure complete tumor coverage and efficacy.

Such a realistic and accurate computer simulation model, embedded in the patient-specific computational MMP, is a predictive tool which provides drug-particle release maps for optimal targeting of predetermined sites. These quantitative results are coupled to the availability of suitable drugs and medical devices.

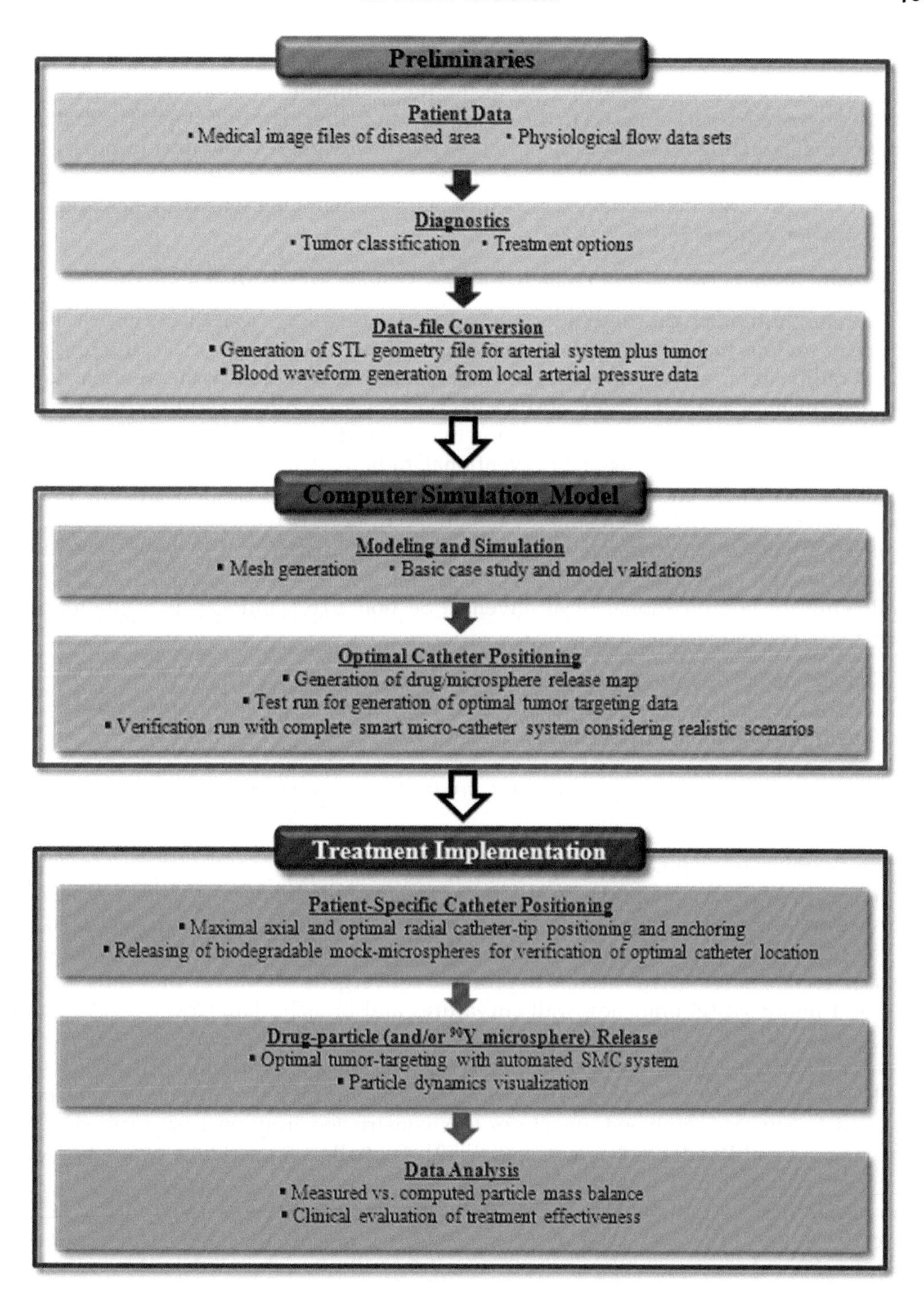

FIGURE 10.5 Computational medical management program.

10.5.2 Applications of Optimal Micro-Drug Delivery

In order to illustrate the optimal TDD technique outlined in Section 10.5.1, two applications are discussed, both being still in the laboratory and clinical testing stage. The first example of optimal targeting is the delivery of radioactive microspheres onto liver tumors, using a smart microcatheter (SMC) connected to a microsphere supply apparatus. The second application deals with targeting predetermined sites in the human lung with drug-aerosols, using a smart inhaler system (SIS).

10.5.2.1 Microsphere Delivery to Liver Tumors by a Smart Microcatheter System

The preferred therapy for solid (liver) tumors is of course surgery. If that is impossible, the next line of options includes radiation, chemotherapy (see also nanoparticle targeting), ablation and radioembolization. Radioembolization (RE) is a form of brachytherapy that consists of permanent implantation of low dose radiation in a solid tumor residing in the liver. A barrier to the adoption of RE is the perception that it is rudimentary and requires exceptional individual skill, particularly with interventional radiology placement of micro-catheters. Thus, the major challenge to successful RE treatment is the need for a targeting system to reliably deliver radioactive microspheres *only* to the tumor. As will be shown, optimal TDD will offer a significant improvement in safety and lower toxicity impact when compared to the currently achievable state-of-the-art methods [51].

Following the computational MMP given in Section 10.5.1 and Fig. 10.5, the following steps have to be implemented:

(i) Conversion of the patient's local (CT/MRI) medical image to arterial/tumor geometry files for flow domain discretization, i.e., generation of a suitable computer mesh;
(ii) Generation of patient-specific blood waveforms from local arterial pressure measurements; and
(iii) Computer code development, model validations and applications to predetermined site targeting with microspheres.

Complications associated with steps (i) and (ii) may include the possible existence of discontinuities in vessel wall boundaries and the problem of distinguishing hepatic vasculature from surrounding tissues. Injection of angiographic contrast will aid in both vessel isolation and best microcatheter positioning as predicted *a priori* by computer simulations. Geometries may be refined or smoothed as needed to promote convergence; however, they must preserve major and minor vessel branches, wall curvatures and branch diameters. Concerning step (iii), Fig. 10.6a depicts a representative arterial system with the inflow waveform to the common hepatic artery (CHA), where 41% of the flow rate is exiting through the major GDA-branch, as well as the measured outlet pressure profile for the four daughter vessels, selectively connected to tumors. The blood waveform was constructed from measured pressure data, employing a novel Windkessel circuit model [52]. That allows for routine evaluation of the influence of changing inlet waveforms.

An application of 'optimal tumor targeting' with microspheres is given in Fig. 10.6b–c. Figure 10.6b shows the *color-coded microsphere-release map with catheter throughout the pulse, which was obtained computationally via backtracking*. Specifically, all particles which selectively depart from one of these five cross-sectional areas exit 100% the associated vessel branch; for

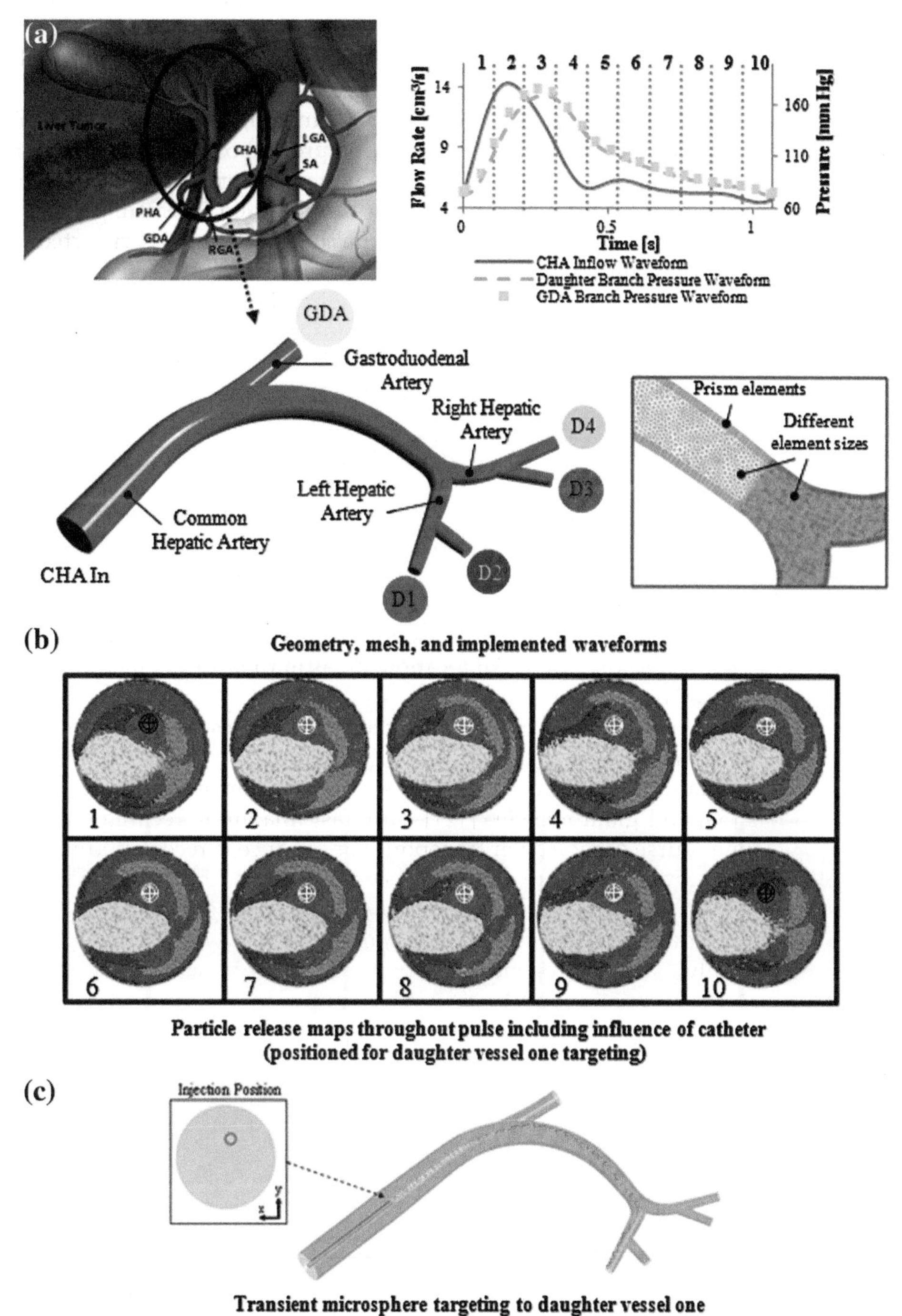

FIGURE 10.6 Representative common hepatic artery system with mesh, inlet and outlet waveforms, and optimal tumor targeting with microspheres. (Sources: [52,53,56].)

example, 'blue' exits daughter vessel 2, while particles released (via injection by a positioned catheter) from the 'red area' exit daughter vessel 1. The latter application is illustrated in Fig. 10.6c. When a catheter is actually deployed and stabilized with a tripod, the particle release map is slightly perturbed compared to that without a catheter [53]. Figure 10.7 depicts a much more complicated case of patient-specific arteries with corresponding flow and pressure waveforms. In this case, the particle release maps throughout one pulse, including the influence of a catheter, are more restrictive than the basic case shown in Fig. 10.6. However, optimal release position and time interval for the specific targeted sites (e.g., right hepatic artery, RHA) still exist; although it needs to be more accurately controlled.

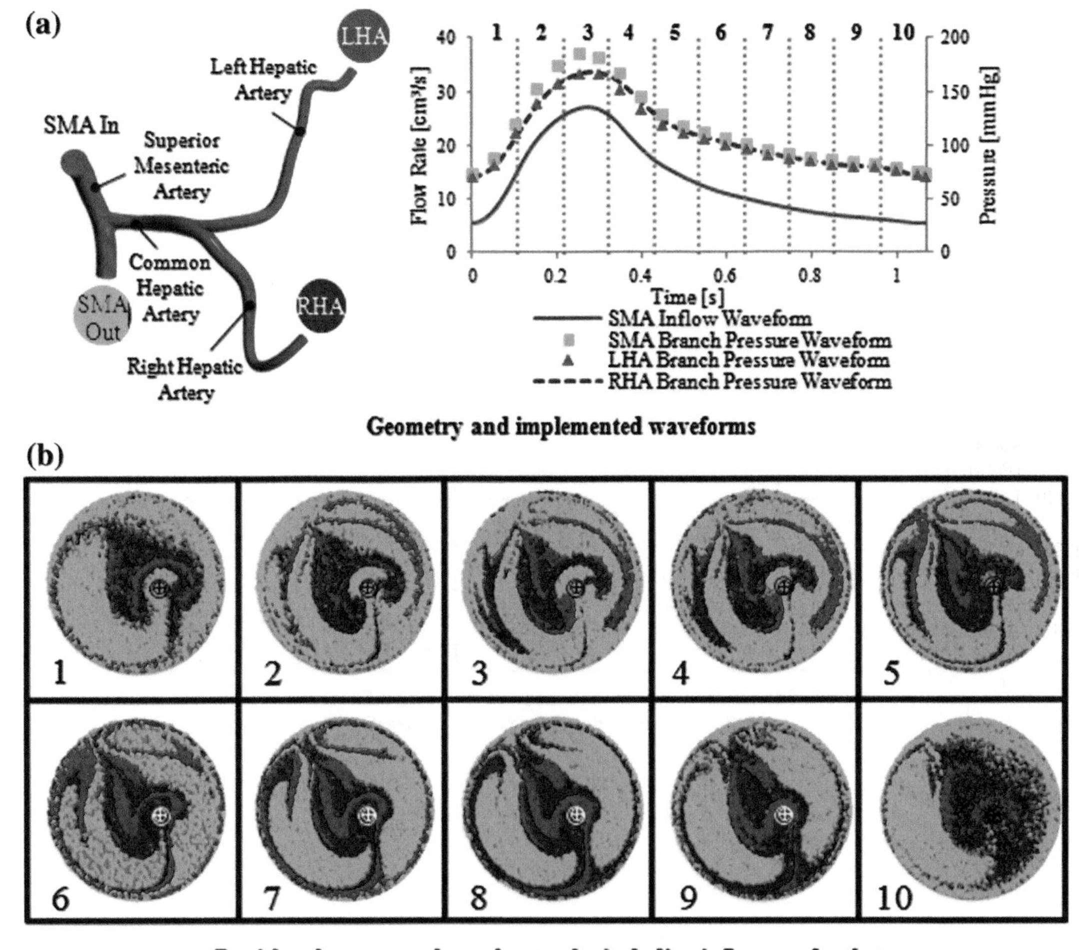

FIGURE 10.7 Patient-specific hepatic artery system with inlet and outlet waveforms and particle release maps throughout pulse including influence of catheter. (Source: [56].)

To validate these findings, a 4x-scaled physical model of the same hepatic artery configuration, matching both the Reynolds and Stokes numbers, was employed to track, visualize and measure microspheres for different injection positions, following the computed particle release map of the associated steady case. The experimental results [54] demonstrated that individual branch targeting is possible for each of the five exits, thereby fully correlating with the predictions of the computer simulation model (see Figs. 10.8a–c and video-link at website http://www.ncsu.mae.edu/cmpl/).

Due to the precise control of the injection speed, position and timing needed for direct targeting, a specially designed catheter (smart microcatheter, SMC) and an associated microsphere supply apparatus (MSA) have been developed. Figure 10.9 shows a schematic of the MSA which includes a programmable syringe pump and supply lines. Best operational conditions for the MSA, i.e., balancing medical need for the right drug dosage and release particle-hemodynamics for optimal targeting, have been determined via computer experiments [55,56].

Because the presence of the (anchored) catheter, the low-shear-rate dependence of blood flow, particle reflux, possibly distensible arterial walls and high dosage, i.e., dense microsphere suspensions, all may influence particle trajectories and hence the accuracy/reliability of drug targeting, further studies are warranted [57–60]. However, it has been demonstrated that direct targeting can be achieved by positioning the smart microcatheter (SMC) in the best axial and *radial* injection location and injecting the microspheres via a programmable microsphere supply apparatus [55,56].

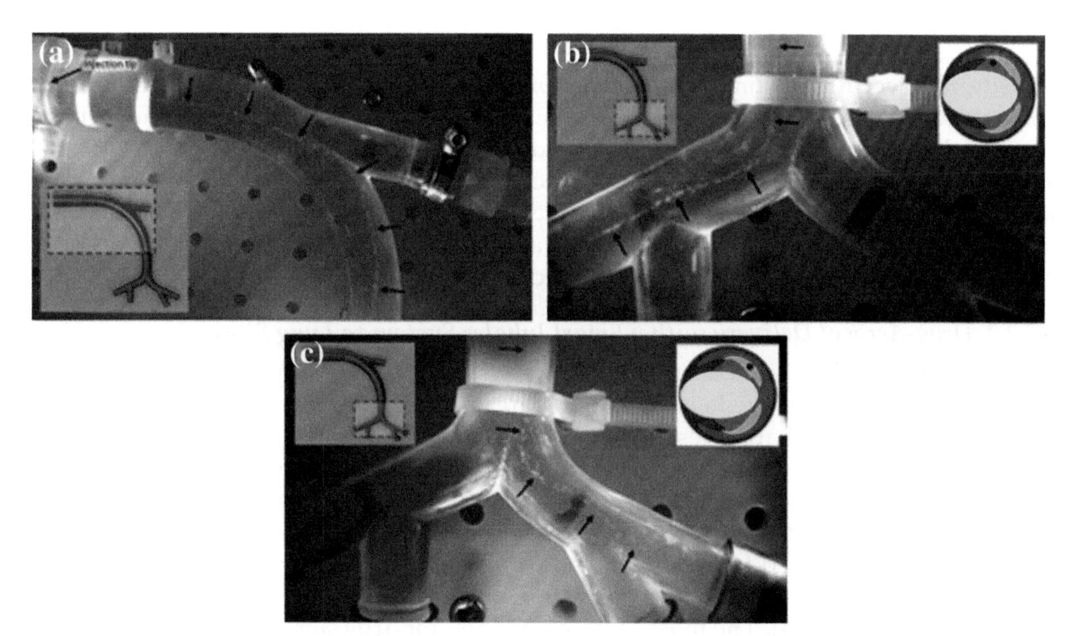

FIGURE 10.8 Video frames depicting a stream of microspheres exiting the injection tube: (a) traveling past the proper hepatic artery (PHA) bifurcation during injection 1; (b) microspheres are shown traveling through the LHA to branch 1 during injection 1; and (c) microspheres are shown traveling through the right hepatic artery (RHA) to branch 4 during injection 2. The insert graphics indicate the location in the model and the injection location. Arrows indicate the microsphere stream. (Source: Electro-Mechanics Research Laboratory [EMRL] at NCSU. Videos are available at http://www.mae.ncsu.edu/cmpl/.)

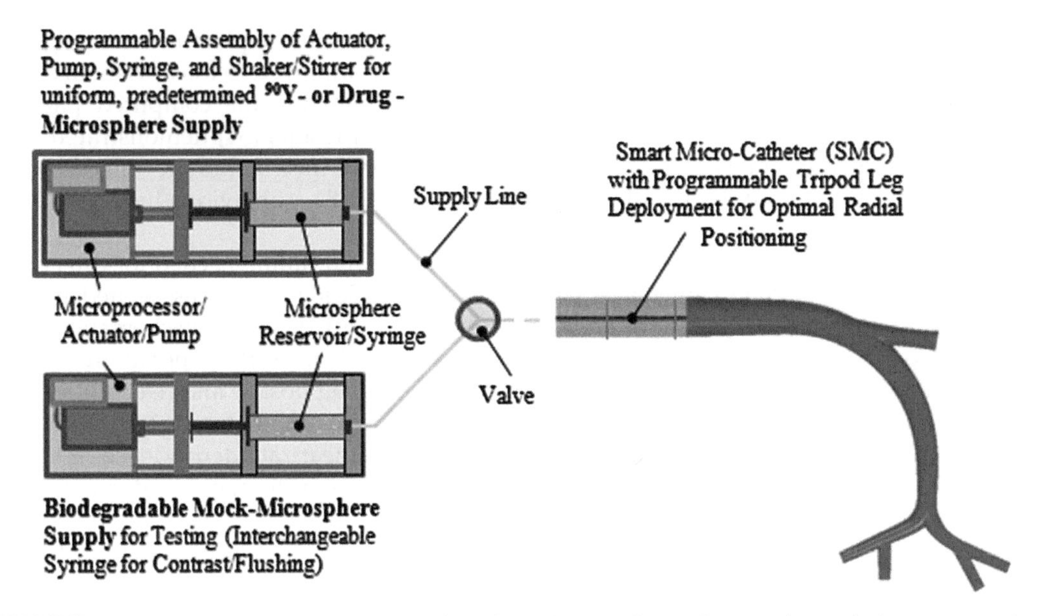

FIGURE 10.9 Microsphere supply apparatus (MSA) supplying well-mixed microsphere solution to smart micro-catheter (SMC) which is optimally positioned in the arterial system.

Lastly, it is worth mentioning that this approach can be readily extended to other direct tumor-targeting cases. For example, it can be used in chemotherapy using loaded microspheres, where no modifications would be needed or expected based on tumor type, location, distribution, prior anti-cancer therapy, etc., because it is patient-specific. Thus, chemo-drugs embedded in (porous) microspheres could be optimally targeted as well. Much has been learned of how to deliver RE with the present approaches already, and no data suggest there are particular issues with the vasculature based on tumor type or prior therapy.

10.5.2.2 Drug-Aerosol Delivery with a Smart Inhaler System to Lung Sites

Of the numerous applications of drug delivery systems, optimally targeted drug-aerosol delivery via inhalation to predetermined lung areas is becoming a very desirable treatment option, not only for combating respiratory disorders but also systemic diseases. Respiratory treatment applications include chronic obstructive pulmonary disease, asthma, cystic fibrosis, bacterial infection and lung tumors. Because of the rapid drug absorption from the deep lung region into the systemic system and ease of administration, the inhalation of medical aerosols has also been used for the treatment of diabetes and cancer.

Common oral drug delivery devices include metered dose inhalers, dry-powder inhalers and nebulizers [61–64]. The most common method for drug-aerosol delivery is the pressurized metered dose inhaler (pMDI). Current versions of these appliances use a hydrofluoroalkane (HFA) propellant and a dose counter. Dry-powder inhalers (DPIs) are breath-actuated, single or multi-dose devices. Nebulizers convert solutions into liquid aerosols of nano-size for nasal inhalation or deeper lung penetration. Typically, inhalers generate rather low particle deposition efficiencies in the lung and they are *non-directional*, i.e., drug-aerosol landing sites cannot be controlled. For example, *Kleinstreuer et al.* [48] analyzed the popular pMDI with a spacer, modified nozzle and HFA propellant. They demonstrated that up to 46.6% of the inhaled polydisperse droplets may reach the lungs, assuming a steady flow rate of Q_{in} equal to 30 liters/minute (see Fig. 10.10). While present inhalers may deliver some of the medicine to the central lung region, most treatments require a high-percentage drug-aerosol deposition at specific lung sites, such as tumors, or in desired lung regions, e.g., inflamed airways of asthma patients. Some drugs are so aggressive and/or expensive that targeted drug-aerosol delivery is imperative, especially for children with asthma to avoid drug delivery to non-inflamed lung areas. Clearly, targeted drug-aerosol delivery to desired lung areas required for the treatment of a specific disease is becoming a more prevalent treatment option because it is fast, convenient, and with reduced topical side-effects [64].

Targeted drug-aerosol delivery can be accomplished with a new, patient-specific 'controlled air-particle stream' methodology. Thus, a new smart inhaler system (SIS) has been patented (US Patent 7,900,625 issued 03/08/2011), which allows for optimal drug-aerosol targeting to patient-specific airway location, as discussed by *Kleinstreuer et al.* [65]. Briefly, the underlying SIS-methodology has been tested, relying on experimentally validated computer simulations

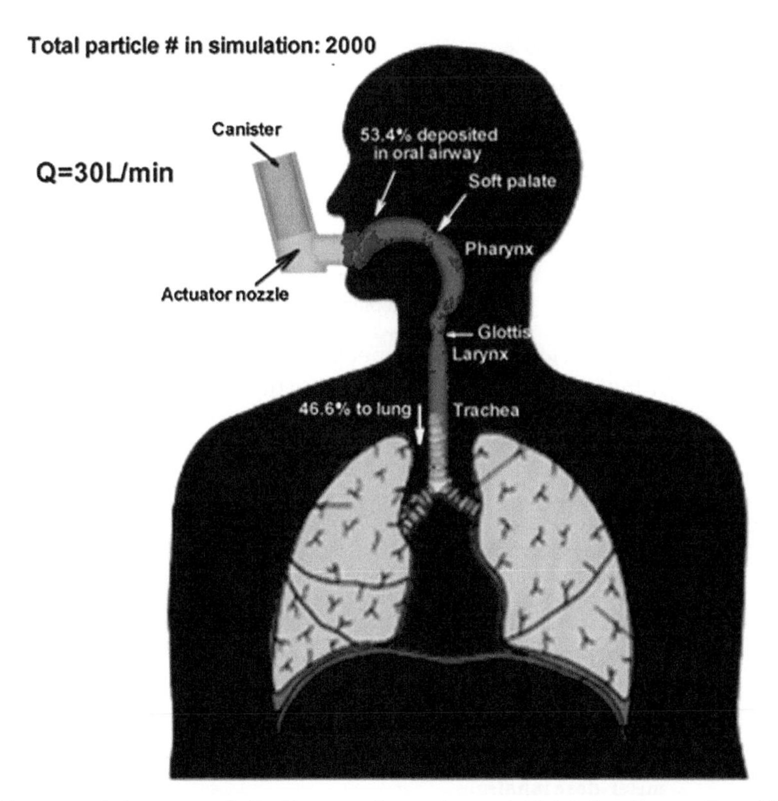

FIGURE 10.10 Typical drug-aerosol distribution when using a pMD inhaler. (Source: http://www.mae.ncsu.edu/cmpl/.)

and laboratory experiments (see Figs. 10.11 and 10.12). Specifically, in Fig. 10.11, the high drug-deposition percentage in the oral cavity and throat using an off-the-shelf inhaler is contrasted with direct targeting of a semi-spherical tumor. The computational/experimental proof-of-concept of optimal drug-aerosol targeting is depicted in Fig. 10.12, where the effect of three different inhaler exit drug release points are illustrated, all exiting 100% at Generation 3.1 (Fig. 10.12a) or the trachea (Fig. 10.12b). This was accomplished as follows. Given *any* respiratory system, all of the inhaled drug-aerosols can be delivered to specific lung sites or regions (presently up to Generation 6 due to still limited MRI/US/CT-scan resolution). In any case, once the patient-specific, CAD-like geometry files have been obtained from medical imaging, the computational tasks to be accomplished for each patient or a representative group of similar patients are:

- Determination of optimal diameter and density of (drug) microspheres;
- Determination of suitable inhalation flow rates, which should be laminar and steady for best air-drug stream control from inhaler-outlet/mouth-inlet to a predetermined target site/region;

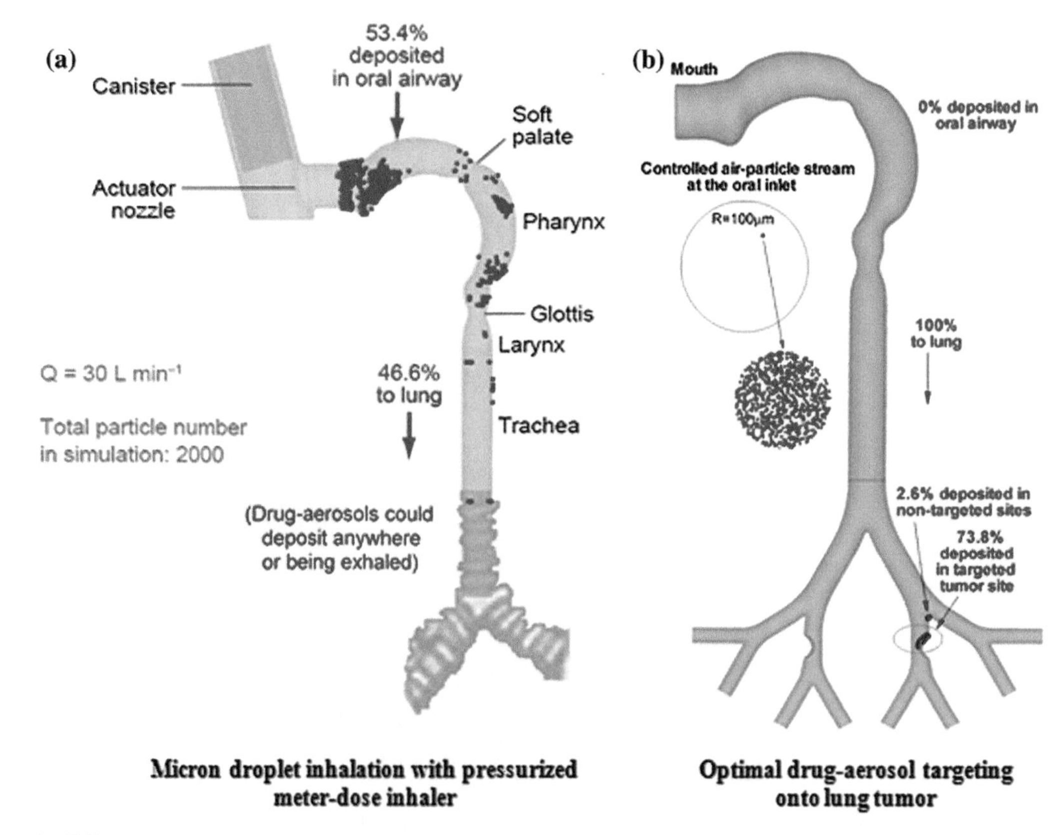

FIGURE 10.11 Comparison of (a) non-directional and (b) more targeted drug-aerosol delivery. (Source: [48].)

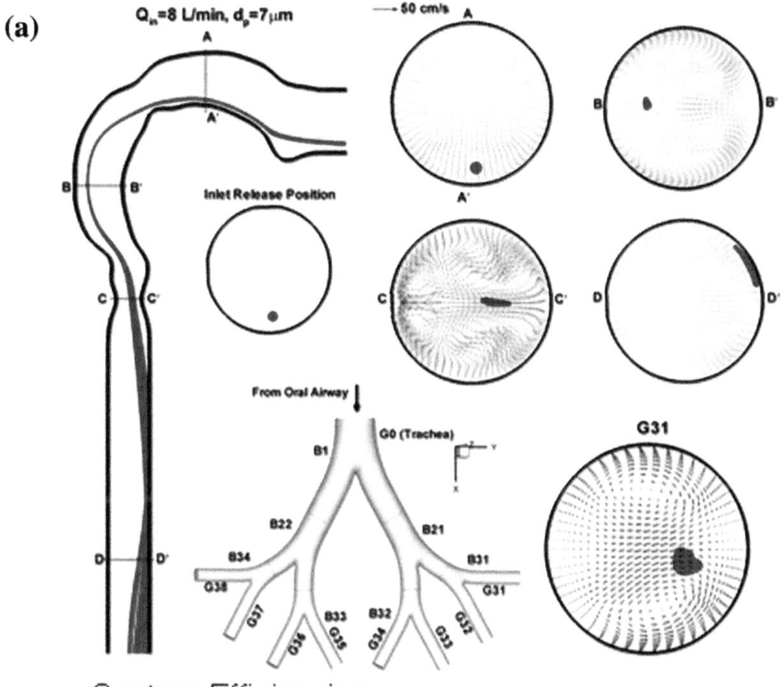

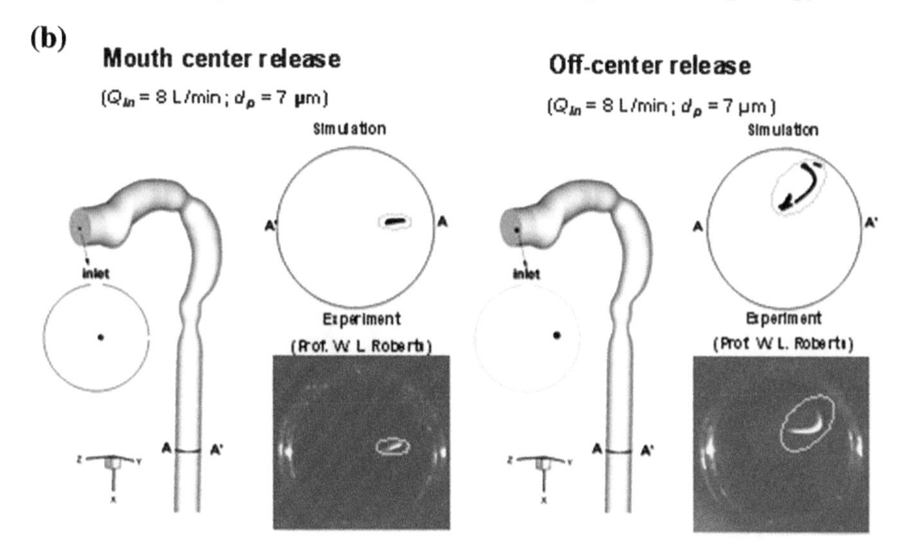

FIGURE 10.12 (a) Controllable particle distributions in the upper airway models when releasing from a given position at mouth inlet with $Q_{in}=8$ L/min and $d_p=7$ μm. The capture efficiency of particles after G3.1 increases from 12.4% (normal, non-directional inhalation) to 100% (targeted inhalation); and (b) comparison between computational and experimental measurements of particle paths through trachea as a function of release point. (Sources: [48,66].)

- Determination of optimal (radial) release position of the drug-aerosols from the (flexible) nozzle tip; and
- Determination of the drug-loading rate to assure deposition of a prescribed dosage onto the lung target area.

These tasks are accomplished via *realistic, experimentally validated* computer simulations as well as computer experiments, finding best microsphere characteristics, inhalation flow rates and drug dosage. Most important is the patient-specific (or group-specific) drug-aerosol release position (i.e., microprocessor-controlled nozzle tip location in the SIS) for optimal targeting. As mentioned, this is obtained via 'backtracking', in which a random particle load (distributed across the inhaler exit tube area) is released and those particles depositing on the target site (or in the target region) are back-tracked. That generates a drug-aerosol release map which correlates specific release zones with specific target sites/regions (see Fig. 10.13).

In order to implement the outlined methodology, the smart inhaler system (SIS) has to feature three key elements: the liquid (or powder) drug-reservoir with aerosol generator, an inhaled flow-waveform modulator to achieve laminar pseudo-steady airflow and a flexible nozzle with controlled radial positioning for optimal drug-aerosol release. It should be noted that the SIS consist from the computational angle only of a 2 cm inhaler tube with an air-particle supply, i.e., a modulated constant airstream and an optimally positioned nozzle from which the 5 ~ 10 micron particles will be released.

In summary, the optimal radial nozzle tip location is determined by patient-specific computer simulation. For the inhalation waveform modulation, computer experiments provide

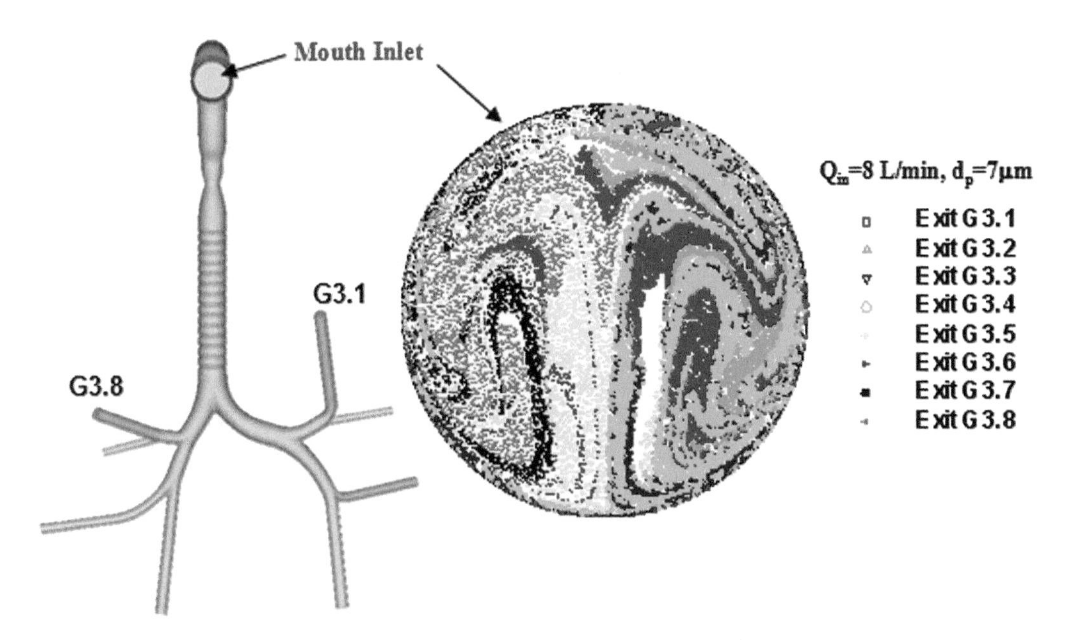

FIGURE 10.13 Regions of optimal particle release positions for targeting at the mouth inlet (or inhaler exit). (Source: [66].)

transient area-parameter values for actuator-controlled flow resistance and air-bleeding area changes. The result is a conversion of a patient's actual transient air flow and pressure profile into a partially steady inhalation flow rate during which the drug-aerosols are released. For most treatments or applications, it is desirable to target not only *spot* locations in the lung, but to selectively cover predetermined target *areas*. In order to accomplish that, the SIS features a flexible nozzle which can be adjusted via two sets of three SMA (shape-memory alloy) wires. With the first set, any pre-determined drug release position can be obtained, while the second set of wires will align the nozzle tip parallel to the air co-flow to avoid premature mixing of the aerosols due to shear layer effects, or even premature wall deposition. The overall device is designed with snap-in panels to allow interior access for precise and reproducible wire mounting and assembly. Electrical leads are embedded into the structure and connect to an external power supply as well as a control and data acquisition system at the bottom with a special minimal-footprint connection to allow simplified mounting and flawless operation.

10.6 CONCLUSIONS

Targeting is a technique which connects a controlled drug release (or injection) point with a pre-determined site. Thus, targeted drug delivery (TDD) is an integral part of drug delivery systems, which include *drug formulation* as well as *the methodology of delivery and the associated devices*. There are two very different TDD approaches in the development and testing phase: injection of multifunctional nanoparticles for treatment of cancer or inflammation, and direct delivery of therapeutic micron particles to point or area targets. In the case of *basic* TDD, the multifunctional nanoparticles in the blood stream allow tumor visualization as well as possible attachment to and destruction of cancer cells. The synthetic nanoparticles most frequently applied include magnetic nanoparticles, polymer drug conjugates and drug-loaded gold nanoparticles, while the drug-carrying biological nanoparticles used are hydrogels, micelles, and dendrimers. These approaches have been selectively applied for gene therapy, cancer-cell targeting and inflammation treatment.

In case of *optimal* TDD, a therapeutic substance is introduced into the body via a controllable device with optimal drug release in order to achieve the goals of high efficacy and safety. Two distinct applications have been reviewed. The first one deals with the controlled delivery of radioactive microspheres and/or chemo-drugs from a smart catheter system to liver tumors. The second application, relying on the same optimal targeting principle, focuses on controlled delivery of inhaled drug-aerosols, using a smart inhaler system, to predetermined lung sites or regions. It is apparent that optimal drug targeting has the potential to eliminate/mitigate severe side-effects caused by nonspecific uptake of aggressive drugs, save cost of expensive drugs by reducing waste and allow for more precise drug dosage to effectively combat the disease.

Acknowledgments

The authors acknowledge the assistance of Dr. Jie Li who generated the images of Figs. 10.1 to 10.3 and formatted the text, as well as Dr. Zhe Zhang for assembling Figs. 10.10 to 10.13. Both were members of the CM-PL Research Team in the MAE Department at NC State University. The insightful comments and suggestions made by Co-editor Dr. Sid Becker are gratefully acknowledged as well.

References

[1] W. Birkfellner, Applied Medical Image Processing, CRC Press Taylor & Francis Group, Boca Raton, FL, 2011.
[2] S. Luo, E. Zhang, Y. Su, T. Cheng, C. Shi, A review of NIR dyes in cancer targeting and imaging, Biomaterials 32 (2011) 7127–7138.
[3] S. Jiang, M.K. Gnanasammandhan, Y. Zhang, Optical imaging-guided cancer therapy with fluorescent nanoparticles, Journal of the Royal Society Interface 7 (2010) 3–18.
[4] P. Carmeliet, R.K. Jain, Angiogenesis in cancer and other diseases, Nature 407 (2000) 249–257.
[5] S.H. Jang, M.G. Wientjes, D. Lu, J.L.-S. Au, Drug delivery and transport to solid tumors, Pharmaceutical Research 20 (2003) 1337–1350.
[6] K. Cho, X. Wang, S. Nie, Z. Chen, D.M. Shin, Therapeutic nanoparticles for drug delivery in cancer, Clinical Cancer Research 14 (2008) 1310–1316.
[7] H. Maeda, The enhanced permeability and retention (EPR) effect in tumor vasculature: the key role of tumor-selective macromolecular drug targeting, Advances in Enzyme Regulation 41 (2001) 189–207.
[8] H. Maeda, Tumor-selective delivery of macromolecular drugs via the EPR effect: background and future prospects, Bioconjugate Chemistry 21 (2010) 797–802.
[9] S. Nie, Understanding and overcoming major barriers in cancer nanomedicine, Nanomedicine (London) 5 (2010) 523–528.
[10] W.H. De Jong, P.J.A. Borm, Drug delivery and nanoparticles: applications and hazards, International Journal of Nanomedicine 3 (2008) 133–149.
[11] E. Blanco, A. Hsiao, A.P. Mann, M.G. Landry, F. Meric-Bernstam, M. Ferrari, Nanomedicine in cancer therapy: innovative trends and prospects, Cancer Science 102 (2011) 1247–1252.
[12] N. Kamaly, Z. Xiao, P.M. Valencia, A.F. Radovic-Moreno, O.C. Farokhzad, Targeted polymeric therapeutic nanoparticles: design, development and clinical translation, Chemical Society Reviews 41 (2012) 2971–3010.
[13] D. Peer, J.M. Karp, S. Hong, O.C. Farokhzad, R. Margalit, R. Langer, Nanocarriers as an emerging platform for cancer therapy, Nature Nanotechnology 2 (2007) 751–760.
[14] Z. Xiao, E. Levy-Nissenbaum, F. Alexis, A. Luptak, B.A. Teply, J.M. Chan, J. Shi, E. Digga, J. Cheng, R. Langer, O.C. Farokhzad, Engineering of targeted nanoparticles for cancer therapy using internalizing aptamers isolated by cell-uptake selection, ACS Nano 6 (2012) 696–704.
[15] K.K. Jain, Drug delivery systems an overview, in: K.K. Jain (Ed.), Methods in Molecular Biology, vol. 437, 2008, pp. 1–50
[16] B. Kateb, K. Chiu, K.L. Black, V. Yamamoto, B. Khalsa, J.Y. Ljubimova, H. Ding, R. Patil, J.A. Portilla-Arias, M. Modo, D.F. Moore, K. Farahani, M.S. Okun, N. Prakash, J. Neman, D. Ahdoot, W. Grundfest, S. Nikzad, J.D. Heiss, Nanoplatforms for constructing new approaches to cancer treatment, imaging, and drug delivery: what should be the policy? NeuroImage 54 (2011) S106–S124.
[17] J.-W. Yoo, N. Doshi, S. Mitragotri, Adaptive micro and nanoparticles: temporal control over carrier properties to facilitate drug delivery, Advanced Drug Delivery Reviews 63 (2011) 1247–1256.
[18] S. Venkataraman, J.L. Hedrick, Z.Y. Ong, C. Yang, P.L.R. Ee, P.T. Hammond, Y.Y. Yang, The effects of polymeric nanostructure shape on drug delivery, Advanced Drug Delivery Reviews 63 (2011) 1228–1246.
[19] A.D. Bangham, M.M. Standish, J.C. Watkins, Diffusion of Univalent Ions across the lamellae of swollen phospholipids, Journal of Molecular Biology 13 (1965) 238–252.
[20] D. Di Paolo, F. Pastorino, C. Brignole, D. Marimpietri, M. Loi, M. Ponzoni, G. Pagnan, Drug delivery systems: application of liposomal anti-tumor agents to neuroectodermal cancer treatment, Tumori 94 (2008) 246–253.
[21] Z. Rao, M. Inoue, M. Matsuda, T. Taguchi, Quick self-healing and thermo-reversible liposome gel, Colloids and Surfaces B: Biointerfaces 82 (2011) 196–202.
[22] Y. Namiki, T. Fuchigami, N. Tada, R. Kawamura, S. Matsunuma, Y. Kitamoto, M. Nakagawa, Nanomedicine for cancer: lipid-based nanostructures for drug delivery and monitoring, Accounts of Chemical Research 44 (2011) 1080–1093.
[23] N.T. Huynh, C. Passirani, P. Saulnier, J.P. Benoit, Lipid nanocapsules: a new platform for nanomedicine, International Journal of Pharmaceutics 379 (2009) 201–209.
[24] Y. Malam, M. Loizidou, A.M. Seifalian, Liposomes and nanoparticles: nanosized vehicles for drug delivery in cancer, Trends in Pharmacological Sciences 30 (2009) 592–599.
[25] M. Rawat, D. Singh, S. Saraf, S. Saraf, Nanocarriers: promising vehicle for bioactive drugs, Biological and Pharmaceutical Bulletin 29 (2006) 1790–1798.

[26] F. Yang, C. Jin, Y. Jiang, J. Li, Y. Di, Q. Ni, D. Fu, Liposome based delivery systems in pancreatic cancer treatment: from bench to bedside, Cancer Treatment Reviews 37 (2011) 633–642.
[27] S.F. Medeiros, A.M. Santos, H. Fessi, A. Elaissari, Stimuli-responsive magnetic particles for biomedical applications, International Journal of Pharmaceutics 403 (2011) 139–161.
[28] T.R. Hoare, D.S. Kohane, Hydrogels in drug delivery: progress and challenges, Polymer 49 (2008) 1993–2007.
[29] Y.B. Patil, U.S. Toti, K. Ayman, L. Ma, J. Panyam, Single-step surface functionalization of polymeric nanoparticles for targeted drug delivery, Biomaterials 30 (2009) 859–866.
[30] S. Parveen, R. Misra, S.K. Sahoo, Nanoparticles: a boon to drug delivery, therapeutics, diagnostics and imaging, Nanomedicine: Nanotechnology Biology and Medicine 8 (2012) 147–166.
[31] S.Y. Madani, N. Naderi, O. Dissanayake, A. Tan, A.M. Seifalian, A new era of cancer treatment: carbon nanotubes as drug delivery tools, International Journal of Nanomedicine 6 (2011) 2963–2979.
[32] Y.-Y. Guo, J. Zhang, Y.-F. Zheng, J. Yang, X.-Q. Zhu, Cytotoxic and genotoxic effects of multi-wall carbon nanotubes on human umbilical vein endothelial cells in vitro, Mutation Research 721 (2011) 184–191.
[33] F. Yang, C. Jin, D. Yang, Y. Jiang, J. Li, Y. Di, J. Hu, C. Wang, Q. Ni, D. Fu, Magnetic functionalized carbon nanotubes as drug vehicles for cancer lymph node metastasis treatment, European Journal of Cancer 47 (2011) 1873–1882.
[34] Q.A. Pankhurst, J. Connolly, S.K. Jones, J. Dobson, Applications of magnetic nanoparticles in biomedicine, Journal of Physics D: Applied Physics 36 (2003) R167–R181.
[35] J.J. Li, S.-L. Lo, C.-T. Ng, R.L. Gurung, D. Hartono, M.P. Hande, C.-N. Ong, B.-H. Bay, L.-Y.L. Yung, Genomic instability of gold nanoparticle treated human lung fibroblast cells, Biomaterials 32 (2011) 5515–5523.
[36] B. Fadeel, A.E. Garcia-Bennett, Better safe than sorry: understanding the toxicological properties of inorganic nanoparticles manufactured for biomedical applications, Advanced Drug Delivery Reviews 62 (2010) 362–374.
[37] M.J.V. Vähä-Koskela, J.E. Heikkilä, A.E. Hinkkanen, Oncolytic viruses in cancer therapy, Cancer Letters 254 (2007) 178–216.
[38] K.W. Lee, B.T. Tey, K.L. Ho, W.S. Tan, Delivery of chimeric hepatitis B core particles into liver cells, Journal of Applied Microbiology 112 (2011) 119–131.
[39] Y. Xiao, H. Hong, A. Javadi, J.W. Engle, W. Xu, Y. Yang, Y. Zhang, T.E. Barnhart, W. Cai, S. Gong, Multifunctional unimolecular micelles for cancer-targeted drug delivery and positron emission tomography imaging, Biomaterials 33 (2012) 3071–3082.
[40] R. Savla, O. Taratula, O. Garbuzenko, T. Minko, Tumor targeted quantum dot-mucin 1 aptamer-doxorubicin conjugate for imaging and treatment of cancer, Journal of Controlled Release 153 (2011) 16–22.
[41] W. Di, X. Ren, H. Zhao, N. Shirahata, Y. Sakka, W. Qin, Single-phased luminescent mesoporous nanoparticles for simultaneous cell imaging and anticancer drug delivery, Biomaterials 32 (2011) 7226–7233.
[42] O.J. Finn, Cancer Immunology, The New England Journal of Medicine 358 (2008) 2704–2715.
[43] L.G. Gouw, K.B. Jones, S. Sharma, R.L. Randall, Sarcoma immunotherapy, Cancers 3 (2011) 4139–4150.
[44] P.J. Cregg, K. Murphy, A. Mardinoglu, Inclusion of interactions in mathematical modeling of implant assisted magnetic drug targeting, Applied Mathematical Modelling 36 (2012) 1–34.
[45] J. Chomoucka, J. Drbohlavova, D. Huska, V. Adam, R. Kizek, J. Hubalek, Magnetic nanoparticles and targeted drug delivering, Pharmacological Research 62 (2010) 144–149.
[46] F. Dilnawaz, A. Singh, C. Mohanty, S.K. Sahoo, Dual drug loaded superparamagnetic iron oxide nanoparticles for targeted cancer therapy, Biomaterials 31 (2010) 3694–3706.
[47] A. Kennedy, D. Coldwell, B. Sangro, H. Wasan, R. Salem, Radioembolization for the treatment of liver tumors: general principles, American Journal of Clinical Oncology 35 (2012) 91–99.
[48] C. Kleinstreuer, H.W. Shi, Z. Zhang, Computational analyses of a pressurized metered dose inhaler and a new drug-aerosol targeting methodology, Journal of Aerosol Medicine-Deposition Clearance and Effects in the Lung 20 (2007) 294–309.
[49] C. Kleinstreuer, Methods and devices for targeted injection of radioactive microspheres, US Patent and PCT Int'l Application (No. PCT/US2010/043552), NC State University, Raleigh, NC, 2011.
[50] J. Wan, B.N. Steele, S.A. Spicer, S. Strohband, G.R. Feijoo, T.J.R. Hughes, C.A. Taylor, A one-dimensional finite element method for simulation-based medical planning for cardiovascular disease, Computer Methods in Biomechanics and Biomedical Engineering 5 (2002) 195–206.
[51] B. Morgan, A.S. Kennedy, V. Lewington, B. Jones, R.A. Sharma, Intraarterial brachytherapy of hepatic malignancies: watch the flow, Nature Reviews Clinical Oncology 8 (2011) 115–120.

[52] C.A. Basciano, Computational particle-hemodynamics analysis applied to an abdominal aortic aneurysm with thrombus and microsphere-targeting of liver tumors. Ph.D. Dissertation, North Carolina State University, Department of Mechanical and Aerospace Engineering, Raleigh, NC, 2010.

[53] C.A. Basciano, C. Kleinstreuer, A.S. Kennedy, W.A. Dezarn, E. Childress, Computer modeling of controlled microsphere release and targeting in a representative hepatic artery system, Annals of Biomedical Engineering 38 (2010) 1862–1879.

[54] A.L. Richards, C. Kleinstreuer, A.S. Kennedy, E. Childress, G.D. Buckner, Experimental microsphere targeting in a representative hepatic artery system, IEEE Transactions on Biomedical Engineering 59 (2012) 198–204.

[55] C. Kleinstreuer, C.A. Basciano, E.M. Childress, A.S. Kennedy, A new catheter for tumor-targeting with radioactive microspheres in representative hepatic artery systems. Part I: impact of catheter presence on local blood flow and microsphere delivery, Journal of Biomechanical Engineering-Transactions of the ASME 134 (2012) 051004.

[56] E.M. Childress, C. Kleinstreuer, A.S. Kennedy, A new catheter for tumor-targeting with radioactive microspheres in representative hepatic artery systems. Part II: solid tumor-targeting in a patient-inspired hepatic artery system, Journal of Biomechanical Engineering-Transactions of the ASME 134 (2012) 051005.

[57] R. Murthy, D.B. Brown, R. Salem, S.G. Meranze, D.M. Coldwell, S. Krishnan, R. Nunez, A. Habbu, D. Liu, W. Ross, A.M. Cohen, M. Censullo, Gastrointestinal complications associated with hepatic arterial Yttrium-90 microsphere therapy, Journal of Vascular and Interventional Radiology, Brief Reports 18 (2007) 553–561.

[58] C.D. South, M.M. Meyer, G. Meis, E.Y. Kim, F.B. Thomas, A.A. Rikabi, H. Khabiri, M. Bloomston, Yttrium-90 microsphere induced gastrointestinal tract ulceration, World Journal of Surgical Oncology 6 (2008) 93.

[59] J. Nair, C. Liu, J. Caridi, R. Zlotecki, T.J. GeorgeJr , Gastroduodenal ulcerations as a delayed complication of hepatic metastasis radioembolization, Journal of Clinical Oncology 28 (2010) e735–e736.

[60] C.H. Cha, M.W. Saif, B.H. Yamane, S.M. Weber, Hepatocellular carcinoma: current management, Current Problems in Surgery 47 (2010) 10–67.

[61] D.A. Edwards, C. Dunbar, Bioengineeiring of therapeutic aerosols, Annual Review of Biomedical Engineering 4 (2002) 93–107.

[62] T.G. O'Riordan, Aerosol delivery devices and obstructive airway disease, Expert Review of Medical Devices 2 (2005) 197–203.

[63] M.J. Telko, A.J. Hickey, Dry powder inhaler formulation, Respiratory Care 50 (2005) 1209–1227.

[64] D.R. Hess, Aerosol delivery devices in the treatment of asthma, Respiratory Care 53 (2008) 699–723.

[65] C. Kleinstreuer, Z. Zhang, J.F. Donohue, Targeted drug-aerosol delivery in the human respiratory system, Annual Review of Biomedical Engineering 10 (2008) 195–220.

[66] C. Kleinstreuer, Z. Zhang, Z. Li, W.L. Roberts, C. Rojas, A new methodology for targeting drug-aerosols in the human respiratory system, International Journal of Heat and Mass Transfer 51 (2008) 5578–5589.

CHAPTER

11

Electrotransport Across Membranes in Biological Media: Electrokinetic Theories and Applications in Drug Delivery

S. Kevin Li, Jinsong Hao, and Mark Liddell

College of Pharmacy, University of Cincinnati, Cincinnati, OH, USA

OUTLINE

11.1 Introduction 418

11.2 Nernst-Planck Theory and Model Simulation Analyses 419

11.3 Electrotransport Under a Constant Electric Field Across Membrane (Symmetric Conditions) 420

11.3.1 Membrane Flux and the Modified Nernst-Planck Equation 421

11.3.2 Membrane Flux and Transference Number 424

11.3.3 Transport Lag Time 424

11.3.4 Electroosmotic Transport 425

11.4 Electrotransport Under Variable Electric Field Across Membrane (Asymmetric Conditions) 426

11.4.1 The Nernst-Planck Equation with Electroneutrality Approximation 426

11.4.2 Electrotransport Under Asymmetric Conditions with Electroosmosis 430

11.5 Electrotransport Across Multiple Barriers/Membranes 430

11.5.1 Electrotransport Across Two-Membrane Systems Under Symmetric and Asymmetric Conditions 432

© 2013 Elsevier Inc. All rights reserved.

http://dx.doi.org/10.1016/B978-0-12-415824-5.00011-4

11.5.2 Effects of Membrane Porosity and Applied Voltage Upon Electrotransport Across Two-Membrane Systems 436

11.6 Electrotransport Under Alternating Current 440

11.7 Electropermeabilization Effect 441

11.8 Electrokinetic Methods of Enhanced Transport Across Biological Membranes 443

11.8.1 Transdermal Iontophoresis 444

11.8.2 Transungual Iontophoresis 447

11.8.3 Transscleral Iontophoresis 448

References 450

11.1 INTRODUCTION

The effective utilization of electric field-assisted transport for drug delivery across biological membranes (i.e., iontophoresis) requires an understanding of the mechanisms and theories behind the process. Most basic electrotransport theories were developed and studied several decades ago (e.g., [1–4]). In the past three decades, a main focus in the field of electrotransport has been the transport behavior of biological membranes. These studies have used diffusion chambers such as side-by-side diffusion cells (e.g., [5,6]), Franz diffusion cells (e.g., [7,8]), Ussing chambers (e.g., [9]), or Transwell plates (e.g., [10]) and constant voltage, constant current or alternating current iontophoresis (e.g., [11,12]) in order to provide insights into the behavior of electrotransport across biological membranes. As a result of better understanding of the electrotransport phenomena, and recent technological advances in electronics and engineering, electric field-assisted drug delivery has become widely available and is now commonly found in both research and clinical settings.

Numerous reviews and book chapters have been published on the theories of the iontophoretic transport process, drug delivery utilizing iontophoresis for different routes of administration (e.g., transdermal, transungual and transscleral iontophoresis), and new drug delivery technologies in these areas (e.g., [13–18]). This chapter covers the topic of electrotransport across membranes in biological media with a focus on its application to drug delivery. Electrotransport theories based on the Nernst-Planck equation are first reviewed. A description of membrane transport under variable electric field and boundary conditions, such as the asymmetric conditions of different total ion concentrations across the membrane, is provided. Electrotransport across multiple membranes during constant current iontophoresis, and electrotransport under alternating current conditions are also discussed. Model simulations are included to further elaborate on the physics underlying the experimental data. The influences of electroosmosis and electropermeabilization in iontophoretic drug delivery are described. Applications of iontophoretic transport in transdermal, transungual and transscleral drug delivery are reviewed. An objective of this chapter is to provide an overview of these topics for graduate students and scientists in pharmaceutics, engineering and biological research.

11.2 NERNST-PLANCK THEORY AND MODEL SIMULATION ANALYSES

The flux of an ionic species (J_i) for one-dimensional mass transfer (along the x-axis) is governed by the modified Nernst-Planck equation [1]:

$$J_i = -D_i \frac{dC_i}{dx} - D_i C_i \frac{z_i F}{RT} \frac{d\psi}{dx} \pm vC_i \tag{11.1}$$

where ψ is the electrical potential, F is the Faraday constant, R is the universal gas constant, T is temperature, v is the average velocity of convective solvent flow, and C_i, x, z_i and D_i are, respectively, the concentration, position in the membrane, valence of charge and diffusion coefficient of the ionic species (i.e., ionic species i). The first term on the right-hand side of Eq. (11.1) describes transport by ion diffusion, the second term by electromigration and the last term by solvent convection. By relating the diffusion coefficient to electrophoretic mobility with the Nernst-Einstein relationship, Eq. (11.1) can be rewritten as:

$$J_i = -D_i \frac{dC_i}{dx} - u_i C_i \frac{d\psi}{dx} \pm vC_i \tag{11.2}$$

where u_i is the electrophoretic mobility and $u_i = \frac{D_i z_i F}{RT}$ according to the Nernst-Einstein relationship. The quantitative measure of the extent to which an ionic species carries the electric current is commonly described as the transference number (or transport number). The transference number is defined as the ratio of the electric current carried by an ionic species of interest (I_i) to the total current carried by all ionic species (I_{total}) in the system:

$$t_i = \frac{I_i}{I_{total}} \tag{11.3}$$

Since the electric current carried by an ion is related to the electrophoretic mobility of the ion, the transference number of the ion is related to its flux across the medium. Eq. (11.3) can be rewritten as a function of the valence of charge and flux of the ionic species:

$$t_i = \frac{|z_i| J_i}{\sum |z_j| J_j} \tag{11.4}$$

where J_i is the flux of species i, J_j is the flux of ionic species j in the system and z_j is the charge of ionic species j. The ionic species j represents all ions including the ion of interest. The flux of the permeant (J_i) is related to the electric current by:

$$J_i = \frac{t_i I_{total}}{AF|z_i|} \tag{11.5}$$

where A is the diffusional surface area. The traditional definition of the transference number is the fraction of the galvanic current carried by the ion in a pure conduction process in a solution. In contrast to this definition, in the context of electrotransport across a membrane, the transference number can be defined as the fraction of the current that is transported by the ion in the membrane with no restriction on its transport mechanisms other than the conditions of steady state and electric field-dominated transport [19].

Thus, the apparent transference numbers under this definition account for the contribution of electric current from all three components in Eq. (11.1): ion diffusion, electromigration and convection.

In this chapter, model analyses of the Nernst-Planck theory were carried out using computer software Comsol (Burlington, MA). Finite-element simulations of electrotransport of ions in a single pore in a membrane were performed using the modified Nernst-Planck equation (Eq. (11.1)) for all ion species in the system and the assumption of charge neutrality:

$$\sum_i z_i C_i = 0 \tag{11.6}$$

For simplicity and only to illustrate the basic concepts, the electrophoretic mobilities and diffusion coefficients of ions in the model simulations were assumed to be the ideal infinite-dilution values given in the literature; i.e., the model did not account for changes in electromobility at high ion concentration. This assumption was used because of the uncertainties in predicting the relationships between the electromobilities and the ion concentrations. This also allows the study to circumvent potential complications that may arise during the simulations resulted from the use of variable electromobility and diffusion coefficients in the membrane.

11.3 ELECTROTRANSPORT UNDER A CONSTANT ELECTRIC FIELD ACROSS MEMBRANE (SYMMETRIC CONDITIONS)

Before the discussion of electrotransport across a membrane, a description is provided of the membrane system configuration adopted here (Fig. 11.1), in which the two sides of the membrane are called the donor and receiver compartments, respectively. The donor compartment contains the ion of interest (or permeant) to be transported across the membrane. Permeant transport occurs along the x-axis from the donor to the receiver compartment. The concentration of the permeant at the donor/membrane interface, i.e., at $x=0$, is defined as the donor concentration C_D (or donor concentrtion of permeant i, $C_{D,i}$). At the receiver/membrane interface where $x=\Delta x$ (the thickness of the membrane), the sink condition for the permeant ($C=0$) is assumed. In anodal electrotransport (i.e., anodal iontophoresis), the anode electrode is on the donor side and the cathode electrode is on the receiver side. In cathodal electrotransport this is

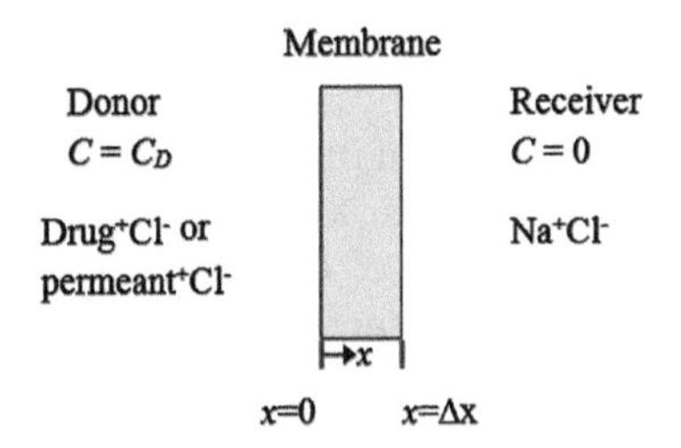

	Donor	Receiver
Anodal iontophoresis	Anode	Cathode
Cathodal iontophoresis	Cathode	Anode

FIGURE 11.1 Membrane system configuration.

reversed, so that the cathode is on the donor side and anode is on the receiver side. Other terminologies commonly used in the present chapter are the symmetric and asymmetric conditions with respect to the total ion concentrations in the donor and receiver. The symmetric condition is defined as the condition in which equal total ion concentrations exist in the donor and receiver compartments (e.g., 0.15 M $Drug^+Cl^-$ in donor and 0.15 M Na^+Cl^- in receiver). The asymmetric condition refers to the condition in which the total concentration of ions in the donor is different from that in the receiver (e.g., 0.15 M Na^+Cl^- in receiver and $Drug^+Cl^-$ concentration in donor $\neq$ 0.15 M).

11.3.1 Membrane Flux and the Modified Nernst-Planck Equation

The steady state flux of an ionic species (J_i) across a membrane can be described by a modified form of Eq. (11.1) [5]:

$$J_i = \varepsilon \left\{ -H_i D_i \left[\frac{dC_i}{dx} + \frac{z_i F C_i}{RT} \frac{d\psi}{dx} \right] \pm W_i C_i v \right\} \tag{11.7}$$

where ε is the combined effective porosity and tortuosity factor of the membrane, H_i is the hindrance factor of diffusion and electrophoretic transport and W_i is the hindrance factor of convective transport due to electroosmosis. H_i and W_i are related to the ratio of solute molecular size to the effective pore size (or pore radius, R_p) of the transport pathway [20]. Under the condition when the conductivity at different positions across the membrane (along the x-axis) is constant, the electric field in the membrane is constant, and Eq. (11.7) becomes:

$$J_i = \varepsilon \left\{ -H_i D_i \left[\frac{dC_i}{dx} - \frac{z_i F C_i}{RT} \frac{\Delta\psi}{\Delta x} \right] \pm W_i C_i v \right\} \tag{11.8}$$

where $\Delta\psi$ is the electrical potential applied across the membrane and Δx is the effective thickness of the membrane. The constant electric field assumption is sometimes called the Goldman approximation. An example of this occurs in the transport of a tracer (or a permeant at relatively low concentration) across a membrane with a high and equal electrolyte concentration on both sides of the membrane (symmetric condition). Since the flux J_i is constant and independent of position x at steady state, the concentration C_i is the only dependent variable in Eq. (11.8). This equation can be rearranged and integrated with the boundary conditions $C_i = 0$ at $x = \Delta x$ and $C_i = C_{D,i}$, at $x = 0$. The integrated form of Eq. (11.8) is:

$$J_i = \frac{\varepsilon H_i D_i C_{D,i}}{\Delta x} (K_{\Delta\psi,i} \pm Pe_i)/(1 - \exp[-K_{\Delta\psi,i} \pm Pe_i]) \tag{11.9}$$

where $K_{\Delta\psi,i} = \frac{z_i F \Delta\psi}{RT}$ and $Pe_i = \frac{W_i v \Delta x}{H_i D_i}$. Pe_i is the Peclet number and is a dimensionless number that describes the extent of the effect of electroosmotic flow to transport – proportional to the ratio of transport by convection to that by diffusion. Equation (11.9) can also be used to estimate the flux of a monovalent permeant under the condition when the permeant and its counterion are the only species in the donor compartment and when the ions in the receiver are monovalent ions with total ion concentration equal to that of the donor (another example of the symmetric condition with respect to donor and receiver). This scenario is illustrated in Fig. 11.2 below.

Figures 11.2a to 11.2d present the steady state concentration profiles of ions in a membrane under the condition of equal total concentration of monovalent ions in the donor and receiver (the symmetric condition) during passive and electrophoretic transport across that membrane. As can be seen in these figures, the diffusion-related, linear concentration profile under the passive transport condition becomes nonlinear in the presence of a moderate electric field, and the curvature of this concentration profile increases with the magnitude of the electric field; the electric field 'drives' the permeant in the membrane from the donor towards the receiver. When electrotransport becomes the dominant transport mechanism and driving force, the permeant (rather than the coion from the receiver) and its counterion are the primary contributors to the conductivity throughout most of the membrane (Fig. 11.2d). This results in relatively constant electric conductivity along the x-axis across the membrane and a constant electric field within the membrane, i.e., a condition described by Eq. (11.8).

Several other points should also be mentioned here. First, the concentration of the permeant in the membrane near the membrane/receiver interface in Fig. 11.2d is close to zero due to the boundary condition of $C_i=0$ in the receiver. This is a region of a very steep permeant concentration gradient where diffusion transport is the dominant transport mechanism and a major contributor to the total flux of the permeant:

$$C_{x\to\Delta x} \approx 0 \text{ and } J_i = \varepsilon\left[-H_i D_i \frac{dC_i}{dx}\right]_{x\to\Delta x} \tag{11.10}$$

Also can be seen in the figure is the small permeant concentration gradient (relatively flat concentration profile) near the donor/membrane interface in the membrane. This is the region of relatively constant and high permeant concentration, corresponding to dominant electromigration of total flux in the membrane. The contribution of diffusion to total flux is negligible in this region:

$$\left[\frac{dC}{dx}\right]_{x\to 0} \approx 0 \text{ and } J_i = \varepsilon\left[-H_i D_i \frac{z_i F C_i}{RT}\frac{d\psi}{dx} \pm W_i C_i v\right]_{x\to 0} \tag{11.11}$$

Third, under steady state the flux of an individual ion (e.g., the permeant) is constant along the x-axis across the membrane, and thus is independent of its position in the membrane: $J_{\text{steady state}}=J$ for all x. In other words, the total transmembrane flux of a permeant can be deduced by its local flux at any axial location along the x-axis in the membrane. For example, when electrotransport is the dominant transport mechanism under high electric fields, the flux of a permeant can be determined at a position near the donor/membrane interface in the membrane without the inclusion of the diffusion term in Eq. (11.8) and the integration of the equation (i.e., the derivation of Eq. (11.9) from Eq. (11.8)) in the calculation of permeant flux. Lastly, similar to permeant flux, the steady state condition dictates that the electric current along the x-axis in the membrane is constant as there cannot be infinite charge build-up over time within the membrane. This implies that the electrical potential gradient is always proportional to the electrical resistance at any x-position in the membrane, or inversely proportional to the

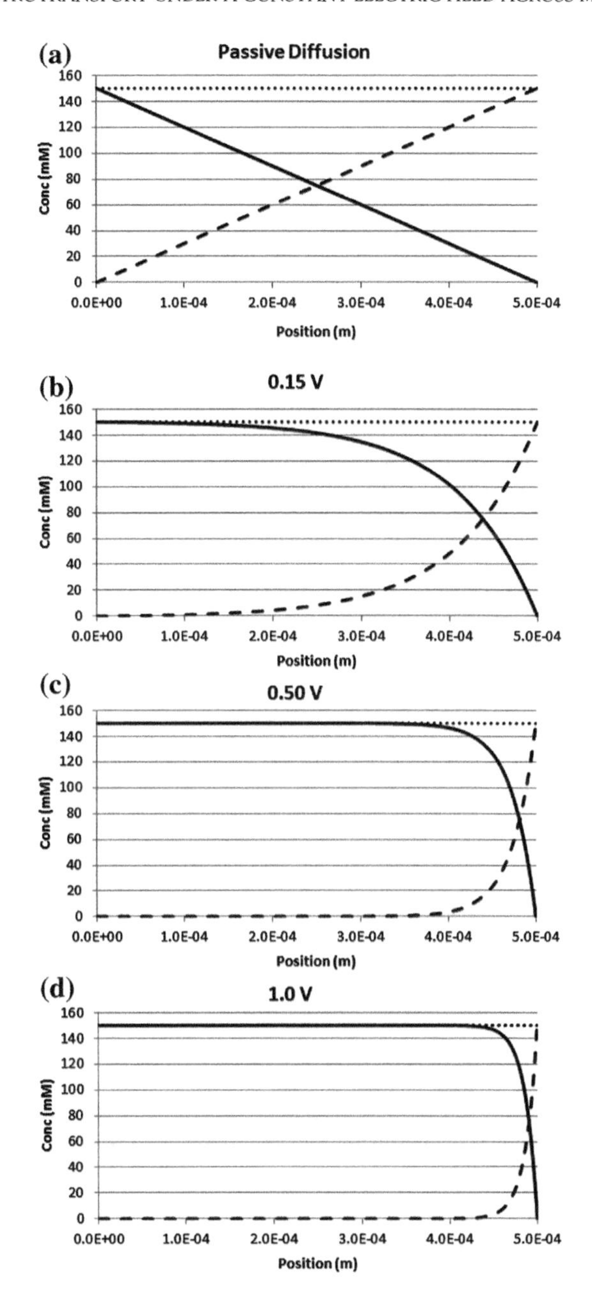

FIGURE 11.2 Steady state concentration profiles of ions in a membrane under the symmetric condition of equal total ion concentration in the donor and receiver during (a) passive transport and electrotransport of (b) 0.15 V, (c) 0.5 V and (d) 1.0 V electrical potential across the membrane. Donor: 0.15 M $Drug^+Cl^-$. Receiver: 0.15 M Na^+Cl^-. Membrane thickness: 0.5 mm. Solid line: permeant concentration profile (cationic drug from donor); dotted line: counterion concentration profile (Cl^- ion from donor and receiver); dashed line: coion concentration profile (Na^+ ion from receiver).

concentrations of the ions at the x-positions in the membrane during steady state electrotransport. The conditions of variable electric fields will be discussed in Section 11.4.

11.3.2 Membrane Flux and Transference Number

In practice during drug delivery, the efficiency of permeant transport across a membrane can be evaluated by the transference number of the permeant during constant current iontophoresis. As pointed out earlier, the apparent transference number for electrotransport across a membrane also accounts for the contribution of ion diffusion and convection to electric current transfer. This allows for the integral characterization of the fraction of current transported in the membranes as a whole (see Eq. (11.5)). Also, this definition of transference number is preferred due to the concentration gradients developed in the membranes with defined boundary conditions as seen in Fig. 11.2b to 11.2d. Under this definition and for iontophoretic drug delivery, it is common practice to simplify Eq. (11.7) when electrophoresis and electroosmosis are the dominant driving forces (i.e., for the case of relatively small passive diffusion contribution):

$$J_i = -\varepsilon \mu_i C_i \frac{d\psi}{dx} \tag{11.12}$$

where μ_i is the effective electrophoretic mobility taking into account transport hindrance and combining the effects of electrophoresis and electroosmosis. Note that $\mu_i = u_i$ when the contribution of electroosmosis to total flux is negligible. From Eqs. (11.4) and (11.12):

$$t_i = \frac{\varepsilon \left| z_i \mu_i C_i \frac{d\psi}{dx} \right|}{\sum_j \varepsilon \left| z_j \mu_j C_j \frac{d\psi}{dx} \right|} \tag{11.13}$$

By simplifying Eq. (11.13), an equation commonly seen in the iontophoresis literature [15,16] is obtained:

$$t_i = \frac{|z_i| \mu_i C_i}{\sum_j |z_j| \mu_j C_j} \tag{11.14}$$

in which C_i and C_j are assumed to be the concentrations of the ionic species in their respective donor chambers. Equation (11.14) provides a good approximation of permeant transference under the condition of dominant electric field-enhanced transport and equal total ion concentration in the donor and receiver (symmetric condition) across the membrane. The validity of Eq. (11.14) under the symmetric condition can be shown using Fig. 11.2d as discussed above.

11.3.3 Transport Lag Time

The lag time of electrotransport ($T_{lag,i}$) from the non-steady state transport equation and Nernst-Planck model has been previously described [21]:

$$T_{lag,i} = \frac{\Delta x^2}{H_i D_i} \frac{(K_{\Delta\psi,i} \pm Pe_i)\coth[0.5(K_{\Delta\psi,i} \pm Pe_i)] - 2}{(K_{\Delta\psi,i} \pm Pe_i)^2} \tag{11.15}$$

In the absence of an electric field ($\Delta\psi=0$ and $v=0$), Eq. (11.15) becomes the well-known transport lag time ($T_{lag,p}$) equation for passive diffusion:

$$T_{lag,p} = \frac{\Delta x^2}{6H_i D_i} \tag{11.16}$$

When the applied electric field across the membrane increases, the lag time of electrotransport decreases.

11.3.4 Electroosmotic Transport

Electroosmosis is the convective solvent flow resulting from the application of an electric field across a solution next to a charged surface. For electroosmosis across a charged porous membrane, the fixed charges on a pore wall in the charged membrane lead to an excess of counterions in the solution adjacent to the surface of the pore (to maintain charge neutrality). This leads to the formation of an electrical double layer in the pores of the membrane. The Debye-Huckel thickness describes the thickness of this double layer, which is related to the ionic strength of the solution in the pores. When an electric field is applied across the membrane (parallel to the charged surface), forces are exerted on the ions in the double layer of the pores. The ions in the double layer move under the influence of the electric field, and as they move, they carry the solvent with them through momentum (Newton's law of action and reaction). This is referred as electroosmotic flow. Electroosmosis assists the transport of both neutral and charged species by this electric field-induced convective solvent flow (see the last term in Eq. (11.7)). For enhanced transport of a non-electrolyte across a membrane due to electroosmosis, the flux can be described by the modification of Eq. (11.7) [22]:

$$J_i = \varepsilon \left\{ -H_i D_i \frac{dC_i}{dx} \pm W_i C_i v \right\} \tag{11.17}$$

Here, the direction of the electroosmotic flow is related to the sign of the net charge of the pores in the membrane. For a net negatively charged pore surface, the convective solvent flow is from the anode (positive electrode) to the cathode (negative electrode) during iontophoresis. Since water is considered incompressible under normal conditions encountered in biological systems, the solvent flow velocity (v) is constant along the x-axis in a homogenous membrane. When the solvent flow across the membrane is in the direction from the donor to the receiver, Eq. (11.17) can be integrated using the boundary conditions $C_i=0$ at $x=\Delta x$ and $C_i=C_{D,i}$ at $x=0$:

$$J_i = \frac{\varepsilon H_i D_i C_{D,i}}{\Delta x} \frac{Pe_i}{1-\exp(-Pe_i)} \tag{11.18}$$

where $Pe_i = \frac{W_i v \Delta x}{H_i D_i}$ as described in Section 11.3.1. Electroosmosis is a function of the surface charge density of the pores in the membrane (related to the fraction of the ionized functional groups on the surface, which is a function of solution pH in the pores), the applied electric field and the ionic strength of the solution in the pores. Under the condition of low surface potential and when the radius of the pores in the membrane is much greater than the electrical double layer thickness ($R_p >> 1/\kappa$), the velocity of the electroosmotic solvent flow can be approximated by:

$$v = \frac{\sigma}{\kappa \eta} \frac{\Delta \psi}{\Delta x} \tag{11.19}$$

where σ is the pore surface charge density, η is the viscosity of the bulk solution and $1/\kappa$ is the thickness of the electrical double layer [23]. Note that κ is proportional to the square root of the solution ionic strength, I_s. Equation (11.19) predicts that the electroosmotic flow velocity increases with the surface charge density, Debye-Huckel double layer thickness, and electric field and decreases with the viscosity and ionic strength of the solution in the membrane. When the electrical double layer thickness is comparable to the radius of the pores in the membrane, the Poisson-Boltzmann and fluid motion equations without the constraints from the magnitude of the surface potential and relative thickness of the electrical double layer should be used to provide more accurate analyses [24–26]. Under this condition, the impact of solution ionic strength (electrical double layer thickness) upon electroosmotic solvent velocity, v, can be less than that predicted by Eq. (11.19).

11.4 ELECTROTRANSPORT UNDER VARIABLE ELECTRIC FIELD ACROSS MEMBRANE (ASYMMETRIC CONDITIONS)

The previous section discusses electrotransport under the symmetric condition in which the total concentrations of ions in the donor and receiver are comparable. When the total ion concentrations in the donor and receiver are unequal under asymmetric conditions, the constant electric field approximation of Eq. (11.8) is no longer valid. Electrotransport under asymmetric conditions is commonly encountered in pharmaceutical research in practice because the receiver chamber usually represents the inside of a cell, tissue or body surface that is under the physiological condition of ~0.15 M ionic strength and generally cannot be manipulated. On the other hand, the ion concentration in the donor chamber is usually optimized to achieve maximum permeant flux in electrotransport. Hence, asymmetric conditions arise in these situations.

11.4.1 The Nernst-Planck Equation with Electroneutrality Approximation

Under the asymmetric conditions and variable electric fields, the Nernst-Planck equation (Eq. (11.7)) can be solved by the electroneutrality approximation (Eq. (11.6)). The electroneutrality approximation can generally be applied to the condition in which the membrane is much thicker than the Debye length of the solution at the membrane boundaries. Under the constraint of electroneutrality, the concentrations of the ions are interrelated to maintain charge neutrality in the membrane. Assuming a negligible contribution from electroosmosis (or convective solvent flow) to electrotransport across an uncharged membrane and sink condition of the permeant in the receiver, the analytical solution of steady state flux across a membrane from the Nernst-Planck equation for an electrolyte solution composed of only monovalent ions was previously discussed [27]:

$$J_i = \frac{\varepsilon H_i D_i C_{D,i}}{\Delta x}\left(1 + \frac{K_{\Delta\psi,i}}{\ln \chi}\right)\frac{(\chi - 1)}{\chi - \exp(-K_{\Delta\psi,i})} \tag{11.20}$$

where χ is the ratio of the total ion concentration in the donor to that in the receiver.

Figures 11.3 and 11.4 present the steady state concentration profiles of monovalent ions in a membrane under the asymmetric conditions, when the total ion concentrations in the donor and receiver are different in the absence of electroosmosis. When the total ion concentration in the donor is different from that in the receiver, the total ion concentration varies along the x-axis across the membrane. This creates a variable electric field within the membrane, a condition that departs from Eq. (11.8). In the figures, the electric field applied across the membrane can be viewed as 'pushing' the permeant and its coion from the donor to the receiver in the membrane. As the electric field increases, it 'pushes' the permeant further towards the receiver. This leads to a steeper permeant concentration profile near the receiver/membrane interface, and enhances the steady state flux of the permeant across the membrane. The counterion maintains a linear concentration gradient from the boundary condition at the donor to that at the receiver under these circumstances. When the total ion concentration of the donor is lower than that of the receiver (Fig. 11.3), the relatively high concentration of counterion in the membrane (due to the concentration in the receiver) will result in enhanced permeant partitioning from the donor into the membrane in order to maintain charge neutrality (between the counterion and permeant). This is a phenomenon similar to ion partitioning due to charge-to-charge interactions. Hence, the concentration of permeant in the membrane is higher than that in the donor. When the total concentration of ions in the donor is higher than that in the receiver (Fig. 11.4), a relatively linear concentration gradient of the permeant across the membrane is observed, in which the permeant concentration in the membrane is lower than that in the donor. According to the concentration profiles in these figures, it is likely that the variable total ion concentration gradient and the variable electric field in the membrane are responsible for the failure of Eq. (11.14) to predict the transference of a permeant under asymmetric conditions. A more appropriate transference number equation for permeant transport under asymmetric conditions has been previously provided and discussed [27]. Another point worth mentioning is the generality of the ion concentration profiles shown in these figures. Although the electrophoretic mobilities of the ions affect the ion fluxes and the electric current passage across the membrane, when electrotransport is operated under the constant voltage condition, the shapes of the ion concentration profiles are relatively independent of the electromobilities of the ions. The ion concentration profiles shown in the figures can be applied to most asymmetric electrotransport settings under similar conditions.

Since diffusion is the dominant transport mechanism in the membrane near the membrane/receiver interface (which also determines the total transmembrane flux of the permeant), the effects of permeant donor concentration upon permeant transference across the

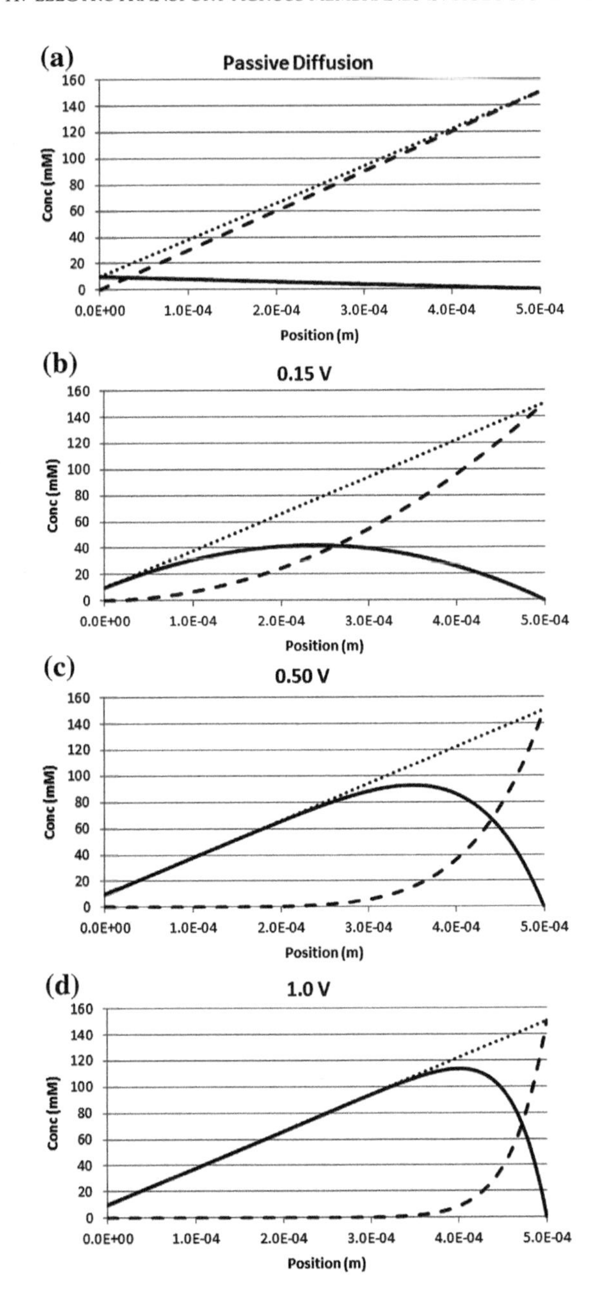

FIGURE 11.3 Steady state concentration profiles of ions in a membrane under the asymmetric conditions of different total ion concentrations in donor and receiver during (a) passive transport and electrotransport of (b) 0.15 V, (c) 0.5 V and (d) 1.0 V electrical potential across the membrane. Donor: 0.01 M $Drug^{+}Cl^{-}$. Receiver: 0.15 M $Na^{+}Cl^{-}$. Membrane thickness: 0.5 mm. Solid line: permeant concentration profile (cationic drug from donor); dotted line: counterion concentration profile (Cl^{-} ion from donor and receiver); dashed line: coion concentration profile (Na^{+} ion from receiver).

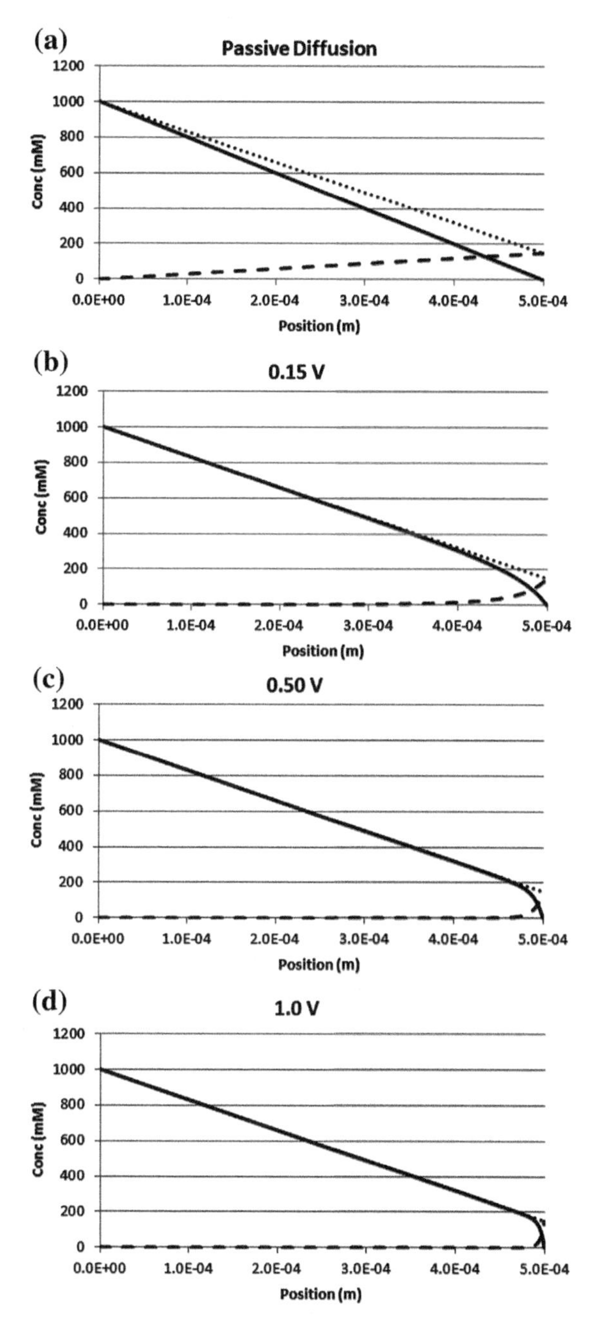

FIGURE 11.4 Steady state concentration profiles of ions in a membrane under the asymmetric conditions of different total ion concentrations in donor and receiver during (a) passive transport and electrotransport of (b) 0.15 V, (c) 0.5 V and (d) 1.0 V electrical potential across the membrane. Donor: 1.0 M $Drug^{+}Cl^{-}$. Receiver: 0.15 M $Na^{+}Cl^{-}$. Membrane thickness: 0.5 mm. Solid line: permeant concentration profile (cationic drug from donor); dotted line: counterion concentration profile (Cl^{-} ion from donor and receiver); dashed line: coion concentration profile (Na^{+} ion from receiver).

membrane can be deduced by comparing the concentration gradients of the permeant in the membrane near this interface (e.g., Figs. 11.3 and 11.4). A comparison of such concentration gradients under the constant current iontophoresis conditions illustrates that there are relatively small effects of permeant donor concentration upon permeant flux across the membrane. This permeant concentration-transport behavior has been observed in constant current transdermal and transscleral iontophoresis studies and is discussed in detail in Section 11.8.

In practice, electrotransport in pharmaceutical research usually occurs under asymmetric conditions, so understanding the interplay of the factors affecting the concentration profiles of ions and permeant flux under the asymmetric conditions can provide insights into strategies to improve electrotransport. These analyses can help pharmaceutical scientists control permeant transport and achieve maximum flux enhancement across cells, tissues or body surfaces. For instance, due to the modest effects of permeant donor concentration on total permeant flux during constant current iontophoresis (as discussed above), the permeant flux of electrotransport does not increase to the same extent as that of passive transport when donor concentration increases under the asymmetric condition. Under this condition, permeant flux enhancement across a membrane due to electrotransport is expected to be smaller from a concentrated solution than that from a dilute solution at the same applied electric current.

11.4.2 Electrotransport Under Asymmetric Conditions with Electroosmosis

Figure 11.5 shows the effects of electroosmosis on the steady state concentration profiles of ions under both the symmetric and asymmetric conditions. If the presence of electroosmosis enhances the transport of the permeant across the membrane, the concentration profiles of the counterions are no longer linear within the membrane. This leads to an increase in permeant flux under symmetric and asymmetric conditions when the total ion concentration in the donor is equal to or higher than that in the receiver. When the total ion concentration in the donor is lower than that in the receiver, electroosmosis from donor to receiver can result in a decrease in permeant flux compared with the case without electroosmosis (at the same applied voltage). This is due to the change in the concentration profile of the counterion by electroosmosis and the decrease in permeant concentration within the membrane. It should be noted that Eq. (11.20) was derived under the assumption of a negligible contribution from electroosmosis to electrotransport (no convective solvent flow) and a linear counterion concentration profile in the membrane, and thus will not be applicable to the situations involving significant electroosmosis.

11.5 ELECTROTRANSPORT ACROSS MULTIPLE BARRIERS/MEMBRANES

Electrotransport that occurs across multiple membranes of different barrier properties when the fluxes of the permeant across the individual membrane barriers in the membrane assembly are different can lead to a permeant concentration build-up in the membranes. To illustrate this, Fig. 11.6 presents a schematic diagram of an assembly of two membranes in series. Here, the intrinsic transference number of the permeant (i.e., the electromobility of the permeant relative to those of the other ions) in the first membrane is unequal to that in the

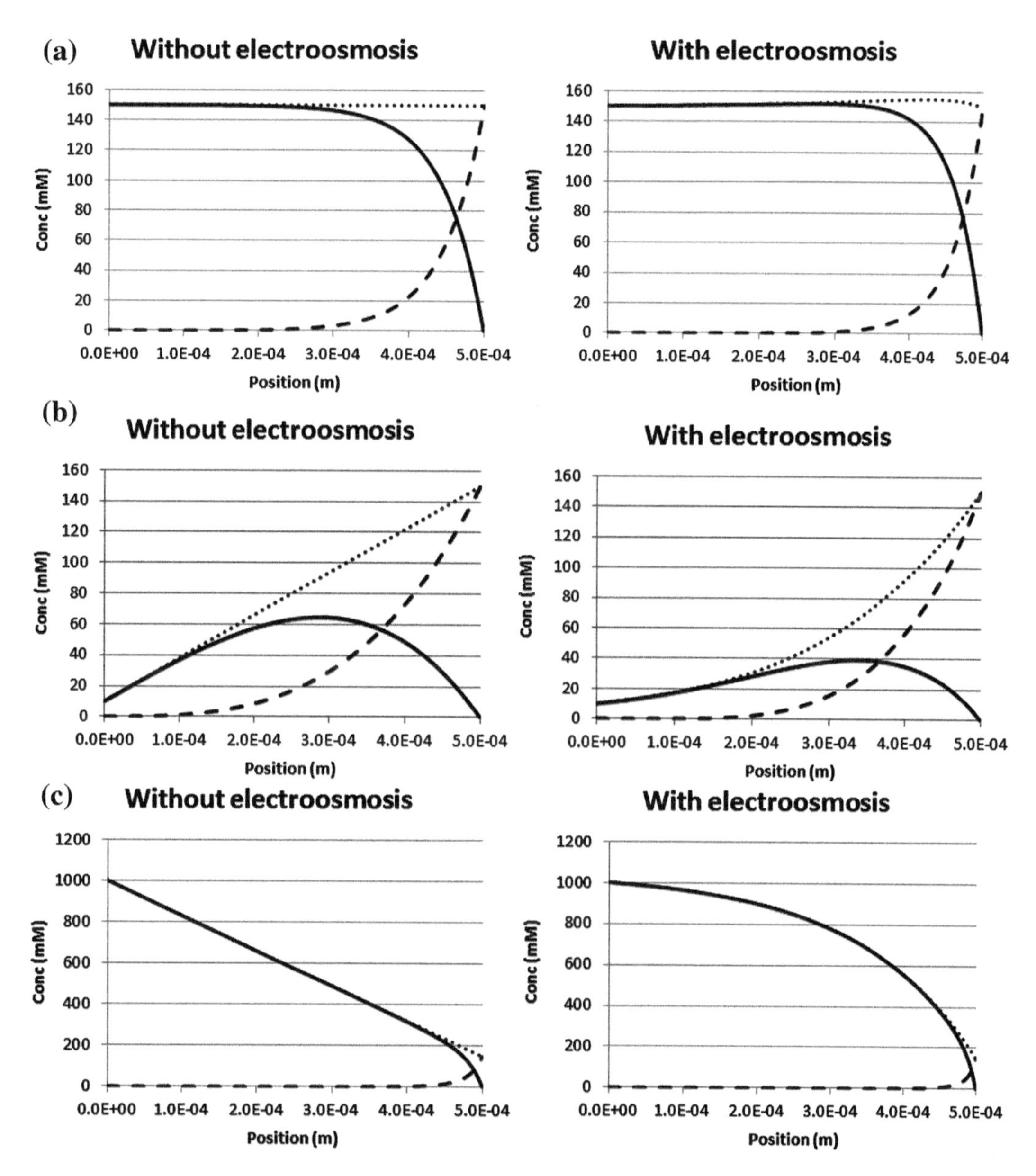

FIGURE 11.5 Steady state concentration profiles of ions in a membrane during electrotransport of 0.25 V electrical potential with and without electroosmosis under the (a) symmetric condition of 0.15 M, (b) asymmetric condition of 0.01 M and (c) asymmetric condition of 1.0 M $Drug^{+}Cl^{-}$ in the donor. The receiver is always 0.15 M $Na^{+}Cl^{-}$. Permeant transport due to electroosmosis contributes to 30% of the total permeant flux under the symmetric condition. Membrane thickness: 0.5 mm. Solid line: permeant concentration profile (cationic drug from donor); dotted line: counterion concentration profile (Cl^{-} ion from donor and receiver); dashed line: coion concentration profile (Na^{+} ion from receiver).

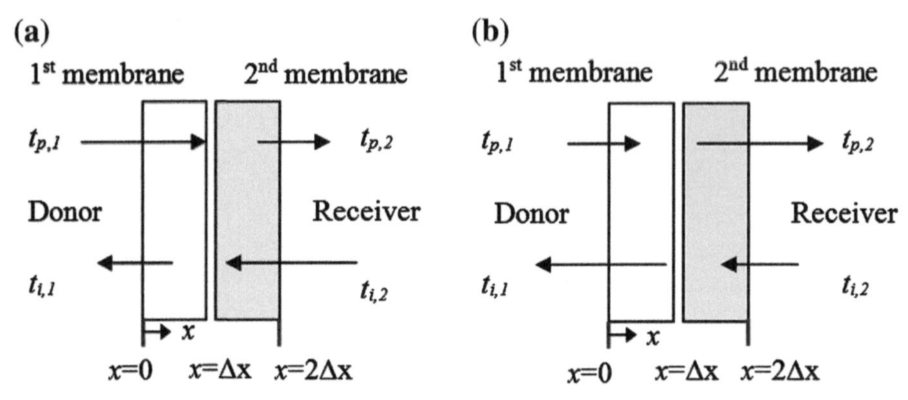

FIGURE 11.6 Schematic diagram of an assembly of two membranes in series of different barrier properties, in which the intrinsic transference number of the permeant in the first membrane ($t_{p,1}$) is (a) higher or (b) lower than that in the second membrane ($t_{p,2}$). $t_{i,1}$ and $t_{i,2}$ are the transference numbers of the counterion in the first and second membranes, respectively. The first membrane is the membrane on the left (white background) facing the donor and the second membrane is the membrane on the right (shaded background) facing the receiver. The length of the arrows indicates the magnitudes of the transference numbers (relative electrophoretic mobilities of the permeant, i.e., permeant electrophoretic mobilities relative to those of the other ions) of each membrane.

second membrane. The intrinsic transference number is defined as the transference number of permeant transport across an individual membrane alone as in a single membrane system. According to Eq. (11.5), the flux of a permeant across an individual membrane is proportional to the transference number and total electric current. Since the intrinsic transference numbers of the first and second membranes are different, the initial fluxes of the permeant across the first and second membranes will also be different. However, the fluxes of the permeant across the first and second membranes are required to be the same under steady state. Furthermore, the total electric current passing through each membrane in the two-membrane system must be equal (no infinite charge build-up in the system). The different intrinsic transference numbers in the two membranes create an interesting scenario, which will be discussed in this section.

11.5.1 Electrotransport Across Two-Membrane Systems Under Symmetric and Asymmetric Conditions

Figure 11.7a shows the steady state concentration profiles of ions across a two-membrane assembly (also see Fig. 11.6) during electrotransport in which the intrinsic transference number of the permeant in the first membrane is higher than that in the second membrane under the symmetric condition. In contrast, Fig. 11.7b shows the steady state concentration profiles in a similar membrane assembly with the intrinsic transference number of the permeant in the first membrane being lower than that in the second membrane. The following description explains the behavior of electrotransport across such two-membrane systems. First, due to the sink condition of permeant concentration in the receiver, the boundary condition at the membrane/receiver interface dictates that only diffusion occurs for the permeant at the interface (as noted in the previous sections). Second, in order to maintain charge neutrality in

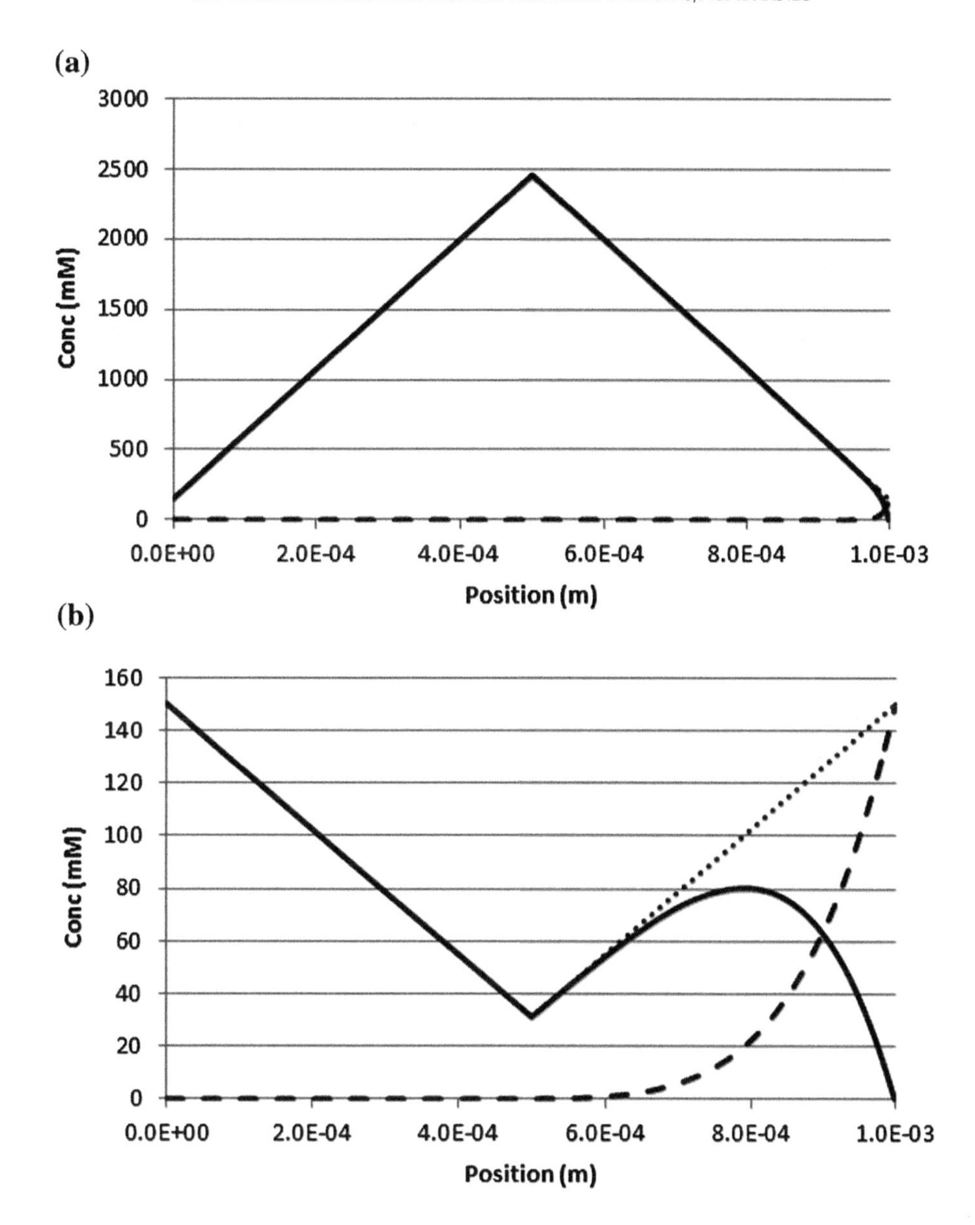

FIGURE 11.7 Steady state concentration profiles of ions across a two-membrane assembly, in which the intrinsic transference number of the permeant in the first membrane is (a) higher or (b) lower than that in the second membrane under the symmetric condition of 0.15 M $Drug^+Cl^-$ in the donor and 0.15 M Na^+Cl^- in the receiver. Electrical potential across the membrane assembly: 0.5 V. Membrane thickness: 0.5 mm each membrane. Solid line: permeant concentration profile (cationic drug from donor); dotted line: counterion concentration profile (Cl^- ion from donor and receiver); dashed line: coion concentration profile (Na^+ ion from receiver).

the membranes, the concentration profiles of the permeant and its counterion overlap in the majority of the membrane assembly (except near the membrane/receiver interface). Third, with variable total ion concentrations along the x-axis across the membranes, the electric field varies along the x-axis in the membranes. The prediction of electrotransport using the Goldman approximation in the Nernst-Planck model would not be appropriate for such a heterogeneous system even under the symmetric condition. Fourth, the different intrinsic transference numbers of the permeant in the two membranes lead to a 'sharp' concentration profile (no longer a smooth function) at the interface between the two membranes and significant contribution of permeant diffusion to the total permeant flux across the membrane assembly. Typically, the contribution of diffusion to total transmembrane flux is comparable to that of electromigration in such multiple membrane systems. This differs from transport in a homogenous membrane system in which the contribution of diffusion to total flux is significant only near the membrane/receiver interface. In Fig. 11.7a, in order to maintain a constant total steady state flux across the membrane assembly, opposing permeant concentration gradients are formed in the two membranes. Diffusion driven by the higher permeant concentration at the membrane/membrane interface enhances permeant transport across the second membrane but retards permeant transport across the first membrane. This balances the effect of the different intrinsic transference numbers of the permeant in the membranes to maintain the same total permeant flux along the x-axis across the membrane system at steady state. The opposite is true in Fig. 11.7b when the difference in permeant electromobilities in the membranes leads to low ion concentration at the membrane/membrane interface. Here, the concentration of the permeant at the interface is lower than those in the donor and second membrane, so permeant diffusion due to the higher permeant concentration in the donor enhances permeant transport across the first membrane and the concentration gradient in the second membrane retards permeant transport across the second membrane. This effectively ensures the same permeant flux across the two membranes during steady state electrotransport.

It should be pointed out that if the barrier properties of the two membranes are only different in porosity or thickness (i.e., they have same intrinsic transference number), the resultant transference number of the permeant across the membrane assembly will be the same as those of the individual membranes. The difference in porosity or thickness between the two membranes (having the same intrinsic transference numbers for the permeant and other conducting ions) only affects the electrical potential gradients across the individual membranes. The electrical potential gradients across the membranes will be different in order to provide the same constant electric current passage across the membranes. The ion concentration profiles along the x-axis observed in Fig. 11.7 are attributed to the different intrinsic permeant transferences of the membranes, which resulted from the different relative electromobilities of the ions in the two membranes. It is also important to point out that these results are obtained under the symmetric condition and the assumption of negligible contribution of electroosmosis. Furthermore, other physical constraints such as the solubility of the permeant and other ions in the membranes (and at the membrane/membrane interface) are not taken into account in these analyses. The discussion of electrotransport across a two-membrane assembly under the asymmetric conditions follows.

Figure 11.8 presents the steady state concentration profiles of ions across the same two-membrane assemblies as those in Fig. 11.7, except that electrotransport takes place under

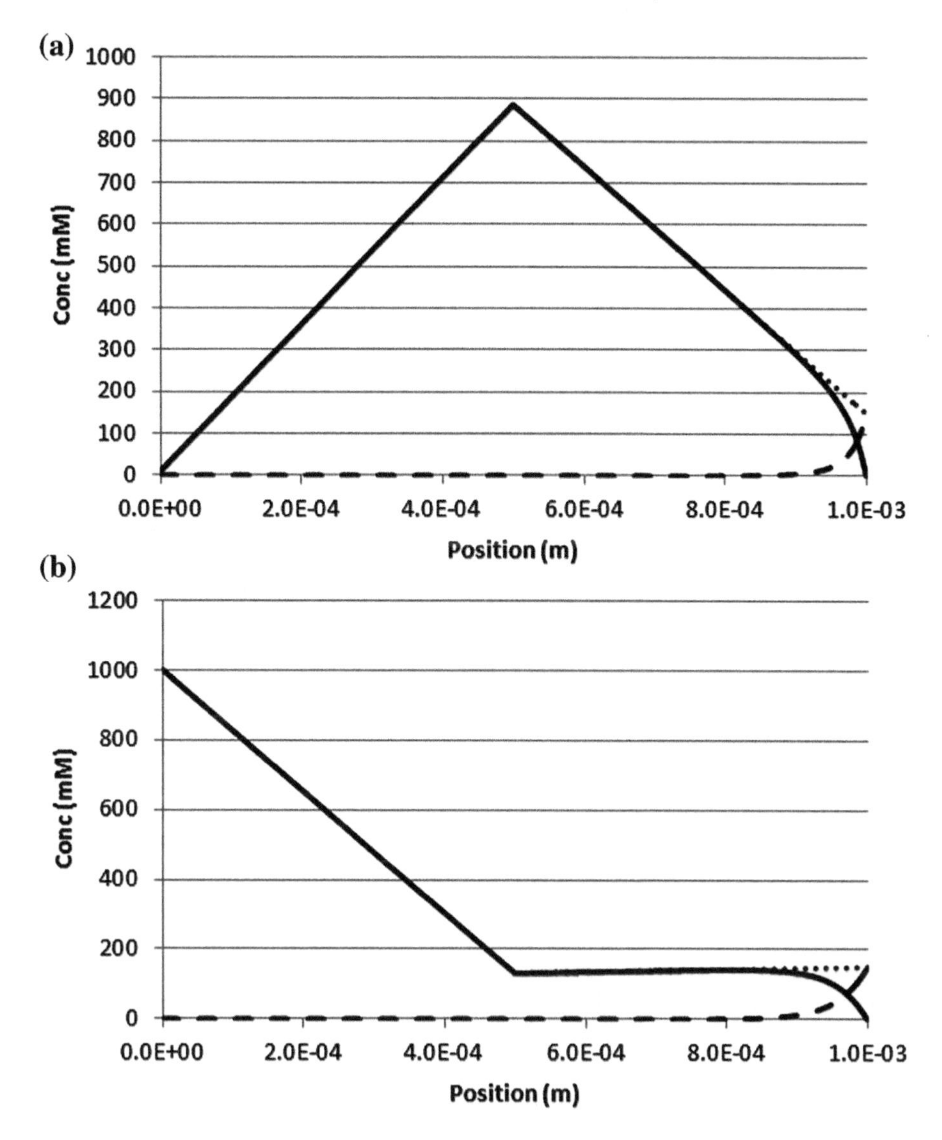

FIGURE 11.8 Steady state concentration profiles of ions across a two-membrane assembly, in which the intrinsic transference number of the permeant in the first membrane is (a) higher and (b) lower than that in the second membrane under the asymmetric condition of (a) 0.01 M and (b) 1.0 M $Drug^{+}Cl^{-}$ in the donor and 0.15 M $Na^{+}Cl^{-}$ in the receiver. Electrical potential across the membrane assembly: 0.5 V. Membrane thickness: 0.5 mm each membrane. Solid line: permeant concentration profile (cationic drug from donor); dotted line: counterion concentration profile (Cl^{-} ion from donor and receiver); dashed line: coion concentration profile (Na^{+} ion from receiver).

asymmetric conditions. A comparison of Figs. 11.7 and 11.8 shows that, while the concentration profiles of ions across the two-membrane assembly are similar under these symmetric and asymmetric conditions, there is a difference in the magnitudes of the ion concentration gradients in the membranes. For example, the higher permeant concentration in the donor in Fig. 11.8b creates a steeper concentration gradient in the first membrane (relative to the symmetric condition that has lower donor concentration in Fig. 11.7b). This concentration gradient is almost sufficient to increase the permeant flux across the first membrane so that a steep concentration gradient in the second membrane is no longer required to balance the different intrinsic transference numbers of the permeant in the two membranes. The figure still shows the 'V-shape' concentration profile (Fig. 11.8b vs. Fig. 11.7b), but the membrane/membrane interface concentration is higher and the concentration gradient in the second membrane is smaller under this asymmetric condition than those under the symmetric condition.

11.5.2 Effects of Membrane Porosity and Applied Voltage Upon Electrotransport Across Two-Membrane Systems

To further analyze the interplay between the concentration profiles and properties of the two-membrane assembly, Fig. 11.9 illustrates the effects of membrane porosity upon the concentration profiles of the permeant in the membranes. When the porosity of the second membrane is decreased relative to the first membrane, the concentration of the permeant at the membrane/membrane interface also decreases. This yields a smaller concentration gradient of the permeant in the second membrane. As the porosity of the second membrane is further decreased, the permeant concentration at the membrane/membrane interface continues to decrease until the interface concentration is equal to that of the donor concentration. Thus, when the porosity of the first membrane (ε_1) is significantly higher than that of the second membrane (ε_2) (i.e., $\varepsilon_1 >> \varepsilon_2$), the two-membrane system effectively acts as a single membrane system with a negligible barrier contribution from the first membrane. This is representative of the simple case of electrotransport across a single membrane in solution; the two-membrane system could be considered as a single membrane system because the first membrane plays such a negligible role that it could be considered as part of the solution itself. An example is that the aqueous unstirred boundary layer in the solution is the first membrane and the biological membrane is the second membrane.

Figure 11.10 summarizes the relationships between the transference numbers of the permeant and the porosities of the membranes in a two-membrane assembly in which one membrane has higher intrinsic transference number for the permeant than the other membrane and in which the lower transference number membrane is facing the receiver. In these simulations, the donor contained the permeant and its counterion, and the receiver contained the counterion and coion at the same concentration as those in the donor (i.e., symmetric condition). The steady state transference numbers of the permeant are plotted against the ratios of the porosities of the two membranes. The membrane porosity ratio is defined as the ratio of membrane porosity of the lower transference number membrane to that of the higher transference number membrane. The figure shows that the resultant transference number of the system ranges between the individual intrinsic transference numbers of the two membranes and is related to the porosity ratio. For a small membrane porosity ratio, the lower transference number membrane becomes the dominant membrane and permeant transference across the membrane assembly approaches the transference number of the membrane with the lower transference number. Conversely, for a high porosity ratio, the transference number of the membrane assembly approaches the value

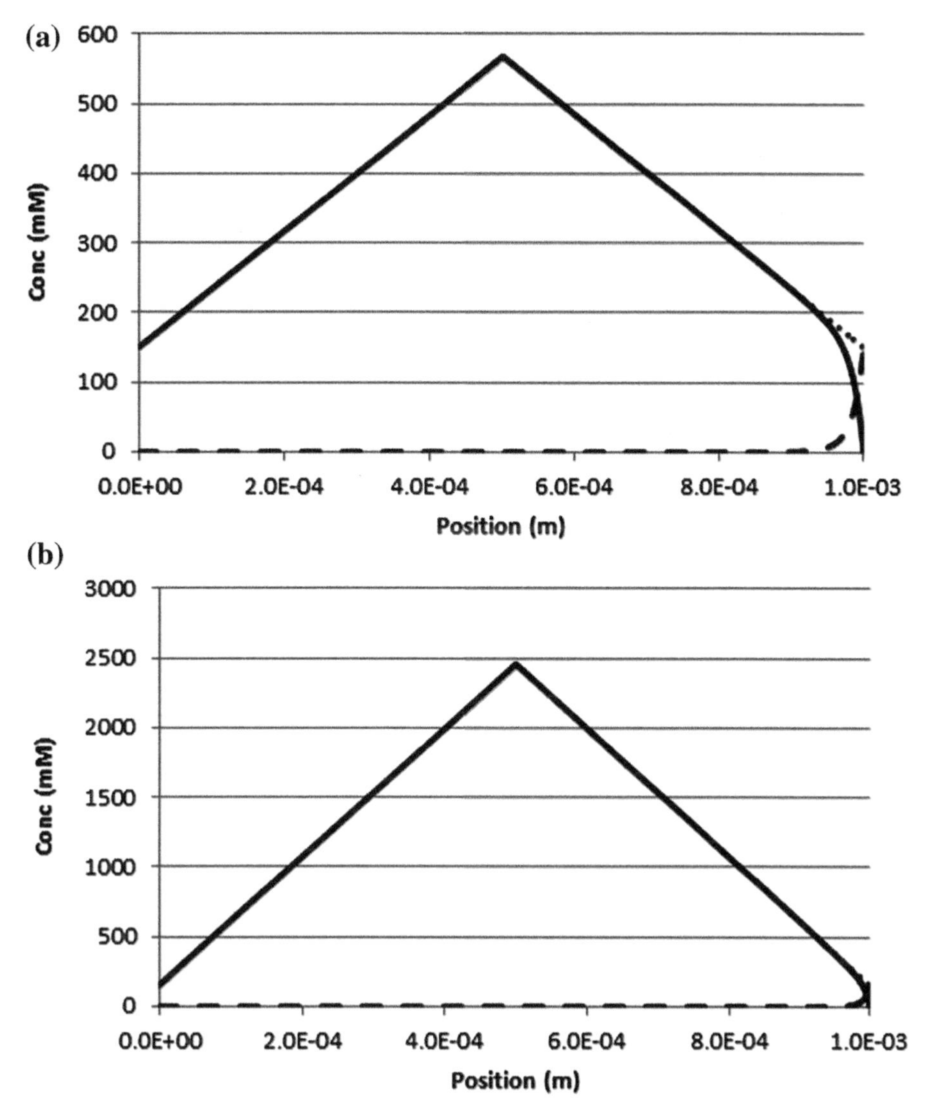

FIGURE 11.9 Effects of membrane porosity of the membranes in the membrane assembly upon the concentration profiles of the ions in the membranes. In Plot (a), membrane porosity of the second membrane is 4x lower than that of the first membrane in the two-membrane assembly. Plot (b) shows a two-membrane assembly with equal membrane porosity of the first and second membranes. Donor: 0.15 M $Drug^+Cl^-$. Receiver: 0.15 M Na^+Cl^-. Electrical potential across the membrane assembly: 0.5 V. Membrane thickness: 0.5 mm each membrane. Solid line: permeant concentration profile (cationic drug from donor); dotted line: counterion concentration profile (Cl^- ion from donor and receiver); dashed line: coion concentration profile (Na^+ ion from receiver). Plot (b) is from Fig. 11.7a.

of the membrane with the higher transference number (the dominant membrane). These results are consistent with the analyses of Fig. 11.9 and the related discussion. Finally, when the respective transference numbers of the two individual membranes approach the same value, the

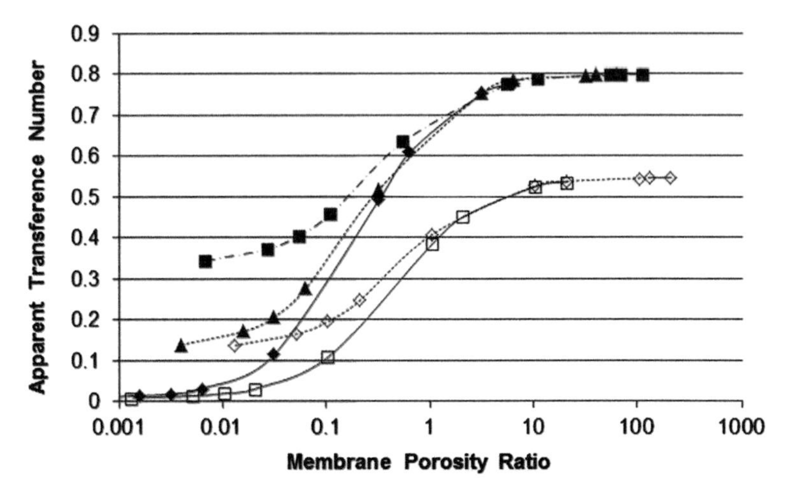

FIGURE 11.10 Relationships between the apparent transference numbers of permeant transport across a two-membrane assembly and the porosity of the membranes in the assembly, with the membrane of lower intrinsic transference number facing the receiver. The total transference number is plotted against the ratio of membrane porosities. Membrane porosity ratio equals the porosity of the lower transference number membrane divided by that of the higher transference number membrane. Symbols and intrinsic transference numbers of the membranes (the first number is the intrinsic transference number of the membrane facing the receiver, the second number is the intrinsic transference number of the membrane facing the donor): solid line with open squares, 0.0069, 0.55; solid line with closed diamonds, 0.0069, 0.80; dotted line with open diamonds, 0.13, 0.55; dotted line with closed triangles, 0.13, 0.80; dash-dot line with closed squares, 0.33, 0.80.

membrane assembly behaves as a single homogenous membrane. It should also be pointed out that changing the membrane thickness of the two membranes in the membrane assembly will have opposite effects as those of membrane porosity: an increase in membrane thickness has the same effect as that of a decrease in membrane porosity.

The concentration of the permeant at the membrane/membrane interface is also related to the applied electric field. Figure 11.11 shows an example of the effects of the applied electric field strength upon the permeant concentration profiles in the membranes. An increase in the electric field (or electric current) will result in an increase in the concentration gradients of the ions in the membranes, i.e., higher diffusion flux, in order to compensate for the increase in flux due to electromigration at the higher electric field. For instance, an increase in electric field leads to a decrease in the concentration at the membrane/membrane interface in the two-membrane assembly in which the intrinsic transference number of the second membrane is higher than that in the first membrane (Fig. 11.11). Conversely, when the intrinsic transference number of the first membrane is higher than that in the second membrane, an increase in electric field will result in an increase in the concentration at the membrane/membrane interface (figure not shown).

The discussion of multiple membrane systems has so far been based on the assumption of a continuous concentration function within a membrane with variable concentration boundary conditions at all the membrane interfaces. In these analyses, the transference number has been defined to account for the contributions of ion diffusion, electromigration and convection to permeant transport (also see Section 11.2) [28]. This differs from the traditional definition of transference number that only accounts for the galvanic current carried by the ion in a

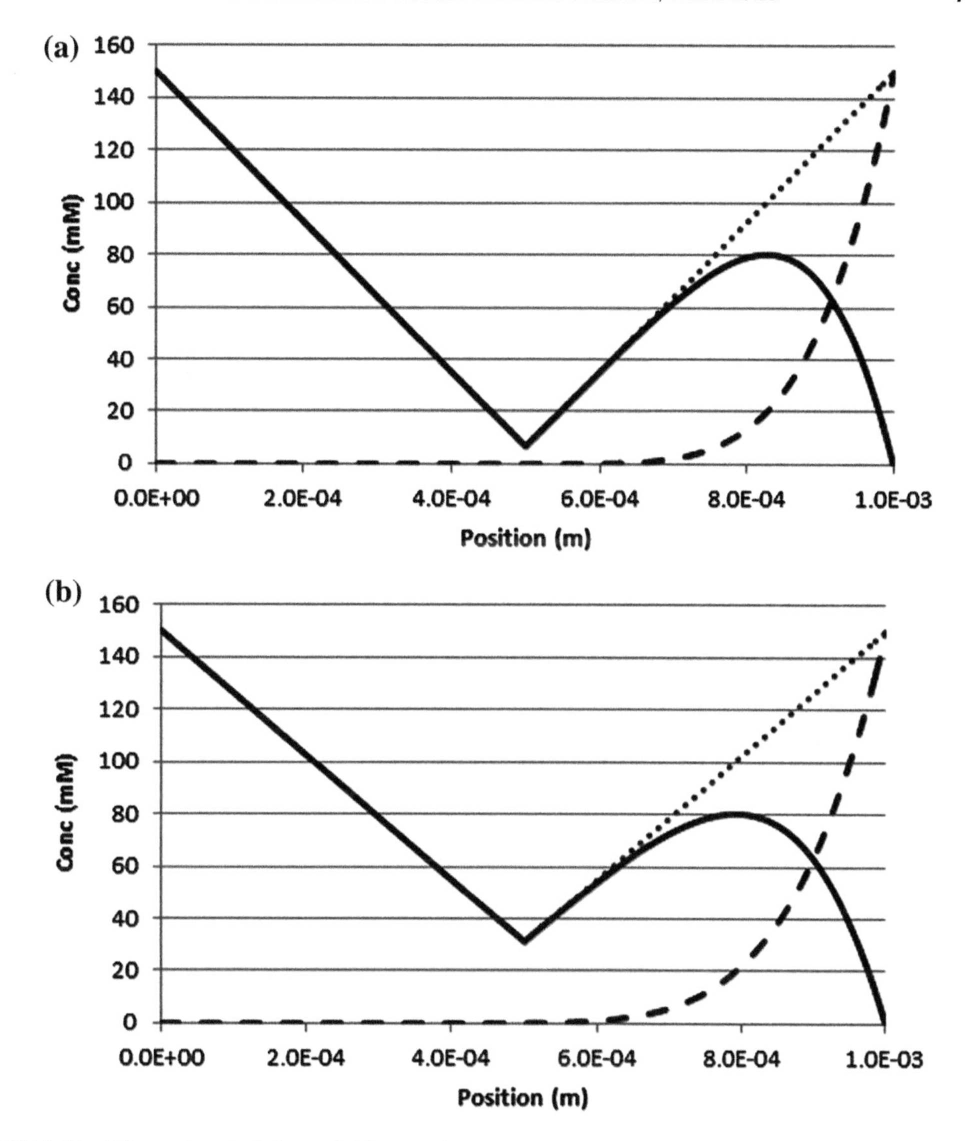

FIGURE 11.11 Effects of applied electric field strength upon the concentration profiles of the ions in the membranes. The applied electric field across the two-membrane assembly in Plot (a) is 2x of that in Plot (b) (1 V and 0.5 V, respectively). Donor: 0.15 M $Drug^+Cl^-$. Receiver: 0.15 M Na^+Cl^-. Membrane thickness: 0.5 mm each membrane. Solid line: permeant concentration profile (cationic drug from donor); dotted line: counterion concentration profile (Cl^- ion from donor and receiver); dashed line: coion concentration profile (Na^+ ion from receiver). Plot (b) is from Fig. 11.7b.

pure conduction process. The traditionally defined transference number has been used to develop an alternative description of permeant transport across multiple membrane systems that relies on transference number discontinuities at the membrane/membrane (or membrane/solution) interface [29]. While this topic will not be discussed in this chapter, readers are referred to the reference for detailed discussion.

11.6 ELECTROTRANSPORT UNDER ALTERNATING CURRENT

Electric field-enhanced drug delivery across biological membranes generally employs the direct current (DC) approach. However, alternating current (AC) has also been investigated for electrotransport in drug delivery. The advantages of AC iontophoresis are as follows. DC iontophoresis alters the electrochemical environment of the solution surrounding the electrodes, leading to electrochemical burns on tissues during long iontophoresis applications or when an inappropriate electrode design is used. AC iontophoresis has been suggested as an alternative to avoid these changes in the solution surrounding the electrodes. In addition, it has been suggested that AC alone, and combinations of DC and AC pulses, can cause less tissue irritation (e.g., skin irritation) compared to DC iontophoresis alone [12]. AC at high frequency also has higher threshold of sensation than DC [30–32] and thus may provide better safety profiles.

The effects of symmetric alternating electric fields (e.g., square-wave AC without DC offset) upon the flux of a charged permeant across a homogenous membrane were investigated under the assumption of Nernst-Planck flux behavior with the Goldman approximation [33]. Other assumptions include an inert membrane whose structure is not altered by the electric field, and a membrane electrical capacitance that is insufficient to cause significant distortion of the resultant AC electric field. The approximate equation for the flux enhancement due to square-wave AC over passive diffusion is:

$$E_{\Delta\psi AC,i} = \frac{1+0.18\sqrt{K_{\Delta\psi,i}/\varphi^3}}{1+1.4\sqrt{K_{\Delta\psi,i}}/\varphi^{14}} + \frac{K_{\Delta\psi,i}}{2(1+0.25\varphi^{10})(1+\varphi+0.2(K_{\Delta\psi,i})^{0.7}\varphi^3)} \tag{11.21}$$

where:

$$\varphi = \frac{f}{f^*} = \frac{2f\Delta x^2}{D_i K_{\Delta\psi,i}}$$

and f is the frequency of the AC. The parameter f^* is defined as the characteristic frequency, in which the value of $0.5/f^*$ is related to the time for the permeant to transport across the entire membrane barrier in one AC half-cycle. This equation describes the relationship between AC frequency and permeant flux under quasi-steady state.

Figure 11.12 presents a plot of the relationship between flux enhancement and AC frequency as described by Eq. (11.21). As can be seen in the figure, low frequency AC electrotransport will result in a maximum flux that is equivalent to half the flux enhancement corresponding to DC electrotransport, i.e., at low AC frequency when $\varphi << 1$, $E_{\Delta\psi AC,i} \approx K_{\Delta\psi,i}/2$ or $\approx z_i F\Delta\psi/(2RT)$. The square-wave AC field is essentially acting as 50% of a DC field to enhance permeant transport across the membrane. In other words, when the period of half an AC cycle provides sufficient time for the permeant to travel across the entire membrane under the electric field during the cycle, maximum utilization of the symmetric AC to assist permeant transport is attained. When the AC frequency increases, flux enhancement decreases. At high AC frequency, the half-cycle time period is too short to enhance permeant movement along the direction of the electric field and the quasi-steady state flux approaches that of passive diffusion.

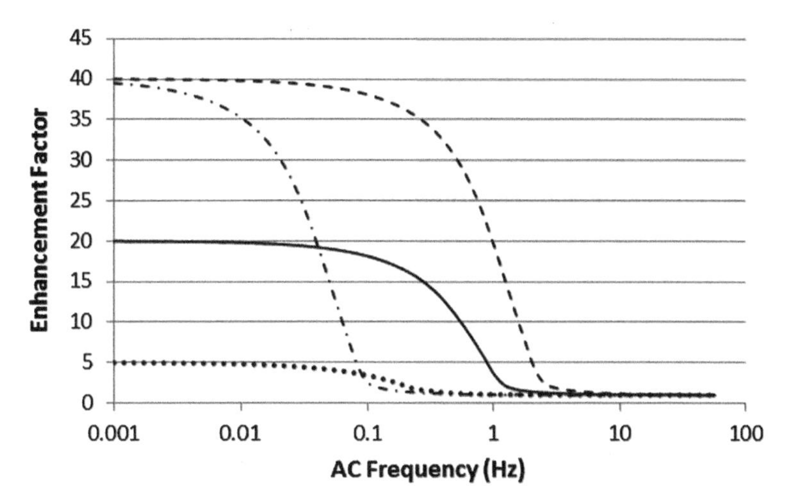

FIGURE 11.12 Relationships between flux enhancement and AC frequency (Eq. (11.21)). Parameters used are: dotted line, 0.25 V AC with membrane thickness of 0.1 mm; solid line, 1.0 V AC with membrane thickness of 0.1 mm; dashed line, 2.0 V AC with membrane thickness of 0.1 mm; dashed dotted line, 2.0 V AC with membrane thickness of 0.5 mm.

It should be pointed out that the derivation of Eq. (11.21) is based on the assumption that the membrane is homogenous. For membranes that are not homogeneous, flux enhancement is expected to deviate from the predictions of this equation. Membrane properties that generally do not affect steady state flux in DC electrotransport can have a significant impact on electrotransport under AC. For example, the loading of a permeant in the reservoirs in a membrane and/or permeant-to-membrane binding can significantly affect AC electrotransport [34,35]. This is due to the dynamic nature of AC electrotransport, which is always under a transit state; the flux in AC electrotransport does not reach steady state and can only be described as quasi-steady state. As a result, there are strong relationships between the AC half-cycle time period, membrane transport lag time and transmembrane flux of AC electrotransport. In general, membrane reservoirs and membrane binding can increase the characteristic 'effective pathlength' of transport across the membrane and decrease the characteristic AC frequency required to achieve maximum flux. Membrane reservoirs also increase the flux enhancement at high AC frequency due to the short transport pathway between the reservoirs (relative to that of the transport pathway across the whole membrane). Therefore, the behavior of AC electrotransport can be complicated for biological membranes as these membranes are typically not homogenous.

11.7 ELECTROPERMEABILIZATION EFFECT

Iontophoresis generally operates by the mechanisms of electrophoresis, electroosmosis and electropermeabilization (or electroporation). Electropermeabilization alters the barrier of a membrane, increasing its intrinsic permeability, and is a major mechanism of iontophoretically enhanced transport for some biological membranes. The mechanisms of membrane electropermeabilization for some biological membranes such as skin are not completely understood. It is generally believed that electropermeabilization is related to the electroporation of lipid bilayer (or lamellae) at the microscopic level. Electroporation is defined as a technique to increase cell membrane permeability to hydrophilic molecules under an applied electric field. The term electroporation is also commonly used to describe the phenomena of electrically induced pore formation (pore induction) in biological membranes. This definition is adopted in this chapter.

Electroporation is the spontaneous formation of a metastable pore in a lipid bilayer due to the energy of an electric field. The minimum threshold voltage to induce electroporation in a lipid bilayer is around 0.2 to 1.0 V per bilayer depending on the bilayer composition. The subject of biomembrane electroporation has been studied and reviewed in the literature. For example, the rates of pore formation (K_f) and destruction (K_d) have been derived [36–38]. K_f can be described by:

$$K_f = \nu_c \exp\left(-\frac{\delta_c - aU^2}{kT}\right) \tag{11.22}$$

where ν_c is the pore creation rate prefactor that is related to the frequency of lateral fluctuations of lipid molecules, k is the Boltzmann constant, δ_c is the energy barrier constant for pore formation, U is the electrical potential and:

$$a = \frac{\pi R_c^2(\varepsilon_w - \varepsilon_m)}{2h} \tag{11.23}$$

where R_c is the critical pore radius, h is pore length, ε_w is dielectric constant of water and ε_m is dielectric constant of the membrane. The rate of pore destruction (K_d) or resealing can be described by:

$$K_d = b_d N \exp\left(-\frac{\delta_d}{kT}\right) \tag{11.24}$$

where N is the pore probability density function, b_d is pore destruction rate prefactor and δ_d is the pore destruction energy barrier constant. According to Eq. (11.22), the rate of pore formation during electroporation increases with the electrical potential across the membrane. The extent of electropermeabilization, which is a function of the rates of pore formation and destruction during iontophoresis, depends on the applied electric field, the properties of the biomembrane and the number of metastable pores in the membrane. Electrotransport during iontophoresis reaches a pseudo steady state when the rate of pore destruction approaches the rate of pore formation. From a macroscopic viewpoint, the pseudo steady state flux of an ionic species across a membrane under electropermeabilization can be described by modifying the porosity term in Eq. (11.7) to take into the account of the newly formed pore pathway in the membrane:

$$J_i = \varepsilon' \left\{ -H_i D_i \left[\frac{dC_i}{dx} + \frac{z_i F C_i}{RT} \frac{d\psi}{dx} \right] \pm W_i C_i v \right\} \tag{11.25}$$

where ε' is the effective porosity of the membrane as a result of electropermeabilization.

Equation (11.25) has been used to analyze electrotransport across electropermeabilized membranes such as skin [39]. The effects of electropermeabilization (or electroporation) will not be discussed here in detail, as this topic is discussed in Chapters 12 and 13.

11.8 ELECTROKINETIC METHODS OF ENHANCED TRANSPORT ACROSS BIOLOGICAL MEMBRANES

Parameters that affect electrotransport across biological membranes can be divided into three main categories: properties of the biological membrane, permeant and solution medium. Table 11.1 lists the different factors that affect electrotransport across biological membranes. The barrier properties of a biological membrane that affect iontophoretic transport include the pore size of the transport pathway in the membrane and the pore charge density of the pathway. In general, membrane porosity and thickness do not affect iontophoretic transport when iontophoresis is operated under the constant current principle. During constant current DC iontophoresis, the electric current is kept constant by controlling the voltage across the membrane for the duration of the iontophoresis application. For example, when the electrical resistance of the tissue decreases during iontophoresis due to an increase in membrane porosity (e.g., electropermeabilization) or a decrease in the effective thickness, the applied voltage across the membrane will decrease and the iontophoretic flux will remain the same (total ion fluxes remain the same at the same electric current).

In practice, the main parameter that indicates the effectiveness of drug delivery across a membrane during constant current iontophoresis is the transference number of the drug according to Eq. (11.5). The transference number of a permeant for a biological membrane is a function of the effective pore size, the charge on the membrane, the physicochemical

TABLE 11.1 Factors That Affect Electrotransport Across Biological Membranes

Factor	Parameter(s) Being Affected
Membrane property	
Membrane pore size	Effective electrophoretic mobility of ions
Membrane pore charge	Permeant-to-membrane partitioning, electroosmosis
Membrane porosity	Membrane electrical resistance
Membrane thickness	Membrane electrical resistance
Permeant property	
Permeant charge	Permeant effective electrophoretic mobility
Permeant size	Permeant effective electrophoretic mobility
Permeant solubility	Highest permeant concentration that can be used
Solution medium property	
Drug concentration	Permeant flux in the presence of background electrolyte
Solution ionic strength	Permeant-to-membrane partitioning, electroosmosis
Solution pH	Fraction of permeant in ionized form, membrane charges
Solution ion composition	Ionic strength, ion competition for electric current
Cosolvent in solution	Ion hydrated radius, medium viscosity, medium dielectric constant

properties of the permeant and the properties of the donor solution. Permeant physicochemical properties which affect electrotransport include the molecular size of the permeant and the charge on the permeant (i.e., electrophoretic mobility). Physical parameters of the solution medium that affect the transference number during iontophoresis include solution pH, permeant concentration and ion composition of the solution in the donor chamber.

The following sections provide a brief review of iontophoretic transport for transdermal, transungual and transscleral drug delivery. Electrotransport across the skin, nail and sclera is discussed from two perspectives: (a) the extent to which the barrier properties of these membranes and physical properties of the solution medium affect iontophoretic transport and (b) the mechanisms of iontophoresis in the context of the electrotransport theories discussed in Sections 11.2. to 11.7. Drug delivery applications of iontophoresis for these administration routes are also reviewed.

11.8.1 Transdermal Iontophoresis

Transdermal iontophoresis is the electrically enhanced transport of molecules across the skin (in particular across the stratum corneum) by an applied electric field. The stratum corneum is the uppermost layer of the epidermis and consists of 10–15 layers of corneocytes embedded in intercellular lipids. This barrier is usually negatively charged, with an isoelectric point of $pI \sim 4$. It has been shown that iontophoresis can effectively enhance the delivery of small molecules such as fentanyl and lidocaine across skin [40,41]. It is also believed that iontophoresis is a promising technology to deliver large molecules such as bioactive peptides and proteins to the body. Iontophoresis can also be used to extract endogenous molecules or drugs from the extracellular fluid in the tissues under the skin, i.e., reverse iontophoresis, for clinical applications such as diagnosis and monitoring of analyte concentration in blood [42–46]. The technology utilizing electrotransport to enhance drug delivery across the skin has been extensively studied since the 1980s [47–49]. For example, transdermal iontophoresis has been examined under various conditions, such as symmetric and asymmetric conditions, constant or alternating current application and constant voltage application. The modified Nernst-Planck (Eq. (11.1)) and transference number (Eq. (11.14)) equations have been used to model transdermal iontophoresis and predict iontophoretic transport enhancement.

In transdermal iontophoresis, the flux enhancing mechanisms include electrophoresis (i.e., the direct interaction of the electric field with the charge of the ionic permeant), electroosmosis (i.e., the convective solvent flow in the preexisting and/or newly created pathways) and electroporation (i.e., electric field induced pore induction). Note that the term electroporation is sometimes used exclusively to describe the technique of high voltage short pulses to permeabilize skin in transdermal delivery [50], which is different from the definition in this chapter. For charged permeants, electrophoresis is usually the dominant transport mechanism and electroosmosis is a secondary mechanism. An exception to this is iontophoresis of macromolecules with low molecular charge-to-mass ratios (i.e., macromolecules with low electrophoretic mobilities). For neutral permeants, electroosmosis is the primary mechanism for enhancing transdermal transport during iontophoresis. Electroporation contributes to transdermal iontophoretic flux enhancement when a moderate to high voltage is applied across the skin.

Previous studies have shown that flux enhancement of monovalent ions across skin during constant low voltage iontophoresis can be predicted by the modified Nernst-Planck equation, e.g., Eq. (11.8) [11]. For transdermal iontophoretic transport of monovalent ions at moderate to high electrical potential in which significant electroporation occurs, the experimental results are in good agreement with the predicted values from the modified Nernst-Planck equation when electroporation is quantified by the change in skin electrical resistance when the background electrolyte ion sizes are the same as those of the permeant ions [39]. However, for molecules of multiple charges ($|z_i| > 1$), a discrepancy between the experimental and theoretical values was observed that is attributed to the problem of using the Einstein relation between electromobilities and diffusion coefficients (see descriptions of Eq. (11.2)) to calculate the electrophoretic mobilities of these molecules in practice (in non-ideal solution) [11]. In this case, an approach using the experimentally determined electromobilities of the permeants to predict flux enhancement and transference number of transdermal iontophoresis is more appropriate. For example, transdermal iontophoretic transport enhancement was shown to be consistent with predictions using the experimental electromobilities of the permeants that were determined by capillary electrophoresis [5]. The experimentally obtained electromobilities of ionic permeants can be used to identify suitable drugs for iontophoretic delivery and to understand the contribution of electrophoresis in transdermal iontophoretic transport [51].

According to Eq. (11.19), the effect of electrososmosis is highly dependent on the pore charge density of a membrane [26,52]. At the physiological pH value, the skin carries a net negative charge. Therefore, electroosmosis enhances the transport of positively charged ions during anodal iontophoresis and retards the transport of negatively charged ions during cathodal iontophoresis [6,53,54]. In transdermal iontophoresis of neutral molecules, anodal iontophoresis enhances the transport of neutral molecules by electroosmosis. The direction of electroosmotic flow can be reversed at pHs below the isoelectric point of skin [55]. *Peck et al.* studied electroosmotic transport behaviors across a synthetic membrane and skin, and showed that flux enhancement due to electroosmosis is related to the molecular size of the permeant [22]. In particular, the electroosmotic flux enhancement increases with the molecular size of the permeant (see the expression of Pe_i in Eq. (11.18)). This is because the velocity of convective solvent transport is relatively constant for permeants of different molecular sizes but the diffusion coefficients of the permeants decrease with permeant molecular sizes, leading to an increase in the ratio of electroosmotic flux to passive flux (i.e., an increase in flux enhancement) with molecular size.

Because electroosmosis is strongly dependent on the effective pore charge density and the pore size of a membrane, the prediction of transdermal electroosmosis is complicated by electroporation during iontophoresis. In transdermal iontophoresis, the effective pore size and pore charge density can be altered due to pore induction and alteration of the existing pore pathway in the skin [56,57]. These changes can also result in flux variability during transdermal iontophoresis [58]. Flux variability is more significant in the transdermal iontophoresis of neutral permeants, for which electroosmosis is the dominant flux enhancing mechanism, in contrast to charged permeants [59]. By controlling the skin electrical resistance and keeping it constant during AC iontophoresis, which is believed to minimize the changes in the effective pore size and pore charge density in skin, the iontophoretic flux variability of neutral permeants was observed to decrease [58].

Electroporation (or pore induction) is believed to be a major transdermal flux enhancing mechanism during constant current iontophoresis and constant voltage iontophoresis at moderate voltage (e.g., > 1 V) [60]. The effect of pore induction on transdermal flux enhancement has been illustrated previously by the observed decrease in skin electrical resistance during iontophoresis and the correlation between transdermal flux and skin electrical resistance. For example, the enhancement factor due to electropermeabilization can be quantified by the change in skin electrical resistance with excellent agreement between theoretical predictions and experimental data [39,61].

A focus of recent transdermal iontophoresis studies was the evaluation of the effects of ion composition and drug concentration in the donor solution upon transdermal iontophoretic delivery and extraction. These studies have shown that the transference number of an ionic drug across skin is a function of molar fraction and electromobility of the drug in the formulation [62–65]. Higher permeant transference number is observed when the molar fraction of the permeant in the formulation is increased, or with an increase in permeant electrophoretic mobility. In the absence of background electrolyte, there was little effect of donor drug concentration upon drug transference in transdermal iontophoretic transport; i.e., transference number becomes independent of drug concentration in the donor [66,67]. This is consistent with the discussion in Section 11.4. For instance, the transference number of lidocaine increased with increasing molar fraction and reached a maximum value of $t_i \sim 0.2$ in the absence of coions in the donor solution [63]. The maximum transference number of lidocaine obtained in the study was relatively low compared to the transference number of sodium ion ($t_i \sim 0.6$) in transdermal iontophoresis [64]. This is due to the low electrophoretic mobility of lidocaine compared with that of sodium ion and the competition of the highly mobile chloride ion transported in the opposite direction for the electric current. The maximum transference numbers were shown to correlate with the electrophoretic mobilities of the permeants and decrease with increasing molecular radii of the permeants up to about 0.8 nm [64]. At higher molecular radii, iontophoretic transport became negligible. Drug-to-skin interactions that affect the pore charge density of the iontophoretic transport pathway in skin (and hence electroosmosis) could also play a major role in the determination of the maximum transference numbers of permeants in iontophoretic drug delivery [67].

In addition to the effects of formulation factors such as ion composition and drug concentration in the donor, the effects of incorporating an ion-exchange membrane in series with the skin were also investigated [68]. It was hypothesized that an ion-exchange membrane in the iontophoresis system would hinder transdermal transport of counterions from the receiver to the donor and therefore could enhance iontophoretic drug delivery. Up to 4-fold flux enhancement in transdermal delivery of salicylate was achieved with the ion-exchange membrane system when the electrical resistance of skin was low during iontophoresis. This finding is consistent with the prediction that decreasing the skin barrier resistance will increase the transference number of a permeant across a two-membrane assembly consisting of skin and an ion-exchange membrane during iontophoresis, in which the ion-exchange membrane in series with the skin has higher intrinsic transference number for the permeant than the skin (see Section 11.5.). Due to this relationship, a subsequent study attempted to reduce the electrical resistance of skin during iontophoresis using AC and chemical permeation enhancers for this ion-exchange membrane enhancement technology [69].

11.8.2 Transungual Iontophoresis

Nails are composed mainly of highly disulfide-linked keratins which have an isoelectric point of pI ~ 4.9–5.4. As a result, the nail plate behaves like a concentrated hydrogel and is negatively charged at a physiological pH of 7.4. Transungual iontophoresis has been studied to enhance drug delivery into and across the nail plates for the treatment of notorious nail diseases such as onychomycosis. In transungual drug delivery, the main mechanisms of iontophoresis are electrophoresis and electroosmosis. Electroporation might contribute to transungual iontophoretic transport but its effect has not been confirmed. Hindered transport in the nail was observed to be critical in transungual delivery.

In the transungual iontophoretic delivery of charged molecules, electrophoresis is the dominant transport mechanism and the effect of electroosmosis is secondary. This is similar to the findings observed in the skin. Previous studies have examined the enhancement effect of iontophoresis on charged molecules and compared the enhancement factors with predictions from the Nernst-Planck equation (Eq. (11.8)). For example, the average enhancement factor of a positively charged permeant, tetraethylammonium, was predicted to be around 26 using the voltage measured in a 0.1 mA (i.e., 0.16 mA/cm^2) anodal iontophoresis study and the Nernst-Planck theory [70]. This value is in close agreement with the experimental value obtained in the same study. In general, the Nernst-Planck theory provides satisfactory predictions of flux enhancement of ionic permeants during transungual iontophoresis.

Although the contribution of electroosmosis to iontophoresis is relatively small compared to electrophoresis for transungual delivery of ionic permeants, electroosmosis is the main flux enhancing mechanism for transungual delivery of neutral permeants. Similar to transdermal electroosmosis, the direction of the electric current and the net charge of the nail determine the direction of the electroosmotic solvent flow in transungual iontophoresis. *Murthy et al.* first reported the pH-dependent iontophoretic transport of uncharged molecules (glucose and griseofulvine) across nails in the pH range of 3–7 [71]. At pH > 6, the electroosmotic solvent flows from the anode to cathode and enhances the iontophoretic transport of neutral permeants, such as mannitol and urea, through the negatively charged nails. At pH < 4, the electroosmotic solvent flows from the cathode to anode and enhances iontophoretic transport of neutral permeants through the positively charged nails in cathodal iontophoresis.

The effects of electroosmosis in transungual iontophoresis were also quantified using urea and mannitol as the model permeants [72]. In this study, flux enhancement of up to 2-fold was observed for mannitol during 0.1 mA anodal iontophoresis at pH 7.4. Higher iontophoretic flux enhancement was observed at pH 9 due to the higher surface charge density of the transport pathway in the nail at this pH. Decreasing the ionic strength of the solution from 0.7 M to 0.04 M increased the iontophoretic transport of mannitol, and increasing the electric current from 0.1 mA (0.16 mA/cm^2) to 0.3 mA (0.5 mA/cm^2) proportionally increased electroosmotic transport across the nail. These observations are consistent with the electrokinetic theory (Eq. (11.19)) which predicts that varying the surface charge density, ionic strength of the solution and electric field can affect the velocity of electroosmotic solvent flow in the nail, and hence electroosmotic transport during transungual iontophoresis.

Chemical enhancers and formulation factors such as solution pH, ion composition and ionic strength have been investigated and found to have significant impacts on transungual iontophoretic delivery. The effects of chemical enhancers upon transungual drug delivery and iontophoresis have been investigated [73]. Thioglycolic acid was shown to increase the effective pore size of the transport pathway in the nail. This would lead to an increase in the effective mobility of the permeant to a greater extent than those of the coions and/or counterions, an increase in the transference number of the permeant, and enhancement in iontophoresis efficiency during constant current transungual iontophoresis (see Eqs. (11.5) and (11.13)). Particularly, the transference number of a positively charged permeant was increased from 0.15 to 0.26 during 0.1 mA anodal iontophoresis when the nail was pretreated with 0.5 M thioglycolic acid in this study.

The effects of pH upon transungual iontophoresis include the alterations of permeant-to-pore charge interactions in the nail and electroosmosis. *Dutet et al.* showed that the transference number of the sodium ion increased from 0.35 to 0.88 when the pH of the solution increased from pH 4 to 7 under the symmetric condition of equal pH in the donor and receiver during constant current transungual iontophoresis [74]. A similar trend of an increase in the transference number of the sodium ion (from 0.25 to 0.6) was observed when the pH on the dorsal side of the nail increased from pH 3 to 11 during constant current iontophoresis under conditions in which the pH in the ventral side of the nail was always pH 7.4 (i.e., to mimic the *in vivo* situation) [75]. The transference numbers at pH 7.4 *in vitro* are consistent with the *in vivo* transungual transference number of sodium ion of approximately $t_i \sim 0.51$ measured in human subjects [76]. The similarity of the values *in vitro* and *in vivo* suggests that the *in vitro* nail plate can be a model for transungual iontophoretic delivery *in vivo*.

Smith et al. investigated the effects of solution ionic strength upon transungual iontophoretic delivery [77]. It was found that the partitioning of an ionic permeant into the nail can be manipulated by changing the solution ionic strength in the donor and receiver. For example, when the ionic strength was decreased, the partitioning of a positively charged permeant increased due to its interaction with the negatively charged nail at physiological pH. In a study of the effects of solution composition upon transungual iontophoresis, the transference numbers of sodium and lithium ions were found to be linearly related to their respective molar fractions in the donor solutions [74]. Enhanced iontophoretic transport efficiency can therefore be achieved by increasing the molar fraction of a permeant in the donor solution or decreasing the concentration of coions in the donor solution. These effects of ionic strength and composition of the solution media upon transungual iontophoretic transport are consistent with the results observed by *Murthy et al.* on the transport of salicylic acid across the nail [78].

11.8.3 Transscleral Iontophoresis

Transscleral iontophoresis is a noninvasive and promising method for ocular drug delivery [14,79]. Although ocular iontophoresis has been known in the ophthalmic community for many years, it had not been studied extensively until the recent emergence of interest in using this technique for the treatments of posterior eye diseases. This approach is supported by recent studies of drug delivery into the eye following iontophoresis and the safety and efficacy of ocular iontophoresis to treat different eye diseases in animals [80–83] and human clinical studies [84–86].

The sclera is a porous membrane composed of collagens, with an isoelectric point pI ~ 3–4 [8,87]. As a result, the sclera carries a net negative charge under physiological conditions [87,88], similar to the skin and nail plate. Compared to skin and nail, the sclera has relatively large pore size and high porosity. The average effective pore radius of sclera was estimated to be around 10–40 nm [89] and the porosity of the sclera was determined to be greater than 50% [90], which allows the penetration of macromolecules [88,89] and nanoparticles [91]. Due to the large pore size, high porosity and collagen nature of the sclera, its the electrical resistance is relatively low, and tissue alteration during transscleral iontophoresis is generally not significant. This suggests that the sclera has higher tolerance of electric current (or electric fields) than the skin. The tolerance threshold of the sclera to electric current can be as high as 20 mA/cm^2, compared to that of the skin (0.5 mA/cm^2). In transscleral iontophoresis, both electrophoresis and electroosmosis contribute to transport enhancement, and their relative contribution to the iontophoretic flux depends on the molecular sizes and charges of the permeants. Electroporation, which has been observed in transdermal iontophoresis, does not seem to be a significant factor in transscleral iontophoresis.

Although the transscleral delivery of ionic permeants can be enhanced via iontophoresis, the transscleral iontophoretic flux enhancement is relatively small when compared with transdermal iontophoresis at the same level of applied electric current (i.e., current density < 0.5 mA/cm^2). Significant flux enhancement in transscleral iontophoresis is observed only when higher electric current density than those of transdermal iontophoresis is employed. This can be attributed to the low electrical resistance of the sclera relative to skin such that a high electric current is required to provide a sufficient level of electrical potential gradient across the membrane for significant flux enhancement during iontophoresis. For example, iontophoretic flux enhancement of small charged permeants (e.g., tetraethylammonium and salicylate) over their passive transport across human sclera were only about 3- to 5-fold during 2 mA (10 mA/cm^2) transscleral iontophoresis [89]. At current densities of 2.5 mA/cm^2 and higher, the iontophoretic flux enhancement of timolol maleate and dexamethasone phosphate was found to be significant, and the fluxes were a function of the applied current in the range of 0.5 to 2.0 mA (2.5 to 10 mA/cm^2) [92]. A similar relationship between the iontophoretic flux and the applied current was also observed for tetraethylammonium and salicylate in transscleral iontophoresis [87].

Electroosmosis in transscleral iontophoresis is similar to that observed in transdermal and transungual iontophoresis: convective solvent flow is from the anode to the cathode for a negatively charged sclera at pH > 5, and is from the cathode to the anode for a positively charged sclera at pH < 3 [92]. At pH 7.4, the convective solvent flow is from the anode to the cathode in transscleral iontophoresis. In addition, the iontophoretic flux enhancement due to electroosmosis depends on the current density (or electric field), molecular size of permeants and ionic strength of solution according to the electrokinetic theory (Eq. (11.19)). *Nicoli et al.* reported insignificant enhancement of acetaminophen (a small molecule) across the sclera in 1.75 mA (2.9 mA/cm^2) anodal iontophoresis experiments [88]. However, the same iontophoresis condition enhanced the transport of dextran (150 kDa) through the sclera by 4-fold. The same study also suggested that the enhancement factors of dextrans across sclera were proportional to their molecular radii in anodal iontophoresis. Thus, for charged macromolecules, the contribution of electroosmosis to transscleral iontophoretic transport can be significant. Bevacizumab has a molecular weight of 149 kDa and is slightly negatively charged [93]. It was found that transscleral transport enhancement of bevacizumab was larger in anodal iontophoresis than in cathodal iontophoresis at pH 7.4 [8,89]. The enhancement for bevacizumab was 32-fold in anodal iontophoresis when 2-mA (10 mA/cm^2) constant current was applied. *Pescina et al.* reported 7.5-fold enhancement of bevacizumab in anodal iontophoresis (3.8 mA/cm^2)

with no enhancement in cathodal iontophoresis [8]. For negatively charged macromolecules of low molecular charge-to-mass ratios such as bevacizumab, the net iontophoretic flux enhancement is related to the sum of the effects of electroosmosis and electrophoresis, in which electroosmosis and electrophoresis drive the flux in opposite directions with electroosmosis from the anode to cathode being the dominant factor. For highly negative charged macromolecules such as polystyrene sulfonic acid of 67 kDa molecular weight, anodal iontophoresis enhancement due to electroosmosis was present but its effect was smaller than that of electrophoresis, and as a result, significant cathodal iontophoresis enhancement of these macromolecules was expected [89]. These observations are in agreement with theoretical predictions.

The ion composition of the solution in the donor affects ion competition for electric current and the velocity of convective solvent flow, and thus influences the iontophoretic flux enhancement due to electrophoresis and electroosmosis in transscleral iontophoresis. *Li et al.* found that the transference numbers of tetraethylammonium and salicylate across the sclera are associated with ion concentration (ionic strength) of the solution in the donor [87]. The iontophoretic transport of timolol maleate, vancomycin and dexamethasone phosphate across sclera was also observed to be related to permeant concentration and ion composition in the donor, consistent with the basic principles of iontophoresis [92]. The flux of mannitol across rabbit sclera increased when the ionic strength in the donor solution decreased from 1 to 0.015 M in anodal iontophoresis [87]. Similar results of formulation parameters such as solution ionic strength upon transscleral iontophoresis of neutral hydrophilic molecules were also observed [94].

Using an ion-exchange membrane assembled in series with the sclera, in which the ion-exchange membrane hinders the transport of the competing counterions and selectively allows the transport of the permeant across the sclera, the ion-exchange membrane can increase the transference number of the permeant in transscleral iontophoretic delivery. In a feasibility study of this method, the steady state flux of salicylate was increased by 3-fold with a positively charged ion-exchange membrane during transscleral iontophoresis; the transference number of salicylate was increased from 0.22 to 0.67 [95]. Similarly, the transference number of tetraethylammonium across the sclera was increased from 0.38 to 0.83 using a negatively charged ion-exchange membrane [28]. Such enhancement was related to the thickness and permeability coefficient of the ion-exchange membrane as well as the electromobility of the permeant across the individual membranes in the membrane assembly. This is consistent with the theoretical predictions of iontophoretic transport across a multiple membrane system as discussed in Section 11.5.

References

[1] A.J. Bard, L.R. Faulkner, Electrochemical Methods: Fundamentals and Applications, first ed., Wiley, New York, 1980.

[2] J.O.M. Bockris, A.K.N. Reddy, Modern Electrochemistry, Plenum, New York, 1970.

[3] J.T.G. Overbeek, Quantitative interpretation of the electrophoretic velocity of colloids, in: H. Mark, E.J.W. Verwey (Eds.), Advances in Colloid Science, Interscience, New York, 1950, pp. 97–134.

[4] R.A. Robinson, R.H. Stokes, Electrolyte Solutions, Academic Press, New York, 1955.

[5] S.K. Li, A.H. Ghanem, C.L. Teng, G.E. Hardee, W.I. Higuchi, Iontophoretic transport of oligonucleotides across human epidermal membrane: a study of the Nernst-Planck model, Journal of Pharmaceutical Sciences 90 (2001) 915–931.

[6] S.M. Sims, W.I. Higuchi, V. Srinivasan, Skin alteration and convective solvent flow effects during iontophoresis. II. Monovalent anion and cation transport across human skin, Pharmaceutical Research 9 (1992) 1402–1409.

[7] A.B. Nair, S.R. Vaka, S.N. Murthy, Transungual delivery of terbinafine by iontophoresis in onychomycotic nails, Drug Development and Industrial Pharmacy 37 (2011) 1253–1258.
[8] S. Pescina, G. Ferrari, P. Govoni, C. Macaluso, C. Padula, P. Santi, S. Nicoli, *In vitro* permeation of bevacizumab through human sclera: effect of iontophoresis application, Journal of Pharmacy and Pharmacology 62 (2010) 1189–1194.
[9] M. Leonard, E. Creed, D. Brayden, A.W. Baird, Iontophoresis-enhanced absorptive flux of polar molecules across intestinal tissue *in vitro*, Pharmaceutical Research 17 (2000) 476–478.
[10] M. Leonard, E. Creed, D. Brayden, A.W. Baird, Evaluation of the Caco-2 monolayer as a model epithelium for iontophoretic transport, Pharmaceutical Research 17 (2000) 1181–1188.
[11] S.K. Li, A.H. Ghanem, K.D. Peck, W.I. Higuchi, Iontophoretic transport across a synthetic membrane and human epidermal membrane: a study of the effects of permeant charge, Journal of Pharmaceutical Sciences 86 (1997) 680–689.
[12] S.K. Li, W.I. Higuchi, H. Zhu, S.E. Kern, D.J. Miller, M.S. Hastings, *In vitro* and *in vivo* comparisons of constant resistance AC iontophoresis and DC iontophoresis, Journal of Controlled Release 91 (2003) 327–343.
[13] M.B. Delgado-Charro, Iontophoretic drug delivery across the nail, Expert Opinion on Drug Delivery 9 (2012) 91–103.
[14] E. Eljarrat-Binstock, A.J. Domb, Iontophoresis: a non-invasive ocular drug delivery, Journal of Controlled Release 110 (2006) 479–489.
[15] M.S. Roberts, P.M. Lai, S.E. Cross, N.H. Yoshida, Solute structure as a determinant of iontophoretic transport, in: R.O. Potts, R.H. Guy (Eds.), Mechanisms of Transdermal Drug Delivery, Marcel Dekker, New York, 1997.
[16] B.H. Sage, Iontophoresis, in: E.W. Smith, H.I. Maibach (Eds.), Percutaneous Penetration Enhancers, CRC Press, Boca Raton, 1995.
[17] S.K. Li, Transdermal delivery: technologies, in: J. Swarbrick (Ed.), Encyclopedia of Pharmaceutical Technology, Informa Healthcare, New York, 2007, pp. 3843–3853.
[18] A.K. Banga, Electrically Assisted Transdermal and Topical Drug Delivery, Taylor and Francis, Bristol, PA, 1998.
[19] J. Manzanares, K. Kontturi, Transport numbers of ions in charged membrane systems, in: T.S. Sorensen (Ed.), Surface Chemistry and Electrochemistry of Membranes, Marcel Dekker, New York, 1999.
[20] W.M. Deen, Hindered transport of large molecules in liquid-filled pores, AIChE Journal 33 (1987) 1409–1425.
[21] G. Kasting, Theoretical models for iontophoretic delivery, Advanced Drug Delivery Reviews 9 (1992) 177–199.
[22] K.D. Peck, V. Srinivasan, S.K. Li, W.I. Higuchi, A.H. Ghanem, Quantitative description of the effect of molecular size upon electroosmotic flux enhancement during iontophoresis for a synthetic membrane and human epidermal membrane, Journal of Pharmaceutical Sciences 85 (1996) 781–788.
[23] S.M. Sims, W.I. Higuchi, V. Srinivasan, Interaction of electric field and electro-osmotic effects in determining iontophoretic enhancement of anions and cations, International Journal of Pharmaceutics 77 (1991) 107–118.
[24] P. Berg, K. Ladipo, Exact solution of an electro-osmotic flow problem in a cylindrical channel of polymer electrolyte membranes, Proceedings of the Royal Society A 465 (2009) 2663–2679.
[25] C.L. Rice, R. Whitehead, Electrokinetic flow in a narrow cylindrical capillary, Journal of Chemical Physics 69 (1965) 4017–4024.
[26] S.M. Sims, W.I. Higuchi, V. Srinivasan, K.D. Peck, Ionic partition coefficients and electroosmotic flow in cylindrical pores: comparison of the predictions of the Poisson-Boltzmann equation with experiment, Journal of Colloid and Interface Science 155 (1993) 210–220.
[27] G.B. Kasting, J.C. Keister, Application of electrodiffusion theory for a homogeneous membrane to iontophoretic transport through skin, Journal of Pharmaceutical Sciences 8 (1989) 195–210.
[28] S.A. Molokhia, Y. Zhang, W.I. Higuchi, S.K. Li, Iontophoretic transport across a multiple membrane system, Journal of Pharmaceutical Sciences 97 (2008) 490–505.
[29] P.H. Barry, Derivation of unstirred-layer transport number equations from the Nernst-Planck flux equations, Biophysical Journal 74 (1998) 2903–2905.
[30] C.F. Dalziel, T.H. Mansfield, Effect of frequency on perception currents, AIEE Trans 69 (1950) 1162–1168.
[31] C.F. Dalziel, F.P. Massoglia, Let-go currents and voltages, AIEE Trans 75 (1956) 49–56.
[32] K. Okabe, H. Yamaguchi, Y. Kawai, New iontophoretic transdermal administration of the beta-blocker metoprolol, Journal of Controlled Release 4 (1986) 79–85.
[33] T.R. Mollee, Y.G. Anissimov, M.S. Roberts, Periodic electric field enhanced transport through membranes, Journal of Membrane Science 278 (2006) 290–300.

[34] G. Yan, S.K. Li, K.D. Peck, H. Zhu, W.I. Higuchi, Quantitative study of electrophoretic and electroosmotic enhancement during alternating current iontophoresis across synthetic membranes, Journal of Pharmaceutical Sciences 93 (2004) 2895–2908.
[35] G. Yan, Q. Xu, Y.G. Anissimov, J. Hao, W.I. Higuchi, S.K. Li, Alternating current (AC) iontophoretic transport across human epidermal membrane: effects of AC frequency and amplitude, Pharmaceutical Research 25 (2008) 616–624.
[36] S.A. Freeman, M.A. Wang, J.C. Weaver, Theory of electroporation of planar bilayer membranes: predictions of the aqueous area, change in capacitance, and pore-pore separation, Biophysical Journal 67 (1994) 42–56.
[37] R.W. Glaser, S.L. Leikin, L.V. Chernomordik, V.F. Pastushenko, A.I. Sokirko, Reversible electrical breakdown of lipid bilayers: formation and evolution of pores, Biochimica Biophysica Acta 940 (1988) 275–287.
[38] G. Saulis, Kinetics of pore disappearance in a cell after electroporation, Biomedical Sciences Instrumentation 35 (1999) 409–414.
[39] H. Zhu, K.D. Peck, S.K. Li, A.H. Ghanem, W.I. Higuchi, Quantification of pore induction in human epidermal membrane during iontophoresis: the importance of background electrolyte selection, Journal of Pharmaceutical Sciences 90 (2001) 932–942.
[40] J.E. Chelly, J. Grass, T.W. Houseman, H. Minkowitz, A. Pue, The safety and efficacy of a fentanyl patient-controlled transdermal system for acute postoperative analgesia: a multicenter, placebo-controlled trial, Anesthesia and Analgesia 98 (2004) 427–433.
[41] J.A. Subramony, A. Sharma, J.B. Phipps, Microprocessor controlled transdermal drug delivery, International Journal of Pharmaceutics 317 (2006) 1–6.
[42] C.C. Bouissou, J.P. Sylvestre, R.H. Guy, M.B. Delgado-Charro, Reverse iontophoresis of amino acids: identification and separation of stratum corneum and subdermal sources *in vitro*, Pharmaceutical Research 26 (2009) 2630–2638.
[43] M.B. Delgado-Charro, R.H. Guy, Transdermal reverse iontophoresis of valproate: a noninvasive method for therapeutic drug monitoring, Pharmaceutical Research 20 (2003) 1508–1513.
[44] B. Leboulanger, M. Fathi, R.H. Guy, M.B. Delgado-Charro, Reverse iontophoresis as a noninvasive tool for lithium monitoring and pharmacokinetic profiling, Pharmaceutical Research 21 (2004) 1214–1222.
[45] S. Nixon, A. Sieg, M.B. Delgado-Charro, R.H. Guy, Reverse iontophoresis of L-lactate: *in vitro* and *in vivo* studies, Journal of Pharmaceutical Sciences 96 (2007) 3457–3465.
[46] R.O. Potts, J.A. Tamada, M.J. Tierney, Glucose monitoring by reverse iontophoresis, Diabetes Metabolism Research and Reviews 18 (Suppl 1) (2002) S49–53.
[47] J. Singh, M.S. Roberts, Transdermal delivery of drugs by iontophoresis: a review, Drug Design and Delivery 4 (1989) 1–12.
[48] R.R. Burnette, B. Ongpipattanakul, Characterization of the permselective properties of excised human skin during iontophoresis, Journal of Pharmaceutical Sciences 76 (1987) 765–773.
[49] J.C. Keister, G. Kasting, Ionic mass transport through a homogeneous membrane in the presence of a uniform electric field, Journal of Membrane Science 29 (1986) 155–167.
[50] A.K. Banga, S. Bose, T.K. Ghosh, Iontophoresis and electroporation: comparisons and contrasts, International Journal of Pharmaceutics 179 (1999) 1–19.
[51] N. Abla, L. Geiser, M. Mirgaldi, A. Naik, J.L. Veuthey, R.H. Guy, Y.N. Kalia, Capillary zone electrophoresis for the estimation of transdermal iontophoretic mobility, Journal of Pharmaceutical Sciences 94 (2005) 2667–2675.
[52] K.D. Peck, J. Hsu, S.K. Li, A.H. Ghanem, W.I. Higuchi, Flux enhancement effects of ionic surfactants upon passive and electroosmotic transdermal transport, Journal of Pharmaceutical Sciences 87 (1998) 1161–1169.
[53] M.J. Pikal, The role of electroosmotic flow in transdermal iontophoresis, Advanced Drug Delivery Reviews 46 (2001) 281–305.
[54] M.B. Delgado-Charro, R.H. Guy, Characterization of convective solvent flow during iontophoresis, Pharmaceutical Research 11 (1994) 929–935.
[55] B.D. Bath, H.S. White, E.R. Scott, Visualization and analysis of electroosmotic flow in hairless mouse skin, Pharmaceutical Research 17 (2000) 471–475.
[56] S.K. Li, A.H. Ghanem, W.I. Higuchi, Pore charge distribution considerations in human epidermal membrane electroosmosis, Journal of Pharmaceutical Sciences 88 (1999) 1044–1049.
[57] S.K. Li, A.H. Ghanem, K.D. Peck, W.I. Higuchi, Characterization of the transport pathways induced during low to moderate voltage iontophoresis in human epidermal membrane, Journal of Pharmaceutical Sciences 87 (1998) 40–48.

[58] H. Zhu, S.K. Li, K.D. Peck, D.J. Miller, W.I. Higuchi, Improvement on conventional constant current DC iontophoresis: a study using constant conductance AC iontophoresis, Journal of Controlled Release 82 (2002) 249–261.

[59] S.K. Li, W.I. Higuchi, R.P. Kochambilli, H. Zhu, Mechanistic studies of flux variability of neutral and ionic permeants during constant current DC iontophoresis with human epidermal membrane, International Journal of Pharmaceutics 273 (2004) 9–22.

[60] H. Inada, A.H. Ghanem, W.I. Higuchi, Studies on the effects of applied voltage and duration on human epidermal membrane alteration/recovery and the resultant effects upon iontophoresis, Pharmaceutical Research 11 (1994) 687–697.

[61] S.K. Li, A.H. Ghanem, K.D. Peck, W.I. Higuchi, Pore induction in human epidermal membrane during low to moderate voltage iontophoresis: a study using AC iontophoresis, Journal of Pharmaceutical Sciences 88 (1999) 419–427.

[62] B. Mudry, R.H. Guy, M. Begona Delgado-Charro, Prediction of iontophoretic transport across the skin, Journal of Controlled Release 111 (2006) 362–367.

[63] D. Marro, Y.N. Kalia, M.B. Delgado-Charro, R.H. Guy, Optimizing iontophoretic drug delivery: identification and distribution of the charge-carrying species, Pharmaceutical Research 18 (2001) 1709–1713.

[64] B. Mudry, P.A. Carrupt, R.H. Guy, M.B. Delgado-Charro, Quantitative structure-permeation relationship for iontophoretic transport across the skin, Journal of Controlled Release 122 (2007) 165–172.

[65] J.B. Phipps, J.R. Gyory, Transdermal ion migration, Advanced Drug Delivery Reviews 9 (1992) 137–176.

[66] A. Luzardo-Alvarez, M.B. Delgado-Charro, J. Blanco-Mendez, Iontophoretic delivery of ropinirole hydrochloride: effect of current density and vehicle formulation, Pharmaceutical Research 18 (2001) 1714–1720.

[67] D. Marro, Y.N. Kalia, M.B. Delgado-Charro, R.H. Guy, Contributions of electromigration and electroosmosis to iontophoretic drug delivery, Pharmaceutical Research 18 (2001) 1701–1708.

[68] Q. Xu, S.A. Ibrahim, W.I. Higuchi, S.K. Li, Ion-exchange membrane assisted transdermal iontophoretic delivery of salicylate and acyclovir, International Journal of Pharmaceutics 369 (2009) 105–113.

[69] Q. Xu, R.P. Kochambilli, Y. Song, J. Hao, W.I. Higuchi, S.K. Li, Effects of alternating current frequency and permeation enhancers upon human epidermal membrane, International Journal of Pharmaceutics 372 (2009) 24–32.

[70] J. Hao, S.K. Li, Transungual iontophoretic transport of polar neutral and positively charged model permeants: effects of electrophoresis and electroosmosis, Journal of Pharmaceutical Sciences 97 (2008) 893–905.

[71] S.N. Murthy, D.C. Waddell, H.N. Shivakumar, A. Balaji, C.P. Bowers, Iontophoretic permselective property of human nail, Journal of Dermatological Science 46 (2007) 150–152.

[72] J. Hao, S.K. Li, Mechanistic study of electroosmotic transport across hydrated nail plates: effects of pH and ionic strength, Journal of Pharmaceutical Sciences 97 (2008) 5186–5197.

[73] J. Hao, K.A. Smith, S.K. Li, Chemical method to enhance transungual transport and iontophoresis efficiency, International Journal of Pharmaceutics 357 (2008) 61–69.

[74] J. Dutet, M.B. Delgado-Charro, Transungual iontophoresis of lithium and sodium: effect of pH and co-ion competition on cationic transport numbers, Journal of Controlled Release 144 (2010) 168–174.

[75] K.A. Smith, J. Hao, S.K. Li, Influence of pH on transungual passive and iontophoretic transport, Journal of Pharmaceutical Sciences 99 (2010) 1955–1967.

[76] J. Dutet, M.B. Delgado-Charro, *In vivo* transungual iontophoresis: effect of DC current application on ionic transport and on transonychial water loss, Journal of Controlled Release 140 (2009) 117–125.

[77] K.A. Smith, J. Hao, S.K. Li, Effects of ionic strength on passive and iontophoretic transport of cationic permeant across human nail, Pharmaceutical Research 26 (2009) 1446–1455.

[78] S. Narasimha Murthy, D.E. Wiskirchen, C.P. Bowers, Iontophoretic drug delivery across human nail, Journal of Pharmaceutical Sciences 96 (2007) 305–311.

[79] M. Halhal, G. Renard, Y. Courtois, D. BenEzra, F. Behar-Cohen, Iontophoresis: from the lab to the bed side, Experimental Eye Research 78 (2004) 751–757.

[80] F.F. Behar-Cohen, A. El Aouni, S. Gautier, G. David, J. Davis, P. Chapon, J.M. Parel, Transscleral Coulomb-controlled iontophoresis of methylprednisolone into the rabbit eye: influence of duration of treatment, current intensity and drug concentration on ocular tissue and fluid levels, Experimental Eye Research 74 (2002) 51–59.

[81] F.F. Behar-Cohen, J.M. Parel, Y. Pouliquen, B. Thillaye-Goldenberg, O. Goureau, S. Heydolph, Y. Courtois, Y. De Kozak, Iontophoresis of dexamethasone in the treatment of endotoxin-induced-uveitis in rats, Experimental Eye Research 65 (1997) 533–545.

[82] M. Voigt, M. Kralinger, G. Kieselbach, P. Chapon, S. Anagnoste, B. Hayden, J.M. Parel, Ocular aspirin distribution: a comparison of intravenous, topical, and coulomb-controlled iontophoresis administration, Investigative Ophthalmology and Visual Science 43 (2002) 3299–3306.
[83] E. Eljarrat-Binstock, F. Raiskup, J. Frucht-Pery, A.J. Domb, Transcorneal and transscleral iontophoresis of dexamethasone phosphate using drug loaded hydrogel, Journal of Controlled Release 106 (2005) 386–390.
[84] A.E. Cohen, C. Assang, M.A. Patane, S. From, M. Korenfeld, Evaluation of dexamethasone phosphate delivered by ocular iontophoresis for treating noninfectious anterior uveitis, Ophthalmology 119 (2012) 66–73.
[85] T.M. Parkinson, E. Ferguson, S. Febbraro, A. Bakhtyari, M. King, M. Mundasad, Tolerance of ocular iontophoresis in healthy volunteers, Journal of Ocular Pharmacology and Therapeutics 19 (2003) 145–151.
[86] M.A. Patane, A. Cohen, S. From, G. Torkildsen, D. Welch, G.W. Ousler, III, Ocular iontophoresis of EGP-437 (dexamethasone phosphate) in dry eye patients: results of a randomized clinical trial, Clinical Ophthalmology 5 (2011) 633–643.
[87] S.K. Li, Y. Zhang, H. Zhu, W.I. Higuchi, H.S. White, Influence of asymmetric donor-receiver ion concentration upon transscleral iontophoretic transport, Journal of Pharmaceutical Sciences 94 (2005) 847–860.
[88] S. Nicoli, G. Ferrari, M. Quarta, C. Macaluso, P. Santi, *In vitro* transscleral iontophoresis of high molecular weight neutral compounds, European Journal of Pharmaceutical Sciences 36 (2009) 486–492.
[89] P. Chopra, J. Hao, S.K. Li, Iontophoretic transport of charged macromolecules across human sclera, International Journal of Pharmaceutics 388 (2010) 107–113.
[90] H. Wen, J. Hao, S.K. Li, Influence of permeant lipophilicity on permeation across human sclera, Pharmaceutical Research 27 (2010) 2446–2456.
[91] E. Eljarrat-Binstock, F. Orucov, Y. Aldouby, J. Frucht-Pery, A.J. Domb, Charged nanoparticles delivery to the eye using hydrogel iontophoresis, Journal of Controlled Release 126 (2008) 156–161.
[92] S. Gungor, M.B. Delgado-Charro, B. Ruiz-Perez, W. Schubert, P. Isom, P. Moslemy, M.A. Patane, R.H. Guy, Trans-scleral iontophoretic delivery of low molecular weight therapeutics, Journal of Controlled Release 147 (2010) 225–231.
[93] S.K. Li, M.R. Liddell, H. Wen, Effective electrophoretic mobilities and charges of anti-VEGF proteins determined by capillary zone electrophoresis, Journal of Pharmaceutical and Biomedical Analysis 55 (2011) 603–607.
[94] S. Pescina, C. Padula, P. Santi, S. Nicoli, Effect of formulation factors on the trans-scleral iontophoretic and post-iontophoretic transports of a 40kDa dextran *in vitro*, European Journal of Pharmaceutical Sciences 42 (2011) 503–508.
[95] S.K. Li, H. Zhu, W.I. Higuchi, Enhanced transscleral iontophoretic transport with ion-exchange membrane, Pharmaceutical Research 23 (2006) 1857–1867.

CHAPTER

12

Mass Transfer Phenomena in Electroporation

Alexander Golberg and Boris Rubinsky

Department of Mechanical Engineering, University of California at Berkeley, Berkeley, CA, USA

OUTLINE

Nomenclature 456

Symbols 457

12.1 Introduction 458

12.2 Electroporation Background and Theory 459

12.3 Applications of Electroporation-Mediated Mass Transport in Biological Systems 463

12.4 Mechanisms of Pulsed Electric Field-Mediated Transport into Cells 464

12.5 Experimental Methods Used to Study Mass Transfer During Electroporation 465

12.6 Mathematical Models Describing Molecular Transport During Reversible Electroporation 467

12.6.1 Physico-Chemical Model for Electroporation 467

12.6.2 Electropermeabilization Model 470

12.6.3 Electrodiffusion Model of DNA Cluster Formation 472

12.6.4 Model of Conductivity Changes During Cell Suspension Electroporation 475

12.6.5 Model of Small Molecule Transport Kinetics Due to Electroporation 477

12.6.6 Two Compartment Pharmacokinetic Model for Molecular Uptake During Electroporation 479

12.6.7 Statistical Model for Cell Electrotransformation 481

12.6.8 A Multiscale Model for Mass Transfer of Drug Molecules in Tissue 481

12.7 Future Needs in Mathematical Modeling of Mass Transport for Electroporation Research 484

References 485

© 2013 Elsevier Inc. All rights reserved.

http://dx.doi.org/10.1016/B978-0-12-415824-5.00012-6

Nomenclature

a	cell radius
C	closed (sealed) state of the membrane
c_m^{out}	concentration of the species of interest on the outer membrane/medium interface
c_m^{in}	concentration of the species of interest on the inner membrane/medium interface
c^{out}	bulk concentration of species outside the cell
c^{in}	bulk concentration of species outside the cell
c_0	infinite external concentration of species
c	local concentration of a chemical of interest
C_{LW}	change in specific capacitance associated with the replacement of lipid molecules with water
C_N	constant that depends on the size of the pores and their growth kinetics
c_B	constant concentration of LY (Lucifer yellow) in the medium
C_o	pore-free lipid bilayer membrane capacitance
D	diffusion coefficient
D_e	Einshtein-Smoluchowski diffusion constant
D_i	diffusion coefficient inside the cell
D_o	diffusion coefficient outside the cell
D_m	diffusion coefficient of a molecule in the membrane phase
d	membrane thickness
E	applied external electric field
E_p	threshold electric field value which causes to electroporation
e_o	elementary charge
F	Faraday constant
f_{per}	fraction of pores which are large enough to cause permeability of ions and molecules through the cell membrane
$f(ADN)$	function which depends on plasmid concentration
f_p	time dependent fraction of porated membrane area
g	parameter that characterizes the fast resealing kinetics of the membrane
J	time dependent flux of molecules
j	thermodynamic current
K	equilibrium distribution constant
k_f^0	cell/tissue characteristic, electric field parameters independent, flow coefficient
k_1	kinetic rate constant for pore formation
k_{-1}	kinetic rate constant for pore resealing
k_B	Boltzmann constant
k_{oi}	coefficient which describes the flow between the outside the cell to the inside of the cell
k_{io}	coefficient which describes the flow between the inside of the cell to the outside
M	flow coefficient
m_0	mass of a substance of interest outside the cell
m_i	mass of a substance of interest inside the cell
N	number of pulses

N	absolute number of molecules absorbed per cell
n	number of molecules
n_c^{in}	amount of transported molecules into one volume compartment
O	pore density
O_0	equilibrium pore density
O_p	number of pores at the end of the electroporation pulse
P_m	permeability coefficient
P	porous state of the membrane
q	electroporation constant
R	gas constant
r	pore radius
S_m	part of membrane surface through which the molecular transport occurs
S_c	total area of outer membrane surface
S_{tot}	represents the total area of cells
S_{por}	total area of all pores of a single cell
S_0	total area of a single cell
T	temperature
t_E	electric pulse duration
t_{obs}	observation time, which is much longer than t_E
Δt_N	interval between two consequent pulses
V	spatially averaged transmembrane voltage ($\langle \Delta\varphi_m \rangle$)
V_c	one volume compartment
V_i	volume of all electropermeabilized cells
V_o	volume of the outside chamber
v	single cell average volume
v	velocity of a molecule in the local electric field
W_s	constant
$\Delta W(r)$	pore formation energy
$\lvert z_{eff} \rvert$	effective charge

Symbols

Γ	energy per area of a flat, pore-free membrane
ε_w	permittivity of pure water and the lipid membrane
ε_m	permittivity of pure lipid membrane
$\Delta\varphi_m$	transmembrane voltage
φ_{ep}	characteristic voltage of electroporation
γ	energy per length edge of the pore
μ	the local electrophoretic mobility
μ_e	mobility of the DNA outside the cell
μ_m	mobility of the DNA inside the membrane
μ_i	mobility of the DNA inside the cell
θ	angle of the radial direction vector
θ_s	pole angle in reference to the direction of the applied electric field

σ_o	conductivity of the suspending medium
σ_i	conductivity of the cell interior
σ_m	conductivity of the cell membrane
$\Delta\sigma_m$	change in the transmembrane conductivity
$\Delta\sigma_{max}$	maximum value of conductivity when the concentrations inside and outside of cells are equal
τ	time constant
ξ	specifies single partition coefficient

12.1 INTRODUCTION

Electroporation is a biophysical phenomenon in which cell membrane permeability is increased through the application of external pulsed electric fields (PEF). Reversible electroporation (RE) occurs when the cell membrane permeabilization is temporary and the treated cells survive. PEFs that lead to cell death are known as irreversible electroporation (IRE). Today, electroporation-based methods are widely used in basic research, biotechnology, medicine and industry. During the past 40 years, IRE has been studied in the food industry as a new non-thermal disinfection method. Given its unique ability to preserve the extracellular matrix undamaged, tissue ablation by non-thermal IRE has become a promising medical technology. RE, a procedure that requires cells to survive the electric field treatment, is an essential tool in the life scientist's, biological engineer's and physician's armamentarium. It is used for cell fusion, gene and drug delivery to cells and tissue samples and for cancer electrochemotherapy. RE is a complex phenomenon which combines cell reversible electropermeabilization with mass transfer of the molecules of interest. The outcome of RE-based molecular delivery is the introduction of cell membrane non-permeant molecules into cells. The molecular mechanisms of cell electropermeabilization and RE mass transfer are still unclear.

Mathematical models are essential for electroporation treatment planning. For example, the mathematical solution of the electric field equation is used to predict the areas of tissue that are permeabilized by PEF. Models also attempt to predict the cell permeabilization efficiency as a function of cell type and treatment parameters.

This chapter will focus primarily on mathematical models of mass transfer related to the application of PEF to cells. The goal of the chapter is to also give the reader an introduction to electroporation-mediated mass transport into cells and tissues to facilitate a fundamental understanding of the models. The chapter consists of six sections. We start with a review of the mechanisms of cell electropermeabilization. The second section briefly discusses mass transfer related applications of PEF. The third section is a summary of the proposed mechanisms of molecular transport across the cell membrane facilitated by PEF. Next, we review experimental methods used for RE mass transfer research. The fifth section is the core of the chapter. It reviews the various mathematical models used to explain and predict molecular transport in RE. We conclude the chapter with recommendations of what we consider to be the more important future needs for modeling of RE-mediated mass transport in life science, biological engineering and medicine.

12.2 ELECTROPORATION BACKGROUND AND THEORY

Certain strong electrical fields, when applied across a cell, can permeabilize the cell membrane, presumably through the formation of nanoscale defects – pores – in the membrane [1]. The use of electroporation can be traced back to 1898, when *Fuller* [2] proposed water sterilization with PEF. PEF entered commercial liquid foods sterilization at the turn of the 20th century, when a process called 'Electropure' was used for milk pasteurization with electricity [3]. Perhaps the most fundamental study on the use of PEF for the inactivation of microorganisms in liquid solutions can be found in a series of three papers by *Sale and Hamilton* [4–6]. These studies drew on the pioneering work of *Doevenspeck* (1961) [7], who reported on the use of PEF in solid food processing. Studies on permeabilization by PEF done in the 1960s and 1970s focused on membrane models and on an empirical understanding of the process. A major advance in the field occurred in 1982, when *Neumann et al.* [8] reported on PEF that facilitated the transfer of genes into a living cell and the consequent gene expression. That paper also coined the term 'electroporation' and started the field of electrotransfection. Today electroporation has become an essential tool in all life science laboratories. Moreover, new clinical applications of both modes of electroporation – RE and IRE – are continuously emerging [1,9].

The empirical studies of the 1960s and 1970s were followed by attempts to establish a theory to describe the effect of PEF on the cell membrane. The first models, which proposed an electromechanical mechanism for membrane breakage [10], could not account for the pulse duration dependence of the critical voltage needed for permeabilization [11], or the dependence of the membrane lifetime on the total membrane area [12]. The current, transient aqueous pore hypothesis argues that the process of electroporation is related to nanoscale pore (defects) formation in the cell membrane [13,14]. Recent molecular dynamics simulations support the theory (Fig. 12.1). Briefly, it is assumed that hydrophobic pores are spontaneously and randomly generated in the cell membrane due to thermal motion of phospholipid molecules. The location and size of these pores vary randomly. The formation energy (Fig. 12.1a) for a single pore is based on classical surface chemistry. Thus, the pore formation energy, $\Delta W(r)$ (in the absence of an external electrical field), is the difference between the gain in energy due to the formation of the outer edge of the pore (γ) and the reduction in energy due to the loss of a circular patch that has been cut out of the membrane when the pore was created ($\pi r^2 \Gamma$) [23]. This approach is based on the much earlier concept of soap film stability proposed by *Deryagin and Gutop* in 1962 [15]. The constant γ is the energy per length edge of the pore and Γ is the energy per area of a flat, pore-free membrane. The expression for the formation energy of a single pore of radius r is shown in Eq. (12.1), as suggested by [13]:

$$\Delta W_p(r) = 2\gamma\pi r - \pi r^2 \Gamma \tag{12.1}$$

In other words, ΔWp describes an activation energy barrier that thermal fluctuations need to overcome in order to create a hydrophobic pore in the absence of a transmembrane voltage ($\Delta\varphi_m$). The activation energy barrier is high enough to prevent cells from frequent self-rupture [16]; however, it is thought that these spontaneous pores provide sites at which phospholipid translocations take place [17].

If the radius *(r)* of the hydrophobic pore exceeds a critical radius *rt* (0.3 to 0.5 nm) [18], it overcomes the energy barrier and turns into a hydrophilic pore (Fig. 12.1a). It is believed that

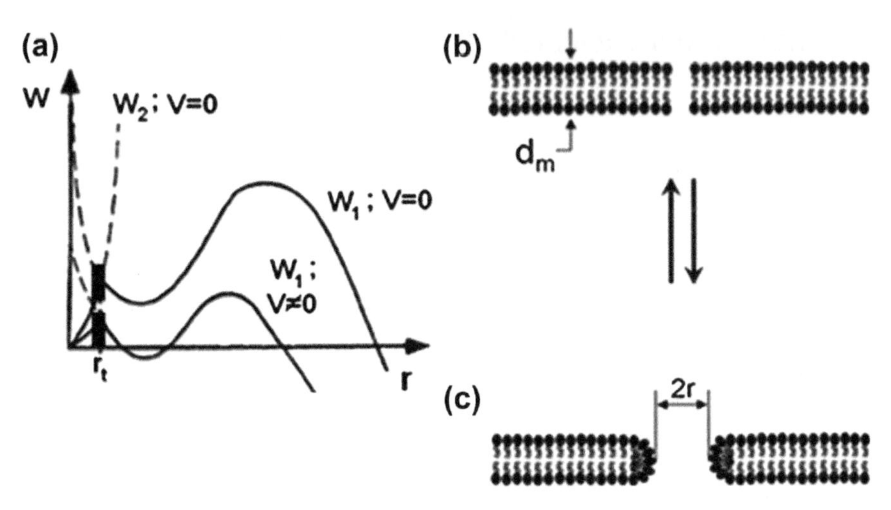

FIGURE 12.1a **Mechanism of pore formation according to the transient aqueous pore hypothesis.** (a) Pore formation energy of a membrane containing a single pore of radius r. W_2 curve (Eq. (12.1)) describes formation energy for a hydrophobic pore. Dashed lines show the theoretical radii that are never achieved due to the formation of a hydrophilic pore at the transition radius r_t. W_1 is the hydrophilic pore formation energy function. V is the average transmembrane potential ($\langle\Delta\varphi_m\rangle$). Below critical radius r_c pores grow uncontrollably, causing membrane rupture. (b) and (c) Schematic of the hydrophobic/hydrophilic pore transition process. d_m is the membrane thickness, r is the pore radius. The figure is based on [13] with permission.

this hydrophilic pore is responsible for the electroporation phenomenon [12,13,19]. Once a hydrophilic pore forms, water molecules get into the inter-membrane space during the pore transformation process (Fig. 12.1a); thus, a nano-hydrophilic environment is created in this usually hydrophobic location. Consequently, nearby lipids reorient in order to form more stable structures (Fig. 12.1a) and a stable hydrophilic pore is formed [18].

It was shown that ions do not enter the small pores because of the high energy barrier [20]. Hence, the pore can be represented by a water-filled, rather than electrolyte-filled, capacitor. Since water molecules in the membrane replace some of the lipids, the hydrophilic pore undergoes an energy change associated with the change in its specific capacitance, *CLW*. As a result, in the presence of transmembrane potential $\Delta\varphi_m$ the energy for hydrophilic pore formation can be described by Eq. (12.2) [13]:

$$\Delta W_p(r, V) = 2\gamma\pi r - \pi r^2\Gamma - 0.5C_{LW}\pi r^2V^2 \tag{12.2}$$

Here, V is the spatially averaged transmembrane voltage ($\langle\Delta\varphi_m\rangle$) and *CLW* is the change in specific capacitance associated with the replacement of lipid molecules with water. The change in the specific capacitance of the pore as water displaces the lipid molecules is given by Eq. (12.3) [13]:

$$C_{LW} = C_0\left(\frac{\varepsilon_w}{\varepsilon_m} - 1\right) \tag{12.3}$$

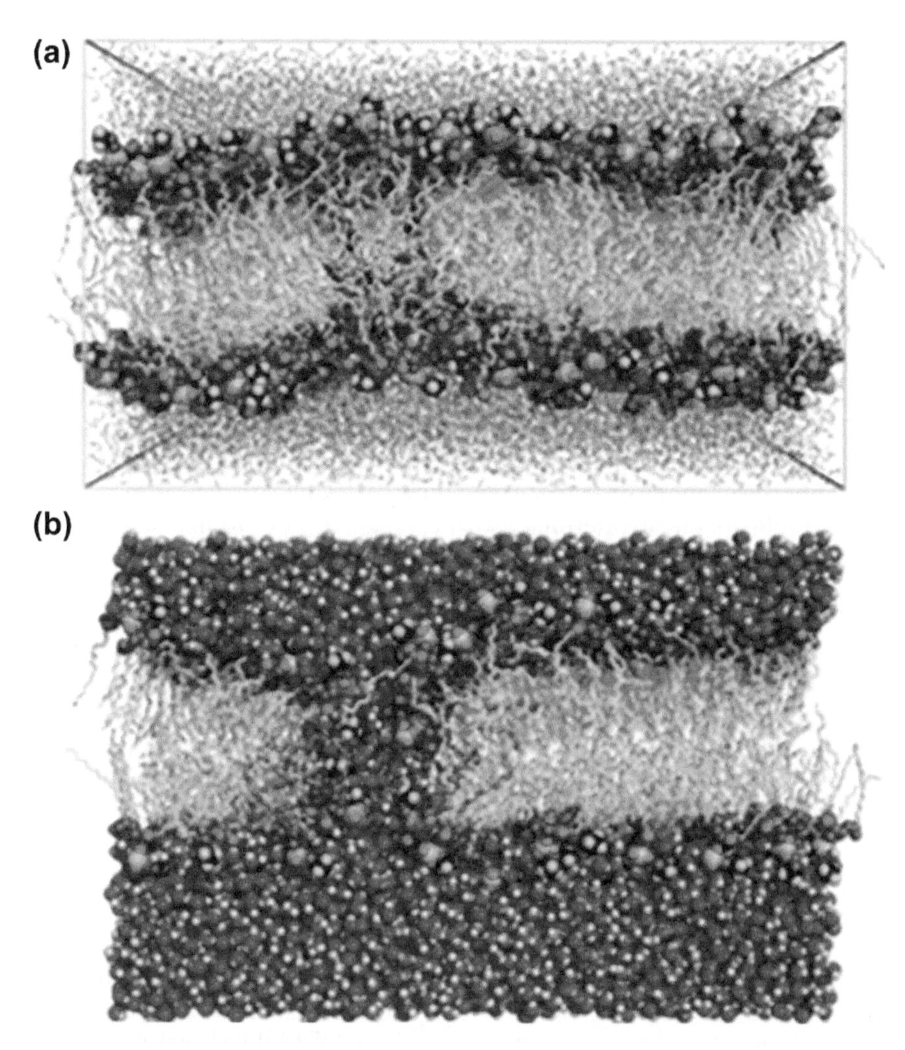

FIGURE 12.1b **Multimolecular stimulation of a model membrane in aqueous media under the application of external electric field.** Lipid tails are described as yellow sticks, the choline groups as blue spheres, the phosphor atoms in green, lipid oxygen atoms in orange, and lipid head carbon atoms in gray. Water is shown in (a) stick representation and in (b) space-filled representation. In panel b, a cut through the center of the pore is shown. Figure adapted from [14] with permission.

where ε_w and ε_m are the permittivity of pure water and the lipid membrane interior respectively, and Co is the pore-free lipid bilayer membrane capacitance. Eq. (12.2) is a parabolic function, for which the maximum $\Delta Wpc(rc, Vc)$ is given by Eq. (12.4) as follows:

$$r_c = \frac{\gamma}{\Gamma + 0.5C_{LW}V^2} \quad \Delta W_{pc} = \frac{\pi \gamma^2}{\Gamma + 0.5C_{LW}V^2} \tag{12.4}$$

Here r_c is the radius associated with the maximum pore formation energy barrier.

The dielectric within the hydrophilic pore (curve *W*2 in Fig. 12.1a) has a much larger permittivity than the lipid of the membrane interior, due to water molecule penetration. Therefore, the function maximum ΔWpc and corresponding pore radius, r_c, both decrease with increasing V (Eq. (12.4)). In addition, increasing the transmembrane potential will rapidly increase the probability of formation (via the Boltzmann probability function) of a pore with radius r_c [20]. The membrane ruptures at stronger applied electric fields, since the pore formation activation energy drops, and the probability of the membrane to acquire pores with $r > r_c$, increases [13]. Consequently, one of these 'super-pores' (pores with $r > r_c$, but $\Delta Wp < \Delta Wpc$) can expand uncontrollably until it eliminates the membrane structure [13].

Studies on bacteria inactivation by PEF revealed the non-deterministic nature of electroporation. Specifically, it was shown that even though all cells were permeabilized by IRE type PEF, not all cells died. It is assumed that the main reason for cell death is the release of intracellular compounds caused by the increased permeability of the membrane, structural membrane changes or osmotic swelling [21]. Cell survival is explained by pore resealing which follows electro-permeabilization [22,23]. This ability of the membrane to rebuild its structure allows a given cell to survive and proliferate after RE treatment. A thermal pore resealing model was developed by *Saulis* (1997) [23], who showed the impact of post-treatment temperature on RE effectiveness. However, it is important to point out that the exact molecular process of pore formation in the cell membrane is not yet understood [13,21].

As shown in the previous section, the formation of pores in the cell membrane depends on the local transmembrane potential $\Delta\varphi_m$ [24]. For cells with a spherical shape, the external electric field induced transmembrane potential can be approximated by an equation derived by *Neumann* [24]:

$$\Delta\varphi_m = -\frac{3}{2}Ef(\sigma)a\cos\theta \tag{12.5}$$

where $\Delta\varphi_m$ is the potential difference at the specific location on the membrane; E is the applied external electric field; a is the cell radius; θ is the angle of the radial direction vector. Note that θ is zero (and $\cos\theta = 1$) when the radial direction vector coincides with the direction of the electrical field. $f(\sigma)$ given by Eq. (12.6) is an explicit function of the electrical conductivities of the suspending medium σ_o, the cell interior σ_i, the cell membrane σ_m, and the ratio of the membrane thickness (d) and the cell radius a [25]. However, since σ_m is in the range of nS m^{-1} for intact membranes, it is readily seen from Eq. (12.6) that $f_{(\sigma)}$ can be approximated by 1, independently of the electrical conductivity of the suspending medium:

$$f_{(\sigma)} = \frac{1}{1 + \frac{(\sigma_m(2+\sigma_i/\sigma_o))}{2\sigma_i d/a}} \tag{12.6}$$

It appears from Eq. (12.5) that $\Delta\varphi_m$ is strongly dependent on position along the cell membrane, through the cosine term in the equation. The maximum $\Delta\varphi_m$ is located at the poles of the cell ($\theta=0,\pi$), where the cosine term has a maximum. According to the transient aqueous pore hypothesis, the largest and most stable pores will form on the poles that are the closest to the electrode locations on the membrane. Experiments with voltage sensitive dyes have

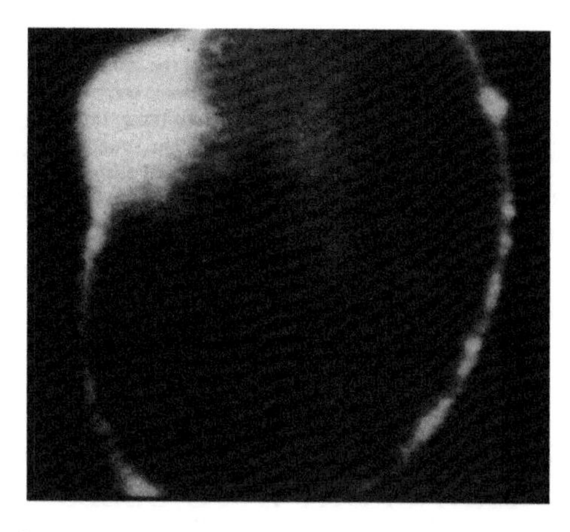

FIGURE 12.2 **Location specific $\Delta\varphi_m$.** The fluorescence response of a single A-431 human carcinoma cell stained with 10 μM di-4-ANEPP (voltage sensitive dye) to an applied electric field of 53.0 V/cm directed from left to right. Image adapted from *D. Gross et al.* [26] with permission.

confirmed that the maximum transmembrane voltage is found on cell poles which face the electrodes (Fig. 12.2) [26].

For cells with a non-spherical shape, an estimate of the transmembrane potential can be obtained by solving Maxwell's equations in ellipsoidal coordinates as in [10]. Numerical solutions are given in [27].

Rapid electrical breakdown and local conformational changes of bilayer structures occur, in excess of a critical $\Delta\varphi_m$ of -1 V, thus causing the electroporation phenomenon [24].

12.3 APPLICATIONS OF ELECTROPORATION-MEDIATED MASS TRANSPORT IN BIOLOGICAL SYSTEMS

The observation by *Neumann* in 1982 [8] that pulsed electric fields facilitate the penetration of DNA molecules into cells led to the development of numerous applications of the technology coined 'electroporation' or 'electropermeabilization'. Applications of electric pulses for cell uptake of small and macromolecules, which intrinsically have very small membrane penetration abilities, were reported for all levels of cell structures: cells, tissues and organs, both *in vitro* and *in vivo*.

Because of its high efficiency, simplicity, consistency and non-chemical and non-viral nature, electroporation has become a 'classical' method for cell transformation in molecular biology [28–31]. Recently cell electroporation was suggested as a tool for *ex vivo* cell preparation for cell based gene therapy [32]. However, an efficient protocol has to be developed for each new type of cells. Therefore, methods for high-throughput optimization and cost reduction have been established [33,34].

Pulsed electric fields are used to deliver drugs, DNA and RNA molecules to cells in tissues in a controlled way [9,35–40]. The electroporation-mediated transfer of cytotoxic drugs into tumors, a procedure coined electrochemotherapy, has become a clinical procedure [41–43].

Furthermore, gene electrotherapy is undergoing clinical trials [44,45]. In addition, research on siRNA electrotransfer has been recently reported [37,46,47].

An additional important application of RE is skin electroporation. Mass transfer in skin takes place due to electroporation of multiple membrane layers, and is beyond the scope of this chapter, but is reviewed in detail in [48–50]. In addition to the non-thermal pore formation mechanism, recent studies have developed a thermodynamic model to describe the structural changes and transport in that occur in skin when the pulse duration is sufficiently long to cause thermal effects [49,51–55].

12.4 MECHANISMS OF PULSED ELECTRIC FIELD-MEDIATED TRANSPORT INTO CELLS

The mechanisms of pulsed electric field enabled molecular transport into cells are complex and not well understood. Observations made during recent decades suggest that the transport mechanisms depend on the size and charge of the molecules involved. The transport processes vary with time during and after the application of the pulse. It was found that the transport mechanisms for the same molecules are different during the application of the electric pulse and after the electric field was removed. Various researchers have used theoretical and experimental studies to propose several mechanisms. Diffusion [56,57], absorption [58], electro-osmosis [59], electrophoresis [60–62], convection [63] and endocytosis [64,65] have all been proposed as possible mechanisms for mass transport across the cell membrane. In this section we will review the different proposed mechanisms for the transport of small molecules, solute, DNA and RNA into cells during the application of pulsed electric fields.

Studies have shown that externally applied electric fields increase the transport of small molecules into the cell, both during and after the application of the electric pulse [66,67]. In these experiments, markers were added to the electropermeabilized cell culture and it was shown that PEF treated cells absorb small molecules during both the application of the electric pulse and after, presumably during the cell membrane defect resealing process. Propidium iodide [68–70], calcein [71–73], calcium ions [69,74,75], Lucifer yellow (LY) [76] and trypan blue [77–80] are the most popular probes used for small molecules mass transfer research. These smaller molecules are also commonly used as experimental tools for membrane resealing kinetics research [80]. Dextrans are used as models for large weight molecules [63].

The electro-osmosis model of *Dimitrov and Sowers* [59], while subjected to numerous challenges, started the field of mechanical models of macromolecule uptake by cells exposed to PEF [57,61,63]. Convection-mediated cellular uptake is possible due to cell movement, cell swelling and electrodeformation. Cell movement due to the application of electric fields was reported in [59,63]. Cell swelling, possibly due to extracellular medium convective transport into the cell, was reported in [59,81,82]. The cell volume could increase because of the osmotic mismatch between the extracellular and intracellular milieu that is accommodated by solute transport from the outside through the cell membrane defects. Furthermore, the reported cell deformations may facilitate connective transport if it increases the total available volume, as shown in [63,83].

In contrast to the simple diffusion of small molecules, DNA transport into cells is a complex process consisting of several steps [66]. Experimental studies suggest that DNA-membrane complexes form during the application of an electric field. It was also found that gene expression was observed only when the DNA molecules are present in solution during the application of the electric field [66]. An absorption model based on these observations was proposed ([58]). Another theory suggests that the DNA is pushed through the membrane by the electrophoretic effect [60–62,84]. In this context additional low amplitude fields applied after the standard electroporation protocol show an increase in the electrotransformation efficiency [60,85,86]. Other proposed mechanisms suggest that the applied pulses lead to an aggregation of ion pumps, which open and permit the free movement of DNA to the cytosol [66]. Later studies challenged the electrophoresis model of DNA electrotransfer, and suggested that DNA and other large molecules penetrate the cell by endocytosis and macropinocytosis routes [87–91]. Indeed, direct measurements showed that a pulsed electric field could stimulate the endocytosis of fluorescein-labeled bovine serum albumin [88–90] and β-galactosidase [64]. Additional evidence for the DNA absorption endocytotic pathway came from a recent study by *Wu and Yaun*, who show in [65] that electroporation efficiency can be reduced by trypsinization of cells post electrotransfection or treatment of cells with (i) pharmacological inhibitors of endocytosis or (ii) anti-dynamin II siRNA prior to electrotransfection. These observations imply that electric field-mediated cellular uptake of DNA molecules must be preceded by binding of DNA to the plasma membrane and is highly dependent upon still unknown molecules implicated in endocytotic vehicle formation.

Paganin-Gioanni et al. [47] showed that externally applied electric fields have two effects on siRNA transport. First, the applied pulsed electric field leads to cell permeabilization, and then the field produces an electrophoretic drag on the negatively charged siRNA molecules from the bulk phase into the cytoplasm. The study suggests a pure electrophoretic mechanism for siRNA PEF-mediated transfer into cells. Direct observation of siRNA transport revealed that it is different from that of DNA, and that no complex is formed between siRNA molecules and the cell membrane. The siRNA molecule moved through the membrane during the application of the external electric field; however, after the electric field was removed the molecular flux stopped, even though the resealing process was not yet finished.

12.5 EXPERIMENTAL METHODS USED TO STUDY MASS TRANSFER DURING ELECTROPORATION

Several methods are currently used for the study of pulsed electric field-mediated transport in cells and tissue. The first method measures electrical conductivity changes in the cell membrane and in tissues, during and after the applications of pulsed electric fields. The first measurements of ion current due to cell membrane leakage under the application of external electric fields were done at the end of the 1970s [92–94]. These experiments tried to explain the cell death due to the application of pulse electric field first reported in a series of seminal papers by *Sale and Hamilton* [4–6]. *Kakorkin et al.* [95] measured conductivity changes of the solution when electric fields were applied on salt filled lipid vesicles. Conductivity changes of the solution during and after the application of external electric fields were correlated to the mass of ions, which left the vesicles to the external solution. A nonlinear current voltage relationship and membrane resealing after the pulse in cell pellets were reported in [81,96]. In addition, the effects of PEF on dense cell suspension during and after electroporation were reported in [97,98]. *Huang and Rubinsky* [29,99] developed a chip with a micro-pore structure in a planar dielectric thin film to trap and isolate single cells for controlled electropermeabilization. Since the increase in cell membrane conductivity will cause an increase in the transport of ions through the cell, the electric current was used as feedback to determine whether electroporation had occurred [100,101]. Measurements of tissue conductivity and impedance were proposed for the real-time control of tissue electropermeabilization [82,102]. *Granot and Rubinsky* [103,104] suggested that electrical impedance tomography be used for monitoring small molecule uptake by tissues exposed to pulse electric fields. Furthermore, galvanic properties of tissue were recently proposed for monitoring tissue permeabilization [105,106]. Current density imaging (CDI), magnetic resonance electrical impedance tomography (MREIT), diffusion-weighted magnetic resonance imaging (DW-MRI) and MRI have been also recently proposed for real-time electroporation process control [107–111].

Several optical methods are used for the direct imaging of electric field-mediated molecular transport into cells. Direct fluorescent monitoring has revealed interesting properties of molecular transport in an externally applied electric field. *Tekle et al.* [112] reported that the transport of molecules is asymptotic. The authors suggested that asymmetric pores are created on both electrode facing poles of the membrane. Higher density, small radius pore sizes occur on the anode side, and fewer pores with larger radiuses are formed on the cathode side. Direct observation using fluorescent markers has shown that the mechanism of DNA electrotransfer is different from the transfer of small molecules. DNA electrotransfer involves cluster formation of DNA molecules on the membrane [113]. Imaging based studies include one of the methods of analyses classified by *Prausnitz et al.* [72] and discussed in the following paragraphs.

First, gene expression analyses are used. In these studies the gene of interest, fused to the gene for a fluorescent protein, is usually transfected into cells. The number of successfully transfected cells is calculated by fluorescent microscopy (histology or whole body imaging are also employed) once the fluorescent protein is expressed. In these techniques, usually done *in vivo*, success is measured by the expression of the gene of interest. This is measured as the level of protein expression, its spatial location, or, as in the case of microorganisms by colony counting [8,36,85,114–120].

Second, total population imaging has been reported in multiple studies; such studies provide information on the average transport of molecules per cell. In these studies the fluorescence, turbidity or radiation of labeled molecules is measured for the whole electroporated population of cells [58] and giant vehicles [121]. Ultrafast imagining techniques have been used to study the time course of molecular transport. For instance, a recent study with 200 ns to 4 ms time resolution revealed a 60 μs delay between the start of the pulse and a first fluorescent signal registered from cells [122]. Although the technique is used often and has led to many discoveries related to electrotransformation mechanisms, it was shown both theoretically and experimentally that a heterogeneous response to electroporation exists within a cell population. Several sub-populations may exist with significantly different responses to electroporation [57,123–127].

Third, single molecular imaging has emerged as an efficient research tool. To overcome the limitation of the population response, single cell imaging analysis was used in [29,84,128–130].

Lastly, flow cytometry has been used for large scale analyses of electroporated cells and tissue models [131,132]. This technique is used for studying molecular uptake and cell damage [72,73,133,134]. Flow cytometry allows the evaluation of the amount of uptake of a given molecule by each cell exposed to a pulsed electric field, in a population of cells in suspension [131]. and in cell monolayers [135]. The evaluation is based on a correlation between the fluorescence emission of electroporated cells and the emission of special calibration beads, which represent a certain number of fluorescent molecules. The technique is described in detail in [72].

To summarize, various methods have been proposed for the study of electroporation associated transport phenomena. Since various molecules apparently have different transport mechanisms, all techniques are essential to study the phenomena. Increased resolution microscopy may improve our understanding of molecular transport during RE.

12.6 MATHEMATICAL MODELS DESCRIBING MOLECULAR TRANSPORT DURING REVERSIBLE ELECTROPORATION

12.6.1 Physico-Chemical Model for Electroporation

A physico-chemical theory for the electroporation-mediated transport of molecules into cells was developed by *Neuman et al.* [136]. This model, based on experimental curves showing DNA and dye uptake by cells, describes electroporation-mediated molecular transport into cells in terms of transport parameters k_f^o, P_m, f_p derived from electrochemical kinetics and thermodynamic analyses of pore creation and resealing [136].

k_f^0 is a cell/tissue characteristic, electric field parameter independent flow coefficient.
P_m is an electric field parameter independent permeability coefficient.
f_p is a time dependent fraction of porated membrane area.

The following paragraphs describe the derivation of an electro-physico-chemical electroporation theory, which describes the molecular transport into cells due to application of electric fields. According to the physico-chemical model, the formation of a pore due to the application of a pulsed electric field can be described by a chemical kinetics approach. Specifically, pore formation and resealing can be viewed as a phase transition according to the following first order kinetic reaction mechanism:

$$C \Leftrightarrow P \tag{12.7}$$

where C describes the closed state and P describes porous state of the membrane.

The degree of membrane electroporation of pore fraction (found experimentally) is defined by:

$$f_p = \frac{[P]}{[P]+[C]} = \frac{K}{1+K} \tag{12.8}$$

where K is an equilibrium distribution constant defined by:

$$K = \frac{[P]}{[C]} = \frac{k_1}{k_{-1}} \tag{12.9}$$

where k_1 is the kinetic rate constant for pore formation and k_{-1} is the kinetic rate constant for pore resealing.

During the application of an electric field ($0 \le t \le t_E$) the following relationship applies:

$$f_p^{C->P}(t, t_E, E) = \frac{K}{K+1}(1 - e^{-t(k_1+k_{-1})}) \tag{12.10}$$

At times after the application of the electric field ($t > t_E$):

$$f_p^{P->C}(t, t_E, E) = f(t_E)e^{-k_{-1}(t-t_E)} \tag{12.11}$$

Experimentally f_P was found from electrooptic and conductometric studies of the electroporative deformation of membranes due to the application of electric field [74,95,137,138]. The formation of pores causes the entrance of water and small ions into the membrane interior, thus decreasing the transmembrane conductivity. Therefore, the upper limit of the fractional increase of transport area is given by:

$$f_P = \frac{\Delta\sigma_m}{\sigma_i} \tag{12.12}$$

where $\Delta\sigma_m$ is the change in the transmembrane conductivity and σ_i is the cell interior conductivity.

According to Fick's first law, the intracellular flow of macromolecules is given by:

$$\frac{dn_c^{in}}{dt} = -D_m S_m \frac{dc_m}{dx} \tag{12.13}$$

where n_c^{in} [M] is the number of molecules transported into one volume compartment V_c; D_m is a diffusion coefficient of a molecule in the membrane phase; and S_m is that part of the membrane surface through which the molecular transport occurs.

The following assumptions were used for the practical application of Eq. (12.13):

1. D_m is a phase specific diffusion coefficient, independent of position within the membrane.
2. The concentration gradient within the membrane is approximated by:

$$\frac{dc_m}{dx} = \frac{(c_m^{out} - c_m^{in})}{d} \tag{12.14}$$

where: c_m^{out} – a concentration of the species of interest on the outer membrane/medium interface, and c_m^{in}– a concentration of the species of interest on the inner membrane/medium interface.

3. Rapid equilibrium is assumed for bulk-membrane interface molecules distribution. The rate limiting step is the transport of species of interest across the membrane. A single partition coefficient is assumed:

$$\xi = \frac{c_m^{out}}{c^{out}} = \frac{c_m^{in}}{c^{in}} \tag{12.15}$$

where c^{out} is the bulk concentration of species outside the cell, $c^{in} = \frac{n_c^{in}}{V_c}$ is the bulk concentration of species outside the cell.

Electric fields applied across the cell membrane lead to the electrodiffusion of charged molecules such as DNA. The electrodiffusion coefficient is given in Eq. (12.16):

$$D_m(E) = D_m \left(1 + \frac{|z_{eff}| e_0 \Delta\varphi_m}{k_B T}\right) \tag{12.16}$$

where $|z_{eff}|$ is the effective charge of the transported molecule, $\Delta\varphi_m$ is the location specific transmembrane potential (Eq. (12.5)). The permeability coefficient P_m for the electroporated parts of the membrane is defined as:

$$P_m = \frac{\xi D_m}{d} \tag{12.17}$$

The experimentally measured flow coefficient k_f for transmembrane transport is defined by:

$$k_f = \frac{P_m S_m}{V_c} \tag{12.18}$$

Substitution of Eqs. (12.14)–(12.19) into Eq. (12.13) leads to Eq. (12.19):

$$\frac{dc^{in}}{dt} = -k_f(c^{out} - c^{in}) \tag{12.19}$$

Since the external volume is much larger than the intracellular volume, we assume that the total external concentration is constant (c_0). Substitution of this additional approximation leads to Eq. (12.20):

$$\frac{dc^{in}}{dt} = -k_f\,(c_0 - c^{in}) \tag{12.20}$$

where c_0 is the initial concentration of the species of interest in the external volume.

Since in RE pores reseal, S_m is time dependent; therefore, the flow coefficient k_f is also time dependent. Given f_p, the time dependent flow coefficient can be expressed as:

$$k_f(t) = k_f^0 f_p(t, t_E, E) \tag{12.21}$$

where k_f^0 is a time independent flow coefficient and is equal to:

$$k_f^0(t) = \frac{P_m S_c}{V_c} = \frac{3P_m}{a} \tag{12.22}$$

Where S_c is the total area of outer membrane surface.

Insertion of Eq. (12.20) into Eq. (12.19) and integration yields the equation for the time dependent internal species concentration:

$$c^{in} = c_0 \left(1 - e^{-k_f^0 \left(\int_{t_0}^{t_E} f_p^{C->p}(t,t,E)dt + \int_{te}^{t_{obs}} f_p^{p->C}(t,t,E)dt\right)}\right) \tag{12.23}$$

where t_{obs} is the observation time, which is much longer than t_e.

The kinetics of yeast cell transformation by electroporation and electroporative dye uptake experimental data was analyzed by the physico-chemical theory described above. Flow (k_f^0) and permeability (P_m) coefficients of Serva blue G [139] and plasmid DNA [140] were calculated for different electroporation protocols providing useful analytical expression for planning future electroporation procedures. In these studies, the upper limit of f_p was used. The applied f_p was calculated from the electrooptic and conductometry studies on membrane conductivity change due to electroporation process [74].

12.6.2 Electropermeabilization Model

The lack of experimental visualization of pores led to a proposal of the 'electropermeabilization theory' developed by *Rols and Teissié* [67,141]. Kinetic studies suggested that a process of electropermeabilization of a biological membrane can be described in the following five steps [56]:

1. 'Induction step' – This step occurs microseconds after the pulse is applied. The externally applied electric field increases the transmembrane potential up to a critical threshold E_p, which causes membrane permeabilization, possibly by the creation of local defects in the membrane's lipid structure. A buffer composition dependent mechanical stress is present.
2. 'Expansion step' – An electrochemical stress and membrane defects expansion take place as long as the electric field is above the threshold value E_p.
3. 'Stabilization step' – For an electric field intensity lower than a particular threshold E_p a millisecond duration recovery of the membrane takes place, leaving the membrane in a permeabilized state for small molecules only.
4. 'Resealing step' – A first order process of membrane leak annihilation occurs on a time scale of minutes and seconds.
5. 'Memory effect' – Cell behavior is restored; however, membrane structure (flip flop) and physiological properties (macropinocytosis) changes can be observed some hours after the pulsed electric field treatment.

These studies revealed that the area of cell surface that is permeabilized is determined by the intensity of E; however, the level of permeabilization depends on the duration and number of pulses [67,68,141,142]. The authors reported an analysis of the electrotransfer into cells of both small molecules [141] and DNA plasmids [84]. Small molecule electrotransfer to cells is described by Eq. (12.24):

$$Q = \int J(t)dt \tag{12.24}$$

where Q is total number of molecules which penetrate a cell in a specific interval of time. J is a time dependent flux of molecules. According to Fick's first law in the form:

$$J = -D_m S_m \frac{dc}{dx} \tag{12.25}$$

or, taking the same assumptions as described previously for Eq. (12.13):

$$J = -P_m S_m (c_0 - c^{in}) \tag{12.26}$$

A quantitative study of calcein incorporated into CHO cells during a defined period after pulsing revealed that membrane resealing obeys a first order kinetic process relationship of the form:

$$c^{in} = c^0_{in} e^{-k_{-1} t_E} \tag{12.27}$$

where c^{in} is the concentration of cancer in a cell exposed to a single pulse and c^0_{in} is the amount of calcein in the cell at time 0. The study showed that c^{in} depends on the electric field, E. However, k_{-1} was found to be field independent [141]. Yet another study proposed the use of a permeabilization probability function to describe the flux of molecules across the cell membrane due to electropermeabilization [137]:

$$dJ = P_m x(N, t_E, E, \theta)(c_0 - c^{in})dS \tag{12.28}$$

where $x(N, t_E, E, \theta)$ is a permeabilization probability function. The function $x(N, t_E, E, \theta)$ describes the progressive transition function between two membrane states.

When the membrane is impermeable to solutes, $x(N, t_E, E, \theta)$ is zero and therefore the flux is zero. The permeabilization probability function is dependent on a particular experimental set-up. Several models for the permeabilization probability function were proposed for the constant N and t_E:

1. x is a constant (y) and is independent of E and θ:

$$\begin{aligned} &x(N, T, E, \theta) = y \\ &J(t) = P_m \cdot y \cdot S/2 \cdot (1 - E_s/E) \cdot (c_0 - c^{in}) \cdot e^{-k(N, t_E)t} \end{aligned} \tag{12.29}$$

2. x is proportional to E:

$$\begin{aligned} &x(N, T, E, \theta) = yE \\ &J(t) = P_m \cdot y \cdot S/2 \cdot (E - E_s) \cdot (c_0 - c^{in}) \cdot e^{-k(N, t_E)t} \end{aligned} \tag{12.30}$$

3. x is proportional to the external electric field induced transmembrane potential difference. The effect depends on the location on the cell surface:

$$\begin{aligned} &x(N, T, E, \theta) = yE\cos\theta, \\ &\text{for } E\cos\theta > E_s \; J(t) = P_m y S/2(E - E_s^2/E)(c_0 - c^{in})e^{-k(N, t_E)t} \end{aligned} \tag{12.31}$$

4. x is proportional to the square of the induced transmembrane potential:

$$\begin{aligned} &x(N, T, E, \theta) = yE^2\cos^2\theta, \\ &\text{for } E\cos\theta > E_s \; J(t) = P_m y S/2(E^2 - E_s^3/E)(c_0 - c^{in})e^{-k(N, t_E)t} \end{aligned} \tag{12.32}$$

In all the described models [137]:

$$E_s = E\cos\theta_s \tag{12.33}$$

where θ_s is the pole angle in reference to the direction of the applied electric field which encloses the part of the cell membrane that becomes permeable [141,143]. In Eqs. (12.29)–(12.32) the factor of 1/2 is relevant for DNA and RNA molecules only. This factor appears because DNA and RNA transport happens only from one pole of the cell. Experimental results for ATP leakage from CHO cells were successfully fitted to Eq. (12.29), while data from *Abidor et al.* [12] were fitted to Eq. (12.32) [12,141,144]. The concepts discussed here are related to the existence of *macro*- and *micro*membrane domains. *Macro domains* are areas of membranes where the permeabilization can take place at the specific external electric field intensity. This area of the membrane is assessed by [56,143] as:

$$S_m = S(1 - E_s/E) \tag{12.34}$$

Within the area of *macro domains* the actual molecular transport takes place through the *micro domains* (defects), the density of which depends on pulse number and duration. References [53,140] also show that the process of small molecules transfer into cells took place mostly after the electric field was removed. Furthermore, the cell membrane permeabilization to small molecules is an asymmetric process (Fig. 12.3) which is enhanced at the anode-facing side of the cell relative to the cathode-facing side [67,113].

It is important to point out that, although for small molecules the uptake takes place mostly after the application of electric pulses, DNA uptake occurs only when DNA is present during the application of pulses [67]. According to the electropermeabilization model, the transfection efficiency (TE) follows the following equation:

$$TE = WNt_E^{2.3}\left(1 - \frac{E_p}{E}\right)f(ADN) \tag{12.35}$$

where W is a constant and $f(ADN)$ is a complex function, dependent on the plasmid concentration. It was experimentally observed that high levels of plasmids are toxic [67,144].

12.6.3 Electrodiffusion Model of DNA Cluster Formation

Microscopy experiments show that negatively charged DNA plasmids flow around the cell towards the anode if the applied electric field is below a certain threshold value (E_p) [113]. However, when fields above the critical permeabilizing value ($E > E_p$) are applied, DNA clusters

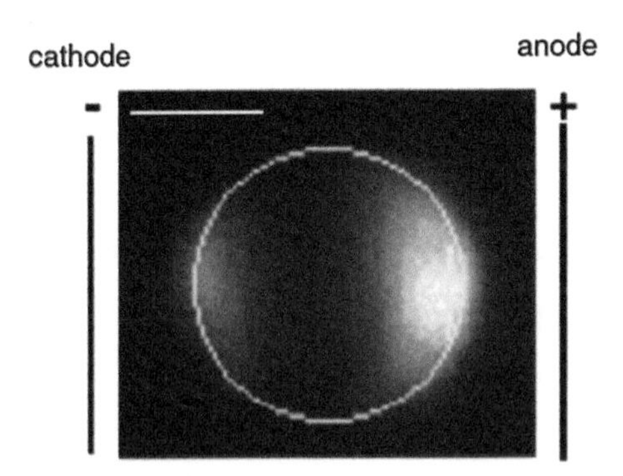

FIGURE 12.3 **The detection of localization and asymmetry of electropermeabilization by propidium iodide.** Die uptake by a CHO cell exposed to 10 pulses, 5 ms, 1 Hz at 0.7 kV/cm. Scale bar: 10 μm. From [67] with permission.

are formed only on the cell pole facing the cathode. The growth of these clusters is conditioned on the presence of the external electric field. The formation of clusters takes place at special membrane *competent sites* in the area of the cathode facing *macro domains* [67]. Figure 12.4 describes the behavior of the DNA plasmid during the application of 10 sequential electric pulses [84].

Following direct optical observations (Fig. 12.4) a multistep model for DNA electrotransport into cells was proposed [67]:

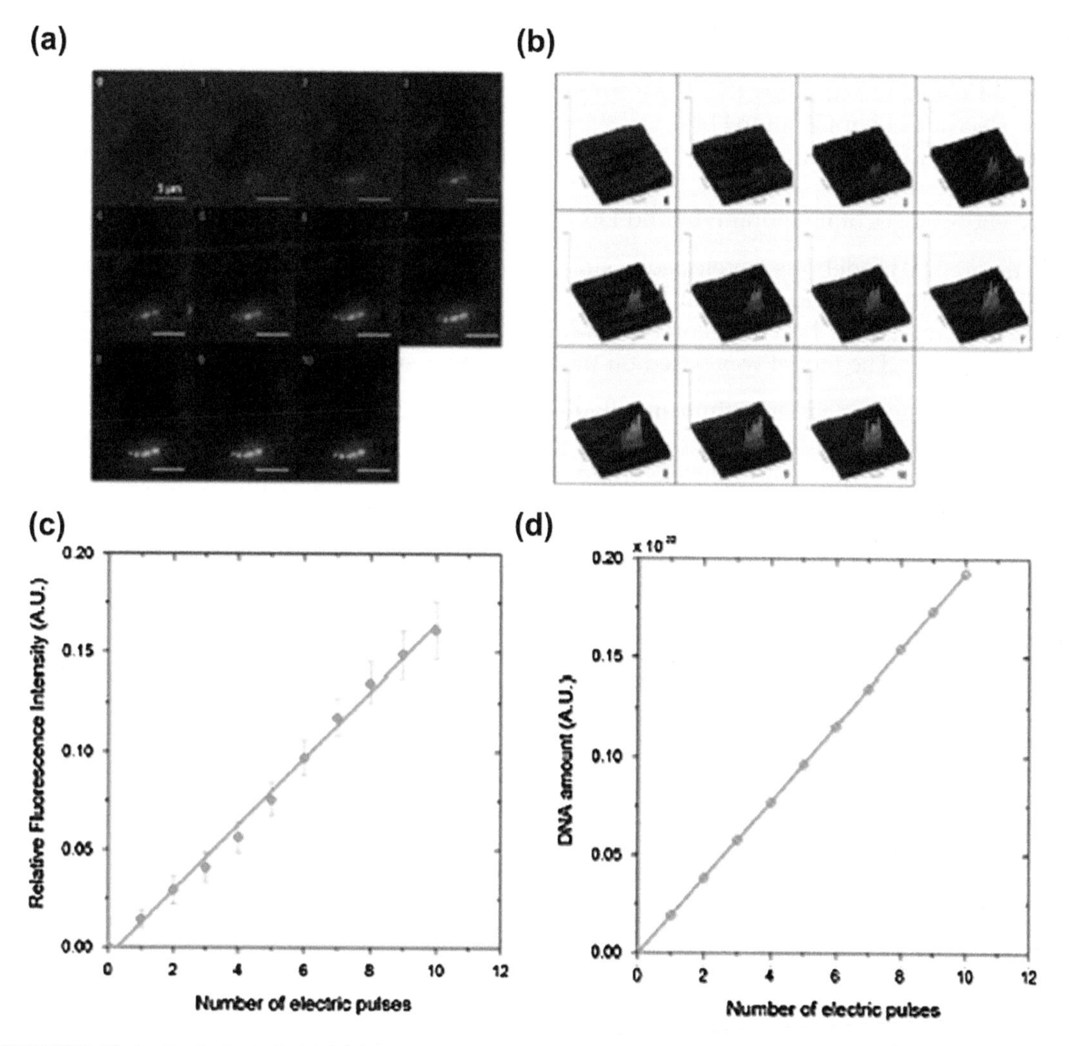

FIGURE 12.4 **Evolution of DNA fluorescence at interaction sites as a function of the number of electric pulses.** (a) Raw imaging data. (b) Quantified increase in DNA fluorescence based on a digital analysis of (a). (c) Experimentally measured increase of total fluorescence due to DNA uptake as a function of the number of applied pulses. (d) Increase in DNA fluorescence as a function of the number of pulses as predicted by the electrodiffusion model. Adapted from [84] with permission.

1. Before the application of the external electric field the membrane acts as a barrier that prevents the entrance of plasmid DNA in the cell.
2. During the application of electric fields above the specific threshold (E_p) the membrane is permeabilized and the negatively charged DNA migrates electrophoretically towards the cathode side of the membrane where it is inserted into *competent sites*, located inside *macro domains*.
3. After the end of electric field application a translocation of a DNA plasmid to cell cytoplasm and nucleus takes place, leading to gene expression. The exact mechanism of the translocation is yet to be elucidated. Several mechanisms for the translocation were proposed. In addition to electrodiffusion [58] they include:

   a. diffusion, increased by the disassembly of the cytoskeleton that may occur during electric field application [145];
   b. aggregation of ion pumps which open and lead to the passage of DNA molecules [146];
   c. endocytoses of membrane-bound DNA [65,67].

The theoretical model was developed to simulate the experimentally observed electrotransfer of 4.7 kbp DNA plasmid into CHO cells [84]. We present the detailed analysis of the model here, to demonstrate how a theoretical framework is developed from direct experimental measurements. The model was based on the following assumptions:

1. The cell membrane is an infinitesimally thin, non-conducting surface.
2. The electric field with magnitude E_0 is applied between two parallel electrodes in the **z** direction, perpendicular to the electrode surfaces.
3. The local charge accumulation due to DNA absorption is neglected.
4. The membrane charging process is much shorter than the duration of the electric pulse and the time scale of the following diffusion processes.
5. Cells have a spherical form.
6. The measured marker fluorescence is proportional to the concentration $c(x,t)$.

In the assumed steady-state regime the local field distribution that causes DNA advection was calculated by the Laplace's equation:

$$\nabla\sigma\nabla\varphi = 0 \tag{12.36}$$

with the boundary conditions:

$$\varphi = -E_0\mathbf{z}, \quad \text{for } \mathbf{z} \to \infty \tag{12.37}$$

The local concentration of DNA, $c(x,t)$, obeys the electrodiffusion equation as follows:

$$\frac{\partial c}{\partial t} = -\nabla j \tag{12.38}$$

where j is the thermodynamic current defined by:

$$j = -D\nabla c + \mu c E \tag{12.39}$$

where D is the local diffusion constant of the DNA, which depends on its location (outside the cell, inside cell membrane and inside the cells) as follows: D_o diffusion coefficient outside the cell; D_m diffusion coefficient inside the membrane; D_i diffusion coefficient inside the cell; μ is the local electrophoretic mobility and is defined as the velocity of a molecule in the local electric field by the following equation:

$$v - \mu E \tag{12.40}$$

where v is the velocity, μ_e is the mobility of the DNA outside; μ_m is the mobility of the DNA inside the membrane; μ_i is the mobility of the DNA inside the cell. The following boundary conditions were used in the study:

$$\begin{aligned} &\nabla c = 0, c \rightarrow \infty \\ &\mu_m = 0 \end{aligned} \tag{12.41}$$

When the electric field is applied, the membrane becomes permeabilized in the *macro domains*. In this study, the authors assumed the formation of a circular pore on each electrode facing cell pole. At the angular region $\theta \in (0, \theta_p)$ facing the cathode and $\theta \in (\pi, \theta_p, \pi)$ facing the anode the conductivity and diffusion constant of the membrane was changed to those of intracellular solutions. θ_p was estimated by direct microscopic observation. The formed pore area is the transfer zone through which species can penetrate into the cell. The velocity of the DNA in the solution under the application of a uniform electric field was measured by direct fluorescence measurement. Using Eq. (12.40), μ_e was estimated. The diffusion coefficient D_e was adopted from literature [147]. μ_i and D_i were assumed to be 3 orders of magnitude smaller than respective outside cell values due to the interaction of DNA and the actin network [147,148]. To validate the model the DNA marker fluorescence was measured with temporal resolution of 2 ms [84].

12.6.4 Model of Conductivity Changes During Cell Suspension Electroporation

Miklavcic et al. [98,149] used a dense cell suspension to analyze how E and N affect transient conductivity changes and the molecular transport. The main focus of these studies is to correlate the experimental findings of electrical field-mediated increase in the membrane conductivity and the related transport of ions during pulses.

The increase in the membrane permeability is described as the fraction of the permeable surface of the cell membrane or the fraction of all transport 'pores' as follows:

$$f_{per} = \frac{S_m}{S_{tot}} = \frac{S_{por}}{S_0} \tag{12.42}$$

where S_{tot} represents the total area of n cells; S_{por} represents the total area of all pores of a single cell; S_0 is the total area of a single cell; f_{per} is the fraction of pores which are large enough to cause permeability of ions and molecules through the cell membrane. The diffusion of ions through the membrane is governed by the Nernst-Planck equation:

$$\frac{dn_e(x,t)}{dt} = -DS_m \frac{dc(x,t)}{dx} - \frac{z_{eff}F}{RT} DSc(x,t) \frac{d\varphi(x,t)}{dx} \tag{12.43}$$

where n_e is the number of molecules in the external medium and F is the Faraday constant. The part of the membrane where the diffusion takes places is defined by Eq. (12.34).

Moreover, the diffusion of ions and small molecules is a slow process and happens mostly after the pulse electric field has ceased. Therefore, the second part of Eq. (12.43) is neglected and the equation is reduced to the following form:

$$\frac{dn_e(x,t)}{dt} = -\frac{(c^{out} - c^{in})}{d} Df_{pc}(E, t_E, N)\left(1 - \frac{E_p}{E}\right) S_0 \tag{12.44}$$

where $f_{per} = f_{pc}(1 - \frac{E_c}{E})$ and $f_{pc} = \frac{S_{pore}}{S_c}$ represent the fraction of pores in the permeabilized region which is calculated in Eq. (12.34). Under the assumption that the volume fraction F′ and surface area of pores S_{por} is approximately constant (resealing kinetics is not taken in consideration in this model), the solution of Eq. (12.44) is an exponential increase of the internal concentration of a species of interest:

$$c_{in}(t) = c_{in}^{max}(1 - e^{-t/\tau}) \tag{12.45}$$

where $c_{in}^{max} = F' c_{in}^{o}$. The time constant τ and flow coefficient k depend on f_{per} in the following way:

$$\tau = \frac{D_m}{f_{per} dr(1 - F')} \tag{12.46}$$

$$k = 1/\tau \tag{12.47}$$

From Eq. (12.45), it appears that the solution of $c_{in}(t)$ increases exponentially to a maximum c_{in}^{max}. Thus, the conductivity between the pulses also increases exponentially due to the ion flux in the membrane. Therefore, the flow coefficient k_N after the Nth pulse can be determined from the measured conductivity between Nth pulse and N+1th pulse (a detailed derivation is given in [98]):

$$k_N = \frac{1}{\Delta t_N} ln\left(\frac{1 - \Delta\sigma_N/\Delta\sigma_{max}}{1 - \Delta\sigma_{N+1}/\Delta\sigma_{max}}\right) \tag{12.48}$$

where Δt_N is the interval between two consequent pulses; $\Delta\sigma_{max}$ is the maximum value of conductivity when the concentrations inside and outside of cells are equal.

Knowing the permeability coefficient k_N, the fraction of pores can be estimated by:

$$f_{per} = k_N \frac{dr\,(1 - F')}{2D_m} \tag{12.49}$$

Based on the recent thermodynamic theories of pore formation energy [13,18,150], the following equation was proposed to link the intensity of pulse electric fields with k defined as:

$$k_N(E) = C_N\left(1 - \frac{E_p}{E}\right) E^2 \tag{12.50}$$

where C_N is a constant that depends on the size of the pores and their growth kinetics and thus on the number of the applied pulses.

12.6.5 Model of Small Molecule Transport Kinetics Due to Electroporation

A study on the kinetics of small molecule transport during and after electroporation was published in [122]. The transport mechanism of ions is different from that of small sized drug-type molecules. The diffusion of ions is much faster than the diffusion of molecules. Therefore, the pore resealing kinetics was neglected in the analyses of ion diffusion assessments through conductivity measurements [98]. Molecular transport was detected and monitored milliseconds, seconds and even minutes after the application of the electric pulses [59,69,128,142,151,152]. The flow of ions and molecules continued until the membrane is fully recovered, or until equilibrium is achieved between the concentration inside and outside the cells [122]. The authors performed an ultra-high speed imaging study of propidium iodine (PI) transport into electropermeabilized CHO cells. The kinetics of dye transport was investigated with time resolutions in the range of from 200 ns to 4 ms. The research revealed new high time resolution curves of the dye uptake by the cells. It was assumed that the fluorescence signal intensity is correlated with the PI accumulation by cells. The work revealed the interesting fact that PI transport into the cells has a delay relative to the application of the pulse. The study shows that the fluorescence signal becomes detectable 60 to 500 μs after the start of the pulse. These results mean that for typical electroporation pulse durations (70–100 μs) the transport of small molecules starts during the pulse. A theoretical model was proposed to describe the behavior of PI transport into CHO cells during and after the pulse.

The transport of PI molecules is quantitatively characterized by the following form of the Nernst–Planck equation:

$$\frac{dc^{in}(t)}{dt}\frac{V_c}{S_m(t)} = -D_e\frac{zF}{RT}c^{in}(t)E - D_e\frac{dc^{in}(t)}{dx} \tag{12.51}$$

where $c(t)$ is the molar concentration of the species of interest passing through the permeabilized part of the membrane; $S_m(t)$ is the surface of permeable structure defects; R is the gas constant; T is the absolute temperature and D_e is the diffusion constant defined by the Einshtein-Smoluchowski relation:

$$D_e = \mu k_B T \tag{12.52}$$

where μ is the mobility; k_B is the Boltzmann constant.

To solve Eq. (12.51) the following assumptions were made:

1. The extracellular concentration c_o of PI is constant. The volume of extracellular solution is much larger than the total volume of cells.
2. Due to the fact that the first term of the right side of Eq. (12.51) describes the PI molecules brought into motion by electrophoresis in the extracellular solution, $c^{in}(t)$ in the first term was approximated by c_o.

3. If the membrane thickness d is known $\frac{dc^{in}(t)}{dx}$ is approximated by $\frac{(c_0-c^{in})}{d}$. Eq. (12.51) can be rewritten as:

$$\frac{dc^{in}(t)}{dt}\frac{V_c}{S_m(t)} = -D_e\frac{zF}{RT}c_0E - D_e\frac{c^{in}(t)-c_0}{d} \tag{12.53}$$

Diffusion processes are slower than the pulse duration during typical electroporation protocols [59,153]; therefore, the contribution of diffusion to the total mass transfer during the pulse is neglected in a first approximation. In addition, the authors assumed that $S_p(t)$ is constant during the pulse application.

The new equation, which describes the mass transport of species c during the application of an electric field, is given by:

$$\frac{dc^{in}(t)}{dt}\frac{V_m}{S_m} = -D_e\frac{zF}{RT}c_0E \tag{12.54}$$

The exact analytical solution of this equation is as follows:

$$c^{in} = \frac{D_eS_m}{V_c}\frac{zF}{RT}c_0Et \tag{12.55}$$

Eq. (12.55) predicts a linear increase in the molecular concentration inside the cells during the application of an electric field. This prediction was proven by fluorescence measurements during the pulse, which did indeed show a linear increase in the fluorescent signal after the first 60 to 500 μs, when the signal was not detectable.

After the end of the pulse, electrophoresis ceases and the diffusion component of Eq. (12.51) is dominant. The transport equation becomes:

$$\frac{dc^{in}(t)}{dt}\frac{V_c}{S_m(t)} = -D_e\frac{c^{in}(t)-c_0}{d} \tag{12.56}$$

The shift from electrophoresis to diffusion driven transport is reflected in the decreasing rate of fluorescence accumulation. The signal achieves saturation by two mechanisms:

1. Equilibrium of concentrations of PI;
2. Membrane resealing.

Membrane resealing kinetics were investigated in [11,18,74,122,154] and were shown to have several phases starting from a short (μs) time scale to a time scale of minutes. A physico-kinetic analysis of pore resealing was developed in [23].

As a first approximation of resealing kinetics, the time dependent permeabilized area of the cell $S_m(t)$ was described by the following equation:

$$S_m(t) = S_{m1}e^{-t/\tau_1} \tag{12.57}$$

This approximation leads to a closed form solution for the intracellular PI concentration:

$$c^{in}(t) = v + b(1 - e^{ce^{-t/\tau_1}}) \tag{12.58}$$

where:

$$
\begin{aligned}
v &= c_0 - (c_0 - c^{in0})(1 - e^{-(DS_{m1}\tau_1/V_c d)e^{-1/\tau_1}}) \\
b &= (c_0 - c^{in0})e^{-(DS_{m1}\tau_1/V_c d)e^{-1/\tau_1}} \\
c &= D_e S_{m1}\tau_1/V_c d
\end{aligned}
\tag{12.59}
$$

Eq. (12.58) was fitted to the experimental data and a good correlation was observed for observation times of $t > 1$ sec. Amendments to Eq. (12.58) include the incorporation of additional exponential terms with a shorter time constant. Eq. (12.60) describes the kinetics of pore resealing in three steps:

$$S_m(t) = S_{m1}e^{-t/\tau 1} + S_{m2}e^{-t/\tau 2} + S_{m3}e^{-t/\tau 3} \tag{12.60}$$

A closed form solution of Eq. (12.56) with $S_p(t)$ described by Eq. (12.60) was not found and a numerical analysis was performed. Fitting Eq. (12.60) to the experimental data revealed that $\tau 1 = 14$ sec, $\tau 2 = 380$ ms, $\tau 3 = 12$ ms. The experimental observations of the fluorescence curve behavior were explained as follows. First, at the very small time scales all three components of Eq. (12.60) play a role. For $t \gg \tau 3$ the third term $S_{m3}e^{-t/\tau 3}$ becomes insignificant. For $t \gg \tau 2, S_{m2}e^{-t/\tau 2}$ becomes insignificant also. Therefore, a good correlation was observed between the experimental curves and Eq. (12.58) for times $t > 1$ sec. This also reveals the nature of c^{in0} in Eq. (12.59). It is found that c^{in0} contains molecules contributed by the transport processes described by $S_{m2}e^{-t/\tau 2} + S_{m3}e^{-t/\tau 3}$.

12.6.6 Two Compartment Pharmacokinetic Model for Molecular Uptake During Electroporation

A two compartment pharmacokinetic model was developed by *Puk et al.* [76] (Fig. 12.5). The experimental data on LY uptake after the application of a single pulse with different amplitudes and duration was used to construct a model which predicts the quantity of molecules introduced by diffusion into the cell after the treatment with a specific pulse electric field parameter. The flow coefficient M was derived from the model. Permeabilization area defines the relative part of the cell surface through which the molecular exchange between the cell and the environment takes place.

Mass transfer in the two compartment model (Fig. 12.5) is described by the following equation [155]:

$$
\begin{aligned}
\frac{dm_0}{dt} &= -k_{oi}m_0 + k_{io}m_i \\
\frac{dm_i}{dt} &= k_{oi}m_o - k_{io}m_i
\end{aligned}
\tag{12.61}
$$

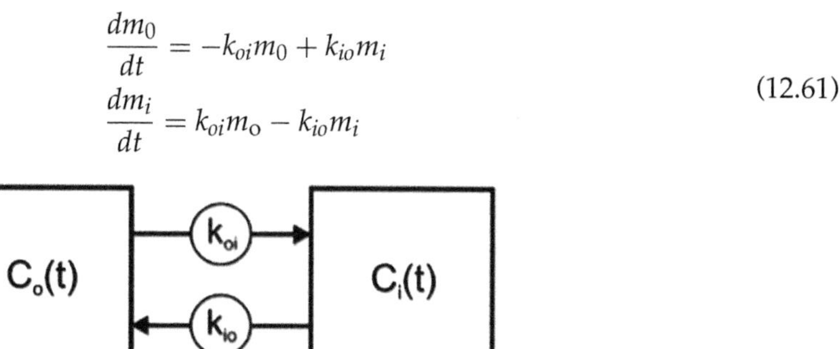

FIGURE 12.5 **Two compartment pharmacokinetic model scheme.** o – denotes the cell exterior and i – the cell interior.

where m_0 and m_i are the masses of a substance of interest outside the cell and inside the cell, respectively; k_{oi} and k_{io} are the coefficients which describe the flow between the outside of the cell to the inside and from the inside of the cell to the outside, respectively.

Written in the form of mass concentration, Eq. (12.61) transforms into Eq. (12.62) in the following way:

$$\begin{aligned} \frac{dc^{out}}{dt} &= -k_{oi}c^{out} + k_{io}\frac{V_o}{V_i}c^{in} \\ \frac{dc^{in}}{dt} &= k_{oi}\frac{V_i}{V_o}c^{out} - k_{io}c^{in} \end{aligned} \tag{12.62}$$

where V_i is the volume of all electropemeabilized cells and V_o is the volume of the outside chamber. At equilibrium, $c^{out} = c^{in}$; therefore:

$$\frac{dc^{out}}{dt} = 0 \Rightarrow k_{oi} = k_{io}\frac{V_i}{V_o} \tag{12.63}$$

Consequently:

$$\begin{aligned} \frac{dc^{out}}{dt} &= k_{io}\frac{V_o}{V_i}(c^{in} - c^{out}) \\ \frac{dc^{in}}{dt} &= k_{io}(c^{out} - c^{in}) \end{aligned} \tag{12.64}$$

However, as discussed earlier, mass transport processes during electropermeabilization can be divided in two phases: during the application of the pulse, and after the pulse during the membrane resealing [58,156,157]. To describe the dynamic behavior of the system an expression for the time dependence k_{io} as is given by:

$$\begin{aligned} k_{io} &= M, \quad 0 < t \le t_E \\ k_{io} &= gMe^{k_{-1}(t_E - t)}, \quad t_E < t \le t_{obs} \end{aligned} \tag{12.65}$$

where M [s^{-1}] is the flow coefficient; k_{-1} [s^{-1}] is the time constant for resealing; g is a parameter that characterizes the fast resealing of the membrane, which takes place immediately after the end of the pulse; t_p is the pulse duration; t_{obs} is the total observation time.

In reference [76], the experiment was performed in two parts. In the first part, cells were immersed in the solution with LY in which the pulse treatment was done. Ten minutes after the pulse field treatment, the cells were transferred to the growth medium. The following simplifying assumptions were made. During the pulse treatment, the concentration of LY (c_B) in the media did not change, even though some molecules were taken up by electropermeabilized cells. After the resuspension in the medium, the outside concentration c_0 is zero and does not change. The additional assumption is that leakage from cells does not happen. The equations describing this process take the following form:

$$\begin{aligned} \frac{dc^{in}}{dt} &= k_{io}(c_B - c^{in}), \quad 0 < t \le t_{10} \\ \frac{dc^{in}}{dt} &= -k_{io}c^{in}, \quad t_{10} < t \le t_{20} \end{aligned} \tag{12.66}$$

The solution of Eq. (12.66), which gives the concentration of molecules inside the cell is:

$$\begin{aligned} c^{in}(t) &= c_B(1 - e^{-Mt}), \quad 0 < t \leq t_E \\ c^{in}(t) &= c_B + (c^{in}(t_E) - c_B)e^{-\frac{Mg}{k_{-1}}(1 - e^{k_{-1}(t_E - t)})}, \quad t_E < t \leq t_{10} \\ c^{in}(t) &= c^{in}(t_{10})e^{-\frac{Mg}{k_{-1}}(e^{k_{-1}(t_E - t_{10})} - e^{k_{-1}(t_E - t)})}, \quad t_{10} \leq t \leq t_{20} \end{aligned} \tag{12.67}$$

Parameter M in this equation was found from fitting the experimental results to the numerical analysis.

12.6.7 Statistical Model for Cell Electrotransformation

A statistical approach to cell electrotransformation analysis was proposed by *Canatella and Prausnitz* [73]. Usually, the developments of electroporation protocols are based on a trial-and-error approach for a specific experimental set-up. A statistical approach, based on more than 200 different experimental conditions for electrotransfection of DU145 prostate cancer was described in [73]. The authors used a nonlinear regression approach and introduced the following equation, which describes the absorption of an absolute number of molecules (N) per cell:

$$N = 7.0 \cdot 10^7 c^{out} t_E^{0.31} n^{0.12} v \left(1 - e^{-1.4.10^{-3} t_E^{2.2} n^{2.1} \Delta\varphi_m^{4.8}}\right) \tag{12.68}$$

where n is the number of pulses, t_E is the pulse length and is the cell volume. Eq. (12.68) was tested for its ability to predict the uptake of a range of different molecules (mostly DNA) in over 900 experimental conditions, including data from 60 different cell lines from 33 different literature studies [73].

12.6.8 A Multiscale Model for Mass Transfer of Drug Molecules in Tissue

When electroporation is done *in vitro*, the cells are usually immersed in a solution containing the chemical species that is to be introduced into the cell. Under these conditions, all cells are exposed to a rather uniform, large, external, extracellular concentration of the solution. However, during *in vivo* electroporation in tissues, the molecule of interest (drugs or DNA) is injected into the tissue; hence the extracellular distribution of the molecules of interest is not homogenous. A basic multiscale model for mass transfer of drug molecules in tissue was introduced in reference [103]. An important thing to consider in the case of electroporation in tissue is the degree of cell packing density in the tissue [158]. The degree of packing and the shape of cells will affect the electrical potential that the cells experience. Therefore, in tissue, the local transmembrane potential at various parts of each cell's membrane changes considerably with the geometry of the cell and the geometry of the tissue [159]. Also, when chemical species (drugs or DNA) are injected in tissue, the extracellular concentration is also location specific and can vary from site to site.

Realistic biophysical models of electroporation should take into account the variation of the transmembrane potential in solid tissue and the variation in the distribution of the molecule of interest in the tissue after the injection. However, due to the limited analyses of mass transport even for a single spherical cell in solutions, the analysis of a complex tissue system is left to future work, and in this text, the authors have focused on developing the multiscale model system now described.

Granot and Rubinsky [103] proposed a basic design for a class of multiscale models that may be used to describe a mass transfer in tissues *in vivo*. The model deals with mass transfer at the single cell and tissue bulk levels. For the tissue bulk level, the model describes the generally non-uniform electric field in the treated region. The electric field is affected by the tissue geometry, temperature, the local conductivity at different regions of the tissue if the tissue is not homogenous, the electroporation electrode configuration and by the applied treatment protocol [160–164]. The statistical effect of applied pulse electric fields in tissues is discussed in [165]. Mass transport of molecules across the tissue is modeled at the bulk tissue scale to evaluate how the concentration changes as the molecule enters the electropermeabilized cells. The single cell scale model describes the electroporation effects on the cell membrane. The cell membrane permeability dynamics due to the applied electric fields are considered and correlated to the transfer of drug molecules across the membrane into the cell interior.

First, a single cell level analysis of molecular transport was performed. The following simplifying assumptions were made:

1. Cells are exposed to a homogeneous electric field. Even though heterogeneous fields arise on the large scale tissue level, the changes in the electric field at the small scale cell level are generally negligible. Therefore, the authors assumed that the electric field in the vicinity of the cell is uniform, although the field strength will vary in different regions of the tissue.
2. Cells have a spherical form. Figure 12.4 shows that cells in tissues have a variety of structures. This fact, consequently, will have an impact on the distribution of the electric field. Nevertheless, the authors made the simplifying assumption of uniform spherical cell structure for the sake of clarity in the description of a general class of models.

The transmembrane potential over a single cell was calculated as described in Eq. (12.36). A single cell model proposed by *Krassowka and Filev* [166] was employed to evaluate the time dependent size of pores in different regions of a cell, both during the application of the electric field and after the field ceased. Using this model the number of pores is computed as a function of $\Delta\varphi_m$ using the pore density function:

$$\frac{dO}{dt} = k_1 e^{(\varphi_m/\varphi_{ep})^2}\left(1 - \frac{O}{O_0 e^{q(\varphi_m/\varphi_{ep})^2}}\right) \tag{12.69}$$

where O is the pores density, O_0 is the equilibrium pore density, k_1 is the pore creation rate, φ_m is the local transmembrane potential, φ_{ep} is the characteristic voltage of electroporation (threshold voltage) and q is an electroporation constant ($q = 2.46$).

An important parameter for molecular transport into cells is pore (membrane defect) size. First, larger pores will allow large molecules to cross the membrane and effectively render the cell permeable to molecules that would not be able to enter the cell if only smaller pores existed, regardless of the number of small pores.

Second, the area of a pore depends on its radius, and thus larger pores contribute more effective area through which molecules can travel than smaller sized pores. In this model, the authors assumed that all pores have the same size and that at the end of the pulse the pore radius shrinks to the minimal pore radius r and begins to reseal as described in [166].

Even though mass transfer occurs during the application of the electric field, the pulse duration is short in comparison with the total time in which molecules penetrate the cell. In this model, the mass transfer that takes place after the application of electric pulses was calculated.

The total area of pores S_m through which the molecular transport occurs is given by Eq. (12.70):

$$S_m = \pi r^2 O_p \tag{12.70}$$

where O_p is the number of pores at the end of the electroporation pulse and is calculated using Eq. (12.71):

$$O_p = \int N dS \tag{12.71}$$

According to the model of pore dynamics [167] and as previously discussed, the pores start to reseal as their number decreases exponentially. This is given by Eq. (12.72):

$$S_m = \pi r^2 O_p e^{-t/\tau} \tag{12.72}$$

Eq. (12.72) describes the relevant area of the cell through which the transport of small molecules takes place after electroporation. With no obstructions to impede their motion, molecules that are found in large concentrations in the extracellular medium will now diffuse into the cell. When no molecules are added or removed from the system the diffusion equation describes the concentration of the molecules (c) as a function of space and time, which can be calculated through Fick's diffusion law (Eq. (12.25)).

The goal of the work in [103] was to build a model for calculating the number of molecules that enter a cell by diffusing through the aqueous pores across the cell membrane. To obtain that, the authors took the diffusion flux J, which is given in dimensions of amount of molecules per unit area per unit time, and multiplied it by the total electroporated area S_m. This results in the number of molecules that enter the cell per unit time. According to Fick's first law of diffusion for an external molecule concentration c^{out} and an internal concentration c^{in}, the flux is given by the following equation:

$$J = -P_m(c^{in} - c^{out}) \tag{12.73}$$

where P_m is the permeability of the molecules through the membrane's pores. The permeability is often determined experimentally and is a measure of how well molecules may flow through the pore under certain conditions. Actual values of the permeability depend on a number of parameters, such as the size of the molecule and the interaction between the molecules. For large pores P_m can be approximated as $P_m = D/d$, where D is Fick's diffusion coefficient and d is the membrane thickness.

The next step in the analysis combines the single cell part of the model and the bulk tissue model. The authors examined single cell electroporation-mediated transport in every cell in the tissue. For the bulk tissue analysis, the authors assumed that cells are infinitesimally small, and modeled them as a distributed reaction rate element. At the single cell level this assumption means that molecules which enter the cell do not leave it again, and may be treated as if they are removed from the system. At the tissue bulk scale this assumption means that a distributed mass sink exists in tissue, through which the molecules leave the system in the regions where electroporation took place.

The extent of electroporation in each region of tissue S_m prescribes the strength of the flux J and thus the number of molecules which disappear in that specific region. The authors assumed that the cells are uniformly packed in the tissue so that each spherical cell is contained in a cube, with edges equal to $2a$ (a-cell radius). In this system the reaction rate of molecular disappearance is described by the following equation:

$$R = \frac{JS_m}{(2a)^3} \tag{12.74}$$

To summarize, in the single cell model a single spherical cell is surrounded by an extracellular medium containing a high concentration of a molecule of interest. The concentration of these molecules in the cell is zero. By inducing an above-threshold transmembrane potential, pores are created in the cell's membrane and some of the molecules diffuse into the cytoplasm. These molecules bind to internal structures in the cell and are effectively removed from the system [103].

For the bulk tissue model, two aspects of the analysis are combined. First, the local electric field values are computed for different points in the tissue. The electric field is defined by the configuration of the electrodes and the tissue's properties. The computed electric field is used for the analyses at the single cell level for the calculation of reaction rate as described in Eq. (12.74). Further, the concentration throughout the extracellular space is calculated using the following equation:

$$\frac{\partial c}{\partial t} - \nabla(D\nabla c) = R \tag{12.75}$$

Eq. (12.75) is solved using the diffusion coefficient D and is subject to initial conditions as well as boundary conditions. The initial concentration of molecules in the extracellular space is assumed to be uniform throughout the tissue, neglecting the volume of cells in which there are no molecules. The boundary conditions can often be modeled as no mass flux at the boundary of the tissue.

12.7 FUTURE NEEDS IN MATHEMATICAL MODELING OF MASS TRANSPORT FOR ELECTROPORATION RESEARCH

The use of electric fields for delivering molecules of different size and charge into cells and tissues is becoming increasingly important in biotechnology, medicine and research.

Electrochemotherapy has become a clinical method, and gene electrotransfer technique is being researched clinically. However, most of the research and the applications on delivering new molecules into cells employ methodologies based on past experience, educated guesses and 'trial-and-error' type principles. The number of parameters involved in the mass transfer process is very large, and the current methodologies require a significant time and labor investment. The mass transfer process depends on pulsed electric field parameters (such as field intensity, pulse number, frequency, duration and shape), medium properties (viscosity and conductivity) and properties of cells and tissue (size, shape, density, growth rate membrane structure and composition). In addition, environmental conditions (such as temperature) also affect the efficiency of the electroporation procedure. Comprehensive mathematical models of electric field facilitated mass transfer in tissue could significantly reduce the time and effort needed for developing new clinical and research protocols. Therefore, we believe that it is of great importance to develop comprehensive and cohesive models of electric field facilitated mass transfer in tissues, which incorporate all the parameters listed above.

Another area in which electric field facilitated mass transfer modeling would be of use is the emerging discipline of Synthetic Biology. Synthetic Biology deals with the design and construction of new biological components, such as enzymes, genetic circuits and cells, or the redesign of existing biological systems. One of the major obstacles in accomplishing the goals of Synthetic Biology is related to the assembly of new metabolic pathways in a host. The challenge is to precisely and reliably deliver large numbers of desirable molecules of different types, most of which are membrane impermeant, into specific targeted cells of interest and to perform large DNA fragment assembly. We believe that mathematical models to predict molecular transport into cells will significantly improve the transfection efficiency, and decrease the time and labor needed for protocol optimization of *in vivo* construction of metabolic pathways. For example, DNA electrotransport was shown to take place only at very specific locations on the membrane. The nature of these locations, their impact on DNA absorption into cells and the impact of external electric fields can be included in models. Furthermore, the kinetics of DNA absorption onto these specific domains and the following translocation into the cell cytosol and especially to the nucleus (in eukaryotic cells) requires new mathematical models of mass transfer, of a type that currently does not exist.

References

[1] B. Rubinsky, Irreversible electroporation in medicine, technology, Cancer Research and Treatment 6 (2007) 255–260.

[2] G.W. Fuller, Report on the investigations into the purification of the Ohio River water at Louisville, Kentucky, D. Van Nostrand Company, New York, 1898.

[3] A. Anderson, R. Finkelshtein, A study of the electro-pure process of treating milk, Journal of Dairy Science 2 (1919) 331–434.

[4] A.J. Sale, W.A. Hamilton, Effect of high electric field on micro-organisms. I. Killing of bacteria and yeast, Biochimica et Biophysica Acta – Biomembranes 148 (1967) 781–788.

[5] A.J. Sale, W.A. Hamilton, Effect of high electric field on micro-organisms. II. Mechanism of action of the lethal effect, Biochimica et Biophysica Acta – Biomembranes 148 (1967) 788–800.

[6] A.J. Sale, W.A. Hamilton, Effects of high electric fields on microorganisms. III. Lysis of erythrocytes and protoplasts, Biochimica et Biophysica Acta 163 (1968) 37–43.

[7] H. Doevenspeck, Influencing cells and cell walls by electrostatic impulses, Fleischwirtschaft 13 (1961) 968–987.

[8] E. Neumann, M. Schaefer-Ridder, Y. Wang, P.H. Hofschneider, Gene transfer into mouse lyoma cells by electroporation in high electric fields, EMBO Journal 1 (1982) 841–845.

[9] L. Mir, Nucleic acids electrotransfer-based gene therapy (electrogenetherapy): past, current, and future, Molecular Biotechnology 43 (2009) 167–176.

[10] U. Zimmerman, J. Vienken, G. Pilwat, Dielectric breakdown of cell membranes, Biophysical Journal 14 (1974) 881–899.

[11] K. Kinosita Jr., T.Y. Tsong, Voltage-induced pore formation and hemolysis of human erythrocytes, Biochimica et Biophysica – Biomembranes 471 (1977) 227–242.

[12] I.G. Abidor, V.B. Arakelyan, L.V. Chernomordick, Y.A. Chizmadhev, V.F. Pastushenko, M.R. Tarasevich, Electric breakdown of bilayer membranes. I. The main experimental facts and their qualitative discussion, Journal of Electroanalytical Chemistry and Interfacial Electrochemistry 6 (1979) 37–52.

[13] J.C. Weaver, Y.A. Chizmadzhev, Theory of electroporation: a review, Bioelectrochemistry and Bioenergetics 41 (1996) 135–160.

[14] R.A. Bockmann, B.L. de Groot, S. Kakorin, E. Neumann, H. Grubmuller, Kinetics, statistics, and energetics of lipid membrane electroporation studied by molecular dynamics simulations, Biophysical Journal 95 (2008) 1837–1850.

[15] B.V. Deryagin, Y.V. Gutop, Theory of the breakdown (rupture) of free films, Kolloidn Zhurnal 24 (1962) 370–374.

[16] J.D. Litster, Stability of lipid bilayers and red blood cell membranes, Physics Letters 53A (1975) 193–194.

[17] C. Taupin, M. Dvolaitzky, C. Sauterey, Osmotic pressure induced pores in phospholipid vesicles, Biochemical Pharmacology 14 (1975) 4771–4775.

[18] R.W. Glaser, S.L. Leikin, L.V. Chernomordik, V.F. Pastushenko, A.I. Sokirko, Reversible electrical breakdown of lipid bilayers: formation and evolution of pores, Biochimica et Biophysica Acta – Biomembranes 940 (1988) 275–287.

[19] J.C. Weaver, Electroporation of biological membranes from multicellular to nano scales, IEEE Transactions on Dielectrics and Electrical Insulation 10 (2003) 754–768.

[20] H.G.L. Coster, The Physics of Cell Membranes, Journal of Biological Physics 29 (2003) 363–399.

[21] H.L.M. Lelieved, S. Notermans, S.W.H. de Haan, Food Preservation by Pulsed Electric Fields. From Research to Application, CRC, Cambridge, England, 2007.

[22] M. Somolinos, P. Mañas, S. Condón, R. Pagán, D. Garcá, Recovery of *Saccharomyces cerevisiae* sublethally injured cells after pulsed electric fields, International Journal of Food Microbiology 125 (2008) 352–356.

[23] G. Saulis, Pore disappearance in a cell after electroporation: theoretical simulation and comparison with experiments, Biophysical Journal 73 (1997) 1299–1309.

[24] E. Neumann, Gene delivery by membrane electroporation, in: P.T. Lynch, M.R. Davet (Eds.), Electrical Manipulation of Cells, Chapman and Hall, New York, 1996, pp. 157–184.

[25] E. Neumann, The relaxation hysteresis of membrane electroporation, in: E. Neumann, A.E. Sowers, C. Jordan (Eds.), Electroporation and Electrofusion in Cell Biology, Plenum Press, New York, 1989, pp. 61–82.

[26] D. Gross, L.M. Loew, W.W. Webb, Optical imaging of cell membrane potential changes induced by applied electric fields, Biophysical Journal 50 (1986) 339–348.

[27] G. Pucihar, D. Miklavcic, T. Kotnik, A time-dependent numerical model of transmembrane voltage inducement and electroporation of irregularly shaped cells, IEEE Transactions on Biomedical Engineering 56 (2009) 1491–1501.

[28] , Transfection of mammalian cells by electroporation, Nature Methods 3 (2006) 67–68.

[29] Y. Huang, B. Rubinsky, Micro-electroporation: improving the efficiency and understanding of electrical permeabilization of cells, Biomedical Microdevices 2 (1999) 145–150.

[30] Z. Fei, S. Wang, Y. Xie, B.E. Henslee, C.G. Koh, L.J. Lee, Gene transfection of mammalian cells using membrane sandwich electroporation, Analytical Chemistry 79 (2007) 5719–5722.

[31] A. Valero, J.N. Post, J.W. van Nieuwkasteele, P.M. ter Braak, W. Kruijer, A. van den Berg, Gene transfer and protein dynamics in stem cells using single cell electroporation in a microfluidic device, Lab on a Chip 8 (2008) 62–67.

[32] T. Geng, Y. Zhan, J. Wang, C. Lu, Transfection of cells using flow-through electroporation based on constant voltage, Nature Protocols 6 (2011) 1192–1208.

[33] M. Flanagan, J.M. Gimble, G. Yu, X. Wu, X. Xia, J. Hu, S. Yao, S. Li, Competitive electroporation formulation for cell therapy, Cancer Gene Therapy 18 (2011) 579–586.

[34] F. Fernand, L. Rubinsky, A. Golberg, B. Rubinsky, Variable electric fields for high throughput electroporation protocol design in curvilinear coordinates, Biotechnology and Bioengineering (2012), http://dx.doi.org/10.1002/bit.24479.

[35] C. Weaver James, R. Langer, O. Potts Russell, Tissue electroporation for localized drug delivery, Electromagnetic Fields, American Chemical Society, 1995, pp. 301–316.

[36] F.M. André, J. Gehl, G. Sersa, V. Préat, P. Hojman, J. Eriksen, M. Golzio, M. Cemazar, N. Pavselj, M.P. Rols, D. Miklavcic, E. Neumann, J. Teissié, L.M. Mir, Efficiency of high- and low-voltage pulse combinations for gene electrotransfer in muscle, liver, tumor, and skin, Human Gene Therapy 19 (2008) 1261–1271.

[37] M. Golzio, L. Mazzolini, A. Ledoux, A. Paganin, M. Izard, L. Hellaudais, A. Bieth, M.J. Pillaire, C. Cazaux, J.S. Hoffmann, B. Couderc, J. Teissie, *In vivo* gene silencing in solid tumors by targeted electrically mediated siRNA delivery, Gene Therapy 14 (2007) 752–759.

[38] T. Kishida, H. Asada, S. Gojo, S. Ohashi, M. Shin-Ya, K. Yasutomi, R. Terauchi, K.A. Takahashi, T. Kubo, J. Imanishi, O. Mazda, Sequence-specific gene silencing in murine muscle induced by electroporation-mediated transfer of short interfering RNA, The Journal of Gene Medicine 6 (2004) 105–110.

[39] S. Li, M. Benninger, Applications of muscle electroporation gene therapy, Current Gene Therapy 2 (2002) 101–105.

[40] L.C. Heller, K. Ugen, R. Heller, Electroporation for targeted gene transfer, Expert Opinion on Drug Delivery 2 (2005) 255–268.

[41] R. Heller, R. Gilbert, M.J. Jaroszeski, Clinical applications of electrochemotherapy, Advanced Drug Delivery Reviews 35 (1999) 119–129.

[42] J. Gehl, S. Li, Electroporation for drug and gene delivery in the clinic: doctors go electric, in: J.M. Walker (Ed.), Electroporation Protocols, Humana Press, 2008, pp. 351–359.

[43] L.M. Mir, J. Gehl, G. Sersa, C.G. Collins, J.-R. Garbay, V. Billard, P.F. Geertsen, Z. Rudolf, G.C. O'Sullivan, M. Marty, Standard operating procedures of the electrochemotherapy: Instructions for the use of bleomycin or cisplatin administered either systemically or locally and electric pulses delivered by the Cliniporator^{TM} by means of invasive or non-invasive electrodes, European Journal of Cancer 4 (2006) 14–25.

[44] A.I. Daud, R.C. DeConti, S. Andrews, P. Urbas, A.I. Riker, V.K. Sondak, P.N. Munster, D.M. Sullivan, K.E. Ugen, J.L. Messina, R. Heller, Phase I trial of interleukin-12 plasmid electroporation in patients with metastatic melanoma, Journal of Clinical Oncology 26 (2008) 5896–5903.

[45] L. Low, A. Mander, K. McCann, D. Dearnaley, T. Tjelle, I. Mathiesen, F. Stevenson, C.H. Ottensmeier, DNA vaccination with electroporation induces increased antibody responses in patients with prostate cancer, Human Gene Therapy 20 (2009) 1269–1278.

[46] A. Paganin-Gioanni, E. Bellard, B. Couderc, J. Teissie, M. Golzio, Tracking *in vitro* and *in vivo* siRNA electrotransfer in tumor cells, Journal of RNAi and Gene Silencing 4 (2008) 281–288.

[47] A. Paganin-Gioanni, E. Bellard, J.M. Escoffre, M.P. Rols, J. Teissie, M. Golzio, Direct visualization at the single-cell level of siRNA electrotransfer into cancer cells, Proceedings of the National Academy of Sciences of the United States of America (2011) ahead of print.

[48] C.A. Gonzalez, B. Rubinsky, Electroporation of the skin, in: J.J. Escobar-Chaves (Ed.), Current Technologies to Increase the Transdermal Delivery of Drugs, Bentham Science Publishers, 2010.

[49] S.M. Becker, Skin electroporation with passive transdermal transport theory: a review and a suggestion for future numerical model development, Journal of Heat Transfer 133 (2011) 1–9.

[50] S.M. Becker, A.V. Kuznetsov, Skin electroporation: modeling perspective, in: K. Vafai (Ed.), Porous Media: Applications in Biological Systems and Biotechnology, CRC Press/Taylor & Francis Group, Boca Raton, FL, 2011, pp. 331–364.

[51] S.M. Becker, A.V. Kuznetsov, Thermal *in vivo* skin electroporation pore development and charged macromolecule transdermal delivery: a numerical study of the influence of chemically enhanced lower lipid phase transition temperatures, International Journal of Heat and Mass Transfer 51 (2008) 2060–2074.

[52] S.M. Becker, A.V. Kuznetsov, Numerical assessment of thermal response associated with *in vivo* skin electroporation: the importance of the composite skin model, Journal of Biomechanical Engineering-Transactions of the ASME 129 (2007) 330–341.

[53] S.M. Becker, A.V. Kuznetsov, Local temperature rises influence *in vivo* electroporation pore development: a numerical stratum corneum lipid phase transition model, Journal of Biomechanical Engineering-Transactions of the ASME 129 (2007) 712–811.

[54] S.M. Becker, A.V. Kuznetsov, Numerical modeling of *in vivo* plate electroporation thermal dose assessment, Journal of Biomechanical Engineering 128 (2006) 76–85.

[55] S.M. Becker, A.V. Kuznetsov, Thermal damage reduction associated with *in vivo* skin electroporation: a numerical investigation justifying aggressive pre-cooling, International Journal of Heat and Mass Transfer 50 (2007) 105–116.

[56] J. Teissié, M. Golzio, M.-P. Rols, Mechanisms of cell membrane electropermeabilization: a minireview of our present (lack of?) knowledge, Biochimica et Biophysica Acta 1724 (2005) 270–280.

[57] J.C. Weaver, A. Barnett, Progress towards a theoretical model for electroporation mechanism: membrane electrical behavior and molecular transport, in: D.C. Chang, B.M. Chassy, J.A. Saunders, A.E. Sowers (Eds.), Guide to Electroporation and Electrofusion, Academic Press, NY, 1992, pp. 91–118.

[58] E. Neumann, S. Kakorkin, K. Toesing, Fundamentals of electroporative delivery of drugs and genes, Bioelectrochemistry and Bioenergetics 48 (1999) 3–16.

[59] D.S. Dimitrov, A.E. Sowers, Membrane electroporation – fast molecular exchange by electroosmosis, Biochimica et Biophysica Acta 1022 (1990) 381–392.

[60] S. Satkauskas, M.F. Bureau, M. Puc, A. Mahfoudi, D. Scherman, D. Miklavcic, L.M. Mir, Mechanisms of *in vivo* DNA electrotransfer: respective contributions of cell electropermeabilization and DNA electrophoresis, Molecular Therapy 5 (2002) 133–140.

[61] V.A. Klenchin, S.I. Sukharev, S.M. Serov, L.V. Chernomordik, Y.A. Chizmadzhev, Electrically induced DNA uptake by cells is a fast process involving DNA electrophoresis, Biophysical Journal 60 (1991) 804–811.

[62] S.I. Sukharev, V.A. Klenchin, S.M. Serov, L.V. Chernomordik, Yu A. Chizmadzhev, Electroporation and electrophoretic DNA transfer into cells. The effect of DNA interaction with electropores, Biophysical Journal 63 (1992) 1320–1327.

[63] D.A. Zaharoff, J.W. Henshaw, B. Mossop, F. Yuan, Mechanistic analysis of electroporation-induced cellular uptake of macromolecules, Experimental Biology and Medicine 233 (2008) 94–105.

[64] M.P. Rols, P. Femenia, J. Teissie, Long-lived macropinocytosis takes place in electropermeabilized mammalian cells, Biochemical and Biophysical Research Communications 208 (1995) 26–35.

[65] M. Wu, F. Yuan, Membrane binding of plasmid DNA and endocytic pathways are involved in electrotransfection of mammalian cells, PLoS ONE 6 (2011) e20923.

[66] M.-P. Rols, S. Li, Mechanism by which electroporation mediates DNA migration and entry into cells and targeted tissues, in: J.M. Walker (Ed.), Electroporation Protocols, Humana Press, 2008, pp. 19–33.

[67] J.-M. Escoffre, T. Portet, L. Wasungu, J. Teissié, D. Dean, M.-P. Rols, What is (still not) known of the mechanism by which electroporation mediates gene transfer and expression in cells and tissues, Molecular Biotechnology 41 (2009) 286–295.

[68] B. Gabriel, J. Teissie, Time courses of mammalian cell electropermeabilization observed by millisecond imaging of membrane property changes during the pulse, Biophysical Journal 76 (1999) 2158–2165.

[69] E. Tekle, R.D. Astumian, P.B. Chock, Selective and asymmetric molecular-transport across electroporated cell-membranes, Proceedings of the National Academy of Sciences of the United States of America 91 (1994) 11512–11516.

[70] C.S. Djuzenova, U. Zimmermann, H. Frank, V.L. Sukhorukov, E. Richter, G. Fuhr, Effect of medium conductivity and composition on the uptake of propidium iodide into electropermeabilized myeloma cells, Biochimica et Biophysica Acta – Biomembranes 1284 (1996) 143–152.

[71] E.A. Gift, J.C. Weaver, Simultaneous quantitative determination of electroporative molecular uptake and subsequent cell survival using gel microdrops and flow cytometry, Cytometry 39 (2000) 243–249.

[72] M.R. Prausnitz, B.S. Lau, C.D. Milano, S. Conner, R. Langer, J.C. Weaver, A quantitative study of electroporation showing a plateau in net molecular transport, Biophysical Journal 65 (1993) 414–422.

[73] P.J. Canatella, M.R. Prausnitz, Prediction and optimization of gene transfection and drug delivery by electroporation, Gene Therapy 8 (2001) 1464–1469.

[74] M. Hibino, H. Itoh, K.J. Kinosita, Time courses of cell electroporation as revealed by submicrosecond imaging of transmembrane potential, Biophysical Journal 64 (1993) 1789–1800.

[75] S.p. Orlowski, L.M. Mir, Cell electropermeabilization: a new tool for biochemical and pharmacological studies, Biochimica et Biophysica Acta – Reviews on Biomembranes 1154 (1993) 51–63.

[76] M. Puc, T. Kotnik, L.M. Mir, D. Miklavcic, Quantitative model of small molecules uptake after *in vitro* cell electropermeabilization, Bioelectrochemistry 60 (2003) 1–10.

[77] H. Wolf, M.P. Rols, E. Boldt, E. Neumann, J. Teissié, Control by pulse parameters of electric field-mediated gene transfer in mammalian cells, Biophysical Journal 66 (1994) 524–531.

[78] M. Golzio, J. Teissie, M.-P. Rols, Cell synchronization effect on mammalian cell permeabilization and gene delivery by electric field, Biochimica et Biophysica Acta – Biomembranes 1563 (2002) 23–28.

[79] J. Teissie, N. Eynard, B. Gabriel, M.P. Rols, Electropermeabilization of cell membranes, Advanced Drug Delivery Reviews 35 (1999) 3–19.

[80] J. Zhen, S.M. Kennedy, J.H. Booske, S.C. Hagness, Experimental studies of persistent poration dynamics of cell membranes induced by electric pulses, IEEE Transactions on Plasma Science 34 (2006) 1416–1424.

[81] I.G. Abidor, L.H. Li, S.W. Hui, Studies of cell pellets. II. Osmotic properties, electroporation, and related phenomena: membrane interactions, Biophysical Journal 67 (1994) 427–435.
[82] A. Ivorra, B. Rubinsky, *In vivo* electrical impedance measurements during and after electroporation of rat liver, Bioelectrochemistry 70 (2007) 287–295.
[83] J. Henshaw, B. Mossop, F. Yuan, Enhancement of electric field-mediated gene delivery through pretreatment of tumors with a hyperosmotic mannitol solution, Cancer Gene Therapy 18 (2010) 26–33.
[84] J.-M. Escoffre, T. Portet, C. Favard, J. Teissié, D.S. Dean, M.-P. Rols, Electromediated formation of DNA complexes with cell membranes and its consequences for gene delivery, Biochimica et Biophysica Acta 108 (2011) 1538–1543.
[85] T. Stroh, U. Erben, A.A. Kuhl, M. Zeitz, B. Siegmund, Combined pulse electroporation a novel strategy for highly efficient transfection of human and mouse cells, PLoS ONE 5 (2010) e9488.
[86] M. Kanduser, D. Miklavcic, M. Pavlin, Mechanisms involved in gene electrotransfer using high- and low-voltage pulses an *in vitro* study, Bioelectrochemistry 74 (2009) 265–271.
[87] S. Satkauskas, M.F. Bureau, A. Mahfoudi, L.M. Mir, Slow accumulation of plasmid in muscle cells: supporting evidence for a mechanism of DNA uptake by receptor-mediated endocytosis, Molecular Therapy 4 (2001) 317–323.
[88] M. Glogauer, W. Lee, C.A.G. McCulloch, Induced endocytosis in human fibroblasts by electrical fields, Experimental Cell Research 208 (1993) 232–240.
[89] Y. Antov, A. Barbul, H. Mantsur, R. Korenstein, Electroendocytosis: exposure of cells to pulsed low electric fields enhances adsorption and uptake of macromolecules, Biophysical Journal 88 (2005) 2206–2223.
[90] A. Barbul, Y. Antov, Y. Rosenberg, R. Korenstein, M. Belting, Enhanced delivery of macromolecules into cells by electroendocytosis macromolecular drug delivery, in: J.M. Walker (Ed.), Humana Press, 2009, pp. 141–150.
[91] Y. Antov, A. Barbul, R. Korenstein, Electroendocytosis: stimulation of adsorptive and fluid-phase uptake by pulsed low electric fields, Experimental Cell Research 297 (2004) 348–362.
[92] K.J. Kinosita, T.Y. Tsong, Hemolysis of human erythrocytes by transient electric field, Proceedings of the National Academy of Sciences of the United States of America 74 (1977) 1923–1927.
[93] K. Kinosita, T.Y. Tsong, Formation and resealing of pores of controlled sizes in human erythrocyte membrane, Nature 268 (1977) 438–441.
[94] K. Kinosita Jr., T.Y. Tsong, Voltage-induced changes in the conductivity of erythrocyte membranes, Biophysical Journal 24 (1978) 373–375.
[95] S. Kakorin, E. Redeker, E. Neumann, Electroporative deformation of salt filled lipid vesicles, European Biophysics Journal 27 (1998) 43–53.
[96] I.G. Abidor, A.I. Barbul, D.V. Zhelev, P. Doinov, I.N. Bandarina, E.M. Osipova, S.I. Sukharev, Electrical properties of cell pellets and cell fusion in a centrifuge, Biochimica et Biophysica Acta 1152 (1993) 207–218.
[97] M. Pavlin, M. Kanduser, M. Rebersek, G. Pucihar, F. Hart, R. Magjarevic, D. Miklavcic, Effect of cell electroporation on the conductivity of a cell suspension, Biophysical Journal 88 (2005) 4378–4390.
[98] M. Pavlin, V. Leben, D. Miklavcic, Electroporation in dense cell suspension – theoretical and experimental analysis of ion diffusion and cell permeabilization, Biochimica et Biophysica Acta 1770 (2007) 12–23.
[99] Y. Huang, B. Rubinsky, Microfabricated electroporation chip for single cell membrane permeabilization, Sensors and Actuators A 89 (2001) 242–249.
[100] R. Davalos, Y. Huang, B. Rubinsky, Electroporation: bio-electrochemical mass transfer in the nano scale, Microscale Thermophysical Engineering 4 (2000) 147–159.
[101] R.E. Diaz-Rivera, B. Rubinsky, Electrical and thermal characterization of nanochannels between a cell and a silicon based micro-pore, Biomedical Microdevices 8 (2006) 25–34.
[102] D. Cukjati, D. Batiuskaite, F. André, D. Miklavcic, L.M. Mir, Real time electroporation control for accurate and safe *in vivo* non-viral gene therapy, Bioelectrochemistry 70 (2007) 501–507.
[103] Y. Granot, B. Rubinsky, Mass transfer model for drug delivery in tissue cells with reversible electroporation, International Journal of Heat and Mass Transfer 51 (2008) 5610–5616.
[104] Y. Granot, B. Rubinsky, Mathematical models of mass transfer in tissue for molecular medicine with reversible electroporation, in: K. Vafai (Ed.), Porous Media. Applications in Biological Systems and Biotechnology, CRC Press, Boca Raton, FL, 2011.
[105] A. Golberg, H.D. Rabinowitch, B. Rubinsky, Galvanic apparent internal impedance: an intrinsic tissue property, Biochemical and Biophysical Research Communications 389 (2009) 168–171.
[106] A. Golberg, S. Laufer, H.D. Rabinowitch, B. Rubinsky, *In vivo* non-thermal irreversible electroporation impact on rat liver galvanic apparent internal resistance, Physics in Medicine and Biology 56 (2011) 951–963.

[107] M. Kranjc, F. Bajd, G. Sersa, D. Miklavcic, Magnetic resonance electrical impedance tomography for monitoring electric field distribution during tissue electroporation, IEEE Transactions on Medical Imaging (2011), http://dx.doi.org/10.1109/TMI.2011.2147328.

[108] M. Hjouj, B. Rubinsky, Magnetic resonance imaging characteristics of nonthermal irreversible electroporation in vegetable tissue, The Journal of Membrane Biology 236 (2010) 137–146.

[109] F. Mahmood, R. Hansen, B. Agerholm-Larsen, K. Jensen, H. Iversen, J. Gehl, Diffusion-weighted MRI for verification of electroporation-based treatments, Journal of Membrane Biology 240 (2011) 131–138.

[110] Y. Zhang, Y. Guo, A.B. Ragin, R.J. Lewandowski, G.-Y. Yang, G.M. Nijm, A.V. Sahakian, R.A. Omary, A.C. Larson, MR imaging to assess immediate response to irreversible electroporation for targeted ablation of liver tissues: preclinical feasibility studies in a rodent model, Radiology 256 (2010) 424–432.

[111] Y. Guo, Y. Zhang, G.M. Nijm, A.V. Sahakian, G.-Y. Yang, R.A. Omary, A.C. Larson, Irreversible electroporation in the liver: contrast-enhanced inversion-recovery MR imaging approaches to differentiate reversibly electroporated penumbra from irreversibly electroporated ablation zones, Radiology 258 (2010) 461–468.

[112] E.R. Tekle, D. Astumian, P.B. Chock, Selective and asymmetric molecular transport across electroporated cell membranes, Proceedings of the National Academy of Sciences of the United States of America 91 (1994) 11512–11516.

[113] M. Golzio, J. Teissie, M.P. Rols, Direct visualization at the single-cell level of electrically mediated gene delivery, Proceedings of the National Academy of Sciences of the United States of America 99 (2002) 1292–1297.

[114] O. Hibbitt, K. Coward, H. Kubota, N. Prathalingham, W. Holt, K. Kohri, J. Parrington, *In vivo* gene transfer by electroporation allows expression of a fluorescent transgene in hamster testis and epididymal sperm and has no adverse effects upon testicular integrity or sperm quality, Biology of Reproduction 74 (2006) 95–101.

[115] J.D. Vry, P. Martinez-Martinez, M. Losen, G.H. Bode, Y. Temel, T. Steckler, H.W.M. Steinbusch, M.D. Baets, J. Prickaerts, Low current-driven micro-electroporation allows efficient *in vivo* delivery of nonviral DNA into the adult mouse brain, Molecular Therapy 18 (2010) 1183–1191.

[116] A.M. Bodles-Brakhop, R. Heller, R. Draghia-Akli, Electroporation for the delivery of DNA-based vaccines and immunotherapeutics: current clinical developments, Molecular Therapy 17 (2009) 585–592.

[117] C. Boutin, S. Diestel, A.I. Desoeuvre, M.-C. Tiveron, H. Cremer, Efficient *in vivo* electroporation of the postnatal rodent forebrain, PLoS ONE 3 (2008) e1883.

[118] T. Matsuda, C.L. Cepko, Controlled expression of transgenes introduced by *in vivo* electroporation, Proceedings of the National Academy of Sciences of the United States of America 104 (2007) 1027–1032.

[119] S.A. Kera, S.M. Agerwala, J.H. Horne, The temporal resolution of *in vivo* electroporation in zebrafish: a method for time-resolved loss of function, Zebrafish 7 (2010) 97–108.

[120] W. Aung, S. Hasegawa, M. Koshikawa-Yano, T. Obata, H. Ikehira, T. Furukawa, I. Aoki, T. Saga, Visualization of *in vivo* electroporation-mediated transgene expression in experimental tumors by optical and magnetic resonance imaging, Gene Therapy 16 (2009) 830–839.

[121] T. Portet, C. Favard, J. Teissié, D. Dean, M.P. Rols, Insights into the mechanisms of electromediated gene delivery and application to the loading of giant vesicles with negatively charged macromolecules, Soft Matter 7 (2011) 3872–3881.

[122] G. Pucihar, T. Kotnik, D. Miklavcic, J. Teissié, Kinetics of transmembrane transport of small molecules into electropermeabilized cells, Biophysical Journal 95 (2008) 2837–2848.

[123] J.C. Weaver, G.I. Harrison, J.G. Bliss, J.R. Mourant, K.T. Powell, Electroporation: high frequency of occurrence of a transient high permeability state in erythrocytes and intact yeast, FEBS Letters 229 (1988) 30–34.

[124] H. Liang, W.J. Purucker, D.A. Stenger, R.T. Kubiniec, S.W. Hui, Uptake of fluorescence-labeled dextrans by 10T 1/2 fibroblasts following permeation by rectangular and exponential-decay electric field pulses, Biotechniques 6 (1988) 556–558.

[125] L.M. Mir, H. Banoun, C. Paoletti, Introduction of definite amounts of nonpermeant molecules into living cells after electropermeabilization: direct access to the cytosol, Experimental Cell Research 175 (1988) 15–25.

[126] H. Lambert, R. Pankov, J. Gauthier, R. Hancock, Electroporation mediated uptake of proteins into mammalian cells, Biochemistry and Cell Biology 68 (1990) 729–734.

[127] G. Pucihar, T. Kotnik, J. Teissié, D. Miklavcic, Electropermeabilization of dense cell suspensions, European Biophysics Journal 36 (2007) 173–185.

[128] W. Mehrle, U. Zimmermann, R. Hampp, Evidence for a symmetrical uptake of fluorescent dyes through electro-permeabilized membranes of Avena mesophyll protoplasts, FEBS Letters 185 (1985) 89–94.

[129] A.E. Sowers, M.R. Lieber, Electropore diameters, lifetimes, numbers, and locations in individual erythrocyte ghosts, FEBS Letters 205 (1986) 179–184.
[130] D. Takao, S. Kamimura, Single-cell electroporation of fluorescent probes into sea urchin sperm cells and subsequent FRAP analysis, Zoological Science 27 (2010) 279–284.
[131] D.C. Bartoletti, G.I. Harrison, J.C. Weaver, The number of molecules taken up by electroporated cells: quantitative determination, FEBS Letters 256 (1989) 4–10.
[132] P.J. Canatella, M.M. Black, D.M. Bonnichsen, C. McKenna, M.R. Prausnitz, Tissue electroporation: quantification and analysis of heterogeneous transport in multicellular environments, Biophysical Journal 86 (2004) 3260–3268.
[133] J. Michie, D. Janssens, J. Cilliers, B.J. Smit, L. Böhm, Assessment of electroporation by flow cytometry, Cytometry 41 (2000) 96–101.
[134] M.R. Prausnitz, C.D. Milano, J.A. Gimm, R. Langer, J.C. Weaver, Quantitative study of molecular transport due to electroporation: uptake of bovine serum albumin by erythrocyte ghosts, Biophysical Journal 66 (1994) 1522–1530.
[135] E.B. Ghartey-Tagoe, J.S. Morgan, K. Ahmed, A.S. Neish, M.R. Prausnitz, Electroporation-mediated delivery of molecules to model intestinal epithelia, International Journal of Pharmaceutics 270 (2004) 127–138.
[136] E. Neumann, S. Kakorkin, K. Toesing, Fundamentals of electroporative delivery of drugs and genes, Bioelectrochemistry and Bioenergetics 48 (1998) 3–16.
[137] S. Kakorin, E. Neumann, Kinetics of the electroporative deformation of lipid vesicles and biological cells in an electric field, Berichte der Bunsengesellschaft für physikalische Chemie 102 (1998) 670–675.
[138] S. Kakorin, S.P. Stoylov, E. Neumann, Electra-optics of membrane electroporation in diphenylhexatriene-doped lipid bilayer vesicles, Biophysical Chemistry 58 (1996) 109–116.
[139] E. Neumann, K. Toensing, S. Kakorin, P. Budde, J. Frey, Mechanism of electroporative dye uptake by mouse B cells, Biophysical Journal 74 (1998) 98–108.
[140] E. Neumann, S. Kakorin, I. Tsoneva, B. Nikolova, T. Tomov, Calcium-mediated DNA adsorption to yeast cells and kinetics of cell transformation by electroporation, Biophysical Journal 71 (1996) 868–877.
[141] M.-P. Rols, J. Teissié, Electropermeabilization of mammalian cells: quantitative analysis of the phenomenon, Biophysical Journal 58 (1990) 1089–1098.
[142] B. Gabriel, J. Teissie, Direct observation in the millisecond time range of fluorescent molecule asymmetrical interaction with the electropermeabilized cell membrane, Biophysical Journal 73 (1997) 2630–2637.
[143] K. Scwister, B. Deuticke, Formation and properties of aqueous leaks induced in human erythrocytes by electrical breakdown, Biochimica et Biophysica Acta – Biomembranes 816 (1985) 322–348.
[144] M.P. Rols, D. Coulet, J. Teissié, Highly efficient transfection of mammalian cells by electric field pulses. Application to large volumes of cell culture by using a flow system, European Journal of Biochemistry 206 (1992) 115–121.
[145] M.P. Rols, J. Teissie, Experimental evidence for the involvement of the cytoskeleton in mammalian cell electropermeabilization, Biochimica et Biophysica Acta 1111 (1992) 45–50.
[146] C. Favard, D.S. Dean, M.P. Rols, Electrotransfer as a non viral method of gene delivery, Current Gene Therapy 7 (2007) 67–77.
[147] G. Lukacs, P. Haggie, O. Seksek, D. Lechardeur, N. Freedman, A. Verkman, Size dependent DNA mobility in cytoplasm and nucleus, Journal of Biological Chemistry 275 (2000) 1625–1629.
[148] E. Dauty, A. Verkman, Actin cytoskeleton as the principal determinant of size dependent DNA mobility in cytoplasm, Journal of Biological Chemistry 280 (2005) 7823–7828.
[149] M. Pavlin, D. Miklavcic, Theoretical and experimental analysis of conductivity, ion diffusion and molecular transport during cell electroporation – relation between short-lived and long-lived pores, Bioelectrochemistry 74 (2008) 38–46.
[150] I.P. Sugar, E. Neumann, Stochastic model for electric field-induced membrane pores electroporation, Biophysical Chemistry 19 (1984) 211–225.
[151] E. Tekle, R.D. Astumian, P.B. Chock, Electro-permeabilization of cell membranes: effect of the resting membrane potential, Biochemical and Biophysical Research Communications 172 (1990) 282–287.
[152] E.H. Serpersu, K. Kinosita Jr., T. Tian Yow, Reversible and irreversible modification of erythrocyte membrane permeability by electric field, Biochimica et Biophysica Acta – Biomembranes 812 (1985) 779–785.
[153] M.R. Prausnitz, J.D. Corbett, J.A. Gimm, D.E. Golan, R. Langer, J.C. Weaver, Millisecond measurement of transport during and after an electroporation pulse, Biophysical Journal 66 (1994) 1522–1530.

[154] M.P. Rols, F. Dahhou, K.P. Mishra, J. Teissié, Control of electric-field induced cell-membrane permeabilization by membrane order, Biochemistry 29 (1990) 2960–2966.
[155] M. Gibaldi, D. Perrier, Pharmacokinetics, Marcel Dekker, New York, 1975.
[156] M.P. Rols, J. Teissie, Electropermeabilization of mammalian cells to macromolecules: control by pulse duration, Biophysical Journal 75 (1998) 1415–1423.
[157] M. Bier, S.M. Hammer, D.J. Canaday, R.C. Lee, Kinetics of sealing for transient electropores in isolated mammalian skeletal muscle cells, Bioelectromagnetics 20 (1999) 194–201.
[158] A.T. Esser, K.C. Smith, T.R. Gowrishankar, J.C. Weaver, Towards solid tumor treatment by irreversible electroporation: intrinsic redistribution of fields and currents in tissue, Technology in Cancer Research and Treatment 6 (2007) 261–274.
[159] T. Kotnik, G. Pucihar, D. Miklavcic, Induced transmembrane voltage and its correlation with electroporation-mediated molecular transport, Journal of Membrane Biology 236 (2010) 3–13.
[160] M. Phillips, E. Maor, B. Rubinsky, Nonthermal irreversible electroporation for tissue decellularization, Journal of Biomechanical Engineering 132 (2010) 091003.
[161] C. Daniels, B. Rubinsky, Electrical field and temperature model of nonthermal irreversible electroporation in heterogeneous tissues, Journal of Biomechanical Engineering 131 (2009) 071006.
[162] C. Daniels, B. Rubinsky, Temperature modulation of electric fields in biological matter, PLoS One 6 (2011) e20877.
[163] S. Corovic, A. Zupanic, D. Miklavcic, Numerical modeling and optimization of electric field distribution in subcutaneous tumor treated with electrochemotherapy using needle electrodes, IEEE Transactions on Plasma Science 36 (2008) 1665–1672.
[164] D. Miklavcic, B. Katarina, D. Semrov, M. Cemazar, F. Demsar, G. Sersa, The importance of electric field distribution for effective *in vivo* electroporation of tissues, Biophysical Journal 74 (1998) 2152–2158.
[165] A. Golberg, B. Rubinsky, A statistical model for multidimensional irreversible electroporation cell death in tissue, Biomedical Engineering Online 9 (2010) 1–13.
[166] W. Krassowska, P.D. Filev, Modeling electroporation in a single cell, Biophysical Journal 92 (2007) 404–417.
[167] J.C. Neu, W. Krassowska, Asymptotic model of electroporation, Physical Review E 59 (1999) 3471–3482.

CHAPTER

13

Modeling Cell Electroporation and Its Measurable Effects in Tissue

Nataša Pavšelj[a], *Damijan Miklavčič*[a], *and Sid Becker*[b]

[a]University of Ljubljana, Faculty of Electrical Engineering, Ljubljana, Slovenia,
[b]University of Canterbury, Department of Mechanical Engineering, Christchurch, New Zealand

OUTLINE

13.1 Introduction – Electroporation 493

13.2 Skin Electroporation 495

13.3 Physical Changes in Biological Tissue Following Electroporation 497
- *13.3.1 Electrical Conductivity Increase* 497
- *13.3.2 Tissue Heating During Electroporation* 499
- *13.3.3 Molecular Transport During Skin Electroporation* 502

13.4 Modeling of Skin Electroporation Transport 504
- *13.4.1 Modeling Non-Thermal Electroporation (Short Pulses)* 506
- *13.4.2 Modeling Thermal Electroporation (Long Pulse)* 507

13.5 Conclusions 514

Acknowledgments 515

References 515

13.1 INTRODUCTION – ELECTROPORATION

Recent advances in pharmacy and biotechnology have yielded a number of drugs that have shown promise in achieving good therapeutic results. However, one of the biggest challenges to successful delivery, which often remains unresolved, is finding the most effective method of delivering molecules to targeted sites inside the human body. Depending on the

http://dx.doi.org/10.1016/B978-0-12-415824-5.00013-8
© 2013 Elsevier Inc. All rights reserved.

nature or the location of the targeted delivery site, these molecules must pass through highly impermeable regions, such as the cell membrane or the skin. One of the methods used to increase cell membrane permeability is electroporation. Electroporation is found in many molecular biology and medical applications, such as electrochemotherapy, DNA transfection for DNA vaccination and gene therapy, transdermal drug delivery, cell fusion and tissue ablation [1–9]. The concept of reversible electroporation involves the use of electric pulses (in a non-destructive manner) in order to establish an increased transmembrane voltage across the phospholipid bilayer of a biological cell to increase its permeability. This transient permeability increase facilitates uptake of drugs or DNA by the cell. Reversible electroporation of the cell membrane can be used on different types of cells and does not affect cell survival nor does it permanently disrupt the cell's functions.

The cell membrane is permeabilized when a sufficiently high transmembrane voltage is reached. The critical value for the reversible cell electropermeabilization is reported to be between 200 mV and 1 V [10,11]. The induced transmembrane voltage – which results from exposing cells to electric field – is superimposed on the resting transmembrane voltage, the origin of which arises from the exchange of (primarily) Na+ and K+ ions across the cell membrane. The phenomenon can be understood in terms of an equilibrium potential: the membrane voltage at which the voltage force precisely balances the concentration gradient force of an ion. When a cell is placed in an electric field, the differences in the conductivities of the cell membrane, the intracellular and the extracellular medium cause a large gradient in the electric field in the membrane, inducing a transmembrane voltage that is superimposed on the resting potential of the membrane [12]. However, if the electric field is further increased, the transmembrane potential may cause irreversible membrane permeabilization, which has the potential to lead to cell death.

The electropermeabilization of cells and the exogenous molecule cell uptake depends on different pulse and physiological factors: the cell and tissue parameters (tissue conductivity, cell size, shape, density, distribution and interactions between them [13–17], and pulse parameters (pulse duration, amplitude and number of pulses) [18–20]. Since the clinician can only directly influence the electric pulse parameters, a wide range of studies has been conducted in order to find the most successful pulsing protocols for different cell types and applications. Roughly speaking, experiments have shown that the number of pulses and their duration determine the molecular transport across the permeabilized membrane, while the pulse amplitude determines the fraction of the membrane surface that is permeabilized [21,22]. Since both the electric field strength and its duration are important determining factors in cell electropermeabilization, protocols combining pulses of different amplitudes and durations have been investigated, especially in gene delivery applications [23–25].

Numerous experiments have to be performed before a biomedical application is put to practical use in a clinical environment. As a complementary work to *in vitro* and *in vivo* experiments, analytical and numerical models can be used to theoretically describe a therapy, in our case, electroporation. In this way we can evaluate different electrical parameters in advance, such as pulse amplitude, duration, number of pulses, or different electrode geometries. Biological effects on cells and tissues are accompanied by different measurable physical phenomena during therapy. These include electrical conductivity changes, thermal effects and molecular transport, all of which can be used quantitatively when constructing theoretical models of electroporation.

13.2 SKIN ELECTROPORATION

Electroporation has been widely investigated as a method to introduce molecules into cells, through the cell membrane which, in general, is highly impermeable to larger molecules. The method has also been extended to the intra- and transdermal delivery of therapeutic molecules, such as drugs or DNA.

Transdermal drug delivery is an attractive delivery route for a number of reasons. First, it is less invasive for the patient than, for example, hypodermic injection. Second, it offers the possibility of steady permeation of drug across the skin, allowing for more consistent serum drug levels, which are often highly desired in therapy. In addition, it can be used as an alternative administration route for patients who cannot tolerate oral dosage forms. Good candidates for transdermal delivery are drugs that may cause gastrointestinal upset and drugs that are degraded by the enzymes and acids in the gastrointestinal system. The intradermal route has great potential when a highly localized response is needed instead of a systemic one, for instance in the treatment of skin conditions (psoriasis) or for cosmetic purposes. The skin is also a very good target organ for DNA immunization because of the large number of potent antigen presenting cells, critical to an effective immune response [26–29]. Also, the accessibility and facilitated monitoring of possible unwanted side effects makes the skin an attractive organ for drug delivery.

However, despite these advantages, only a small number of drug types can currently be delivered through the transdermal delivery route. Mostly these are drugs of relatively low molecular weight (<500 Da), low melting point (<200°C), moderate lipophilicity (logP 1–3), moderate aqueous solubility and high pharmacological potency. Three potential routes through the epidermis are possible: via appendages (shunt route), through the intercellular lipid domains or by a transcellular route [30].

Several enhancement techniques are used to overcome the skin barrier and facilitate intra- and transdermal drug delivery. Chemical penetration enhancers and drug encapsulation techniques can successfully enhance the permeation of small molecules, but their application can sometimes be limited by side effects such as skin irritation. Also, the amount of molecular delivery is often too low to provide any significant therapeutic effect [31–35]. Another approach involves the use of direct physical methods of penetration, such as microneedles or jet injectors [36–41]. Laser or radiofrequency energy can be used to increase the temperature and induce thermal microablation of the stratum corneum [42–46]. When ultrasound is applied to the skin, drug permeation is enhanced by disrupting the stratum corneum lipid bilayers due to microcavitation in the skin [47–49]. Furthermore, some electrically assisted methods are used to deliver substances through skin. Iontophoresis involves the use of a weak direct electric current that promotes the movement of electrically charged drug molecules into the skin, predominantly at skin appendages [50–52]. In contrast, electroporation uses shorter but much more intense electrical pulses in order to transiently and directly create new pathways for drug delivery [53–56]. In combination with electroporation, longer electric pulses of lower voltage can be used to electrophoretically drive the electrically charged drug into and/or through the skin (Fig. 13.1).

It has been shown that high voltage pulses create aqueous pathways in the lipid bilayer membranes of the stratum corneum. Pulses of durations as short as 1 ms can cause an up to

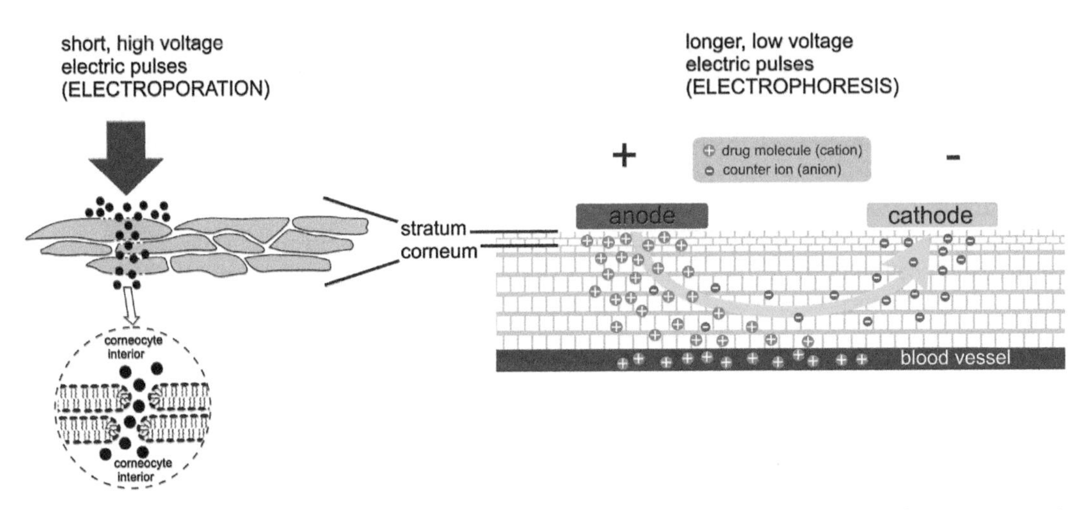

FIGURE 13.1 Schematics of a combination of stratum corneum electroporation – creation of aqueous pores in lipid bilayers of corneocytes – and electrophoresis. The latter provides a driving force to deliver an electrically charged molecule into or through the skin.

1000-fold increase in transdermal transport. The electroporation, and consequently the increase in the conductivity of the stratum corneum, is not homogeneous throughout the electroporated area. Ionic and molecular transport across the skin subjected to high voltage pulses is highly localized in sites termed 'local transport regions' (LTRs). Their size depends on pulse duration, while pulse amplitude dictates the number of LTRs that appear on the affected skin surface [54,55,57–59].

In addition to creating new transport routes, electroporation can also be used when the targeted sites of delivery are the viable cells of the epidermis, the dermis or subcutaneous tissue. Electrically permeabilized cell membranes facilitate cell uptake thus increasing the efficacy of the therapeutic molecules. This delivery method can be used for the treatment of skin disorders or, being a secretor of a variety of important systemic (blood borne, body-wide) proteins, it can be used to elicit a systemic effect. Skin is also a very good target organ for DNA immunization because of the large number of potent antigen presenting cells, critical to an effective immune response [26–29]. If necessary, large areas of skin can be treated and can easily be monitored for potential complications, such as infection and neoplasia, by observation or biopsy. Electroporation seems particularly effective for improving DNA transfection after intradermal and topical delivery without any significant alteration of skin structure. Transdermal gene delivery by electroporation has been tried by topically applying DNA on the skin's surface before electroporation [60]. Unfortunately, due to the barrier function of the skin, the transfection rate is rather low and restricted to the epidermis. However, by using intradermal injection in combination with electric pulses, the transdermal barrier is avoided so that electroporation is then used for successful permeabilization of targeted cells for gene transfection [24].

13.3 PHYSICAL CHANGES IN BIOLOGICAL TISSUE FOLLOWING ELECTROPORATION

13.3.1 Electrical Conductivity Increase

Predictions of the response of any biological material to an applied electrical stimulation require data on the conductivities and relative permittivities of the tissues making up the media. Because tissue types perform specific physiological functions, their make-up can be remarkably different, so at all organizational levels each biological material can demonstrate a number of tissue specific characteristics that have to be considered. Firstly, many tissues exhibit strongly inhomogeneous tendencies. The cell itself is comprised of an insulating membrane enclosing a conductive cytosol. A suspension of cells can be regarded at low frequencies simply as insulating inclusions in a conducting fluid, the insulation being provided by the cell membrane. At frequencies in the MHz range, capacitive coupling across this membrane becomes more important. Going up from microscopic to macroscopic descriptions, the cells are surrounded by an extracellular matrix, which can be extensive, as in the case of bone, or minimal, as in the case of epithelial tissue. Furthermore, many tissues do not contain cells of a single size and function. Another macroscale consideration is that there can be large thermo-electrical influences related to the presence and influence of the vascular system's local perfusion of blood, the extent of which is also tissue and site specific. It is thus difficult to extrapolate from the dielectric properties of a cell suspension to those of an intact tissue. The described differences are clearly reflected in very different bulk properties of biological materials [61–64]. Secondly, some biological materials are distinctly anisotropic. For example, muscles are composed of fibers that are very large individual cells which are aligned in the direction of muscle contraction. Electrical conduction along the length of the fiber is thus significantly easier than conduction between the fibers in the extracellular matrix because the extracellular matrix is less conductive than the cell. Therefore, muscle tissue manifests typical anisotropic electrical properties. The longitudinal conductivity is significantly higher than the transverse conductivity, even when path differences in the charge transport are taken into account; this is especially marked in the low frequency range. Thirdly, nonlinear changes in the electrical properties of biological material exposed to intense electric fields have been observed.

As the electric current passes through a biological tissue, it is distributed through different parts of the tissue, depending on local tissue electrical conductivity. Because blood is electrically conductive, tissues such as liver and muscle that are highly perfused have higher electrical conductivities, while poorly perfused tissues such as bone and fatty tissue have much lower conductivities. While for more conductive tissues the current will flow more easily, for the same current the local electric field in these tissues will be lower than in tissues with low conductivity. *In vivo* or clinical applications of electroporation rarely only deal with a single homogeneous tissue. Instead, electric pulses are normally applied to a composite tissue that is made of individual parts or layers of different dimensions and electrical conductivities (inverse to electrical resistivities). Upon applying electric pulses, the voltage is divided among these layers in a distribution that is proportional to their electrical resistances, as in a voltage divider. An example of such a voltage divider is skin tissue, composed of different layers, as depicted in Fig. 13.2. Similarly, where objects of different electrical conductances are in parallel configuration, the current is divided among them in a manner that is proportional to their electrical conductances, as in a current divider. This leads to a more complex electric field distribution, meaning that some parts of the tissue whose electrical conductivities are disproportionally lower than the rest of the tissue are exposed to a much stronger local electric field. The electric field is the highest in the layer with the highest resistivity (lowest conductivity).

One of the most identifiable results of tissue electroporation is a marked increase in tissue electrical conductivity (σ). This is due to the microscale effects of the increased electric field (E) that causes cell membrane electropermeabilization [65–67]. This nonlinear property of biological tissue is considered a threshold phenomenon, meaning that the local electric field has to reach a certain value, termed the reversible electropermeabilization threshold E_{rev} in order to cause conductivity changes. Further, for the duration of the pulse, this conductivity change is an irreversible phase transition process. More specifically, once the conductivity is increased in a given tissue volume, it cannot be changed back to its lower value during pulse delivery, even if the local electric field strength drops below the threshold value due to the dynamic increases in the conductivities of the surrounding tissue and global system response. It should be noted that, while the long-term effects of reversible electroporation ensure that at times long after the pulse application the tissue will return to its original state, at very short time scales (times *during* the pulse), the effects are non-reversible, so that the electrical conductivity of the tissue can only increase. It is important to determine the required amplitude of electric pulses at a given electrode-tissue set-up, in order to achieve an electric field distribution in the tissue that is adequate for a specific application. The reversible and irreversible electric field thresholds are both inherent characteristics of the tissue type, and are independent of the specific electrode type or whether the tissue is electroporated in isolation or in conjunction with surrounding tissue.

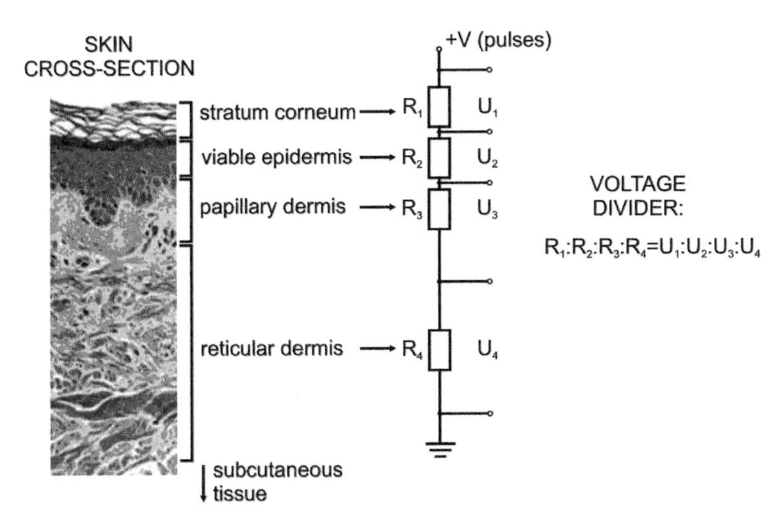

FIGURE 13.2 Layered skin tissue represented as a voltage divider: The voltage of the applied pulses is distributed between layers according to their resistance ratios. Consequently, the electric field is the highest in the layer with the highest resistivity (the stratum corneum).

Skin is a highly inhomogeneous, composite tissue. As already mentioned, electroporation has been used on skin tissue with the aim of increasing the transdermal delivery of different molecules with cosmetic or clinical therapeutic effect, or the delivery of such molecules into viable skin cells. Skin can be conceptualized as being composed of at least three different layers: the epidermis, dermis and the subcutaneous tissue. The epidermis is further divided into different sublayers, but the one that has the greatest influence on its transport and electric properties is the outermost layer, the stratum corneum (SC). The SC is composed of layers of dead, flat skin cells that are constantly being shed and renewed (the process takes about two weeks). Although it is very thin (typically around 20 μm), the SC contributes the primary resistance to electric current and molecular transport of the skin. Its high resistivity makes the SC one of the most electrically resistive tissues in the human body. Deeper layers, including the rest of the epidermis, the dermis and the subcutaneous tissue (fat, connective tissue, larger blood vessels and nerves), all have much lower resistivities [68–73].

High voltage pulses applied on the skin cause a significant rise in skin conductivity of up to three orders of magnitude due to the creation of pathways for ionic current [74–78]. This is reflected in increased electric current during pulse delivery (see Fig. 13.3). If the applied voltage is not too high, partial or full recovery of the skin resistance can be observed within a post-pulse period that can range from microseconds up to several hours [53,57,65,75]. At the onset of electric pulses, most of the applied voltage rests across the highly resistive stratum corneum. Consequently, the electric field in that layer is likely to rise above its critical strength, the reversible electropermeabilization threshold. This in turn causes a breakdown of the SC's protective barrier function and a drop in its resistance. As a result, the barrier to molecular transport is greatly reduced, which facilitates transdermal drug delivery, especially when used in combination with longer but lower electrophoretic pulses. Also, due to the drop in the resistance of the stratum corneum, the skin layers below it can now be subjected to an electric field whose intensity is sufficiently large to initiate the permeabilization of cell membranes in those viable tissues. This is particularly beneficial when the cell interior is the target of delivery (such as in gene therapy).

13.3.2 Tissue Heating During Electroporation

One of the concerns associated with the exposure of living tissue to intense electric fields is the potential of the generation of resistive heating. Excessive heating is unwanted, not only in order to ensure patient safety by avoiding burns and damage to viable cells, but also in order to avoid damage to the delivered molecules. This is especially true when we are dealing with sensitive substances such as plasmid DNA. A lot of work, both theoretical and experimental, has been done to evaluate different pulse parameters and electrode design and their influence on the outcome of the treatment [79]. Although potentially excessive resistive heating has been demonstrated during electroporation [80–87], this aspect is mostly overlooked in treatment planning and theoretical analysis of specific applications of electroporation-based treatments. Since the amount of resistive heat is specific to the conditions of pulse application, thermal effects should be considered, and these can be predicted in theoretical models.

The equation most commonly used in the modeling of heat transfer in living tissue is the Pennes' bioheat equation [88], which includes contributions from blood flow and metabolic heat production. This model was developed primarily to predict the thermal effects of blood

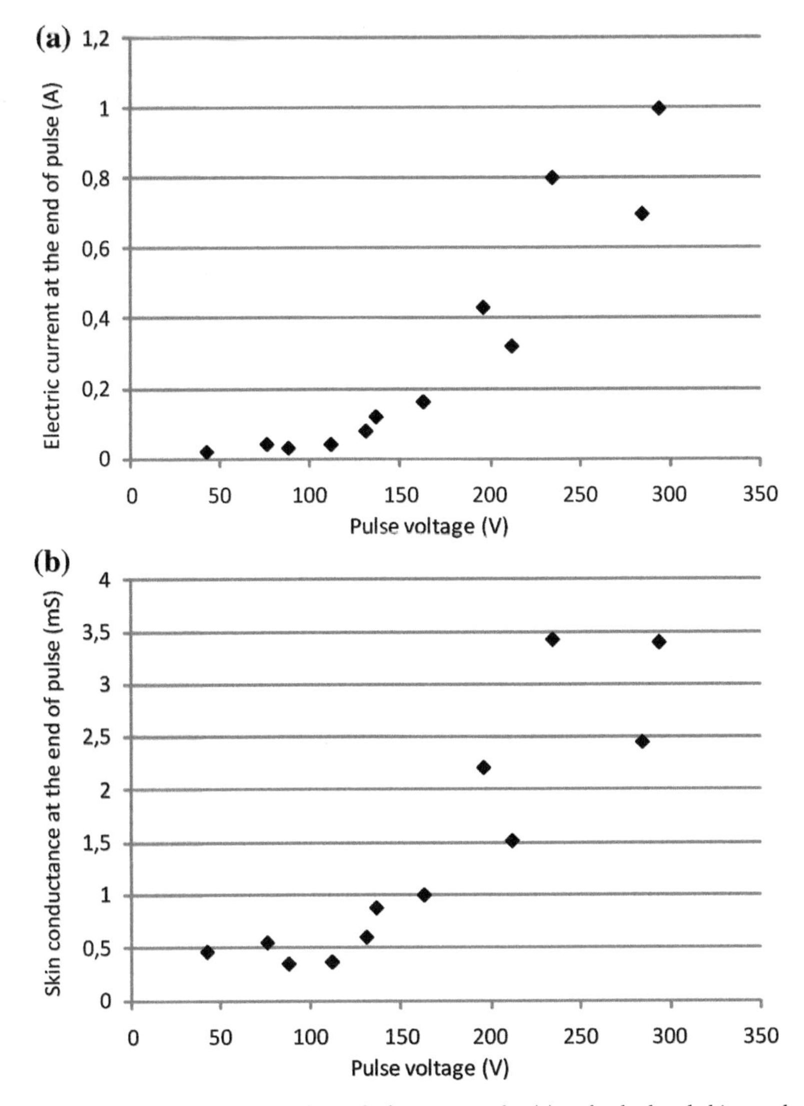

FIGURE 13.3 Measured electric current at the end of a 100 μs pulse (a) and calculated skin conductance (b), as a function of pulse amplitude. Measurements were performed on rat skin fold. Pulses were delivered through parallel plate electrodes of 2.8 cm distance.

perfusion in living tissue, and it was derived from measurements of temperature distribution in the forearm of human subjects:

$$\rho \cdot c \frac{\partial T}{\partial t} = \nabla \cdot (k \cdot \nabla T) - w \cdot c_b(T - T_b) + Q_M + Q_{DC} \tag{13.1}$$

where ρ is the mass density of the material (tissue), c is its specific heat capacity, k is its thermal conductivity and T is the temperature. The second term on the right side of the equation describes tissue perfusion, where c_b is the specific heat capacity of blood and T_b is the arterial blood temperature. Q_M is the metabolic heat generation and Q_{DC} is the resistive heat that is generated during the pulse(s).

In the applied settings of electroporation, some terms in the Pennes' bioheat equation can be neglected. In this situation, the cooling effect of the blood in the skin during the short electric pulses is very small compared to the contribution of the resistive heating. Moreover, it has been shown that the application of electric pulses during electroporation induces a rapid and significant reduction in blood flow in the treated area [89–91]. Secondly, the rate of metabolic heat generation during the pulse protocol is, again, negligible compared to the contribution of the resistive heating. Both are especially true in the epidermis (almost no vascularization or metabolic heat production), where most of the heat dissipation takes place during the electroporation of stratum corneum. However, for applications of electroporation on tissues other than skin, especially when pulses are delivered to highly perfused tissues or in the vicinity of large blood vessels (e.g. electroporation of liver [92]), blood flow cannot be neglected, as it contributes significantly to the cooling of the tissue.

An example of analysis of the amount of heating in and around an electrically created pore in the stratum corneum can be seen in Fig. 13.4. This depicts the result of a study [87] whose aim was to evaluate possible thermal damage of skin, as well as the delivered substance, for a pulse protocol consisting of one short high voltage pulse, followed by a longer low voltage pulse.

Various nonlinear processes take place when biological material is heated. First, the heat capacity and thermal conductivity value are often temperature dependent. The reported approximate rise in heat capacity between 60 and 85°C amounts to 17%, while the rise in thermal conductivity between physiological temperature (37°C) and 85°C is about 25%

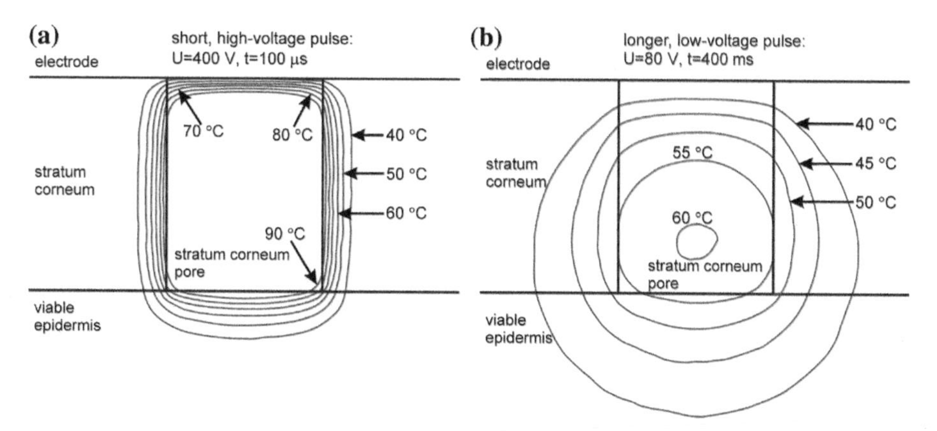

FIGURE 13.4 Resistive heating in and around the electrically created pore in the stratum corneum (diameter = 15 μm) during pulse protocol combining a short high voltage pulse (electroporation), followed by a longer, low voltage pulse (electrophoresis). (a) Isotherms at the end of the high voltage (HV) pulse. Pulse duration was 100 μs, pulse amplitude was 400 V. (b) Isotherms at the end of the low voltage (LV) pulse. Pulse duration was 400 ms, pulse amplitude was 80 V.

[93–95]. In the case of the skin, the stratum corneum undergoes several thermally induced phase transitions at elevated temperatures, lasting in the order of minutes. These temperatures may differ slightly for different animal species; however, in the range of 30–120°C, roughly four phase transitions have been identified: at around 45°C (lipid packing phase transition), at 70°C (lipid melting, disappearance of the bilayer structure, increased water permeation), at 85°C (protein-associated lipid transition, irreversible), and at 100°C (irreversible protein denaturation) [96–100]. Therefore, it is reasonable to expect changes in thermal properties of skin as the temperature increases during pulse delivery.

Furthermore, a strong coupling between the thermal effects and the electrical effects exists through the temperature dependence of the electrical conductivity of the tissue. The tissue's electrical conductivity increases with temperature, and since the local temperature rise is a result of the resistive heating, which is a function of the local electric field, a mutual dependence results. Once a part of a tissue is permeabilized, it becomes more conductive and the current density increases by several times, increasing resistive heating. In turn, tissue conductivity increases even more, as most biological materials exert a positive temperature coefficient of electrical conductivity in the range of 1–3%$°C^{-1}$ [101]. However, this increase in electrical conductivity is significantly smaller than the increase in tissue conductivity due to electroporation (several fold). Further discussion of the coupling of thermal and electrical responses of the skin is continued in Section 13.4.

13.3.3 Molecular Transport During Skin Electroporation

One method which uses electric current to deliver substances through the skin is iontophoresis. Sweat gland ducts and hair follicles provide pre-existing aqueous pathways that potentially allow the passage of water-soluble molecules upon the application of low voltages across the skin. However, the permeation flux provided by iontophoresis is often much lower than desirable. Another way to temporarily breach the barrier function of the skin is electroporation, which actually creates new aqueous pathways across lipid-based structures [102–106]. It has been shown that pulses of high voltage create aqueous pathways penetrating the multi-lamellar lipid bilayer membranes of the stratum corneum. In this process, adjacent corneocytes may become connected, resulting in pathways spanning the stratum corneum. Pulses lasting for as short a time as 1 ms have been shown to cause an increase of up to four orders of magnitude in the transdermal transport of charged molecules of molecular weights of up to 1000 grams/mol (most pharmaceutical compounds fall within this range). Even larger molecules (e.g. heparin), as well as highly charged ones (antisense oligonucleotides) can be transported across the skin effectively using electroporation techniques. Transport is generally localized to small LTRs [53–55,58,107]. When describing the response of the skin to electrical stimulation, it is common practice to categorize the pulse intensity as low, medium or high voltage [54]. In low voltage stimulation (the electric potential drop across the skin $U_{skin} < 5$ V), the mechanism responsible for ionic and molecular transport through the skin is iontophoresis, and movement takes place mainly through pre-existing pathways (skin appendages). In moderate voltage stimulation (5 V $< U_{skin} <$ 50 V), two parallel pathways are involved: one involving the appendages and the other one crossing the multilamellar lipid bilayers and corneocytes within the stratum corneum. In high voltage stimulation ($U_{skin} >$ 50 V), the decrease in R_{skin} by up to three orders of magnitude is very rapid, typically occurring within a few microseconds. Also the onset of molecular transport associated with these higher intensity pulses is much faster than in iontophoresis. In relation to pulse parameters, it needs to be emphasized that in addition to the amplitude of the pulse(s), the pulse duration strongly influences the nature and the size of the created pathways.

A number of *in vitro* experimental studies have been conducted on skin in order to study the formation of aqueous pores through the stratum corneum as a result of electroporation [53,55,58]. The main question was whether the increase in the permeability of the stratum corneum is fundamentally an electrical or a thermal phenomenon. A widely accepted answer is that while the electroporation is responsible for the creation of small pores in the stratum corneum, it is the secondary thermal effect which causes the rapid expansion of these nanoscale pathways which in turn leads to the formation of the much larger LTRs. Experimental *in vitro* electroporation studies have used a number of exponentially decaying pulses, having time constants ranging from as little as 1 ms up to 300 ms. These high voltage pulses resulted in mass transport that was highly localized and that was generally focused in the LTRs. For short pulses, typical minimum LTR size was found to be about 100 μm in diameter; for longer pulses, the sizes increase up to 300 μm. An increase in pulse voltage results in an increase in the density of the LTR distribution (number of LTRs per cm^2) but does not significantly influence their size. Alternately, longer duration pulses produce larger LTRs with almost no changes in their density of distribution [53,55]. Also after applying short, high voltage pulses, the skin's electrical resistance drops by up to three orders of magnitude, while for longer, medium voltage pulses, the drop is only up to two orders of magnitude. However when more conservative pulse protocols are used [87], large LTRs due to thermal expansion are not

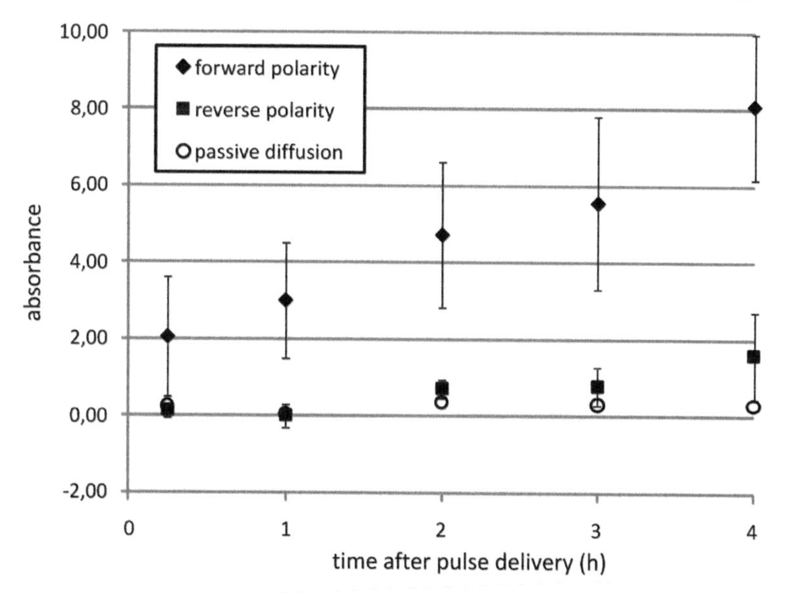

FIGURE 13.5 The delivery of negatively charged calcein through dermatomed pig skin. Absorbance was measured in the Franz cell receiver compartment, after the delivery of three high voltage pulses (1000 V, 1 ms, 1 kHz), followed by one low voltage pulse (200 V, 500 ms). For 'forward polarity', the negative electrode was in the donor compartment, and the positive electrode was in the receiver compartment (and vice versa for the 'reverse polarity').

observed (neither under a microscope, nor in theoretical analysis), for instance, when one short high voltage pulse (causing the electroporation of lipid bilayers in the stratum corneum) is followed by a longer low voltage pulse (providing the driving force), instead of trains of 50 pulses or more (see Fig. 13.4). Still, the electrophoretic contribution associated with the low voltage pulses plays an important role in the intra- or transdermal delivery of molecules, by providing the driving force that carries the delivered substance through the created aqueous pathways. This is illustrated in Fig. 13.5, which shows the delivery of negatively charged calcein through dermatomed pig skin after a pulse protocol including both an electroporation pulse and a low voltage electrophoretic pulse. The latter is a very important part of the protocol, as the reversal of pulse polarity reduces the amount of the delivered molecule to background value.

13.4 MODELING OF SKIN ELECTROPORATION TRANSPORT

Because the local transport regions evolve during an applied skin electroporation pulse of relatively high voltage (in which the transdermal voltage difference, $U_{skin} > 50$ V) and long duration (tens to hundreds of milliseconds), the transport of solute into the skin during the pulse application(s) is transient in nature. The skin's transport characteristics are dynamically altered during the application of the electric pulse [106,108,109], and the primary contributor to the transport of large charged molecules is electrophoresis [53]. For these reasons, traditional models of transdermal transport consider only passive transport (which does not capture the changes in the SC structure) or they only consider steady transport (which do not capture the transient effects). Therefore these models cannot adequately describe transport associated with the creation of the LTRs. Transport modeled in this study is derived from the modified Nernst-Planck equation which describes the transient transport of a charged solute in the presence of an electric field:

$$\frac{\partial C}{\partial t} = \nabla \cdot [(D_i \nabla C) + (m_i C \nabla V)] \tag{13.2}$$

The subscript i is used to denote a specific layer of the skin (recall the composite nature of the tissue in Fig. 13.2) in which the thermo-physical parameter values may vary between tissue layers. The parameter D is the effective diffusion coefficient, m is the effective electrophoretic mobility, V is the electric potential and C is the dimensionless solute concentration which has been normalized by the relation: $C = C^*/C_O$ where C^* is the local solute concentration and C_O is some reference concentration, usually that of the initial homogenous concentration within the applicator gel. Even longer electroporation pulse times are usually only tens to hundreds of ms in duration, so that the electro-osmotic effects (which can only occur during the pulse) are negligible compared to the electrophoretic contributions [106,108,109]. The contributions of electro-osmotic flow have been neglected from Eq. (13.2). During the application of the pulse the contribution of diffusion is also much smaller than that of electrophoresis (for large molecules this has been shown to be negligible at short time scales [110]). However, since diffusion governs transport between pulses, this term should be retained for studies including times between or long after electroporation pulse application. In Eq. (13.2) electrophoretic transport is dependent on the gradient of the electric potential, V. This is typically evaluated from the Laplace equation:

$$\nabla \cdot (\sigma_i \nabla V) = 0 \tag{13.3}$$

where σ_i is the tissue electrical conductivity. Because the electric field's transient and non-Ohmic behavior occurs at very short time scales (relative to those of thermal and mass transport), the capacitive charging times are neglected and as the local stratum corneum electrical conductivity increases with the degree of electroporation, a new solution of Eq. (13.3) is sought for each time step of Eq. (13.2). Again, note that the subscript i in Eqs. (13.2) and (13.3) implies the composite nature of the skin that is depicted in Fig. 13.2. The subscript may refer to any of the composite layers of the skin: stratum corneum (SC), epidermis, dermis or subcutaneous tissue, each of which has its own associated thermo-physical parameter values. While the dermis and subcutaneous tissues can experience electroporation at the cellular level, it is the outermost layer of the skin, the stratum corneum, which experiences the most dramatic architectural alterations during skin electroporation and thermal expansion of the created pores. Because electric pulses applied on skin can result in local transport regions within the SC, the following discussion considers only the thermo-electrical response of the SC and focuses on treating transport within the local SC architecture:

$$\frac{\partial C}{\partial t} = \nabla \cdot [(D_{SC} \nabla C) + (m_{SC} C \nabla V)] \tag{13.4}$$

and:

$$\nabla \cdot (\sigma_{SC} \nabla V) = 0 \tag{13.5}$$

where D_{SC} and m_{SC} correspond to the transport coefficients of the stratum corneum, and σ_{SC} the SC electrical conductivity. It is the increase in these coefficients that reflects the alteration of the SC's permeability to mass transport, and therefore the challenge in modeling the SC's increase in permeability lies in linking the increases of the SC transport coefficients to the electric field. Because the skin's electrical characteristics can be directly measured experimentally, studies often use the order of magnitude increases in SC electrical conductivity, σ_{SC}, as the physical parameter to quantify electroporation related changes in the SC microstructure.

As stated in Section 13.3.3, the response of the SC microstructure depends on the intensity and length of the applied pulse: for short, high voltage pulses the routes of transport are much smaller and much more evenly distributed throughout the SC compared to the transport routes associated with longer medium voltage pulses. This is best described in the conclusive remarks of the experimental study by Pliquet and Gusbeth in [55]:

> For small ions, the transport surface area for charged species is orders of magnitude less, i.e. 0.001. It is unlikely that the pathways initiated by electroporation are the same pathways for large charged molecule transport. The skin can become permeable for such large charged species only if a synergistic effect between electroporation and Joule heating produces an alteration of the stratum corneum structure. The primary pathways created by electroporation recover fast, while the LTR skin lipid phase transition recovery takes much longer.

Because the size and evolution of these two pathway types are entirely different, the models used to describe the transport and the degree of electroporation are also different. The following discussion of transport models is categorized accordingly.

13.4.1 Modeling Non-Thermal Electroporation (Short Pulses)

Electroporation pulse times classified as short duration are typically in the order of 100 μs. It has been observed that when the drop in electric potential across the skin reaches some critical value (30–100 V) [57], the skin's permeability increases by orders of magnitude. In the short pulse regime, the resulting transport routes consist of nanometer sized pores that traverse the SC and the barrier function of the SC is only inhibited for solutes of very small molecular size. In this situation, the electroporation could be considered from a macroscopic perspective so that the transport coefficients, D_{SC}, m_{SC} and σ_{SC} of Eqs. (13.4) and (13.5) could be modeled empirically from experimental observation. Computational model development incorporating the electro-physical parameter changes associated with skin electroporation has been described previously [111–113]. In these models the degree of cellular electropermeabilization is directly linked to the local gradient in electric potential.

Novel empirical approaches have been developed that directly relate the degree of permeability (represented by increases in the electrical conductivity) to the local electric field [66,113]. The implication is that the degree of electroporation is related to the intensity of the gradient in electrical potential. These studies use data from skin electroporation experiments which monitor the relationship between electrical conductivity changes and the electric field magnitude. The electric field is determined from the gradient in potential:

$$E = \nabla V \tag{13.6}$$

In the study [113], a finite element analysis was conducted of a skin fold undergoing electroporation, and used experimentally determined relations between SC electrical conductivity and local electric field magnitude. The increase in electrical conductivity with electric field is modeled by a series of step functions:

$$\sigma_{SC}(S/m) = \begin{cases} E < 600 & : \sigma_{SC} = 0.0005 \\ 600 < E < 800 & : \sigma_{SC} = 0.0165 \\ 800 < E < 1000 & : \sigma_{SC} = 0.06 \\ 1000 < E < 1200 & : \sigma_{SC} = 0.178 \\ 1200 < E & : \sigma_{SC} = 0.5 \end{cases} \tag{13.7}$$

The study allows a macroscopic representation of the current–applied voltage relationship that closely resembles that of the experimental data, and compares well with experimental findings relating the electrical behavior of the skin fold at various applied voltages.

The concept of using the magnitude of the electric field to define degree of permeability of the skin and underlying tissues is used in a detailed model in reference [66]. The basis of this study is that, experimentally, it has been shown that electroporation occurs at or near a specific electric field threshold, E_O, but that when a certain electric field value is exceeded, E_1, irreversible cellular damage begins. The study relies on experimentally derived values of E_O and E_1 to represent the threshold values of various tissues undergoing electroporation. The electric field dependent electrical conductivity, $\sigma(E)$, is related to degree of electroporation by some relationship between electrical conductivity before permeabilization, σ_o, and a maximum value of electrical conductivity due to electroporation, σ_1. The parameter values corresponding to skin are: $E_O = 400\ V/cm, E_1 = 900\ V/cm, \sigma_o = 0.002\ \text{S/m}$ and $\sigma_1 = 0.16\,\text{S/m}$. The study finds that the empirical model allows for computational results that agree very well with experiments also conducted in that study.

A search through existing literature yields no studies using the transport Eqs. (13.2) and (13.4) to directly model transport associated with short pulse electroporation. However, it is postulated here that the same empirically based method used to relate the electric field to the breakdown in the SC electrical resistance could be used to describe the SC's increase in permeability to mass transport. If from experiment, the pre-breakdown coefficients are found to be $m_{SC,U}$ and $D_{SC,U}$ and once some critical value in the electric field is reached, E_C, the coefficients are measured to as $m_{SC,M}$ and $D_{SC,M}$, then a simple step function could be used to represent the increase in SC permeability increases:

$$m_{SC} = \begin{cases} E < E_C : m_{SC,U} \\ E \geq E_C : m_{SC,M} \end{cases} \tag{13.8}$$

and:

$$D_{SC} = \begin{cases} E < E_C : D_{SC,U} \\ E \geq E_C : D_{SC,M} \end{cases} \tag{13.9}$$

In this way the SC transport coefficients and electrical conductivity of Eqs. (13.4) and (13.5) could be evaluated iteratively or simply updated at each time step. The challenge of such an empirically based method lies in the experimental evaluation of the transport coefficients. The evaluation of the post-breakdown diffusion coefficient could require a time scale that exceeds the life of the transport routes. The experimental evaluation of the solute's electrophoretic mobility in the post-electroporated SC would require measurements on a timescale equal to or less than that of the applied electric pulse.

13.4.2 Modeling Thermal Electroporation (Long Pulse)

Longer duration pulses (which may last several hundred ms) amplify two important secondary effects: electrophoresis and resistive heating (often referred to as Joule heating). Experimental skin electroporation studies have shown that electro-osmosis (the electrically induced movement of fluid) plays a negligible role, while electrophoresis (the electrically induced movement of charged particles) is the primary contributor in transdermal delivery of large charged molecules [106,108,109]. Resistive heating resulting from the applied electric pulses also plays an important secondary role in the increases in SC permeability. Early researchers of skin electroporation used the phrase 'Local Transport Region' (LTR) to describe a phenomenon in which the applied pulses result in localized regions of increased SC permeability [58,114]. The results of *in vitro* studies show that within the LTR, the electrical and mass permeabilities may be orders of magnitude greater than outside the LTR [53,55], and that the development of LTRs is always associated with thermal effects [115]. Experimental observations of 300 ms pulses have shown high temperature contours exceeding 60°C that originate near skin appendages (sweat glands) and grow radially outwards [116]. Direct evidence of the localized moving heat front and LTRs is found in *in vitro* studies in which human SC is removed and electroporated under observation [53,55,57,59,116–118]. Depending on pulse intensity, the scale in size of these high temperature fronts is on the order of 100 μm occurring on a time scale of 10–100 ms [53,116].

The relationship between resistive heating and increased permeability can be linked through experimental studies, which document that the SC lipid structure experiences a structural rearrangement at elevated temperatures. Evidence of lipid chain melting and increased permeability is given in [96,119], where it is shown that heated porcine SC becomes more permeable to water flux abruptly at temperatures approaching 70°C. It is generally accepted that the SC experiences four primary endothermic transitions over the temperature range 40–130°C. These have been independently confirmed by experiment [96,98,120,121]. Of great interest to the thermally assisted destabilization of the SC's lipid barrier function is the phase change which occurs within the temperature range 60–70°C, and it is this transition that is attributed to the disordering of the lamellar lipid phase [121].

The thermodynamic models of skin electroporation are based on these experimental observations. The idea is that while the electric field is applied to the skin, the current attempts to cross the electrically resistive SC. Any SC location with a lower electrical resistance (as would be associated with a sweat gland or hair follicle, as well as any electrically created pores in the lipid bilayers of the SC) will experience a much higher current density. This high current density results in the resistive heating that provides the necessary power for the sudden intense local temperature rises. With these increases in local temperatures, the SC lipids' internal energy exceeds some critical value and the highly structured lipid organization is destroyed. With this change in lipid architecture, the SC becomes much more permeable to transport of mass and electrical charge.

Recently, a feasibility evaluation was presented in a study of skin electroporation in which the electrically induced temperature rises within an LTR contribute to variations in the thermo-electrical property values of the skin [87]. Models of transdermal mass transport have been developed which focus on the electro-transfer of large charged solutes through the SC during the thermally related increases in SC permeability which arise during the electroporation pulse [85,86,110,122].

It has been postulated that the transport route associated with thermal effects of electroporation occurs through the lipid filled spaces as they become more disorganized, and it is noteworthy to mention that when the thermal effects of resistive heating are modeled, the SC architecture should be considered. The thin SC corneocytes-lipid matrix architecture may be conceptualized as a brick and mortar structure [123], in which the highly impermeable corneocytes are represented by the bricks and the slightly less impermeable lipid sheets are represented by the mortar filled spaces between the bricks (see Fig. 13.6). To account for the lipid path through the SC (in which transport occurs only within the lipid filled spaces and not through the corneocytes) the transport is modeled in a porous media context:

$$\frac{\partial C}{\partial t} = \frac{\varepsilon_{SC}}{\tau_{SC}}[\nabla \cdot (m_L C \nabla V) + \nabla \cdot (D_L \nabla C)] \tag{13.10}$$

The porosity, ε, and tortuosity, τ, in Eq. (13.10) are given the subscripts SC to emphasize that these terms account for the SC's structure as a whole. Using the asymmetric brick and mortar model described in [123], the porosity and tortuosity values can be approximated from estimations of the local SC structure and a detailed parametric account is given in [124]. Previous

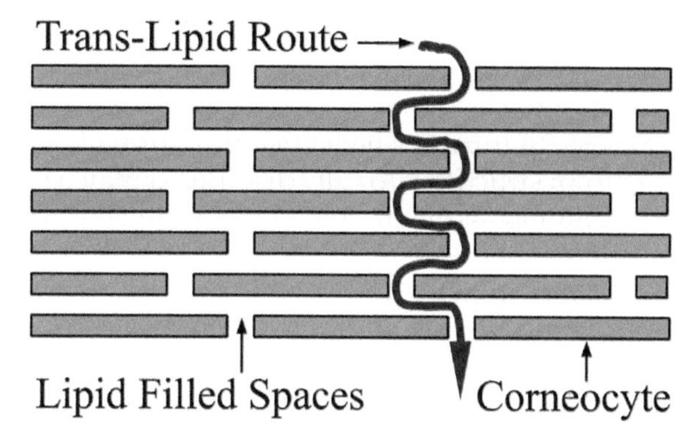

FIGURE 13.6 Brick and mortar representation of the SC structure, depicting the trans-lipid route of transport.

studies have used the porous media concept to describe the SC during electroporation and have used the ratio of SC porosity to tortuosity in the range [$1.5 \times 10^{-4} \leqslant \varepsilon_{SC}/\tau_{SC} \leqslant 5.8 \times 10^{-4}$].

In Eq. (13.10) the transport coefficients refer to those of the solute within the SC lipid microstructure where m_L is the electrophoretic mobility of the solute in the SC lipids and D_L is the effective diffusion coefficient in the SC lipids. What remains to be shown is a link between the increase in the SC lipid transport coefficients, m_L and D_L, and the increase in temperature (and lipid melting) that result from the applied electric field.

Recalling the earlier description of the skin as being composed of distinct layers, skin electroporation is modeled by a composite system so that each composite layer has its own thermo-electrical parameter values. To capture the sudden temperature rises, a modified Pennes' bioheat heat equation is used to describe the transient distribution of thermal energy within each layer of the system:

$$\rho_i c_i \frac{\partial T}{\partial t} = \nabla \cdot (k_i \nabla T) - \omega_i c_b (T - T_b) + Q_{M,i} + Q_{DC} \tag{13.11}$$

As was discussed in Section 13.3.2, the perfusion term and the metabolic heat generation may be neglected. The Joule heat, Q_{DC}, generated from the induced electric field occurs only during the applied electric pulse, and it is defined as:

$$Q_{DC} = \sigma_i |\nabla V|^2 \tag{13.12}$$

The subscript i in Eqs. (13.11) and (13.12) is used to indicate that the electrical conductivity of each of the layers have different values reflecting the composite nature of the skin. In order for a localized high current density to evolve, the SC electrical conductivity must be modeled to have some initial lateral variation. This would be associated with the much higher relative local electrical conductivity of a sweat gland or pre-existing pore. The pre-existing pore sizes used in the computations of previous studies have varied in the range $1.5\mu m \leqslant R_P \leqslant 15\ \mu m$ and are assigned the same electrical properties as the gel layer (roughly the same as those of water) [86,87,110,122,125].

In recent studies [86,125], the authors have proposed a thermodynamically based model to describe the structural changes associated with skin electroporation by adapting methods that have been traditionally designed to model melting and solidification processes occurring over a temperature range [126]. The analogy being that the thermally influenced transition of the SC lipid microstructure from structured to fully disordered may be represented by a similar function to the ones used to model the thermal phase change of a material from solid to liquid.

The function referred to as the lipid melt fraction is used to describe the degree of lipid disorder. It is assumed that the disordering of the lipid structure occurs over the temperature range corresponding to the lipid phase transition $[T_{E1} < T \leqslant T_{E2}]$. Here T_{E1} corresponds to the temperature at which the SC lipid phase transition is initiated and T_{E2} corresponds to the temperature at which the lipids have become fully disrupted. The lipid melt fraction is defined as:

$$\varphi = \frac{(H - c_{SC}(T - T_{E1}))}{\Delta H_E} \tag{13.13}$$

where c_{SC} is the SC specific heat capacity and ΔH_E is the latent heat associated with this phase transition. The total enthalpy, H, is defined as:

$$H = \int_{T_{E1}}^{T} c_{SC,APP} d\tau \quad [T_{E1} < T \leqslant T_{E2}] \tag{13.14}$$

To simplify the description considerably, a rectangular shaped specific heat vs. temperature curve is used to model phase transitions, so that the apparent SC specific heat, $c_{SC,APP}$, is represented by:

$$c_{SC,APP} = c_{SC} + c_{SC,L} \tag{13.15}$$

where $c_{SC,L}$ is the latent SC lipid specific heat:

$$c_{SC,L} = \frac{\Delta H_E}{T_{E2} - T_{E1}} \tag{13.16}$$

Under the approximation of rectangular shaped enthalpy vs. temperature curves, Eqs. (13.14)–(13.16) may be substituted into Eq. (13.13) to provide a much simpler expression of the lipid melt fraction:

$$\varphi = \begin{cases} 0 & : \quad T \leqslant T_{E1} \\ \frac{T - T_{E1}}{T_{E2} - T_{E1}} & : \quad T_{E1} \leqslant T \leqslant T_{E2} \\ 1 & : \quad T_{E2} \leqslant T \end{cases} \tag{13.17}$$

The lipid melt fraction of Eq. (13.17) is a thermodynamic ratio of thermal energy added to the lipids to thermal energy required to complete lipid phase transition. The melt fraction value represents the degree of lipid disorder ($0 \leqslant \varphi \leqslant 1$) where $\varphi = 0$ corresponds to the unaltered lamellar lipid structure that exists below phase transition temperatures ($T \leqslant T_{E1}$) and $\varphi = 1$ corresponds to much more permeable, fully disrupted SC lipid architecture that exists above temperatures associated with the lipid phase transition ($T_{E2} \leqslant T$). The parametric values used are referenced from experimental data provided in [121] and are: $T_{E1} = 65^{\circ}C$, $T_{E2} = 75^{\circ}C$, $\Delta H_E = 5300$ J/kgK.

The next step of the thermodynamic model is to relate the lipid melt fraction to the electrical and molecular permeability changes of the SC. The electrical conductivity of the SC has been shown to experience increases during the disruption of its architecture. Experimental observations of the thermal dependence of SC electrical resistance have shown that the magnitude of the electrical resistance can drop by two orders of magnitude at lipid thermal phase change temperatures [127]. Large changes in the SC electrical conductivity will strongly influence the voltage distribution of Eq. (13.5) and subsequently the electric field of Eq. (13.6), which in turn is coupled to the resistive heating in Eq. (13.11) and the electrophoretic driving contribution in Eq. (13.4). It is very important to attempt to correctly describe the electrical conductivity increases in terms of lipid thermal-structural behavior. Consider that it is possible to experimentally measure $\sigma_{SC,U}$ which is the SC electrical conductivity associated with the unaltered lipid structure ($\varphi = 0$) as well as $\sigma_{SC,M}$ which is the electrical conductivity associated with the SC after full lipid melting ($\varphi = 1$). For instance the results suggested in reference [127] have been interpreted in the studies [86,110,125] so that the conductivity values $\sigma_{SC,U} = 10^{-5}$ S/m and $\sigma_{SC,M} = 10^{-3}$ S/m represent the two order of magnitude increase in electrical conductivity of the SC. To relate the increase in electrical conductivity, different functional dependencies have been proposed [110], a linear function:

$$\sigma_{SC} = \sigma_{SC,U} + \varphi(\sigma_{SC,M} - \sigma_{SC,U}) \tag{13.18}$$

and an exponential function:

$$\sigma_{SC} = \sigma_{SC,U} + (\sigma_{SC,M} - \sigma_{SC,U}) \exp\left[(\varphi - 1)/\varphi\right]. \tag{13.19}$$

The conductivity values $\sigma_{SC,U} = 10^{-5}$ S/m and $\sigma_{SC,M} = 10^{-3}$ S/m have been chosen to represent the two order of magnitude increase in electrical conductivity with lipid restructuring as suggested by the results in [127].

The experimentally observed increases in permeability to mass transport can be satisfied by relating the transport coefficients directly to the lipid melt fraction $m_L = f(\varphi)$ and $D_L = g(\varphi)$. In studies [125,122] a linear dependence of transport coefficients on the lipid melt fraction has been assumed so that:

$$m_L = m_{L,U} + \varphi(m_{L,M} - m_{L,U}) \tag{13.20}$$

and:

$$D_L = D_{L,U} + \varphi(D_{L,M} - D_{L,U}) \tag{13.21}$$

The transport coefficients $m_{L,U}$ and $D_{L,U}$ correspond to mobility and diffusion of the solute in the SC lipids that have not experienced any thermal transition ($\varphi = 0$) and $m_{L,M}$ and $D_{L,M}$ represent the highest values of transport coefficient after full lipid melting ($\varphi = 1$).

As an alternative to the linear description of transport coefficient behavior, consider the passive transport studies [127,124], in which the diffusion coefficient within the SC lipids is exponentially dependent upon the lipid state. Those studies are not concerned with the dynamic thermal alteration of the lipid microstructure, and their description of the lipid state is based on the molecular interactions between the solute and a non-thermally altered lipid structure. Despite these differences, an exponential dependence on lipid state could be implemented in this model such that the transport coefficients are represented as:

$$m_L = m_{L,U} + \left(m_{L,M} - m_{L,U}\right) \exp\left[(\varphi - 1)/\varphi\right] \tag{13.22}$$

$$D_L = D_{L,U} + \left(D_{L,M} - D_{L,U}\right) \exp\left[(\varphi - 1)/\varphi\right] \tag{13.23}$$

Correctly estimating the solute flux relies explicitly on an accurate estimation of the transport coefficients. The values of the diffusion coefficient and the mobility of large molecules within tissues can be estimated experimentally or semi-empirically [128–130]. Estimates that have been used in the literature to approximate large DNA mobility in the lipids correspond to $m_{L,U} = 2.67 \times 10^{-14}$ and $m_{L,M} = 5.96 \times 10^{-8}$ m^2 s^{-1} V^{-1}. A detailed description of the implications of these values is provided in [110].

Parametric studies that have compared the linear dependencies of Eqs. (13.18), (13.20) and (13.21) to the exponential dependencies of Eqs. (13.19), (13.22) and (13.23) have shown that, when compared over the same temperature range, there is a relative difference in the total solute transported or the LTR region of influence.

It is also worth mentioning that some studies consider the possibility that these dramatic increases in permeability do not occur continuously over the temperature range ΔT_E at all. Instead they might be considered to occur suddenly, at a specific threshold temperature which lies within the range of lipid phase transition temperatures. This is precisely the approach that is used in the recent study [87] in which the development of the LTR is represented by local SC thermo-electrical property values that experience sudden changes at a discrete threshold temperature, T_C. While that study focused on modeling the alteration of the SC's thermo-electrical characteristics, such a step function approach could be implemented into the current transport study by representing the increases in transport coefficients as follows:

$$\sigma_{SC} = \begin{cases} T < T_{EC} : \sigma_{SC,U} \\ T \geqslant T_{EC} : \sigma_{SC,M} \end{cases} \tag{13.24}$$

$$m_L = \begin{cases} T < T_{EC} : m_{L,U} \\ T \geqslant T_{EC} : m_{L,M} \end{cases} \tag{13.25}$$

$$D_L = \begin{cases} T < T_{EC} : D_{L,U} \\ T \geqslant T_{EC} : D_{L,M} \end{cases} \tag{13.26}$$

In [87] a representative threshold temperature of $T_{EC} = 60$ °C is used. It seems that any temperature near the experimentally observed lipid phase transition temperature range would be applicable. From a strictly thermodynamic perspective, it does seem physically unlikely that the SC's thermally associated increases in permeability occur at a discrete temperature; however, it may be that the temperature ranges over which these increases in permeability occur are so tightly grouped that a step function is an adequate representation. It is important to consider that if the step function is implemented into a model of mass transport, the solution is critically sensitive to the choice of threshold temperature; in [110], an order of magnitude increase in total solute delivered is shown when this temperature is increased from $T_{EC} = 65^{\circ}$C to $T_{EC} = 75^{\circ}$C. The implications of this finding were discussed in detail in [125], in which it was postulated that lowering the lipid phase transformation temperatures using chemical enhancers will result in greater effectiveness, and allow skin electroporation with a lower risk of unintentional thermal damage to the tissue.

Parametric investigations in [110] have been conducted which compare the total solute transported and the region of influence of lipid thermal transition, and show clearly that the total solute transported is more sensitive to parameters influencing the electrical conductivity than to factors affecting the effective diffusion coefficient and mobility. This is because any variation in the transport coefficients, $m_{L,M}$ and $D_{L,M}$, will not influence the evolution of the LTR, while a variation in electrical conductivity will strongly affect the evolution of the LTR. The effects of SC porosity $\varepsilon_{SC}/\tau_{SC}$ on total solute transport are much less intensive than the effects of the SC thickness. The influence of the sensitivity of transport to the lipid phase transition temperature range is depicted in Fig. 13.7. In order to make direct comparisons between parameter influences on the quantities of solute transported into the skin, a normalized concentration ratio is introduced. This parameter is used to represent the ratio of total solute

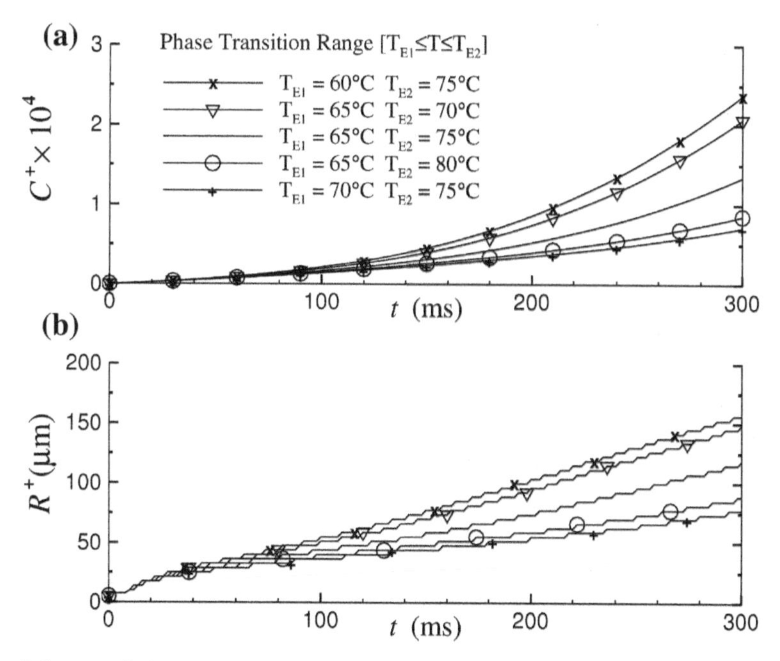

FIGURE 13.7 Influence of phase transition temperature range on: (a) total solute transported below the SC surface, and (b) development of the LTR size.

transported below the SC surface to the total available solute initially contained in the gel layer. It is represented mathematically by the expression:

$$C^{+}(t) = \frac{\text{Solute transported below the SC surface}}{\text{Total initial available solute}} = \frac{\int_{r=0}^{R_{MAX}} \int_{z=0}^{Z_{MAX}} C(r,z,t)dzdr}{\int_{r=0}^{R_{MAX}} \int_{z=th_{GEL}}^{0} C(r,z,0)dzdr} \tag{13.27}$$

In order to quantify parameter effects on LTR development during the pulse, an effective LTR radius ($R^{+}(t)$) is introduced. It is defined as the minimum radial location within the SC at which the SC lipids have not experienced any thermal restructuring ($\varphi = 0$). Thus $R^{+}(t)$ is used to characterize the LTR size at any time during the applied pulse. The variables C^{+} and R^{+} are used to compare parametric influences in all subsequent discussion.

The onset temperature, T_{E1}, represents the minimum temperature at which the SC lipid structure is affected and T_{E2} represents the end of the phase transition temperature range, above which no increase in permeability of the SC is associated. The onset temperature is the most influential on solute transport and lipid melting. If the SC lipids experience the onset of thermal phase transition at lower temperatures, less energy in the form of resistive heat is required to initiate the growth of the LTR. While slightly less influential, the temperature at which lipid melting ends, T_{E2}, is also important to total solute transported.

13.5 CONCLUSIONS

In this chapter, the physical phenomena associated with electroporation of the skin are reviewed. It is well documented that the physical response of the skin to electroporation depends on the intensity and duration of the applied electric pulses. The current understanding of the field is that there exist two primary pulsing regimes that result in a dramatic restructuring of the SC architecture. Short intense pulses are believed to result in nanometer-sized pathways that are sufficient for the delivery of very small solutes. Longer duration pulses provide two secondary effects that greatly increase the success of skin electroporation. Resistive heating has been shown to raise the internal energy of the SC lipids to such an extent that their microstructure experiences a dramatic alteration. The restructuring of the SC lipids results in order of magnitude increases in SC permeability to mass transport and electrical conductivity. Another important secondary effect of long pulse skin electroporation is electrophoresis, which greatly increases the success of the transport of large charged solute. In fact it is experimentally confirmed that in some cases the transport of solute is negligible in the absence of the electrophoretic contribution.

The field's computational and theoretical descriptions of skin electroporation have been summarized here as well. The phenomena that the models capture can be divided into two categories: the degree of electroporation and the transport of solute. By experimental observation, the degree to which the SC's structure has been altered by the electroporation pulses is conveyed through the measured increases in electrical conductivity and permeability to mass transport. Models of electroporation attempt to represent these alterations in the SC structure through increases in local electrical conductivity and increases in local mass permeability. Models of short intense pulse electroporation rely entirely on empirical models to relate the increase in electrical conductivity to the magnitude of the local electric field. Long medium pulse electroporation models rely on a thermodynamic description of the SC lipid behavior to describe increases in electrical conductivity and increased permeability.

While the field can currently offer theoretical models that provide a structure that can be used to describe LTR development and associated mass transport through the SC, future work should include the influences of subcutaneous cellular electroporation. The next generation of models should extend to include the effects of the variations in parameter values of the underlying tissues that result from the electroporation of cells associated with the electric field within the dermis.

Acknowledgments

Research was performed in the scope of LEA EBAM and was in part financed by the European Regional Development Fund (Biomedical Engineering Competence Center, Slovenia), COST TD1104, the Slovenian Research Agency (L2-2044), and the Alexander von Humboldt Foundation.

References

[1] G. Sersa, D. Miklavcic, M. Cemazar, Z. Rudolf, G. Pucihar, M. Snoj, Electrochemotherapy in treatment of tumors, Ejso 34 (2008) 232–240.

[2] L. Mir, Nucleic acids electrotransfer-based gene therapy (electrogenetherapy): past, current, and future, Molecular Biotechnology 43 (2009) 167–176.

[3] J. Escoffre, T. Portet, L. Wasungu, J. Teissie, D. Dean, M. Rols, What is (still not) known of the mechanism by which electroporation mediates gene transfer and expression in cells and tissues, Molecular Biotechnology 41 (2009) 286–295.

[4] J. Escobar-Chavez, D. Bonilla-Martinez, M. Villegas-Gonzalez, A. Revilla-Vazquez, Electroporation as an efficient physical enhancer for skin drug delivery, Journal of Clinical Pharmacology 49 (2009) 1262–1283.

[5] R. Davalos, L. Mir, B. Rubinsky, Tissue ablation with irreversible electroporation, Annals of Biomedical Engineering 33 (2005) 223–231.

[6] C. Ramos, J. Teissie, Electrofusion: a biophysical modification of cell membrane and a mechanism in exocytosis, Biochimie 82 (2000) 511–518.

[7] D. Hannaman, Electroporation for DNA immunization: clinical application, Expert Review of Vaccines 9 (2010) 503–517.

[8] A. Gothelf, J. Eriksen, P. Hojman, J. Gehl, Duration and level of transgene expression after gene electrotransfer to skin in mice, Gene Therapy 7 (2010) 839–845.

[9] M. Usaj, K. Trontelj, D. Miklavcic, M. Kanduser, Cell-cell electrofusion: optimization of electric field amplitude and hypotonic treatment for mouse melanoma (B16-F1) and Chinese hamster ovary (CHO) cells, Journal of Membrane Biology 236 (2010) 107–116.

[10] K. Kinosita Jr., T.Y. Tsong, Voltage-induced pore formation and hemolysis of human erythrocytes, Biochimica et Biophysica Acta (BBA) – Biomembranes 471 (1977) 227–242.

[11] J. Teissié, M.P. Rols, An experimental evaluation of the critical potential difference inducing cell membrane electropermeabilization, Biophysical Journal 65 (1993) 409–413.

[12] E. Tekle, R.D. Astumian, P.B. Chock, Electro-permeabilization of cell membranes: effect of the resting membrane potential, Biochemical and Biophysical Research Communications 172 (1990) 282–287.

[13] R. Susil, D. Semrov, D. Miklavcic, Electric field induced transmembrane potential depends on cell density and organization, Electro- and Magnetobiology 17 (1998) 391–399.

[14] T. Kotnik, D. Miklavčič, Analytical description of transmembrane voltage induced by electric fields on spheroidal cells, Biophysical Journal 79 (2000) 670–679.

[15] M. Pavlin, N. Pavselj, D. Miklavcic, Dependence of induced transmembrane potential on cell density, arrangement, and cell position inside a cell system, IEEE Transactions on Biomedical Engineering 49 (2002) 605–612.

[16] B. Valič, M. Pavlin, D. Miklavčič, The effect of resting transmembrane voltage on cell electropermeabilization: a numerical analysis, Bioelectrochemistry 63 (2004) 311–315.

[17] M. Essone Mezeme, G. Pucihar, M. Pavlin, C. Brosseau, D. Miklavčič, A numerical analysis of multicellular environment for modeling tissue electroporation, Applied Physics Letters 100 (2012) 143701–143701-4.

[18] H. Wolf, M.P. Rols, E. Boldt, E. Neumann, J. Teissié, Control by pulse parameters of electric field-mediated gene transfer in mammalian cells, Biophysical Journal 66 (1994) 524–531.

[19] A. Macek-Lebar, D. Miklavcic, Cell electropermeabilization to small molecules *in vitro*: control by pulse parameters, Radiology and Oncology 35 (2001) 193–202.

[20] G. Pucihar, J. Krmelj, M. Reberšek, T.B. Napotnik, D. Miklavčič, Equivalent pulse parameters for electroporation, IEEE Transactions on Biomedical Engineering 58 (2011) 3279–3288.

[21] P.J. Canatella, J.F. Karr, J.A. Petros, M.R. Prausnitz, Quantitative study of electroporation-mediated molecular uptake and cell viability, Biophysical Journal 80 (2001) 755–764.

[22] M.-P. Rols, J. Teissié, Electropermeabilization of mammalian cells to macromolecules: control by pulse duration, Biophysical Journal 75 (1998) 1415–1423.

[23] L.M. Mir, M.F. Bureau, J. Gehl, R. Rangara, D. Rouy, J.M. Caillaud et al., High-efficiency gene transfer into skeletal muscle mediated by electric pulses, Proceedings of National Academy of Science 96 (1999) 4262–4267.

[24] N. Pavselj, V. Preat, DNA electrotransfer into the skin using a combination of one high- and one low-voltage pulse, Journal of Controlled Release 106 (2005) 407–415.

[25] F. Andre, J. Gehl, G. Sersa, V. Preat, P. Hojman, J. Eriksen et al., Efficiency of high- and low-voltage pulse combinations for gene electrotransfer in muscle, liver, tumor, and skin, Human Gene Therapy 19 (2008) 1261–1271.

[26] L. Zhang, G. Widera, D. Rabussay, Enhancement of the effectiveness of electroporation-augmented cutaneous DNA vaccination by a particulate adjuvant, Bioelectrochemistry 63 (2004) 369–373.

[27] G. Vandermeulen, E. Staes, M.L. Vanderhaeghen, M.F. Bureau, D. Scherman, V. Preat, Optimisation of intradennal DNA electrotransfer for immunisation, Journal of Controlled Release 124 (2007) 81–87.

[28] J.W. Hooper, J.W. Golden, A.M. Ferro, A.D. King, Smallpox DNA vaccine delivered by novel skin electroporation device protects mice against intranasal poxvirus challenge, Vaccine 25 (2007) 1814–1823.

[29] J. Drabick, J. Glasspool-Malone, S. Somiari, A. King, R. Malone, Cutaneous transfection and immune responses to intradermal nucleic acid vaccination are significantly enhanced by *in vivo* electropermeabilization, Molecular Therapy 3 (2001) 249–255.

[30] H.A.E. Benson, S. Namjosh, Proteins and peptides: strategies for delivery to and across the skin, Journal of Pharmaceutical Sciences 97 (2008) 3591–3610.

[31] M. Foldvari, M. Baca-Estrada, Z. He, J. Hu, S. Attah-Poku, M. King, Dermal and transdermal delivery of protein pharmaceuticals: lipid-based delivery systems for interferon alpha, Biotechnology and Applied Biochemistry 30 (1999) 129–137.

[32] B. Magnusson, P. Runn, Effect of penetration enhancers on the permeation of the thyrotropin releasing hormone analogue pGlu-3-methyl-His-Pro amide through human epidermis, International Journal of Pharmaceutics 178 (1999) 149–159.

[33] P. Gupta, V. Mishra, A. Rawat, P. Dubey, S. Mahor, S. Jain et al., Non-invasive vaccine delivery in transfersomes, niosomes and liposomes: a comparative study, International Journal of Pharmaceutics 293 (2005) 73–82.

[34] A.C. Williams, B.W. Barry, Penetration enhancers, Advanced Drug Delivery Reviews 56 (2004) 603–618.

[35] I.B. Pathan, C.M. Setty, Chemical penetration enhancers for transdermal drug delivery systems, Tropical Journal of Pharmaceutical Research 8 (2009) 173–179.

[36] J. Matriano, M. Cormier, J. Johnson, W. Young, M. Buttery, K. Nyam et al., Macroflux (R) microprojection array patch technology: A new and efficient approach for intracutaneous immunization, Pharmaceutical Research 19 (2002) 63–70.

[37] J. Schramm, S. Mitragotri, Transdermal drug delivery by jet injectors: energetics of jet formation and penetration, Pharmaceutical Research 19 (2002) 1673–1679.

[38] H. Dean, D. Fuller, J. Osorio, Powder and particle-mediated approaches for delivery of DNA and protein vaccines into the epidermis, Comparative Immunology, Microbiology, & Infectious Diseases 26 (2003) 373–388.

[39] D. McAllister, P. Wang, S. Davis, J. Park, P. Canatella, M. Allen et al., Microfabricated needles for transdermal delivery of macromolecules and nanoparticles: fabrication methods and transport studies, Proceedings of the National Academy of Sciences USA 100 (2003) 13755–13760.

[40] M.R. Prausnitz, Microneedles for transdermal drug delivery, Advanced Drug Delivery Reviews 56 (2004) 581–587.

[41] L. Daugimont, N. Baron, G. Vandermeulen, N. Pavselj, D. Miklavcic, M.-C. Jullien et al., Hollow Microneedle Arrays for Intradermal Drug Delivery and DNA Electroporation, Journal of Membrane Biology 236 (2010) 117–125.

[42] A.C. Sintov, I. Krymberk, D. Daniel, T. Hannan, Z. Sohn, G. Levin, Radiofrequency-driven skin microchanneling as a new way for electrically assisted transdermal delivery of hydrophilic drugs, Journal of Controlled Release 89 (2003) 311–320.

[43] J. Fang, W. Lee, S. Shen, H. Wang, C. Fang, C. Hu, Transdermal delivery of macromolecules by erbium: YAG laser, Journal of Controlled Release 100 (2004) 75–85.

[44] J. Birchall, S. Coulman, A. Anstey, C. Gateley, H. Sweetland, A. Gershonowitz et al., Cutaneous gene expression of plasmid DNA in excised human skin following delivery via microchannels created by radio frequency ablation, International Journal of Pharmaceutics 312 (2006) 15–23.

[45] G. Levin, Advances in Radio-Frequency Transdermal Drug Delivery, Pharmaceutical Technology, supplement (2008) S12–S14, S16–S19.

[46] C. Gómez, A. Costela, I. García-Moreno, F. Llanes, J.M. Teijón, D. Blanco, Laser treatments on skin enhancing and controlling transdermal delivery of 5-fluorouracil, Lasers in Surgery and Medicine 40 (2008) 6–12.

[47] R. Alvarez-Roman, G. Merino, Y. Kalia, A. Naik, R. Guy, Skin permeability enhancement by low frequency sonophoresis: lipid extraction and transport pathways, Journal of Pharmaceutical Sciences 92 (2003) 1138–1146.

[48] S. Mitragotri, J. Kost, Low-frequency sonophoresis – a review, Advanced Drug Delivery Reviews 56 (2004) 589–601.

[49] I. Lavon, J. Kost, Ultrasound and transdermal drug delivery, Drug Discovery Today 9 (2004) 670–676.

[50] J. Raiman, M. Koljonen, K. Huikko, R. Kostiainen, J. Hirvonen, Delivery and stability of LHRH and Nafarelin in human skin: the effect of constant/pulsed iontophoresis RID B-5491-2008, European Journal of Pharmaceutical Sciences 21 (2004) 371–377.

[51] S. Mayes, M. Ferrone, Fentanyl HCl patient-controlled iontophoretic transdermal system for the management of acute postoperative pain, The Annals of Pharmacotherapy 40 (2006) 2178–2186.

[52] L. Hu, P. Batheja, V. Meidan, B.B. Michniak-Kohn, Iontophoretic transdermal drug delivery, in: V.S. Kulkarni (Ed.), Handbook of Non-Invasive Drug Delivery Systems, William Andrew Publishing, Boston, 2010, pp. 95–118.

[53] R. Vanbever, U.F. Pliquett, V. Preat, J.C. Weaver, Comparison of the effects of short, high-voltage and long, medium-voltage pulses on skin electrical and transport properties, Journal of Controlled Release 69 (1999) 35–47.

[54] J.C. Weaver, T.E. Vaughan, Y.A. Chizmadzhev, Theory of electrical creation of aqueous pathways across skin transport barriers, Advanced Drug Delivery Reviews 35 (1999) 21–39.

[55] U. Pliquett, C. Gusbeth, Surface area involved in transdermal transport of charged species due to skin electroporation, Bioelectrochemistry 65 (2004) 27–32.

[56] U. Pliquett, J.C. Weaver, Feasibility of an electrode-reservoir device for transdermal drug delivery by noninvasive skin electroporation, IEEE Transactions on Biomedical Engineering 54 (2007) 536–538.

[57] U. Pliquett, R. Langer, J.C. Weaver, Changes in the passive electrical properties of human stratum corneum due to electroporation, Biochimica et Biophysica Acta 1239 (1995) 111–121.

[58] U.F. Pliquett, R. Vanbever, V. Preat, J.C. Weaver, Local transport regions (LTRs) in human stratum corneum due to long and short 'high voltage' pulses, Bioelectrochemistry and Bioenergetics 47 (1998) 151–161.

[59] U. Pliquett, S. Gallo, S.W. Hui, C. Gusbeth, E. Neumann, Local and transient structural changes in stratum corneum at high electric fields: contribution of joule heating, Bioelectrochemistry 67 (2005) 37–46.

[60] N. Dujardin, P. Van Der Smissen, V. Preat, Topical gene transfer into rat skin using electroporation, Pharmaceutical Research 18 (2001) 61–66.

[61] C. Gabriel, S. Gabriel, E. Corthout, The dielectric properties of biological tissues: I. Literature survey, Physics in Medicine and Biology 41 (1996) 2231–2249.

[62] S. Gabriel, R.W. Lau, C. Gabriel, The dielectric properties of biological tissues: II. Measurements in the frequency range 10 Hz to 20 GHz, Physics in Medicine and Biology 41 (2004) 2251–2269.

[63] C. Gabriel, Dielectric properties of biological tissue: variation with age, Bioelectromagnetics Supplement 7 (2005) S12–S18.

[64] C. Gabriel, A. Peyman, E.H. Grant, Electrical conductivity of tissue at frequencies below 1 MHz, Physics in Medicine and Biology 54 (2009) 4863.

[65] U. Pliquett, J.C. Weaver, Electroporation of human skin: simultaneous measurement of changes in the transport of two fluorescent molecules and in the passive electrical properties, Bioelectrochemistry and Bioenergetics 39 (1996) 1–12.

[66] N. Pavselj, Z. Bregar, D. Cukjati, D. Batiuskaite, L. Mir, D. Miklavcic, The course of tissue permeabilization studied on a mathematical model of a subcutaneous tumor in small animals, IEEE Transactions on Biomedical Engineering 52 (2005) 1373–1381.

[67] D. Cukjati, D. Batiuskaite, F. Andre, D. Miklavcic, L.M. Mir, Real time electroporation control for accurate and safe *in vivo* non-viral gene therapy RID A-9497-2008, Bioelectrochemistry 70 (2007) 501–507.

[68] T. Suchi, Experiments on electrical resistance of the human epidermis, The Japanese Journal of Physiology 5 (1955) 75–80.

[69] J.C. Lawler, M.J. Davis, E.C. Griffith, Electrical characteristics of the skin. The impedance of the surface sheath and deep tissues, The Journal of Investigative Dermatology 34 (1960) 301–308.

[70] J.F. Lane, Electrical impedances of superficial limb tissues: epidermis, dermis, and muscle sheath, Annals New York Academy of Science 170 (1970) 812–825.

[71] T. Yamamoto, Y. Yamamoto, Electrical properties of the epidermal stratum corneum, Medical and Biological Engineering 14 (1976) 151–158.

[72] T. Yamamoto, Y. Yamamoto, Dielectric constant and resistivity of epidermal stratum corenum, Medical and Biological Engineering 14 (1976) 494–499.

[73] Y.A. Chizmadzhev, A.V. Indenbom, P.I. Kuzmin, S.V. Galichenko, J.C. Weaver, R.O. Potts, Electrical properties of skin at moderate voltages: contribution of appendageal macropores, Biophysical Journal 74 (1998) 843–856.

[74] M.R. Prausnitz, B.S. Lau, C.D. Milano, S. Conner, R. Langer, J.C. Weaver, A quantitative study of electroporation showing a plateau in net molecular transport, Biophysical Journal 65 (1993) 414–422.

[75] S.A. Gallo, A.R. Oseroff, P.G. Johnson, S.W. Hui, Characterization of electric-pulse-induced permeabilization of porcine skin using surface electrodes, Biophysical Journal 72 (1997) 2805–2811.

[76] P. Pawlowski, S.A. Gallo, P.G. Johnson, S.W. Hui, Electrorheological modeling of the permeabilization of the stratum corneum: theory and experiment, Biophysical Journal 75 (1998) 2721–2731.

[77] A. Jadoul, J. Bouwstra, V. Preat, Effects of iontophoresis electroporation on the stratum corneum. Review of the biophysical studies, Advanced Drug Delivery Reviews 35 (1999) 89–105.

[78] U. Pliquett, M.R. Prausnitz, Electrical impedance spectroscopy for rapid and non-invasive analysis of skin electroporation, in M.J. Jaroszeski, R. Heller, R. Gilbert (Eds.), Electrically Mediated Delivery of Molecules to Cells, Electrochemotherapy, Electrogenetherapy and Transdermal Delivery by Electroporation, Humana Press, Totowa, NJ, 2000, 377–406.

[79] A. Gothelf, J. Gehl, Gene electrotransfer to skin; review of existing literature clinical perspectives, Current Gene Therapy 10 (2010) 287–299.

[80] R.V. Davalos, B. Rubinsky, L.M. Mir, Theoretical analysis of the thermal effects during *in vivo* tissue electroporation, Bioelectrochemistry 61 (2003) 99–107.

[81] R.V. Davalos, B. Rubinsky, Temperature considerations during irreversible electroporation, International Journal of Heat and Mass Transfer 51 (2008) 5617–5622.

[82] I. Lacković, R. Magjarević, Three-dimensional finite-element analysis of joule heating in electrochemotherapy and *in vivo* gene electrotransfer, IEEE Transactions on Dielectrics and Electrical Insulation 16 (2009) 1339.

[83] U. Pliquett, Joule heating during solid tissue electroporation, Medical and Biological Engineering and Computing 41 (2003) 215–219.

[84] C. Daniels, B. Rubinsky, Electrical field and temperature model of nonthermal irreversible electroporation in heterogeneous tissues, Journal of Biomechanical Engineering 131 (2009) 071006.

[85] S.M. Becker, A.V. Kuznetsov, Numerical assessment of thermal response associated with *in vivo* skin electroporation: the importance of the composite skin model, Journal of Biomechanical Engineering 129 (2007) 330–340.

[86] S.M. Becker, A.V. Kunetsov, Local temperature rises influence *in vivo* electroporation pore development: a numerical stratum corneum lipid phase transition model, Journal of Biomechanical Engineering – Transactions of the Asme 129 (2007) 712–721.

[87] N. Pavselj, D. Miklavcic, Resistive heating and electropermeabilization of skin tissue during *in vivo* electroporation: a coupled nonlinear finite element model, International Journal of Heat and Mass Transfer (2011)

[88] H.H. Pennes, Analysis of tissue and arterial blood temperatures in the resting human forearm, Journal of Applied Physiology 1 (1948) 93.
[89] G. Sersa, T. Jarm, T. Kotnik, A. Coer, M. Podkrajsek, M. Sentjurc et al., Vascular disrupting action of electroporation and electrochemotherapy with bleomycin in murine sarcoma, British Journal of Cancer 98 (2008) 388–398.
[90] G. Sersa, M. Cemazar, C.S. Parkins, D.J. Chaplin, Tumor blood flow changes induced by application of electric pulses, European Journal of Cancer 35 (1999) 672–677.
[91] T. Jarm, M. Cemazar, D. Miklavcic, G. Sersa, Antivascular effects of electrochemotherapy: implications in treatment of bleeding metastases, Expert Review of Anticancer Therapy 10 (2010) 729–746.
[92] I. Edhemovic, E.M. Gadzijev, E. Brecelj, D. Miklavcic, B. Kos, A. Zupanic et al., Electrochemotherapy: a new technological approach in treatment of metastases in the liver, Technology in Cancer Research and Treatment 10 (2011) 475–485.
[93] D. Haemmerich, I. Santos, D.J. Schutt, J.G. Webster, D.M. Mahvi, *In vitro* measurements of temperature-dependent specific heat of liver tissue, Medical Engineering & Physics 28 (2006) 194–197.
[94] D. Haemmerich, D.J. Schutt, I. Santos, J.G. Webster, D.M. Mahvi, Measurement of temperature-dependent specific heat of biological tissues, Physiological Measurement 26 (2005) 59–67.
[95] A. Bhattacharya, R.L. Mahajan, Temperature dependence of thermal conductivity of biological tissues, Physiological Measurement 24 (2003) 769–783.
[96] G. Golden, D. Guzek, A. Kennedy, J. Mckie, R. Potts, Stratum-corneum lipid phase-transitions and water barrier properties, Biochemistry 26 (1987) 2382–2388.
[97] S.M. Al-Saidan, B.W. Barry, A.C. Williams, Differential scanning calorimetry of human and animal stratum corneum membranes, International Journal of Pharmaceutics 168 (1998) 17–22.
[98] H. Tanojo, J.A. Bouwstra, H.E. Junginger, H.E. Boddé, Thermal analysis studies on human skin and skin barrier modulation by fatty acids and propylene glycol, Journal of Thermal Analysis and Calorimetry 57 (1999) 313–322.
[99] C.L. Silva, S.C.C. Nunes, M.E.S. Eusébio, A.A.C.C. Pais, J.J.S. Sousa, Thermal behavior of human stratum corneum, Skin Pharmacology and Physiology 19 (2006) 132–139.
[100] C.L. Silva, S.C.C. Nunes, M.E.S. Eusébio, J.J.S. Sousa, A. Pais, Study of human stratum corneum and extracted lipids by thermomicroscopy and DSC, Chemistry and Physics of Lipids 140 (2006) 36–47.
[101] F.A. Duck, Physical Properties of Tissue: A Comprehensive Reference Book, Academic Press, London, 1990.
[102] R. Vanbever, V. Preat, Factors affecting transdermal delivery of metoprolol by electroporation, Bioelectrochemistry and Bioenergetics 38 (1995) 223–228.
[103] M.R. Prausnitz, The effects of electric current applied to skin: A review for transdermal drug delivery, Advanced Drug Delivery Reviews 18 (1996) 395–425.
[104] M.R. Prausnitz, A practical assessment of transdermal drug delivery by skin electroporation, Advanced Drug Delivery Reviews 35 (1999) 61–76.
[105] A.R. Denet, V. Preat, Transdermal delivery of timolol by electroporation through human skin, Journal of Controlled Release 88 (2003) 253–262.
[106] A.R. Denet, R. Vanbever, V. Preat, Skin electroporation for transdermal and topical delivery, Advanced Drug Delivery Reviews 56 (2004) 659–674.
[107] U. Pliquett, Mechanistic studies of molecular transdermal transport due to skin electroporation, Advanced Drug Delivery Reviews. 35 (1999) 41–60.
[108] V. Regnier, N. De Morre, A. Jadoul, V. Préat, Mechanisms of a phosphorothioate oligonucleotide delivery by skin electroporation, International Journal of Pharmaceutics 184 (1999) 147–156.
[109] S. Satkauskas, F. Andre, M. Bureau, D. Scherman, D. Miklavcic, L.M. Mir, Electrophoretic component of electric pulses determines the efficacy of *in vivo* DNA electrotransfer, Human Gene Therapy 16 (2005) 1194–1201.
[110] S. Becker, Transport modeling of skin electroporation and the thermal behavior of the stratum corneum, International Journal of Thermal Sciences 54 (2012) 48–61.
[111] N. Pavselj, D. Miklavcic, Numerical models of skin electropermeabilization taking into account conductivity changes and the presence of local transport regions, IEEE Transactions on Plasma Science 36 (2008) 1650–1658.
[112] N. Pavselj, D. Miklavcic, A numerical model of permeabilized skin with local transport regions, IEEE Transactions on Biomedical Engineering 55 (2008) 1927–1930.

[113] N. Pavselj, V. Preat, D. Miklavcic, A numerical model of skin electropermeabilization based on *in vivo* experiments, Annals of Biomedical Engineering 35 (2007) 2138–2144.
[114] U.F. Pliquett, T.E. Zewert, T. Chen, R. Langer, J.C. Weaver, Imaging of fluorescent molecule and small ion transport through human stratum corneum during high voltage pulsing: localized transport regions are involved, Biophysical Chemistry 58 (1996) 185–204.
[115] U. Pliquett, C. Gusbeth, R. Nuccitelli, A propagating heat wave model of skin electroporation, Journal of Theoretical Biology 251 (2008) 195–201.
[116] U.F. Pliquett, C.A. Gusbeth, Perturbation of human skin due to application of high voltage, Bioelectrochemistry 51 (2000) 41–51.
[117] M.R. Prausnitz, Do high-voltage pulses cause changes in skin structure? Journal of Controlled Release 40 (1996) 321–326.
[118] R. Vanbever, V. Preat, *In vivo* efficacy and safety of skin electroporation, Advanced Drug Delivery Reviews 35 (1999) 77–88.
[119] R.O. Potts, M.L. Francoeur, Lipid biophysics of water loss through the skin, Proc Natl Acad Sci USA 87 (1990) 3871–3873.
[120] G.M. Golden, D.B. Guzek, R.R. Harris, J.E. McKie, R.O. Potts, Lipid thermotropic transitions in human stratum corneum, Journal of Investigative Dermatology 86 (1986) 255–259.
[121] P. Cornwell, B. Barry, J. Bouwstra, G. Gooris, Modes of action of terpene penetration enhancers in human skin differential scanning calorimetry small-angle x-ray diffraction and enhancer uptake studies, International Journal of Pharmaceutics 127 (1996) 9–26.
[122] S.M. Becker, Skin electroporation with passive transdermal transport theory: a review and a suggestion for future numerical model development, Journal of Heat Transfer 133 (2011) 011011.
[123] H.F. Frasch, A.M. Barbero, Steady-state flux and lag time in the stratum corneum lipid pathway: results from finite element models, Journal of Pharmaceutical Sciences 92 (2003) 2196–2207.
[124] J. Kushner 4th, D. Blankschtein, R. Langer, Evaluation of the porosity, the tortuosity, and the hindrance factor for the transdermal delivery of hydrophilic permeants in the context of the aqueous pore pathway hypothesis using dual-radiolabeled permeability experiments, Journal of Pharmaceutical Sciences 96 (2007) 3263–3282.
[125] S.M. Becker, A.V. Kuznetsov, Thermal *in vivo* skin electroporation pore development and charged macromolecule transdermal delivery: a numerical study of the influence of chemically enhanced lower lipid phase transition temperatures, International Journal of Heat and Mass Transfer 51 (2008) 2060–2074.
[126] M.N. Ozisik, Heat Conduction. second ed., Wiley-Interscience, 1993.
[127] W.H.M. Craane-van Hinsberg, J.C. Verhoef, H.E. Junginger, H.E. Boddé, Thermoelectrical analysis of the human skin barrier, Thermochimica Acta 248 (1995) 303–318.
[128] D.A. Zaharoff, F. Yuan, Effects of pulse strength and pulse duration on *in vitro* DNA electromobility, Bioelectrochemistry 62 (2004) 37–45.
[129] N.C. Stellwagen, DNA mobility anomalies are determined primarily by polyacrylamide gel concentration, not gel pore size, Electrophoresis 18 (1997) 34–44.
[130] S.K. Li, A.H. Ghanem, C.L. Teng, G.E. Hardee, W.I. Higuchi, Iontophoretic transport of oligonucleotides across human epidermal membrane: a study of the Nernst-Planck model, Journal of Pharmaceutical Sciences 90 (2001) 915–931.

CHAPTER

14

Modeling Intracellular Transport in Neurons

Andrey V. Kuznetsov

Department of Mechanical and Aerospace Engineering, North Carolina State University, Raleigh, NC, USA

OUTLINE

Nomenclature 521

Subscripts 522

Abbreviations 522

14.1 Introduction 522

14.2 A Model of Axonal Transport Drug Delivery 523

14.3 Effect of Dynein Velocity Distribution on Propagation of positive Injury Signals in Axons 532

14.4 Simulation of Merging of Viral Concentration Waves in Retrograde Viral Transport in Axons 540

14.5 Conclusion 547

References 547

Nomenclature

- D diffusivity of particles ($\mu m^2/s$)
- $H(\eta)$ Heaviside step function
- $I_1(\eta)$ modified Bessel function of the first kind of order 1
- k first order rate kinetic constant (s^{-1})
- n particle number density ($1/\mu m^3$)
- N Laplace transform of the function $n(x,t)$
- t time (s)
- t_c duration of exposure at the axon terminal (s)
- v velocity of dynein motors ($\mu m/s$)

http://dx.doi.org/10.1016/B978-0-12-415824-5.00014-X

© 2013 Elsevier Inc. All rights reserved.

x linear coordinate that starts at the axon terminal and is directed toward the neuron soma (μm)

Subscripts

a accumulated particles
0 free particles
– dynein-driven particles

Abbreviations

MT microtubule
PAC pharmaceutical agent complex
PIS positive injury signal
PNS peripheral nervous system

14.1 INTRODUCTION

Neurons have a relatively small body (soma) and are characterized by long processes, dendrites and axons (also called neurites). Dendrites receive signals while axons transmit them. The total volume of the processes in one neuron cell can be 1000 times larger than the volume of the body of the neuron (*Goldstein and Yang* [1]). In the human body, axons can be up to one meter in length. Most organelles are synthesized in the neuron soma, and they have to be transported to a particular location in the neurite where they are needed. Since organelles are typically large vesicles, in accordance with the Stokes-Einstein equation they have small diffusivity. In order to transport them in a timely fashion, cells contain a sophisticated 'railway' system, in which organelles are pulled by molecular motors running on microtubules (MTs). Fast anterograde axonal transport is powered by kinesin molecular motors, while retrograde axonal transport is powered by dynein motors [2–6].

Molecular motor-driven transport is used not only by intracellular organelles; many harmful viruses utilize the axonal transport machinery to spread in the neural system. Also, various chemical signals (for example, trophic factors) are transported by molecular motors. Various authors have discussed similarities in trafficking of trophic factors and pathogenic proteins and have also reviewed research on the utilization of axonal transport for targeted drug delivery for treating neurological diseases [7–12].

Investigating axonal trafficking is important because its defects are linked to various neurodegenerative disorders, such as Alzheimer's disease and Down's syndrome [13,14]. Retrograde axonal transport can be used for targeted drug delivery to neurons [15]. There is also a link between the transport of chemical injury signals in an axon and its ability to regenerate [16,17]. Also, many neurotropic viruses rely on retrograde axonal transport to attack neurons [18–22].

This provides sufficient motivation for developing mechanistic models of intercellular trafficking. This chapter reviews various minimal models of intracellular trafficking in axons that can be solved utilizing either purely analytical or hybrid numerical and analytical techniques. In addition to bringing a physical insight to the processes, the obtained solutions are expected to be useful for testing numerical codes that would be based on extended versions of the models reviewed in this chapter.

14.2 A MODEL OF AXONAL TRANSPORT DRUG DELIVERY

Possible applications of axonal transport for targeted drug delivery have been discussed in [23–27]. Potential benefits include the reduction of toxicity of the drug and an increase in its half life. *Filler et al.* [15] chemically synthesized a tripartite complex that consisted of an axon transport facilitator molecule, a polymer linker and a large number of drug molecules for targeted drug delivery to neurons. These authors [15] reported that they were able to load up to 100 drug molecules per complex; the obtained results indicated a tenfold increase in the drug's half life and a 300 fold decrease in the necessary dose compared to systemic administration of the same drug.

The research of *Filler et al.* [15] indicates that a significant amount of the transported drug (in the form of pharmaceutical agent complexes, or PACs) en route to the neuron soma accumulated in the axon and then was re-released. The explanation of this phenomenon, as proposed in [15], is based on the fact that the material from the axoplasm can be endocytosed by paranodal complexes of Schwann cells at the Nodes of Ranvier ([28], but see also [29]). *Filler et al.* [15] also reported that the accumulated drug was subsequently re-released (rather than entering the degenerative pathway [28]). The goal of this section is to review mathematical models developed in [30–34] that account for drug accumulation as it is transported retrogradely down the axon and then is re-released. The chapter will also discuss analytical solutions of model equations for the transient situation. It will be shown that by comparing the qualitative behavior of the obtained solutions with the results reported in *Filler et al.* [15], it is possible to gain some information about reasonable values of kinetic parameters involved in the model.

A sketch of the problem is displayed in Fig. 14.1a. In order to have an analytically tractable problem that captures main features of axonal transport drug delivery it is assumed that during the exposure to the drug, $0 < t < t_c$, the concentration of dynein-driven PACs at the axon terminal is equal to a constant value n_b and it drops to zero for $t > t_c$. This generates a pulse of dynein-driven PACs that propagates toward the neuron soma. Dynein-driven PACs can be absorbed at the Nodes of Ranvier and thus transition to the accumulated state; PACs in the accumulated state can be re-released and re-enter dynein-driven transport, in accordance with a kinetic diagram displayed in Fig. 14.1b. (Due to a large number of Nodes of Ranvier the absorption and re-release of PACs are assumed to occur continuously along the axon length). PACs in the dynein-driven state are also subject to diffusion. The situation is thus different from that analyzed in *Smith and Simmons* [35] where no population of accumulated particles was considered.

Utilizing the above problem description, the conservation equations for the accumulated and dynein-driven PACs are presented by the following coupled equations. Equation (14.1) below expresses the conservation of PACs accumulated in the axon, while Eq. (14.2) expresses the conservation of PACs that are transported by dynein motors. These two equations are coupled through the kinetic terms that describe transitions between these two populations of PACs (see Fig. 14.1b); the kinetic terms are represented by the two terms on the right-hand side of Eq. (14.1) and the second and third terms on the right-hand side of Eq. (14.2).

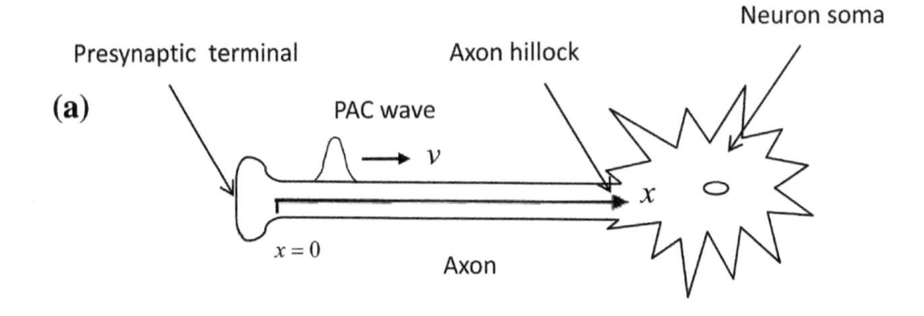

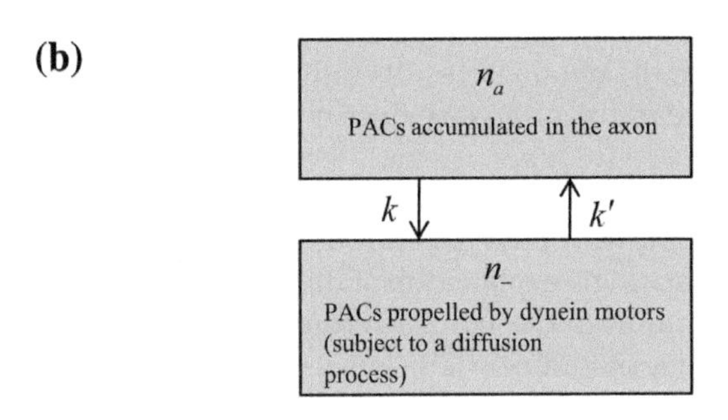

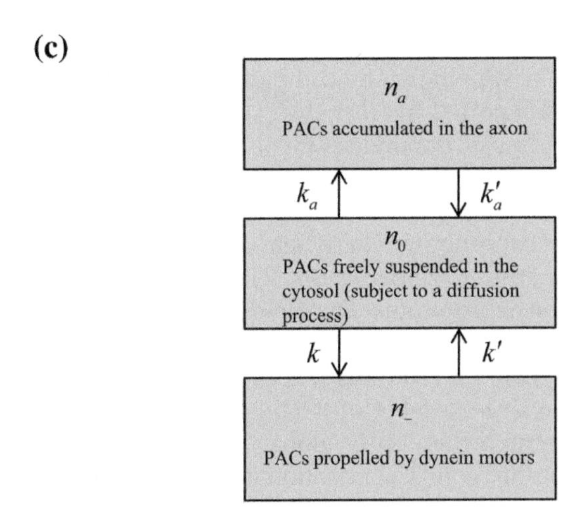

FIGURE 14.1 (a) A sketch of axonal transport drug delivery in a neuron; (b) Kinetic diagram for a two-population PAC model; (c) Kinetic diagram for a three-population PAC model.

The first term on the right-hand side of Eq. (14.2) describes PAC diffusion, and the fourth term on the right-hand side of Eq. (14.2) represents dynein-driven PAC transport:

$$\frac{\partial n_a}{\partial t} = -kn_a + k'n_- \tag{14.1}$$

$$\frac{\partial n_-}{\partial t} = D\frac{\partial^2 n_-}{\partial x^2} + kn_a - k'n_- - v\frac{\partial n_-}{\partial x} \tag{14.2}$$

where D is the diffusivity of dynein-driven PACs, k is the first order rate constant characterizing the rate at which PACs are re-released from the accumulated state, k' is the first order rate constant characterizing the rate at which PACs are absorbed at the Nodes of Ranvier (see Fig. 14.1b), n_a is the number density of PACs accumulated at a particular location in the axon, n_- is the number density of PACs transported retrogradely by dynein motors, t is the time, v is the average velocity of dynein motors and x is the linear coordinate that starts at the axon terminal and is directed toward the neuron soma (see Fig. 14.1a).

It is assumed that PAC absorption at the axon terminal produces a PAC concentration pulse of duration t_c. All PACs entering the axon are initially attributed to the dynein-driven state (this particular assumption does not have any significant effect on the solution since the kinetic processes quickly redistribute PACs between dynein-driven and accumulated states). Mathematically, the initial pulse is described as a stepwise variation of dynein-driven PAC concentration at $x=0$ (a constant PAC concentration, n_b, when $0 < t < t_c$, and a zero PAC concentration for $t > t_c$), as follows:

$$n_-(0,t) = n_b\left[1 - H(t - t_c)\right] \tag{14.3}$$

where $H(\eta)$ is the Heaviside step function.

It is further assumed that the neuron soma acts as a perfect absorber of PACs (no wave reflection at the axon hillock [36]), which means that the solution is identical to that obtained for a semi-infinite axon. This is an acceptable approximation since axon length is much larger than its diameter.

Equations (14.1) and (14.2) with boundary condition (14.3) and zero initial conditions (it is assumed that at $t=0$ there are no PACs in the axon, which is always the case as long as no treatment is administered) are solved by the Laplace transform. The subsidiary equations are:

$$sN_a = -kN_a + k'N_- \tag{14.4}$$

$$sN_- = D\frac{\partial^2 N_-}{\partial x^2} + kN_a - k'N_- - v\frac{\partial N_-}{\partial x} \tag{14.5}$$

where $N_a(x,s)$ and $N_-(x,s)$ are the Laplace transforms of the functions $n_a(x,t)$ and $n_-(x,t)$, respectively.

The Laplace transform of boundary condition (14.3) is:

$$N_-(0,s) = n_b\frac{1 - e^{-st_c}}{s} \tag{14.6}$$

The solutions of subsidiary Eqs. (14.4) and (14.5) subject to boundary condition (14.6) and the condition that $N_-(x,s)$ remains finite as $x \to \infty$ are:

$$N_a(x,s)/n_b = \frac{k'}{s(k+s)}\exp\left[-st_c + \frac{v - \sqrt{\frac{4Ds(k+k'+s)}{k+s} + v^2}}{2D}x\right]\left(-1 + \exp[st_c]\right) \tag{14.7}$$

$$N_{-}(x,s)/n_b = \frac{1}{s}\exp\left[-st_c + \frac{v - \sqrt{\frac{4Ds(k+k'+s)}{k+s} + v^2}}{2D}x\right]\left(-1+\exp\left[st_c\right]\right) \tag{14.8}$$

Equations (14.7) and (14.8) become singular as $D \to 0$. Therefore, the case of zero diffusivity is analyzed separately. $D=0$ is substituted in Eq. (14.5) and then Eqs. (14.4) and (14.5) are solved subject to boundary condition (14.6). This results in the following:

$$N_a(x,s)/n_b = \frac{k'}{s(k+s)}\exp\left[-st_c + \frac{k+k'+s}{(k+s)v}sx\right]\left(-1+\exp\left[st_c\right]\right) \tag{14.9a}$$

$$N_{-}(x,s)/n_b = \frac{1}{s}\exp\left[-st_c + \frac{k+k'+s}{(k+s)v}sx\right]\left(-1+\exp\left[st_c\right]\right) \tag{14.9b}$$

The inverse Laplace transforms of the right-hand sides of Eqs. (14.7) and (14.8) were found numerically by utilizing the method and algorithm described in [37] and [38]. The precision of the method is also discussed in these references. The inverse Laplace transforms of Eqs. (14.9a) and (14.9b) (for the $D=0$ case) were found analytically by using convolution integral [39] with a help of Mathematica 8.0© package (Wolfram Research, Inc.), following [32], where a solution was obtained for a different boundary condition (the result presented below is thus new):

$$\begin{aligned} n_a(x,t)/n_b = {} & \frac{k'}{k}\exp\left[-\frac{k'}{v}x\right]\left(1-\exp\left[-k\left(t-\frac{x}{v}\right)\right]\right)H\left[t-\frac{x}{v}\right] \\ & + \frac{k'}{k}\int_0^t \exp\left[-k\tau-\frac{k'}{v}x\right]\left(1-\exp\left[-k\left(t-\tau-\frac{x}{v}\right)\right]\right) \\ & \times H\left[t-\tau-\frac{x}{v}\right]\sqrt{\frac{kk'x}{\tau v}}I_1\left[2\sqrt{\frac{\tau kk'x}{v}}\right]d\tau \\ & - H\left[t-t_c\right]\left\{\frac{k'}{k}\exp\left[-\frac{k'}{v}x\right]\left(1-\exp\left[-k\left(t-t_c-\frac{x}{v}\right)\right]\right)H\left[t-t_c-\frac{x}{v}\right]\right. \\ & + \frac{k'}{k}\int_0^{t-t_c} \exp\left[-k\tau-\frac{k'}{v}x\right]\left(1-\exp\left[-k\left(t-t_c-\tau-\frac{x}{v}\right)\right]\right) \\ & \left.\times H\left[t-t_c-\tau-\frac{x}{v}\right]\sqrt{\frac{kk'x}{\tau v}}I_1\left[2\sqrt{\frac{\tau kk'x}{v}}\right]d\tau\right\} \end{aligned} \tag{14.10}$$

$$\begin{aligned} n_{-}(x,t)/n_b = {} & \exp\left[-\frac{k'}{v}x\right]H\left[t-\frac{x}{v}\right] + \int_0^t \exp\left[-k\tau-\frac{k'}{v}x\right]H\left[t-\tau-\frac{x}{v}\right]\sqrt{\frac{kk'x}{\tau v}}I_1\left[2\sqrt{\frac{\tau kk'x}{v}}\right]d\tau - H\left[t-t_c\right] \\ & \times\left\{\exp\left[-\frac{k'}{v}x\right]H\left[t-t_c-\frac{x}{v}\right] + \int_0^{t-t_c}\exp\left[-k\tau-\frac{k'}{v}x\right]H\left[t-t_c-\tau-\frac{x}{v}\right]\sqrt{\frac{kk'x}{\tau v}}I_1\left[2\sqrt{\frac{\tau kk'x}{v}}\right]d\tau\right\} \end{aligned} \tag{14.11}$$

where $I_1(\eta)$ is a modified Bessel function of the first kind, of order 1.

According to [40] and [41], the velocity of cytoplasmic dynein walking on MTs, v, is approximately 1 μm/s. The duration of the PAC exposure, t_c, is estimated as 60 s. Estimating the two kinetic constants involved in the model, k and k', is difficult since the dynamics of PAC absorption/release at the Nodes of Ranvier has not been quantified in the published literature. In fact, one of the goals of the reviewed research is to estimate reasonable values of the kinetic constants by analyzing model predictions, so that model predictions are in agreement with what was observed in [15]. Two cases are considered: of large and small values of the kinetic constants. k and k' are not likely to exceed 1 s^{-1}; these are the values used by *Smith and Simmons* [35] in their model describing the dynamics of organelle attachment/detachment to/from MTs, and the dynamics of particle absorption/release at the Nodes of Ranvier is expected to be slower than that. Hence 1 s^{-1} is used as a large value for k and k'. 10^{-4} s^{-1} is then used as a small value for k and k'.

Diffusivities of organelles can be estimated using Einstein-Stokes relation; based on this Smith and Simmons [35] estimated the diffusivity of a 1 μm sphere in water as 0.4 $\mu m^2/s$. To show the effect of diffusivity the results are displayed for two cases: $D = 1$ $\mu m^2/s$ and $D = 0$ $\mu m^2/s$.

Figures 14.2–14.5 are computed using $t_c = 60$ s and $v = 1$ μm/s. In order to check the accuracy of the numerical inverse Laplace transform of Eqs. (14.7) and (14.8) (which was used to obtain data presented in Figs. 14.2 and 14.4) the problem given by Eqs. (14.1), (14.2) and (14.3) was also solved by Comsol Multiphysics®. The obtained solutions were identical to those displayed in Figs. 14.2 and 14.4.

Figure 14.2 displays the case of small values of the kinetic constants and large diffusivity. This figure is computed for $k = 10^{-4}\,s^{-1}$, $k' = 10^{-4}\,s^{-1}$ and $D = 1$ $\mu m^2/s$. It is evident that when kinetic constants are small, the concentration of PACs accumulated at the Nodes of Ranvier, displayed in Fig. 14.2a, propagates as a front (a front is characterized by a constant PAC concentration followed by a rapid decrease to zero), while the concentration of dynein-transported PACs, displayed in Fig. 14.2b, propagates as a pulse. There is no destruction of PACs in the model, so as the pulse propagates toward the neuron soma, the total number of PACs (accumulated and dynein-driven) remains constant. It can be seen in Fig. 14.2b that as the pulse of dynein-driven PACs propagates, its amplitude decreases. One reason for this is that the pulse spreads out as it propagates, and the other reason is the transition of PACs from the dynein-driven state to the accumulated state. Indeed, as the pulse of dynein-driven PACs in Fig. 14.2b propagates, the number of accumulated PACs (shown in Fig. 14.2b) increases; PACs are accumulated all the way between the axon terminal and the edge of the pulse.

Figure 14.3 demonstrates the effect of decreased diffusivity of PACs; it is similar to Fig. 14.2, but is computed assuming zero diffusivity of PACs, $D = 0$ $\mu m^2/s$. A comparison between Figs. 14.2 and 14.3 clearly shows the effect of diffusivity: diffusivity makes the pulse and the front much smoother; the pulse in Fig. 14.3b (computed with zero diffusivity) retains its rectangular shape as it propagates; its amplitude decreases very slowly due to PAC transition to the accumulated state (due to a small rate of PAC absorption at the Nodes of Ranvier the accumulation of PACs in the axon occurs very slowly).

Figure 14.4 presents the case of large values of the kinetic constants and large diffusivity. This figure is computed for $k = 1\,s^{-1}$, $k' = 1\,s^{-1}$ and $D = 1$ $\mu m^2/s$. The situation is markedly different from that displayed in Fig. 14.2: now both accumulated (displayed in Fig. 14.4a) and

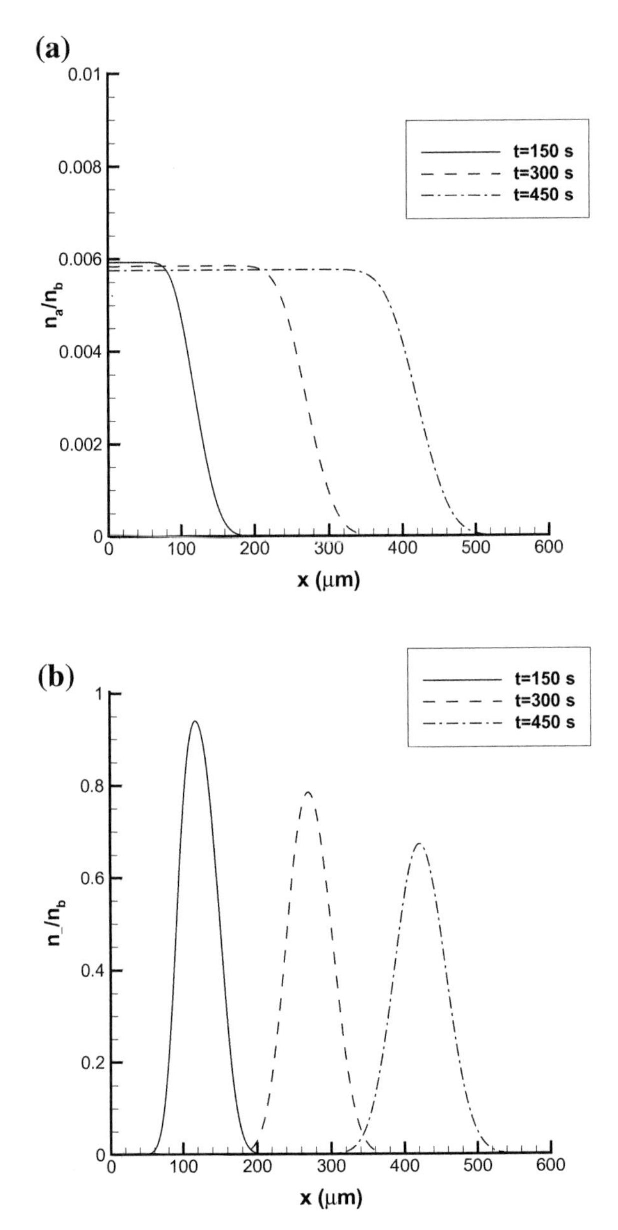

FIGURE 14.2 The case of small values of kinetic constants: (a) Number density of PACs accumulated at the Nodes of Ranvier; (b) Number density of PACs transported retrogradely by dynein motors. ($D=1\ \mu m^2/s$, $k=10^{-4}\,s^{-1}$, $k' = 10^{-4}\ s^{-1}, t_c = 60$ s, $v=1\ \mu m/s$).

dynein-driven (displayed in Fig. 14.4b) PACs form almost identical pulses that spread out as they propagate away from the axon terminal. The reason for this is the large rates of PAC absorption/release at the Nodes of Ranvier: as the pulse of dynein-driven PACs propagates,

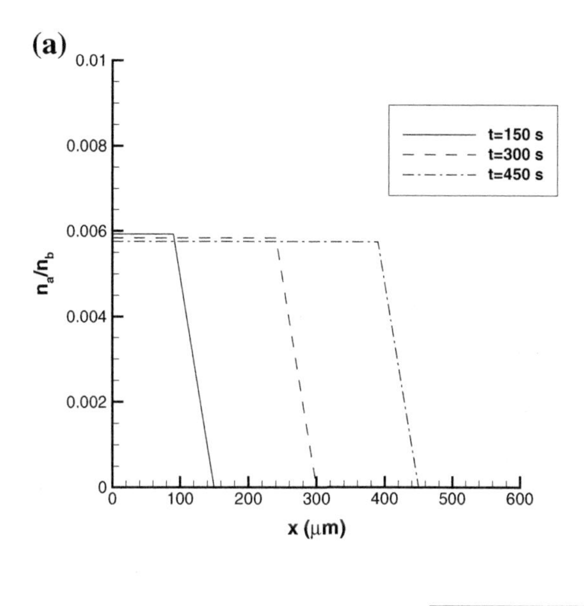

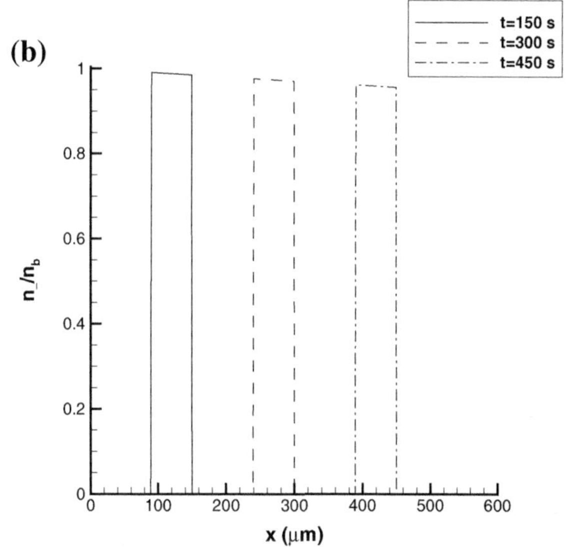

FIGURE 14.3 Similar to Fig. 14.2, but now for $D = 0\,\mu m^2/s$.

PACs are absorbed at the Nodes of Ranvier, but once the pulse moves away from a particular location, they immediately detach and re-enter dynein-driven transport. There is another interesting insight that can be gained by comparing Figs. 14.4b and 14.2b: the waves in Fig. 14.4b move at almost half the speed of those in Fig. 14.2b. This is because in the situation displayed in Fig. 14.4, a typical PAC spends only half of its time in the dynein-driven state while the other half it spends in the accumulated state (the latter is similar to the pausing

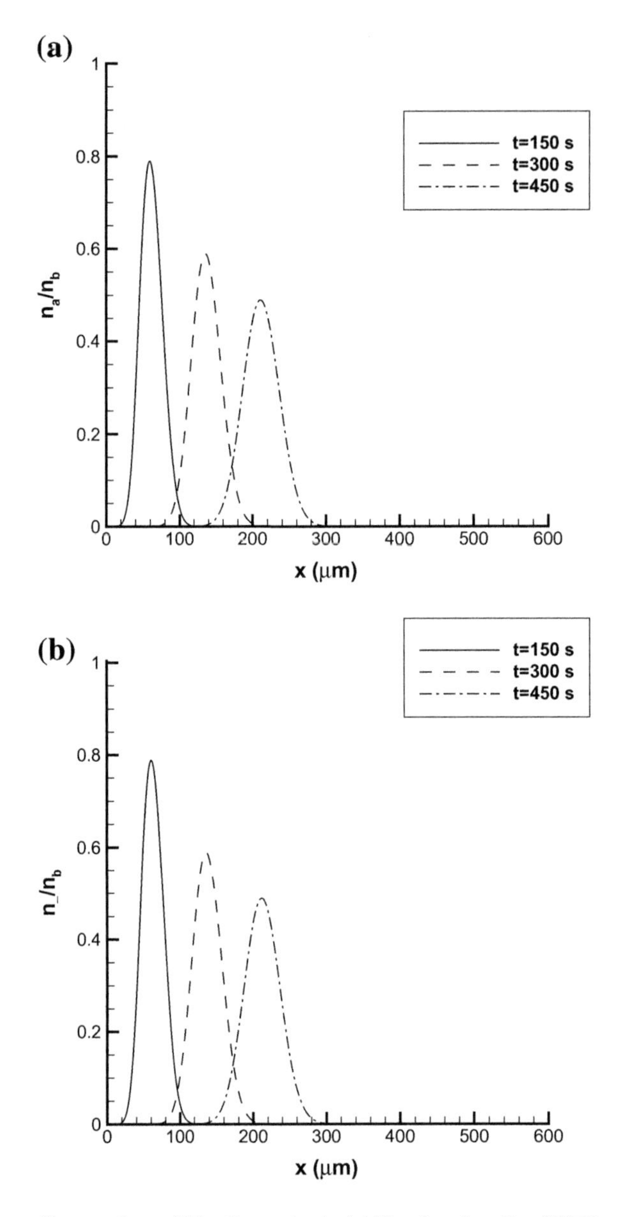

FIGURE 14.4 The case of large values of kinetic constants: (a) Number density of PACs accumulated at the Nodes of Ranvier; (b) Number density of PACs transported retrogradely by dynein motors. ($D=1\,\mu m^2/s$, $k=1\,s^{-1}$, $k' = 1\,s^{-1}, t_c = 60$ s, $v=1\,\mu m/s$).

state in the *Jung and Brown* model [42]). Since the velocity of a PAC in the accumulated state is zero, its average velocity is half of the dynein velocity. In the situation displayed in Fig. 14.2, a PAC spends only a small portion of its time in the accumulated state; therefore, its velocity is close to that of a dynein motor.

Figure 14.5 again demonstrates the effect of decreased diffusivity of PACs. It is similar to Fig. 14.4, but is computed assuming zero diffusivity of PACs, $D=0\ \mu m^2/s$. The velocities of the waves in Figs. 14.5a and 14.5b are close to those in Figs. 14.4a and 14.4b (and are equal to

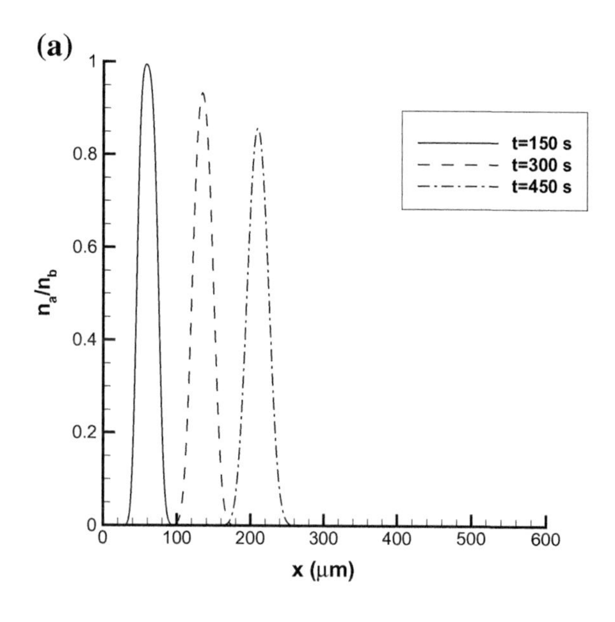

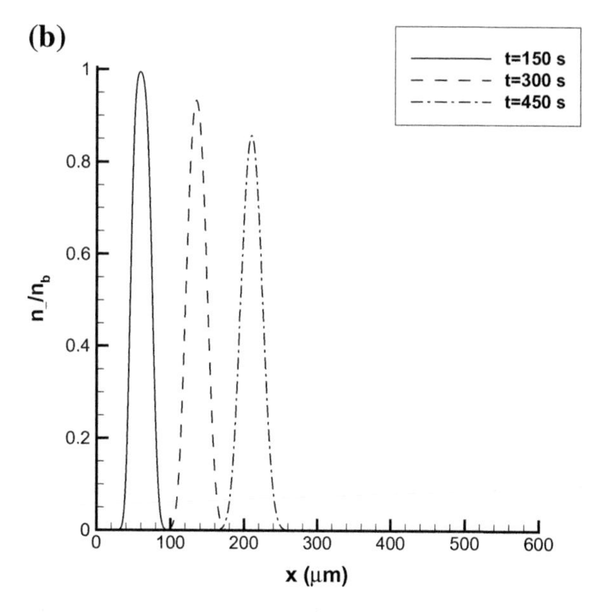

FIGURE 14.5 Similar to Fig. 14.4, but now for $D=0\,\mu m^2/s$.

half of the dynein motor velocity), but the waves are narrower now, which is due to the lack of diffusion. The pulses in Fig. 14.5b are different from those in Fig. 14.3b due to large rates of PAC exchange with the accumulated state.

A more sophisticated model of axonal transport drug delivery that includes three populations (kinetic states) of PACs, namely, PACs transported by dynein motors, PACs freely suspended in the cytosol, and PACs accumulated at the Nodes of Ranvier (see Fig. 14.1c), is developed in [34]. In that model, conservations of the accumulated, freely suspended and dynein-driven PACs, respectively, are expressed by the following equations:

$$\frac{\partial n_a}{\partial t} = -k_a' n_a + k_a n_0 \tag{14.12}$$

$$\frac{\partial n_0}{\partial t} = D\frac{\partial^2 n_0}{\partial x^2} - k_a n_0 - k n_0 + k_a' n_a + k' n_- \tag{14.13}$$

$$\frac{\partial n_-}{\partial t} = -k' n_- + k n_0 - v\frac{\partial n_-}{\partial x} \tag{14.14}$$

where D is now the diffusivity of PACs freely suspended in the cytosol, k_a' is the first order rate constant characterizing the rate at which PACs are re-released from the accumulated state to become free PACs, k_a is the first order rate constant characterizing the rate at which PACs from the free state are absorbed at the Nodes of Ranvier, k' is now the first order rate constant characterizing the rate at which dynein-driven PACs detach from MTs to become PACs freely suspended in the cytosol, k is now the first order rate constant characterizing the rate at which free PACs attach to MTs to become dynein-driven PACs and n_0 is the number density of free PACs suspended in the cytosol (see Fig. 14.1c).

This model, however, does not allow an analytical solution and has to be solved numerically.

14.3 EFFECT OF DYNEIN VELOCITY DISTRIBUTION ON PROPAGATION OF POSITIVE INJURY SIGNALS IN AXONS

It is well known that neurons of the peripheral nervous system (PNS) possess some limited regenerative ability, which enables them to repair themselves after an injury, whereas neurons of the central nervous system do not have such ability [17,43]). Understanding the mechanisms that enable PNS neurons to regenerate is important for developing treatments for neurodegenerative diseases [44]. In injured neurons, there are three types of signaling through which information about the location and the extent of injury is transmitted to the neuron body. The first type of signal is electrophysiological, related to the discharge of axonal potentials. The second is the negative regulation related to the interruption of normal supply of tropic factors. The third is the positive regulation related to transport of various positive injury signals (PISs), such as cytokines, importin and members of mitogen-activated protein kinase family, which are transported from the injury site toward the neuron soma by dynein motors [17,43–45]. Since PISs are large molecular complexes, their retrograde transport is accomplished by dynein motors.

The apparent redundancy in signaling mechanisms may be necessary to ensure the reliability of information transmission, but it also may be necessary to enable a neuron to detect the distance between the cell body and the lesion site. Indeed, neuronal response to injury depends on the distance from the soma to the lesion site and the exact location of the lesion [16]. The particular response also depends on the neuron type. For example, primary sensory neurons with cell bodies in the dorsal root ganglion have two axonal branches: peripheral and central. The peripheral branch regenerates when injured and the central branch does not. However, if two injuries are inflicted, firstly on the peripheral branch and then on the central branch, then the central branch does regenerate [44,46–48].

Kam et al. [16] considered two mechanisms that neurons may use to determine the distance to the injury site. The first possible mechanism is based on the assumption that the injury site initiates two signals, fast (electrophysiological) and slow (chemical, composed of PISs transported retrogradely by dynein motors), and that sensors located in the neuron body (soma) can detect the time delay between the arrival of these two signals. The second possible mechanism is based on the assumption that by measuring the degree of spreading of the chemical signal alone it is possible to determine the distance to the lesion site.

In what follows, the problem considered in [16] is revisited; the problem is simplified such that an analytical solution can be obtained. This section reviews the analysis and results originally reported in [49,50].

A schematic diagram of PIS transport in an axon is displayed in Fig. 14.6a. The PISs are released from the axon injury site. The origin of the coordinate system, $x=0$, is located at the injury site. It is assumed that the PISs are released for a limited time, t_c, resulting in a constant flux of PISs, j_0, between $t=0$ and $t=t_c$. For $t > t_c$ the flux of PISs from the injury site is assumed to be zero. The wave of PISs generated this way at the axon injury site travels toward the neuron soma via association with dynein molecular motors, which pull their cargo with velocity v. It is also assumed that while the PISs travel toward the neuron soma, there is some randomness in their motion, which may be caused by motor navigation around obstacles as well as by Brownian motion of free PISs in the cytoplasm if they detach from MTs. This random component of the motion of PISs is modeled by a diffusion process with an effective diffusivity D. This modeling approach is somewhat similar to that developed in [35], but it accounts only for one population of transported molecular complexes (in this case PISs). This assumption is made in order to make the problem analytically tractable. It is also assumed that on their way to the neuron soma some PISs may be destroyed; this is modeled by the first order decay rate, k_d. The above assumptions result in the following advective diffusion equation, as proposed in [49,50], that governs the propagation of the concentration wave of PISs from the axon injury site toward the neuron soma:

$$\frac{\partial n}{\partial t} = D\frac{\partial^2 n}{\partial x^2} - v\frac{\partial n}{\partial x} - k_d n \tag{14.15}$$

where n is the PIS number density.

Since the total flux of PISs, j, is the sum of diffusion and motor-driven fluxes, it is given by:

$$j = -D\frac{\partial n}{\partial x} + vn \tag{14.16}$$

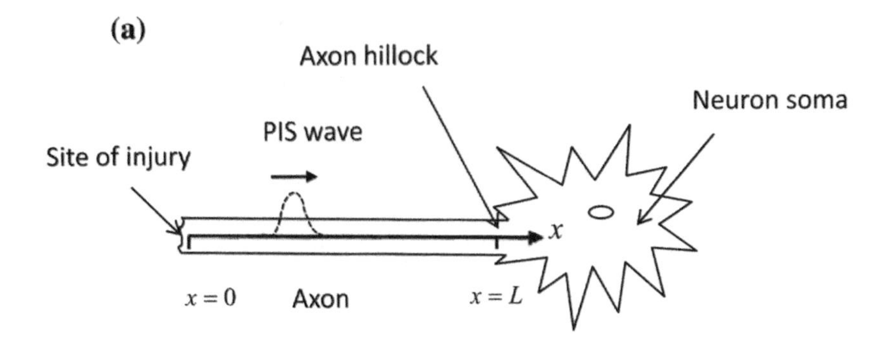

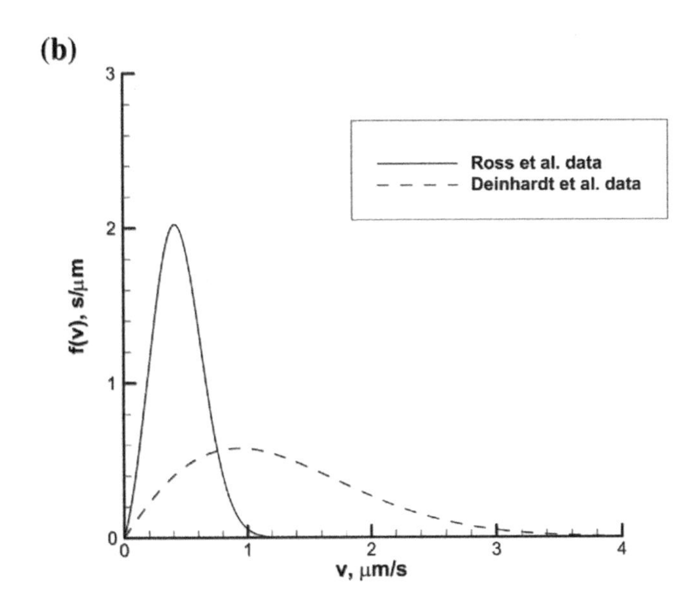

FIGURE 14.6 (a) Schematic diagram of a PIS wave traveling in an axon; (b) Pdfs of dynein velocity distributions based on data reported in [54] and [55]. These results are obtained by normalizing analytical curve fits of these data presented in [16] such that $\int_0^\infty f(v)dv = 1$.

The boundary condition at $x=0$ describing a PIS flux of duration, t_c, then is:

$$j(0,t) = -D\frac{\partial n}{\partial x}(0,t) + vn(0,t) = j_0\left[1 - H(t - t_c)\right] \tag{14.17}$$

where $H(\tau)$ is the Heaviside step function. Again, it is assumed that the neuron soma acts as a perfect absorber of PISs, which means that the solution is identical to that obtained for a semi-infinite domain.

Equation (14.15) with boundary condition (14.17) and zero initial condition (it is assumed that initially, prior to the injury, there are no PISs in the axon) is solved by Laplace transform. The subsidiary equation is:

$$sN = D\frac{\partial^2 N}{\partial x^2} - v\frac{\partial N}{\partial x} - k_d N \tag{14.18}$$

where $N(x, s)$ is the Laplace transform of the function $n(x, t)$.

The Laplace transform of boundary condition (14.17) is:

$$-D\frac{\partial N}{\partial x}(0, s) + vN(0, s) = j_0\frac{1 - e^{-st_c}}{s} \tag{14.19}$$

The solution of subsidiary Eq. (14.18) subject to boundary condition (14.19) and the condition that the solution remains finite as $x \to \infty$ is:

$$N(x, s) = \frac{2j_0}{s\left(v + \sqrt{4D(k_d + s) + v^2}\right)}\left(-1 + \exp\left[st_c\right]\right)\exp\left[-st_c - \frac{-v + \sqrt{4D(k_d + s) + v^2}}{2D}x\right] \tag{14.20}$$

Calculating the inverse Laplace transform [51] of the right-hand side of Eq. (14.20), the following solution for the viral concentration wave that propagates from the axon terminal toward the neuron soma was obtained:

$$\begin{aligned} n(x, t; v) = {} & \frac{j_0}{4Dk_d} \\ & \times \exp\left[-k_d t - \frac{\left(-v + \sqrt{4Dk_d + v^2}\right)x}{2D}\right]\left\{2v\exp\left[\frac{\left(v + \sqrt{4Dk_d + v^2}\right)x}{2D}\right]\operatorname{erfc}\left[\frac{vt + x}{2\sqrt{Dt}}\right] + e^{k_d t}\left\{\left(-v + \sqrt{4Dk_d + v^2}\right)\right.\right. \\ & \times \operatorname{erfc}\left[\frac{-t\sqrt{4Dk_d + v^2} + x}{2\sqrt{Dt}}\right] - \left(v + \sqrt{4Dk_d + v^2}\right)\exp\left[\frac{x\sqrt{4Dk_d + v^2}}{D}\right]\operatorname{erfc}\left[\frac{t\sqrt{4Dk_d + v^2} + x}{2\sqrt{Dt}}\right]\Bigg\} \\ & + \left\{\left(v + \sqrt{4Dk_d + v^2}\right)\exp\left[t + \frac{x\sqrt{4Dk_d + v^2}}{D}\right]\operatorname{erfc}\left[\frac{\sqrt{4Dk_d + v^2}(t - t_c) + x}{2\sqrt{D(t - t_c)}}\right]\right. \\ & - 2v\exp\left[k_d t_c + \frac{\left(v + \sqrt{4Dk_d + v^2}\right)x}{2D}\right]\operatorname{erfc}\left[\frac{v(t - t_c) + x}{2\sqrt{D(t - t_c)}}\right] \\ & \left.\left. - \left(-v + \sqrt{4Dk_d + v^2}\right)e^{k_d t}\operatorname{erfc}\left[\frac{\sqrt{4Dk_d + v^2}(-t + t_c) + x}{2\sqrt{D(t - t_c)}}\right]\right\}H(t - t_c)\right\} \end{aligned} \tag{14.21}$$

Here $\operatorname{erfc}(\tau)$ is the complementary error function.

For the case when PISs are not destroyed, $k_d = 0$, Eq. (14.21) simplifies to:

$$\begin{aligned} n(x, t; v) = {} & \frac{j_0}{2v}\left\{\operatorname{erfc}\left[\frac{-vt + x}{2\sqrt{Dt}}\right] - \operatorname{erfc}\left[\frac{-v(t - t_c) + x}{2\sqrt{D(t - t_c)}}\right]H(t - t_c)\right. \\ & \left. + \exp\left(\frac{vx}{D}\right)\left[-\operatorname{erfc}\left[\frac{vt + x}{2\sqrt{Dt}}\right] + \operatorname{erfc}\left[\frac{v(t - t_c) + x}{2\sqrt{D(t - t_c)}}\right]H(t - t_c)\right]\right\} \end{aligned} \tag{14.22}$$

Equations (14.21) and (14.22) give PIS concentrations under the assumption that all PISs move with the same velocity v (in these equations v is constant). The situation when the dynein velocity distribution is characterized by a pdf $f(v)$ can be approached by the method proposed in [49]. Imagine that signal particles and dynein motors are characterized by color, depending on the velocity of the dynein motor that transports a particle; green particles are transported only by green dynein motors that move with velocity v_1, red particles are transported only by red dynein motors that move with velocity v_2, etc. The particles do not switch the type of dynein motors they are riding: green particles always ride on green motors, and red particles always ride on red motors. At $x=0$ the flux of green particles is κj_0 and the flux of red particles is $(1-\kappa)j_0$, where $\kappa < 1$. If one is color-blind and is interested only in the total concentration of signal particles, one needs to calculate the sum with respect to different velocities to obtain the total concentration, that is $\kappa n\,(x,t;v_1)+(1-\kappa)\,n\,(x,t;v_2)$. Now imagine that one has a large number of colors, and that the fluxes of these colored particles at $x=0$ are distributed according to a stepwise version of $f(v)$, with the total flux of injected particles (over the whole velocity range) still being j_0. Physically one can imagine that once signal particles are released from the injury site, they are picked up by dynein motors whose velocity distribution is characterized by a pdf $f(v)$. The larger the number of dynein motors in a certain velocity range, the larger is the probability that an injected particle will be picked up by a dynein motor moving in this velocity range. In the limit of an infinite number of colors, the total concentration is given by the following integral, which expresses a superposition of concentrations corresponding to different dynein velocities:

$$n_{f(v)}\,(x,t)=\int_0^{\infty} n\,(x,t;v)\,f(v)dv \tag{14.23}$$

It should be noted that although Eq. (14.15) is valid separately for each single-colored species of particles (with corresponding velocities), the total concentration $n_{f(v)}$ defined by Eq. (14.23) does not have to satisfy Eq. (14.15). This is because when one considers the total concentration, one calculates the number of particles in a unit volume, all of which are moving with different velocities. The total concentration is thus found as a post-processing step rather than by solving a fundamental conservation equation. In practice, the integral on the right-hand side of Eq. (14.23) only needs to be calculated over the interval of v where $f(v)$ is non-zero (see Fig. 14.6b). If v is a constant (all dynein motors move with one and the same velocity v_0) then $f(v)=\delta\,(v-v_0)$, where δ is a Dirac delta function.

The value $D=0.01\ \mu m^2/s$ is used in computations. There is no published data known to the author concerning the rate of chemical injury signal destruction in the axon. Two values for the constant k are used in computations: $k_d=0$, which assumes that all signal complexes released at the injury site reach the neuron soma intact, and $k_d=10^{-4}\,s^{-1}$, which is taken to be a representative value for the cases illustrating the effect of PIS destruction.

Following [16], it is assumed that a PIS is composed of 500 moving particles. In order to estimate the flux of PIS particles from the injury site, j_0, one needs to know the duration of the PISs release at the axon injury site, t_c. The latter parameter is hard to estimate. *Lukas et al.* [52] reported that in retinal ganglion cells it takes about 30 min for the cell soma to detect the axonal damage; this gives the upper limit for the duration of the chemical signal, but the signal itself is probably much shorter – it just takes a long time for it to travel from the injury site to the neuron soma. Using for example a 60 s signal duration, and estimating the average axonal diameter, d, as 1.1 μm based on [53] one obtains an estimation for j_0 of 8.8 particles/(μm²s). It has been established by numerical simulations conducted by the author (not shown) that for $t \gg t_c$ results are not sensitive to t_c; what matters is the number of particles released from the injury site. Consequently, one can increase t_c and reduce j_0 proportionally; at large times this will have a minimal effect on the solution.

Kam et al. [16] utilized experimental data for dynein velocity obtained in [54,55] (the difference between velocity distributions reported in the two studies is explained by the fact that different systems and different test conditions were used, and the behavior of dynein motors depends on those) and produced two correlations for the distributions of dynein velocities based on these results. Normalizing the results given in [16] so that $\int_0^\infty f(v)dv = 1$ yields the following pdf of the dynein velocity distribution that is based on data reported in [54]:

$$f_1(v) = 5.8145\, v \exp\left[-\left(\frac{v - 0.2738}{0.3363}\right)^2\right] \tag{14.24}$$

where the velocity of dynein motors must be given in μm/s. The expected value of dynein velocity in this case is $E(v) = 0.445$ μm/s.

The normalized pdf of the dynein velocity distribution that is based on data reported in [55] is:

$$f_2(v) = 1.4616\, v \exp\left[-\left(\frac{v + 0.6657}{1.7231}\right)^2\right] \tag{14.25}$$

The expected value of the dynein velocity in this case is $E(v) = 1.272$ μm/s. Pdfs given by Eqs. (14.24) and (14.25) are illustrated in Fig. 14.6b.

Figures 14.7a and 14.7b illustrate PIS waves for the cases when there is a distribution of dynein velocities. In computing Fig. 14.7a the dynein velocity distribution was based on data presented in [54] while in computing Fig. 14.7b the dynein velocity distribution was based on data presented in [55]. In computing these figures, the kinetic constant characterizing the rate of destruction of PISs in an axon is assumed to be zero ($k_d = 0$), so that the number of signal complexes remains constant until the signals reach the neuron soma. The peak of the PIS wave moves faster in Fig. 14.7b because the expected velocity, $E(v)$, is much larger for the dynein velocity distribution based on data presented in [55]. Also, in the case displayed in Fig. 14.7b the PIS wave spreads out much faster because the range of dynein velocities for the distribution based on [55] data is much wider.

Figure 14.8a illustrates the performance of a two-signal model suggested in [16]. (In computing Figs. 14.8a and 14.8b, k_d is also assumed to be zero.) This figure depicts a time delay between the fast and slow signals, Δt, versus the distance between the injury site and axon hillock, L. The sensitivity of the slow signal detector is set to 50%. This means that the detector responds when 50% of molecular signaling complexes arrive to it. Figure 14.8a shows that for dynein velocity distributions based on both *Ross et al.* [54] and *Deinhardt et al.* [55] data, the correlation between Δt and L is linear, indicating that this principle can indeed be used for measuring L. The accuracy of measuring L is expected to be better for the velocity distribution based on [54] because the slope is larger.

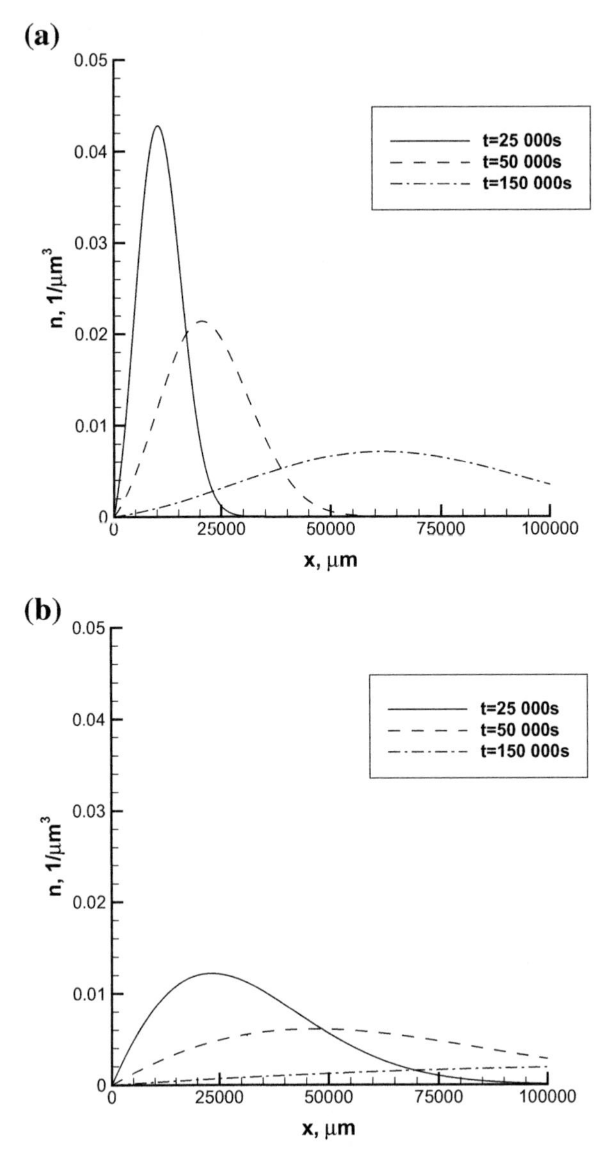

FIGURE 14.7 Chemical signal waves at various times (it is assumed that there is a distribution of dynein velocities). (a) The dynein velocity distribution is based on [54]; (b) The dynein velocity distribution is based on [55]. (The rate of destruction of PISs in the axon is zero: $k_d = 0$).

Figure 14.8b illustrates the performance of a two-detector model suggested in [16]. This figure depicts a time delay between the response times to the same slow signal by two different detectors versus the distance between the injury site and axon hillock. The sensitivity of the first detector is set to 5% and the sensitivity of the second detector is set to 80%. This means that the first detector responds when 5% of molecular signaling complexes arrive to it

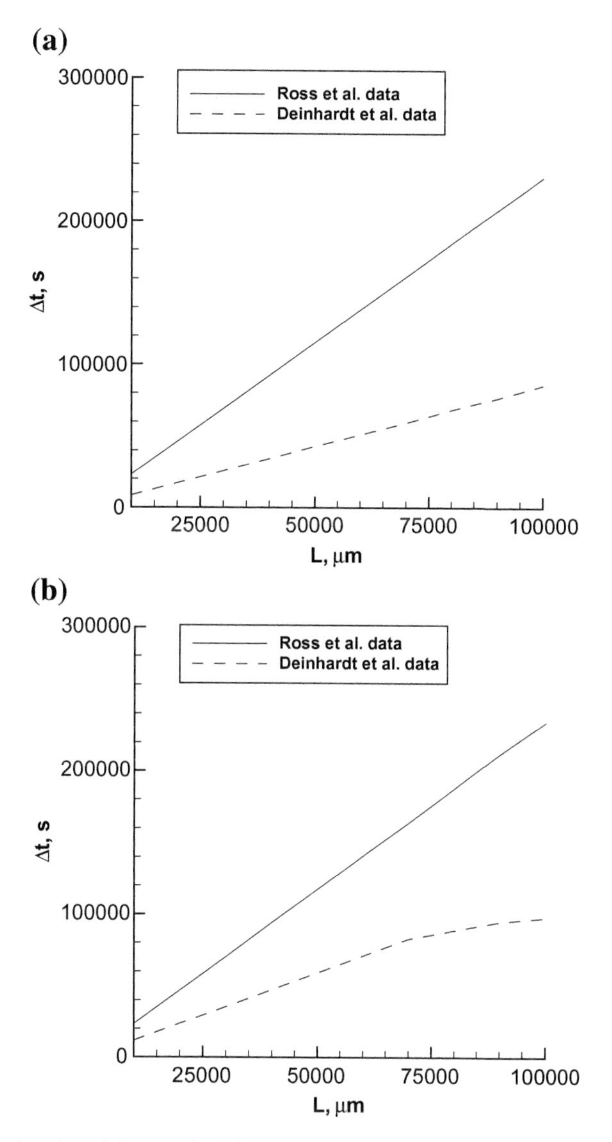

FIGURE 14.8 (a) Two-signal model: time delay between the fast and the slow signals versus the distance between the injury site and axon hillock. The sensitivity of the slow signal detector is set to 50% (which means that the detector responds when 50% of molecular signaling complexes arrive to it). (b) Two-detector model: time delay between the response times to the same slow signal by two different detectors versus the distance between the injury site and axon hillock. The sensitivity of the first detector is set to 5% and the sensitivity of the second detector is set to 80%. (The rate of destruction of PISs in the axon is zero: $k_d = 0$.)

and the second detector responds when 80% of molecular signaling complexes arrive there. Again, this figure demonstrates that this principle can also be used for measuring L, although its accuracy, especially for the dynein velocity distribution based on [55], is much better for small L since the dependence of Δt on L becomes flat as L increases.

14.4 SIMULATION OF MERGING OF VIRAL CONCENTRATION WAVES IN RETROGRADE VIRAL TRANSPORT IN AXONS

Equation (14.15) can also be utilized to model the retrograde motor-driven transport of neurotropic viruses (viruses that infect nerve cells). Many such viruses enter the axons at their presynaptic terminals [56]. Since viruses have no means to propel themselves independently, and because due to their size their diffusivity is small, after being internalized they have to rely on cellular transport machinery in order to reach the neuron soma [57]. Some viruses, such as the West Nile virus, can spread in both retrograde and anterograde directions [58]; however, most neurotropic viruses, such as the rabies virus, herpes virus and polio virus enter the neuron via endocytosis at the presynaptic terminal of the axon, and then utilize dynein motors in order to be transported in the retrograde direction toward the neuron soma [18–21].

Another area where understanding retrograde viral transport is important is gene therapy, since this relies on the delivery of special genes and various growth factors by viral vectors, such as a modified adeno-associated viral (AAV) vector. This has potential in retrograde delivery of the anti-apoptotic genes for neuronal protection [59], treating of spinal injury [60–62], glaucoma treatment [63] and promoting central nervous system repair [64].

When virus enters an axon, its concentration forms a wave that propagates toward the neuron soma. If such an entry occurred twice within a short period of time (the situation when viral waves would enter a cell in a temporally coordinated fashion can be easily created in a lab), the two induced viral concentration waves will interact. The study of the interaction of such waves is physically interesting; the positions of the peaks of the waves can be used as markers in an experiment, and this information can be useful for measuring certain parameters related to viral transport, such as the velocity of viral propagation, the rate of viral degradation and the viral diffusivity. A comparison of simulation results with a future experiment can provide insight into mechanisms of intracellular viral trafficking.

The solutions reviewed in this section were obtained in [30,50,65,66].

A schematic diagram of the problem is displayed in Fig. 14.9a. The virus enters the axon of the PNS at the synapse (located at $x=0$). It is assumed that the synapse is exposed to a constant viral flux, j_0, for a limited time, t_c, and then, after time t_d, to another viral pulse of the same intensity, also of duration t_c (the boundary condition at the synapse is illustrated in Fig. 14.9b). The viral diffusivity models transport of a free virus in the cytoplasm of the cell, as well as the situation when an endosome containing viral particles detaches from an MT. Diffusivity can also be caused by cargo navigation around obstacles during motor-driven transport. Degradation of the virus as it travels from the axon terminal toward the neuron soma is accounted for by a first order decay rate (such degradation is important, for example, in the case of the poliovirus [18]). Under these assumptions, retrograde transport of viruses in an axon is governed by Eq. (14.15) (where n is now the viral concentration) and the total viral flux (by diffusion and motor-driven transport) is given by Eq. (14.16).

The boundary condition at $x=0$ now is (see Fig. 14.9b):

$$j(0,t) = -D\frac{\partial n}{\partial x}(0,t)+vn(0,t) = j_0\{1 - H[t - t_c] + H[t - (t_d + t_c)] - H[t - (t_d + 2t_c)]\} \tag{14.26}$$

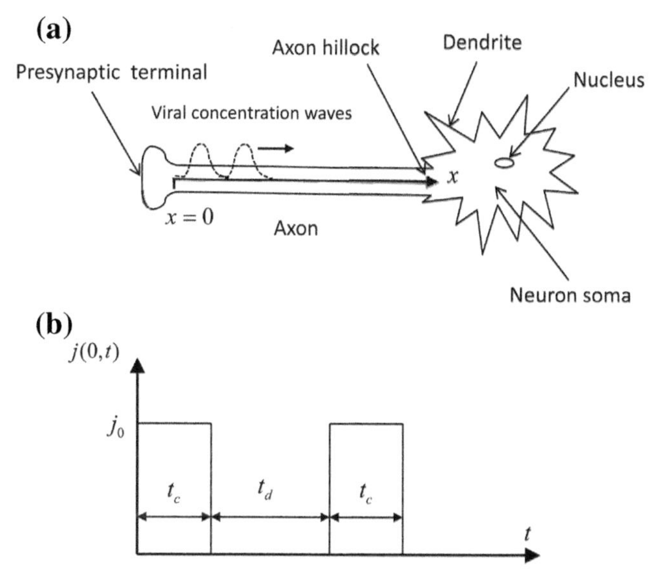

FIGURE 14.9 (a) Schematic diagram of retrograde viral trafficking (two viral concentration waves are displayed); (b) Viral flux at the axon terminal versus time (the boundary condition).

where the Heaviside step functions are used to produce two initial waves (pulses) displayed in Fig. 14.9b.

Since it is assumed that the neuron soma acts as a perfect absorber of viruses, the solution is identical to that obtained for a semi-infinite domain. Assuming that initially there are no viruses in the axon, the subsidiary equation is given by Eq. (14.18).

The Laplace transform of boundary condition (14.26) is:

$$-D\frac{\partial N}{\partial x}(0,s)+vN(0,s)=j_0\frac{1-e^{-s\,t_c}+e^{-s(t_d+t_c)}-e^{-s(t_d+2t_c)}}{s} \tag{14.27}$$

The solution of subsidiary Eq. (14.18) subject to boundary condition (14.27) and the condition that the solution remains finite as $x\to\infty$ is:

$$\begin{aligned}N(x,s)=&\frac{2j_0}{s\left(v+\sqrt{v^2+4D(k_d+s)}\right)}\left\{\exp\left[\frac{v-\sqrt{v^2+4D(k_d+s)}}{2D}x\right]\right.\\&-\exp\left[-st_c+\frac{v-\sqrt{v^2+4D(k_d+s)}}{2D}x\right]+\exp\left[-s(t_c+t_d)+\frac{v-\sqrt{v^2+4D(k_d+s)}}{2D}x\right]\\&\left.-\exp\left[-s(2t_c+t_d)+\frac{v-\sqrt{v^2+4D(k_d+s)}}{2D}x\right]\right\}\end{aligned} \tag{14.28}$$

The solution for the viral concentration can be obtained by calculating the inverse Laplace transform of the right-hand side of Eq. (14.28); the result is given in [30]. Alternatively, the inverse Laplace transform can be found numerically.

For the case when dynein velocity is not constant but is rather characterized by a pdf $f(v)$, the total viral concentration is again given by Eq. (14.23).

It is assumed that a single pulse contains 500 viral particles (1000 particles in two pulses). Since the problem is linear, this assumption does not affect the generality of the trends displayed in the figures. In order to estimate the flux of viral particles from the injury site, j_0, one needs to know the duration of one viral pulse, t_c. Using, for example, a 60 s duration of the pulse and estimating the average axonal diameter, d, as 1.1 μm based on [53] one again obtains $j_0 = 8.8\,\text{particles}/(\mu\text{m}^2\text{s})$. The time between the pulses, t_d, is assumed to be twice the pulse duration, 120 s.

For plasmids, the disintegration rate, k_d, was estimated as $1/3600\,\text{s}^{-1}$ [57,67]. Two values are used here for this parameter: $k_d = 0\,\text{s}^{-1}$, which corresponds to the assumption that no degradation of the virus occurs, and $k_d = 10^{-3}\,\text{s}^{-1}$ which corresponds to a relatively large rate of viral degradation.

Figure 14.10 displays viral concentration waves at three times: 300, 600 and 900 s. Figure 14.10a is computed assuming that all dynein motors move with the same velocity ($v_1 = 0.445\ \mu\text{m/s}$) and Fig. 14.10b is computed assuming that there is a distribution of dynein velocities, which is based on data reported in [54], see Eq. (14.24) (v_1 used for Fig. 14.10a is obtained as an expected value of this velocity distribution). Other parameter values utilized for this figure are $D = 1\ \mu\text{m}^2/\text{s}$, $j_0 = 8.8\ 1/\mu\text{m}^2\text{s}$, $t_c = 60$ s, $t_d = 120$ s and $k = 0\ \text{s}^{-1}$.

In Fig. 14.10a one can see how two waves merge (the merging is complete by approximately $t = 900$ s). It should be noted that merging of the waves is observed only if dynein velocity is assumed constant; in Fig. 14.10b, which is computed for the case when dynein velocity is distributed, the two pulses do not induce two distinct waves, which is due to a broad range of dynein velocities, so the waves merge right at the beginning. Observing this situation in a future experiment can thus suggest how nonuniform the dynein velocity distribution is in a particular cellular system.

Figure 14.11 is similar to Fig. 14.10, but it is computed for the dynein motor velocity distribution based on [55], see Eq. (14.25). This distribution gives a much larger expected motor velocity ($v_2 = 1.272\ \mu\text{m/s}$, which is 2.86 times larger than the value used for Fig. 14.10a). Since the waves in Fig. 14.11a merge much more slowly than those displayed in Fig. 14.10a, the waves are shown at larger times (2000, 4000 and 6000 s). To explain why the time when the waves merge is affected to such a degree by the motor velocity, Eq. (14.15) is converted into a dimensionless form:

$$\frac{\partial \hat{n}}{\partial \hat{t}} = \widehat{D}\frac{\partial^2 \hat{n}}{\partial \hat{x}^2} - \frac{\partial \hat{n}}{\partial \hat{x}} - \hat{n} \tag{14.29}$$

where the dimensionless variables are defined as follows:

$$\widehat{D} = \frac{Dk}{v^2}, \quad \hat{x} = \frac{xk}{v}, \quad \hat{n} = n\frac{v}{j_0}, \quad \hat{t} = tk \tag{14.30}$$

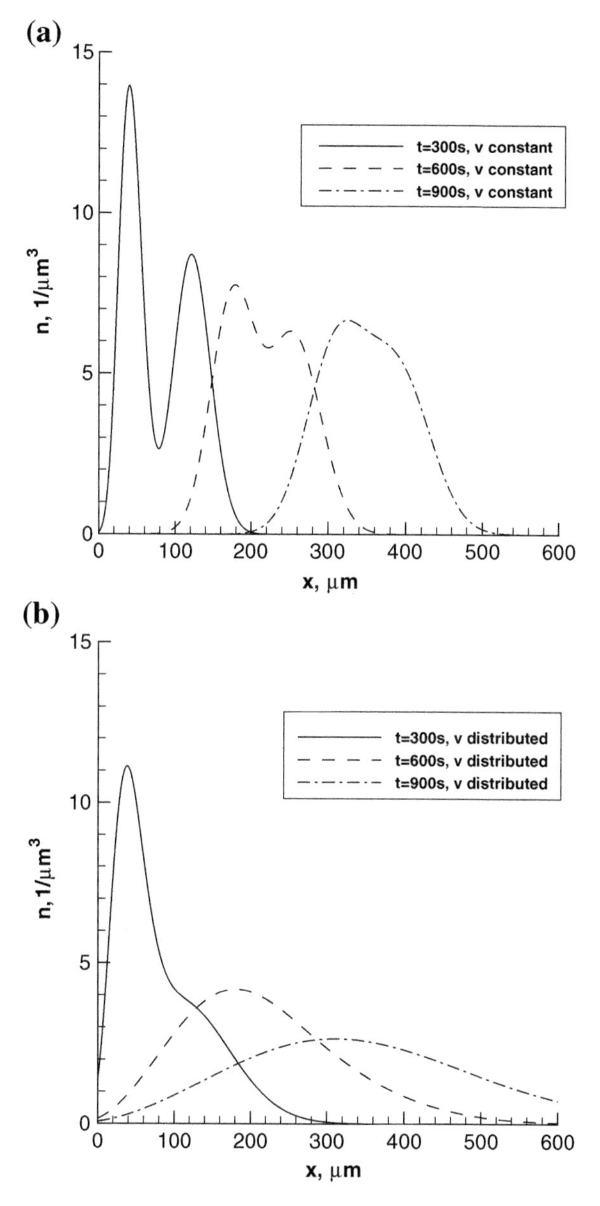

FIGURE 14.10 Viral concentration waves at various times. (a) Constant velocity of all dynein motors is assumed (calculated as an expected value of a corresponding velocity distribution); (b) A pdf describing the dynein motor velocity distribution based on [54] is assumed. ($D = 1\,\mu\text{m}^2/\text{s}$, $j_0 = 8.81\,\mu\text{m}^2/\text{s}$, $t_c = 60$ s, $t_d = 120$ s, $k = 0\,\text{s}^{-1}$).

One can check that if one recomputes the results utilizing the same dynein velocity that was used for Fig. 14.11a ($v_2 = 1.272\,\mu\text{m/s}$) but modifies the rest of the parameters as follows:

$$D^* \to \left(\frac{v_2}{v_1}\right)^2 D, \; n^* \to \frac{v_2}{v_1} n, \; x^* \to x \Big/ \left(\frac{v_2}{v_1}\right) \tag{14.31}$$

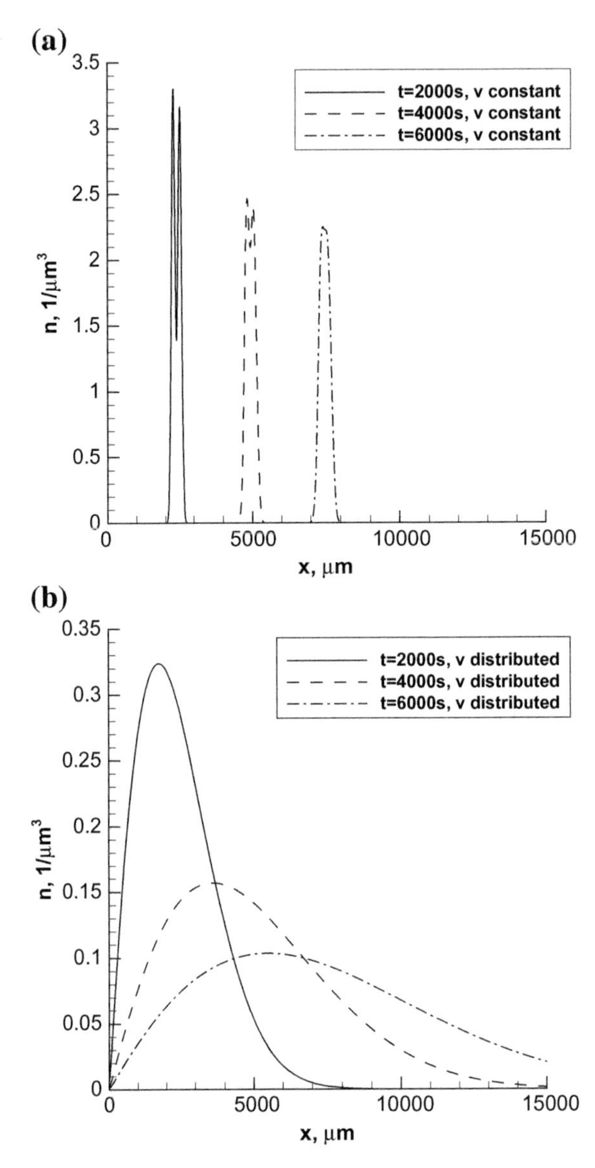

FIGURE 14.11 Similar to Figure 14.10, but now for the dynein motor velocity distribution based on [55].

and uses $t = 300$, 600 and 900 s, one obtains the results displayed in Fig. 14.10a. This means that the results for a smaller velocity v_1 can be recovered from the results for a larger velocity v_2 by increasing the viral diffusivity by a factor of $\left(\frac{v_2}{v_1}\right)^2$, increasing the wave amplitude by a factor of $\frac{v_2}{v_1}$, and shrinking x by a factor of $\frac{v_2}{v_1}$.

Figure 14.12 displays values of x where the function $n(x)$ takes its maximum values. In physical terms, Fig. 14.12 shows the positions of the two peaks of the two concentration

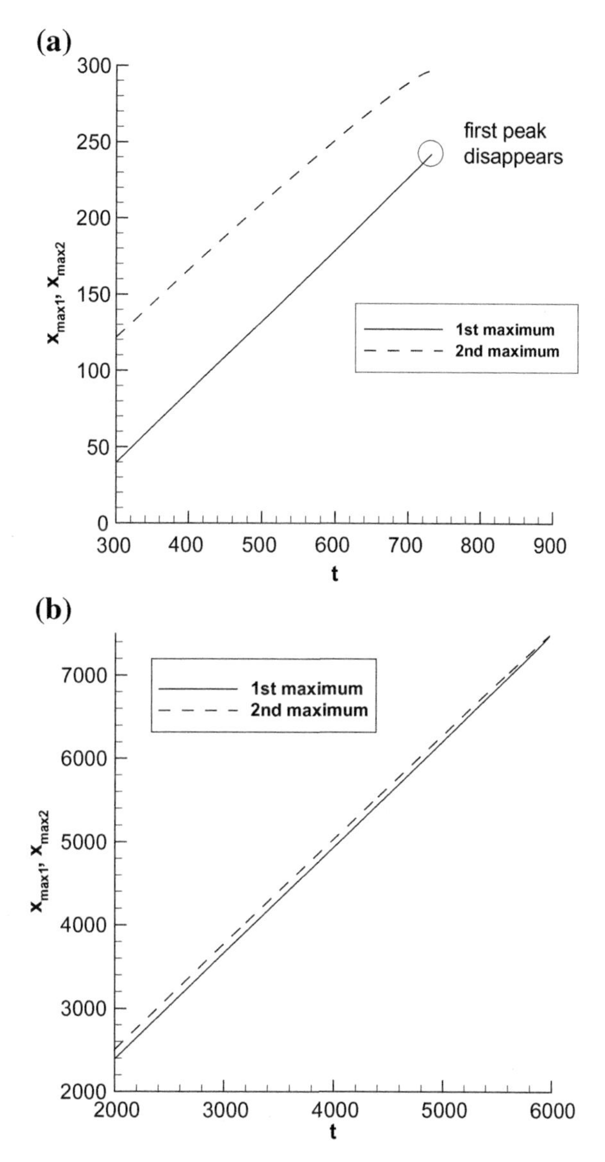

FIGURE 14.12 Positions of the waves' peaks versus time computed for the case when all dynein motors move with the same velocity. (a) The dynein velocity equals the expected value that follows from [54] (0.4453 μm/s); (b) The dynein velocity equals the expected value that follows from [55] (1.272 μm/s). (D=1 μm²/s, $j_0 = 8.81/\mu m^2 s, t_c = 60$ s, $t_d = 120$ s, $k = 0$ s^{-1}).

waves. Constant velocity of dynein motors is assumed. Fig. 14.12a is computed using the expected value of dynein velocity that follows [54] ($v_1 = 0.445$ μm/s), and Fig. 14.12b is computed using the expected value of dynein velocity that follows from [55] ($v_2 = 1.272$ μm/s). Figure 14.12 is computed for the case with no viral degradation, $k = 0$ s^{-1}. Figure 14.12a thus corresponds to the waves displayed in Fig. 14.10a (in this case the waves merge at $t_m = 732\,s$) and Fig. 14.12b corresponds to those displayed in Fig. 14.11b (in this case the waves merge at $t_m = 5{,}978\,s$). As long as the waves are distinct, there are two peaks. Eventually, the first peak disappears (see Fig. 14.12a), and from that moment on one can say that the two waves have merged. It is evident that as the waves propagate, the two peaks become closer, but only slightly, so merging of the two waves is not a process involving the coalescence of two peaks; rather, the first peak becomes less and less pronounced and eventually becomes a region with a horizontal slope.

14.5 CONCLUSIONS

Three new models of intracellular dynein-driven axonal transport have been reviewed. The models are based on the continuum mechanics approach to intracellular transport and allow for a large-scale view of vesicular trafficking in axons.

1. A model of axonal transport drug delivery accounts for PAC exchange between accumulated and dynein-driven states and PAC diffusivity. The obtained results indicate that if diffusivity of PACs is not zero, the pulse of dynein-driven PACs spreads outs as it propagates away from the axon terminal. Even if diffusivity is zero, the pulse spreads out (although not as fast) unless the rate of PAC absorption at the Nodes of Ranvier is negligible. The velocity of the pulse propagation is unaffected by diffusivity but depends on the kinetic constants characterizing the rates of PAC transition between the dynein-transported and accumulated states. If the rate of dynein absorption at the Nodes of Ranvier is large, a typical PAC spends a large portion of its time in the accumulated state (not moving at all) and the velocity of the pulse of dynein-driven PACs is small. If the rate of dynein absorption at the Nodes of Ranvier is small, the velocity of the pulse approaches the velocity of a dynein motor. Possible extensions of this model include accounting for a population of PACs freely suspended in the cytosol.
2. A model of transport of chemical signals from the axon injury site to the neuron soma makes it possible to investigate the feasibility of different mechanisms that neurons may use to measure the distance between the injury site and the axon hillock. The results indicate that both the two-signal and two-detector methods suggested in [16] are feasible; the accuracy depends on the dynein velocity distribution and on the distance to the injury site. The accuracy generally decreases if the distance to the injury site becomes too large.
3. Utilizing a model of retrograde viral transport in an axon, the situation when two waves of viral concentration are induced in the axon by subjecting the presynaptic terminal to two viral impulses has been investigated. Two cases are simulated: when all dynein motors move with the same velocity and when there is a distribution of dynein motor velocities. Merging of the waves is observed only if dynein velocity is assumed constant; when dynein velocity is distributed, the two pulses do not induce two distinct waves, the concentration disturbances in this case merge right in the beginning.

References

[1] L.S.B. Goldstein, Z.H. Yang, Microtubule-based transport systems in neurons: the roles of kinesins and dyneins, Annual Review of Neuroscience 23 (2000) 39–71.

[2] S.P. Gross, Hither and yon: a review of bi-directional microtubule-based transport, Physical Biology 1 (2004) R1–R11.

[3] M.A. Welte, Bidirectional transport along microtubules, Current Biology 14 (2004) R525–R537.

[4] P.E. Gallant, Axonal protein synthesis and transport, Journal of Neurocytology 29 (2000) 779–782.

[5] A.D. Pilling, D. Horiuchi, C.M. Lively, W.M. Saxton, Kinesin-1 and dynein are the primary motors for fast transport of mitochondria in drosophila motor axons, Molecular Biology of the Cell 17 (2006) 2057–2068.

[6] S. Ally, A.G. Larson, K. Barlan, S.E. Rice, V.I. Gelfand, Opposite-polarity motors activate one another to trigger cargo transport in live cells, Journal of Cell Biology 187 (2009) 1071–1082.

[7] T.L. Deckwerth, R.M. Easton, C.M. Knudson, S.J. Korsmeyer, E.M. Johnson, Placement of the BCL2 family member BAX in the death pathway of sympathetic neurons activated by trophic factor deprivation, Experimental Neurology 152 (1998) 150–162.

[8] M. Deshmukh, E.M. Johnson, Evidence of a novel event during neuronal death: development of competence-to-die in response to cytoplasmic cytochrome c, Neuron 21 (1998) 695–705.

[9] A. Riccio, S. Ahn, C.M. Davenport, J.A. Blendy, D.D. Ginty, Mediation by a CREB family transcription factor of NGF–dependent survival of sympathetic neurons, Science 286 (1999) 2358–2361.

[10] M.A. Vogelbaum, J.X.X. Tong, K.M. Rich, Developmental regulation of apoptosis in dorsal root ganglion neurons, Journal of Neuroscience 18 (1998) 8928–8935.

[11] L.S. Zweifel, R. Kuruvilla, D.D. Ginty, Functions and mechanisms of retrograde neurotrophin signaling, Nature Reviews Neuroscience 6 (2005) 615–625.

[12] C.S. von Bartheld, Axonal transport and neuronal transcytosis of trophic factors, tracers, and pathogens, Journal of Neurobiology 58 (2004) 295–314.

[13] G.B. Stokin, C. Lillo, T.L. Falzone, R.G. Brusch, E. Rockenstein, S.L. Mount, R. Raman, P. Davies, E. Masliah, D.S. Williams, L.S.B. Goldstein, Axonopathy and transport deficits early in the pathogenesis of Alzheimer's disease, Science 307 (2005) 1282–1288.

[14] D.D. Hurd, W.M. Saxton, Kinesin mutations cause motor neuron disease phenotypes by disrupting fast axonal transport in drosophila, Genetics 144 (1996) 1075–1085.

[15] A.G. Filler, G.T. Whiteside, M. Bacon, M. Frederickson, F.A. Howe, M.D. Rabinowitz, A.J. Sokoloff, T.W. Deacon, C. Abell, R. Munglani, J.R. Griffiths, B.A. Bell, A.M.L. Lever, Tri-partite complex for axonal transport drug delivery achieves pharmacological effect, BMC Neuroscience 11 (2010) 8.

[16] N. Kam, Y. Pilpel, M. Fainzilber, Can molecular motors drive distance measurements in injured neurons? Plos Computational Biology 5 (2009) e1000477.

[17] N. Abe, V. Cavalli, Nerve injury signaling, Current Opinion in Neurobiology 18 (2008) 276–283.

[18] K.Z. Lancaster, J.K. Pfeiffer, Limited trafficking of a neurotropic virus through inefficient retrograde axonal transport and the type I interferon response, Plos Pathogens 6 (2010) e1000791.

[19] N.D. Mazarakis, M. Azzouz, J.B. Rohll, F.M. Ellard, F.J. Wilkes, A.L. Olsen, E.E. Carter, R.D. Barber, D.F. Baban, S.M. Kingsman, A.J. Kingsman, K. O'Malley, K.A. Mitrophanous, Rabies virus glycoprotein pseudotyping of lentiviral vectors enables retrograde axonal transport and access to the nervous system after peripheral delivery, Human Molecular Genetics 10 (2001) 2109–2121.

[20] M. Miranda-Saksena, R.A. Boadle, A. Aggarwal, B. Tijono, F.J. Rixon, R.J. Diefenbach, A.L. Cunningham, Herpes simplex virus utilizes the large secretory vesicle pathway for anterograde transport of tegument and envelope proteins and for viral exocytosis from growth cones of human fetal axons, Journal of Virology 83 (2009) 3187–3199.

[21] G.S. Tan, M.A.R. Preuss, J.C. Williams, M.J. Schnell, The dynein light chain 8 binding motif of rabies virus phosphoprotein promotes efficient viral transcription, Proceedings of the National Academy of Sciences of the United States of America 104 (2007) 7229–7234.

[22] S.E. Encalada, L.S.B. Goldstein, Prion transport, in: Larry R. Squire (Ed.), Encyclopedia of Neuroscience, Academic Press, Oxford, 2009, pp. 1071–1075.

[23] B. Bizzini, P. Grob, M.A. Glicksman, K. Akert, Use of the B-iib tetanus toxin derived fragment as a specific neuro-pharmacological transport agent, Brain Research 193 (1980) 221–227.

[24] A.G. Filler, Axonal-transport and MR-imaging – prospects for contrast agent development, JMRI-Journal of Magnetic Resonance Imaging 4 (1994) 259–267.

[25] A.G. Filler, B.A. Bell, Axonal-transport, imaging, and the diagnosis of nerve compression, British Journal of Neurosurgery 6 (1992) 293–295.

[26] A.G. Filler, H.R. Winn, L.E. Westrum, P. Sirrotta, K. Krohn, T.W. Deacon, Intramuscular injection of wga yields systemic distribution adequate for imaging of axonal transport in intact animals, Society for Neuroscience Abstracts 17 (1991) 1480.

[27] R.H. Haschke, J.M. Ordronneau, A.H. Bunt, Preparation and retrograde axonal-transport of an anti-viral drug horseradish-peroxidase conjugate, Journal of Neurochemistry 35 (1980) 1431–1435.

[28] K.P. Gatzinsky, C.H. Berthold, Lysosomal activity at nodes of ranvier during retrograde axonal-transport of horseradish-peroxidase in alpha-motor neurons of the cat, Journal of Neurocytology 19 (1990) 989–1002.

[29] K.P. Gatzinsky, C. Berthold, C. Fabricius, A. Mellström, Lysosomal activity at nodes of Ranvier in dorsal column and dorsal root axons of the cat after injection of horseradish peroxidase in the dorsal column nuclei, Brain Research 566 (1991) 131–139.

[30] A.V. Kuznetsov, Merging of viral concentration waves in retrograde viral transport in axons, Central European Journal of Physics 9 (2011) 1493–1502.

[31] A.V. Kuznetsov, A model of axonal transport drug delivery: effects of diffusivity, International Journal for Numerical Methods in Biomedical Engineering 8 (2012) 1083–1092.

[32] A.V. Kuznetsov, A model of axonal transport drug delivery, Central European Journal of Physics 10 (2012) 320–328.

[33] A.V. Kuznetsov, Effect of pharmaceutical agent degradation on axonal transport drug delivery: an analytical solution for a transient situation, International Communications in Heat and Mass Transfer 38 (2011) 1317–1321.

[34] A.V. Kuznetsov, A three-kinetic-state model of axonal transport drug delivery, Journal of Mechanics in Medicine and Biology 12 (2012) 1250044.

[35] D.A. Smith, R.M. Simmons, Models of motor-assisted transport of intracellular particles, Biophysical Journal 80 (2001) 45–68.

[36] B. Engquist, A. Majda, Absorbing boundary-conditions for numerical-simulation of waves, Mathematics of Computation 31 (1977) 629–651.

[37] P.P. Valko, S. Vajda, Inversion of noise-free laplace transforms: towards a standardized set of test problems, Inverse Problems in Engineering 10 (2002) 467–483.

[38] P.P. Valko, J. Abate, Comparison of sequence accelerators for the Gaver method of numerical Laplace transform inversion, Computers & Mathematics with Applications 48 (2004) 629–636.

[39] M. Abramowitz, I.A. Stegun, Handbook of Mathematical Functions, with Formulas, Graphs, and Mathematical Tables, Dover Publications, Mineola, NY, 1965.

[40] S.J. King, T.A. Schroer, Dynactin increases the processivity of the cytoplasmic dynein motor, Nature Cell Biology 2 (2000) 20–24.

[41] S. Toba, T.M. Watanabe, L. Yamaguchi-Okimoto, Y.Y. Toyoshima, H. Higuchi, Overlapping hand-over-hand mechanism of single molecular motility of cytoplasmic dynein, Proceedings of the National Academy of Sciences of the United States of America 103 (2006) 5741–5745.

[42] P. Jung, A. Brown, Modeling the slowing of neurofilament transport along the mouse sciatic nerve, Physical Biology 6 (2009) 046002.

[43] I. Rishal, M. Fainzilber, Retrograde signaling in axonal regeneration, Experimental Neurology 223 (2010) 5–10.

[44] F. Sun, V. Cavalli, Neuroproteomics approaches to decipher neuronal regeneration and degeneration, Molecular & Cellular Proteomics 9 (2010) 963–975.

[45] R. Ohara, K. Hata, N. Yasuhara, R. Mehmood, Y. Yoneda, M. Nakagawa, T. Yamashita, Axotomy induces axonogenesis in hippocampal neurons by a mechanism dependent on importin beta, Biochemical and Biophysical Research Communications 405 (2011) 697–702.

[46] P.M. Richardson, V.M.K. Issa, Peripheral injury enhances central regeneration of primary sensory neurons, Nature 309 (1984) 791–793.

[47] D.S. Smith, J.H.P. Skene, A transcription-dependent switch controls competence of adult neurons for distinct modes of axon growth, Journal of Neuroscience 17 (1997) 646–658.

[48] S. Neumann, C.J. Woolf, Regeneration of dorsal column fibers into and beyond the lesion site following adult spinal cord injury, Neuron 23 (1999) 83–91.

[49] A.V. Kuznetsov, An analytical solution describing propagation of positive injury signals in an axon: effect of dynein velocity distribution, Computer Methods in Biomechanics and Biomedical Engineering, in press, http://dx.doi.org/10.1080/10255842.2011.632376.

[50] A.V. Kuznetsov, Analytical modelling of retrograde transport of nerve growth factors in an axon: a transient problem, Computer Methods in Biomechanics and Biomedical Engineering 16 (2013) 95–102.

[51] H.S. Carslaw, J.C. Jaeger, Conduction of Heat in Solids. second ed., Clarendon Press, Oxford, 1959.

[52] T.J. Lukas, A.L. Wang, M. Yuan, A.H. Neufeld, Early cellular signaling responses to axonal injury, Cell Communication and Signaling 7 (2009) 5.

[53] E. Bergers, J.C.J. Bot, C.J.A. De Groot, C.H. Polman, G.J.L.A. Nijeholt, J.A. Castelijns, P. van der Valk, F. Barkhof, Axonal damage in the spinal cord of MS patients occurs largely independent of T2 MRI lesions, Neurology 59 (2002) 1766–1771.

[54] J.L. Ross, K. Wallace, H. Shuman, Y.E. Goldman, E.L.F. Holzbaur, Processive bidirectional motion of dynein-dynactin complexes *in vitro*, Nature Cell Biology 8 (2006) 562–570.

[55] K. Deinhardt, S. Salinas, C. Verastegui, R. Watson, D. Worth, S. Hanrahan, C. Bucci, G. Schiavo, Rab5 and Rab7 control endocytic sorting along the axonal retrograde transport pathway, Neuron 52 (2006) 293–305.

[56] E.L. Bearer, M.L. Schlief, X.O. Breakefield, D.E. Schuback, T.S. Reese, J.H. LaVail, Squid axoplasm supports the retrograde axonal transport of herpes simplex virus, Biological Bulletin 197 (1999) 257–258.

[57] T. Lagache, E. Dauty, D. Holcman, Physical principles and models describing intracellular virus particle dynamics, Current Opinion in Microbiology 12 (2009) 439–445.

[58] M.A. Samuel, H. Wang, V. Siddharthan, J.D. Morrey, M.S. Diamond, Axonal transport mediates West Nile virus entry into the central nervous system and induces acute flaccid paralysis, Proceedings of the National Academy of Sciences of the United States of America 104 (2007) 17140–17145.

[59] B.K. Kaspar, D. Erickson, D. Schaffer, L. Hinh, F.H. Gage, D.A. Peterson, Targeted retrograde gene delivery for neuronal protection, Molecular Therapy 5 (2002) 50–56.

[60] Y. Liu, B.T. Himes, J. Moul, W.L. Huang, S.Y. Chow, A. Tessler, I. Fischer, Application of recombinant adenovirus for *in vivo* gene delivery to spinal cord, Brain Research 768 (1997) 19–29.

[61] D.D. Pearse, M.B. Bunge, Designing cell- and gene-based regeneration strategies to repair the injured spinal cord, Journal of Neurotrauma 23 (2006) 438–452.

[62] X.N. Bo, D.S. Wu, J. Yeh, Y. Zhang, Gene therapy approaches for neuroprotection and axonal regeneration after spinal cord and spinal root injury, Current Gene Therapy 11 (2011) 101–115.

[63] K.R.G. Martin, H.A. Quigley, D.J. Zack, H. Levkovitch-Verbin, J. Kielczewski, D. Valenta, L. Baumrind, M.E. Pease, R.L. Klein, W.W. Hauswirth, Gene therapy with brain-derived neurotrophic factor as a protection: Retinal ganglion cells in a rat glaucoma model, Investigative Ophthalmology & Visual Science 44 (2003) 4357–4365.

[64] M. Berry, L. Barrett, L. Seymour, A. Baird, A. Logan, Gene therapy for central nervous system repair, Current Opinion in Molecular Therapeutics 3 (2001) 338–349.

[65] A.V. Kuznetsov, An analytical solution describing retrograde viral transport in an axon, International Communications in Heat and Mass Transfer 38 (2011) 1313–1316.

[66] A.V. Kuznetsov, Analytical investigation of various regimes of retrograde trafficking of neurotropic viruses in axons, Central European Journal of Physics 9 (2011) 1372–1378.

[67] D. Lechardeur, K.J. Sohn, M. Haardt, P.B. Joshi, M. Monck, R.W. Graham, B. Beatty, J. Squire, H. O'Brodovich, G.L. Lukacs, Metabolic instability of plasmid DNA in the cytosol: a potential barrier to gene transfer, Gene Therapy 6 (1999) 482–497.

Index

^{90}Y microsphere deposition efficiency, 401
$\Delta\Psi$-dependency of mitochondrial Ca^{2+} uniporter
  alternative formulation, 201–204
  free energy formalism, 197–200
  reparameterization, 201

A

Absorption, 464
Alternating current, in electrotransport, 440–441
  and flux enhancement, 440
  iontophoresis, advantages of, 440
  membrane properties, 441
  symmetric, effects of, 440
Aneurysm, 294
Anzelius's solution, 85–86
Aristotle, Metaphysic [1045a.10], 7
Arterial wall
  anisotropic mechanisms, 313–314
  and collagen, 283–291
  compliance of, 281–282
  constitutive models, 309–324
  continuous damage, 321
  damage models, 320–321
  design requirements, 281–283
  discontinuous damage, 321
  distribution of fiber orientations, 316–317
  distribution of fiber recruitment stretch, 317–319
  elastin, 291–292
  enzymatic damage, 321
  fibril-forming collagens, 286–287
  functioning, 276
  isotropic mechanism, 312–313
  layers of, 279–281
  mechanical integrity, 276–277
  multi-mechanism models (GR&D), 319–324
  multiple mechanism models, 311–312
  N-fiber anisotropic models, 314–316
  positive and negative growth, 319–320
  remodeling, 320
  transition region, 278
  tubular segment of, 279–281
  tunica adventitia, 308–309
  tunica intima, 301–306
  tunica media, 306–308
  vascular cells, 279
Articular cartilage
  theory of interacting continua, 17–18
Autoregulated cerebral perfusion, 258–259
  arteriolar models, 264–267
  local autoregulation, 258–259
  neurovascular coupling process, 265
  organ autoregulation, 258
  simple models, 259–264
  smooth muscle and endothelial cell
      processes, 265
Axonal transport drug delivery
  accumulated and dynein-driven PACs, 523–525
  assumptions of, 525
  benifits, 523
  diffusivities of organelles, 527
  Heaviside step function, 525
  inverse Laplace transforms, 526
  kinetic states of PACs, 532
  Laplace transform of boundary condition, 525
  problems in, 523
  research of, 523

B

Basal laminae, 289
Basement membrane, 302–303
Beaded filament forming, 289–290
Bell's model, 163–165
Biological cells, 11–12
  mechanical properties, 25–29
  mechanics of red blood cell, 11–12
  structural transition, 12
Biological membrane, in carrier mediated
      transport
  antiport, 208
  cotransport, 208
  experimental transport research, 190–191
  kinetic modeling of simple carrier, 183–187
  kinetic treatment of simple pore, 187–188
  passive carrier-mediated transport, 182
  physicochemical principles, 183–190
  of solute transport, 188–190
  thermodynamics of solute transport, 183
  transport process, 182
Biomacromodules, 8–9
  coarse-grained methods, 9

molecular dynamics method, 8
protein actin, 8–9
Biot's theory of poroelasticity, 16
Blood (blood flow)
cell properties and microvessel structure, 350
see also Red blood cells (RBCs)
Blood-brain barrier, 124–129
cultured endothelial cell monolayers, 135
fiber matrix model, 137–141
hydraulic conductivity and reflection coefficient, 130–132
measurement, 133–135
permeability measurements, 130
single perfused microvessel, 130–135
solute permeability, 132–133
transport coefficients, 129–130
transport pathways, 128–129

C

Ca^{2+} uniporter, 192
limitation of, 201
Calcein, 464
Calcium ions, 464
Capillary blood flow
continuum formulation, 214
geometrical model of, 215–216
homogeneous networks, 226–235
in microcirculation, 214
numerical simulations in discrete model, 215
steady capillary blood flow, 216–222
theoretical models, 214
tree models of capillary networks, 222–226
unsteady blood flow, 216
unsteady capillary blood flow, 235–243
Carbon-based particles, 397
Cell electrotransformation, statistical model for, 481
Cell electroporation, transformation
complex process, 465
conductivity on cell suspension electroporation, 475–477
number of molecules calculation, 483
pore (membrane defect) size, 482–483
single cell and tissue bulk levels, 482
single cell model and bulk tissue model, combination on, 483–484
Cell membrane electropermeabilization, 498
Cell membrane, 10–11
bacterial cells, 10
mechanism of transport, 10
Cell partitioning law
emprical relation, 221
general expression, 221
Cell swelling, 464
Cellular automata, 33–34
Cellular biological media, 2–4
asymptotic homogenization, 19–21
Biot's theory of poroelasticity, 16
calulation of constitutive parameters, 23–32
conservation of linear momentum, 12–13
conservation of mass, 12–13
conservation of solute, 12–13
continuum-based model, 33
development of sharpened experimental tools, 4
diffusion coefficients, 31–32
diffusion/chemotaxis process, 33
discrete-based models, 33–36
engineering principles, 6–7
equation-free approaches, 23
fluid shear, 5
formulation of laws and constitutive relations, 12–23
generic framework for theoretical calculation, 23–24
growth and pattern formation, 32–36
hierarchical structure, 3
interactions, 7–8
mass transport mechanisms, 5
mechanical properties, 29
mechanotransduction, 5
microbial biofilm, hierarchial structure of, 4
multiscale bottom-up approaches, 18–23
multiscale computational, 23
protein-mediated mechanisms, 5
remarks on experimental determination, 24–25
scales of observation, 3
single-scale, single phase approches, 12–16
spatial averaging method, 21–23
stress tensor, 14
theory of interacting continua, 16–17
transport phenomena, 4–6
transport processes in macroscopic scale, 6
upscaling methods, 18–19
viscoelastic behavior, 17–18
Ceramic nanoparticles, 398
Cerebral arterial structure, 254–256
Cerebral capillary beds, 267
behavior, 267
functional reactivity, 267
Green's function, 269
hemodynamic parameters, 267
Henry's law, 268–269
infinite three-dimensional domain, 268
Michaelis-Menten equation, 267–268
oxygen transport, 267–272
Cerebral perfusion, 254–257
arterial blood vessels, 257–258
arterial network and cortex, 256–257

autoregulation, 258–259
capillary beds, 267–272
cerebral arterial structure, 254–256
circle of Willis, 256
complex models, 264–272
local autoregulation, 258–259
organ autoregulation, 258
simple models of autoregulation, 259–264
vascular network and its numerical simulation, 259
Circle of Willis, 256
blow flow, 259–261
variability of, 256
Collagen
anisotropic mechanisms, 313–314
collagen IV, 289
collagen V, 287–289
collagen VI, 289–290
collagen VIII, 290
collagen XII, 290–291
collagens types I and III, 287
to extracellular matrix, 300–301
Fiber Attachment Stretch λfa, 323
fibril-forming collagens, 286–287
GR&D, 323–324
growth/atrophy, 324
remodelling, 323–324
and supramolecular assemblies, 284–291
types and subfamilies, 285–286
platelet adhesion, 290
Collagens types I and III, 287
Comsol Multiphysics®, 527
Constant electric field (symmetric conditions), 420–426
donor compartment, 420–421
receiver compartments, 420–421
Convection, 464
Cytoplasm, 350
Cytoskeleton, 11
functions, 11
structural specific proteins, 11

D

Debye-Huckel thickness, 425
Dembo et al.'s model, 165
Dentinal fluid flow (DFF), 42
analysis of, 46–47
shear stress, 54
thermal expansion/contraction, 46
tooth thermal pain, implications for, 53–54
Dentinal microtubules, 42
analysis of thermomechanics, 43–46
deformation, 46
in dentinal fluid flow (DFF), 42
displacement, 46
Dextrans, 464
Diffusion, 464
Diffusivities of organelles, 527
using Einstein-Stokes relation, 527
Direct drug-particle transport
applications of optimal micro-drug delivery, 404–413
implementation of optimal targeted drug delivery, 401–404
Discharge hematocrit, 217–218
DNA cluster formation
electrodiffusion model, 472–475
Doxil®, 396
Drug release, in biological tissues
analytical solutions for local mass non-equilibrium, 78–87
Anzelius's solution, 85–86
averaging concentration, 67–68
conservation of, 73–78
continuity equation, 69–71
continuum modeling, 65–73
Darcy's Law, 68
drug-eluting stent, 92–115
extended continuity equation, 71–72
Fick's equation, 73
first-order consumption rate coefficient, 76
in fluid phase, 74–77
fluid-saturated porous media, 67
governing equations, 78
mathematical difficulties of analytical solution, 78–87
microscopic approach, 65
molecular dynamics (MD) simulations, 64
Nusselt's solution, 79–83
partial differential equations, 78
permeability, 68
phases of, 65
Schumann's solution, 83–85
in solid phase, 77–78
temperature of fluid mixture, 78
tortuosity, 72–73
and volume-averaged variables, 66–68
Drug-aerosol delivery, 408–413
in smart inhaler system to lung, 408–413
Drug-eluting stent (DES), 61
Dynein-driven PACs, 523–525
boundary conditions, 535
complementary error function, 535
computations, 536
duration, 534
inverse Laplace transform, 535
on positive injury signals in axons, 532–540
PIS transport, 533
velocity of, 537

E

Elastin molecules, 291
  damages, 322–323
  degradation of, 322
  GR&D for, 321–323
Electrochemotherapy, 484–485
Electrokinetic methods, electrotransport in biological membrane, 443–450
  effectiveness, 443–444
  factors affecting, 443
  transdermal iontophoresis, 444
  transscleral Iontophoresis, 448–450
  transungual iontophoresis, 447–448
Electro-osmosis, 464
Electroosmotic transport, 425–426
Electropermabilization model, 470–472
  biological membrane, 470–471
Electropermeabilization
  pulse and physiological factors, 494
Electropermeabilization, 441–443
  in electrotransport, 442
  mechanisms of, 441–442
Electrophoreses, 464
Electroporation
  realistic biophysical models, 481–482
Electroporation, 493–495
  electrical conductivity increase, 497–499
  physical changes in biological tissue, 497–504
  tissue heating, 499–502
Electropure', 459
Electrotransport in biological membrane, 418–419
  alternating current, 440–441
  constant electric field (symmetric conditions), 420–426
  diffusion chambers, 418
  donor/membrane interface, 422
  electrokinetic methods, 443–450
  electroosmosis, 425–426, 430
  electropermeabilization effect, 441–443
  independent positioning, 422–424
  lag time, 424–425
  membrane flux and transference number, 424
  membrane/receiver interface, 422
  multiple barriers/membranes, 430–440
  Nernst-Planck theory, 418–420
  steady state concentration, 422
  variable electric field (asymmetric conditions), 426–430
Endocytosis, 464
Endothelial cell monolayers, 135
Endothelial cells, 293
  computational fluid dynamic, 294
  local hemodynamic conditions, 294
  mechanical stimuli sensor, 294
  monolayer, 293
  morphology of, 294
  regulation of permeability, 294
  viscous drag force, 293
Endothelial glycocalyx layer, 301–302
Endothelial surface glycocalyx, 146–152
  composition, thickness, and structure, 146–148
  effect of charge, 151–152
  mechanosensor, 150–151
  molecular sieve, 148
  role of, 148–151
  Starling's hypothesis
Exponential model
  isotropic mechanism, 312–313
Extracellular matrix (ECM), 2–3
  components of mammalian tissues, 3
  transport properties, 25

F

FACIT, 290–291
Fahraeus–Lindqvist effect, 11–12
Fenestrae, 152
Fiber matrix model, 137–141
Fibril-forming collagens, 286–287
  collagen IV, 289
  collagen V, 287–289
  collagen VI, 289–290
  collagen VIII, 290
  collagen XII, 290–291
  collagens types I and III, 287
  morphology, 287
  non-centrosymmetric nature, 287
Fibroblasts, 297
  heterogeneous phenotypes, 297
  procollagen molecules, 297
  vessel injury, 297
Fibropositor model, 297–298
Fick's equation, 73
Fick's first law, 468
Fluid shear, 5
Fluid wall model
  advection-reaction-diffusion equation, 102–106
  concentration solutions and results, 99–102
  general physiological and mathematical description, 96–99
  pure diffusion approximation, 96–102
Free transport, in trageted drug delivery, 394–400
  nanodrug carriers, 395–400
  passive and active targeting, 395
Functional hyperemia, 258–259

G

Glycocalyx, 9–10
  fluid shear stress, 9–10
Glycosaminoglycans, 279
Goldman approximation, 421
Gram-positive bacteria, 10

H

Hodgkin-Huxley (H-H) model, 49
Homogeneous capillary blood flow, 226–235
  area- or space-filling networks, 230
  dimensions and parameters, 229
  effective hydraulic permeability, 229–230
  geometrical construction, 228
  interior nodes, 227
  numerical method, 228
  pristine network, 228–229
  significance of bifurcation law, 231–233
  theoretical model, 226
  viscosity correlation, significance of, 233–234
Hopf bifurcation theory, 248
Human heart, 281

I

Image-guided drug delivery (IGDD), 394
Immersed-boundary method (IBM)
  for fluid-membrane interaction, 354–355
Immunotherapy, 399–400
Interendothelial (intercellular) cleft, 122–124
Inter-endothelial cleft
  water and solutes transportation model, 135–146
  pore-slit theory, 135–137
  paracellular pathway of peripheral microvessel wall, 141–144
  1-D models, 135–141
  3-D models, 141–146
Internal elastin lamina, 304–306
Intracellular matrix, 11
Intracranial aneurysms, 324–333
  cerebral artery bifurcations, 325–326
  CFD analyses, 327–328
  computational modeling, 326–329
  prevalance rates of, 325
  risk factors, 326–327
Intramural stress state, 294
  and wall shear stress vector, 294
Iontophoresis, 441–442

K

Klitzman and Johnson cell partitioning law, 220–221

L

Lattice-Boltzmann Method (LBM)
  for fluid dynamics, 351–352
Lidocaine, 446
Liposomes, 396
Liver tumours
  in Smart Microcatheter System, 404–408
Local Transport Region (LTR), 495–496, 507–508
Lucifer yellow (LY), 464

M

M. tuberculosis, 10
Magnetic drug targeting, 392
Magnus-Keizer uniporter model, 193
Marle's formulation, 21
Mass transfer phenomenon, in electroporation, 458–459
  bacteria inactivation, 462
  direct imaging, 466
  effect of PEF on cell membrane, 459
  electrochemotherapy, 484–485
  electro-osmosis model, 464
  experimental methods, 465–467
  galvanic properties, 465–466
  mass transport in biological systems, 463–464
  mass transport in cells, 464–465
  physico-chemical model, 467–470
  reversible electroporation, 467–484
  on siRNA transport, 465
  small molecule transport kinetics, 477–479
  Synthetic Biology, 485
  two compartment pharmacokinetic model for molecular uptake, 479–481
  use of, 459
Mechanobiology, of arterial wall, 276–277
  on human tissues, 278
  mathematical models, 277
  vascular cells, 292–301
Mechanosensor, 150–151
Mechanotransduction, 5
Metal-based nanoparticles, 398
Mg^{2+} inhibition
  kinetic scheme, 204–206
  and mitochondrial Ca^{2+} uniporter model, 191–208
  and Pi regulation, 204
Microcirculation
  Bell's model, 163–165
  decreased microvessel permeability, 156–158
  Dembo et al.'s model, 165
  effect of curvature, 166–168
  effect of wall shear stress, 168–170
  under flow conditions, 162–163
  general cell adhesion models, 163–165
  increased microvessel permeability, 154–156
  mathematical models for, 163–166
  model predictions for, 166–170

modulation, 153–162
transportation modulation, 153–162
tumor cell adhesion, 162–170
Microvascular hyperpermeability, 158–162
vascular endothelial growth factors (VEGFs), 158–161
Microvascular transport, 122–153
blood-brain barrier, 124–129
interendothelial (intercellular) cleft, 122–124
peripheral microvessels, 122–124
transvascular pathways, 122–129
Mitochondria, 191
Mitochondrial ATP/ADP antiporter, 208
Mitochondrial Ca^{2+} uniporter, 191–208
and Ca^{2+} uniporter, kinetic scheme for, 193–195
derivation of, 195–197
dissociation constants, 199–200
equilibrium constant, 198–199
free energy barrier formalism, 197–200
full cooperative binding, 196
IMM $\Delta\Psi$ dependency, 192
kinetics of, 192–193
limitation of Ca^{2+} uniporter model, 201
mathematical modeling, 192–193
measurements of, 192
Mg^{2+} inhibition and Pi regulation, 204
no cooperative binding, 196–197
parameterization, 200–201
partial cooperative binding, 196
rate constants, 200
simulations, 201
$\Delta\Psi$-dependency of, 197–200
Mitochondrial inner membrane Na^+/Ca^{2+} exchanger, 208
Molecular sieve, 148
Mooney-Rivlin model
isotropic mechanism, 312–313
Multifunctional nanodrugs (passive targeting), 392
Multi-layered wall model
concentration solutions and results, 111–115
general physiological and mathematical description, 107–111
Multiple barriers/membranes, electrotransport in, 430–440
behavior of, 432–434
concentration profiles and properties, 436
membrane porosity and applied voltage, effects of, 436–440
membrane/membrane interface, 438
symmetric and asymmetric conditions, 432–436
transference numbers, of permeant and porosities, 436–438
Mycobacteria, 10

N

Na^+/H^+ exchanger, 208
Nails, 447
pH-dependent iontophoretic transport, 447
in transungual iontophoresis, 447–448
Nanodrug, in trageted drug delivery, 395–400
carbons, 397–398
ceramics, 398
immunotherapy, 399–400
lipids, 396–397
metals, 398
multifunctional nanoparticles, 398–399
polymers, 397
viruses, 398
Neo-Hookean model
isotropic mechanism, 312–313
Nernst-Planck theory, 419–420
and electroneutrality approximation, 426–430
finite-element simulations, 420
flux of ionic species, 419
membrane flux, 421–424
and model simulation analyses, 419–420
one-dimensional mass transfer, 419
Network forming, 290
Neurons, modeling intracellular transport
dynein velocity distribution on positive injury signals, 532–540
gene therapy, 540
investigating axonal trafficking, 522
merging viral concentration waves in retrograde viral transport, 540–546
nolecular motor-driven transport, 522
pharmaceutical agent complexes, 523
Nociceptor transduction, 47–51
modeling transduction, 49–51
shear stress modeling, 47–49
Nodes of Ranvier, 527
Non-thermal electroporation (long pulse), in skin electroporation, 506–507
Nusselt's solution, 79–83

O

Optimal micro-drug delivery, 404–413
solid (liver) tumors, 413–414
Optimal micro-drug delivery, 404–413
computational MMP, 404
computational tasks, 409–412
liver tumors in Smart Microcatheter System, 404–408
patient-specific arteries, 404–406
smart inhaler system (SIS), 412
validation, 407
with microspheres, 404–406

P

Pharmaceutical agent complexes, 523
 diffusivities of organelles, 527
 effect of decreased diffusivity, 531–532
pH-dependent iontophoretic transport, 447
Piola–Kirchhoff stress tensor, 14–15
Plasma membrane, 208
Plasma, 350
Polymer drug, 397
Pore induction
 effect of, 446
 see also Electroporation
Pore-slit theory, 135–137
Porous cellular biological media
 hydraulic permeability of, 29–31
Porous media, 65–73
 analytical solutions for local mass non-equilibrium, 78–87
 Anzelius's solution, 85–86
 averaging concentration, 67–68
 conservation of, 73–78
 continuity equation, 69–71
 continuum modeling, 65–73
 Darcy's Law, 68
 drug-eluting stent, 92–115
 extended continuity equation, 71–72
 Fick's equation, 73
 first-order consumption rate coefficient, 76
 in fluid phase, 74–77
 fluid-saturated porous media, 67
 governing equations, 78
 mathematical difficulties of analytical solution, 78–87
 microscopic approach, 65
 molecular dynamics (MD) simulations, 64
 Nusselt's solution, 79–83
 partial differential equations, 78
 permeability, 68
 phases of, 65
 in solid phase, 77–78
 Schumann's solution, 83–85
 temperature of fluid mixture, 78
 tortuosity, 72–73
 and volume-averaged variables, 66–68
Propidium iodide, 464
Protein actin, 8–9
Proteoglycans, 9
Pulsed electric fields, 463–465
 molecular transport in cells, 464

R

Radioembolization (RE), 404
Red blood cells (RBCs), 350
 immersed-boundary method (IBM), 354–355
 intercellular aggregation, 353–354
 and membrane mechanics, 352–353
 plasma viscosity, 350–351
 shear stress variation, 351
Reversible cell electropermeabilization, 494
Reversible electroporation
 application of, 468
 assumptions, 468
 electro physico-chemical electroporation theory, 467
 electrodiffusion coefficient, 469
 electropermabilization model, 470–472
 mathematical models, 467–484
 permeabilization probability function, 471–472
 pore fraction, 467
 time dependent flow coefficient, 469
 time dependent internal species concentration, 469
 yeast cell transformation, 470
Reversible electroporation, 493–494
 cell membrane is, 494

S

Schumann's solution, 83–85
Sclera, 448–449
 ion-exchange membrane, 450
Skin electroporation, 495–497
 cell membrane electropermeabilization, 498
 cell membranes delivery, 496
 electrical conduction, 497–498
 electrophoresis, 504–505
 feasibility evaluation, 508
 high voltage pulses, 499
 intra- and transdermal drug delivery, 495
 magnitude of the electric field, 506–507
 melting and solidification processes, 510
 microscopic to macroscopic descriptions, 497
 modeling, 504–514
 molecular transport, 502–504
 non-thermal electroporation (short pulses), 506–507
 pulses of durations, 495–496
 resistive heating, 499–501
 thermal electroporation (long pulse), 507–514
 tissue electroporation, 498
 transdermal drug delivery, 495
Skin electroporation, transformation, 464
 multiscale model of drug molecules in tissue, 481–484
Skin, 447
 in transungual iontophoresis, 447–448
Solulte permeability, 132–133
Starling's hypothesis
 capillary wall, 148–150

Steady capillary blood flow, 216–222
bifurcation node, 217
cell partitioning, 219–220
converging bifurcations, 220
discharge hematocrit, 217–218
diverging bifurcations, 220
effective viscocity, 218–219
Klitzman and Johnson cell partitioning law, 220–221
Poiseuille's law, 216
suspension flow, 221–222
tree networks, 222–226
tube hematocrit, 218
volumetric flow rate, 216
Stent-based drug delivery, 61
Stokesian-dynamics' method, 139
Stratum corneum (SC), 499
Stratum corneum, 444, 495
Structural organization, in bone, 3
Subarachnoid hemorrhage (SAH), 325
Subendothelium, 303–304
Synthetic Biology, 485

T

Targeted drug-aerosol delivery, 409–412
Thermal electroporation (long pulse), in skin electroporation, 507–514
Thioglycolic acid, 447–448
Tissue electroporation, 498
Tissue heating
on electroporation, 499–502
intense electric fields, 499–501
resistive heating, 499–501
Tissue space (Interstitium), 152–153
Tooth, thermal pain in
1D Fourier heat transfer, 44
cold stimulation, 51–52
dentinal microtubules, 42
difference on hot and cold pain, 54–56
enamel and dentine layers, in-plane thermal stress, 45
exposed dentine, 52–53
external noxious stimuli, 42–43
hot and cold stimuli, 42–43
physical properties of, 44
physiological parameters, 48
thermally induced dentinal fluid flow, 43–47
thermomechanical model, 43
Trageted drug delivery (TDD), 392
active targeting, 392
development of smart drugs, 392
direct transport, 401–413
forced transport, 400–401
free transport, 394–400
image-guided drug delivery (IGDD), 394
imaging techniques, 392–394
multifunctional nanodrugs (passive targeting), 392
nanodrugs, 395–400
passive targeting, 392
Transdermal drug delivery, 62–64
Transdermal iontophoresis, 444–447
anodal iontophoresis, 445
effective pore size and pore charge density, 445
flux enhancing mechanisms, 444–445
ion composition and drug concentration, effects of, 446
monovalent ions, 445
stratum corneum, 444
transference number, 446
Transport lag time, 424–425
Transscleral iontophoresis, 448–450
for ocular iontophoresis, 448
electroosmosis in, 449–450
sclera, 448–449
significant flux enhancement, 449
Transungual iontophoresis, 447–448
chemical enhancers and formulation factors, 447–448
electroosmosis in, 447
nail plates, 447
pH, effects of, 447–448
solution ionic strength, 448
Tree models capillary blood flow, 222–226
geometrical construction, 222–223
numerical method, 223
Trypan blue, 464
Tube hematocrit, 218
Tumor metastasis, 158–162
and integrin signaling, 161–162
vascular endothelial growth factors (VEGFs), 158–161
Tunica adventitia, 279–281
Tunica intima, 257, 279–281, 301–306
basement membrane, 302–303
endothelial glycocalyx layer, 301–302
internal elastin lamina, 304–306
subendothelium, 303–304
Tunica media, 279–281, 306–308
of elastic arteries, 306–308
of muscular arteries, 308

U

Unsteady capillary blood flow, 235–243
balances at bifurcations, 237–238
circular capillaries, 236–237
correlations, 237

microvascular network, 235
numerical methods, 238–240
oscillations in velocity of blood flow, 235
single-node dynamics, 240–243
state space, 248
in straight capillary, 235–236
for subcritical exponents, 244–246
supercritical exponents, 246
in tree networks, 243–249

V

Vascular cells, 292–301
endothelial cells, 293
fibroblasts, 297
matrix assembly, 297–301
vascular smooth muscle cells, 295–297
Vascular smooth muscle cells, 295–297
environmental factors, 296–297
peak force development, 295–296
phenotypes, 296
principal function, 295–296
Viruses, 398

W

Wall shear stress, 168–170
West Nile virus, 540
Wolff's law, 4–5